11/20 857

75,00

# CP VIOLATION
## in Particle Physics and Astrophysics

XXVth Anniversary of CP Violation Discovery

*Château de Blois, France - May 22-26, 1989*

**"CP Violation in Particle Physics and Astrophysics"**

**EDITIONS FRONTIERES**
B. P. 33
91192 Gif-sur-Yvette Cedex - France

ISBN 2-86332-071-8


Printed in Singapore by Kim Hup Lee Printing Co. Pte. Ltd.

# CP VIOLATION
## in Particle Physics and Astrophysics

edited by
J. TRAN THANH VAN

*Editions Frontières*

XXVth Anniversary of CP Violation Discovery

**"CP Violation in Particle Physics and Astrophysics"**

## International Advisory Committee

Banner M. *(Saclay)*
Cabibbo N. *(Roma)*
Darriulat P. *(CERN)*
Davier M. *(Orsay)*
Frisch H. *(Chicago)*
Gilman F. *(SLAC)*
Glashow S. *(Harvard)*
Harari H. *(Rehovot)*
Kobayashi M. *(KEK)*
Lee T. D. *(Columbia)*
Lehmann P. *(Paris)*
Mannelli I. *(Pisa)*
Michel L. *(Bures)*
Okun L. *(Moscow)*
Rubbia C. *(CERN)*
Sakharov A. *(Moscow)*
Salam A. *(Trieste)*
Schatzman E. *(Paris)*
Steinberger J. *(Austin)*
Weinberg S. *(CERN)*
Willis W. *(CERN)*
Yang C. N. *(Stony Brook)*

## International Program Committee

Adair R. K. *(Yale)*
Audouze J. *(Paris)*
Bloch Ph. *(CERN)*
Fournier D. *(Orsay)*
Jarlskog C. *(Stockholm)*
Lutz A. M. *(Orsay)*
Oliver L. *(Orsay)*
Paschos E. *(Dortmund)*
Peyaud B. *(Saclay)*
Tran Thanh Van J. *(Orsay)*
Truong T. N. *(Palaiseau)*
Turner M. *(Chicago)*
Wahl H. *(CERN)*
Winstein B. *(Chicago)*
Wolfenstein L. *(Pittsburgh)*

## Sponsors :

Ministère de la Recherche et de la Technologie

Centre National de la Recherche Scientifique (IN2P3, MPB)

Commissariat à l'Energie Atomique

Conseil Régional de la Région Centre

Conseil Général du Loir-et-Cher

Ville de Blois

Rhône Poulenc

Air France

Tektronix

A. Sakharov.
Monday December 11th, 1989, Meeting at Lebedev Institute for the abolition of the article 6 of the USSR Constitution : communist party leadership. →

**We acknowledge the help of :**

Monsieur Didier Hamon, Ministère de la Culture
Mosieur Alain Brunet, Ministère de la Culture
Monsieur Alain Voindrot, Mairie de Blois
Monsieur Frank Trouilloud, Mairie de Blois
Monsieur le Comte du Pavillon et Madame, Château de Beauregard
Madame Martine Tissier de Mallerais, Château de Blois
Madame Martine Rebière, Palais des Congrès
Monsieur Antoine Perragin, Palais des Congrès
Monsieur Charles Mary, Château de Chambord
Monsieur de la Morandière, Château de Chenonceau
Monsieur Jean Saint Bris, le Clos Lucé
Monsieur André de Marco, Rhône Poulenc
Madame Véronique Mare, Rhône Poulenc
Madame Sylvie Gille-Naves, Lenôtre-Paris
Monsieur Gilbert Cotteau
Monsieur et Madame Alain Pasquier and their staff
Madame Claude Lotéguy
Monsieur Christophe Averty

**Andrey Dmitrievich Sakharov**
(1921-1989)

This book is dedicated to the memory of Andrey Dmitrievich Sakharov for
his vision of the universe and his exceptional scientific contributions to many fields of physics,
his vision of human beings and his strong conviction in the human heart, spirit and intelligence,
his unconquerable force of spirit, his noble and pure soul,
and his courageous devotion to humanity.

## *Message of Professor A. Sakharov*

Moscow May 17, 1989

Dear Professor J. Tran Thanh Van,

To my greatest regret, I have to inform you of the latest events that have made it impossible for me to keep my promise to attend the CP Conference.

Until yesterday, I had been hoping that my being elected as a member of our new Parliament would not deprive me of honorable and pleasant opportunity to take part in the work of this conference even though I had been forced to reduce my visit to a few days, since the session of the Parliament is to begin on May 25. Yesterday, however, I was appointed a member of the organizing committee which is to establish the program and the procedure of the forthcoming session. This work, beginning on May 18, will keep me occupied completely. Because of the critical situation of my country, I feel it is my unavoidable duty and responsibility to cancel my visit.

I trust you understand how deep is my regret to miss the chance of meeting my colleagues I wanted so much to see. I bring forth my excuses for all the trouble caused by the late and unexpected change in my plans. I would appreciate it if I could have a chance to respond to your kind and generous invitation sometime later when my time will be at my disposal.

Sincerely yours,

A. Sakharov

*Discours de Monsieur Jack Lang,*
*Ministre de la Culture et de la Communication et*
*Maire de Blois*
*à l'inauguration de la conférence, le 22 Mai 1989*
*au Château de Blois*

Cher Monsieur Trân,
Mesdames, Messieurs,

Je serai bref.

Vous êtes certainement impatients d'aborder les sujets pour lesquels vous êtes réunis au cours de ces journées.

Je voudrais d'abord, Mesdames et Messieurs, en votre nom, si j'ose dire, féliciter Monsieur Trân, Madame Trân et toute l'équipe d'organisation qui, avec une pugnacité et un enthousiasme qui renversent toutes les montagnes, ont réussi cet exploit d'abord de vous réunir, de vous arracher à vos universités et à vos centres de recherches, puis de trouver aussi des financements pour assurer la tenue de cette manifestation importante.

Naturellement, personne ne peut résister à Monsieur et Madame Trân, personne. Vous en êtes la preuve vivante, moi aussi, un peu. Et il faut des hommes et des femmes comme eux, capables de faire bouger les choses, pour que précisément la Science dont vous êtes les représentants puisse aller de l'avant.

Je voudrais aussi vous donner lecture d'un message du Président de la République, M. François MITTERRAND. Un message adressé à votre réunion et à vous-même. Je vous le lis, il est court :

*Mesdames,*
*Messieurs,*

*Vous êtes réunis au Château de Blois pour célébrer le 25ème anniversaire d'une grande découverte en physique des particules.*

*C'est grâce à vous, M. Turlay et à vos amis américains que nous savons depuis 25 ans que deux grandeurs qui se conservent dans la nature lorsqu'elles sont séparées ne se conservent pas quand elles sont associées.*

*Beaucoup de livres de science fiction évoquent des mondes dans lesquels l'antimatière existe autant que la matière. C'est grâce à cette découverte dans l'infiniment petit que l'académicien Andrei Sakharov, que je suis heureux de saluer aussi, a démontré que nous vivons dans un univers presque exclusivement de matière. Ceci démontre, une fois de plus, l'existence d'une relation intime entre le comportement de la particule élémentaire et celui de l'univers dans son ensemble.*

*Je profite de cette occasion pour saluer , au nom de mon pays, la communauté scientifique internationale que vous représentez et qui est aujourd'hui rassemblée, à l'occasion de cette manifestation.*

*François MITTERRAND*

J'ajouterai 2 mots et j'en resterai là. Pour l'ignorant que je suis, très impressionnné de parler devant des femmes et des hommes de Science, les mots mêmes que vous utilisez: l'infiniment petit, la particule élémentaire, la structure de la matière, ces mots eux-mêmes, indépendamment de leur contenu, ont une vertu, je dirais, poétique, qui réjouit l'homme de culture artistique et littéraire que je suis. Je me souviens d'avoir lu avec beaucoup de passion le livre de l'un de vos collègues, Ilya Prigogine; ce livre, qu'il a écrit avec une jeune femme très brillante, Isabelle Stenger, a le beau nom de "La Nouvelle Alliance", la nouvelle alliance, comme au fond l'alliance de Dieu avec les hommes, la nouvelle alliance des hommes avec la nature. Et ce qui m'a beaucoup plu dans ce livre, c'est, lorsque citant le poète Saint John Perse, Prigogine insistait sur ce qu'il appelait la lecture poétique du monde.

Il expliquait - je ne sais si vous les savants êtes en accord avec la pensée de votre confrère - il expliquait que l'homme de Science, quel que soit son savoir, quelle que soit la rigueur de son raisonnement, ne pouvait pas ne pas avoir eu en lui une intention poétique du monde.

C'est que je suis convaincu qu'indépendamment de la méthodologie, de l'épistémologie qui vous gouvernent, le flair, le "nez", doit certainement jouer un grand rôle dans vos travaux, dans vos recherches et d'une certaine façon, cela vous rend plus humains, à nos yeux, nous qui attendons beaucoup de vous et de vos découvertes.

Voilà ce que je désirais vous dire, en vous souhaitant un grand succès dans vos travaux et en espérant que vous serez ici assez heureux pour aimer y revenir; comme nous en avons parlé avec Monsieur et Madame Trân, après tout, pourquoi ne pas institutionnaliser vos Rencontres sous la forme, par exemple, des entretiens de Blois, qui permettraient, année après année, de prendre le pouls de l'avancée de vos recherches et des débats scientifiques.

Merci à tous, Merci.
Beau succès et à bientôt.

*Discours de Monsieur Hubert Curien,*
*Ministre de la Recherche et de la Technologie*
*à la Conférence de Presse, le 25 Mai 1989*
*au Château de Chambord*

Merci beaucoup Monsieur Trân.
Merci de m'avoir invité à cette séance tout à fait sympathique.
Il est assez rare de siéger à la même table qu'un si grand nombre de lauréats du prix Nobel: on se sent vraiment honoré dans des circonstances telles que celles là.

Les Rencontres de Moriond ont acquis, depuis longtemps déjà, une réputation très grande et notre reconnaissance va à Monsieur Trân qui, aidé par Madame Trân depuis de nombreuses années déjà, organise ces Rencontres dans des endroits, j'allais dire divers, d'hiver, puisque j'ai quelques témoignages de physiciens célèbres qui se réjouissaient de venir parler de physique et de faire un peu de ski. Aujourd'hui, on ne fait pas de ski mais on profite de quelques merveilles de notre passé.

Un auditoire de très grande qualité et la chance, vous l'avez dit, d'arriver à un état de développement de la science des particules où l'on peut apporter des réflexions vraiment originales et l'espoir de résoudre quelques problèmes essentiels grâce aux machines qui vont très prochainement être mises en service et à d'autres qui sont d'ores et déjà dans l'imagination de nos collègues physiciens.

Il est tout à fait intéressant de voir comment la science se développe avec une vitesse prodigieuse en deux secteurs pourtant extrêmement difficiles et très différents l'un de l'autre :

- Le secteur qui vous intéresse, c'est à dire la nature intime, ultime, de la matière,
- Le secteur des Sciences de la Vie, c'est à dire la combinaison la plus complexe possible de cette matière que vous traquez dans ses derniers détails.

Et cependant la Science reste une, ce sont bien les même types de raisonnement, c'est la même formation qui est nécessaire à nos chercheurs et parlant de la recherche qui nous intéresse ici, je voudrais souligner combien il est nécessaire que les pays les plus avancés s'intéressent à la Science la plus difficile. C'est en s'attaquant aux problèmes extrêmement coriaces, qui sont ceux qui nous préoccupent quotidiennement, qu'on fait évidemment les découvertes les plus extraordinaires. Les problèmes simples ont peut-être des solutions simples, voire compliquées. Il est rare que des problèmes très simples nous amènent à des coups de génie. Les affaires difficiles sont évidemment la matière de réflexion de nos savants les plus habiles. Nous allons parler tout à l'heure de coopération et là aussi je voudrais insister sur l'extraordinaire solidarité internationale des chercheurs et comme ici, des physiciens.

Vous avez besoin naturellement pour vos travaux d'une intuition extrêmement poussée, mais vous avez aussi besoin d'expériences cruciales qui vous montrent si oui ou non vous êtes sur la bonne voie et qui vous permettent de vous relancer dans des travaux nouveaux ou d'abandonner ceux qui s'avèrent moins fructueux. On ne peut pas faire de physique uniquement avec un crayon. Le crayon est peut-être l'instrument le plus précieux de tous les scientifiques mais, s'il était le seul, arriverait à s'user !

La coopération internationale est, tout particulièrement aujourd'hui, brillamment illustrée par la présence de collègues parmi les plus réputés dans le monde entier dans cette discipline et là aussi, c'est le très grand mérite de notre ami Trân d'avoir cette force de conviction, cette autorité qu'il a acquise par sa propre compétence scientifique mais aussi par la gentillesse de ses propos et l'aménité de son accueil.

Mais je ne veux pas faire un discours beaucoup plus long. Je voudrais dire simplement que le gouvernement de notre pays est très intéressé par la recherche. Il est très intéressé par la recherche pour de multiples raisons dont les deux plus évidentes sont les suivantes :

1 - On ne peut pas avoir une économie prospère si l'on ne fait pas de recherche,

2 - On n'est pas très intelligent si l'on ne fait pas de recherches.

Ces deux motivations, la motivation culturelle et la motivation économique, sont très profondément liées dans notre esprit; mes compatriotes qui sont ici, entendant les déclarations de notre Président de la République savent qu'il n'est guère de discours où le Président de la République Française ne rappelle que pour un pays avancé, la recherche et tout ce qui lui est lié : la formation, l'éducation, la promotion de l'innovation dans l'industrie, sont très évidemment les devoirs du gouvernement. Or ces priorités, nous les avons non seulement au coeur mais aussi à l'esprit, quand nous avons à prendre des décisions à caractère budgétaire. Je ne vais pas ennuyer mes collègues étrangers avec nos propres discussions budgétaires, vous avez les vôtres, mais je sais que cette solidarité internationale est une très grande force de la science, une des très grande joies des chercheurs et encore une fois je voudrais remercier au nom des autorités françaises tous nos collègues qui ont bien voulu se déplacer et venir discuter - avec aménité certes, mais quelquefois sans doute, avec âpreté - des problèmes qui les intéressent. Je suis sûr que vous arriverez à résoudre ces problèmes là, pas tout de suite, mais vous y arriverez.

# AVANT PROPOS

La conférence internationale sur la violation de la symétrie CP s'est tenue du 22 au 26 mai 1989 au Château de Blois, chef d'oeuvre de la Renaissance Française.

L'idée nous en est venue en 1984 lors d'une discussion avec Jim Cronin : il nous est apparu qu'il serait très difficile de rassembler les 4 "mousquetaires" de la découverte de la violation de la symétrie CP à Moriond et que le meilleur moyen d'y parvenir serait d'organiser une conférence pour célébrer le 25ème anniversaire de la découverte. En même temps, l'idée de lier science et culture en faisant découvrir à nos amis étrangers l'un des trésors du patrimoine français nous tenait de plus en plus à coeur : en 1985, nous avons débuté au Château de Blois les "Blois Workshops" consacrés à la diffusion élastique et diffractive; puis en 1988, nous avons organisé à Arles et au Château de Tarascon le XIXème "Multiparticle Dynamics Symposium". Les Châteaux de Blois et Chambord nous semblèrent alors les lieux privilégiés pour célébrer les 25 ans de cette importante découverte et honorer nos amis Jim Christenson, Jim Cronin, Val Fitch et René Turlay.

A l'époque de cette découverte, la physique expérimentale et la physique théorique étaient en effervescence : les résultats des diverses expériences ne cadraient pas avec la théorie, ce qui donna naissance à plusieurs hypothèses. Le groupe expérimental de Princeton arriva au bon moment, avec l'équipement adéquat, et découvrit la violation de la symétrie CP. D'autres groupes à Brookhaven, à Columbia, en Illinois, à Princeton et à Yale ont aussi contribué de façon importante à la découverte. Tous méritent notre reconnaissance et notre respect et je suis heureux que toutes ces collaborations aient pu être présentes à Blois.

La découverte de la violation de la symétrie CP suscita une explosion de travaux expérimentaux et théoriques pendant au moins une dizaine d'années. Récemment, les progrès de la recherche, expérimentale aussi bien que théorique, ont conduit à un renouveau d'intérêt pour la violation de la symétrie CP. Deux collaborations internationales - NA31 au CERN et E731 à FNAL - ont déployé des efforts considérables pour la mesurer lors des désintégrations K. Quelques-uns de leurs résultats furent annoncés pour la première fois au cours de cette conférence. Les résultats concernant la différence de phase $\Phi_{+-} - \Phi_{00}$ sont cohérents avec la conservation de la symétrie CPT et les valeurs $\varepsilon'/\varepsilon$ concordent avec le modèle standard, sans pour autant disqualifier le modèle superfaible.

En plus de tous ces résultats complémentaires qui montrent que la recherche dans ce domaine est encore très fructueuse, même 25 ans après, la trentaine d'orateurs invités ont relaté avec intérêt l'historique et l'état actuel des connaissances sur la violation de la symétrie CP. Les thèmes suivants ont été exposés par les physiciens les plus compétents dans leur domaine : violation de la symétrie CP dans le cadre et au delà du modèle standard, violation de la symétrie CP à LEAR, dans les désintégrations rares, dans le système B et en cosmologie, ainsi que la violation de la symétrie T dans la mesure du moment électrique dipolaire du neutron.

Je voudrais remercier les membres du Comité International pour leurs suggestions concernant les thèmes à aborder au cours de la conférence ainsi que les membres du Comité de Programme pour le travail considérable qu'ils ont fourni dans le choix et la coordination du programme. Mes remerciements vont tout particulièrement à Anne-Marie Lutz, Bernard Peyaud et Bruce Winstein pour l'aide qu'ils ont apportée, dans l'organisation tant scientifique que matérielle du colloque.

La conférence a été honorée par les plus hautes autorités du pays : le Président François Mitterrand a félicité les auteurs de la découverte dans son message aux participants. Monsieur Jack Lang, Ministre de la Culture et Maire de Blois, a apporté son soutien à la conférence, présidé la session d'ouverture et offert aux participants et à leurs familles une réception dans la magnifique Salle des Etats Généraux du Château de Blois. Monsieur Hubert Currien, Ministre de la Recherche et de la Technologie, présida la conférence de presse à Chambord et a tenu à souligner à cette occasion tout l'intérêt qu'il porte à la collaboration scientifique internationale. La présence des ministres, ainsi que celle de Monsieur Maurice Dousset, Président du Conseil Régional de la Région Centre et de Monsieur Michel Fromet, Député du Loir-et-Cher, est le signe de l'importance qu'ils accordent à la recherche; cette présence est appréciée par les participants comme un encouragement aux efforts de notre communauté scientifique. Je voudrais remercier Jean Audouze, Daniel Cribier, Michel Davier, Pierre Lehmann, Pierre Petiau, et Daniel Thoulouze pour l'aide précieuse qu'ils m'ont apportée pendant l'organisation de cette conférence.

L'organisation de cette conférence n'aurait pas été possible sans le support financier et logistique du Ministère de la Recherche et de la Technologie, du Centre National de la Recherche Scientifique (IN2P3, MPB), du Commissariat à l'Energie Atomique (IRF), du Ministère de la Culture et de la Communication, du Conseil Régional de la Région Centre, du Conseil Général du Loir et Cher et de la Mairie de Blois. La contribution de la National Science Foundation (NSF), qui a permis à 15 jeunes physiciens de participer à la conférence a été particulièrement appréciée. Je remercie aussi Air France et Tektronix pour leur soutien.

Mes remerciements vont aussi à Rhône Poulenc dont l'aide financière et le concours technique furent déterminants.

J'exprime ma profonde gratitude à Madame Monique Lang pour l'intérêt et l'attention particulière qu'elle a portée à cette conférence et à Monsieur Jean-René Fourtou, Président de Rhône Poulenc, pour ses encouragements au développement de la recherche scientifique.

Enfin, l'organisation matérielle a été prise en charge par le Laboratoire de l'Accélérateur Linéaire d'Orsay, le Département de Physiques de Particules Elémentaires de Saclay et les Rencontres de Moriond. Je remercie vivement Monique Bonamy, Aïda Cherkaoui, Rolande Dosbioz, Eliane Frassati, Odile Lebey, Nathalie Lesbre, Nathalie Leman, Lê Van Suu, Laurence Massiot, Nicole Mathieu, Lucienne Norry, Nicole Ribet, Clémentine Rollet, Trân Dai Thanh Binh, Emmanuelle Tran Thanh Tam et Anne-Claire Trân Thanh Binh Minh qui ont consacré beaucoup d'énergie et de temps à la réussite et au caractère chaleureux de cette conférence.

J. Trân Thanh Vân

# FOREWORD

The International Conference on CP Violation was held in the Château de Blois, pearl of the French Renaissance, from May 22 to May 26, 1989.

The idea for such a conference arose from a discussion with Jim Cronin back in 1984, when it turned out that to bring together in Moriond the "Four Musketers" of the CP Violation discovery would be a difficult enterprise. We agreed to try again, for the 25th anniversary of the event. Meanwhile, we became increasingly enthusiastic about the idea of linking Science and Culture, while promoting the historic and cultural treasures of France. In 1985, we started in the Château of Blois the "Blois Workshops" on Elastic and Diffractive Scattering; in 1988, we organized the Multiparticle Dynamics Symposium at Arles and the Château of Tarascon. The Châteaux of Blois and Chambord seemed to us the natural settings to celebrate the silver anniversary of this important discovery and to honour our friends : Jim Christenson, Jim Cronin, Val Fitch and René Turlay.

As noted by both A. Pais and R. Adair, experimental and theoretical physics alike were in a state of ferment twenty-five years ago: results from various experiments did not fit into the theoretical framework, giving rise to various hypotheses. The Princeton group's experiment arrived at the right time and with the adequate equipment and succeeded in discovering CP violation. Other groups at Brookhaven, Columbia, Illinois, Princeton and Yale have also made important contributions : all deserve our respect and recognition. I am happy that all these collaborations were present at this meeting.

The discovery of CP Violation led to an explosion of experimental and theoretical work which did not subside for at least a decade. In the last few years, progress in experimentation as well as in theoretical research has led to a renewed interest in CP Violation. Two international collaborations : NA31 at CERN and E731 at FNAL devoted considerable effort to measure CP Violation in K decays and some of their results were announced for the first time at this conference. The results on the phase difference $\Phi_{+-} - \Phi_{00}$ are consistent with CPT conservation and the values of $\varepsilon'/\varepsilon$ agree with the Standard Model without however disqualifying the superweak model.

In addition to these new results which clearly indicate that the field is very active even after 25 years of hard work, about thirty very interesting invited talks covered both the history of the discovery and the state of the art in CP Violation. CP Violation in and beyond the Standard Model, CP Violation at LEAR, in the rare decays, in the B system and in Cosmology, as well as T Violation through the measurement of the electric dipole moment of the neutron were reviewed by the most talented physicists in the field.

I would like to thank the members of the International Advisory Committee for many suggestions concerning the topics to be discussed and the members of the Program Committee for the considerable work expended in choosing and coordinating the program. I wish to address my special thanks to Anne Marie Lutz, Bernard Peyaud and Bruce Winstein for their help extended to the practical organization.

The Conference was favoured by the highest authorities of the country. President François Mitterand sent a message to the participants and his praise to the authors of the discovery. Professor Jack Lang, Minister of Culture and Mayor of Blois strongly supported the Conference, chaired the opening session and offered a welcome reception to the participants and their families in the magnificent *"Salle des Etats Généraux"* of the Château de Blois. Professor Hubert Curien, Minister of Scientific Research and Technology, chaired the press conference and expressed his views on international scientific collaboration at the Château de Chambord. Their presence, and that of Mr. Maurice Dousset, President of the "Conseil Régional de la Région Centre" and of Mr. Michel Fromet, deputy of Loir et Cher, on many occasions during the meeting were recognized by the participants to be a sign of their interest in scientific research and constituted strong encouragement to our community.

The organization of this conference would not have been possible without the support and logistic help of the "Ministère de la Recherche et de la Technologie", the "Centre National de la Recherche Scientifique" (IN2P3, MPB), the "Commissariat à l'Energie Atomique" (IRF), the "Ministère de la Culture et de la Communication" the "Conseil Régional de la Région Centre", the" Conseil Général du Loir et Cher" and the town authorities of Blois. The contribution of the National Science Foundation (NSF) which helped fifteen young physicists to attend the conference is particularly appreciated. I would like to thank Jean Audouze, Daniel Cribier, Michel Davier, Pierre Lehmann, Pierre Petiau and Daniel Thoulouze for their constant support during the organization of the conference.

The sponsoring of Rhône Poulenc which strongly supported the conference, of Air France and of Tektronix are gratefully acknowledged.

I would like to express my gratitude to Madame Monique Lang for her interest and the special care she devoted to the conference and to Mr Jean-René Fourtou for his encouragement and generous support.

The practical organization of the conference was handled by the "Laboratoire de l'Accélérateur Linéaire" at Orsay , the "Département de Physique des Particules Elémentaires" at Saclay and the "Rencontres de Moriond" and I am happy to express my special thanks to Monique Bonamy, Aida Cherkaoui, Rolande Dosbioz, Eliane Frassati, Odile Lebey, Nathalie Lesbre, Nathalie Leman, Lê van Suu, Laurence Massiot, Nicole Mathieu, Lucienne Norry, Nicole Ribet, Clémentine Rollet, Trân Dai Thanh Binh, Emmanuelle Trân Thanh Tâm and Anne Claire Trân Thanh Binh Minh who devoted much of their time and energy to bring a special touch to the conference.

J. Trân Thanh Vân

# SCIENTIFIC PROGRAM

## *Monday May 22, 1989*

*Morning : Opening Session*

## 25th ANNIVERSARY CELEBRATION

***Chairman*** **: J. Lang** (Minister of Culture and Mayor of Blois)

| | |
|---|---|
| J. Tran Thanh Van | Welcome |
| J. Lang | Message of François Mitterrand, President of the Republic<br>Welcome Address |

***Chairman*** **: P. Lehman**

| | |
|---|---|
| A. Pais | Discovery of the CP Violation |

***Chairman*** **: V. Soergel**

| | |
|---|---|
| R. Adair | Early Experiments on CP Violation |
| J. Steinberger | Current Experimental Status of CP Violation |

*Afternoon session*

## Guest Lectures

***Chairman*** **: H. Lipkin**

| | |
|---|---|
| J. Christenson | Dø : Progress and Promise |
| J. Cronin | High Energy Gamma Ray Astronomy |

***Chairman*** **: L. Lederman**

| | |
|---|---|
| V. Fitch | The Fifth Force, Fact or Fiction ? |
| R. Turlay | LEP Experimental Program |

## *Tuesday May 23, 1989*

*Morning Session*

## THEORETICAL INTRODUCTION AND RESULTS OF CURRENT EXPERIMENTS

***Chairman* : R. Dalitz**

C. Jarlskog — CP Violation in the Standard Model
R. Sachs — T or CP Violation

***Chairman* : I. Mannelli**

K. Peach — $\varepsilon'/\varepsilon$ : (NA31)
B. Winstein — $\varepsilon'/\varepsilon$ : (E731)
H. Bech Nielsen — Strong CP Conservation from quantum Gravity Wormholes

*Afternoon Session*

***Chairman* : A. Wattenberg**

P. Debu — $\phi_{oo} - \phi_{+-}$ : (E731)
V. Gibson — $\phi_{oo} - \phi_{+-}$ : (NA31)
K. Smith — Electric Dipole Moment of the Neutron : (Experiment)
O. Pene — Electric Dipole Moment of the Neutron : (Theory)

Public lecture : La Vie dans l'Univers (in french) : A. Vidal Madjar

## *Wednesday May 24, 1989*

*Morning Session*

***Chairman* : M. Holder**

H. Schroder — New ARGUS Results on B Mesons
P. Drell — BBbar Mixing and b $\rightarrow$ u Transition : (CLEO)
H. Harari — Generation Mixing and the Parameters of the Standard Model

***Chairman* : A. Morel**

P. Sikivie — Strong CP Violation and the Axion
S. Barr — Strong CP Without an Axion
J. Hagelin — Flavour Changing Neutral Currents, CP Violation in Supersymmetry, ...

## *Vednesday May 24, 1989*

*Afternoon Session*

***Chairman* : G. Kane**

| | |
|---|---|
| G. Martinelli | Heavy and Light Meson Decays in Lattice QCD |

## $\phi$ Factories

| | |
|---|---|
| J. Bernabeu | $\varepsilon'/\varepsilon$ and Rare Decay at $\phi$ Factories |
| N. T. Truong | The background in $\Phi$ Factory |
| A. Skrinsky | (Experiment) |
| K. Kleinknecht | (Discussion) |

## CP Violation at LEAR

***Chairman* : E. Gabathuler**

| | |
|---|---|
| L. Golowich | Detecting CP Violation with Triple Product Correlations |
| Ph. Bloch | CP LEAR Experiment |
| G. Bassompierre | Testing CP Invariance in Hyperon Decays |
| G. Gollin | $\phi_{oo} - \phi_{+-}$ : (E773) |
| Y. Yamaguchi | Origin of CP Violation and Neutral K System |
| C. Chardin | CP Violation : a Matter of Gravity |

## *Thursday May 25, 1989*

*Morning Session*

## THEORETICAL AND EXPERIMENTAL TRENDS

## Rare Decays

***Chairman* : R. Rau**

| | |
|---|---|
| F. Gilman | CP Violation in Rare K Decays |
| E. de Raphael | Long Distance Effects in Rare K Decays |
| A. J. S. Smith | BNL 787 - Neutral Current $K^+$ decays |
| J. Ritchie | BNL 791 and 77 : Lepton Flavour Violation |

***Chairman* : E. Paschos**

| | |
|---|---|
| E. Auge | NA31 - Studies on rare K Decays |
| G. Block | K Facility at FNAL |
| R. Patterson | E 731 upper Limits on $K^\circ\overline{K}^\circ \rightarrow \pi^\circ\gamma\gamma$ |
| M. Schmidt | BNL 845 - Progress on a Search for $K \rightarrow \pi^\circ$ e+e- |

## *Thursday May 25, 1989*

*Afternoon Session*

***Chairman :*** **P. K. Kabir**

| | |
|---|---|
| A. Sanda | CP Violation in the B System |

## B Factories

| | |
|---|---|
| K. Schubert | BETA - The B Meson Factory |
| D. Hitlin | A High Luminosity Asymetric B Factory |
| C. T. Murphy | Comments on B Physics in fixed Target at FNAL and SSC |
| G. Wormser | Beauty Physics at $e^+e^-$ Colliders |
| A. Martin | Reconstruction of the Kobayashi-Maskawa Matrix |

## *Friday May 26, 1989*

*Morning Session*

## COSMOLOGY AND PERSPECTIVES

## Cosmology

***Chairman :*** **J. Cronin**

| | |
|---|---|
| J. Audouze | Cosmology and Particles. An Introduction |
| E. Kolb | Review of Cosmological Baryogenesis |

***Chairman :*** **V. Fitch**

| | |
|---|---|
| R. Peccei | CP Violation for Baryogenesis |
| L. McLerran | Is it possible to generate the Baryon Asymmetry at the E.W. Transition ? |

*Afternoon Session*

***Chairman :*** **J. Tran Thanh Van**

| | |
|---|---|
| A. Linde | Non-Euclidean Quantum Cosmology |
| L. Wolfenstein | Conclusions |

*End of the Conference*

# *Contents*

## V. GENERATION MIXING



## VI. CP VIOLATION BEYOND THE STANDARD MODEL



## VII. PHI FACTORIES

## VIII. CP VIOLATION AT LEAR



## IX. RARE DECAYS



## X. CP VIOLATION IN THE B SYSTEM



## XI. CP VIOLATION AND COSMOLOGY

## XII. CONCLUSIONS AND PERSPECTIVES

*In front of the Conference Building. Château de Blois.*

*From left : Val Fitch, René Turlay, Jim Cronin and Jim Christenson.*
*Château de Beauregard. Courtesy of Count du Pavillon.*

*Conference room in the Gaston d'Orléans Wing. Château de Blois.*

*May 21st, 1989. Double celebration at the Château de Blois of the 25th anniversary of the CP Violation discovery and Andrei Sakharov's sixty-eighth birthday. Welcome address to participants by Jack Lang, Minister of Culture and Mayor of Blois. Cake offered by Lenôtre and presented by Chef Marcel Lenôtre.*

*Reception offered by Jack Lang in the famous "Salle des Etats Généraux", Château de Blois.*

*Welcome addresses by Jack Lang, Leon Lederman and Jean Trân Thanh Vân.*

*From left : Val Fitch, Leon Lederman, Jack Steinberger and Jim Cronin.*

*Mr Maurice Dousset "Président du Conseil Régional" and Count du Pavillon at the Château du Beauregard.*

*At the Clos Lucé where Leonardo da Vinci spent the last part of his life.*
*From left : Anne Marie Lutz, Leon Lederman, Orrin Fackler and Val Fitch.*

*Hubert Curien, Minister of Research and Technology and Mrs Anne Perrine Curien.*
*Château de Chambord May 26, 1989.*

*Press Conference in the "Salle de Chasse" of the Château de Chambord on May 26, 1989.
From left : Daniel Cribier, Jean Teillac, Leon Lederman, Jean Trân Thanh Vân, Hubert Curien, Jim Cronin, Val Fitch and Pierre Lehmann.*

*From left : Antoine Perragin, Kim and Jean Trân Thanh Vân and Martine Tissier de Mallerais.*

# CP VIOLATION CELEBRATION
# PAST AND PRESENT

*from left : Daisy Fitch, Ellen Lederman, Annette Cronin, Anne-Perrine Curien, Barry James, Jean-François Seiller, Jean Audouze, Jacques Haïssinski, Pierre Petiau, Michel Davier, Bernard Peyaud, René Turlay.*

# CP-violation: the first 25 years

A. Pais

(Rockefeller University, New York, New York, 10021)

It was a very good year, 1964, both in theoretical and in experimental high energy physics.

Higgs produced his mechanism for spontaneous symmetry breaking. It would take several years before it became clear how important that concept is for the current version of electroweak unification. It was also the year in which it was proposed that all hadrons known till then are composites of three species of quarks and antiquarks, and in which that idea led in turn to the beginnings of current algebra.

Also in 1964 a fourth quark species was introduced, carrying a new quantum number, charm. The motivations were various: to avoid quarks with fractional charge; to find new constraints in hadron spectroscopy; and to implement a baryon-lepton symmetry, four leptons versus four quarks. Charm did not draw much attention until 1970, when the GIM mechanism was proposed as a means for eliminating highly unwanted contributions to neutral currents.

Again in 1964, we received word of electron-positron collisions obtained in a single-ring collider, the first accelerator of that type, built in Frascati. Two-ring colliders had been proposed and studied earlier. Then Bruno Touschek pointed out that a single ring with counterrotating particles of opposite charge suffices. Such was the Frascati device, AdA, Anelli di Accumulazione, which had generated $e^+e^-$ interactions at 200 + 200 Mev[1]. Since then we have gained a factor nearly 5000 in collider energies -- not bad.

1964 was most especially a very good year for Brookhaven National Laboratory.

On February 24 a paper appeared[2] in which the Samios group reported the first observation of an $\Omega^-$-particle, seen in the 80 inch hydrogen bubble chamber exposed to a 5 GeV separated $K^-$-beam produced at the AGS.

On July 15 the first papers on static SU(6) symmetry appeared, written by some of BNL's theoretician summer visitors.[3]

On July 27 a paper appeared authored by Christenson, Cronin, Fitch and Turlay, entitled "Evidence for the $2\pi$-decay of the $K_2^o$ meson." The result of their experiment, done with a $K_2$-beam produced at the AGS, was a branching ratio $(2,0\pm0,4)\times10^{-3}$ for decay into $\pi^+\pi^-$, quite close to the current more precise value. Their conclusion: "The presence of a two-pion decay mode implies that the $K_2^o$ meson is not a pure eigenstate of CP."

I had been aware of that result since one morning in early June when I was having breakfast in the old, rather dilapidated, Brookhaven cafeteria which some of us remember with slight nostalgia. That morning Jim and Val had walked up to me and had asked could they talk to me. Naturally, I said. They proceeded to tell me that they had found $K_2$-decays into $\pi^+\pi^-$. How can that be I asked; it violates CP-invariance. They knew that, they said, but there it was. Why was the effect not due to regeneration of short-lived K's, I wanted to know. Because, they said, that effect was far too small in the helium bag where the $2\pi$ events had been found. I asked many more questions, why were they not seeeing $2\pi\gamma$ decays with a soft photon, or $\pi\mu\nu$ decays with a soft neutrino and perhaps some confusion about mass. They had thought long and hard about these and other alternatives (the actual experiment had been concluded the previous July) and had ruled them out one by one. After they left I had another coffee. I was shaken by the news. I knew quite well that a small amount of CP-violation would not drastically alter the earlier discussions, based on CP-invariance, of the neutral K-complex. Also, the experience of seeing a symmetry fall by the wayside was not new to anyone who had lived through the 1956-7 period. At that time, however, there had at once been the consolation that P- and C-violation could be embraced by a new and pretty concept, CP-invariance. What shook all concerned now was that with CP gone there was nothing elegant to replace it with. The very smallness of the $2\pi$ rate, CP-invariance as a near miss, made the news even harder to digest.

It so happened that during the same month of June I had begun preparing a rapporteur's talk on weak interactions for the Dubna conference to be held that August. I was of course quite content to have such a juicy item as CP-violation for inclusion in my report. I never made it to Dubna, however. Within weeks I had my hands full with SU(6). I was fortunate enough to find my good friend Sam Treiman willing to take over my Russian assignment.

I shall return in a while to the year 1964 but will next lay out what I plan to discuss in the rest of this lecture.

When I was invited, quite a while ago, to give this talk I felt I did not want to refuse this opportunity to pay my respects to good friends. In addition, le chateu de Blois vaut bien une messe. As time grew nearer I began to feel uncomfortable, however. Why should it be me to explain what the gang of four had done when they are right here in the audience? Also, much of the past material had already been extensively discussed, for example in Kabir's monograph on the CP-puzzle,[4] and in Sach's book on time reversal.[5] Review articles of high quality continue to appear periodically.[6] Conference reports are widely available. Even historians have tried to get into the act.[7] So what should I say at this time?

I decided that it might perhaps be of interest to backtrack a little and include remarks on the beginnings of K-particle physics and on the origins of discrete symmetries. I do that next, after which I shall move forward from 1964 on.

Accordingly I turn back to the years immediately following the Second World War, the time when experimental particle physics still fell under the more general heading cosmic ray physics. Those were the years when two British groups made major discoveries: the Blackett group in Manchester, and the Powell group in Bristol.

Blackett's first important work dates back to his Cavendish days in the early 1930s when he, a cloud chamber expert, joined forces with Occhialini who knew everything about Geiger counters. "According to Blackett, at that time the Geiger counter was a very delicate instrument. As he put it: 'In order to make it work you had to spit on the wire on some Friday evening in Lent.' They decided to pool their instrumental knowledge and investigate cosmic rays."[8] The result was the invention of the counter controlled cloud chamber, with which in 1933 they discovered the electron-positron formation process in cosmic ray showers. In 1937 Blackett was appointed professor in Manchester. He brought along a cloud chamber and an 11-ton magnet, with which he continued his cosmic ray research. That was the equipment with which the first neutral K particle was observed.

That, however, came to pass only after major interruptions caused by the second World War. In the summer of 1939 most of the U.K. members of Blackett's group were sent to work at a secret radar station. After six weeks it was realized that the station did not require highly trained physicists who were then assigned other war duties. One of these, George Dixon Rochester, was sent back to Manchester to teach honours students and cadets, run a University fire brigade, and help with civil defense. He had also sufficient free time to continue cosmic ray research.[9]

Immediately after the War Blackett returned to Manchester. With his blessings Rochester, together with "a bright young colleague," Clifford Charles Butler, "completely modernized the Blackett cloud chamber."[10]

That was the apparatus with which, on October 15, 1946 Rochester and Butler saw the first $K^o$ -- or $V^o$, as it was called then -- an inverted fork as seen looking upward. They could rapidly exclude conventional interpretations, such as scattering, electron-positron pair production, nuclear disintegrations. Here was something new, a two-body decay of a never seen particle.

Decay into what?

At this point it is quite significant to note that in October 1946 pions were yet to be discovered! We are still in the confused period when what now are called muons -- with mass known to be about 200 m (m=electron mass) -- were thought to be Yukawa's nuclear force mesons. All Rochester and Butler could conclude was that if the secondaries have mass 200 m then the primary has mass 870 ± 200 m.

"These results were not published immediately in the hope that other V particles would appear. Patience was rewarded ... on May 23, 1947."[9] A new particle was found, charged this time, a $K^+$ or $K^-$ decaying into $\pi^+$ or $\pi^- + \pi^o$, with mass 1080 ± 100 m, again assuming that the secondaries have mass 200 m.*

It was on the day after this discovery that the discovery of the charged pion by the Powell group appeared in print[13] -- but the neutral pion was not discovered until 1950![14] So we have an odd situation: neutral kaons are older than charged pions, charged kaons older than neutral pions. Rochester and Butler's paper on their two events finally appeared in print in December 1947.[15]

I must have heard of the V-particles in 1948 but do not recall their causing a stir at that time. Rochester later wrote: "The two years following 1947 were tantalising and embarrassing to the Manchester group because no more V-particles were found."[9] Nor, apparently, did others at once re-examine earlier cloud chamber pictures for V-particle evidence.[16] Years later I raised that issue with Wilson Powell, a cosmic ray expert

---

*The existence of a charged particle with mass about 1000 m had been claimed earlier.[11] However, "till today [that] event remains something of a puzzle."[12]

from Berkeley, asking him in particular why V-particles had not been discovered earlier as well they might have. He promised an answer by the following day. When next we met he had with him a handful of cloud chamber pictures he had taken in the thirties, each one showing either a fork or a kink. He told me that some time in the early fifties he had convinced himself that these were V-particles, adding with a smile that, because of the great interest in showers in the thirties, experimentalists would load their cloud chambers with many metal plates in order to obtain high particle multiplicities, and would pay little attention to individual tracks. Rochester and Butler had noticed their two events in a chamber containing only a single plate ...

As said, in 1948 there was hardly any reaction to the news. The discussion on cosmic rays at the Pocono Conference (30 March-1 April 1948) does not contain[17] any mention of V-particles. Rochester's own report[18] on cosmic ray work to the symposium in honor of Millikan's eightieth birthday does not include V-particles in the list of main results. When I recently queried George Rochester about this he replied: "It is true that there is meagre reference in the printed version ... I did, however, say more about the V-particles in my talk and I spoke privately with a number of the participants, notably, Anderson, Brode and Rossi."[19]

Then, later in 1948, Blackett received a letter[20] from Carl Anderson at Cal Tech who wrote:

"Rochester and Butler may be glad to hear that we have about 30 cases of forked tracks similar to those they described in their article in Nature about two years ago, and so far as we can see now their interpretation of these events as caused by new unstable particles seems to be borne out in our experiments."

The paper[21] by the Anderson group, published in May 1950 marked the turning point. Physicists now began to pay attention. My own active interest in the subject also began only after having read the Cal Tech paper.

There followed a rapid expansion of experimental activities. Alternative K decay modes were found. Hyperons were discovered. The best summary of those years are the proceedings of a cosmic ray conference held in July 1953 in Bagnères-de-Bigorre, in the French Pyrenees. It was at that meeting that the name K-particle was adopted. It is a real pity that those proceedings were never printed. (They do exist in dittoed form.)

During the closing remarks at Bagnères C. F. Powell said: "Gentlemen we have been invaded ... the accelerators are here."

In April 1953 two V-events were reported[22] found at Brookhaven's Cosmotron. At the 1955 Rochester Conference the good news[23] was that a dozen V's had been found at the Berkeley bevatron. That was just the beginning. Beam intensities at the accelerators have improved, energies have gone up, new detectors have been developed, beam builders have created electromagnetically separated beams with high enrichment factors. Today the AGS generates about $1,5 \times 10^{13}$ protons per pulse every 2 1/2 secs., with a 40% duty factor. A partially separated beam now produces $10^{10} K^+$ per 1000 hours. The Christenson-Cronin-Fitch-Turlay experiment lasted for 40 days and nights with over 200 hours net running time. According to my friend Nick Samios, the current director of BNL, one single pulse of the AGS in its present mode would suffice to yield the same number of $\pi^+\pi^-$ decays of $K_2^o$ as they found 25 years ago. So let us all pay our respects to the builders of machines, beams and detectors.

By the mid 1950s theorists had gotten into the act.

In 1952 the associated production mechanism was proposed as a means for reconciling the copious production of the new particles with their long life times.[24] In 1953 the strangeness scheme was put forward.[25] In 1954 the $\Delta I = 1/2$ rule for strangeness changing decay was suggested.[26] In 1955 it was shown that the strongly produced $K^o$, with strangeness $S = +1$, and $\overline{K^o}$, with $S = -1$, are not particles but particle mixtures.[27]

The reason given at that time was based on charge conjugation (C-) invariance. You know the argument. Consider the decay of a $K^o$, with strangeness $S = +1$ into $\pi^+\pi^-$. The operation C transforms the final state into itself but $K^o$ into $\overline{K^o}$ with $S = -1$. How can C transform the final but not the initial state into itself? It cannot do so as long as S is a good quantum number (strong interactions) but it can, and does, when S is not a good quantum number (weak interactions). Leave aside for a moment the small mixing $K^o \rightleftarrows \pi^+\pi^- \rightleftarrows \overline{K^o}$ and introduce new states

$$K_1 = \frac{K^o+\overline{K^o}}{\sqrt{2}}, \quad K_2 = \frac{K^o-\overline{K^o}}{\sqrt{2}} .$$

$K_1$ can decay into $\pi^+\pi^-$ but $K_2$ cannot, since the amplitudes for $K^o \to \pi^+\pi^-$ and $\overline{K^o} \to \pi^+\pi^-$ are equal. More generally $K_1(K_2)$ can only decay into states even(odd) under C. Thus $K_1$ and $K_2$ have different life times $\tau_1, \tau_2$. Since we should properly reserve the name "particle" for an object with a unique life time, $K_1$ and $K_2$ are true particles, $K^o$ and $\overline{K^o}$ particle mixtures. Prior to any experimental evidence for such behavior the consequences of such mixing for the e3 and μ3 modes of neutral K decay were analyzed by Zeldovich[28] and by Treiman and Sachs[29].

It had at once been suspected[27] that $\tau_2 >> \tau_1$ simply because the rates of $\pi^+\pi^-$- and $2\pi^o$-decay (both inaccessible to $K_2$) have to be so much larger than the rates for alternative modes. The first evidence fo a long lived neutral K was found in 1956 in Brookhaven by the Lederman group from Columbia, in one of the last cloud chamber experiments at accelerators.[30] In 1957 it was observed that no more than half of the particles in a beam consisting only of $K^o$'s can decay into two pions, another prediction[27] of particle mixing.

After I had given a seminar on this subject at Columbia University sometime early in 1955, Piccioni came up to talk to me. Out of our long discussion about the fate of a freshly created $K^o$ beam in the course of time grew the idea of regeneration.[32] You know the point. Follow a $K^o$ beam for a time long compared with $\tau_1$. It now consists almost entirely of $K_2$. Let it hit an absorber. The $K_o$ component of $K_2$ is selectively removed. The emergent beam has a restored $K_1$ component, about 1/4 the intensity of the original $K^o$ beam if the absorber is nearly $K^o$-transparent, hence $K_1$ decays should reappear. The effect was first observed in 1960[33], regeneration by $K_2$-diffraction was first seen in 1961.[34]

These observations illustrate how the neutral K-system serves as a testing ground, not only for particle properties but also for such quite general concepts as the quantum mechanical superposition principle. As Feynman later put it: "It is not based on an elegant hocus-pocus. ... We have taken the superposition principle to its ultimate logical conclusion ... It works."[35]

All that was fine and dandy. Meanwhile, however, the assumption of C-invariance, underlying particle mixing in its original version, had been found wanting. In late 1956 experiments revealed parity (P)- and C-violation in certain weak interaction processes. On January 16, 1957, the New York Times carried on its front page an article headlined: "Basic concept in physics is reported upset in tests/Conservation of parity in nuclear theory challenged by scientists at Columbia and Princeton Institute."

We had been prepared for the possibility of such violations ever since public discussions of this issue at the sixth Rochester conference (April 1956). For example, before the news became known, Treiman and I had submitted a paper[36] on $K_{e3}$ decay in which we had been careful to note that our results were independent of the validity of P-, C-, and T- (time reversal) invariance. I remember very well how nevertheless I was thoroughly shaken when I first heard the news. The face of physics had changed.

I shall not discuss here the marvelous experimental and theoretical contributions on C, P, and T made around 1956[37], but have thought it might be of some interest to this audience to say a few words about how these symmetries had entered physics in the first place.

T-invariance is the oldest known of these three principles. I do not know who stated it first, but it is implicit in Newton's mechanical laws. It is certain that it played a prominent role in the classical physics of the 1870's, to wit in the reconciliation of apparently irreversible thermodynamic principles (the second law) with the T-invariant laws of classical mechanics.* (In quantum mechanics unitarity actually suffices.) The extension of T-invariance to classical electrodynamics is due to Boltzmann who noted[39] in 1897 that the Maxwell equations remain unchanged "if one at a given moment leaves all electric forces and polarizations unchanged but reverses the time direction and the magnetic forces and polarizations," this in refutation of an argument to the contrary by Planck.[40]

T-invariance in quantum mechanics finds its roots in a paper of 1930 by Kramers, entitled "General theory of paramagnetic rotations in crystals,"[41] in which he proved his "degeneracy theorem": the energy states of an odd number of spin-1/2 particles are at least doubly degenerate in the absence of an external magnetic field. Kramers used specific dynamical considerations. In 1932 Wigner reduced the theorem to

---

*Sachs has noted that a theorem due to Painlevé in 1904 illustrates how T-invariance may be used directly in the solution of dynamical problems.[38]

its essence, a consequence of T-invariance.[42] The theorem exemplifies a general feature of T-invariance implications: these always involve relations between different states, never an intrinsic property of one state, quite unlike the situation for parity.

Next, parity. Early in 1924, Otto Laporte, a student of Sommerfeld, discovered[43] that the energy levels of iron atoms consist of two subsets that do not intercombine. Later this result turned out to be a special case of a general quantum mechanical rule according to which all spectral lines produced by electric dipole radiation are generated by transitions between atomic (or molecular) states of opposite parity. In 1924, quantum mechanics was not yet there, however, hence parity could not even be defined. Heisenberg, in his last paper[44] prior to his discovery of quantum mechanics, attempted to formulate a rationale for the 'Laporte rule.' That was still too early for a correct interpretation. In May 1927, two years after quantum mechanics, Wigner found the right answer.[45] He divided atomic states into 'normal terms' (we would say even-parity states) and 'reflected (gespiegelte) terms' (odd-parity) and noted that for electric dipole radiation only transitions normal ↔ reflected are allowed. In February 1928, he returned to this problem in his seminal paper 'On the conservation laws of quantum mechanics.'[46] There he introduced the parity opeator P, with eigenvalues, the parity quantum number or parity for short, equal to +1 or -1 and stated the law of parity conservation. He also commented: "Only rarely will one be able to use [parity] since it has only two eigenvalues and has therefore too little predictive power." He further stressed that parity has no analog in classical mechanics.

Thirdly, C. In 1934 Heisenberg had demanded "that the symmetry of nature with respect to positive and negative charges is expressed from the outset in the fundamental equations of the theory."[47] Three years later Kramers cast Heisenberg's requirement in the general form of invariance under charge conjugation, a term he introduced at that time.[48] The great novelty about C-invariance was that it can only be consistently formulated in quantum field theory.

Shortly before the discovery of P- and C-violation a theorem linking the operations P, C, and T had been come upon. This "CPT-theorem"[49] states that combined CPT invariance holds true under quite general conditions: (1) invariance under proper (no space reflections) orthochronous (no time reversal) inhomogeneous (space-time translations included) Lorentz transformations; (2) there exists a state of lowest (finite) energy; (3) the metric in Hilbert space is positive definite; and (4) distinct fields either commute or anticommute for space-like separations (this last condition can be slightly weakened). This theorem became acutely relevant once the violations of P and C had been seen. No violations of CPT have ever been found. The best test by far of this principle is once again provided by the neutral K's: The $K^o - \overline{K^o}$ mass difference, zero by CPT, is less than $10^{-18}$ times the $K^o$-mass. CPT is assumed to hold in what follows. It would mean a drastic revision of the foundations of quantum field theory if it were ever to occur.

Also the issue of CP invariance had been raised before the detection of C and P violation. In the course of their study of super selection rules, Wick, Wightman, and Wigner had remarked as early as 1952: "The disturbing possibility remains that C and P are both approximate and CP is the ony exact symmetry law."[50] Next, in 1956 when P-violation had become a distinct possibility but before its discovery, Landau[51] and Lee and Yang[52] had posed the question: suppose P is violated; is there a weaker but nevertheless elegant symmetry that could take its place? Yes, they answered, look in a mirror and flip the charge and all will be symmetric.

After this detour I return to the neutral K particles.

In 1957 it was noted[53] that CP invariance suffices to maintain the main features of the neutral K-system as initially discussed on the assumption of C invariance. Redefine $K_1(K_2)$ as states with CP = +1(-1) instead of C = +1(-1). Since zero angular momentum states $\pi^+\pi^-$ and $2\pi^o$ are even under P and C, they have CP = +1. Thus it remains true that $K_1$ can, $K_2$ cannot, decay into $2\pi$, so that the earlier argument for a long-lived $K_2$ continued to hold.

The switch from C to CP does affect other decay modes, however, As Triman and I noted[54], if C were good then $3\pi$-decays would be much more probable for $K_1$ than for $K_2$. If CP holds then the situation is just the opposite.

In 1957 and for a number of years thereafter CP-invariance seemed to work quite well. It was clear of course that the issue was an experimental one. The neutral K complex now acquired new significance as a testing ground for CP invariance. Was it indeed true that $K_2$ never decayed into $2\pi$? In 1961 a Dubna group reported[55] an upper limit of 0,3 percent for the branching ratio of $K_2 \to \pi^+\pi^-$. Another important test concerned the difference in rates of $K_2 \to \pi^-\ell^+\nu$ and $K_2 \to \pi^+\ell^-\nu$ ($\ell$=e or $\mu$). Any deviation from zero of this difference would violate C and T, hence CP by the CPT theorem.[56] By 1964 experiment showed no difference from zero.[57] Nor had earlier tests[58] for T-violation using polarized neutrons shown any effect.

Looking back on these years, Cronin reminisced in 1982: "Even though parity had been overthrown a few years before, one was quite confident about CP-symmetry."[59]

That was the situation in early 1963, when Ronald Rau, secretary of the hgigh energy advisory committee at Brookhaven, received a research proposal of which I copy the first few lines.

PROPOSAL FOR $K^o_2$ DECAY AND INTERACTION EXPERIMENT

J. W. Cronin, V. L. Fitch, R. Turlay

(April 10, 1963)

I. INTRODUCTION

The present proposal was largely stimulated by the recent anomalous results of Adair et al., on the coherent regeneration of $K^o_1$ mesons. It is the purpose of this experiment to check these results with a precision far transcending that attained in the previous experiment. Other results to be obtained will be a new and much better limit for the partial rate of $K^o_2 \to \pi^+ \pi^-$, a new limit for the presence (or absence) of neutral currents as observed through $K_2 \to \mu^+ + \mu^-$. In addition, if time permits, the coherent regeneration of $K_1$'s in dense materials can be observed with good accuracy.

II. EXPERIMENTAL APPARATUS

Fortuitously the equipment of this experiment already exists in operating condition. We propose to use the present $30^{o}$ neutral beam at the A.G.S. along with the di-pion detector and hydrogen target currently being used by Cronin, et al. at the Cosmotron. We further propose that this experiment be done during the forthcoming μ-p scattering experiment on a parasitic basis.

The di-pion apparatus appears ideal for the experiment.

. . . . . . . . . . . . . . . . .

I should explain.

"Adair et al." refers to a Brookhaven-Yale hydrogen bubble chamber experiment on the regeneration of $K_1$'s by a $K_2$-beam. The effect observed was "far greater than that anticipated from conventional mechanisms."[60] A new coherent regeneration mechanism was conjectured resulting from a new weak long range proton-K interaction.

Fitch has recalled how "In view of my previous experience with K-mesons I responded to this news like a fire horse at the sound of a bell. And my colleague Cronin and I had many discussions about this over lunch tables at Brookhaven."[61] There is no doubt that this anomaly was the chief motivation for their experiment. In fact half of their about 140,000 recorded events were of the regeneration type, in carbon, copper, and lead. Another 23000 were regenerations in hydrogen. One of their good but less remembered results[62] was that regeneration was smaller by a factor 15 than the Brookhaven-Yale data implied.

The dipion detector mentioned in the proposal refers to a spark chamber arrangement used in an experiment on dipion production in π-p collisions, performed at the Cosmotron in Brookhaven in the winter and spring of 1963 by Clark, Christenson, Cronin, and Turlay.[63] The main interest was in the region of production of ρ mesons which decay into 2π. "The modifications to convert from ρ mesons to neutral K mesons were minor."[61]

Receipt of the research proposal was acknowledged on April 12. On May 6 Rau wrote to Cronin and Fitch: "Dear Jim and Val, The High Energy Advisory Committee has asked me to inform you that they have approved your experiment ... for approximately 200 hours."[64] Because of shutdowns the allotted time had to be used between May 27 and July 22. "We immediately moved the ρ-spectrometer from the cosmotron and started to debug the gear on June 2 at the A.G.S."[61]

They used their time like this. First, regeneration in heavy metals. Then the regenerator material was removed and replaced by a helium bag for the "CP invariance run" which started June 20 and took about 100 running hours. Then regeneration again, in hydrogen. Analysis of their photographs began shortly after the end of July, when the run was over. Cronin has told me that they did not look at the CP film until January 1964, that it was definitely a lower priority item, and that, when they finally did look, "2π decays stared us in the face." Fitch has said: "What we saw was enough to perk up the adrenalin supply."[61] They had obtained $45 \pm 10$ $K_2$ decays into $\pi^+\pi^-$.

Next, in Fitch's words: "It was now time to kill the effect in every conceivable way."[61] They took long and hard looks at all possible ways that might have fooled them but "these data survived. After almost six months of this examination we published. And I must say we were totally confident that the data would survive the critical look they were going to get."[61]

On August 6 the New York Times, apparently aware of the CPT theorem, caried on page 68 a news item entitled: "High energy physics experiment finds time reversal may affect physics laws." The face of physics had changed again.

What were the early responses?

Fitch's characterization was: "This discovery was not expected, it was not wanted when found, and the early explorations were directed to finding a way through and around it."[61] I have no recollections of expressions of doubt about the quality of the experiment. The perpetrators were widely regarded as real pros.

The best way I can think of introducing my own comments on the early reactions is to display the following plot prepared by Cronin in 1965.

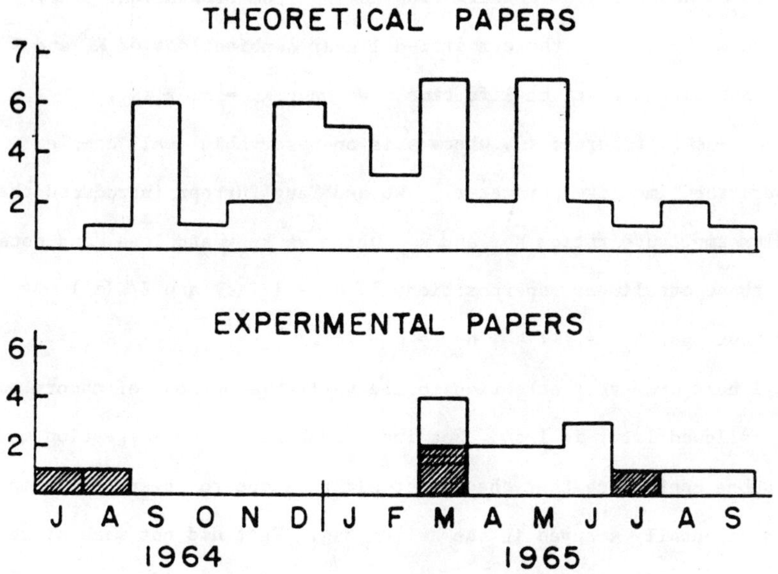


Fig. 1. Cronin's plot[65] of the time distribution of theoretical and experimental papers on CP-violation. Cross hatched experiments are those specifically measuring $K_2^o \rightarrow 2\pi$.

The July 1964 paper is the one that started it all. In August the Abashian group from Urbana reported[66] on an experiment which carried "at least a suggestion that CP may actually be violated" by a small amount. It took until March 1965 before the first confirmations of $2\pi$ decay of $K_2$ were reported, at Cern[67] and at Harwell[68] -- which goes to show that these experiments were no kid stuff.

The contents of the single and important theoretical paper from August is taught in every CP-violation 101 class. It is by Wu and Yang[69] and is yet another 1964 contribution coming from Brookhaven. They introduced $K_S$ and $K_L$, those modified linear combinations of $K^o$ and $\overline{K^o}$ that each have a distinct life time. We have $K_S = K_1 + \varepsilon K_2$, $K_L = K_2 + \varepsilon K_1$, to order $\varepsilon^2$, where $\varepsilon$ is an observable small complex number, the "impurity parameter." Wu and Yang further introduced the complex amplitude ratios $\eta_{+-}$ and $\eta_{oo}$ which we know and love, and noted that these are linear superpositions of $\Delta I = 1/2(\varepsilon)$ and $3/2(\varepsilon')$ contributions, $\eta_{+-} = \varepsilon + \varepsilon'$, $\eta_{oo} = \varepsilon - 2\varepsilon'$.

I next draw your attention to the veritable outpour of theory papers that followed later in 1964. Not included here is the suggestion made at the Dubna conference that the effect might be due to regeneration in a fly accidentally trapped in the helium bag. That did not wash since the fly's density would have to be far in excess of that of uranium.[70]

The papers in question reflect a wide-spread unease initially caused by the new effect. They include a variety of efforts at explaining it in a CP invariant way. A list of such inevitably exotic though, I wish to stress, by no means frivolous alternatives includes the introduction of new very weakly-coupled long-range fields; of <u>ad hoc</u> new particles; of a shadow universe coupled to ours by weak interactions; of deviations from the exponential decrease of decay rates with time; of deviations from the quantum mechanical superposition principle: and others.[71] Most of these proposals could be experimentally disproved; others faded away. CP-violation has turned out to be inevitable.

One paper in this 1964 group is of a quite different nature. That is Wolfenstein's suggestion from August[72] that CP-violation is attributable to a "superweak" interaction, about $10^{-9}$ times weaker than weak, that allows strangeness changes equal to 2. That idea "has never been regarded as more than a phenomenological stopgap pending a more comprehensive theory."[73] The superweak description makes one important prediction, however, to wit that $\varepsilon' = 0$.

Having related at some length how experimental and theoretical beachheads in this new territory were established, I now turn to later developments. Since these are the subject of this conference, I shall be quite brief, and in particular will leave the quotation of numbers to others more expert. Here then is my short list of highlights. First, experiment.

1) In 1967 the first evidence for CP-violating charge asymmetries was reported, from Stanford[74] for $K_{\mu 3}$, from Columbia[75] for $K_{e3}$.

2) The first reports on the detection of $K_L \rightarrow 2\pi^o$ also date from that year.[76] In his 1980 Nobel lecture[70] Cronin noted that by then he had already spent six years of his professional life in pursuit of that decay mode.

3) In 1970 Cern data[77] on $K_L - K_S$ interference were analyzed without imposing CPT invariance from the outset, resulting in direct evidence for T-violation.[78]

4) Already in 1957 Landau had noted[51] that the neutron could have an electric dipole moment if and only if P and T are violated. The search for such a moment has gone on for many years now, mainly by a Leningrad group and by an international consortium directed by Ramsey. Their last

published papers, dating from 1986[79] and 1984[80] respectively, report no evidence for the existence of such a moment. Precision has improved meanwhile, so a month ago I called Ramsey to check on the present status. He gave me his presently known upper bounds, adding that there still was no conclusive evidence for a dipole moment. At this conference Ken Smith will tell us in technical detail about the current state of affairs.

5) Since 1964 more than twenty years of hard labor had seemed to indicate that all CP data could be reduced to the single complex number $\varepsilon$, in other words that $\varepsilon' = 0$. Then, in the spring of 1988, came forth a claim from Cern[81] for a non-zero $\varepsilon'$, while Fermilab[82] reported that there $\varepsilon'$ measurements had much improved. In the next few days we shall undoubtedly hear more on this subject.

As to theoretical developments:

1) In 1967 Sakharov pointed out[83] that an initially baryon-symmetric universe might turn asymmetric provided there is a baryon-nonconserving interaction, C and CP violation, and a deviation from thermal equilibrium. This may explain why after baryon-antibaryon annihilations the currently observed one baryon per $10^{10}$ photons survives. His seminal paper marks the beginning of the interplay between particle physics and cosmology, so actively pursued these days.[84] This is yet another topic about which we shall hear more later.

2) In 1975 Treiman and I noted[85] that decays of the charmed $D^o$, $\overline{D^o}$ need not be a repeat performance of the kaon story, since here there are many decay options both for $D_1$ and $D_2$, and since there is no single dominant channel like $2\pi$ for neutral kaons. Life times of $D_1$ and $D_2$ are probably

comparable. It was also noted that the accessibility of many decay channels provides new CP tests, the comparison of decay rates in pairs of exclusive conjugate reactions for charged D's. Similar remarks apply to B mesons, discovered later. Subsequent more detailed theoretical analyses indicate that CP effects in the neutral D-system will be small.[86] By contrast the B-case looks very promising. Quite recent experiments at ARGUS[87], confirmed at CLEO[88], give evidence for considerable $B^o - \overline{B^o}$ mixing. That is most encouraging for CP-effects in B decays, which have not as yet been seen, however.

3) What have theorists contributed in the course of 25 years toward a quantitative understanding of CP-violation?

Nothing.

We have understood that there are several ways of accommodating the effect within the framework of gauge theories: either by extending the Higgs system of the standard model; or by extending the standard gauge group; or -- and that is the prettiest result -- the Kobayashi-Maskawa way, where one notes[89] that for the standard model with more than two quark doublets there is room for CP violation in the quark mass matrix.[90] None of the above is of any help, however, in explaining why CP invariance is such a near miss.

Superstrings are no help either. It is true that, as I am told, compactification to a Calabi-Yao manifold leaves room[91] for generic CP violation, but quantitative predictions are lacking -- as in fact they are for any physics question you may ask stringers.

Much new physics has emerged since 1964. Think of the experimental discoveries of hard deep inelastic lepton scattering, of the discoveries of the $J/\Psi$, the upsilon, the $\tau$-lepton, jet events, the W and Z. I would not say we have fully digested all these facts but they do fit neatly into the ongoing dialog between theory and experiment.

Not so with CP.

Have we ever seen anything analogous in physics? Once, perhaps. In 1859 Leverrier wrote of his discovery of a tiny anomalous precession of the perihelion of Mercury, "cui theoriae lumen nundum accesserit," on which no light has been thrown. There it lay, discussed yes, but not understood, until about sixty years later Einstein would explain it from general relativity "without the need [he wrote] of any special [further] hypothesis."[92] That was not a frontal attack on a tiny effect but a byproduct, one might say, of a fundamentally new world view. Is CP somehow not amenable to frontal attack either?

Finally, what are the prospects?

We pray for the second coming of CP violation, in the dipole moment of the neutron, or the decays of heavier neutral mesons, preferably in both.

We pray for a quantitative theory of CP violation. How long will that take? Who can say?

Recently I read a popular best seller -- I found it neither popular nor best -- in which it was said that we are near to finding out God's design for the physical universe. That attitude reminds me of the well-known story of the little bird that almost froze to death in the Russian steppes. My own views coincide with what Cronin has said[70] some years ago:

"We must continue to seek the origin of the CP symmetry violation by all means at our disposal. We know that improvements in detector technology and quality of accelerators will permit even more sensitive experiments in the coming decades. We are hopeful, then, that at some epoch, perhaps distant, this cryptic message from nature will be deciphered."

References


1. For the history of AdA see C. Bernardini, Scientia 113, 27, 1978.
2. V. E. Barnes et al. Phys. Rev. Lett. 12, 204, 1964.
3. F. Gürsey and L. A. Radicati, Phys. Rev. Lett. 13, 173, 1964; A. Pais, ibid. 13, 175, 1964; F. Gürsey, L. A. Radicati and A. Pais, ibid. 13, 299, 1964.
4. P. K. Kabir, "The CP puzzle," Ac. Press, New York 1968.
5. R. G. Sachs, "The physics of time reversal," Univ. of Chicago Press 1987.
6. The most recent one is by L. Wolfenstein, Ann. Rev. Nucl. Ptcle Sci. 36, 137, 1986.
7. A. Franklin, Hist. St. Ph. Sci. 13, 207, 1983.
8. J. L. Heilbron, summary of interview with P. M. S. Blackett, December 17, 1962, copy in N. Bohr Library, Am. Inst. of Phys. New York.
9. Here I quote, in part verbatim, from G. D. Rochester, "The earliest history of strange particles," in "Early history of cosmic ray studies," Y. Sekido and H. Elliot Eds., Reidel, Boston 1985.
10. G. D. Rochester, Coll. hist. particle phys., J. de Phys. 43, suppl. to no. 12, p. 169, 1982.
11. L. Leprince-Ringuet and M. Lhéritier, Compt. Rend. Paris 219, 618, 1944; see also Rev. Mod. Phys. 21, 42, 1949.
12. Ch. Peyron, ref. 10, p. 7.
13. C. M. G. Lattes et al. Nature 159, 694, 1947.

14. See J. Steinberger, W. K. H. Panofsky, and J. Steller, Phys. Rev. 78, 802, 1950.

15. G. D. Rochester and C. C. Butler, Nature 160, 855, 1947.

16. See however ref. 15, footnotes 3 and 4.

17. Dittoed notes of the Pocono Conference, unpublished.

18. G. D. Rochester, Rev. Mod. Phys. 21, 20, 1949.

19. G. D. Rochester, letter to A. Pais, November 25, 1984.

20. C. D. Anderson, letter to P. M. S. Blackett, November 28, 1948, repr. in ref. 9.

21. A. J. Seriff et al. Phys. Rev. 78, 290, 1950.

22. W. B. Fowler et al. Phys. Rev. 90, 1126, 1953; 91, 1287, 1953.

23. G. Goldhaber, Proc. 5th Rochester Conf., p. 104, Interscience, New York 1955.

24. A. Pais, Phys. Rev. 86, 663, 1952.

25. M. Gell-Mann, Phys. Rev. 92, 833, 1953; T. Nakano and K. Nishijima, Progr. Theor. Phys. 10, 581, 1953.

26. M. Gell-Mann and A. Pais, Proc. Glasgow Conf. on nuclear and meson phys., p. 342; E. H. Bellamy and R. G. Moorhouse Eds., Pergamon, Oxford 1955.

27. M. Gell-Mann and A. Pais, Phys. Rev. 97, 1387, 1955.

28. Y. B. Zeldovich, JETP 30, 1168, 1956.

29. S. B. Treiman and R. G. Sachs, Phys. Rev. 103, 1545, 1956.

30. K. Landé et al., Phys. Rev. 103, 1901, 1956.

31. F. Eisler et al., Nuov. Cim. 5, 1700, 1957.

32. A. Pais and O. Piccioni, Phys. Rev. 100, 1487, 1955.

33. F. Muller et al., Phys. Rev. Lett. 4, 418, 1960.

34. R. H. Good et al., Phys. Rev. 124, 1221, 1961.

35. R. P. Feynman, "Theory of fundamental processes," p. 50, Benjamin, New York 1961.

36. A. Pais and S. B. Treiman, Phys. Rev. 105, 1616, 1957.

37. For some of my impressions of that time see A. Pais, Nucl. Phys. 5, 297, 1958.

38. Ref. 5, p. 14.

39. L. Boltzmann, Berl. Ber. 1897, p. 660, 1897, p. 1016.

40. M. Planck, Berl. Ber. 1897, pp. 55, 715.

41. H. A. Kramers, Proc. Kon. Ak. Wet. Amsterdam 33, 959, 1930; repr. in H. A. Kramers, "Collected scientific papers," p. 522, North-Holl. Cy, Amsterdam 1956.

42. E. P. Wigner, Goett. Nachr. 1932, p. 546.

43. O. Laporte, Zeitschr. f. Phys. 23, 135, 1924.

44. W. Heisenberg, Zeitschr. f. Phys. 32, 841, 1925, section 5.

45. E. P. Wigner, Zeitschr. f. Phys. 43, 624, 1927, sections 15, 25.

46. E. P. Wigner, Goett. Nachr. 1927, p. 375.

47. W. Heisenberg, Zeitschr. f. Phys. 90, 209, 1934.

48. H. A. Kramers, Proc. Kon. Ak. Wet. Amsterdam 40, 814, 1937, repr. in Collected papers (ref. 41), p. 697.

49. For the history and proofs of the theorem see R. Jost, in "Theoretical Physics in the twentieth century," p. 107, M. Fierz and V. Weisskopf Eds., Interscience, New York 1960; R. F. Streater and A. S. Wightman, "PCT, spin and statistics, and all that," Benjamin, New York 1964.

50. G. C. Wick, A. S. Wightman, and E. P. Wigner, Phys. Rev. 88, 101, 1952.

51. L. D. Landau, JETP 32, 405, 1957.

52. C. N. Yang, Rev. Mod. Phys. 29, 231, 1957.

53. L. D. Landau, Nucl. Phys. 3, 127, 1957; R. Gatto, Phys. Rev. 106, 168, 1957.

54. A. Pais and S. B. Treiman, Phys. Rev. 106, 1106, 1957.

55. D. Neagu et al., Phys. Rev. Lett. 6, 552, 1961.

56. T. D. Lee, R. Oehme and C. N. Yang, Phys. Rev. 106, 340, 1957; H. W. Wyld and S. B. Treiman, Phys. Rev. 106, 169, 1957; also R. G. Sachs and S. B. Treiman, Phys. Rev. Lett. 8, 137, 1962; R. G. Sachs, Phys. Rev. 129, 2280, 1963.

57. D. Luers et al., Phys. Rev. 133, 1276, 1964.

58. M. A. Clark et al., Phys. Rev. Lett. 1, 100, 1958; M. T. Burgy et al., Phys. Rev. Lett. 1, 324, 1958.

59. J. Cronin, Phys. Today 35, July 1982, p. 38.

60. L. B. Leipuner et al., Phys. Rev. 132, 2285, 1963.

61. V. L. Fitch, in "Symmetries in physics (1600-1980)," p. 452, Proc. Conf. at San Felice de Guixols, Spain, Sept. 1983, M. Doncel chief ed. Univ. of Barcelona, 1987.

62. J. H. Christenson, J. W. Cronin, V. L. Fitch, and R. Turlay, Phys. Rev. Lett. 13, 138, 1964.

63. A. R. Clark, J. H. Christenson, J. W. Cronin, and R. Turlay, Phys. Rev. 139B, 1556, 1965.

64. The letters of April 10, 12 and May 6 are reproduced in full in ref. 61.

65. J. W. Cronin, Proc. Int'l. Conf. Weak Ints. at Argonne Lab, October 25-27, 1965, p. 17.

66. A. Abashian et al., Phys. Rev. Lett. 13, 243, 1964.

67. X. de Bouard et al., Phys. Lett. 15, 58, 1965.

68. W. Galbraith et al., Phys. Rev. Lett. 14, 383, 1965.

69. T. T. Wu and C. N. Yang, Phys. Rev. Lett. 13, 380, 1964.

70. J. W. Cronin, Rev. Mod. Phys. 53, 373, 1981.

71. For a review see J. Prentki, Proc. Int'l. Conf. on Elem. Ptcles, Oxford 1965, p. 47, Rutherford Laboratory 1966; also ref. 7.

72. L. Wolfenstein, Phys. Rev. Lett. 13, 562, 1964.

73. B. Schwarzschild, Phys. Today, October 1988, p. 17.

74. J. Dorfan et al., Phys. Rev. Lett. 19, 987, 1967.

75. S. Bennett et al., Phys. Rev. Lett. 19, 993, 1967.

76. J. M. Gaillard et al., Phys. Rev. Lett. 18, 20, 1967; J. W. Cronin et al., Phys. Rev. Lett. 18, 25, 1967.

77. J. C. Chollet et al., Phys. Lett. 31B, 658, 1970.

78. K. R. Schubert et al., Phys. Lett. 31B, 662, 1970.

79. I. S. Altarev et al., JETP Lett. 44, 460, 1986.

80. J. M. Pendlebury et al., Phys. Lett. 136B, 327, 1984.

81. H. Burkhardt et al., Phys. Lett. 206, 169, 1988.

82. M. Woods et al., Phys. Rev. Lett. 60, 1695, 1988.

83. A. D. Sakharov, JETP Lett. 5, 24, 1967.

84. For a review see E. W. Kolb and M. S. Turner, Ann. Rev. Nucl. Part. Sci. 33, 645, 1983.

85. A. Pais and S. B. Treiman, Phys. Rev. D12, 2744, 1975; also L. B. Okun, V. I. Zakharov, and B. M. Pontecorvo, Lett. Nuov. Cim. 13, 218, 1975.

86. A. C. Carter and A. I. Sanda, Phys. Rev. D23, 1567, 1981.

87. H. Albrecht et al., Phys. Lett. 192B, 245, 1987.

88. A. Jawahery, Proc. 24th High Energy Int'l Conf. Munich, August 1988, p. 545, Springer, New York 1989.

89. M. Kobayashi and T. Maskawa, Progr. Theor. Phys. 49, 652, 1973.

90. For more details on these three methods see ref. 6.

91. See A. Strominger and E. Witten, Comm. Math. Phys. 101, 341, 1985; M. B. Green, J. Schwarz, and E. Witten, "Superstring theory," Vol. 2, p. 511, Cambridge Univ. Press 1987.

92. For more on the perihelion of Mercury see A. Pais, "Subtle is the Lord," p. 253 ff. Oxford Univ. Press 1982.

*cake offered by Lenôtre-Paris. Height : 3 meters - Weight : 120 kg.*

# CP-Nonconservation – The Early Experiments

Robert K. Adair
Yale University

**Abstract**

The history of the early experiments relevant to the discovery of CP-invariance violation, as well as the revolutionary experiment of the Princeton group of Christenson, Cronin, Fitch, and Turlay, are discussed here from a personal view. All of this work was conducted at Brookhaven National Laboratory when I was closely associated with the laboratory. In particular, I recount the discovery of the $K_L^0$ by Chinowsky, Lederman, Lande, Booth, and Impeduglia; I discuss the anomaly in $K_L^0 \to \pi^+ + \pi^-$ transitions in hydrogen reported by Leipuner, Chinowsky, myself and others which excited special interest in the character of $K_L^0$ transitions; and I review the Illinois experiments of Abashian, Smith, Wattenberg, and others that saw evidence of CP-violation soon after the Princeton results were known.

## Prologue

As I understood the charge given me by the Organizing Committee, I was to talk this morning only about the precursor – and postcursor – experiments related to the discovery of CP-violation. I assumed then that a review of the remarkable central experiment of the Princeton group – the experiment that we are celebrating – would be covered by others. Hence, I saw my talk as a kind of parallel to a possible talk by an Art Historian who was to devote a lecture to De Vinci's treatment of the backgrounds and clothing in his Mona Lisa – leaving the discussion of the face and smile to others. To those who love art and the history of art, such a talk may not be without interest. Similarly, I supposed that you who have come here to celebrate the 25th anniversary of the detection of the flaw in CP invariance at Brookhaven, might find my reminiscences of related experiments in that golden age of Brookhaven National Laboratory, interesting also. However, since no speaker actually discussed the character of the Princeton experiment in any detail, I have added in this written document, such a description – a description of the face and enigmatic smile of the portrait to complete my description of the landscape that forms the Mona Lisa background.

With rare exceptions science is evolutionary rather than revolutionary and even the revolutions, such as CP non-invariance, have evolutionary precursors and postcursors. I have chosen to talk, primarily, about three such experiments; one, the discovery of the $K_L^0$ meson (then $K_2^0$) by Willy Chinowsky, Leon Lederman, and others (a major discovery in itself); two, a curious precursor measurement by Leipuner, Chinowsky, myself and others that suggested anomalous transitions of the $K_L^0$ to two pions; and three, a most interesting experiment – or program – by the University of Illinois group of Abashian, Smith, Wattenberg and others that was both precursor and postcursor to the Princeton Experiment. The first two measurements were conducted at the BNL Cosmotron, arguably the first truly high energy accelerator, while the Illinois and Princeton experiments took place at the BNL AGS, in concept at least, the first of the strong focusing machines.

## The Discovery of the $K_L^0$

Since the only solid evidence we have even now of CP-invariance violation stems from observations of nature made with the marvelous vernier of the neutral $K$ system, it seems right and proper that I begin with the discovery of the $K_L^0$ predicted by Gell-Mann and Pais.[1] Chinowsky, then at BNL, and Lederman, Lande, Booth, and Impeduglia noted that the long-lived neutral $K$ sibling should be observable and proceeded to look for it late in 1955.

---

[1] M. Gell-Mann and A. Pais, *Phys. Rev.* **97**, 1387 (1955) After it was discovered that neither $P$ nor $C$ were conserved, the prediction of Pais and Gell-Mann of a short- and a long-lived neutral $K$-meson where the long-lived particle would not decay to two pions, which depended upon C-conservation, was extended by Lee, Oehme, and Yang to hold also if CP were conserved; T.D. Lee, R. Oehme, and C.N. Yang, *Phys. Rev.* **106**, 340 (1957). These authors noted also that one might expect a long-lived $K^0$ *even if CP were not conserved* but that the longer lived sister would then occasionally decay into two pions.

**Observation of Long-Lived Neutral *V* Particles***

K. Lande, E. T. Booth, J. Impeduglia, and L. M. Lederman,
*Columbia University, New York, New York*
and
W. Chinowsky, *Brookhaven National Laboratory, Upton, New York*
(Received July 30, 1956)

To search for these particles, the BNL-Columbia group placed their 36" diameter expansion chamber, set in a magnetic field, in a neutral beam far from a copper target bombarded by 3 GeV protons from the Cosmotron. They then looked for $K^0$-decays in the cloud chamber. The chamber was actually outside of the Cosmotron building in a temporary galvanized iron cylindrical structure, similar to structures called 'cribs' used by American midwestern farmers to store corn ('maize' in British English), and was then known as the 'corn-crib'. The diagram of Figure 1 shows the experimental setup.

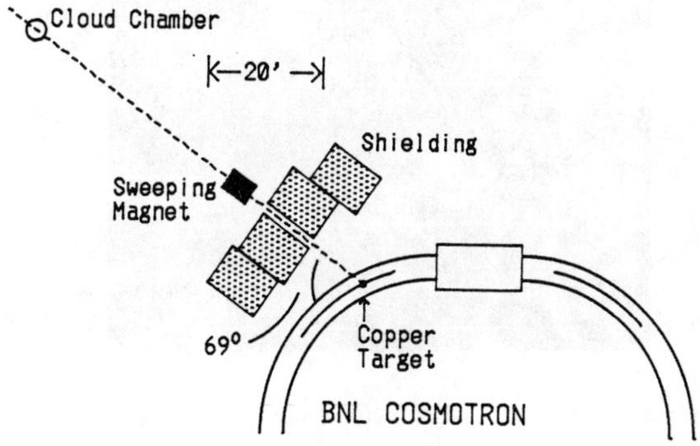


Figure 1: Experimental arrangement for the BNL-Columbia discovery of the $K_2^0$ at the BNL Cosmotron.

The decays were to be known by their characteristic $V$-shaped track in the chamber. But the pictures of the expanded chamber showed hundreds of fine tracks from the general background generated by the Cosmotron. Could a $V$-shape track be identified in that sea of background? Yes! In their initial publication, they described 23 such tracks.[2] But the recognition was not easy.

At BNL Willy Chinowsky had set up his picture-scanning apparatus near an experiment Leipuner, Larsen, and I were running at the Cosmotron. (It seems that the only room Willy could find to work was on the Cosmotron floor.) One afternoon, Willy brought me over to look at the second decay he had found. I thought that he had a vivid imagination as, even after searching a few minutes, I saw nothing. But I did see it after – but only after – Willy pointed it out to me. And then it was beautiful. I recall that in the initial scanning conducted at Columbia and BNL, the group was encouraged that both scanning groups had found 4 events in a perusal of a roll of film. But they were somewhat less enthusiastic when

[2] K. Lande, E.T. Booth, J. Impeduglia, L.M. Lederman, and W. Chinowsky, *Phys. Rev.* **103**, 1901 (1956)

they found that they had each seen four different events for a total of eight. But the difficult scanning problem was solved and they reported 23 decays in their initial publication – and 186 eventually. Moreover, not one these events was kinematically consistent with a $K^0 \to 2\pi$ decay and CP-invariance violation.

The picture shown in Figure 2 shows a typical decay from a second experiment.[3] in which one can clearly see the the $K_2^0$ decay tracks. The decay is of the type $K_L^0 \to \pi^+ + \pi^- + \pi^0$; the $\pi^+$ is marked $P^+$ in the photograph.

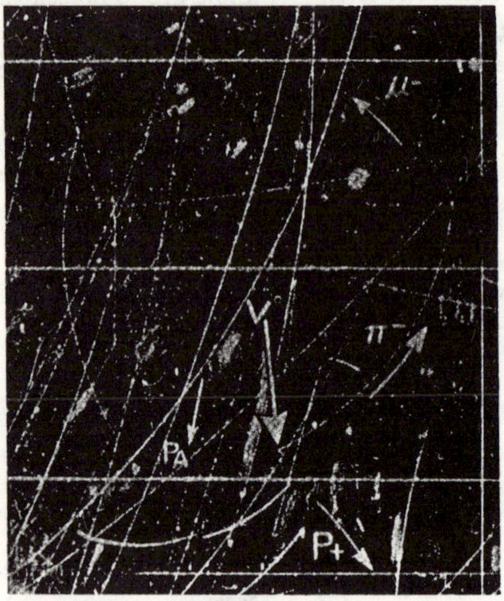


Figure 2: A picture from the Columbia-BNL cloud chamber experiment in which one can see a typical $K_L^0$ decay at the point indicated by the arrow.

Eventually, after a second run, a total of 186 decays were identified. And, strikingly, none could be fit kinematically under the assumption that the $K_L^0$ decayed to two pions. Clearly, CP was conserved to a high degree.

**DECAY PROPERTIES OF $K_2{}^0$ MESONS***

D. Neagu, E. O. Okonov, N. I. Petrov, A. M. Rosanova, and V. A. Rusakov
Joint Institute of Nuclear Research, Moscow, U. S. S. R.
(Received April 20, 1961)

A few years later a JINR cloud chamber group expanded[4] on the evidence that the $K_L^0$ rarely, if ever, decayed to two pions with a similar experiment where they saw 225 decays of which none fit the assignment $K^0 \to 2\pi$. From the two experiments, it seemed CP must then be conserved in the $K$-decays at least to the degree illustrated by a $2\pi$ branching ratio not much larger than 1/400.

Playing the game of *If*; if these groups had been able to collect four times as many events, they should have seen three or four $K_L^0 \to \pi^+ + \pi^-$ events and they might have discovered the violation of CP-invariance.

[3]M. Bardon, K. Lande, L.M. Lederman, and William Chinowsky, *Ann. of Phys.*, **5**, 156 (1958)

[4]D. Neagu, E.O. Okonov, N.I. Petrov, A.M. Rosanova, and V.A. Rusakov, *Phys. Rev. Letts.* **6**, 552 (1961)

# Anomalous $K_L^0 \rightarrow \pi^+ + \pi^-$ Events

## Anomalous Regeneration of $K_1^0$ Mesons from $K_2^0$ Mesons*

L. B. Leipuner, W. Chinowsky,† and R. Crittenden
*Brookhaven National Laboratory, Upton, New York*

and

R. Adair,‡ B. Musgrave,§ and F. T. Shively†
*Yale University, New Haven, Connecticut*
(Received 13 March 1963; revised manuscript received 27 August 1963)

About 1956 Larry Leipuner, Rich Larsen, and I built a liquid hydrogen bubble chamber 14" in diameter and 8" deep that operated in a magnetic field of about 17 kgauss. At the time the chamber first ran successfully, only Luis Alvarez's 15" chamber was larger – though very shortly the BNL 20" chamber came on line and soon Luis changed the whole game with his 72" chamber. In those dear, departed days of highly personal physics, Larry and Rich and I designed and built the bubble chamber, designed and built the measuring machines, wrote the programs that handled the data, designed and ran the experiments – and even produced the phenomenological theory that purported to explain our results. (Regrettably, in retrospect – except possibly for Larry's remarkable measuring machines that had no more descendants than ancestors – we did none of this as superbly well as the larger groups; but we had a good time.)

In 1960 Larry and I became interested in putting the chamber in a neutral beam where we might reasonably expect to accept much larger fluxes than from the charged beams we had been using and see a larger density of interesting events. In discussing this with Chinowsky, Willy pointed out that the neutral beams we were considering should be rich in $K_L^0$ mesons and agreed to join us in an experiment designed to look at interactions and decays of neutral particles – primarily K-mesons – in the bubble chamber. We were joined in this by Ray Crittenden, a post-doc from Wisconsin (now at the University of Indiana), Brian Musgrave, a Yale post-doc from Birmingham (now at Argonne), and Frank Shively my first Yale graduate student.

The experiment was simple; we placed the chamber in a neutral beam at the Cosmotron produced in the forward direction by the interactions of an external 3 GeV proton beam on a steel target. Figure 3 shows an artist's (RKA) view of what chamber and events were to look like. I do not include the "paintings" of aurochs and sabre-tooth tigers we scrawled on the walls of the concrete cave enclosing the chamber nor the inscribed slogans "Yanqui go home!" that amused a group of Soviet visitors in those pre-perestroika days.

We took about 90,000 pictures in the neutral beam in the spring of 1961 and then we looked at the film – or I looked at the film (along with our professional scanners). Indeed, for reasons that I cannot wholly recollect but had something to do with the fact that everyone else found more useful things to do preparing for our next experiment, I did a large part of the scanning myself.

In the course of that scanning, I seemed to see too many events that looked to me like $K_L^0 \rightarrow 2\pi$ decays. (Larry tells me that he remembers me saying almost every day, "I found another one.") I now suspect that there was an element of illusion involved, as such decays are hardly self-evident but it did trigger our interest in that mode and led to our decision to run all of the decays through the programs with the $2\pi$ identity assignment. Such a decision

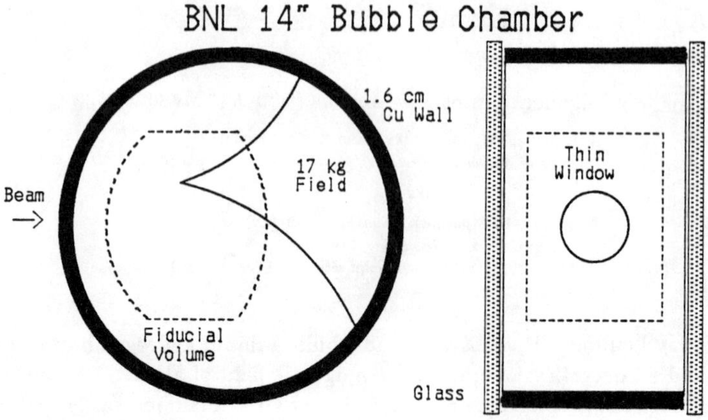


Figure 3: An artist's view of the chamber and a decay event. The dashed lines show the limits of the chamber fiducial region.

constitutes a real intellectual jump since we well knew that almost all of the events we were formally treating as $2\pi$ decays must, in fact, have at least one track that was not a pion.

When all the decay events were analyzed as $2\pi$ decays, the angular distribution showed a sharp peak (of 19 events of a total of 252 charged decays) in the forward direction as shown in Figure 4. In the section of the paper[5] devoted to the interpretation of the data (a section that I wrote), it says that the probability of the peak being due to chance is about $10^{-6}$ – which is not so much wrong as disingenuous in the neglect of known processes – evaluated in the paper – that contributed to 9.5 events to the 19 event peak. (Shakespeare has Hamlet admit that he "is indifferent honest." I am afraid that the young author of that part of the paper was also "indifferent honest.") The proper evaluation of the probability that 19 or more events would show up if only 9.5 were expected is more like 1:500.

Then from where did the extra 9.5 events come? Certainly not from violation of CP-invariance where a $2\pi$ decay had not been seen in 411 decays at BNL and Dubna. So I invented a weak, long-range, fifth force between the $K$-mesons and protons that was designed to coherently regenerate $K_1^0$ mesons from the $K_2^0$ flux with an intensity that fit the experimental results while escaping contradiction by other experiments. It was a clever enough idea that turned out to have been considered by Steve Weinberg also: indeed we agreed to write a joint paper on the subject but the Princeton results presented at Dubna in 1964 saved us from that effort.

We know now that there may be other fifth forces, but not this one. But the extra events were real. If these events played a role in the decision by Fitch and Cronin to look for $K \to 2\pi$ decays, from where did they come? Was it simply some kind of experimental error, a gross statistical fluctuation, or was the "Old One" generous enough to provide a hint of CP-invariance violation for they who would listen?

It seems that there was such a hint. *After* the Princeton result was announced, we came

[5] L.B. Leipuner, W. Chinowski, R. Crittenden, R.K. Adair, B. Musgrave, and F.T. Shively, *Phys. Rev.* **132**, 2285 (1963) The analysis presented in the paper was largely completed by the summer of 1962 before I left in August for a year at CERN and the results were related in talks. The paper, slow to be written, was received by the *Physical Review* March 13, 1963 but was not published until early fall.

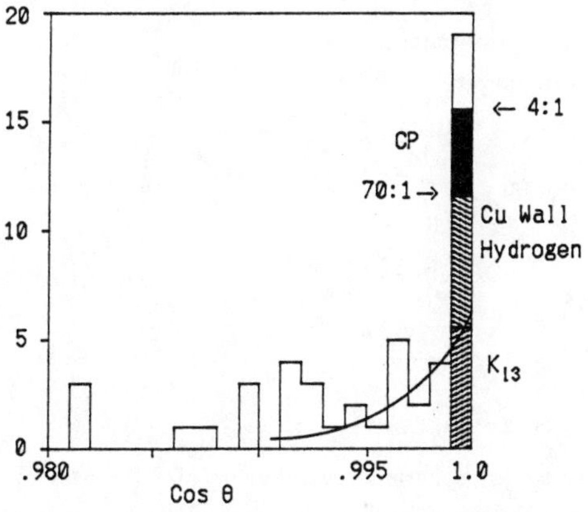


Figure 4: Angular distribution of the Yale-BNL events that fit $K \to 2\pi$ kinematics.

to understand the peak as an intensity generated from the coherent sum of three amplitudes of different origin, all nearly in phase: (1) the regenerated $K_1^0$ amplitude from the copper in the chamber walls, (2) the coherent $K_1^0$ amplitude from the hydrogen in the chamber, and (3) the CP-invariance violating $K_2^0 \to K_1^0 \to 2\pi$ amplitude – augmented by the Old One's gift of almost one standard deviation fluctuation.[6]

The magnitude of the $K_L^0 \to K_s^0$ regeneration amplitude for hydrogen, though not the phase, is (and was in 1962) known directly through our[5] measurements of the incoherent regeneration cross sections in the forward direction. And, from the $K^+, p$ and $K^-, p$ total cross sections, it was clear that the regeneration amplitude was largely imaginary. Similarly, the imaginary part of the regeneration cross section for copper could be estimated adequately from direct measurements augmented by estimates of the total cross sections for $K^0$ and $\overline{K}^0$ derived from measurements of $K^+$ and $K^-$ total cross sections. Moreover, at $K$ energies as high as 1 GeV, an energy characteristic of all of the experiments considered here, the imaginary parts of the amplitudes can be expected to be appreciably larger than the real parts – which are not known as well. It is the imaginary parts of the *initial* regeneration amplitudes for copper and hydrogen that tend to be in phase with each other and with the CP-invariance violating $K_1^0$ part of the $K_L^0$ and it is these regeneration amplitudes, derived from the imaginary parts of the forward scattering amplitudes, that are important to our considerations.

The constructive interference of the three amplitudes, which results in the large effect of CP-violation, is suggested by the diagram of Figure 5. Here the three amplitudes are shown as vectors summed in the complex plane for different positions in the chamber. When it is noted that the intensity is the square of the total amplitude, the importance of the CP-violation part of the amplitude is evident.

The coherent "intensities" from the wall, taken separately, would add up to about 2.3 events; about 1.2 events from the wall, about 0.4 events from the hydrogen, and 0.7 events from the CP-even part of the $K_L$ intensity. In the approximation that these are all in

[6]This is discussed by Chris Hawkins who joined our group from Oxford for a few years: C.J.B. Hawkins, *Phys. Rev.* **156**, 1444 (1967). The results presented here, recalculated by me using more modern parameters, differ only slightly from Hawkins' 1967 values.

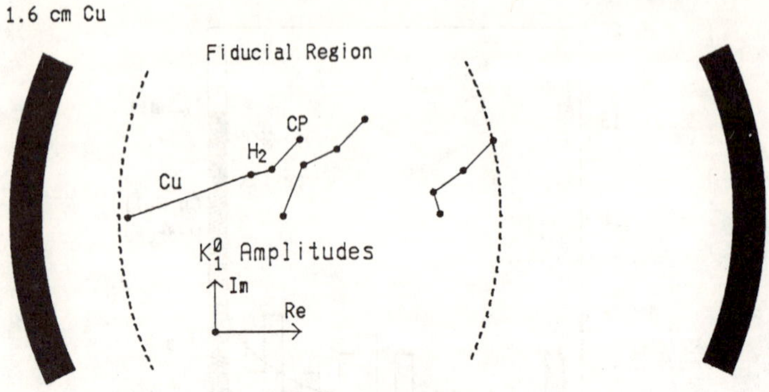


Figure 5: Vector amplitudes for coherent regeneration of $K_1^0$ states by the copper chamber wall, coherent regeneration by the hydrogen, and the $K_1^0$ part of the $K_L^0$, at three positions in the chamber.

phase, the total intensity from these processes would be about 6.6 events. Using the same approximation, without the CP-violating amplitude, the intensity would be only about 3 events. More complete calculations give total coherent intensities of 2.6 events without a CP-violating contribution and 6.3 events with the CP violation. We estimate that the – poorly known – contributions of the real parts of the forward scattering amplitudes are much less important.

In summary, we should have expected 15.25 events of which the existence of the CP violating amplitude contributed about 3.7 events. Or, if we were ignorant of CP violating amplitudes – as of course, we were – we should have expected only 11.55 events. The probability of getting 19 or more events when one expects 11.55, is about 1 in 70. The probability of getting 19 or more events when one expects 15.25 is about 1 in 4; that difference is a little less than one standard deviation.[7]

In this sense, the CP-invariance violating amplitude contributed almost 4 events through the *coherent* contribution of the amplitude (the incoherent intensity would have been only about 0.7 events). This, together with the gift of a one deviation statistical fluctuation (of almost 4 events) and the enthusiasm and optimism of youth, convinced us that we were seeing something odd – and we were right.

Why were we so struck by so small an effect (two standard deviations today – but about three as calculated in 1962 using more primitive mass differences, cross sections, and lifetimes). As I said, we were younger and less wise. Yesterday Larry Leipuner and I were talking about the experiment and we decided that with our greater maturity (read senility) we would never crawl out so far on a limb today. But there is also a kind of confidence in visual techniques that it is difficult to match with more abstract methods. In the maturity of these methods, there was something substantial about bubble chamber pictures, cloud chamber pictures, and emulsion tracks, that one doesn't quite meet with today. We *saw* the events and we believed them.

[7] In the original paper, I calculated an expectation of but 9.5 events, about two events less than the number I arrived at now. A reduction of about one event follows from my use in 1962 of the value of $\Delta m = 1.5$ incorrectly accepted at that time rather than the value of 0.5 known today. The source of the rest of the discrepancy (of about one event) is not obvious though other parameters have changed less dramatically.

The Princeton group took our result seriously – and so did we.[8] Although Maurice Goldhaber told us that we should use spark chambers to look further and better at whatever we thought we saw, we decided that it would be easier and quicker for us to do a second, better and cleaner, bubble chamber experiment.[9] We recognized that the copper wall, at best, contributed unwanted backgrounds in our experiment so we rebuilt the chamber so as to have a large thin window and operated the chamber in a 600 MeV/c monoenergetic $K_L^0$ beam produced by the interaction of a pion beam from the Cosmotron with a hydrogen target. We also ran a longer exposure that should have been be quite sufficient to firmly establish the anomalous hydrogen regenerated that we suspected *according to the specific model of a new, weak, long range interaction that we had postulated.* We had estimated the effect to be $2.5\% \pm 0.9\%$ of all charged decays for 1000 MeV/c $K$-mesons.

However, at the lower energy, and without the copper, we did not see a large effect. Though we calculated that we had an excess of about 1%, that was hardly outside of the statistical uncertainty in the experiment.

Before our run with the monoenergetic 600 MeV/c $K_L^0$ beam at the Cosmotron, Fred Eisler, who had worked with us briefly, borrowed our chamber, which had been fitted with the large thin window, and with T.C. Bacon and H.W.K. Hopkins ran a similar exposure in a beam at the AGS which provided a high intensity of $K_L^0$-mesons over a broad energy range near 1000 MeV/c. With better statistics, they were able to show definitely that there was no effect as large as 2.5%.[10] However, again they seemed to have a significant excess of the order of $0.8\% \pm 0.33\%$ of events that fitted $K_1^0$ regeneration kinematics, as we pointed out[11], but again there were both statistical and systematic uncertainties. The results from Eisler's run, shown in Figure 5 (taken from our paper)[11] show our evaluation of the excess of events in their forward peak.

Today we know that the constructive interference of the hydrogen regeneration amplitude and the CP-violating amplitude will give a substantial contribution to the forward $K_1^0$ intensity. The rate of generation (per length of path) of the imaginary part of the $K_1^0$ amplitude regenerated coherently from the interaction of the $K_L^0$ beam with the hydrogen is almost exactly equal to the rate of spontaneous generation of $K_1^0$ from the $K_2^0$ part of the $K_L^0$ beam – and that phase is also imaginary. Asymptotically, the phase shifts from the $K_1^0 - K_2^0$

[8]The set of high energy physicists – not so large today – was smaller then and we all knew each other well. When Jim Cronin came to Brookhaven after getting his degree with Sam Allison at Chicago working in low energy nuclear physics, he moved into an office in the Cosmotron building next door to that used by Larry and I. And before the experimental run of the Princeton group, the consequences of which occupy us today, Val Fitch and Dave Cassels joined with Larry Leipuner and Ray Crittenden (while I went to CERN for a year) to measure the electromagnetic form factor of the pion at BNL; an experiment that Dave used for his Ph.D. thesis.

[9]I review the chronology: After our bubble chamber run, we began preparation of an experiment designed to look at the decays of $\Lambda^0$s using a small spark chamber constructed by Brian Musgrave to run in the magnetic field of our bubble chamber magnet. The apparatus worked well in a test run in the late spring of 1962, but an analysis of that data – with Lynn Stevenson – suggested that we had been too optimistic concerning event rates and we withdrew the experiment. By late summer, when we had completed enough of the analysis of the bubble chamber run to convince us that we had an anomaly, Larry and Val had agreed to work together on the pion form factor and I left for CERN. At that time, we decided to put a large thin window in our bubble chamber in preparation for another, cleaner, look at the anomaly we had seen. When I returned from CERN in June, 1963, Fitch and Cronin were just beginning their spark chamber experiment and we began preparation for our second bubble chamber run which took place – belatedly – the next summer.

[10]F. Eisler, T.C. Bacon, and H.W.K. Hopkins, *Proc. Int. Conf. Fund. Aspects of Weak Interactions* (Brookhaven National Laboratory, 1963), p. 82.

[11]L.B. Leipuner and R.K. Adair, *Phys. Letts.* **12**, 67 (1964). This paper also contains a more complete analysis of the data described in reference 5.

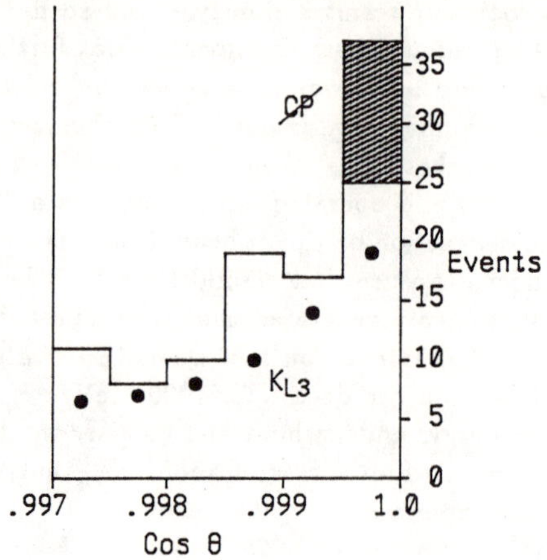


Figure 6: Angular distributions of events, from Eisler *et al.*, that fit the kinematics of $K^0 \rightarrow 2\pi$ decays. Our estimate[11] of the excess of events in the forward direction is shown by cross hatching.

mass difference and the attenuation of the amplitude by decay, conspire to create coherent amplitudes from a long region of path that differ in phase from the incident beam by the angle $\phi = tan^{-1}[2(M_1 - M_2)c^2\tau_1/\hbar] \approx 45^o$. Since the final phase (of $45^o$) is from the same origin, asymptotically the part of the initial regeneration amplitude derived from the imaginary parts of the forward scattering amplitudes will be *exactly* in phase with the $K_1^0$ part of the $K_L^0$. Accidently, the asymptotic (after about two $K_s^0$ mean-lives) magnitude of the $K_1^0$ regeneration amplitude from the imaginary part of the forward scattering amplitudes for 1000 MeV/c $K_L^0$, $p$ interactions is about equal to the $K_1^0$ amplitude part of the $K_L^0$. Hence the total intensity of $K_1^0$ in the beam will be more than four times that which might have been expected from CP-violation alone. At 600 MeV/c, the $K$-momentum in our second experiment, the hydrogen regeneration amplitude is only about 0.6 times the $K_2^0 \rightarrow K_1^0$ generation amplitude and the anomalous regeneration from the constructive interference of these two amplitudes is appreciably smaller.

For the Eisler experiment, and our second experiment (when certain biases that follow from energy cuts are considered), the $K \rightarrow 2\pi$ decays will be near 1% of the total charged $K$-decay intensity. This is rather smaller than the excess stated[11] to be $2.5\% \pm 0.9\%$, observed in our initial experiment[5] – which was assisted by the coherent amplitude from the copper wall and the Old One's nudge of a one standard deviation statistical fluctuation – but it is, nevertheless, a significant effect and quite accessible by bubble chamber measurements. However, measurements in the small bubble chamber were probably not quite sufficient to firmly establish so small an anomaly.[12]

[12]The date of reception of our *Physics Letters* paper was July 23, 1964, just before the July 27 publication of the Princeton result. After some reflection, it then became evident to Leipuner and I that we had been observing a CP-violation effect constructively interfering with normal coherent regeneration. It is interesting to note, that nowhere in our conjectures as to the origin of the anomaly that so concerned us, is CP-invariance violation mentioned. I, at least, believed in CP-invariance so strongly that I was not able to entertain the idea of its violation.

# The Princeton Experiment and CP-Invariance Violation

EVIDENCE FOR THE $2\pi$ DECAY OF THE $K_2^0$ MESON*†

J. H. Christenson, J. W. Cronin,‡ V. L. Fitch,‡ and R. Turlay§
Princeton University, Princeton, New Jersey
(Received 10 July 1964)

Professor Pais has spoke of his deep reluctance to believe that CP-invariance was flawed. He tells me that he does not remember – thank goodness – a conversation with me shortly after he had discussed the Princeton data with Fitch and Cronin. But I remember it clearly. We had left the Cosmotron building by a door near my office at the end of the corridor, and were walking towards the Physics Building. Bram said to me that it looked like Fitch and Cronin had seen $K_2^0$ decays to two pions and hence had evidence for CP-invariance violation and asked what I thought of such a possibility. (I had heard, less directly, of their result.) I remember saying that while they [*i.e.* Fitch and Cronin] were excellent physicists, they must be in error; CP *must* be invariant.

Such complete faith in CP-invariance seems foolish now, but I suspect that most physicists would have answered as I did; we didn't believe that the Old One could play such games with us and build us a cock-eyed universe.[13] Then why was the conversion so complete after the Princeton result was announced? Why were there no experimental caveats at all? (Of course, there were theoretical suggestions for evasions though these were as radical as CP-violation.) There were no caveats because the Princeton experiment was *seamless and complete.* That experiment gave CP-invariance such a blow that it did not need another.

Although you can read the paper[14] – and I hope you have read it recently, it is only three pages long – I will try to show you succinctly something of the power of the result.

The decay of a $K^0$ moving in a well defined direction to two charged pions is kinematically triply over constrained; it is a 3C fit in old bubble chamber jargon. This means that the signal-to-background for detection of such a process is a strong function of the accuracy of the kinematic measurements. Even a small signal *must* stand out if the measurements can be made with sufficient accuracy and if the flux of particles is sufficient. With this in mind, the Princeton group designed a state-of-the-art detection system tuned in its design to observe $K_L^0 \rightarrow \pi^+ + \pi^-$ decay events with maximum sensitivity. A schematic diagram of their double-arm spectrometer is shown in Figure 7. This detector was then set up in a neutral beam produced at an angle of 30° by the interaction of 30 GeV protons on a beryllium internal target at the BNL AGS. At this large angle, the mean momentum of the $K_L^0$ mesons was known to be about 1000 MeV/c (In their measurements, the Princeton group determined the mean momentum of the $2\pi$ events they selected to be 1100 MeV/c. This is about the same as the mean $K_L^0$ momentum in the Cosmotron beam produced in the forward direction but the backgrounds were smaller at the AGS.)

For some measurements, material targets were placed in the beam; for the set-up shown here, directed towards observing decays, a helium bag was placed about the region covered by the spectrometer arrays. The long lever arm of the magnetic-spark chamber combinations allowed precise measurements of the particle momenta and direction. The trigger, which

[13] To me, at least, the Mach's Principle arguments against CP-violation seemed unassailable. Remember this was long before we appreciated the disruptive power of spontaneous symmetry breaking. Even now, I would be upset if it turned out the fundamental equations were not invariant under CP though I know the solutions – the world we live in – are not so invariant.

[14] J. Christenson, J. Cronin, V. Fitch, and R. Turlay, *Phys. Rev. Letts.* **13**, 138 (1964)

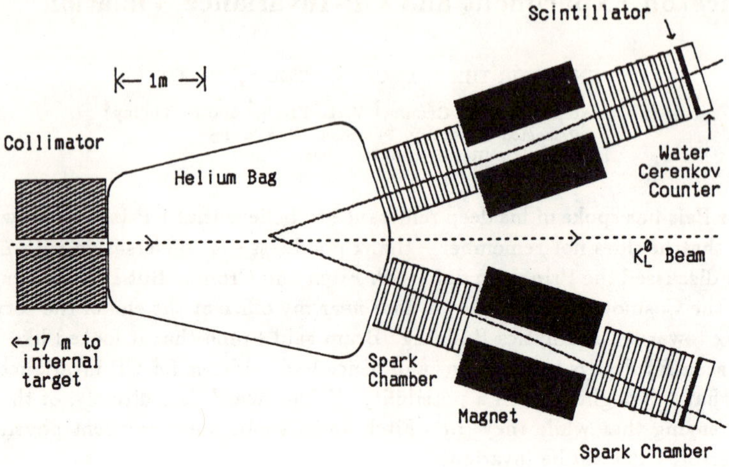


Figure 7: Plan view of the Princeton spark chamber magnetic spectrometer.

required that a particle in each arm trip both a scintillator and a water Ĉerenkov counter was simple and effective in eliminating backgrounds from slow particles such that $v/c < 3/4$. The acceptance was not large, but very much greater for two-body decays than three-body decays of the $K^0$.

At that time, the coherent forward regeneration of $K_1^0$ states from the interaction of a $K_2^0$ beam was well known and the group used the decay $K_1^0 \rightarrow \pi^+ + \pi^-$ produced in this manner by the interactions of the $K_2^0$ beam with a 47 gm/cc tungsten target, placed in various positions along the beam, to test and calibrate their apparatus. Using these decays, they were able to show that the angular resolution of their apparatus corresponded to $\sigma_\Theta \approx 3.4$ milliradians while the invariant mass resolution was found to be $\sigma_m \approx 3.6\ MeV/c^2$.[15]

Then, armed with this highly precise, accurately calibrated, spectrometer, the Princeton group looked at the "regeneration" in the vacuum (considering helium as close enough to a vacuum as far as regeneration is considered). The results are best expressed in a two-dimensional plot of the number of events, interpreted as $\pi^+ + \pi^-$ decays, plotted as a function of $\cos\Theta$, the cosine of the angle the reconstructed momentum makes with the beam direction), and the invariant mass, $m$. Real $K_L^0 \rightarrow \pi^+ + \pi^-$ events should show up on such a plot at $\cos\Theta = 1$ and at $m = 498\ MeV/c^2$. The sharp peak at $\cos\Theta = 1$, and at $m_{\pi\pi} = m_K = 498$ MeV, then established the existence of that decay.

The plot of Figure 8, together with a brief perusal of the Princeton protocol just about ends the argument; there are no experimental caveats. A particle with almost exactly the mass of the $K_2^0$ meson has decayed into two charged particles, each with almost exactly the mass of the pion. Moreover, the probability of this process is that which could be expected if the $K_L^0$ decays in this mode with a branching ratio near 0.002. And after 25 years of further study of CP-invariance violation, we know no other non-zero number. This number

[15]The corresponding resolutions for the Yale-BNL 14" bubble chamber were reported[5] to be $\sigma_\Theta \approx 26$ mr and $\sigma_m \approx 20\ MeV/c^2$. From these values, one should expect that the Princeton signal-to-background would be as much as 300 times better than for the bubble chamber! That ratio actually seems to have been about 150 as the resolution of the vacuum (decay) events was slightly worse than for the tungsten calibration events.

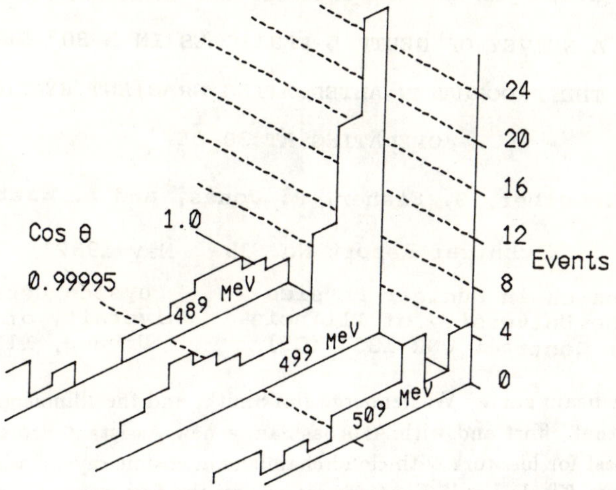


Figure 8: The number of events from the Princeton experiment in intervals of $\Delta \cos \Theta = 10^{-5}$ and $\Delta m = 10\ MeV/c^2$ is shown.

*completely* describes all that we know of CP-invariance violation.[16]

## The Illinois Program

By the spring of 1961, a new $30^o$ beam had been opened in "Inner Mongolia" at the AGS; the informal but universally adopted name followed from the fact that the beam was directed to the *inside* of the AGS ring. That spring (1961) Al Wattenberg, with Gerber, Fisher, and Jones, all from the University if Illinois, undertook the responsibility of conducting a survey of the intensities of neutrons, gamma-rays and $K_2^0$ mesons in that beam. Scintillation counters, a lead-glass Ĉerenkov counter, and two spark chambers were use in the course of these measurements. In a report to BNL Associate Director Rod Cool[17], they reported, "If we assume a vector interaction, we obtain that there are about $1.2 \times 10^8$ $K_2^0$ mesons with energies above 1 BeV [*sic*] per steradian per $10^{11}$ circulating protons; this latter figure is probably good only to a factor of 3." The report continues on page 6 to say, "We were joined by Professor J.H. Smith ...[for the *K*-meson measurements]." This was the beam used by

[16]More precisely, the $K_2^0 \rightarrow K_1^0$ transition rate $= |\epsilon|/(2\tau_s sin\phi)$ – one real number – is the only measure of CP-invariance violation we know. We note the value of $\phi$ is determined by CP-invariant factors. Of course the validity of this (superweak-like) formulation follows from the absence of other numbers and then from many invaluable negative results.

[17]H-J. Gerber, G. Fisher, R. Jones, and A. Wattenberg, *University of Illinois Technical Report No. 39*, May 1962

the Princeton group a year later to conduct the experiment we are now celebrating.

A SURVEY OF NEUTRAL PARTICLES IN A 30° BEAM

FROM THE BROOKHAVEN ALTERNATING GRADIENT SYNCHROTRON

OPERATING AT 30 BEV

H-J. Gerber, G. Fisher, R. Jones, and A. Wattenberg

Technical Report No. 39 May 1962

Research in Nuclear Physics at the University of Illinois under Contract ONR 1834 (05)

Physics Department University of Illinois Urbana, Illinois

After their beam survey Wattenberg, Jim Smith, and the Illinois group were interested in exploiting that effort and with Al Abashian, a new Assistant Professor, and Bob Sard, known then best for his work with cloud chambers in cosmic rays, made a proposal in April 1962 to measure $K^0_{l3}$ decays.[18] I note a sentence on the first page of the proposal, "In the $\mu$ meson polarization runs we hope to test time reversal invariance by measuring the up-down asymmetry of the electron." If I could not accept the possibility of a flaw in CP-invariance (and through the CPT theorem, a breakdown in time-reversal invariance), the Illinois group did not share that blindness.[19]

The laboratory approved the proposal and the group, sans Wattenberg who had arranged to spent the year in Italy, ran their experiment, primarily designed to look at the three-body correlations in $K^0_L3$ decays, beginning in November 1962 and running through the first week of January, 1963. Then they dismantled their apparatus to make way for the Princeton group who were scheduled to look for anomalous $K^0_L \to \pi^+ + \pi^-$ transitions in that beam.

The diagram of Figure 9 shows a schematic representation of the Illinois apparatus which is largely a magnetic spectrometer where spark chambers are to do the tracking. Muons from $K^0_2 \to \pi + \mu + \nu$ decays were to be identified by their passage through the lead block and their range in the heavy plates making the last set of spark chambers. The lead was removed for the part of the run devoted to an examination of $K^0_L \to \pi + e + \nu$ decays and the electrons were identified through the showers they induced in the lead-and-spark-chamber sandwich. Since the object of the experiment was the measurement of three-body modes, it was necessary that the spectrometer have a large acceptance. Hence, the spark chambers were placed between the poles of the magnet giving a much larger acceptance than for the Princeton two-arm spectrometer at the cost of a reduced accuracy in angle and momentum measurement. Moreover, the triggers were designed to accept only events with muons, or

[18] *Proposal to study $K^0_2$ Decays in a Neutral Beam from the AGS: Abashian, Sard, Smith, and Wattenberg; submitted April 17, 1962.* Again, the principal players were old friends. I had first met Al Wattenberg when I gave a talk at Argonne National Laboratory about 1950 when I was still a Wisconsin Graduate student and Al was head of the Physics Department there. And I used to joke about a grudge I supposedly held against Al Abashian. Within two weeks after we heard of the result of C.S. Wu and Eric Ambler which ended illusions of parity non-conservation, a group of us at BNL did a very nice experiment in a little hydrogen bubble chamber that Larry Leipuner, Rich Larsen and I had built, showing parity non-conservation in the $\pi \to \mu \to e$ decay chain. We agreed to list the names on the paper alphabetically. As that is a game I do well in, I had visions of this nice paper being known as *Adair et al.*. It was then a rude shock to realize that I was out alphabetized by Abashian and I was to be only an *et al.* of: Abashian, *et al.*, *Phys. Rev.* **105**, 1927 (1957)

[19] In this suggestion the proposal was premature as well as prescient. That experiment was too difficult for the techniques the Illinois group was using. Twenty years later Larry and I and our colleagues spent several years measuring this transverse muon polarization.

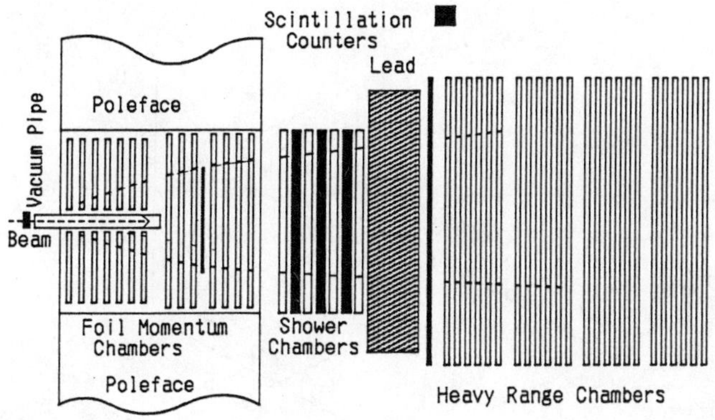


Figure 9: Plan view of the Illinois spark-chamber spectrometer. The event is a symbolic reconstruction of a $K$ decay where both decay particles are identified as pions through their range in the range chambers.

events with electrons. A sentence in the proposal reads, "We will trigger our spark gaps only if we have an electron event or a $\mu$ event." Fortunately (?), the discrimination against $2\pi$ decays (which were of course not expected) was not complete.

Since spark chamber data was recorded photographically, the work of analyzing a typical 100,000 picture run was labor-intensive and slow. Much of the first year after the experiment was spent in measuring the photographs and reducing the data to a form that could be run through computers. And then, the first analyses were naturally devoted to those events in which electrons, or muons, were identified visually through their characteristic spark chamber signatures.

So it was much later, perhaps after a May 1963 meeting at Argonne in Chicago where René Turlay talked about preliminary Princeton results (but did not announce CP violation), that the Illinois group – and largely Abashian – examined their events (especially those that did not pass the electron or muon signature tests) on the assumption that the two charged particles could be pions. When they did this they seemed to see an excess of events in the forward direction with masses near 505 MeV.

SEARCH FOR *CP* NONCONSERVATION IN $K_2^0$ DECAYS*

A. Abashian, R. J. Abrams, D. W. Carpenter, G. P. Fisher, B. M. K. Nefkens, and J. H. Smith
Department of Physics, University of Illinois, Urbana, Illinois
(Received 27 July 1964)

Figure 10 shows the data from the "electron" run (without the lead block and with a trigger set to accept events with at least one electron.) where only events where neither particle can be identified visually as a muon or electron are considered. From todays vantage, the sharp peak in the 505 MeV data suggests a presence of the $K_L^0 \to \pi^+ + \pi^-$ decay mode where the measurement of the field integral of the magnet was in error by about 1%. A similar analysis of the "muon" run seemed to show – albeit marginally – a smaller peak of four events – with a peak of three events from the sample of certified muon events with a

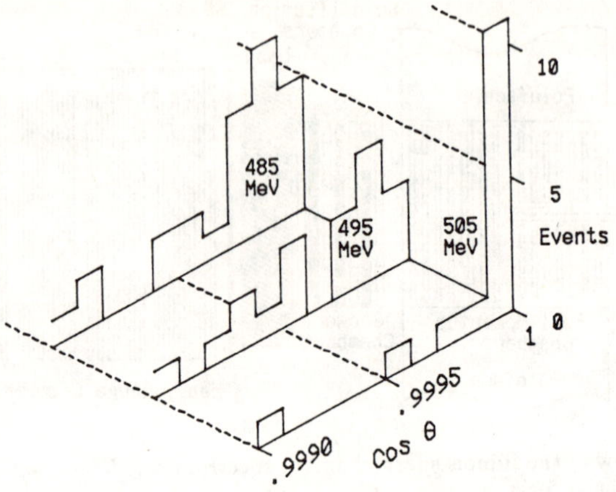


Figure 10: The number of events from the Illinois experiment are plotted in intervals of $\Delta \cos \Theta = 10^{-4}$ and $\Delta m = 10MeV/c^2$.

$\pi - \pi$ invariant mass of 500 MeV $< m <$ 510 MeV.

Perhaps, most interesting, two events of the eleven making up the peak in Figure 10 were such that both decay particles seemed to interact in the spectrometer in a manner that could not be expected if the particle were a muon or electron but was characteristic of a pion. In one case, one particle stopped in plate 5 of the range chamber and one in plate 11 (as suggested by the sketch in Figure 9) where, from the measured momenta, each would have passed through the whole set if it had been a muon. Assuming both particles were pions, their invariant mass was measured to be 501 MeV. In a second example, one particle stopped in plate 15 (with a momentum that would have carried a muon through all of the plates) and the second made a 35 degree scattering in one of the lead plates in the shower detector. The invariant mass of this pair was measured to be 506 MeV. None of the four particles displayed any showering in the shower chambers.

Although, according to Abashian, they knew that the Princeton group was seeing evidence for CP-violation, in their paper, entitled ***Search for CP Nonconservation in $K_2^0$ Decays***[20] they concluded only that, "this [the peak in Figure 10] places an upper limit on on the violation of CP-invariance of $5 \times 10^{-6}$ [*i.e.* $\epsilon^2$], with at least a suggestion that CP may be actually violated by this amount." Abashian says that in title and conclusion, the Illinois group attempted to proceed as they would have if they had not heard of the Princeton result – which had not been published at the time they had prepared their paper – though, knowing that result they were confident that they had seen CP-noninvariance.

## Alternate Universes

American writers have explored in their fiction the hypothesis, "What if the South had won the American Civil War?" If Lee's orders, wrapped around three cigars, had not been

[20]A. Abashian, R.J. Abrams, D.W. Carpenter, G.P. Fisher, B.M.K. Nefkins, and J.H. Smith, *Phys. Rev. Letts.* **13**, 243 (1964). The paper was received at PRL July 27, 1964, the same day that Princeton paper was published.

dropped on the battle field at Antietam to be picked up by a Union soldier; if Lee and Longstreet had let Hood get to Little Roundtop before Lawrence Chamberlain and the 20th Maine at Gettysburg. Surely there must be French – or English – novels describing Europe if Napoleon had prevailed in the campaign that ended at Waterloo. We can play that game also with the discovery of CP-noninvariance – and by enrolling those who endow Quantum Mechanics with parallel universes, we have better credentials.

I have mentioned that if the cloud chamber measurements at BNL and at Dubna had been carried out with a total of 1500 events, rather than 400, the breakdown of CP-invariance would have been observed in 1957. From such a number of events, one could expect to see about 4 $K_L^0 \rightarrow \pi^+ + \pi^-$ events and the with the accuracy of measurement possible with such cloud chambers, this would have been a clean signal. But to collect these 4 events, either experiment, *as designed*, would have to take, scan, and measure about 7 times as many pictures changing a difficult enough experiment into a *tour de force* to be entertained only if very important results were almost certain. I think one would have to shuffle through many, many alternate universes to find one where such a brute force approach would have been conducted.

However, the extended experiment would not be quite so impractical if a beam very much richer in low energy $K_L^0$ mesons were available. But, I don't believe that in this universe or any other, any such far better beams can be created at either the Cosmotron or at the JINR synchrophasotron.

But how would physics have progressed in an alternate universe where there was no Princeton experiment because the gang-of-four had found other occupations. I consider universes where, Val *meta*-Fitch, instead of working in physics, would follow an old American midwestern tradition and became a bank robber like Jesse James, Butch Cassidy, and the Sundance Kid – only Val would specialize in Nebraska banks. Similarly, I see Jim *meta*-Cronin as the ruthless Texas oil-billionaire, J.C. – *a la* J.R. in *Dallas*. I shall leave Jim *meta*-Christenson and René *meta*-Turlay to find their own congenial occupations outside of physics.

Then would the Yale-BNL program, unimpeded by the Princeton success, reach the same conclusion in a year or two. I think not. That road is not plain and clear. With one further good experiment, in a moderately larger bubble chamber such as Ralph Shutt's BNL 20" chamber where more accurate measurements could be made, the anomaly in the hydrogen regeneration (which we now know follows from the constructive interference of regeneration in hydrogen and the CP-nonconserving $K_1^0$ impurity in the $K_L^0$) would be established firmly. But I do not see the anomaly as being interpreted very quickly as a CP-nonconservation effect. At least, I think one would have to look through a very large number of universes to find a *meta*-Adair, much smarter than the *ortho* variety, who would have worked out such a conclusion.

Then, in that faerie universe, would the Illinois group discover CP-noninvariance? And announce their conclusion in October 1964, after a more careful data analysis than that described in their paper published in Middle Earth? Perhaps, but I would bet against it. It is hardly clear that without the ambience created by the Princeton effort, the Illinois group would have looked in their data for $2\pi$ decays of the $K_2^0$. Such a special search does not fall near the main line of their basic inquiry into the three-body decay distributions.

But they might have looked in that direction. I have noted that they showed themselves open to the idea of CP-noninvariance in their proposal. And Al Wattenberg told me that he had a graduate student looking at possible $K_2^0 \rightarrow \mu + \mu$ decays thinking such events might indicate CP-noninvariance.[21] But if they had carried out the kind of analysis that they did in this realm, what would their conclusions be?

Their data, as published in this universe in August 1964, might have excited more interest (in themselves and others) in the possibility of CP-noninvarant $K_2^0 \rightarrow \pi^+ + \pi^-$ transitions, but their results, standing alone, would not have convinced good physicists – or themselves, as they are very good physicists. The conclusions that they published are as strong as it is possible to make responsibly from the data that they present.

Could they have made a much more convincing case through a more complete data analysis? Perhaps. But the paper they published is not a shoddy job. Hence, there was probably little to be gained by more extensive analyses. The difficulties are fundamental; the effect they would examine is seen only in backgrounds in an experiment designed to do quite different things – and biased towards rejection of the $2\pi$ signal.

If neither the Yale-BNL or the Illinois programs could have been expected to uncover CP-noninvariance in a short time in the absence of the Princeton effort, how long would it take to reveal that asymmetry in nature – and who would make the discovery? Here I would guess that in our perusal of alternate universes without the Princeton group, we would find a mean discovery date in the summer of 1966 – about two years later than it was seen in our universe where we are fortunate enough to have our four friends as physicists. And in some of these other universes, the discovery would have been made by the Illinois group – though in a second experiment. (I haven't found any universe in which *I* discovered CP-noninvariance, but I am still looking.)

## Epilogue

I suppose that it is a part of the human condition to look back with regret at lost opportunities – and some have so looked back at the discovery of CP-nonconservation. Following the Bible verse I took at my Confirmation in the Lutheran Church when I was 13, "No man having put his hand to the plow, and looking back, is fit for the Kingdom of God" [Luke 9:62], I have thought that I do less bewailing of the past than most. However I have wondered about the possibility that Leipuner and I might have taken a wiser path after we noted the anomalous $K_2^0 \rightarrow \pi + \pi$ events in the bubble chamber. Indeed I often tell my undergraduate physics students at Yale in jest that, "One must be lucky and clever to make an important discovery in experimental physics. And I am in the unenviable position of having demonstrably been lucky enough, but most evidently not clever enough." Then I add, "I have noted also that the luckiest people I know are also the cleverest." Good people make their own luck.

But though history looks at the past from the present, we live traveling through time in the other direction – and that is not so easy. In preparing this talk – and trying to review the remarkable discovery that took place 25 years ago in the context of that time, I found that I only renewed my respect for the people involved at Brookhaven, Columbia, Illinois, Yale, and, above all, Princeton; and I include myself – as I was 25 years ago – in that respect.

[21] Such decays are, of course not forbidden by CP-invariance but we were all more naive about the consequences of CP-conservation at that time.

# EXPERIMENTAL STATUS OF CP VIOLATION

J. Steinberger
CERN, Geneva, Switzerland
and
Scuola Normale Superiore, Pisa, Italy

**ABSTRACT**

The experiments on CP violation as well as the phenomenology of CP violation in K decay are reviewed.

The two most recent experiments searching for direct CP violation — the University of Chicago-Saclay experiment at FNAL, as well as the CERN experiment which has given a positive result — are discussed in some detail.

# 1. EXPERIMENTS ON CP VIOLATION

In the unfolding of the story of the elementary particles and their interactions, underlying symmetries have often played an important role. An excellent example is the discovery of parity violation in 1957 [1]. This stimulated a number of vital experiments, and within a year the V − A current structure of the weak interaction was known [2, 3].

The discovery of CP violation has not had a similar impact, and the few manifestations of CP violation subsequently discovered had no immediate effect on other areas of particle physics; instead, the way in which CP violation enters the general framework of particle theory has remained obscure.

The advent of the electroweak theory, and especially the realization [4] that CP violation could quite naturally occur in it—through the quark mixing matrix—has generated a new experimental incentive to test this hypothesis, in particular the experiments searching for 'direct' CP violation that are reviewed in Sections 3 and 4.

## 1.1 Discovery of CP violation

The experiment of Christenson et al. [5] was probably motivated by the perplexing result (later understood to be erroneous) of 'anomalous regeneration' [6]. The experiment consisted of a double-armed spectrometer looking at a He-filled region for decays taking place in a neutral beam sufficiently far from its target of origin so that the $K_S$ component had decayed (see Fig. 1). The $K_{e3}$ decays were rejected by means of threshold Cherenkov counters. The signal was dominated by the $\pi^+\pi^-\pi^0$ and $\mu^\pm\pi^\mp\nu$ decays of the long-lived kaon. The invariant mass distribution, assuming that both dectected particles are pions, shows no clear peak at the $K^0$ mass above the three-body background, but the mass slice at the $K^0$ has a peak in the forward direction (in contrast to neighbouring mass bins) (Fig. 2), which demonstrated the $\pi^+\pi^-$ decay of the long-lived kaon, and therefore CP violation. The branching ratio was found to be small

$$\frac{\Gamma_{L,+-}}{\Gamma_{L,all}} = (2 \pm 0.4) \times 10^{-3} ,$$

corresponding to

$$|\eta_{+-}| \equiv \frac{\langle + - |T|L\rangle}{\langle + - |T|S\rangle} \simeq 2.3 \times 10^{-3} .$$

## 1.2 $K_L^0 \to 2\pi^0$

In addition to the decay $K_L \to \pi^+\pi^-$, CP violation has been observed in only two channels: the decay $K_L \to \pi^0\pi^0$ and the charge asymmetry in $K_L$ semileptonic decay.

The $2\pi^0$ decay presented considerable difficulty at the time, and the first observation [7, 8] yielded results for the partial width that were in disagreement with the values accepted at present. The first results in agreement are due to the bubble chamber experiment of Budagov et al. [9] and the counter experiment of Banner et al. [10]. The apparatus of the

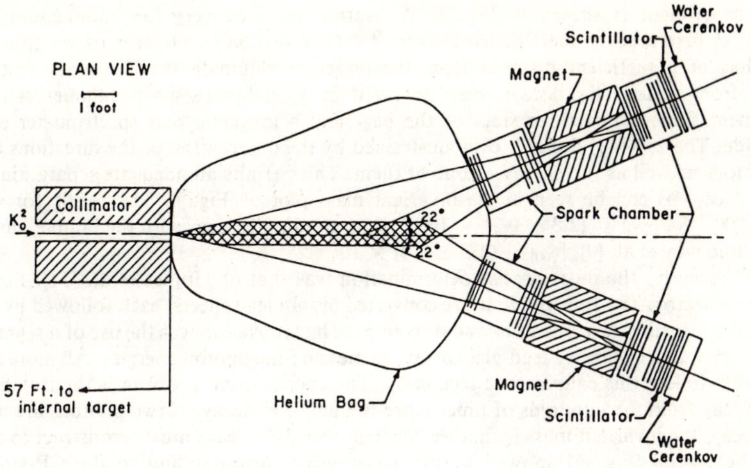


Fig. 1 Apparatus of the experiment in which CP violation was discovered [5].

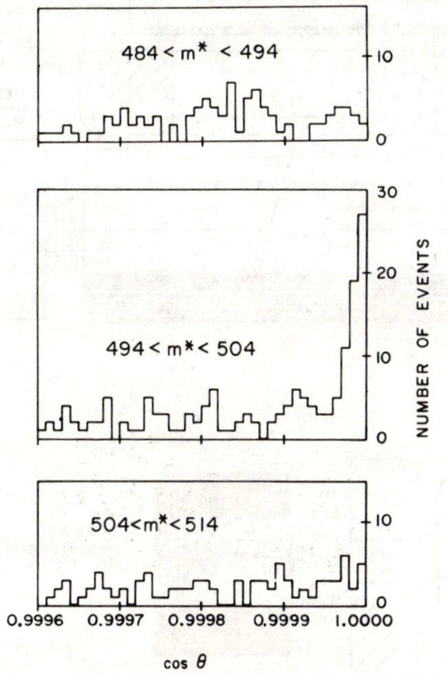


Fig. 2 Angular distributions of the observed events near the forward direction in three mass bins. The forward peak at the $K^0$ mass demonstrated CP violation [5].

counter experiment is shown in Fig. 3. A neutral beam of very low momentum (200–300 MeV/c) produced at the Princeton–Penn 1.3 GeV proton accelerator passes through a helium bag at a sufficient distance from the target to eliminate the $K_S$ component. The photons from neutral $K_L$ decays were observed in a lead-plate/spark-chamber sandwich arrangement surrounding three sides of the bag, and a magnetic pair spectrometer on the fourth side. The $2\pi^0$ ($4\gamma$) decay is overconstrained by the observation of the directions of the four photons as well as the energy of one of them. This permits an adequate separation from the $3\pi^0$ decay, as can be seen in the invariant mass plot of Fig. 4. The result for $\eta_{00} \equiv \langle 00|T|L\rangle / \langle 00|T|S\rangle$, $|\eta_{00}| = (2.3 \pm 0.3) \times 10^{-3}$, is in good agreement with the bubble chamber result of Budagov et al. [9], $|\eta_{00}| = (2.2 \pm 0.4) \times 10^{-3}$.

Until recently, the most precise determination was that of Holder et al. [11]. Figure 5 shows the detector. The four photons are converted in thin lead sheets, each followed by spark chambers to measure their positions and directions. The innovation was the use of a segmented calorimeter, in this instance a lead-glass array, to measure the photon energies. All more recent experiments follow this calorimetric technique. The experimental problem is the isolation of the $2\pi^0$ decay from the hundreds of times more frequent $3\pi^0$ decay. If two photons are missed in this decay, the invariant mass is smaller, whereas true $2\pi^0$ decays must reconstruct to the $K^0$ mass. The results (Fig. 6) showed a very clean peak corresponding to the CP-violating $K_L \rightarrow 2\pi^0$ decay, and the value for the relative rates of CP violation in the $\pi^0\pi^0$ and $\pi^+\pi^-$ decays, $|\eta_{00}/\eta_{+-}|^2 = 1.0 \pm 0.12$.

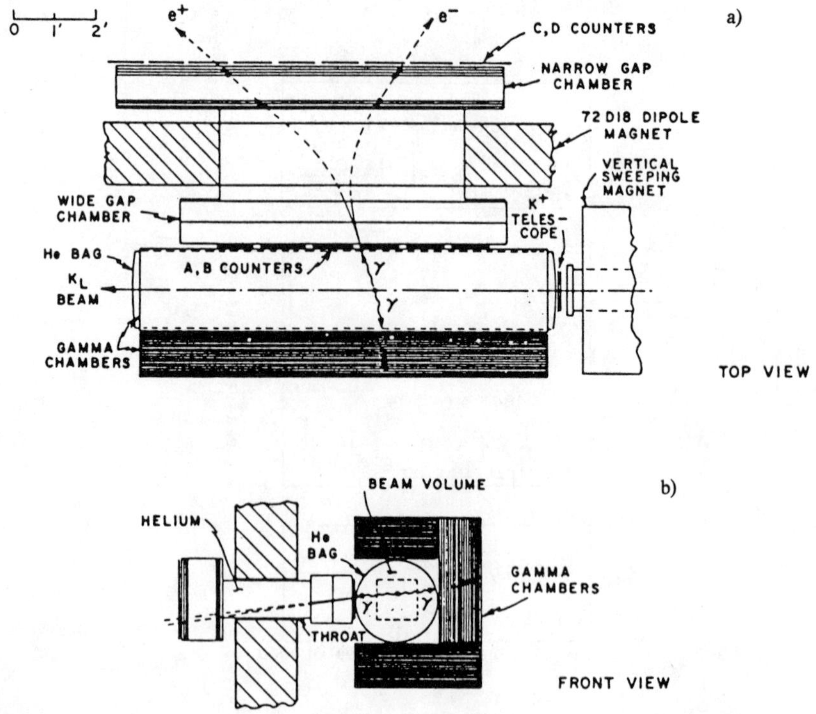


Fig. 3 Apparatus of Banner et al. for the measurement of $\Gamma_{L,00}$. At least one of the four photons is required to be detected in the pair spectrometer and its momentum vector is therefore well measured. The other photons are converted and observed in the lead/spark-chamber ('gamma chamber') stacks on the other three sides.

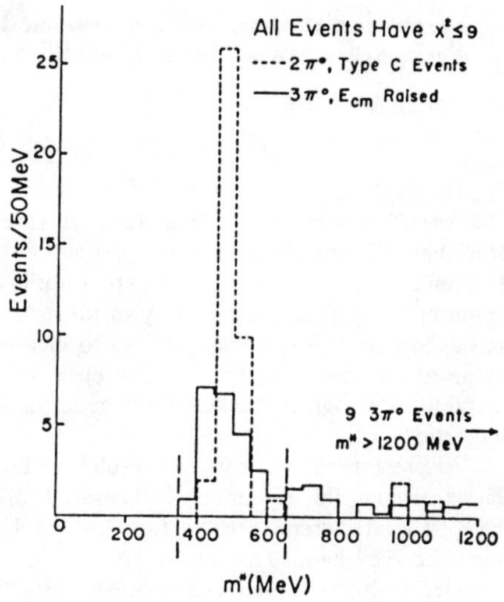


Fig. 4 Invariant mass distributions for events identified as $2\pi^0$ decays and for the $3\pi^0$ background. Experiment of Banner et al. [10].

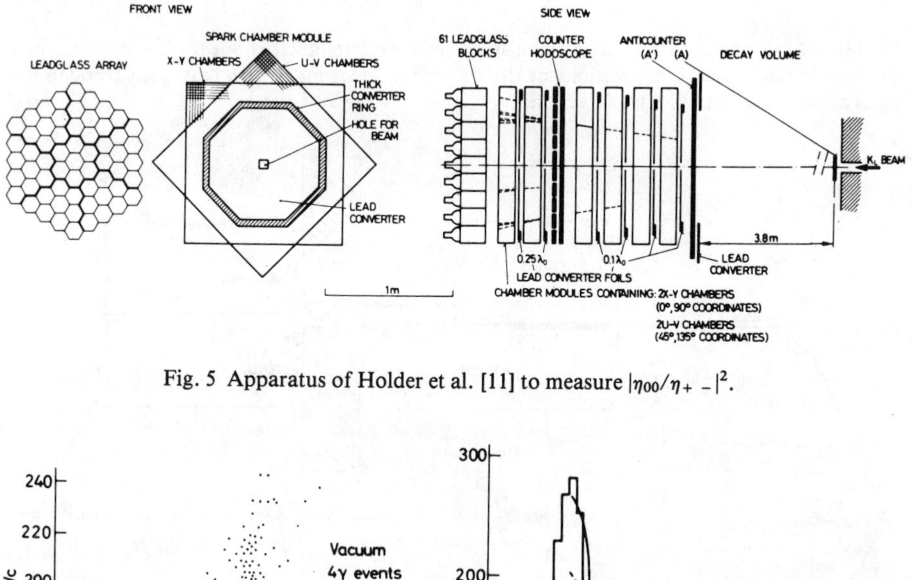


Fig. 5 Apparatus of Holder et al. [11] to measure $|\eta_{00}/\eta_{+-}|^2$.

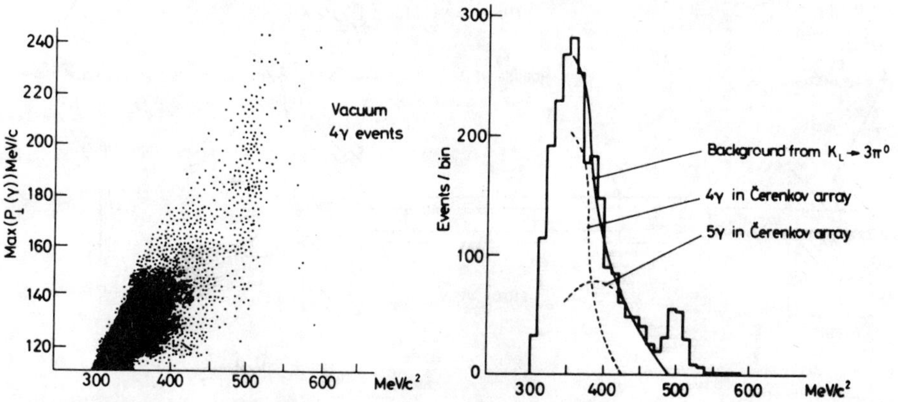


Fig. 6 Invariant mass distributions for the decays of $K_S$ and $K_L$ into four photons, from Ref. [11]. In $K_S$ decay the $2\pi^0$ mode is completely dominant. The $K_L$ decay shows a small $2\pi^0$ peak at the $K^0$ mass, containing 167 events, above the dominant background of $3\pi^0$ decays.

**1.3 The charge asymmetry in $K_L^0$ semileptonic decay**

The charge asymmetry in the decays $K_L^0 \to \ell^{\pm} + \pi^{\mp} + \nu$,

$$\delta_e = \frac{N_{e^+} - N_{e^-}}{N_{e^+} + N_{e^-}} ,$$

violates CP symmetry. It was first observed in two simultaneous experiments: one at Brookhaven in the electron decay (Bennett et al. [12]), the other at SLAC in the muon decay (Dorfan et al. [13]). The problem here is not in the identification of the decay mode—both are common decay channels and easy to identify—but rather in obtaining adequate statistics as well as low enough systematic biases to measure the small asymmetry of ~ 0.003. The BNL apparatus is shown in Fig. 7. The electrons are identified in a gas threshold Cherenkov counter. Their sign is selected in a magnet, and the asymmetry is determined by reversing the magnetic field.

At present, the most precise results for both decays are those of Geweniger et al. [14]. In this experiment the asymmetry is measured sufficiently close to the K-producing target so that the $K_S$–$K_L$ interference (see Section 2.8) can be observed as well and therefore the $\Delta S = \Delta Q$ rule also checked. The layout of the apparatus is shown in Fig. 8. The observed time-dependent asymmetries are shown in Fig. 9. The results for $\delta_\ell$ are

$$\delta_e = (3.41 \pm 0.18) \times 10^{-3} \quad \text{and} \quad \delta_\mu = (3.13 \pm 0.29) \times 10^{-3} .$$

The biggest systematic correction in these asymmetry measurements is due to the difference in the nuclear interaction cross-sections of the $\pi^+$ and $\pi^-$ mesons, which must pass through a small amount of matter before detection.

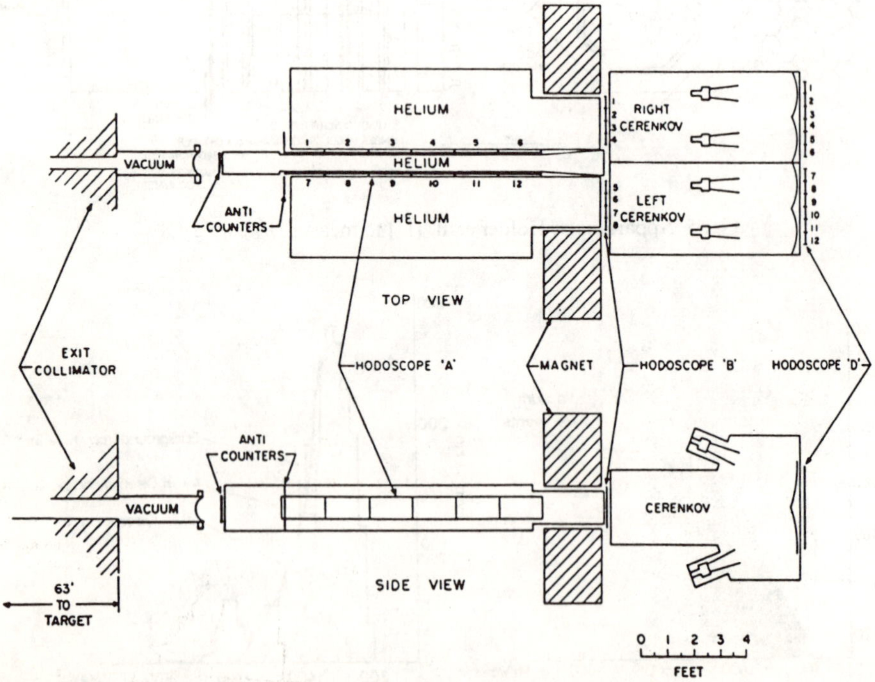


Fig. 7 Apparatus of Bennett et al. [12] for determining the charge asymmetry in $K_{e3}$ decay. The electrons and positrons are charge selected by the magnet and identified in the gas (ethylene) threshold Cherenkov counter.

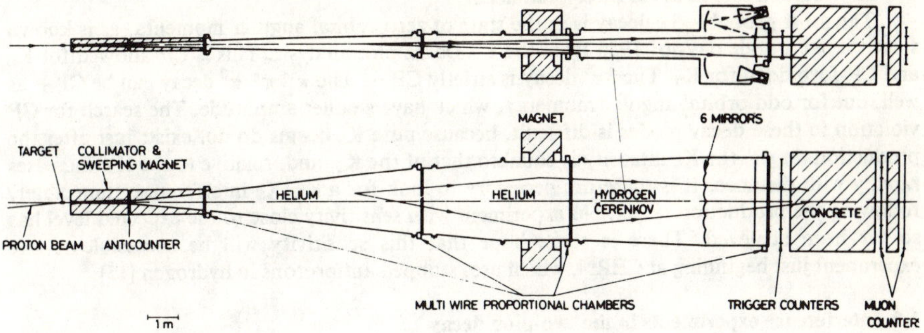


Fig. 8 Apparatus of Geweniger et al. [14] for the determination of $K_S$–$K_L$ interference in the charge symmetry of the decays $K^0 \to e^{\pm} + \pi^{\mp} + \nu$ and $K^0 \to \mu^{\pm} + \pi^{\mp} + \nu$.

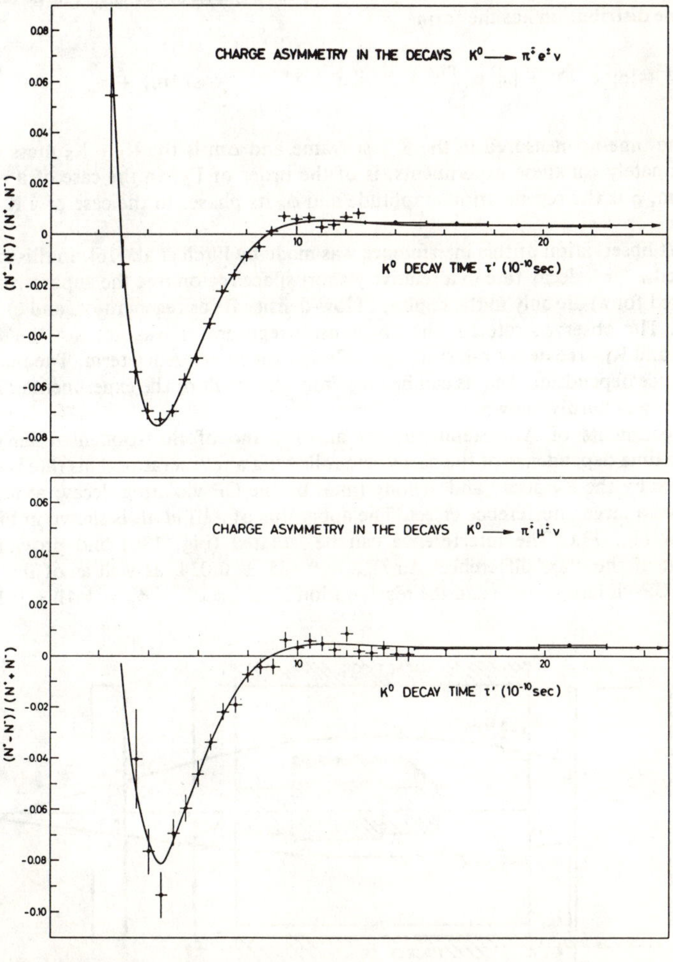


Fig. 9 Charge asymmetries in $K_{e3}$ and $K_{\mu3}$ decays as function of $\tau$, the time in the kaon rest frame since its production. The asymmetry at long times is the CP-violating asymmetry $\delta_\ell$. The oscillating asymmetry at short times is due to $K_S$–$K_L$ interference and does not violate CP (Geweniger et al. [14]).

**1.4 CP violation search in the three-pion decay**

The most probable $3\pi$ decay is in the state of zero orbital angular momenta, as is known since the $\tau$–$\theta$ puzzle days of 1956 from the $\tau^+$ Dalitz plot analysis. This is CP-allowed for $K_L$ and CP-forbidden for $K_S$. The $3\pi^0$ decay is strictly CP –. The $\pi^+\pi^-\pi^0$ decay can be CP + as well, but for odd orbital angular momenta, which have smaller amplitude. The search for CP violation in these decay modes is difficult, because pure $K_S$ beams do not exist; just after the production target, the $K_L$ intensity is equal to that of the $K_S$, and, relative to the $K_S$, increases rapidly with distance. It is therefore necessary to look for a $K_L$–$K_S$ interference immediately following the production target. No experiment with sensitivity close to the expected level has so far been achieved. There is some hope that this sensitivity will be accessible to the experiment just beginning at CERN, which uses stopped antiprotons in hydrogen [15].

**1.5 Interference experiments in the two-pion decay**

A coherent mixture of $K_S$ and $K_L$ will, in its time evolution, show a CP-violating interference, which can be used to measure the phases of $\eta_{+-}$ and $\eta_{00}$. The mixed state can be produced by passing a $K_L$ beam through a 'regenerator' or by observing the decay sufficiently close to the neutral-beam production target so that the $K_S$ amplitude is still large enough. The expected time distribution has the form

$$\frac{dN_{\pi\pi}}{d\tau} \propto |\varrho|^2 e^{-\Gamma_S \tau} + |\eta|^2 e^{-\Gamma_L \tau} + 2|\varrho\eta| e^{-(\Gamma_S+\Gamma_L)\tau/2} \cos(\Delta m \tau + \phi_\varrho - \phi_\eta) \ .$$

Here $\tau$ is the time as measured in the K rest frame and $\Delta m$ is the $K_L - K_S$ mass difference, which, fortunately for these experiments, is of the order of $\Gamma_S$. In the case of a regenerator in a $K_L$ beam, $\varrho$ is the regeneration amplitude and $\phi_\varrho$ its phase. In the case of a K ($\overline{K}$) beam, $\varrho = 1\,(-1)$.

The first observation of this interference was made by Fitch et al. [16]. In this experiment, the integrated $\pi^+\pi^-$ decay rate in a relatively short space region (see the apparatus in Fig. 10) was compared for a) air only in the region, b) low-density Be as regenerator, and c) solid Be as regenerator. The observed rate for the low-density regenerator was not just the sum of the regenerated and $K_L$-free decay rates; it also reflected the interference term. The magnitude of the interference depends on $\Delta m$, as can be seen from the result of the experiment in Fig. 11. At that time, $\Delta m$ was hardly known.

The experiments of Alff-Steinberger et al. [17] and of Bott-Bodenhausen et al. [18] observed the time dependence of the decay rate following a regenerator. This rate is dominated at short times by the $K_S$ decay and at long times by the CP-violating decay; at intermediate times it shows a large interference effect. The apparatus of Alff et al. is shown in Fig. 12, and the result in Fig. 13a. The interference can be isolated (Fig. 13b) and provides a good measurement of the mass difference, $\Delta m/\Gamma_S = 0.455 \pm 0.034$, as well as of the difference between the CP-violating phase and the regeneration phase, $\phi_{+-} - \phi_\varrho = 1.41 \pm 0.18$.

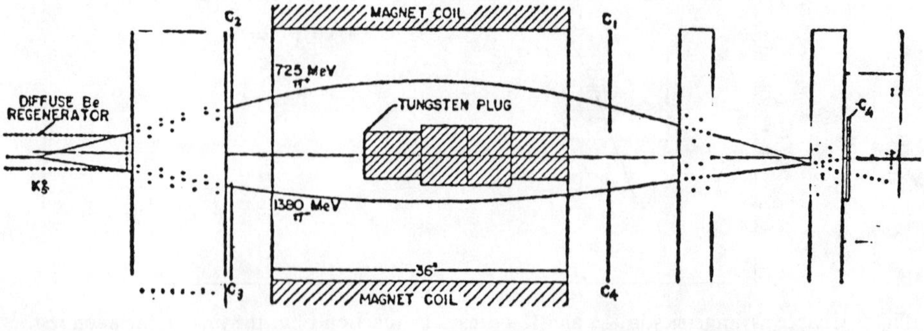


Fig. 10 Spark-chamber spectrometer of Fitch et al. for the demonstration of $K_S$–$K_L$ interference in $\pi^+\pi^-$ decay [16].

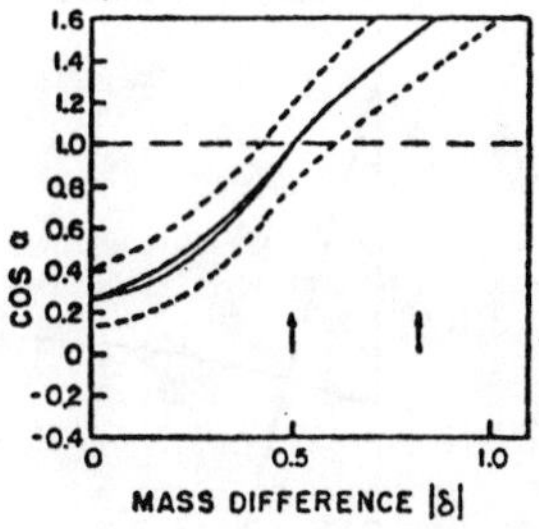


Fig. 11 Result of Fitch et al. [16] for the phase $\alpha = \phi_{\varrho}-\phi_{\eta}$ as a function of the mass difference $\Delta m/\Gamma_S$, which at the time was poorly known.

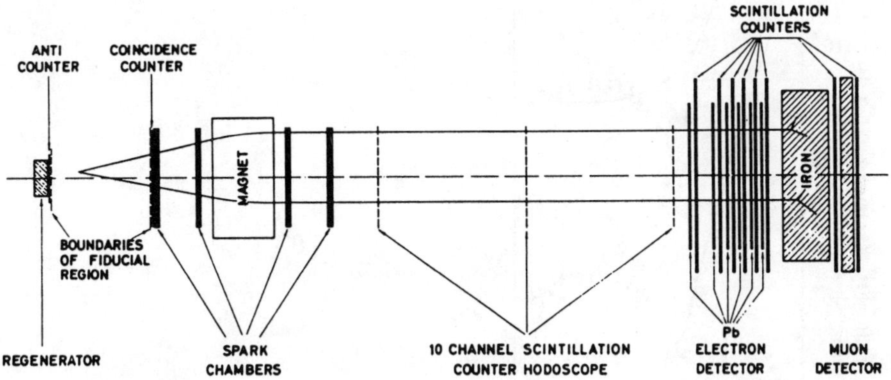


Fig. 12 Spark-chamber spectrometer of Alff-Steinberger et al. [17] used in the observation of the $K_S$-$K_L$ interference in $\pi^+\pi^-$ decay, and its time dependence.

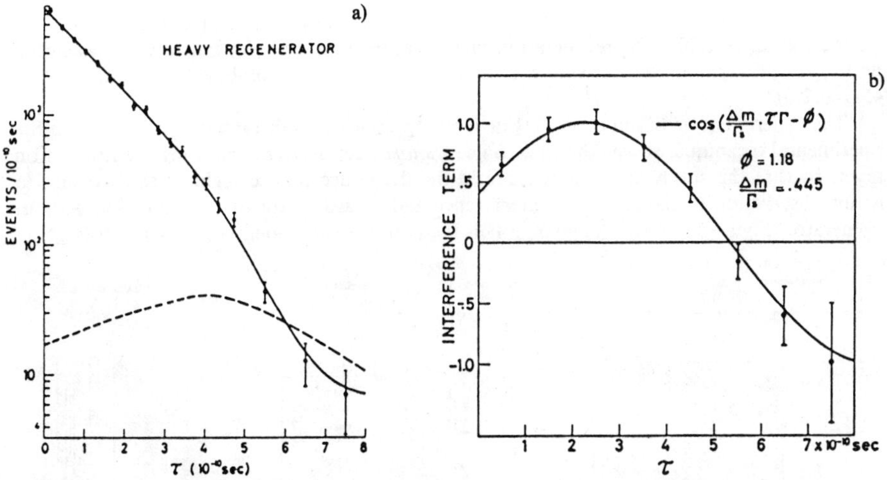


Fig. 13 a) Proper time distribution following a 12 cm thick copper regenerator. The dotted line is the acceptance of the apparatus. At short times the $K_S$ decay dominates; at longer times the interference is very clear. The observed times are too short to reach the region ($\tau > 12 \times 10^{-10}$ s) dominated by $K_L$ decay. b) The isolated interference term permits a measure of $\phi_{\eta}-\phi_{\varrho}$ and $\Delta m/\Gamma_S$.

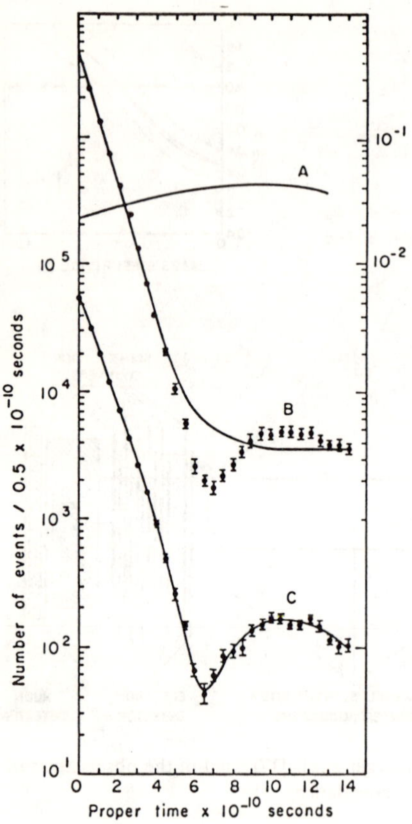


Fig. 14 $K_S$–$K_L$ interference in the $\pi^+\pi^-$ decay following a carbon regenerator (Carithers et al. [19, 20]).

Figure 14 shows the result of the more recent and much more precise experiment of Carithers et al. [19, 20]. The regeneration phase was determined independently by measuring the $K_{\ell 3}$ charge asymmetry following the same regenerator. The combined result is $\phi_{+-} = (45.5 \pm 2.8)°$.

The experiment of Böhm et al. [21] in 1969 represented a substantial step forward in the experimental technique. It was the first to use a *short* neutral beam following the production target, so that the surviving $K_S$ amplitude in the decay acceptance region was sufficient to permit observation of the $K_S - K_L$ interference and measurement of $\phi_{+-}$ directly, without regenerator. Figure 15 shows the layout. This method was subsequently used by Geweniger et

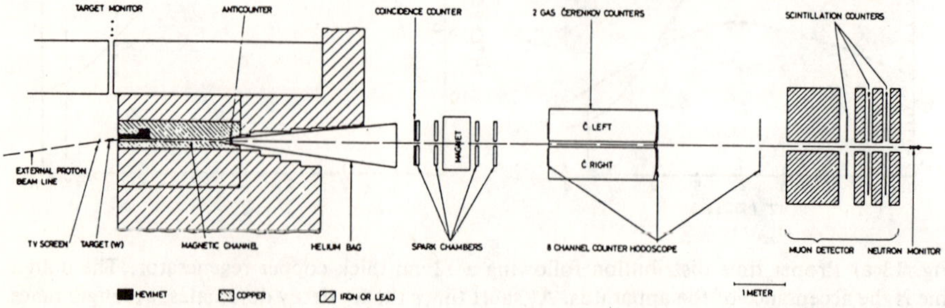


Fig. 15 Apparatus of Böhm et al. [21] for the measurement of the time-dependent $K_S$–$K_L$ interference in the $\pi^+\pi^-$ decay.

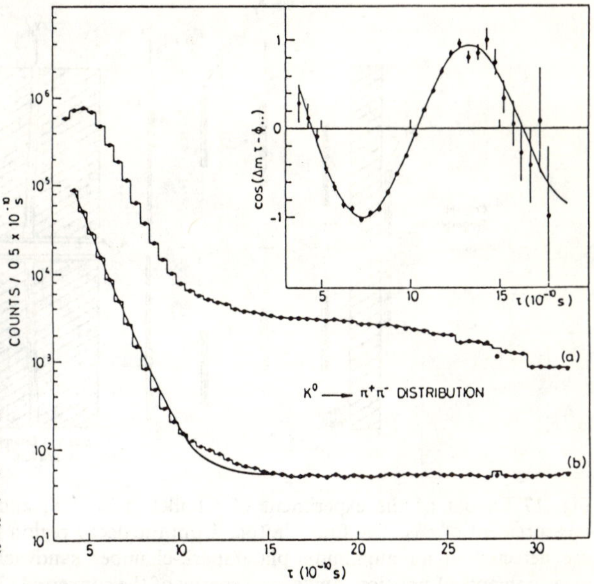


Fig. 16 Proper time distribution of $\pi^+\pi^-$ decay following a target in which mostly K (rather than $\bar{K}$) are produced (Geweniger et al. [22]). The upper histogram is raw data. In the lower histogram the data are corrected for acceptance, and the solid curve is a best fit neglecting interference. The inset shows the isolated interference term which gives a precise value of $\phi_{+-}$.

al. [22, 23] in the most precise measurements available up to now. The experimental arrangement was basically that of Böhm et al. [21]; however, the experiment exploited the multiwire proportional chambers (MWPCs) invented in the mean time by Charpak [24]. It was the first application of this powerful new technique, and large chambers and associated circuitry had first to be developed [25]. The MWPCs permit rates that are hundreds of times higher than those possible with spark chambers, and they can be self-triggered, thus avoiding trigger scintillators, so that there can be less material in the detector. The apparatus is the same (Fig. 8) as that used also in the $K_{\ell 3}$ charge asymmetry measurements already cited [14, 16]. The $K_S - K_L$ interference in the $\pi^+\pi^-$ decay is shown in Fig. 16. The results were:

$$\phi_{+-} = (45.9 \pm 1.6)^\circ \ ,$$

$$|\eta_{+-}| = (2.30 \pm 0.035) \times 10^{-3} \ .$$

In addition, the experiment yielded a more precise measurement of $\Gamma_S$:

$$\Gamma_S = (1.119 \pm 0.006) \times 10^{10}\, s^{-1} \ .$$

Similar experiments were performed to measure $\phi_{00}$, but this proved more difficult. The first results were those of Chollet et al. [26] and of Wolff et al. [27]. The four photons from a decay region following a copper regenerator are detected in the aluminium-plate/spark-chamber arrangement shown in Fig. 17, which permits the measurement of position and direction of each of the four $\gamma$-rays, as well as a crude measurement of each energy. From the proper time distribution of Fig. 18, $\phi_{00}$ was found to be: $\phi_{00} = (38 + 25)^\circ$.

Using the apparatus (Fig. 5) of Holder et al. [11], Barbiellini et al. [28] compared the $K \to 2\pi^0$ intensities following dilute and dense copper regenerators; they found $\phi_{00} - \phi_\varrho = (99.6 \pm 17)^\circ$. Making a comparison with previous results on $\pi^+\pi^-$ decay, $\phi_{+-} - \phi_\varrho = (92 \pm 5.5)^\circ$, the phase difference is $\phi_{00} - \phi_{+-} = (7.6 \pm 18)^\circ$.

Until very recently, the most precise experiment on $\phi_{00}$ and the phase difference was that of Christenson et al. [29], using a short neutral beam, a proportional wire spectrometer for the charged mode, and lead-glass Cherenkov detector blocks for the neutral mode, as shown in Fig. 19. The efficiency-corrected proper time distribution and the isolated interference term are shown in Fig. 20. The results are the following:

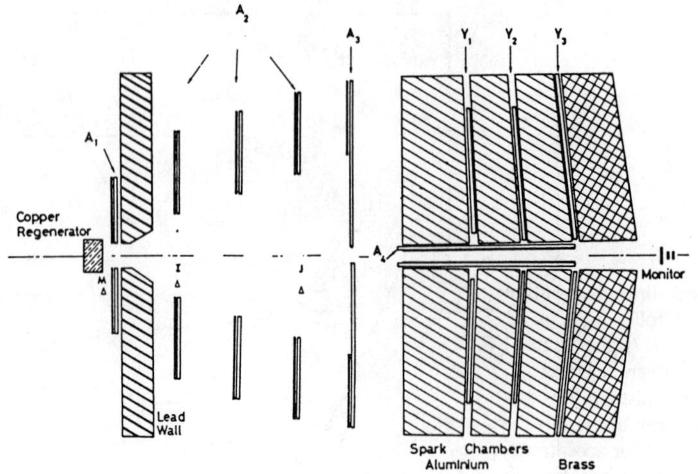


Fig. 17 Layout of the experiment of Chollet et al. [26] and of Wolff et al. [27] for the measurement of $\phi_{00}$. The four photons from the decay region following a copper regenerator are detected in an aluminium-plate/spark-chamber sandwich arrangement which permits measurements of position, angle, and energy of the converted photons.

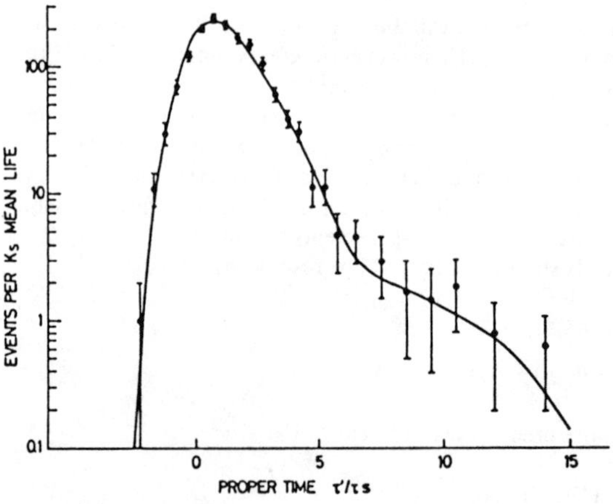


Fig. 18 Proper time distribution of the $K_S \to \pi^0\pi^0$ decay following a copper regenerator, as observed by Chollet et al. [26].

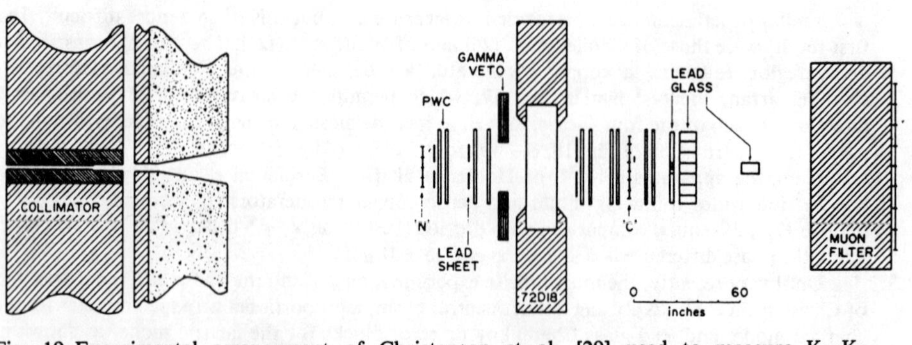


Fig. 19 Experimental arrangement of Christenson et al. [29] used to measure $K_S$-$K_L$ interference in the neutral and charged two-pion decay modes.

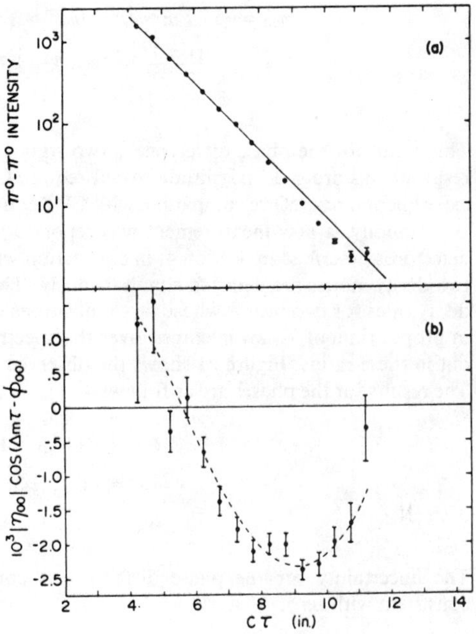


Fig. 20 Result of Christenson et al. 1979 [29]: a) time dependence of decay rate. The solid line is the expected rate without interference; b) isolated interference term.

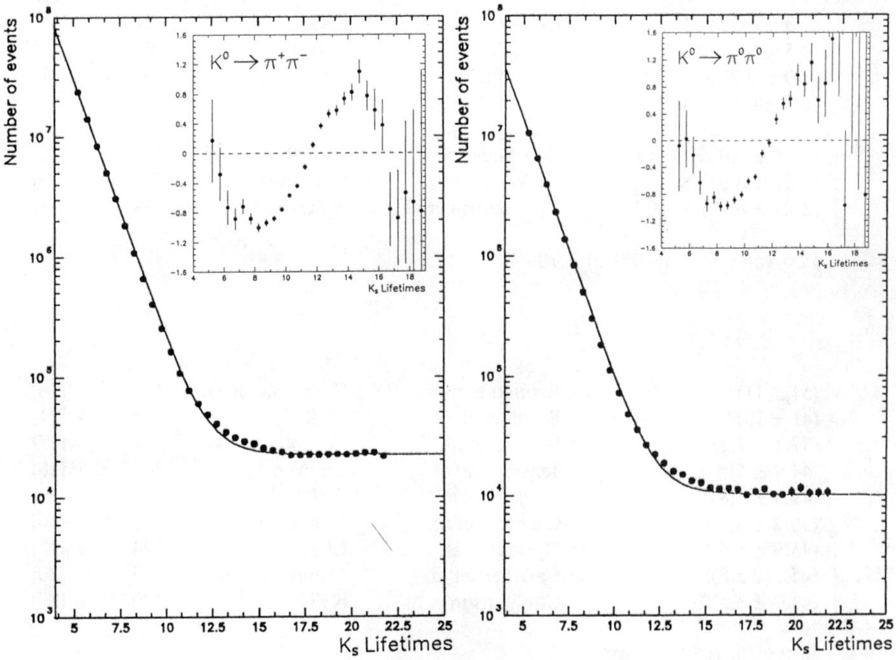


Fig. 21 The rate of decay of kaons to charged and neutral pions as a function of the $K_S$ lifetime. Superimposed are the fitted lifetime distributions without interference. The insets show the interference terms extracted from the data. (From Ref. [30].)

$$\phi_{00} = (55.7 \pm 5.8)^\circ \ , |\eta_{00}| = (2.33 \pm 0.18) \times 10^{-3} \ ,$$
$$\phi_{+-} = (41.7 \pm 3.5)^\circ \ , |\eta_{+-}| = (2.27 \pm 0.12) \times 10^{-3} \ ,$$
$$\phi_{00} - \phi_{+-} = (12.6 \pm 6.3)^\circ \ , |\eta_{00}/\eta_{+-}| = 1.00 \pm 0.09 \ .$$

This result for the phase difference is two standard deviations from a null result. A non-zero result of this order of magnitude would require CPT violation (see Section 2). Since all other experimental results are compatible with CPT conservation, this result was of great interest.

Happily, a new measurement was reported very recently by the CERN group [30]. The detector is described in Section 4, in connection with the result of this experiment for $\epsilon'/\epsilon$. The two decay modes are detected simultaneously. The phase is determined from the ratios of the decay rates for two beams whose target positions differ by 14 m, corresponding to a difference in proper time of $\sim 2\tau_s$ averaged over the spectrum. The acceptance of the apparatus drops out in these ratios. Figure 21 shows the observed decay time distributions for the two modes. The results for the phases are as follows:

$$\phi_{+-} = (46.8 \pm 1.5)^\circ \ ,$$
$$\phi_{00} = (47.1 \pm 2.3)^\circ \ ,$$
$$\phi_{00} - \phi_{+-} = (0.2 \pm 2.9)^\circ \ .$$

The uncertainty for the phase difference is considerably reduced, and the value itself is consistent with zero.

### 1.6 Summary of experimental results on CP violation

$|\eta_{+-}|$

| | | | | Ref. |
|---|---|---|---|---|
| $(2.3 \pm 0.2) \times 10^{-3}$ | Christenson et al. | BNL | '64 | [5] |
| $(2.5 \pm 0.3) \times 10^{-3}$ | Abashian et al. | BNL | '64 | [31] |
| $(3.0 \pm 0.7) \times 10^{-3}$ | de Bouard et al. | CERN | '65 | [32] |
| $(2.3 \pm 0.2) \times 10^{-3}$ | Galbraith et al. | Nimrod | '65 | [33] |
| $(2.23 \pm 0.05) \times 10^{-3}$ | Messner et al. | SLAC | '73 | [34] |
| $(2.30 \pm 0.035) \times 10^{-3}$ | Geweniger et al. | CERN | '74 | [22] |
| $(2.25 \pm 0.05) \times 10^{-3}$ | De Voe et al. | Chicago | '77 | [35] |
| $(2.27 \pm 0.12) \times 10^{-3}$ | Christenson et al. | NYU | '79 | [29] |

Best value $(2.27 \pm 0.025) \times 10^{-3}$

$\phi_{+-}$

| | | | | |
|---|---|---|---|---|
| $(51 \pm 11)^\circ$ | Bennett et al. | Columbia–BNL | '68 | [36] |
| $(41 \pm 15)^\circ$ | Böhm et al. | CERN | '69 | [21] |
| $(37.1 \pm 7.1)^\circ$ | Bennett et al. | Columbia–BNL | '69 | [37] |
| $(44.7 \pm 5)^\circ$ | Jensen et al. | Chicago | '69 | [38] |
| $(49.3 \pm 6.8)^\circ$ | Faissner et al. | CERN | '69 | [39] |
| $(36.2 \pm 6.1)^\circ$ | Carnegie et al. | Princeton–BNL | '72 | [40] |
| $(45.9 \pm 1.6)^\circ$ | Gjesdal et al. | CERN | '74 | [23] |
| $(45.5 \pm 2.8)^\circ$ | Carithers et al. | Columbia–BM | '75 | [20] |
| $(41.7 \pm 3.5)^\circ$ | Christenson et al. | NYU | '79 | [29] |

Best value $(45.2 \pm 1.4)^\circ$

$|\eta_{00}|$

| | | | | |
|---|---|---|---|---|
| $(2.2 \pm 0.4) \times 10^{-3}$ | Budagov et al. | CERN | '68 | [9] |
| $(2.3 \pm 0.3) \times 10^{-3}$ | Banner et al. | Princeton | '68 | [10] |
| $(3.6 \pm 0.6) \times 10^{-3}$ | Gaillard et al. | CERN | '69 | [41] |
| $(2.0 \pm 0.23) \times 10^{-3}$ | Barmin et al. | Serpukhov | '70 | [42] |

$|\eta_{00}/\eta_{+-}|$

| | | | | |
|---|---|---|---|---|
| $(1.03 \pm 0.07)$ | Banner et al. | Princeton | '72 | [43] |
| $(1.00 \pm 0.06)$ | Holder et al. | CERN | '72 | [11] |
| $(1.00 \pm 0.09)$ | Christenson et al. | NYU | '79 | [29] |

Best value $(1.01 \pm 0.04)$

$\phi_{00}$

| | | | | |
|---|---|---|---|---|
| $(51 \pm 30)°$ | Chollet et al. | CERN | '70 | [26] |
| $(38 \pm 25)°$ | Wolf et al. | CERN | '71 | [27] |

$\phi_{00} - \phi_{+-}$

| | | | | |
|---|---|---|---|---|
| $(7.6 \pm 18)°$ | Barbiellini et al. | CERN | '73 | [28] |
| $(12.6° \pm 6.2)°$ | Christenson et al. | NYU | '79 | [29] |
| $(0.2 \pm 2.9)°$ | Barr et al. | CERN | '89 | [30] |

Best value $(2.4 \pm 2.6)°$

$\delta_\ell$

| | | | | |
|---|---|---|---|---|
| $(2.24 \pm 0.36) \times 10^{-3}$ | Bennett et al. | Columbia–BNL | '67 | [12] |
| $(4.05 \pm 1.35) \times 10^{-3}$ | Dorfan et al. | SLAC | '67 | [13] |
| $(3.46 \pm 0.33) \times 10^{-3}$ | Marx et al. | Columbia–BNL | '70 | [44] |
| $(2.8 \pm 0.5) \times 10^{-3}$ | Piccioni et al. | SLAC | '72 | [45] |
| $(3.33 \pm 0.5) \times 10^{-3}$ | Williams et al. | Yale–BNL | '73 | [46] |
| $(3.18 \pm 0.38) \times 10^{-3}$ | Fitch et al. | Princeton | '73 | [47] |
| $(3.41 \pm 0.18) \times 10^{-3}$ | Geweniger et al. | CERN | '74 | [14] |
| $(3.13 \pm 0.29) \times 10^{-3}$ | Geweniger et al. | CERN | '74 | [14] |

Best value $(3.33 \pm 0.15) \times 10^{-3}$
Search in $3\pi$ decay

$\eta_{000}$

| | | | | |
|---|---|---|---|---|
| $0.08 \pm 0.18 + i(0.05 \pm 0.27)$ | Barmin et al. | Serpukhov | '83 | [48] |

$\eta_{+-0}$

| | | | | |
|---|---|---|---|---|
| $0.13 \pm 0.2 + i(0.17 \pm 27)$ | Metcalf et al. | CERN | '72 | [49] |
| $0.05 \pm 0.17 + i(0.15 \pm 0.33)$ | Barmin et al. | Serpukhov | '85 | [50] |
| $0.04 \pm 0.03 + i(0.015 \pm 0.07)$ | Border et al. | Serpukhov | '86 | [51] |

## 2. PHENOMENOLOGY OF CP VIOLATION IN K DECAY

### 2.1 Early history

In the beginning [52] there were '$V^0$'s'—new neutral particles decaying to two charged. The $K^0$ and its decay into two pions emerged clearly in 1952 [53]. Following the discovery of the isotopic spin–electric charge relationships of the strange particles in 1953 by Gell-Mann [54], Gell-Mann and Pais [55] were the first to notice that there must be two $K^0$'s, with opposite assignment of isospin projection and strangeness, $K^0$ and $\bar{K}^0$. This was beautifully confirmed by Landé et al. in 1956 [56].

It follows that if CP were conserved in the decay interaction, the two states $|K_1\rangle = |K\rangle + |\bar{K}\rangle$ and $|K_2\rangle = |K\rangle - |\bar{K}\rangle$ would be eigenstates of the weak interaction, with well-defined but different lifetimes, slightly different (second order in the weak interaction) masses, and with the $K_1$ decaying to CP-even and the $K_2$ to CP-odd final states. Both $K_1$ and $K_2$ evolve exponentially with time:

$$\frac{d|K_1\rangle}{d\tau} = -iM_S|K_1\rangle\,, \qquad \frac{d|K_2\rangle}{d\tau} = -iM_L|K_2\rangle\,,$$

with

$$\begin{aligned} M_S &= m_S - i\Gamma_S/2 & \qquad M_L &= m_L - i\Gamma_L/2 \\ m_S &= 497.7\ \text{MeV} & m_L &= 497.7\ \text{MeV} \\ \Gamma_S &= 1.121 \pm 10^{10}\ \text{s}^{-1} & \Gamma_L &= 1.93 \pm 10^{7}\ \text{s}^{-1}\,. \end{aligned}$$

The dominant branching ratios are

| | | | |
|---|---|---|---|
| $\pi^+\pi^-$ | 69% | $\pi^+\pi^-\pi^0$ | 12.4% |
| $\pi^0\pi^0$ | 31% | $\pi^0\pi^0\pi^0$ | 21.5% |
| | | $\pi^\pm\mu^\mp\nu$ | 27.1% |
| | | $\pi^\pm e^\mp\nu$ | 38.7% |

The mass difference is

$$\Delta m \equiv m_L - m_S = 0.477\Gamma_S \simeq 10^{-16}\, m_K\,.$$

The $K_1$ and $K_2$ are orthogonal, and there can be no interference between $K_1$ and $K_2$ in decays to any CP eigenstate.

### 2.2 $K_1$–$K_2$ mixing

CP violation changes all this. The phenomenology of CP violation in $K^0$ decay was first discussed by Wu and Yang [57]. There are still two exponentially decaying eigenstates of the weak interaction, but these are now mixtures of $K_1$ and $K_2$:

$$|K_S\rangle = |K_1\rangle + (\tilde{\epsilon} + \delta)\,|K_2\rangle \qquad \text{and} \qquad |K_L\rangle = |K_2\rangle + (\tilde{\epsilon} - \delta)\,|K_1\rangle\,.$$

The observed CP violation is small, so we anticipate that $\tilde{\epsilon}$ and $\delta$ are small and neglect higher-order terms.

$K_S$ and $K_L$ are no longer orthogonal:

$$\langle L|S\rangle = 2\,\text{Re}\,\tilde{\epsilon} + 2i\,\text{Im}\,\delta\,.$$

### 2.4 Mass matrix

The time evolution of a general neutral-kaon state $\alpha|K\rangle + \beta|\bar{K}\rangle$ can also be expressed by the mass matrix:

$$\frac{d}{d\tau}\begin{pmatrix}\alpha\\ \beta\end{pmatrix} = -i\begin{pmatrix}M & A\\ B & \bar{M}\end{pmatrix}\begin{pmatrix}\alpha\\ \beta\end{pmatrix}.$$

The elements of this matrix, in terms of $M_S$, $M_L$, and the mixing parameters $\tilde{\epsilon}$ and $\delta$ are then:

$$M = \frac{M_S + M_L}{2} + \delta(M_S - M_L)\,,$$

$$\bar{M} = \frac{M_S + M_L}{2} - \delta(M_S - M_L)\,,$$

$$A = \frac{M_S - M_L}{2}(1 + 2\tilde{\epsilon})\,,$$

$$B = \frac{M_S - M_L}{2}(1 - 2\tilde{\epsilon})\,.$$

**2.4 CPT conservation**

If CPT is conserved—and therefore T is violated—then

$$\langle K|T|K\rangle = \langle \bar{K}|T|\bar{K}\rangle\,.$$

Therefore

$$M = \bar{M}$$

and

$$\delta = 0\,.$$

**2.5 T conservation**

If, instead, T were conserved—and therefore CPT violated—then

$$\langle K|T|\bar{K}\rangle = \langle \bar{K}|T|K\rangle\,.$$

Therefore

$$A = B$$

and

$$\tilde{\epsilon} = 0\,.$$

As we will see, the experiments show clear T violation and are consistent with CPT conservation.

**2.6 Unitarity [58]**

Unitarity here is the statement of probability conservation: the decay of the kaon is compensated by the appearance of the decay product,

$$-\frac{d}{d\tau}|\langle\psi^*|\psi\rangle|^2 = \sum_F |\langle F|T|\psi\rangle|^2\,.$$

Here $|\psi\rangle$ is a general $K^0$ state: $|\psi\rangle = \alpha|S\rangle + \beta|L\rangle$, and F is the final state. Since $\alpha$ and $\beta$ are arbitrary, this results in the two intuitively immediate relations

i) $\Gamma_S = \Sigma_F |\langle F|T|S\rangle|^2$

and

ii) $\Gamma_L = \Sigma_F |\langle F|T|L\rangle|^2$,

but also in the less obvious result from the cross-term:

$$\text{iii)}\quad \left(-i\Delta m + \frac{\Gamma_S + \Gamma_L}{2}\right)\langle L|S\rangle = \sum_F \langle F|T|L\rangle^* \langle F|T|S\rangle\,. \qquad (2)$$

Both sides of Eq. (2) are CP-violating: the left, because $\langle L|S\rangle = 0$ if CP were conserved; the right, because the final states of S and L could not be the same.

Let $\eta_F$ be the CP-violating amplitude ratio:

$$\eta_F = \frac{\langle F|T|L\rangle}{\langle F|T|S\rangle} \qquad \text{if F is CP +}\,,$$

and

$$\eta_F = \frac{\langle F|T|S\rangle^*}{\langle F|T|L\rangle^*} \qquad \text{if F is CP} - .$$

Then Eq. (2) becomes

$$\left(-i\Delta m + \frac{\Gamma_S + \Gamma_L}{2}\right) \langle L|S\rangle = \sum_F \eta_F^* \Gamma_F . \tag{2'}$$

The right-hand side is dominated by the two-pion decay. Present upper limits on CP violation in other channels ensure that these contribute at most several per cent. It is therefore an adequate approximation for the following argument to write Eq. (2′):

$$\left(-i\Delta m + \frac{\Gamma_S + \Gamma_L}{2}\right) \langle L|S\rangle = \eta_{+-}^* \Gamma_{S,+-} + \eta_{00}^* \Gamma_{S,00} . \tag{2''}$$

Experimentally, $\eta_{+-} = (2.27 \pm 0.025) \times 10^{-3}$, $\phi_{+-} = (45.2 \pm 1.3)°$, $(\eta_{00}/\eta_{+-}) = 1.01 \pm 0.04$, and $\phi_{00} - \phi_{+-} = (2 \pm 2.5)°$, so that Eq. (2″) becomes:

$$\left(-i\frac{\Delta m}{\Gamma_S} + \frac{1}{2}\right) \langle L|S\rangle = \eta_{\pi\pi}^* ,$$

where $|\eta_{\pi\pi}| = (2.27 \pm 0.03) \times 10^{-3}$ and $\phi_{\pi\pi} = (45.8 \pm 1.6)°$.

*2.6.1 Unitarity, CPT conservation, and T violation*

Since CP is violated, either CPT is violated, or T is violated, or both

| **If CPT is good** | **If T is good** |
|---|---|
| $\delta = 0$ | $\tilde{\epsilon} = 0$ |
| $\langle L\|S\rangle = 2\,\text{Re}\,\tilde{\epsilon}$ | $\langle L\|S\rangle = 2i\,\text{Im}\,\delta$ |
| $2\left(\frac{i\Delta m}{\Gamma_S} + \frac{1}{2}\right) \text{Re}\,\tilde{\epsilon} = \eta_{\pi\pi}$ , | $2\left(\frac{\Delta m}{\Gamma_S} - \frac{i}{2}\right) \text{Im}\,\delta = \eta_{\pi\pi}$ , |
| $\therefore \phi_{\pi\pi} = \tan^{-1}\frac{2\Delta m}{\Gamma_S} = (43.7 \pm 0.3)°$ | $\therefore \phi_{\pi\pi} = \tan^{-1}\frac{-\Gamma_S}{2\Delta m} = (138.7 \pm 0.3)°$ |

The experiment is in good agreement with CPT conservation. *T-violation is clearly demonstrated.* From now on we therefore take $\delta = 0$.

## 2.7 Isospin analysis of the two-pion decay

The states of well-defined charge can be resolved into states of well-defined isospin:

$$|+-\rangle = \sqrt{\frac{2}{3}}\,|I = 0\rangle + \sqrt{\frac{1}{3}}\,|I = 2\rangle ,$$

$$|00\rangle = -\sqrt{\frac{1}{3}}\,|I = 0\rangle + \sqrt{\frac{2}{3}}\,|I = 2\rangle .$$

Let $A\,e^{i\delta_0} = \langle I = 0|T|K\rangle$, where $\delta_0$ is the $\pi\pi$ scattering phase; then

$$A_0^*\,e^{i\delta_0} = \langle I = 0|T|\bar{K}\rangle .$$

This is a consequence of CPT conservation and unitarity (Watson's theorem).

Similarly, let

$$A_2 e^{i\delta_2} = \langle I = 2|T|K\rangle ,$$

then

$$A_2^* e^{i\delta_2} = \langle I = 2|T|\bar{K}\rangle ;$$

$A_0$ and $A_2$ are real if CP is conserved. Let

$$\epsilon = \frac{\langle I = 0|T|L\rangle}{\langle I = 0|T|S\rangle} = i\,\frac{\text{Im}\, A_0}{\text{Re}\, A_0} + \tilde{\epsilon} ,$$

and

$$\epsilon' = \frac{1}{\sqrt{2}}\left[\frac{\langle I = 2|T|L\rangle}{\langle I = 0|T|S\rangle} - \tilde{\epsilon}\,\frac{\langle I = 2|T|S\rangle}{\langle I = 0|T|S\rangle}\right]$$

$$= \frac{i}{\sqrt{2}}\, e^{i(\delta_2 - \delta_0)}\left(\frac{\text{Im}\, A_2}{\text{Re}\, A_0} - \frac{\text{Im}\, A_0}{\text{Re}\, A_0}\,\frac{\text{Re}\, A_2}{\text{Re}\, A_0}\right) .$$

Then

$$\eta_{+-} = \epsilon + \epsilon' \qquad \text{and} \qquad \eta_{00} = \epsilon - 2\epsilon' ,$$

where we have ignored terms that are smaller by $\text{Re}\, A_2/\text{Re}\, A_0$. This ratio measures the admixture of $I = 2$ in the decay amplitude, and is obtained from the $K_S$ branching ratios:

$$\text{Re}\, A_2/\text{Re}\, A_0 \simeq \left[\frac{\Gamma_{S,+-}}{\Gamma_{S,00}}\,\frac{1 - (2m_{\pi^0}/m_K)^2}{1 - (2m_{\pi^+}/m_K)^2} - 2\right] \div [\cos(\delta_2 - \delta_0)\, 6\sqrt{2}] .$$

$$\simeq 0.049 .$$

The measurements of the $\pi\pi$ scattering phase $\delta_2 - \delta_0$,

| | | |
|---|---|---|
| $\delta_2 - \delta_0$ | $= (41 \pm 8)°$ | Devlin and Dickey 1979 [59] |
| | $= (29 \pm 3)°$ | Biswas et al. 1981 [60] |
| | $= (56 \pm 3)°$ | Kamal 1987 [61] |

are not in agreement within their errors. Here it will be adequate to use $\delta_2 - \delta_0 = (45 \pm 15)°$ as proposed in a recent review by Wahl [62].

The quantity $\eta_{+-}\Gamma_{+-} + \eta_{00}\Gamma_{00}$ occurring in the unitarity relation, is equal to $\epsilon\Gamma_S$ with high precision. The phase of $\epsilon$ is therefore predicted by unitarity. Since the phase of $\epsilon'$ is also known, the phenomenology doubly constrains the two complex amplitudes $\eta_{+-}$ and $\eta_{00}$; $\epsilon$ and $\epsilon'$ are expected to be nearly collinear, as illustrated in Fig. 22.

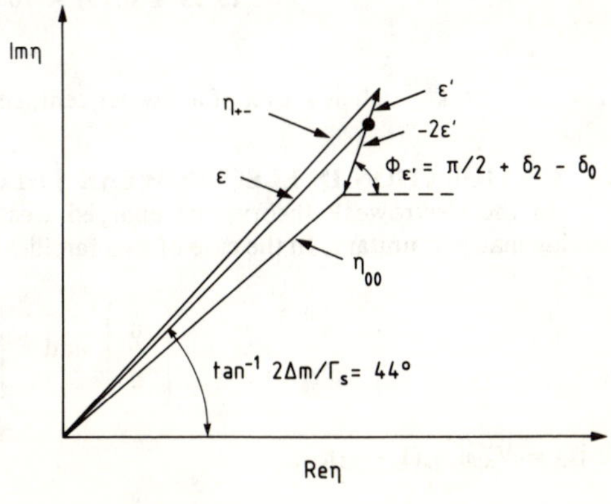


Fig. 22 Relationship between the amplitude ratios $\eta_{+-}$, $\eta_{00}$, $\epsilon$ and $\epsilon'$.

**2.8 Charge asymmetry in semileptonic decay**

If x is the ratio of the $\Delta S = \Delta Q$ violating amplitude to the $\Delta S = \Delta Q$ conserving amplitude in the $K^0$ semileptonic decay:

$$x = \frac{\langle \ell^- |T| K\rangle}{\langle \ell^+ |T| K\rangle},$$

then the time-dependent charge asymmetry for a state that is initially $|K\rangle$ is

$$\delta_\ell(\tau) = \frac{2(1-|x|^2)\,[e^{-\Gamma_S \tau} + \mathrm{Re}\,\epsilon\, e^{-\Gamma_L \tau} + e^{(\Gamma_S+\Gamma_L)\tau/2}\cos \Delta m\tau]}{|1+x|^2 e^{-\Gamma_S\tau} + |1-x|^2 e^{-\Gamma_L\tau} + 4\,\mathrm{Im}\,x\, e^{-(\Gamma_S+\Gamma_L)\tau/2}\sin \Delta m\tau}.$$

For a state that is initially $|\bar{K}\rangle$ the sign of the interference in the numerator changes. In the quark model, $x = 0$, and the asymmetry simplifies:

$$\delta_\ell(\tau) = \frac{2\,[e^{-\Gamma_S\tau} + \mathrm{Re}\,\epsilon\, e^{-\Gamma_L\tau} + e^{-(\Gamma_S+\Gamma_L)\tau/2}\cos \Delta m\tau]}{e^{-\Gamma_S\tau} + e^{-\Gamma_L\tau}}.$$

The oscillating term here is CP-conserving. The CP-violating asymmetry is the asymmetry left over at long times:

$$\delta_\ell = 2\,\mathrm{Re}\,\epsilon\,.$$

The interference (not CP-violating) at shorter times has had an important application in the determination of the regeneration phase for those experiments measuring $\phi_{+-}$ and $\phi_{00}$ following a regenerator [19, 37].

**2.9 Mixing only. The superweak model [63]**

If CP violation in K decay is manifested by the mixing parameter $\tilde{\epsilon}$ only, then $\eta_F = \tilde{\epsilon}$ for all decay channels F of well-defined CP. In particular, $\epsilon' = 0$. There is then only one parameter, the strength of the mixing, since the phase is fixed by unitarity. There are five measured quantities and therefore four checks on this model:

| | **Experiment** | **Mixing only** |
|---|---|---|
| $\lvert\eta_{+-}\rvert$ | $(2.27 \pm 0.025) \times 10^{-3}$ | – |
| $\lvert\eta_{00}/\eta_{+-}\rvert$ | $1.01 \pm 0.04$ | 1 |
| $\phi_{+-}$ | $(45.8 \pm 1.4)^\circ$ | $(43.7 \pm 0.3)^\circ$ |
| $\phi_{00} - \phi_{+-}$ | $(2.4 \pm 2.6)^\circ$ | 0 |
| $\delta_\ell$ | $(3.33 \pm 0.15) \times 10^{-3}$ | $(3.30 \pm 0.04) \times 10^{-3}$ |

The four checks, each at a level of a few per cent, support the model.

## 3. CP VIOLATION IN ELECTROWEAK THEORY

In the electroweak theory, the charged weak quark currents are flavour-mixed. The mixing matrix is unitary. In the case of two families [64, 65],

$$\begin{pmatrix} u \\ d \end{pmatrix} \text{ and } \begin{pmatrix} c \\ s \end{pmatrix},$$

if $j^\mu_{\alpha\beta} = V_{\alpha\beta}\bar{\psi}_\alpha \gamma_\mu (1-\gamma_5)\psi_\beta$,

$$V_2 = \begin{pmatrix} V_{ud} & V_{us} \\ V_{cd} & V_{cs} \end{pmatrix}$$

With the proper choice of the arbitrary phases of the quark states,

$$V_2 = \begin{pmatrix} \cos\theta_C & \sin\theta_C \\ -\sin\theta_C & \cos\theta_C \end{pmatrix}$$

There is a single mixing angle and the currents are CP-conserving.

In the case of three families, the unitary mixing matrix has three mixing angles and one complex phase, the latter giving rise to CP violation within the model. This was first noted by Kobayashi and Maskawa (KM) in 1973, long before the third family had been discovered [4]. For three families, the KM mixing matrix is

$$V_3 = \begin{pmatrix} V_{ud} & V_{us} & V_{ub} \\ V_{cd} & V_{cs} & V_{cb} \\ V_{td} & V_{ts} & V_{tb} \end{pmatrix} = \begin{pmatrix} c_1 & c_3 s_1 & s_1 s_3 \\ -c_2 s_1 & c_1 c_2 c_3 - s_2 s_3 e^{i\delta} & c_1 c_2 s_3 - c_3 s_2 e^{i\delta} \\ s_1 s_2 & -c_1 c_3 s_2 - c_2 s_3 e^{i\delta} & -c_1 s_2 s_3 + c_2 c_3 e^{i\delta} \end{pmatrix}$$

where $s_1 = \sin\theta_1$, $c_1 = \cos\theta_1$, etc.; $s_1$ is essentially the Cabibbo angle, determined from hyperon and kaon semileptonic decay: $s_1 = 0.22$; $s_3$ is known with modest precision from the B lifetime, $s_3 = 0.05 \pm 0.01$; and an upper limit on $s_2$ can be put from the experimental upper limit on the $b \to u$ branching ratio, $s_2 < 0.01$.

The phase $\delta$ gives rise to CP-violation effects. It has been shown by Jarlskog [66], Greenberg [67], and Wu [68] that these effects are proportional to $J = \mathrm{Im}\,(V_{us}V_{cb}V_{ub}^*V_{cs}^*) = \mathrm{Im}\,(V_{ud}V_{cs}V_{us}^*V_{cd}^*) = c_1c_2c_3s_1^2s_2s_3 \sin\delta$.

The smallness of the observed CP-violation effects is immediately understood in terms of the smallness of J. For instance, $\tilde{\epsilon}$ would be expected to be of the order of $J/s_1^2 \gtrsim 0.0025$; the agreement is much too good.

Attempts at a real calculation of $\tilde{\epsilon}$ on the basis of the box diagram

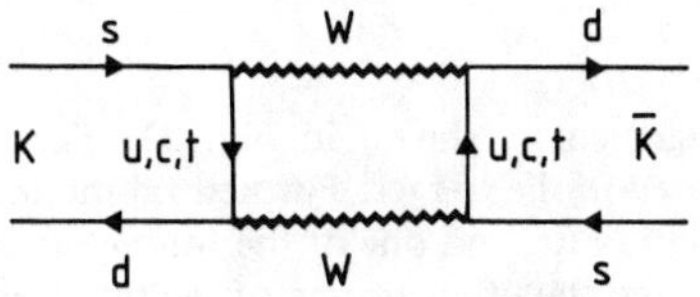


meet with difficulties associated with the problems of dealing with mesons rather than quarks, the long-distance or dispersive contributions, and the unknown top-quark mass (see, for example, the review of Wolfenstein [69]). The model definitively predicts non-zero $\epsilon'$, but again the calculations, based on the 'penguin' diagram

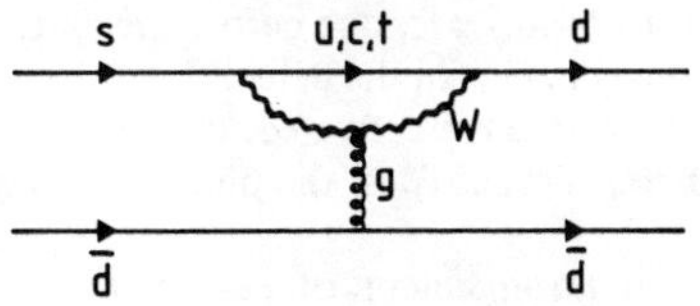


suffer from the same uncertainties. A recent summary of these calculations (Donoghue et al. [70]) gives the result

$$1 \times 10^{-3} < |\epsilon'/\epsilon| < 7 \times 10^{-3}.$$

Such small values of $\epsilon'/\epsilon$ were not accessible to the older experiments.

Since the phase of $\epsilon'$ is within $\sim 15°$ of the phase of $\epsilon$ (Section 2), the magnitude of $\epsilon'$ is best seen as a difference in the magnitudes of $\eta_{+-}$ and $\eta_{00}$ rather than of their phases. With adequate accuracy $|\eta_{+-}| = |\epsilon| + |\epsilon'|$ and $|\eta_{00}| = |\epsilon| - 2|\epsilon'|$, so that $\epsilon'/\epsilon = {}^1\!/_6(1 - |\eta_{00}|/|\eta_{+-}|^2)$.

The summary result $|\eta_{+-}/\eta_{00}| = 1.01 \pm 0.04$ corresponds to $\epsilon'/\epsilon = (3 \pm 12) \times 10^{-3}$, and is not sensitive to the levels expected theoretically. This has encouraged new experimental efforts to reach the required sensitivity.

## 4. $\epsilon'/\epsilon$. DIRECT CP VIOLATION

Since $|\eta_{+-}|^2 = \Gamma_{L,+-}/\Gamma_{S,+-}$ and $|\eta_{00}|^2 = \Gamma_{L,00}/\Gamma_{S,00}$, the measurement of $\epsilon'/\epsilon$ requires the measurement of the double ratio $(\Gamma_{L,00}/\Gamma_{S,00})/(\Gamma_{L,+-}/\Gamma_{S,+-})$ with adequate systematic and statistical accuracy. To achieve the systematic accuracy, it is desirable to observe all four reactions simultaneously, or at least two at a time, so that beam intensities and detection efficiencies cancel. This was not the case for the earlier experiments. The rates should be such that at least $10^5$ events are observed in each channel. The most difficult of the four channels is $K_L \to 2\pi^0$, where the earlier experiments were limited to $\sim 100$ events. The required statistical factor of $\sim 1000$ represented a great challenge, not so much for the kaon beam as for the detector.

The first of the 'new wave' experiments were published in 1985. The sensitivity was considerably improved, but not sufficiently. The Brookhaven–Yale experiment of Black et al. [71] found $\epsilon'/\epsilon = (2 \pm 8) \times 10^{-3}$, and the Chicago–Saclay experiment of Bernstein et al. [72] at Fermilab found $\epsilon'/\epsilon = (-5 \pm 5) \times 10^{-3}$. For the rest of this review we give a brief account of the two most recently published experiments: the second round of the Fermilab experiment [73], and the CERN experiment of Burkhardt et al. [74].

### 4.1 Second Fermilab experiment

Some of the details of this description are taken from the thesis of Jarry [75]; the result is that of Woods et al. [73].

In this experiment the $K_L$ and $K_S$ decays are measured simultaneously. There are two parallel beams of $K_L$, and in one of them a regenerator is inserted ahead of the sensitive decay volume of the detector. The $2\pi$ decays from this beam are essentially $K_S$. The regenerator is alternated from one beam to the other between bursts. The $K_L$ and $K_S$ decays are observed simultaneously. However, the detector is not exactly the same for neutral and charged detection, so the two decay modes are observed alternately.

#### *4.1.1 The detector*

The experimental arrangement is shown in Fig. 23. The 17 m long decay volume is terminated by a scintillator used in the trigger. For neutral-mode observation, 0.5 mm of lead is inserted in front of this scintillator, and one of the four $\gamma$'s is required to convert in it. The materialized pair serves to assign the event to one of the two beams. This piece of lead is the essential difference in the detector for the two decay modes.

The evacuated decay region is extended by 20 m to permit the decay particles to separate; at this point it is terminated by a thin Kevlar–Mylar window.

This is followed by an 18 m long spectrometer to measure the charged tracks. The spectrometer consists of a pair of drift chambers and a magnet, followed by another pair of drift chambers. Two scintillator hodoscopes, one before the magnet and one after the last drift chamber, serve to trigger charged decays. In the detection of the neutral mode, a compensating magnet just following the lead converter is activated. It is adjusted in such a way that the pair, after traversing the two magnets, is reunited in the photon detector, so that to this detector it looks like a photon.

The photon detector is an arrangement of 804 lead-glass blocks (see Fig. 24), each 5.7 cm × 5.7 cm laterally and 20 radiation lengths long. Calibration is effected by means of electrons and of $\pi^0$'s whose origin is known. The spatial resolution obtained is $\approx 3.5$ mm in x and y. The energy resolution is found to be

$$(\Delta E/E)^2 = \surd(6\%/\sqrt{E})^2 + (3\%)^2 \ .$$

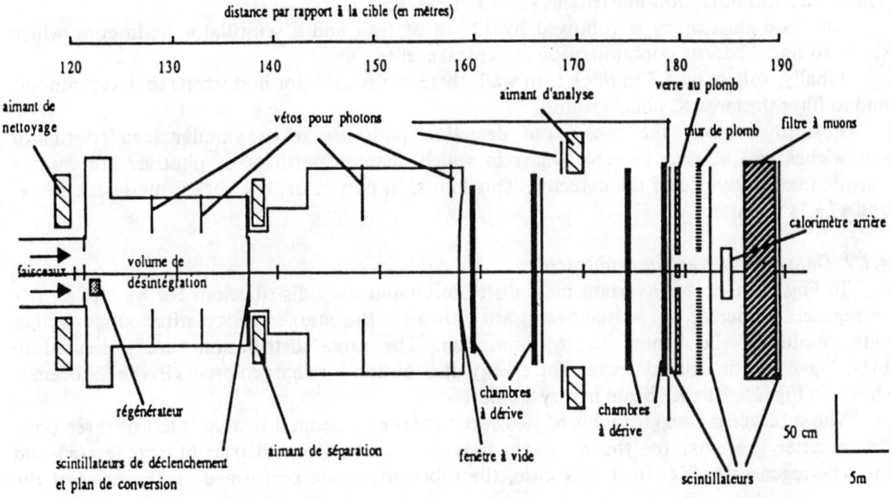


Fig. 23 Experimental arrangement for the experiment of Woods et al. [73].

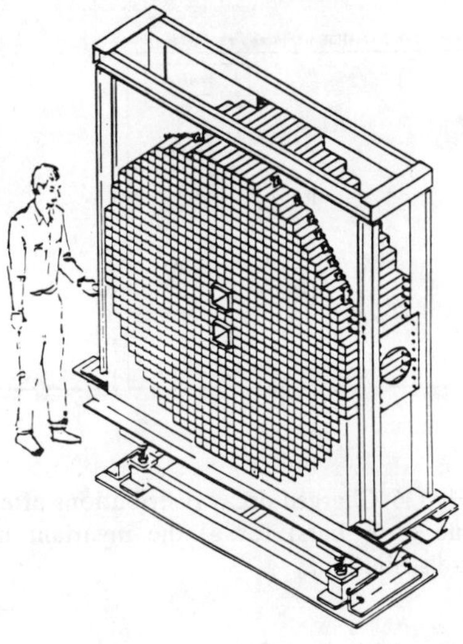

Fig. 24 The photon detector consisting of 804 lead-glass blocks.

Fig. 25 Linearity of the lead-glass detector for electrons and photons.

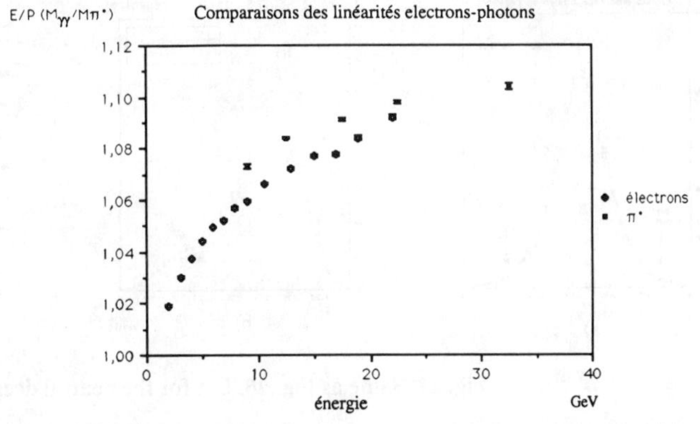

There is a substantial non-linearity as shown in Fig. 25.

The lead-glass array is followed by 12 cm of lead and a scintillator hodoscope which serves to signal hadron contamination in neutral events.

Finally, following a 3 m thick iron wall, there is a scintillator hodoscope to detect muons, and to filter the large $K_{\mu 3}$ background.

Positioned along the decay and detection path are seven annular lead/scintillator sandwiches which serve to veto events in which charged particles or photons are emitted outside the acceptance of the detector. This limits, in particular, the backgrounds due to $3\pi^0$ and $\pi^+\pi^-\pi^0$ decays.

### *4.1.2 Data and background subtraction*

In Figs. 26a–c, the invariant mass distribution and the z distributions for $K_S$ and $K_L$ ($K_S$ = regenerated beam, $K_L$ = free beam) are shown for the charged decays after cuts, together with results of the Monte Carlo simulation. The same distributions are presented in Figs. 27a–c for the neutral decay. The energy distribution for charged events in the $K_S$ beam is shown in Fig. 28. The $K_L$ beam is very similar.

The subtracted background is of two sorts: for the $K_L$ beam it is what is left of three-body decays after the cuts; for the $K_S$ beam it is the decays from diffraction-regenerated and inelastic-regenerated $K_S^0$. In the $K_S$ case, the subtractions are performed on the basis of the

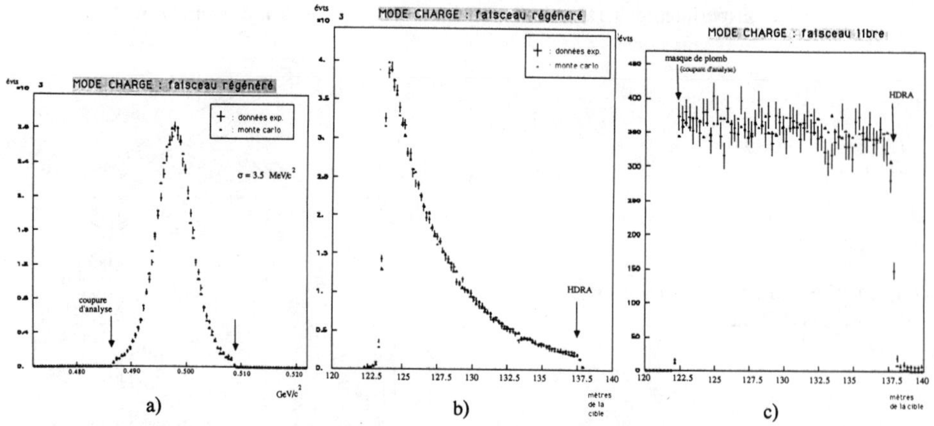


Fig. 26 Charged decay distributions after cuts (error bars) as well as Monte Carlo simulation (no error bars) for a) the invariant mass distribution, b) $K_S$ z-distribution, and c) $K_L$ z-distribution.

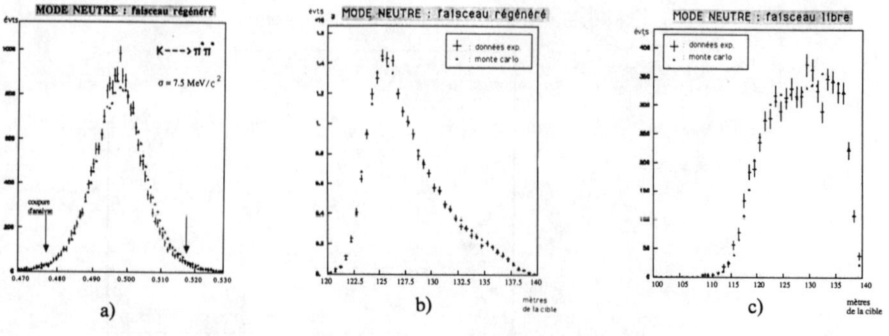


Fig. 27 Same as Fig. 26, but for the neutral decay.

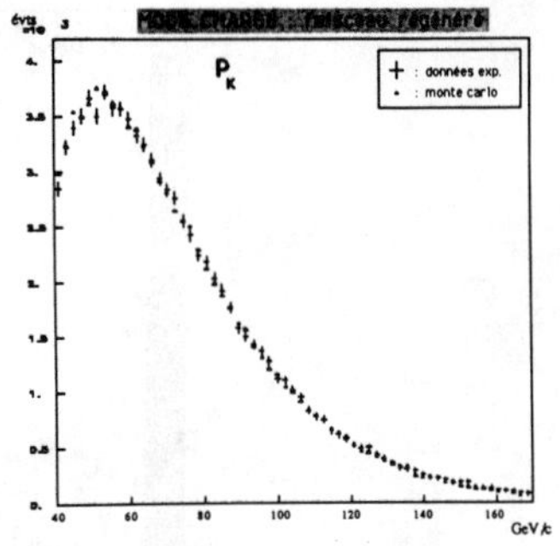


Fig. 28 Energy spectrum of $K_S$ charged events, as well as Monte Carlo simulation.

distributions in the square of the transverse momentum. The unwanted decays give tails in this distribution, and these form the basis for extrapolation to the region of small $p_T^2$ that is accepted. In the $K_L$ case, the observed mass distributions are used to estimate the backgrounds, although in the $K_L \to 2\pi^0$ channel the $p_T^2$ distribution is used as well.

For the $K_S$ subtractions, the $p_T^2$ dependence of the diffraction regeneration is calculated according to a model; the $p_T^2$ dependence of the inelastic regeneration is assumed to be of the form $\alpha\, e^{-\beta p_T^2}$, where $\alpha$ and $\beta$ are fitted to the data. These are shown in Fig. 29 for the neutral decays and in Fig. 30 for the charged decays. The neutral background is evaluated at $(3 \pm 0.3)\%$, the charged at $(0.56 \pm 0.1)\%$. The difference between the two is due to the fact that the charged $p_T^2$ cut for acceptance is 400 $(\text{MeV/c})^2$ whilst the neutral cut is ten times higher.

For the $K_L \to 2\pi^0$ channel, the background is the $3\pi^0$ mode, where one or both extra photons overlay the other four. On the basis of the mass distribution of Fig. 31, the background is estimated to be $(2.9 \pm 1)\%$. Essentially the same result is obtained if one extrapolates in $p_T^2$. For the charged $K_L$ channel, the background is calculated on the basis of the wings of the mass distribution, interpolating to the accepted mass interval with a Monte Carlo calculation, assuming the background to be $K_{e3}$ and $K_{\mu 3}$ (see Fig. 32). A background of $(1.9 \pm 0.3)\%$ is found.

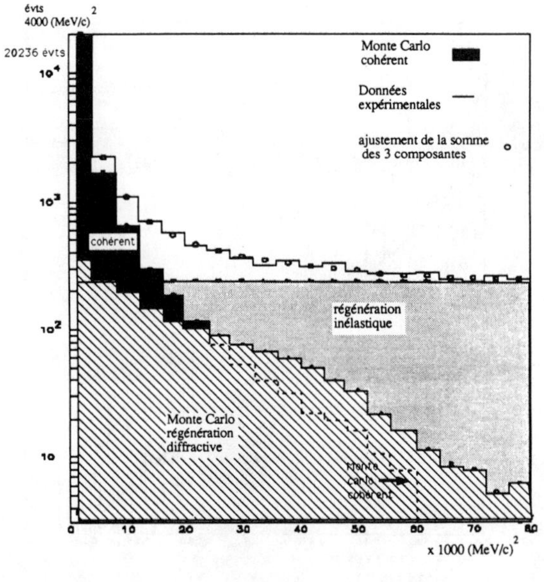


Fig. 29 The $p_T^2$ distribution for $K_S$ neutral decay. The fitted diffractive and inelastic backgrounds are shown. The $p_T^2$ cut for good events is 4000 $(\text{MeV/c})^2$.

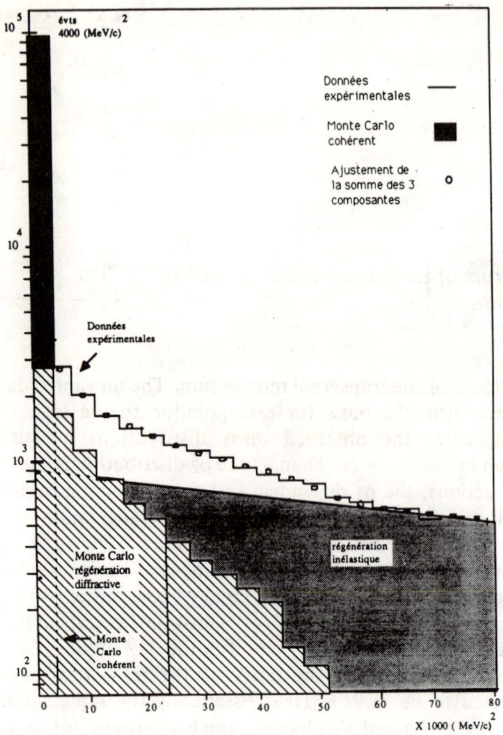


Fig. 30 Same as Fig. 29 but for $K_S$ charged decay. The $p_T^2$ cut for good events is 400 $(MeV/c)^2$.

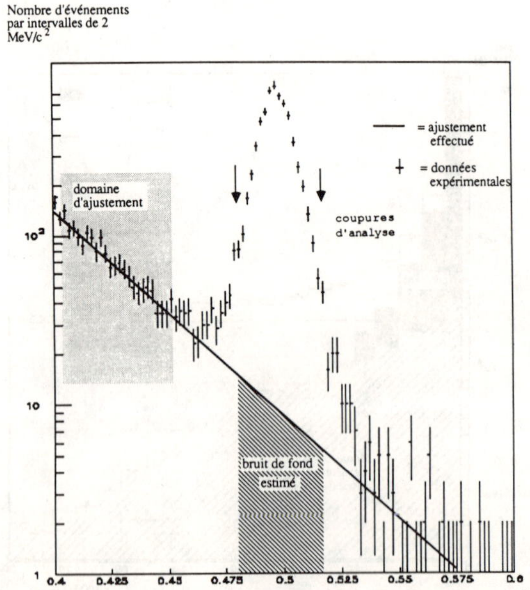


Fig. 31 Invariant mass distribution in $K_L$ neutral decay, and the method of background subtraction.

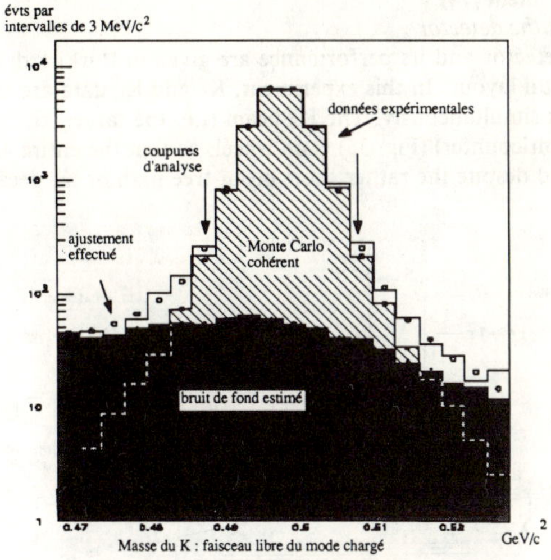


Fig. 32 Mass distribution in $K_L$ charged decay, with indication of how the background is estimated.

### *4.1.3 Results*

The event numbers, after background subtraction, are

$$K_S \to \pi^+\pi^-: \quad 88{,}696$$
$$K_S \to \pi^0\pi^0: \quad 19{,}629$$
$$K_L \to \pi^+\pi^-: \quad 20{,}242$$
$$K_L \to \pi^0\pi^0: \quad 6{,}230$$

The statistics are limited by the $K_L \to 2\pi^0$ channel, as is always the case.

The double ratio is obtained in energy bins but not in z bins, because the z region is inadequately scanned by the $K_S$ decays, as can be seen from Fig. 26. Instead, the z-distributions are integrated. This necessitates a correction for the fact that neutral and charged acceptances have different z-dependences. This correction is performed on the basis of a Monte Carlo calculation, which can be seen in Fig. 33.

The result is $\epsilon'/\epsilon = 0.0032 \pm 0.0028$ stat. $\pm$ 0.0012 syst.

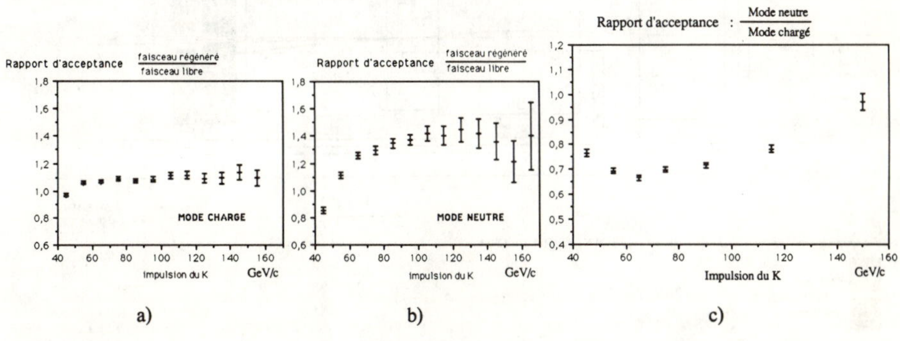


Fig. 33 Monte Carlo calculations of acceptance ratios as a function of kaon momentum: a) charged mode, $K_S/K_L$; b) neutral mode, $K_S/K_L$; c) double ratio neutral/charged, $K_S/K_L$.

## 4.2 The CERN experiment [74]

### *4.2.1 The beams and the detector*

Details of the detector and its performance are given in Burkhardt et al. [76]. Figure 34 shows the experimental layout. In this experiment, $K_S$ and $K_L$ data are taken alternately, and the two decay modes simultaneously. The $K_S$ beam (i.e. the target, the sweeping magnet, the collimator, and the anticounter) (Fig. 35) is on wheels so that the entire 48 m decay region can be uniformly explored despite the rather short mean-free path of $K_S$ decay. The decay region,

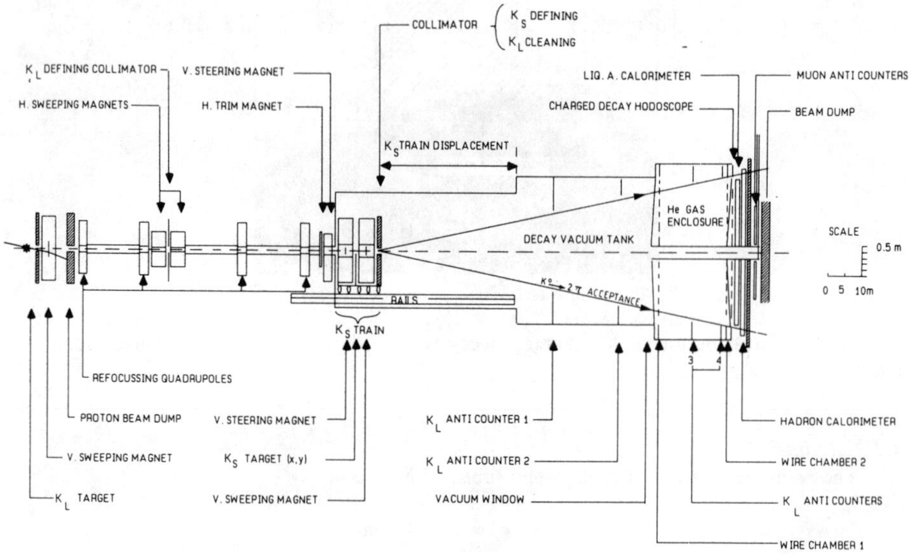


Fig. 34 Layout of CERN experiment.

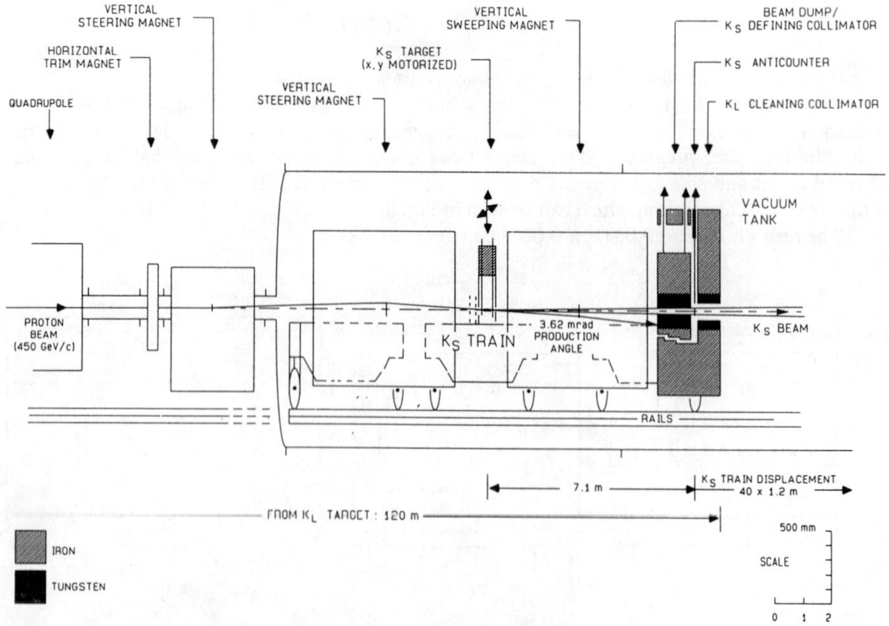


Fig. 35 The $K_S$ beam train.

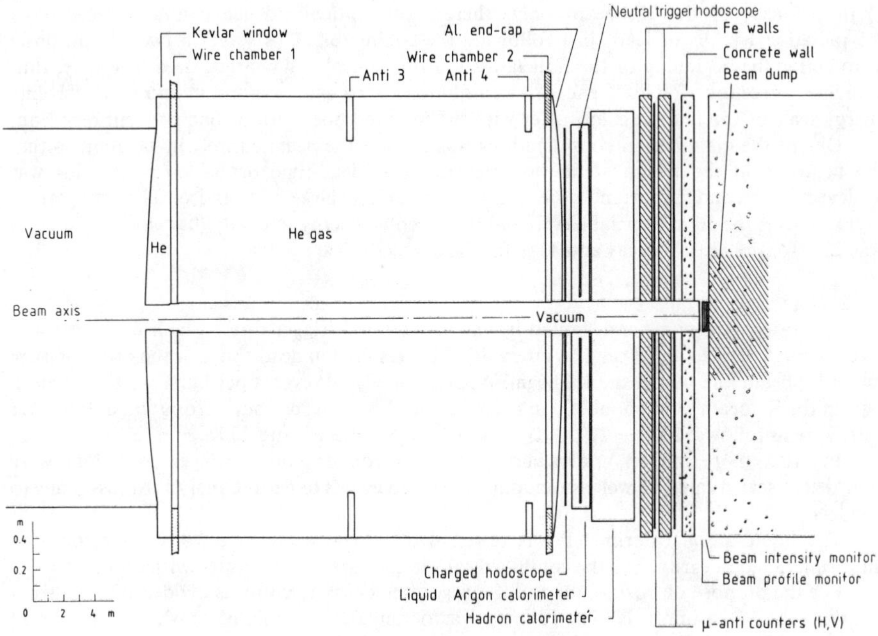


Fig. 36 Detector layout.

and the following 50 m region that serves to let the decay products spread sufficiently to enter the sensitive region of the detector, are evacuated. The neutral beam then continues through an evacuated pipe, and the decay products leave the vacuum region by a 1 mm hemispherical Kevlar window and then enter the helium-filled detector region.

The detector consists of the following elements (see Fig. 36).

1) Minidrift chambers. These are two proportional wire chambers, each with four wire planes, horizontal, vertical, 53°, and 143°. The wire spacing is 6 mm; the position is interpolated by means of drift-time measurement. The precision is $\approx$ 300 $\mu$m in each projection. These chambers serve to measure the tracks of the charged decay products. Reconstruction precision of the decay vertex along the decay axis is $\approx$ 35 cm.

2) A scintillator hodoscope that serves as the trigger for charged decays.

3) A lead/liquid-argon calorimeter. There are 80 layers, each consisting of a 1.5 mm lead sheet clad on both sides with 0.2 mm aluminium, a liquid-argon gap of 2 mm, and a printed circuit board covered on both sides with copper-strip anodes with a pitch of 12.5 mm, and again 2 mm of liquid argon. In succeeding gaps the strips alternate from horizontal to vertical. The total thickness corresponds to 25 radiation lengths.

The energy resolution for photons is $7.5\%/\sqrt{E(\mathrm{GeV})}$. The space resolution is $\approx$ 0.5 mm in x and y.

The stack is read out independently in quadrants and in two groups of 40 layers in depth each. In the central plane of the calorimeter there is a scintillator hodoscope that serves to trigger the neutral decays.

4) A hadron calorimeter consisting of 48 sandwiches of 25 mm iron plate and 5 mm scintillator. The scintillators are arranged in strips, 12 cm wide, alternately horizontal and vertical. The stack is read out in quadrants and in two depth layers. The energy resolution for pions of the two calorimeters is $0.65/\sqrt{E}$.

5) A muon anticoincidence arrangement, consisting of two layers, each with a 70 cm thick iron absorber and a scintillator hodoscope.

6) Four photon anticoincidence annular rings along the decay and detector length, to eliminate some of the $3\pi^0$ and $\pi^+\pi^-\pi^0$ background. They consist of lead sheets and scintillation counters.

7) In the case of the $K_S$ beam only, there is an anticoincidence counter preceded by 1.5 radiation lengths of lead, just following the collimator. This serves a twofold purpose: a) to define the beginning of the fiducial region for neutral and charged decays equally, thus avoiding corrections for differences in z resolution; b) to calibrate and monitor the absolute energy scale of the photon calorimeter with the required one tenth of one per cent precision.

One of the main design considerations was to have the beam entirely in vacuum, so that the neutrons in the beam would not contribute to dead-time or background. This was achieved. A second important objective was to have a large acceptance. The acceptance, averaged over the 50 m fiducial length and the accepted energy interval, $70 < E_K < 170$ GeV, was 22% for the neutral decay and 43% for the charged decay.

*4.2.2 Operation*

The rates were in general limited by the background trigger rates, which were kept at a level of $\approx 10^5$ per 2 s burst (at 12 s intervals). This resulted in dead-times of the order of 12% in the $K_L$ beam and 4% in the $K_S$ beam. Approximately $10^3$ events per burst were written on tape in the $K_L$ beam, and about 300 in the $K_S$ beam. The rates per burst, for good events after cuts, were as follows: 1 $K_L \rightarrow 2\pi^0$, 3 $K_L \rightarrow \pi^+\pi^-$, 30 $K_S \rightarrow 2\pi^0$, and 70 $K_S \rightarrow \pi^+\pi^-$.

In order to permit an understanding of backgrounds, looser trigger conditions were recorded at scaled rates, as well as randomly triggered events to permit analysis of losses due to accidentals.

The liquid-argon calorimeter was tested with known energy electrons for resolution, uniformity, and linearity, and the combined calorimetric system was tested with pions.

For the purpose of positioning the $K_S$ target, the decay region was divided into 40 'train positions' at 1.2 m intervals. The running was organized into 'miniperiods', of which there were 36, all in 1986. Each miniperiod consisted of one 24 h exposure in the $K_L$ beam, and ten exposures of ~ 3/4 h each in the $K_S$ beam. These $K_S$ exposures were spaced by four train positions, so that one miniperiod had exposures at, say, positions 2, 6, 10, etc.; the next would use positions 4, 8, 12, etc. In four miniperiods, each position in the decay region was covered.

The $K_L$ and $K_S$ beams were produced by protons striking the target at the same angle, 3.6 mrad. Nevertheless the spectra were appreciably different, owing to the preferential decay of lower-energy kaons in the 8 m distance between $K_S$ target and anticounter. The spectra of the observed decays are shown in Fig. 37 for the two beams.

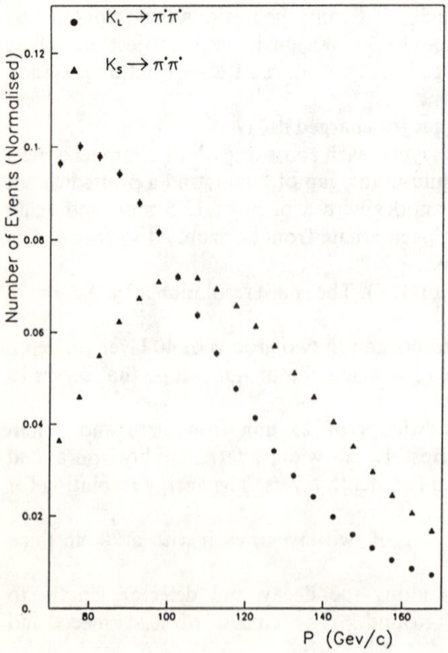


Fig. 37 Observed energy spectra in the $K_S$ and $K_L$ beams.

*4.2.3 Data reduction and background subtraction*

The on-line trigger and off-line analysis were arranged so that for the two decay modes the dead-times were the same, and accidental effects were as nearly the same as possible. For instance, an extra space point in the front chamber invalidated charged events; it therefore did the same for neutrals. Or, an extra photon (the fifth for neutral decays, and the first for charged decays) disqualified both types of events. However, an extra photon is not quite the same when there are already four present, or there are not, because of overlap problems; nor is an extra space point quite the same if two space points are already there, or if they are not. It is therefore necessary to make a correction for the differences in the effects of accidentals on the four event classes.

4.2.3.1 Neutral events

For each event, four photons of more than 5 GeV are required. They must be separated by 5 cm in at least one projection. The centre of energy in the liquid-argon calorimeter must be in the beam; therefore,

$$r = \left[(\Sigma\, E_i x_i)^2 + (\Sigma\, E_i y_i)^2\right]^{1/2} / \Sigma\, E_i$$

is required to be less than 10 cm. Figure 38 shows the distribution in r for the $K_S$ beam.

The decay vertex is reconstructed, using the K mass, according to

$$z_v = 1/m_K \left\{\sum_{i<j}^{4} E_i E_j [(x_i - x_j)^2 + (y_i - y_j)^2]\right\}^{1/2} .$$

The resolution in $z_v$ is $\approx$ 1 m.

Given the decay vertex $z_v$, the invariant masses of photon pairs may be calculated:

$$m_{ij} = 1/z_v \sqrt{E_i E_j [(x_i - x_j)^2 + (y_i - y_j)^2]} .$$

Those of the three combinations for which the two masses are nearest to the $\pi^0$ mass are chosen. For $2\pi^0$ decays, these masses will be close to the $\pi^0$ mass, but not, in general, for $3\pi^0$ decays for which only four of the six photons have been detected. The distribution in these masses is therefore used in the background subtraction.

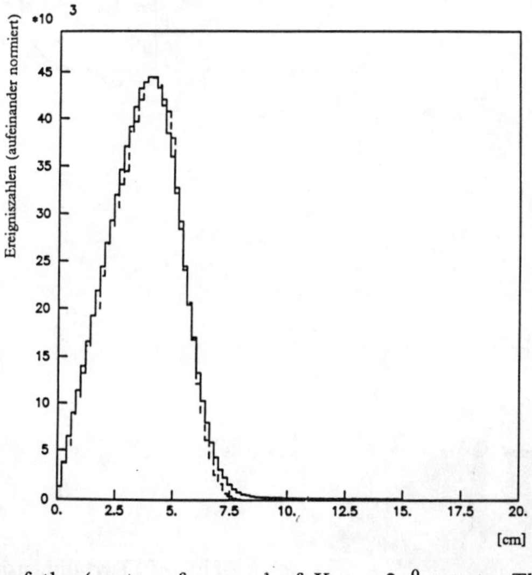


Fig. 38 Distribution of the 'centre of energy' of $K_S \to 2\pi^0$ events. This shows the radial distribution of the $K_S$ beam in the plane of the liquid-argon calorimeter. The dashed histogram is the Monte Carlo simulation.

It is possible to proceed inversely: to reconstruct the decay vertex using the photon pairs, assuming that they have the $\pi^0$ mass, and then to use this vertex, $z_v(\pi^0)$, to find the invariant mass of the event:

$$m^2 = \sum_{i<j}^{4} E_iE_j[(x_i - x_j)^2 + (y_i - y_j)^2]/z_v^2(\pi^0) \ .$$

This mass is not used in the event selection, but illustrates the resolution of the detector. It is shown in Fig. 39 for $K_S$ events.

Event selection is based on the two-dimensional $\gamma_i\gamma_j$ mass distribution. It is shown in Fig. 40 for $K_S$ events as a Lego plot and in Fig. 41 for $K_L$ events as a scatter plot. The $K_S$ events have essentially zero background, whilst the $3\pi^0$ background in the $K_L$ decay is uniformly distributed. In order to select good events and to subtract the $K_L$ background, constant $\chi^2$ ellipses of linearly increasing area are constructed and the event numbers in the annular regions are plotted (see Fig. 42). Of the good events, 98.5% are in the inner ellipse, as seen in the background-free $K_S$ case. In the $K_L$ case, extrapolation to the inner ellipse gives the $3\pi^0$ background of (4.0 ± 0.2)%.

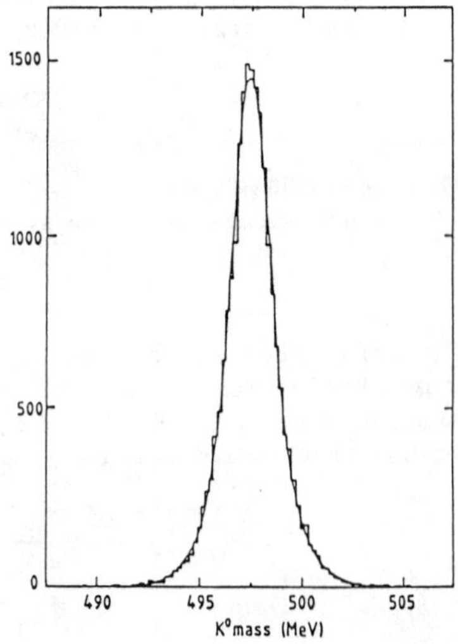


Fig. 39 Mass resolution for $2\pi^0$ decay in the $K_S$ beam.

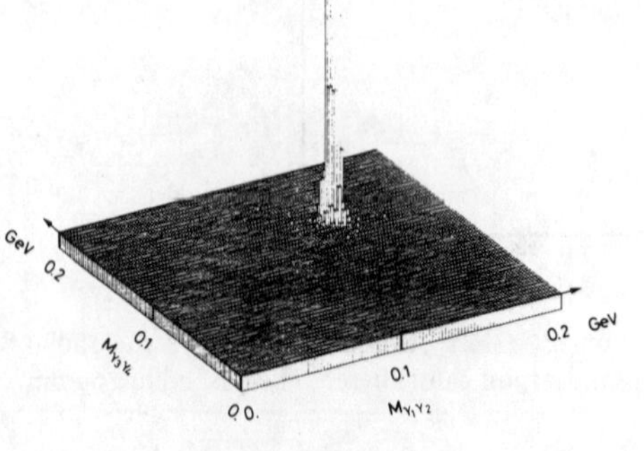


Fig. 40 Two-dimensional histogram of $K_S$ neutral events in the plane $m_{\gamma_1\gamma_2}$ versus $m_{\gamma_3\gamma_4}$.

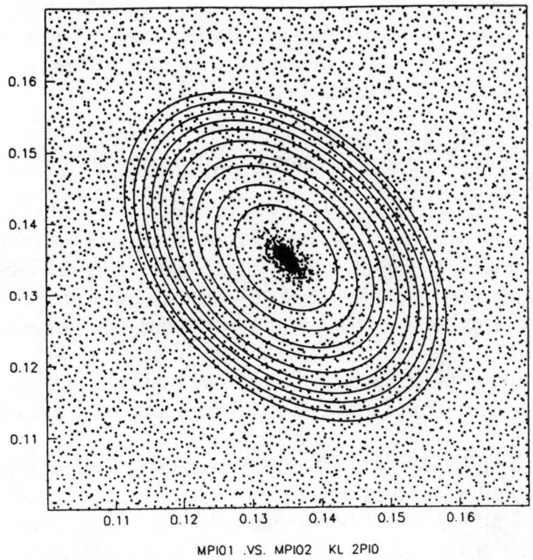


Fig. 41 Scatter plot in the plane $m_{\gamma_1\gamma_2}$ versus $m_{\gamma_3\gamma_4}$ for $K_L$ neutral events. The ellipses correspond to $\chi^2 = 9, 18, 27, \dots$ .

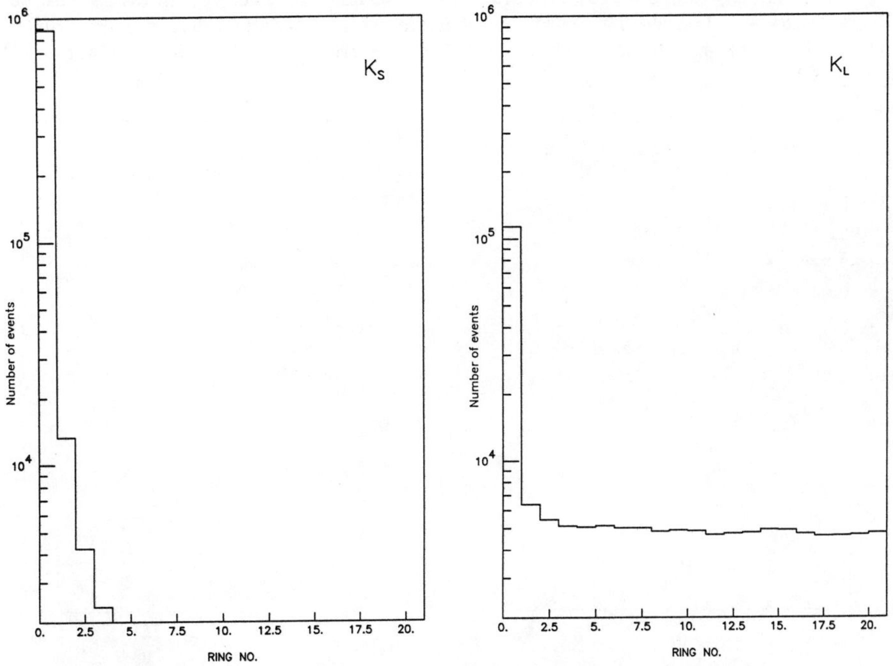


Fig. 42 Event numbers in the annulus between successive ellipses for $K_S$ and $K_L$ events; 98.5% of events are contained in the inner ellipse. Events in rings numbers 5 to 10 are used to extrapolate to zero for background subtraction in the $K_L$ case.

4.2.3.2 Charged events

The energy of an event is not determined calorimetrically, but geometrically from the opening angle $\theta$ and the *ratio* R of the calorimetric energies of the two pions. This is more accurate, and is less dependent on calorimetry calibration:

$$E^2 = \frac{2 + R + (1/R)}{\theta^2}\{m_K^2 - m_\pi^2[2 + R + (1/R)]\} .$$

The resolution deteriorates with R, so the events are restricted to $0.4 < R < 2.5$. This cut is also effective against the $\Lambda \to \pi^- p$ background. The average energy resolution is $\approx$ 1%.

A very important cut requires that at least 20% of the energy of *each* pion be deposited in the hadron calorimeter. This cut loses 50% of the signal, but reduces the $K_{e3}$ background to an acceptable level.

A further cut, again to reduce the three-body decay background, requires that the distance from the beam centre to the line joining the two points in the back chamber be less than 10 cm.

The invariant mass of each event is calculated from the calorimetric pion energies and the opening angle. The resolution depends on the kaon energy, and a $2.1\sigma$ cut is applied.

The $K_S$ sample has no significant background. The $K_L$ background is determined on the basis of the distribution of the events in the distance between the event plane and the target where the kaons are produced. This is shown in Fig. 43 (the $K_S$ distribution is scaled geometrically so that it can be compared directly). The signal region is taken to be less than 5 cm, and the control region for determining the background is between 7 and 12 cm. This background is mainly due to $K_{e3}$ events and, to a lesser extent, to $K_{\mu 3}$ and $K_{\pi 3}$ events, and their relative contributions can be determined from the many auxiliary events in which the decay is identified. It is extrapolated to the signal region by means of a Monte Carlo calculation, as shown. The background subtracted is $(0.65 \pm 0.2)$%, where the systematic error is included.

It has already been pointed out that the energy scales of the two types of events must be the same to $\sim$ 0.1%. We have also seen that the energy in charged decays is determined geometrically. In order to achieve the equality of the energy scales, the overall calibration of

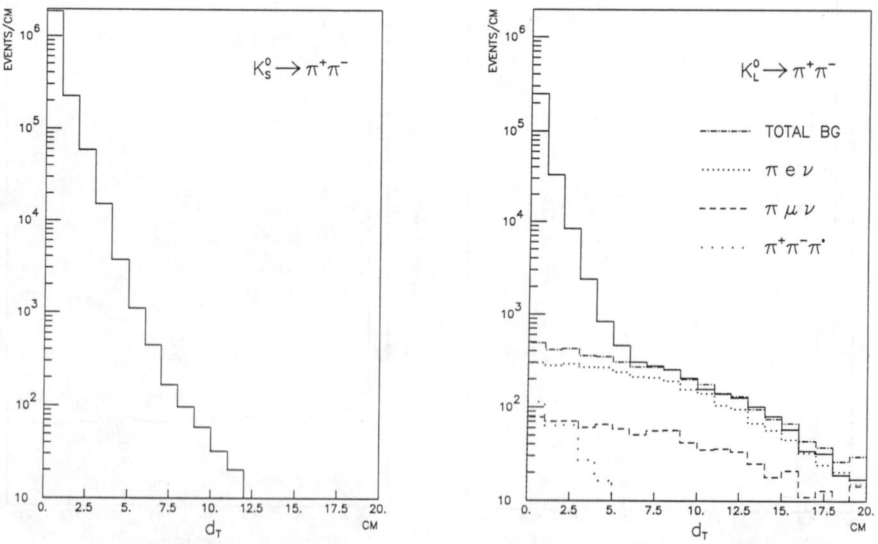


Fig. 43 Distributions in the distance between the production target and the decay plane for charged events. The Monte Carlo extrapolation of the three-body background from the tail to the signal region is indicated.

the liquid-argon calorimeter is also determined geometrically. It is chosen by requiring that the reconstructed anticounter position in the $K_S$ beam is the same for both neutral and charged decay. Since, in the neutral decay, only photon energies and positions in the liquid argon are measured, a 1% change in overall energy scale will translate to approximately 1 m in anticounter position. The 0.1% equality in energy scales is easily achieved.

*4.2.4 Results*

In the final tabulation the energy interval $70 < E < 170$ GeV and the decay region $10.5 < z < 48.9$ m were retained. The z interval at the beginning of the decay region, $z < 10.5$ m, was dropped to avoid complication caused by a low level of regeneration in the final $K_L$ collimator. The results are summarized in Table 1 and the contributions to the systematic error are given in Table 2.

The statistics are limited essentially by the 109,000 events in the $K_L \to 2\pi^0$ channel. The data were analysed in bins of energy and of position of the decay along the beam axis, and for each miniperiod. They were corrected by a Monte Carlo calculation for binning effects, for the slight difference in the $K_S$ and $K_L$ beam shapes, and for the scattering of $K_S$ in the anticounter and collimator. The net Monte Carlo correction is 0.3%.

**Table 1**
Event statistics and corrections

| | Signal events (× 1000) | Background (%) | Scattering (%) | Inefficiencies Pretrigger (%) | Inefficiencies Trigger (%) | Accidental losses (%) |
|---|---|---|---|---|---|---|
| $K_L \to 2\pi^0$ | 109 | 4.0 | < 0.1 | 0.06 ± 0.06 | 0.20 ± 0.10 | 2.6 ± 0.07 |
| $K_L \to \pi^+\pi^-$ | 295 | 0.6 | | 0.37 ± 0.07 | 0.05 ± 0.06 | 2.6 ± 0.05 |
| $K_S \to 2\pi^0$ | 932 | < 0.1 | 0.3 | 0.04 ± 0.02 | 0.12 ± 0.03 | 2.5 ± 0.05 |
| $K_S \to \pi^+\pi^-$ | 2300 | < 0.1 | | 0.48 ± 0.03 | 0.01 ± 0.01 | 2.8 ± 0.05 |
| Effect on R | | | 0.3 | −0.12 ± 0.10 | −0.03 ± 0.12 | −0.34 ± 0.10 |

**Table 2**
Systematic uncertainties on the double ratio R
(in percent)

| | |
|---|---|
| Background subtraction for $K_L \to 2\pi^0$ | 0.2 |
| Background subtraction for $K_L \to \pi^+\pi^-$ | 0.2 |
| $2\pi^0/\pi^+\pi^-$ difference in energy scale | 0.3 |
| Regeneration in the $K_L$ beam | < 0.1 |
| Scattering in the $K_S$ beam | 0.1 |
| $K_S$ anticounter inefficiency | < 0.1 |
| Difference in $K_S/K_L$ beam divergence | 0.1 |
| Calorimeter instability | < 0.1 |
| Monte Carlo acceptance | 0.1 |
| Gains and losses by accidentals | 0.2 |
| Pretrigger and trigger inefficiency | 0.1 |
| Total systematic uncertainty | ±0.5% |

The result for R is

$$R = 0.980 \pm 0.004\ (\text{stat.}) \pm 0.005\ (\text{syst.})\ .$$

This translates to

$$\epsilon'/\epsilon = (3.3 \pm 1.1) \times 10^{-3}\ .$$

This is the first statistically significant evidence for 'direct' CP violation; that is, for an effect other than that due to $K_1$–$K_2$ mixing. It clearly is in disagreement with the superweak model [63], but agrees in sign and order of magnitude with the predictions of the electroweak theory [69, 70], if it is assumed that the electroweak currents are responsible for CP violation [4].

**4.3 Present status of $\epsilon'/\epsilon$ experiments**

The positive result of the CERN experiment for $\epsilon'/\epsilon$ is interesting but, with three standard deviations, near the limit of its sensitivity. The Fermilab and CERN experiments are therefore continuing. At the recent conference on Lepton and Photon Interactions at High Energies (SLAC, Aug. 89), Winstein reported preliminary results based on a small fraction of data obtained at Fermilab during the autumn and winter of 1987. These data were obtained without the thin lead sheet converter for photons, so that neutral and charged modes could be observed together. It is the first experiment in which the four rates are measured simultaneously. On the basis of 52,000 neutral $K_L$, 43,000 charged $K_L$ events, and approximately four times as many $K_S$ events, the result is $\epsilon'/\epsilon = (-0.5 \pm 1.5) \times 10^{-3}$ (preliminary), a null result differing by two standard deviations from the CERN result.

The CERN experiment has continued with small modifications designed to increase the rate modestly and to reduce the systematic uncertainty. The runs in '88 and '89 should quadruple the statistics.

## REFERENCES


[1] C.S. Wu et al., Phys. Rev. **105** (1957) 1413.
[2] R.P. Feynman and M. Gell-Mann, Phys. Rev. **109** (1958) 193.
[3] E.C.G. Sudarshan and R. Marshak, Phys. Rev. **109** (1958) 1860.
[4] M. Kobayashi and K. Maskawa, Prog. Theor. Phys. **49** (1973) 652.
[5] J.H. Christenson, J.W. Cronin, V.L. Fitch and R. Turlay, Phys. Rev. Lett. **13** (1964) 138.
[6] R. Adair et al., Phys. Rev. **132** (1963) 2285.
[7] J.M. Gaillard et al., Phys. Rev. Lett. **18** (1967) 20.
[8] J.W. Cronin et al., Phys. Rev. Lett. **18** (1967) 25.
[9] I.A. Budagov et al., Phys. Lett. **B28** (1968) 215.
[10] M. Banner et al., Phys. Rev. Lett. **21** (1968) 1107.
[11] M. Holder et al., Phys. Lett. **B40** (1972) 141.
[12] S. Bennett et al., Phys. Rev. Lett. **19** (1967) 993.
[13] D. Dorfan et al., Phys. Rev. Lett. **19** (1967) 987.
[14] C. Geweniger et al., Phys. Lett. **B48** (1974) 483.
S. Gjesdal et al., Phys. Lett. **52B** (1974) 113.
[15] L. Adiels et al., CP violation studies at LEAR, presented by P. Bloch at the 25th Anniversary of the Discovery of CP Violation, Blois (France), 1989.

[16] V.L. Fitch et al., Phys. Rev. Lett. **15** (1965) 73.
[17] C. Alff-Steinberger et al., Phys. Lett. **20** (1966) 207 and **21** (1966) 595.
[18] M. Bott-Bodenhausen et al., Phys. Lett. **20** (1966) 212 and **23** (1966) 277.
[19] W.C. Carithers et al., Phys. Rev. Lett. **34** (1975) 1240.
[20] W.C. Carithers et al., Phys. Rev. Lett. **34** (1975) 1244.
[21] A. Böhm et al., Nucl. Phys. **B9** (1969) 605.
[22] C. Geweniger et al., Phys. Lett. **B48** (1974) 487.
[23] S. Gjesdal et al., Phys. Lett. **B52** (1974) 119.
[24] G. Charpak et al., Nucl. Instrum. Methods **62** (1968) 262.
[25] P. Schilly et al., Nucl. Instrum. Methods **91** (1971) 221.
[26] J.C. Chollet et al., Phys. Lett. **B31** (1970) 658.
[27] B. Wolff et al., Phys. Lett. **B36** (1971) 517.
[28] G. Barbiellini et al., Phys. Lett. **B43** (1973) 529.
[29] J.H. Christenson et al., Phys. Rev. Lett. **43** (1979) 1209 and 1212.
[30] G.D. Barr et al., A measurement of the phases of the CP-violating amplitudes in $K^0 \rightarrow 2\pi$ decays and a test of CPT invariance, to be submitted to Phys. Lett. B.
[31] A. Abashian et al., Phys. Rev. Lett. **13** (1964) 245.
[32] X. de Bouard et al., Phys. Lett. **15** (1965) 58.
[33] W. Galbraith et al., Phys. Rev. Lett. **14** (1965) 383.
[34] R. Messner et al., Phys. Rev. Lett. **30** (1973) 876.
[35] R. DeVoe et al., Phys. Rev. **D16** (1977) 565.
[36] S. Bennett et al., Phys. Lett. **B27** (1968) 248.
[37] S. Bennett et al., Phys. Lett. **B29** (1969) 317.
[38] D.A. Jensen et al., Phys. Rev. Lett. **23** (1969) 615.
[39] H. Faissner et al., Phys. Lett. **B30** (1969) 204.
[40] R.K. Carnegie et al., Phys. Rev. **D6** (1972) 2335.
[41] J.M. Gaillard et al., Nuovo Cimento **A59** (1969) 453.
[42] V.V. Barmin et al., Phys. Lett. **B33** (1970) 377.
[43] M. Banner et al., Phys. Rev. Lett. **28** (1972) 1597.
[44] J. Marx et al., Phys. Lett. **B32** (1970) 219.
[45] R. Piccioni et al., Phys. Rev. Lett. **29** (1972) 1412.
[46] H.H. Williams et al., Phys. Rev. Lett. **31** (1973) 1521.
[47] V.L. Fitch et al., Phys. Rev. Lett. **31** (1973) 1524.
[48] V.V. Barmin et al., Phys. Lett. **B128** (1983) 129.
[49] M. Metcalf et al., Phys. Lett. **40B** (1972) 703.
[50] V.V. Barmin et al., Nuovo Cimento **A85** (1985) 67.
[51] P.M. Border et al., Proc. 23rd Int. Conf. on High-Energy Physics, Berkeley, 1986 (World Scientific, Singapore, 1987), Vol. II, p. 845.
[52] C.C. Rochester and G.D. Butler, Nature **160** (1947) 855.
[53] R.W. Thompson et al., Third Rochester Conference, 1952, and Phys. Rev. **90** (1953) 329. R. Armenteros et al., Phil. Mag. **43** (1951) 113.
[54] M. Gell-Mann, Phys. Rev. **92** (1953) 1833.
[55] M. Gell-Mann and A. Pais, Phys. Rev. **97** (1955) 1387.
[56] K. Landé et al., Phys. Rev. **103** (1956) 1901.
[57] T.T. Wu and C.N. Yang, Phys. Rev. Lett. **13** (1964) 380.
[58] The argument presented here is due to J.S. Bell, Proc. Conf. on Elementary Particles, Oxford, 1965 (Rutherford Lab., Chilton, Didcot, 1966), p. 195.
[59] T.T. Devlin and J.O. Dickey, Rev. Mod. Phys. **51** (1979) 237.
[60] N.N. Biswas et al., Phys. Rev. Lett. **47** (1981) 1378.
[61] A.N. Kamal, J. Phys. **G12** (1986) L43.
[62] H. Wahl, preprint CERN–EP/89–86, talk given at the Rare Decay Symposium, Vancouver, 1988.
[63] L. Wolfenstein, Phys. Rev. Lett. **13** (1964) 562.
[64] N. Cabibbo, Phys. Rev. Lett. **10** (1963) 531.
[65] S.L. Glashow, J. Iliopoulos and L. Maiani, Phys. Rev. **D2** (1970) 1285.
[66] C. Jarlskog, Phys. Rev. Lett. **55** (1985) 1039.
[67] D.W. Greenberg, Phys. Rev. **D32** (1985) 1841.

[68] D.D. Wu, Phys. Rev. **D33** (1986) 860.
[69] L. Wolfenstein, Annu. Rev. Nucl. Sci. **36** (1986) 137.
[70] J.F. Donoghue, J.F. Golowich and B.R. Holstein, Phys. Rep. **131** (1986) 320.
[71] J.K. Black et al., Phys. Rev. Lett. **54** (1985) 1628.
[72] R.H. Bernstein et al., Phys. Rev. Lett. **54** (1985) 1631.
[73] M. Woods et al., Phys. Rev. Lett. **60** (1988) 1695.
[74] H. Burkhardt et al., Phys. Lett. **B206** (1988) 163.
[75] P. Jarry, Thèse, Orsay, 1987.
[76] H. Burkhardt et al., Nucl. Instrum. Methods **A268** (1988) 116.

# GUEST LECTURES

*Jim and Victoria Christenson, René Turlay.*

# DO: PROGRESS AND PROMISE

James H. Christenson
Fermi National Accelerator Laboratory

ABSTRACT

The DO detector currently being assembled at the Tevatron at Fermilab to study proton-antiproton collisions at 2 TeV is described. The rationale for the experiment is given and the design goals and major features of the apparatus are presented. A few physics topics are discussed and the status of each of the major detector systems is briefly reviewed.

The DO detector is currently being assembled at the Tevatron at Fermilab and will begin to study proton-antiproton collisions at 2 TeV during the next collider run in the beginning of 1991. The rationale for the experiment is given below, and the design goals and major features of the apparatus are described. A few physics topics are briefly discussed to illustrate the capability of the detector. Finally, the status of each of the major detector systems is briefly reviewed.

The DO project is a collaborative effort of about one hundred eighty physicists and graduate students from twenty-two universities and laboratories around the world. At Fermilab, the development of the experiment is directly supported by more than one hundred sixty engineers, technicians, and programmers.

## Rationale for DO

DO is truly a second generation collider experiment. The detector has been designed to provide optimal response to the events and physics observed at CERN. The DO effort is complementary to the CDF detector that is built and taking data. In DO, great emphasis has been placed on lepton detection, with enhanced electron identification and 4-pi muon detection. Hermetic and homogeneous calorimetry will provide improved missing transverse energy resolution and excellent jet energy and position measurement. By simultaneously taking data with both large facilities, the effective luminosity of the Tevatron will be doubled, and confidence in reports of new physics in one experiment will be strengthened by corroboration from the other.

## Primary Design Goals

The experiment has been designed to search for departures from the

Standard Model, with high sensitivity to short-distance, high transverse-momentum phenomena. To this end, one needs good:

1. Lepton identification over the full solid angle. The leptonic decay modes are relatively clean. The systematics involved in the detection of electrons and muons are quite different and the simultaneous observation of both channels will provide added confirmation of new physics.

2. Missing transverse energy determination. The importance of the detection of energetic neutrinos was amply and indelibly demonstrated in the discovery at CERN of the W boson through its leptonic decays.

3. Spatial resolution of jets and leptons. It is essential to recognize leptons near jets to observe decays of heavy quarks, and to identify photons and neutral pions near charged particles.

To meet these goals, we have designed a detector with the following basic features:

1, No central magnetic field. Particle energies will be measured by calorimetry, a technique whose resolution improves with energy. Neutral and charged particles are thereby detected on the same footing. Tracking and pointing charged particles in the presence of many others is simpler and can be accomplished with a system smaller than that needed in a magnetic field. The resultant detector is more compact, cheaper (in principle!), and amenable to improved calorimetry and muon detection.

2. Uranium - Liquid Argon Calorimetry. This calorimeter technology is conducive to fine transverse and longitudinal segmentation to reveal the electron shower development, and to permit the separation of photons from neutral pions as well as electrons from hadrons. The detectors can be

built with a minimum of cracks and hot spots. The unit gain of the liquid argon provides stability and relative ease of calibration. Excellent energy resolution can be obtained.

The very high density of the uranium absorber plates results in a relatively compact calorimeter. The detector is well 'compensated', responding equally to electromagnetic and hadronic particles, resulting in good energy resolution for jets comprised of a variable mixture of electromagnetic and hadronic particles.

3. Hermetic muon detection with magnetized iron. By surrounding the entire apparatus with a thick magnetized iron wall, muons can be identified by their penetration with little misidentification from hadronic 'punchthrough'. The total number of interaction lengths of the calorimeter and muon systems varies from thirteen at ninety degrees to eighteen near the beam axes. By absorbing hadrons, muons can be observed inside jets. Good spatial resolution can be used to match the muons to minimum ionizing particles in the tracking system and the calorimeters.

4. TRD in front of calorimeters. The identification of electrons can be improved by a factor of fifty by the deployment of transition radiation detectors in front of the calorimeters.

## Layout of the Detector

The overall layout of the D0 detector is shown in Figure 1. The interaction region is immediately surrounded by a tracking system consisting of three cylindrical detectors: a vertex drift chamber, a transition radiation detector, and a conventional drift chamber. In the forward region, particles are tracked by a series of planar drift chambers.

The tracking system is, in turn, surrounded by a set of three large liquid argon calorimeters, one Central (barrel) Calorimeter, and two End (planar) Calorimeters. Each calorimeter contains a finely segmented electromagnetic section some twenty radiation lengths thick, followed by a coarser hadronic section about ten interaction lengths deep. The calorimeters are surrounded on all sides, including underneath, by magnetized iron toroids instrumented before and after with large planes of drift tubes to determine the momentum (and sign) of muons that penetrate the entire detector.

Particles emitted at angles from 1 to 179 degrees are detected by the DO apparatus. The detector weighs 5500 tons, occupies a volume of 65X45X45 cubic feet, and has 100,000 individual detector channels. The calorimeters contain 300 tons of depleted uranium and 15,000 gallons of liquid argon. The detector is supported by a large movable platform that houses all the tracking and calorimetry analog front-end electronics. A farm of fifty MicroVax's will filter triggering information to select events.

The DO Assembly Building is complete and has been occupied for some time. The large iron toroids have been installed on the detector platform and tested, and the Central Calorimeter is being assembled in a clean room adjacent to the detector, prior to being rolled on to the platform. All digitizing electronics, trigger and event processing equipment, and power supplies will be located in a three-story counting house that rolls with the detector toward the collision hall, but remains outside the shielding wall by virtue of an articulated cable bridge. The iron muon toroids can be retracted, and the End Calorimeters rolled back, to provide access to various portions of the compact detector.

## Physics Topics

One example of physics that demonstrates the power of the D0 detector is a precision measurement of the W-Z mass difference in search of deviations from the Standard Model. A single Tevatron run at a luminosity of 10**30 will yield an integrated luminosity of 10 inverse picobarns and 15,000 W -> e+ neutrino and 1,500 Z -> e+ e-. Expected increased luminosity should yield ten times more data. We will measure the mass of both the W and Z in our apparatus. Since the mass of the Z will be measured more accurately by SLC and LEP, we need only determine the mass difference, or the ratio of the two. This quantity is independent of the energy scale and most systematics. We will actually measure the transverse mass, a quantity that D0 is particularly well-suited to determine since it depends on the transverse electron energy that we will measure with good resolution, and on the transverse energy of the neutrino that will be well measured by the hermetic, compensated calorimeter. Monte Carlo simulation of a 10 inverse picobarn run indicates that we will measure the Z mass to +/-120 MeV and the W mass to +/-200 MeV. These results provide a 1% measure of the square of the sine of the Weinberg angle, and a 15% measure of the 'radiative correction' that is sensitive to the masses of the Higgs particle and the top quark. If the mass of the top quark is not known, these measurements will place limits on it. In a 100 inverse picobarn run, a heavy top quark should be discovered, and the W/Z mass measurement should place meaningful limits on the Higgs mass or reveal a departure from the Standard Model.

One of the most exciting opportunities for D0 is the search for the top quark. Preliminary indications from CDF are that the mass is greater than 60 GeV, and a complete analysis of all of their data will likely raise the limit to 80-90 GeV, near the mass of the W. If the top mass exceeds that of the W, the top quark will decay to a W and a bottom quark. If one of the W's decays leptonically, and the other to two jets, the event will be distinguished by an isolated lepton, missing neutrino energy, and four or more jets. If the mass of the top quark is 120 GeV, we would expect some 600 events of this type in a 10 inverse picobarn run. Monte Carlo simulation shows that background from events containing jets and a leptonic W decay can be suppressed by requiring four jets, each with transverse momenta in excess of 10 GeV. Under these conditions, the mass distribution of the W and nearest jet reveals some sixty events peaked at the top mass. The prospect of ten times as many events improves our sensitivity to even heavier top quarks.

## Status of Detector Systems

### Central Tracking System -

The layout of the tracking system is shown in Figure 2. The detectors fill the 1.5 meter diameter cylindrical aperture in the Central Calorimeter. Information from the chambers is processed in 7232 Flash ADC's.

The Vertex Detector is a high precision drift chamber 32 centimeters in diameter, consisting of three layers of staggered jet cells each containing eight longitudinal sense wires along the path of the particle. The coordinate along the wire is determined by charge division. Using carbon dioxide, the drift resolution is 60-70 microns. Fabrication of the detector will be complete late this year, and the device will be tested in a beam at Fermilab at the beginning of next year.

The large Central Drift Chamber consists of thirty-two modular sectors. As in the Vertex Detector, the azimuthal coordinates of tracks are measured in four layers of jet cells. Here, however, the longitudinal coordinate is measured with delay lines placed at the inner and outer radii of each layer. The structural shells are made of Rohacell to minimize the amount of material in front of the calorimeter to aid in the separation of photons and neutral pions from electrons. Fabrication of this detector is complete. The drift resolution for single hits has been measured to be less than 200 microns, and the delay line resolution is 2 millimeters. Pulse height measurements serve to reject two minimum ionizing particles in a single jet cell by a factor of seventy. The first p-pbar collisions at D0 have recently been observed in a test of three modules in the Collision Hall.

The Forward Drift Chambers at each end of the system consist of two separate 'Theta' planes and one 'Phi' plane. Construction of these devices will be finished in September and subsequently beam- tested.

The Transition Radiation Detector is complete and has been thoroughly tested at the CERN PS. The detector consists of three radiator stacks of polypropylene foils, each followed by a xenon/methane filled radial drift chamber. Flash ADC spectra exhibit clusters from individual x-ray conversions in the chamber, and the pion rejection of the device is 50:1 for an electron efficiency of 90%.

Muon System -

Muons are detected by a set of four proportional drift tube planes mounted on the inside or front faces of the magnetized iron toroids, and two sets of three planes 1.2 meters apart, placed behind the iron, as shown in Figure 3. The magnetic field in the iron of 1.9 Tesla results in a momentum resolution of 18% up to 300 GeV. Each of the 176 modules in the system is assembled from interlocking aluminum extrusions to form units eight feet wide, up to twenty feet long, and three or four cells thick. The drift resolution has been measured to be 200 microns. The coordinate along the sense wire is measured with a cathode pad vernier that is formed by a 24 inch long diamond shaped array of strips. Charge division between the strip portion inside the diamond to that outside determines the longitudinal coordinate to 3 millimeters. The struck diamond is found by charge division on the sense wire itself.

All of the modules have been built and tested and installation is in progress.

A Small Angle Muon System is under development to extend coverage in the forward direction from 11 degrees down to 4 degrees. This system will aid in studies of muons from heavy flavors.

Calorimeter System -

The liquid argon calorimeter system is the 'heart' of DO. It features fine segmentation over the full solid angle down to one degree of the beam line. More than 4500 projective towers .1 units in pseudorapidity and .1 radians in azimuth are divided longitudinally in eight or nine depth segments. At the peak depth of an electromagnetic shower, the segmentation is doubled in each direction. Ionization in the liquid argon is measured by a system of 50,000 charge-sensitive preamplifiers mounted on the cryostats, baseline subtractors located beneath the apparatus in the platform, and ADC's in the movable counting house.

The Electromagnetic Section of the three calorimeters is comprised of some twenty radiation lengths of 3-4 millimeter uranium plates. The sampled ionization is readout in four depth segments. The energy resolution is about 14%/sqrt(E) and the spatial resolution is 1-2 millimeters. These Electromagnetic Sections are followed by Fine Hadronic Sections consisting of 6 millimeter uranium plates for a total of 7-9 interaction lengths, divided into three or four depth segments. The resolution for hadrons is about 45%/sqrt(E). Leakage through the Fine Hadronic Section is collected in the Coarse Hadronic Section built of 1.83 inch copper or stainless steel plates with a total thickness of 3-6 interaction lengths.

The deployment of modules in the Central and End Calorimeters is shown in Figure 4. The facing heads of the cryostats are shaped to allow the vessels to be nested close together to optimize the response in the transition region to particles that pass through the Central and one of the End Calorimeters.

All sixty-four modules of the Central Calorimeter have been fabricated and cold-tested at Brookhaven National Laboratory. At least one module of each of the three types has been successfully tested in liquid argon in a test beam at Fermilab. The cryostat has been placed in the clean room at the DO Assembly Building and installation of modules in the cryostat is underway. Once all of the modules are in place, the coaxial cables that route signals from the modules to feedthroughs atop the cryostat will be installed, and the modules and cabling to the preamplifiers will be thoroughly tested. The modules will be enclosed in the cryostat by welding the two pairs of heads on each end of the cylindrical vessel, and the calorimeter will be moved onto the detector platform for final cryogenic hookup, cabling, and testing.

Modules for the two End Calorimeters have been designed and are currently being fabricated. Each End Calorimeter contains an

Electromagnetic module being developed at Lawrence Berkeley Laboratory, followed by a system of Hadronic modules under construction at Fermilab. The Hadronic Section is comprised of a large cylindrical Inner Hadronic module, surrounded by sixteen Middle Hadronic units, in turn encompassed by a ring of sixteen Outer Hadronic modules.

The Electromagnetic module is a very highly segmented disk-shaped device that contains 7488 individual channels. The detector employs 4 millimeter uranium absorbers, is 88 inches in diameter, and weighs 4.9 tons. All components for the first module have been procured, and assembly of the unit has begun. After the module has been cold-tested, it will be shipped to Fermilab to be beam-tested at the beginning of the next year. Because of the length of time required to test the first module, the second unit will actually be installed first at D0.

The Outer Hadronic module stainless steel absorber plate assemblies are being fabricated by Serpukhov in the Soviet Union and will be transported by ship to the United States soon. At Fermilab, signal and readout boards will be installed in the six ton absorber assemblies, wired and cabled, and the completed detectors cold-tested in liquid nitrogen. Prior to being instrumented, sixteen of the bare assemblies will be installed in a ring in the module assembly building to test the mechanical properties of this array that must support the full 210 ton weight of all the modules.

Whereas the large-angle Outer Hadronic modules contain Coarse Sections only, consisting solely of 1.83 inch thick stainless steel plates, the five ton Middle Hadronic modules contain Fine Sections of fifty-nine 6 millimeter uranium plates, followed by Coarse Sections of fourteen stainless steel plates. Assembly of these modules is well underway at Fermilab. Facilities for removing the oxide from the uranium plates, and installing ground pins and spacers, have been developed. The uranium plates and signal boards of the Fine Section are stacked in a clean room, compressed to flatten the plates, and held together by straps welded between stainless steel front and back plates. The Coarse Section absorber weldment, again fabricated by Serpukhov, is then attached to the rear of the module, and the entire unit wired, cabled, and cold-tested. A prototype module was extensively studied in the test beam last year.

The Inner Hadronic module is 68 inches in diameter, 67 inches long, and weighs thirty-two tons. It contains more than six thousand calorimeter channels. A Fine Section of fifty-nine 6 millimeter uranium plates is followed by a Coarse Section of twelve stainless steel plates from Serpukhov. This unit is the most difficult module to construct and its readiness will pace the completion of the End Calorimeters. Components are being prepared and stacking of the first module will begin in the Fall. As for the Electromagnetic module, the first unit will be beam-tested early next year, and the second unit will be installed first at DO.

A few modules from the Central and End Calorimeters have undergone extensive testing at Fermilab with pions and electrons from 10 to 150 GeV. The module response is found to be linear over the full energy range. The energy resolution of the modules as a function of energy is characterized by a sampling term of 16% for electrons, and 44% for pions, close to that expected. From the ratio of energy deposition in adjacent pads in a Central Calorimeter Electromagnetic module, a position resolution of 2 millimeters has been measured. The response of an End Calorimeter Middle Hadronic module to energy deposited by an electron is seen to be the same as that for a pion to within a few percent.

## Summary

Substantial progress has been made in preparing the DO experiment. All detectors are built or in production. Major components of all subsystems have been beam-tested. Software to process triggers and to control and monitor the apparatus is 50% complete. Installation of detectors on the platform in the DO Assembly Building is underway. The magnetic toroids have been installed and tested, cables are being laid, the cryogenic system is being built, and on-line computers put in place. Ten percent of the muon chambers have been mounted. Final assembly of the Central Calorimeter has begun. The full experiment should be ready for collisions in the beginning of 1991.

DO is well-suited for interesting physics. The Tevatron is working superbly and luminosities as high as 5X10**31 will be obtained before long, greatly extending the reach for new physics. The challenge for DO is clear. We must complete the detector prior to the next Tevatron run in early 1991. We must not only commission the detector but take meaningful data during the first run. We must accomplish a future upgrade of the detector without missing any data taking.

DO is the second detector at the Tevatron, but it is surely not second-best!

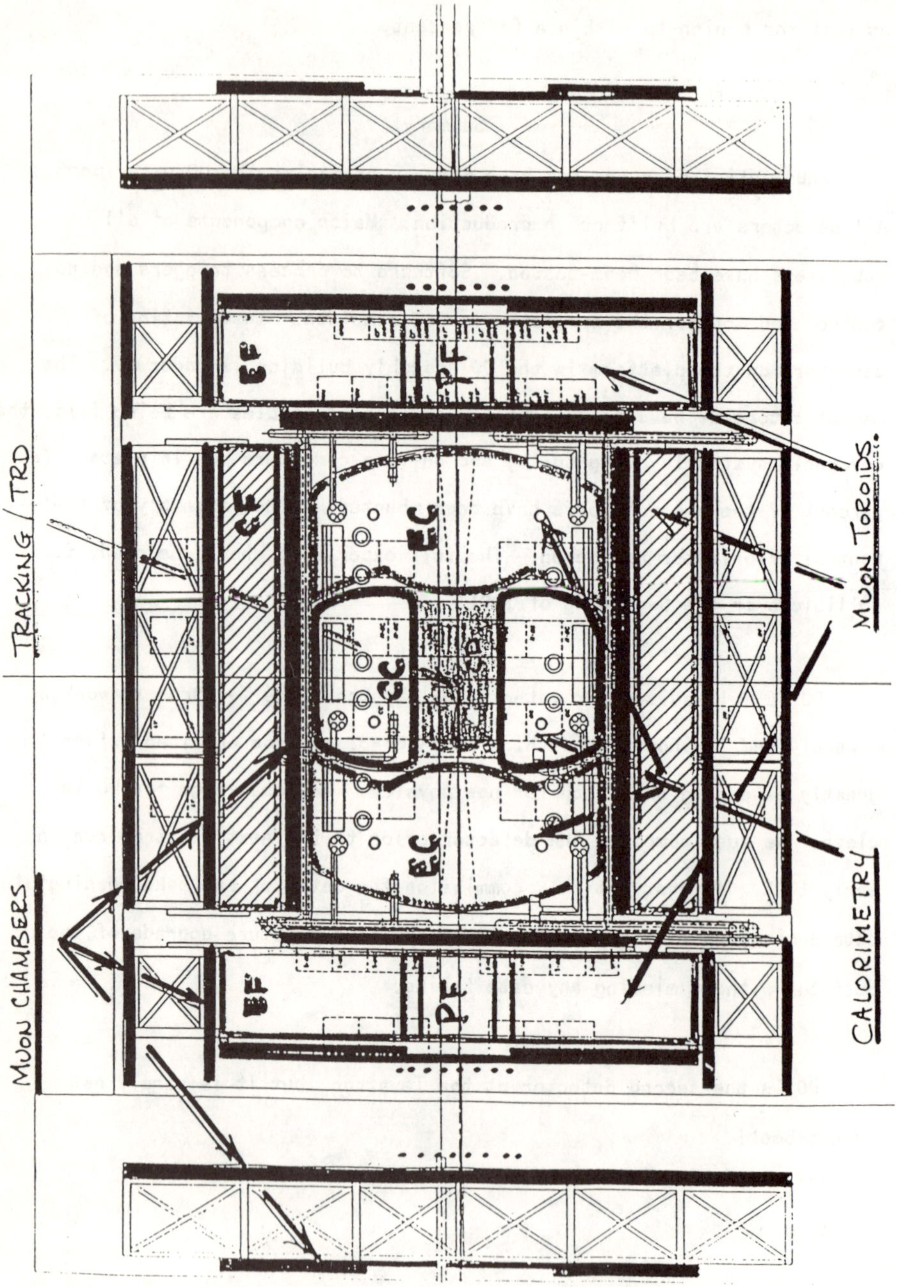


Figure 1
Overall Layout of the D0 Detector.

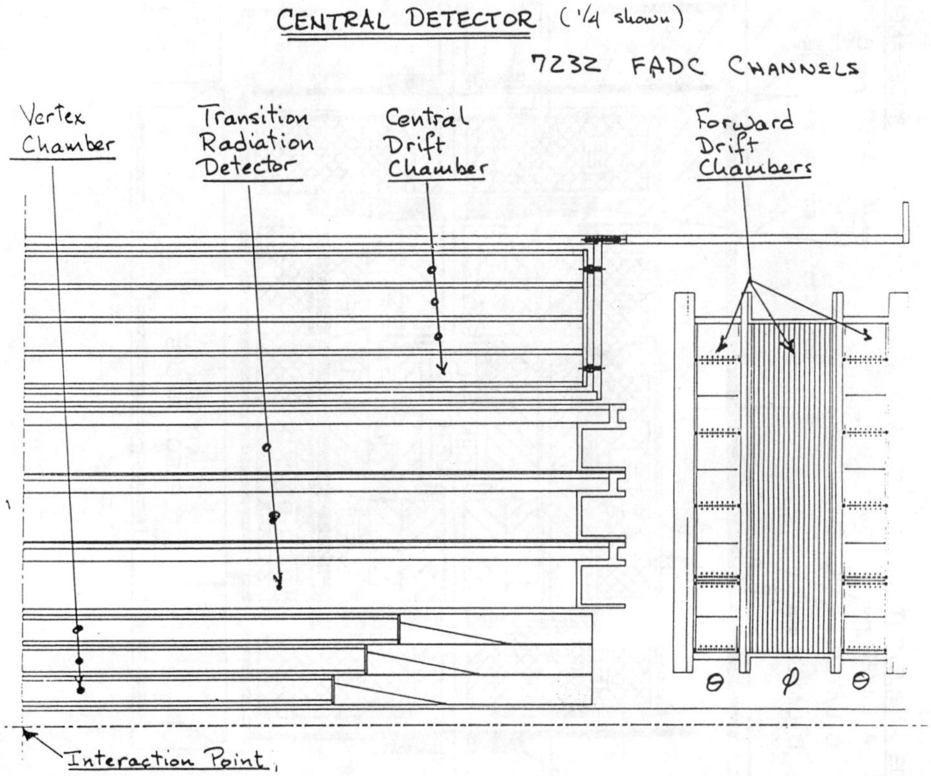


Figure 2
Layout of the Central Tracking System.

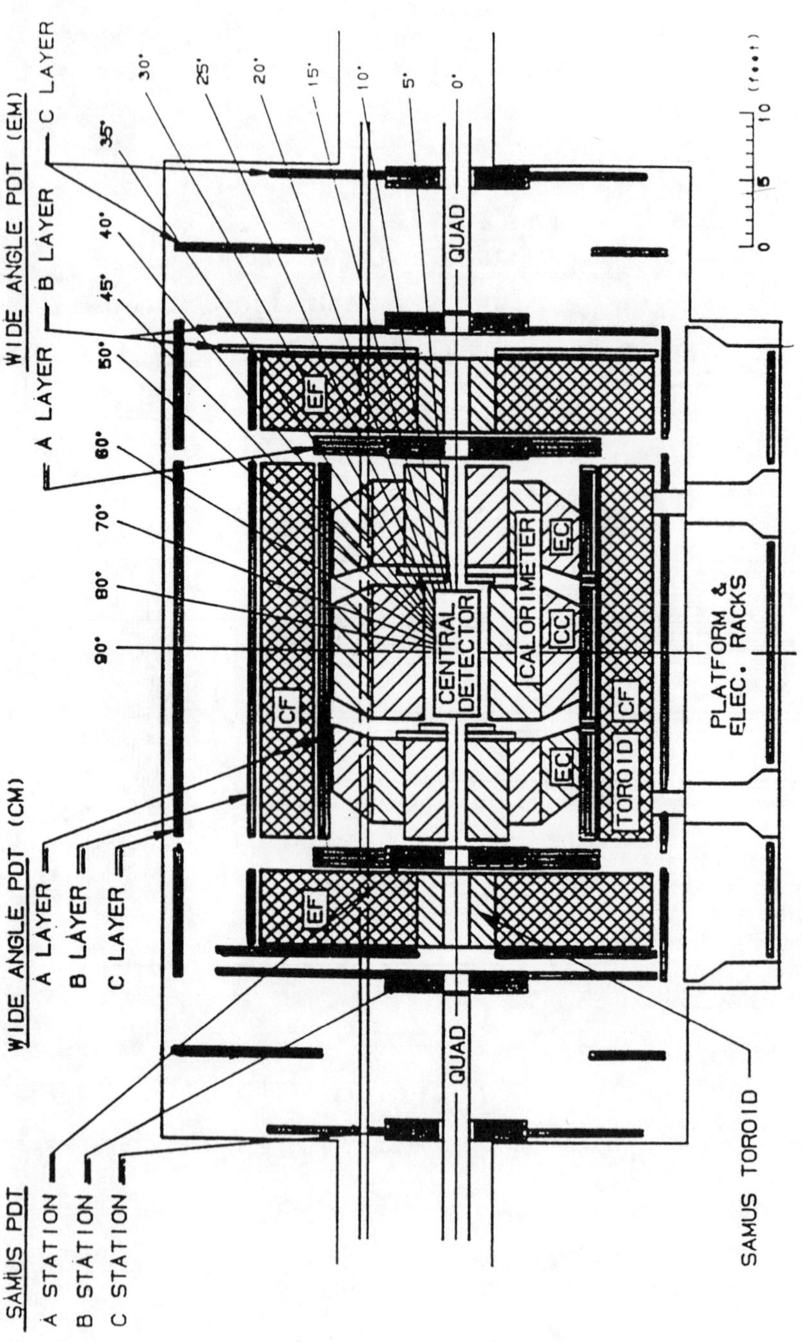


Figure 3
Layout of the Muon System.

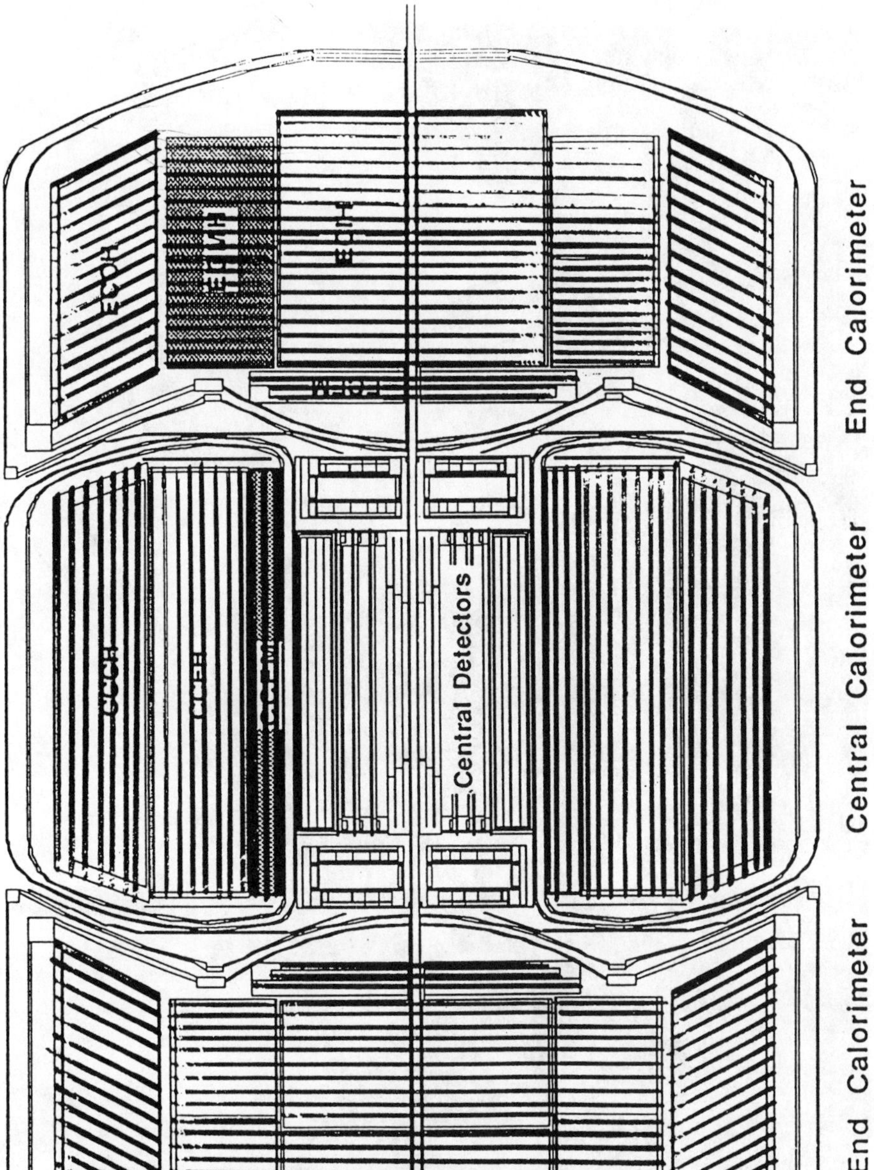


Figure 4
Layout of the Calorimeter System.

*Jim Cronin and René Turlay.*

# HIGH ENERGY GAMMA RAY ASTRONOMY†

James W. Cronin

Enrico Fermi Institute and Department of Physics

The University of Chicago, Chicago, Illinois 60637 USA

When Jim Christensen, Rene Turlay, Val Fitch and I were working in the hot and dusty experimental area inside the ring of the AGS (named affectionately Inner Mongolia), none of us dreamed that there would be an entire conference some twenty-six years later devoted to CP violation discovered by our experiment. I am grateful to Tran and his colleagues for organizing this conference which combines excellent new results concerning CP violation with the magnificent ambience of the Loire Valley. The heat in Blois, however, is reminiscent of that summer in 1963.

I am also pleased that Tran has asked each of us to talk briefly about our present research. I will take this opportunity to tell you a bit about an instrument for high energy gamma ray astronomy that my colleagues[1)] and I are building. Also I will mention how I came to enter a field which is at best on the fringes of what we consider particle physics.

We all know that the important physics of the next decade is going to come principally from the colliders. It is the well-defined problems that set the scale and size of the experiments but, undoubtedly, it will be the unanticipated discoveries at colliders and elsewhere that will lead to the most exciting physics.

By training and temperament I have tried to work on a small scale. My participation in a new measurement of the lifetime of the neutral pion, while on sabattical at CERN is an example.[2)] But I have always wanted to build something large, and I have always been fascinated with cosmic rays. When I was at Princeton, even before the Princeton-Penn Accelerator, I used cosmic ray muons to test scintillation counters for experiments at the Brookhaven Cosmotron. It was a delight to have these minimum ionizing particles coming for free from the heavens.

So when George Cassiday came to Chicago from the University of Utah a few years ago and gave a seminar on the measurement of the upper end of the cosmic ray spectrum ($10^{17}$eV to $10^{20}$eV), I decided to look seriously at the possibility of working in some fashion in cosmic ray physics.

In looking over the possibilities a very clear problem emerged which has been recognized by many people throughout the world. In 1972 air showers of about 1 TeV pointing back to the X-ray binary Cygnus X-3 were observed by detection of the Cerenkov light produced in the atmosphere.[3)] In 1984 air showers of about 1000 TeV pointing back to Cygnus X-3 were observed by a surface air shower array at Kiel.[4)] These air showers were presumed to have been produced by gamma rays incident on the earth's atmosphere. They cannot have been produced by charged particles because bending in the galactic magnetic field decouples the direction of the particle as it strikes the earth from its source. One disturbing factor, however, in the Kiel result was the fact that muons were found to accompany the showers from Cygnus X-3 in a rate far exceeding that expected for gamma rays.

The ratio of muons to electrons (and positrons) for gamma-induced showers is expected to be less than 2% of that expected for hadron-induced showers.[5)] This fact is due to the very small cross section for hadron production by gamma rays. If the Kiel result were correct, then exciting phenomena in both particle physics and astronomy would be the consequence. A number of other experiments seemed to confirm the Kiel result, using mainly the 4.8 hour modulation of Cygnus X-3.[6)] None of these confirming experiments had data on accompanying muons.

The Kiel array was very small. It was at sea level and, consequently had a high threshold. It required four years to collect fifteen events. Subsequent to the Kiel result many groups began to prepare apparatus to look for sources of UHE gamma rays ($E_\gamma > 10^{14}$ eV). Many of these experiments were put together using existing apparatus and were not optimized for this gamma ray astronomy; others were optimized but were in some cases too small to collect data quickly.

At Chicago we have designed and are now building an array that has a much larger area and a much lower threshold than Kiel. Our array, when complete, will be 2500 times more sensitive than the Kiel array. This array is being installed at the Dugway Proving Grounds in Utah where the University of Michigan is installing 2500m$^2$ of muon detectors and the

University of Utah is operating the Fly's Eye detector which observes high energy showers by both atmospheric fluorescence and Cerenkov light. This combination of detectors should prove to be a very effective instrument for high energy gamma ray astronomy.

Figure 1 shows schematically how an air shower array functions. Typically at an altitude of 1500 meters a 100 TeV gamma ray produces about 25000 charged particles (mostly electrons and positrons). Figure 2 shows the lateral distribution of these electrons. A typical particle density is $40/m^2$ at the core dropping to $\sim 1/m^2$ at a distance of 40 meters. The shower particles travel at the speed of light in a disk of thickness ~1 meter which is normal to the shower direction. Measurement of the relative times of arrival of the particles at a grid of scintillators allows the direction of the shower to be reconstructed with a precision of about 1°. A measurement of the pulse heights in the scintillators permits the total number of particles in the shower ($N_e$) to be estimated.

Figure 3 shows the Chicago Air Shower Array (CASA) being built at Dugway, Utah. When complete, it will consist of 1089 detectors placed on a square grid with a 15 meter spacing. Each detector consists of four 61cm × 61cm × 1.2cm scintillators placed in a plastic box. A unique feature of this array is the placement of analog and digital electronics inside each detection station. Each station is controlled by a micro-computer. Data are stored locally and read out periodically by an ethernet local area network system. Time signals are exchanged between nearest neighbor boxes so that the relative shower arrival time is measured between stations. Figure 4 shows an exploded view of the components in each station. A technical description of the detector has been presented elsewhere.[7)]

The placement of the electronics in the field has a number of advantages. It avoids long runs of high quality cable needed to preserve fast timing, and avoids the congestion and expense of thousands of cables coming to a common center. The array is operated by only three 50Ω cables that pass to a central controller, which processes triggers generated in the field, issues commands to encode data, and collects data. The controller is coupled to a $\mu$-VAX computer which logs the data on disk and ultimately writes it on video tape.

As indicated in Fig. 3, a 49-station "engineering" array has been running since December 1988. At present 400 additional stations have been installed, and we expect 529 units to be operating by December 1989. We expect the array to be complete by December 1990. At present 8 muon detectors, each consisting of sixty-four 2.5 $m^2$ scintillation counters have been installed by Michigan and are in operation.

The sensitivity of the detector can be evaluated in two ways. First, for steady sources one can ask how long one must run to observe a signal which is six times larger than a one standard deviation fluctuation of the background of cosmic rays. Or in the case of bursts at an elevated flux, one can ask how many events will be accumulated in a single passage of the source overhead.

In the first case, the time required to see an effect which is N standard deviations above background is given by

$$T = \frac{N^2 F_B \, \Delta\Omega}{F_s^{\,2} A \, R_\mu}$$

where $F_B$ is flux per unit solid angle of isotropic cosmic ray background above the threshold, $\Delta\Omega$ is the angular resolution of the detector [typically $\pi(1°)^2$], $F_s$ is the flux of the source above the threshold of the detector, A is the area of the detector and $R\mu$ is the rejection factor for muon showers using the Michigan muon array.

There is a premium on good angular resolution (small $\Delta\Omega$), large area, and a large $R_\mu$. For our installation $R_\mu$ is about 100 at $10^{14}$ eV and becomes >1000 at $10^{15}$ eV. Figure 5 shows the correlation between the size of the shower detected in the CASA engineering array and the number of muons detected by the Michigan muon detectors. One can see the power of the muon detection to reject hadron showers. If, however, the showers from point sources really do have a muon content similar to hadronic showers, then one will not be able to use the muon detectors to reduce background.

For the CASA array, with integral fluxes typical of the Kiel observation ($10^{-12}$ cm$^{-2}$ sec$^{-1}$ at $10^{14}$ eV), an effect of six standard deviations can be seen in about ten days. If the actual flux is ten times smaller, then 1000 days is required to observe the source with the same significance. One can appreciate the value of the muon rejection (if it works) since for $R_\mu$ ~100 the observation time for a significant measurement is reduced by a factor 100. For sources like Cygnus X-3 of Hercules X-1 which pass overhead, an integral flux of $10^{-14}$ cm$^{2}$ sec$^{-1}$ at $10^{14}$ eV should be detectable.

These sources as observed in radio and X-rays show a propensity to burst. Observations have been reported where the integral flux for $10^{14}$ eV was $\geq 10^{-11}$ cm$^{-2}$ sec$^{-1}$ for periods of hours to days. For the CASA array this would produce more than 300 events in a single pass overhead, an effect which would be easily detected even without muon rejection.

My colleagues and I are eagerly awaiting the completion of this large array. Nature will decide whether or not we will be pursuing both particle physics and astrophysics. If one takes seriously a recent report,[8)] then one is dealing with a very low mass (~50 MeV) neutral particle which has strong interactions. I personally find it difficult to believe that such is the case. The observation of point sources where the muon content is similar to the cosmic ray background make it more likely that the observations are a fluctuation of a background. Of course I would be delighted to be found in error.

†This work is supported by NSF grants PHY 88-23033 AND PHY 87-02822.

References


1) James W. Cronin, Kenneth G. Gibbs, Hans A. Krimm, Nicholas Mascarenhas, Timothy A. McKay, Dietrich Müller, Brian J. Newport, René A. Ong, Leslie J Rosenberg, and Mark E. Wiedenbeck.

2) H.W. Atherton et al., Phys. Lett. 158B, 81 (1985).

3) B.M. Vladiminsky, A.A. Stepanian and V.P. Fomin, Proc. 13th Int. Cosmic Ray Conf. (Denver 1973) 1, 456.

4) M. Samorski and W. Stamm, Astrophys. J. (Lett.) 268, L17 (1983).

5) T. Stanev, T.K. Gaisser, and F. Halzen, Phys. Rev. D 32, 1244 (1985); P.G. Edwards, R.J. Protheroe and J. Rawinski, J. Phys. G 11, L101 (1985).

6) Discussion of ultra-high energy gamma rays from Cygnus X-3, and other sources can be found in the reviews: T.C. Weekes, Physics Reports 160, 1 (1987); R.J. Protheroe, Proc. 20th Int. Cosmic Ray Conf. (Moscow 1987) 8, 21.

7) K.G. Gibbs, Nucl. Instrum. Methods A264, 67 (1988); R.A. Ong, 1989, G. Yodh and D.C. Wold, eds., Proc. Workshop Physics and Experimental Techniques of High Energy Neutrinos and VHE and UHE Gamma Ray Particle Astrophysics, Arkansas, USA, to be published in Nuclear Physics; D. Sinclair, 1989, Nucl. Instrum. Methods A278 583.

8) B.L. Dingus et al., Phys. Rev. Lett. 61, 1906 (1988).

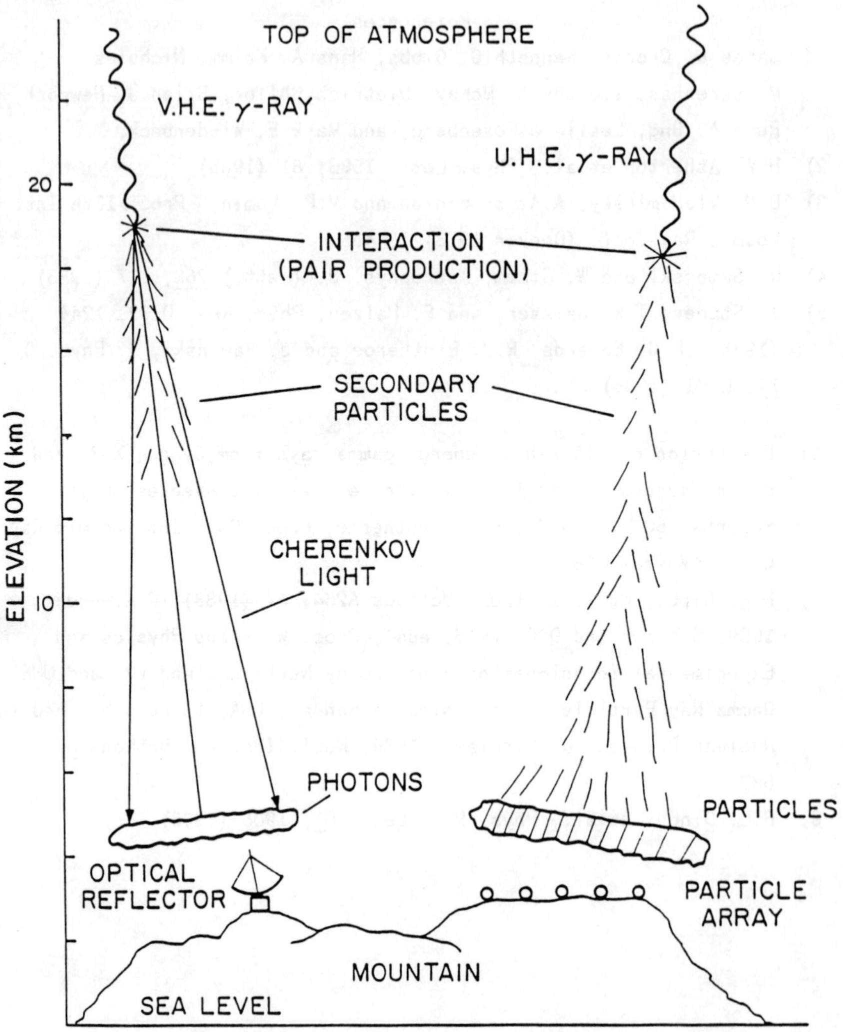


Fig. 1

1. Schematic view of air shower detection techniques. On the left is shown the atmospheric Cerenkov technique, which detects optical photons from ~$10^{12}$ eV showers which do not reach the ground. On the right is a surface array which detects the particles which reach the ground for ~$10^{14}$ eV showers.

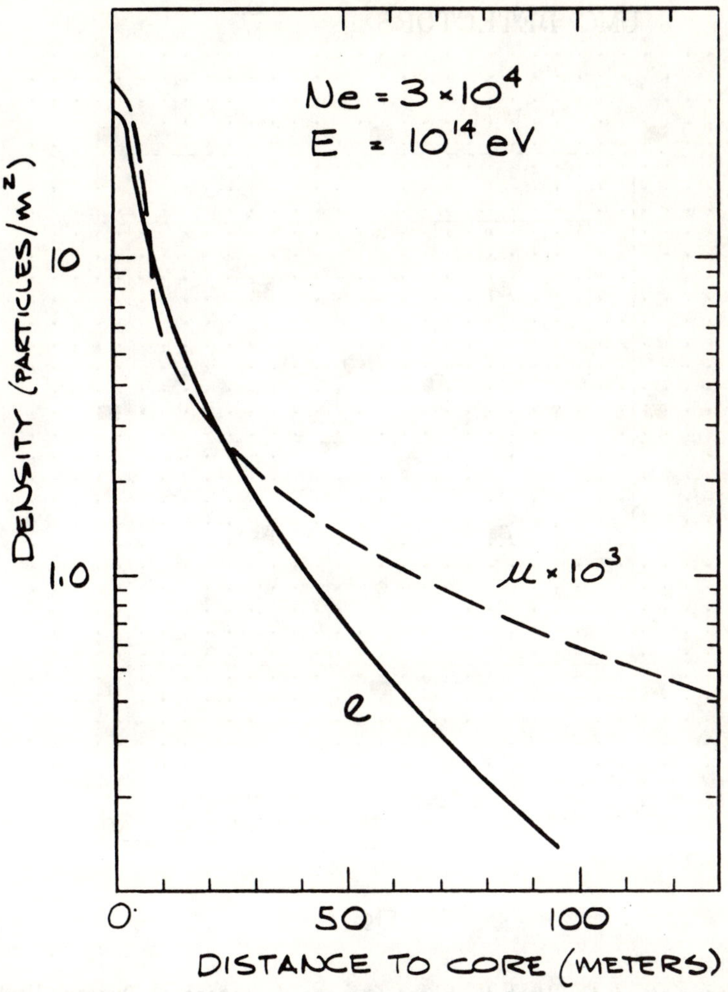


Fig. 2

2. Lateral distribution of electrons (positrons) and muons in a $10^{14}$ eV air shower at an atmospheric depth of 850 gms/cm$^2$. Note that the peak muon density is much lower but the lateral spread of the muons is much wider than the electrons.

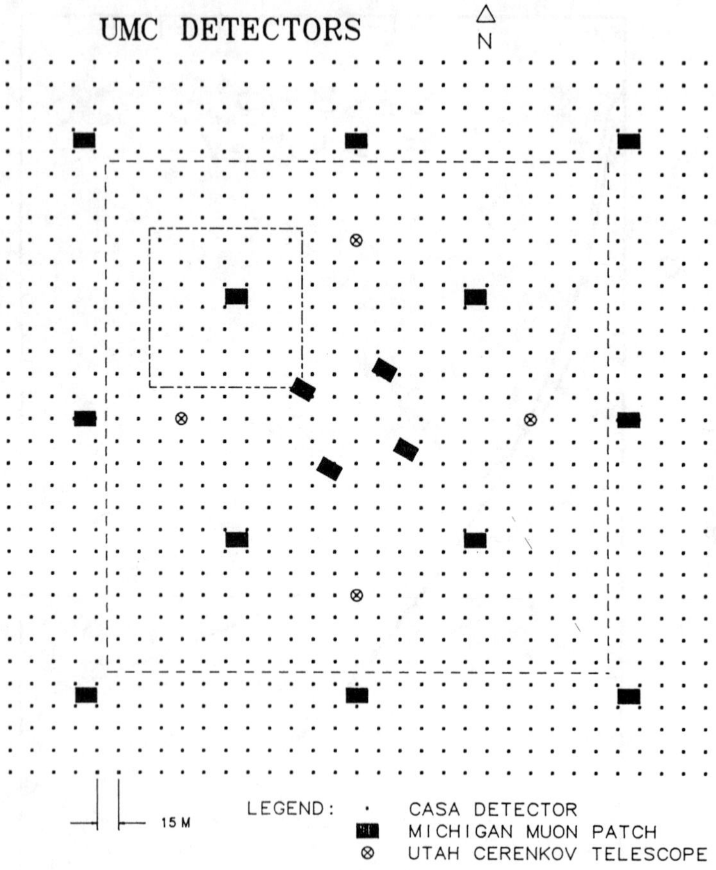


Fig. 3

3. Plan view of the Utah-Michigan-Chicago detectors at Dugway, Utah. The dashed lines indicate the boundaries of the operating 49-station prototype array and the 529-station array (complete December 1989). The full array of 1089 detectors will be complete by December 1990. The Fly's Eye II experiment located at the center of the array is not shown.

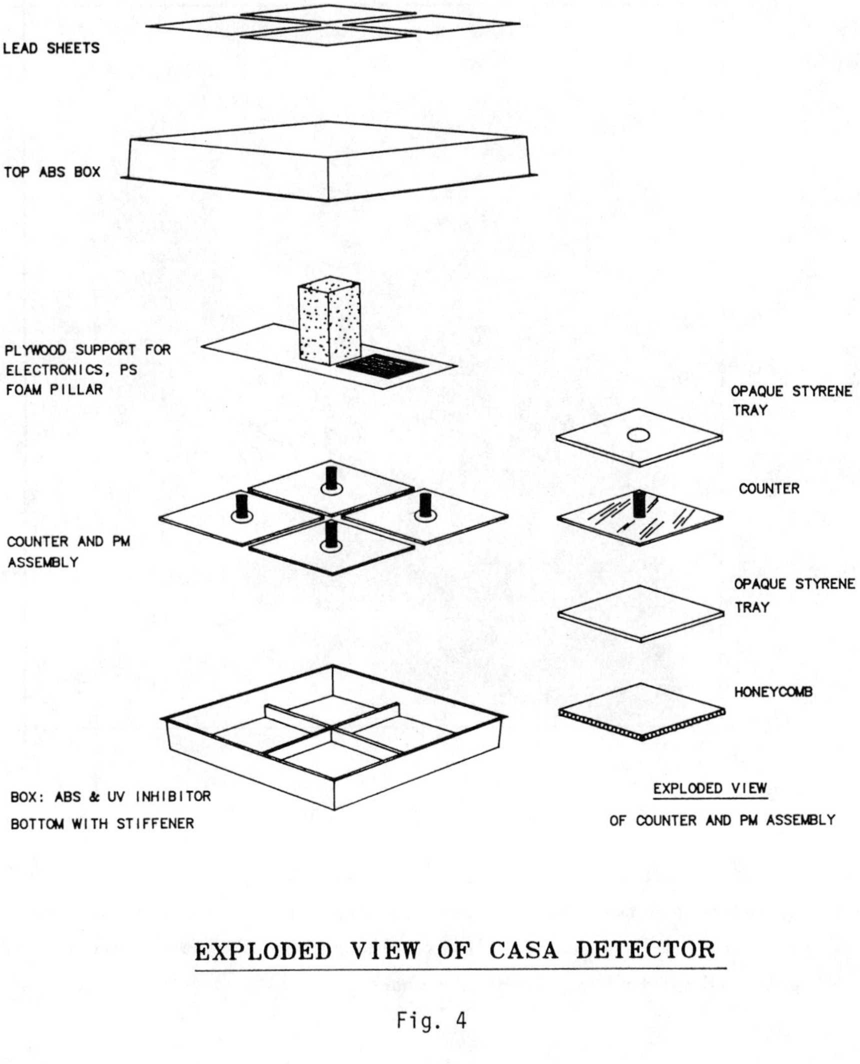


EXPLODED VIEW OF CASA DETECTOR

Fig. 4

4. Exploded view of components in a CASA station.

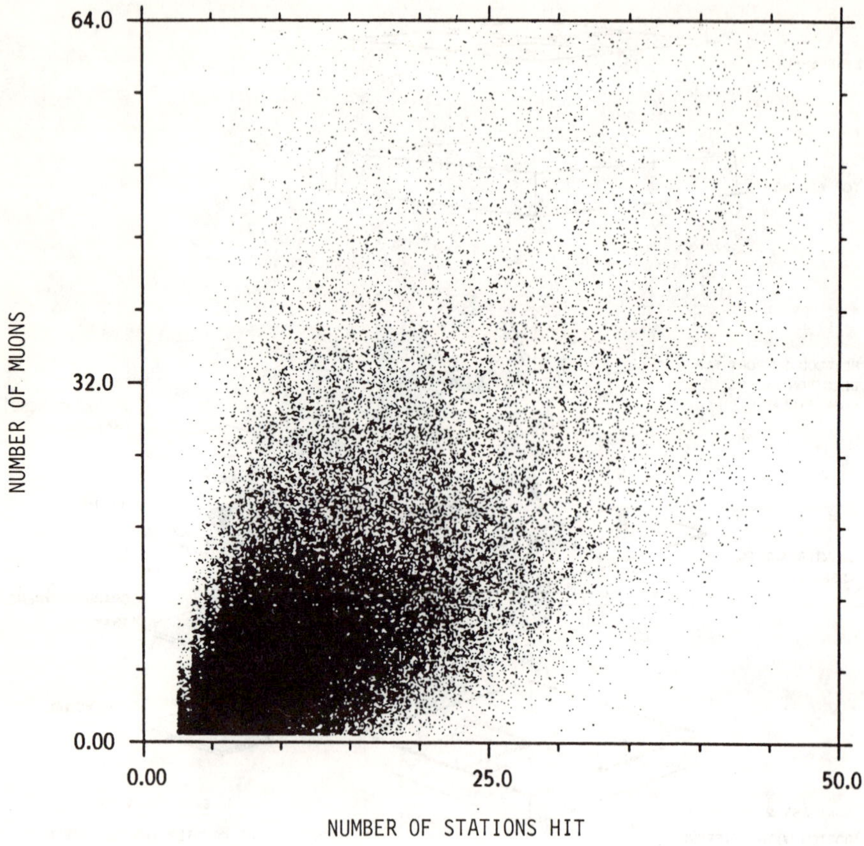


Fig. 5

5. Correlation between number muon counters hit and number of surface stations hit by shower particles. The number of stations hit is proportional to the size of the shower.

# THE FIFTH FORCE, FACT OR FICTION?[1]

Val L. Fitch
Princeton, University
Princeton, N. J. ,USA

Some brief comment on the relevance of the subject of this talk to the principal topic of the conference is appropriate. The particle which has been invented to solve the strong CP problem in quantum chromodynamics, the axion, will produce forces between material objects at macroscopic ranges. However, two characteristics remove the axion from consideration as a carrier of the so-called fifth force. First, it is a pseudoscalar particle and correspondingly couples to the spins of the interacting particles and the spins of the particles which compose the usual macroscopic objects average nearly to zero. Second, from cosmological considerations, the preferred mass of the axion is such that the expected range of any force would be of the order of one centimeter, much shorter that the range which has been associated with any fifth force, i.e., from a few tens of meters to a few kilometers. The strong CP problem will be discussed later in the conference by some of the experts.

There has been a certain amount of theoretical speculation down through the years on the possibility of the existence of new forces which have been add-ons to the usual tensor gravitational force[2]. Indeed, within the framework of various supersymmetric theories there is room for scalar and vector contributions of finite range to the usual gravitational force. There has been nothing compelling about the theoretical speculations, neither for nor against. However, there is room for new forces should they exist. So it becomes a purely experimental question. To the extent that the theoretical speculations could predict the ranges of any new force something in the range of one km was preferred and it was just in this range that ordinary Newtonian gravity was least tested. Starting in the early 1980's a few experiments showed some results which were suggestive of the existence of forces with a strength of around 1 % that of gravity and with a range of a few hundred meters, easily within the bounds of the theoretical conjectures.

The real impetus for the current interest was the reanalysis by Fischbach and his collaborators of the classic Eötvös Experiment[3]. They showed that the Eötvös data, rather than demonstrating an equivalence of inertial and gravitational mass to about one part in $10^8$, actually showed a rather striking dependence of the gravitational acceleration on baryon number. Since the equivalence principle subsequently had been demonstrated to be good to one part in $10^{11}$ by Dicke and Bragzinskii using the sun as the source of inertial effects, Fischbach, et al, concluded that the EÖtvös data was evidence for the existence of a new force of intermediate range. The Eötvös experiment used the spinning earth as the source of inertial effects and, in general, would be sensitive to intermediate range forces. Indeed, Fischbach, et.al. held the apparent lack of equivalence of the gravitational and inertial masses that they observed in the reanalyzed Eötvös data as evidence for a fifth force*[1].

Fischbach, et al, went further and in an attempt to account for some of their own experimental results involving the neutral K-meson system, proposed that the new force was due to a hyperphoton coupling to hypercharge. For those of us who had been in the K-meson business for a long time this suggestion immediately invoked feelings of deja vu since one of the original ideas for accounting for the two pion decay of the $K_L$ without CP violation was to postulate the existence of such a hyperphoton. This idea was ruled out at the time by the absence of such decays as $K^+ \longrightarrow \pi$ + hyperphoton and by the absence of a dependence of the $K_L \longrightarrow 2\pi$ decay on the K momentum. One expects that the branching ratio to go as $E^{2J}$ where E is the energy of the $K_L$ and J is the spin of the quantum involved, equal to unity in this case since it is a vector. At very high momentum Fischbach et al had measured a somewhat smaller branching ratio than the accepted value obtained at low momentum and suggested a tiny bit of hyperphoton interaction superimposed on the CP violation to account for their results. However, their conclusion from their data that the branching ratio was slightly momentum dependent has not been confirmed by other experiments. Furthermore, the absence of $K_L \longrightarrow \pi$ + hyperphoton provides a sufficiently stringent limit to eliminate the hyperphoton as a source of the fifth force. Correspondingly, we dismiss the possibility that the fifth force has anything to do with hypercharge, neutral K mesons and CP violation. Remaining open are the separate questions of the validity of the discrete symmetries in gravity. However, these will not be discussed here.

The experimental results which have been published are in two categories. There are those that are specifically material dependent, the torsion balance and neutral buoyancy experiments are of this kind. Those experiments that measure g as a function of elevation above the earth or depth in the earth that can be either be dependent or totally independent of the nature of the material. Each of the experiments in both categories has been somewhat different in design and execution. This is good and bad. It is good because, invariably, it is the systematic effects which are limiting and they are different is each of

* In fairness to Eötvös, the atomic mass data , which are necessaary to examine the dependence on baryon number, were not available at the time of his work. If they had been and he had seen the effect later uncovered by Fischbach,et al, surely he would have gone back and further examined his experiment for systematic errors. He fully recognized the importance of the issues involved and would have thought long and hard before announcing a departure from the Equivalence Principle. If he had made such an announcement in the early part of this century would that have deterred Einstein from General Relativity?

the experiments. It is bad because a precise comparison of the results is sometimes difficult.

The experiments are interpreted in terms of a Yukawa potential superimposed on a Newtonian, viz.,

$$V(r) = \frac{G\,m_1 m_2}{r}\,(1 + \alpha\, e^{-r/\lambda})$$

where $\alpha$ is a strength factor and $\lambda$ is the range. It is natural to add more terms to account for the possibility of scalar, vector and tensor additions each with a characteristic strength and range. This can lead to some confusion since, presumably, with six parameters to fit, one can account for the divergent results of six different experiments. The strength parameter, $\alpha$, is factored into the product of two "charges" which characterize whatever coupling is being considered (and always normalized to the mass (atomic) of the material) times an intrinsic strength, $\alpha_0$. For example, in the case of coupling to baryon number, B, the charge would be $B/\mu$ where $\mu$ is the atomic mass. In this case, the earth with $B_0/\mu_0$ as a source a fifth force will produce on the ith atom, a horizontal acceleration of

$$a_H = \frac{3}{2}\,\frac{\alpha_o \lambda}{R_o}\,\frac{\rho}{\rho_o}\,\sin(\theta_{tilt})\,g\,\frac{B_o}{\mu_o}\,\frac{B_i}{\mu_i}$$

where $R_0$ is the radius of the earth, $\rho$ is the local rock density, $\rho_0$ is the mean density of the earth, $\theta$ is the local angle of tilt of the terrain appropriately averaged over the range $\lambda$ and g is the local acceleration of gravity. From the above relations one sees that experiments using laboratory sources tend to measure $\alpha$ (since the fifth force ranges under consideration are greater than laboratory distances)whereas experiments using the earth as a source are sensitive to the product $\alpha\lambda$. Of course, the average angle of tilt will be sensitive to the range. At very large ranges, a few tens of kilometers, it must go to zero from considerations of isostasy.

Laboratory sources have an advantage since the constitution is known. However, the sensitivity is very much lower (comparing a 1 ton source at 1 m with the earth the factors are more than 100 for ranges greater than 100 m.)

If a fifth force exists to what is it coupled? There is the strong theoretical prejudice that any fundamental force be coupled to a conserved quantity. As long as the proton is not seen to decay, baryon number is a candidate. But everyone believes the proton does decay so what then? Perhaps the best candidate for a conserved quantity is B -L, baryon minus lepton number. However, keeping an open mind and following the fashion, we summarize the published results in terms of coupling to nonconserved as well as conserved quantities; to isospin, to baryon number and to B - L. In order to account for diverse results obtained in different experiments performed under dissimilar circumstances some authors have suggested various linear combinations of these quantities. Personally, I am inclined to ascribe disparate results to the always lurking subtle systematic effects to which all of these experiments are susceptible. After a fifth force is convincingly demonstrated, we can then worry about the details.

In Table I we summarize the published results from those experiments which are specifically material dependent[4,5,6,7,8,9,10,11,12,13]. In Table II the results from the gravimeter measurements are shown[14,15,16]. As noted above, these later experiments can be sensitive to both material dependent or material independent forces. If an effect is seen in Table II and not in Table I, it would suggest the presence of, e.g., a scaler force coupled to mass.

Several comments are in order. The early results from the neutral buoyancy experiment of Theiberger has not been borne out by subsequent results from torsion balance experiments or by the other neutral buoyancy experiment of Bizzetti, et al. Theiberger observed a definite effect and while many people have speculated on possible sources of systematic effects which could have produced the his result, none have been especially convincing. At the same time it must be noted that in this experiment there were no controls; for example, an investigation under similar conditions of the behavior of a ball made largely of polyethylene. The experiment was never repeated. Only one other experiment (Stubbs, et al) has had sufficient sensitivity to look for the relatively tiny effect reported by Boynton and has not confirmed it. The results in Table I are all published. Stubbs et al are in the luxurious position of having their home laboratory on terrain with a significant slope and, in residence, have been refining their apparatus and measurements for the last few years. At the last Moriond conference they reported results which improve those in Table I by almost a factor of 25 and still no effect is found. That group is approaching the precision needed to improve the Equivalence Principle limits.

The gravimeter experiments all look for deviations from standard gravity under terrestrial conditions. The test mass in the instruments is tungsten. The three published experiments all show an effect. Unfortunately the results are quite different. The latter two even show the opposite sign to the original Holding and Tuck Australian mine experiment. It should be noted that the borehole (mine) experiments are quite insensitive to range, only to strength. It has been shown by Bartlett and Tew[17] that when surface contours are properly taken into account the effect originally observed by Holding and Tuck in the Australian mine shaft can be made to disappear, at least, largely. Bartlett also pointed out that the result of Eckhardt, et al, was not corrected properly for the surface contour. Eckhardt's group has reanalyzed its data taking finer details of the surface topology into account and, as a result, has decreased the magnitude of the published effect by about one third.

In summary, we conclude that there is no evidence for material dependent effects, certainly at the level which was originally suggested. The question of the existence of material independent effects is still somewhat open. Are the current results telling us something new of fundamental importance or are the experiments just sophisticated ways of learning about standard Newtonian gravitational anomalies? New experiments are needed. The issues are clearly of the highest significance.

| Experiment | Type | Source | Detector | N - Z | N + Z | B - L | Range |
|---|---|---|---|---|---|---|---|
| Thieberger | N Buoyancy | Terrestial | Cu-H2O | | - 1.4 ± .3 m | | @ 100 m |
| Stubbs, et al | Torsion Bal | Terrestial | Cu-Be | | 0 ± 2 cm | 0 ± 1 cm | @ 100 m |
| " | | | | | 0 ± 8 cm | 0 ± 4 cm | @ 1 km |
| Boynton, et al | Torsion Bal | Terrestial | Al-Be | | -2.3 ± .6 cm | -.25 ± .08 cm | @ 100 m |
| Niebauer, et al | Free Fall | Terrestial | Cu, Ur | | 1.6 ± 6 m | 3.3 ± 12 cm | 10 m-100km |
| Fitch, et al | Torsion Bal | Terrestial | Cu-CH2 | | -4 ± 7 cm | -.16 ± .3 cm | 25m-1.6 km |
| Speake-Quinn | Beam Balance | Pb and Brass | C-Cu | .008 ± .02 * | | | > 1 m |
| Cowsik, et al | Torsion Bal | Pb | Cu-Pb | 0 ± .0015 | | | |
| Bennett | Torsion Bal | H20 | Pb-Cu | -.0005 ± .001 | | | > 50 m |
| Stubbs, et al | Torsion Bal | Pb | Al-Be | 0 ± .0012 | .0002±.002 | | > .5 m |
| Bizzetti, et al | N Buoyancy | Terrestial | H20-Nylon | | < .3 m | | 60 - 1000m |

TABLE I. Results of material dependent experiments

| Experiment | Detector | Strength | Range |
|---|---|---|---|
| Holding, et al | gravimeter | -0.01 | ≈ 200m |
| Eckhardt et al | " | 0.02 | 300m |
| Ander et al | " | 0.03 | ≈ 1 km |

TABLE II. Results of material independent experiments

## References


1. For the past three years Moriond conferences have been devoted to the development of this subject. The Proceedings give a good account of the history.

2. There are many theoretical references. I give two for the flavor of the arguments. J. Scherk, Unification of ;the Fundamental Particle Interactions, S. Ferrara, et al, ed, Plenum, (1980). R. D. Peccei, J. Sola, and C. Wetterich, Phys. Lett. B 195, 183 (1987)

3. E. Fischbach, D. Sudarsky, A Szafer, C. Talmadge, and S. H. Aaronson, Phys. Rev. Lett. 56, 3 (1986).

4. P. Thieberger, Phys. Rev. Lett. 58, 1066 (1987)

5. C. W. Stubbs, et al, Phys. Rev. Lett. 58, 1070 (1987)

6. P. E. Boynton, et al, Phys. Rev. Lett. 59, 1385 (1987)

7. T. M. Niebauer, M. P. McHugh, and J. R. Faller, Phys. Rev. Lett. 59, 609 (1987)

8. V. L. Fitch, Mr. V. Isaila, and M. A. Palmer, Phys. Rev. Lett. 60, 1801 (1988)

9. C. C. Speake and T. J. Quinn, Phys. Rev. Lett. 61, 1340 (1988)

10. R. Cowsik, N. Krishnan, S. N. Tandon, and C. S. Unnikrishnan, Phys. Rev. Lett. 61, 2179 (1988)

11. W. R. Bennett, Jr., Phys. Rev. Lett. 62, 365 (1989)

12. C. W. Stubbs, et al, Phys. Rev. Lett. 62, 609 (1989)

13. P. G. Bizzetti, et al, Phys. Rev. Lett. 62, 2901 (1989)

14. S. C. Holding and G. J. Tuck, Nature (London) 307, 714 (1984). S. C. Holding, F. D. Stacey and G. J. Tuck, Phys. Rev. D 33, 3487 (1986)

15. D. H. Eckhardt, et al, Phys. Rev. Lett. 60, 2568 (1988)

16. M. E. Ander, et al, Phys. Rev. Lett. 62 985 (1989)

17. D. F. Bartlett and W. L. Tew, Phys. Rev. D40, 673 (1989)

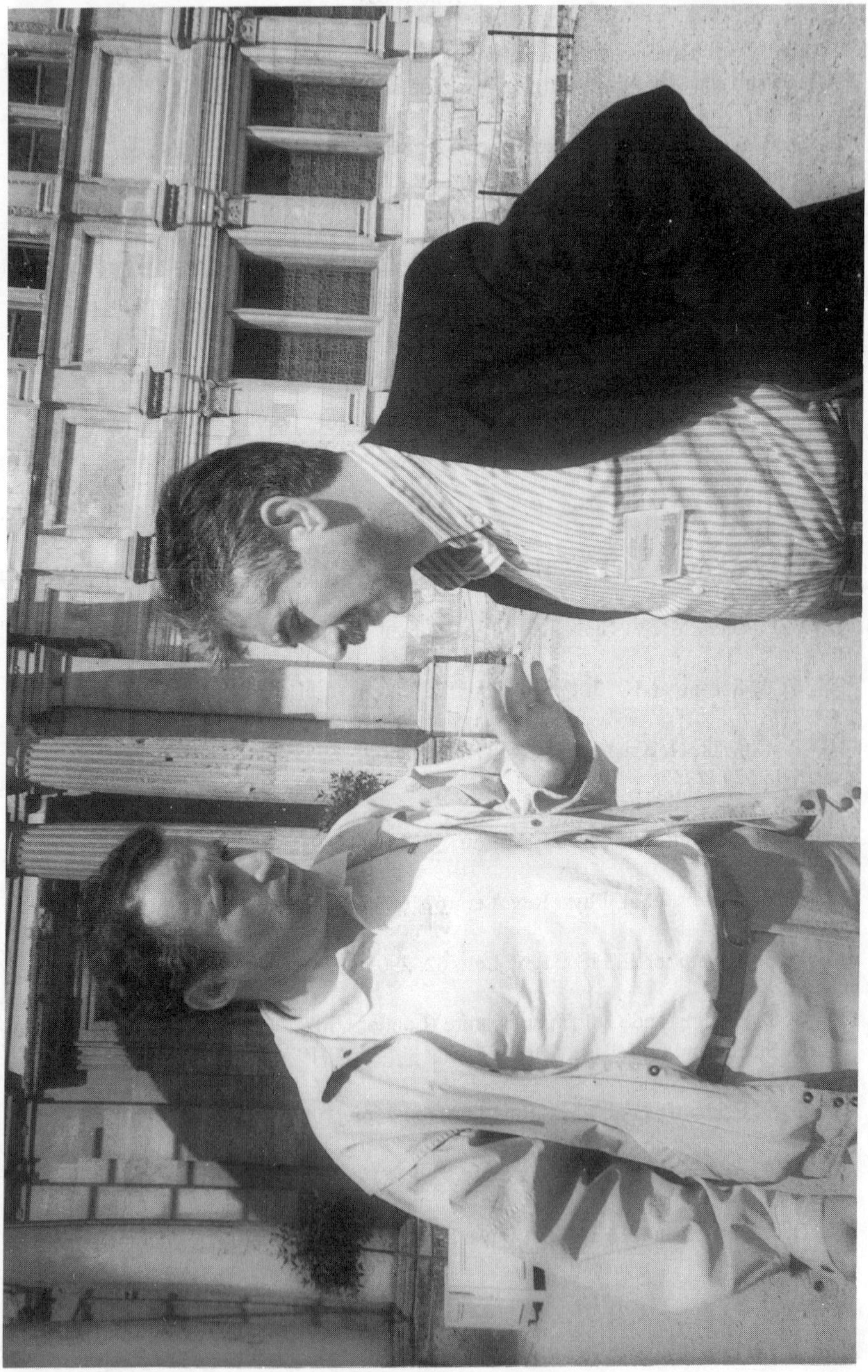

*Bernard Peyaud and Luis Oliver*

## GUEST LECTURE

**R. Turlay**

**D.Ph.P.E., CEN-Saclay, France**

When I asked the organizers of the conference for information on what a guest lecture is and what they expect from me I was given the following answer :

i) you should not say one word about the CP discovery ;

ii) tell us what physics you did during the last 25 years ;

iii) give us a title for your talk !

Quite a programme ! I shall ignore the first advice by expressing how I am grateful to Tran Thanh Van, my old friend, to have had the splendid idea to celebrate the $25^{th}$ anniversary of the discovery of CP violation. It is a great occasion to spend one week of scientific discussion in a good friendly atmosphere. I meet again with great pleasure my three colleagues J. Christenson, J. Cronin, V. Fitch, from whom I have learnt so much.

I shall say a few words on the main aspects of the physics I participated in since 1964, and that will explain why I have chosen as guest lecture title "the LEP experimental programme".

**1964-1989**

In 1964 after two years at Princeton University I came back home and J. Cronin decided to join our laboratory for a sabbatical year. Great help and a good time for us. We decided to look at $K_L^\circ$ decay with the 3 GeV/c proton accelerator "Saturne". The energy was low and the statistics marginal but we could confirm the CP violation at low energy and study the $K_{e3}^\circ$ and $K_{\mu3}^\circ$ form factors. We found a strange result on the $K_{\mu3}^\circ$ form factors but could not find the bug[1].

1968-1972

It was not possible to continue to work on $K$ physics at this energy and with this intensity. I did not want to join any of the CERN groups working on CP but did not want to stop this kind of physics. We decided with Ronald Mermod from Geneva University, (P. Exterman was also one of the main collaborators) to study the validity of the rule $\Delta S = \Delta Q$ by studying the $Ke_4$ decay $K^+ \to \pi^+\pi^-e^+\nu$.

We found a limit : $\Gamma\left(Ke_4 \to e^-\right)/\Gamma\left(Ke_4 \to e^+\right) < 3.6\ 10^{-4}$ at 95% cl. , and many other valuable results came from this experiment :

– $\pi\pi$ phase shift $\delta = \delta_\circ^\circ - \delta_1^1$ which gives[2] an extrapolated value $\delta_\circ^\circ \simeq 37 \pm 5°$ at the $\pi\pi$ invariant mass of 500 MeV/c$^2$ and can be used to determine the phase of $\varepsilon'$ ;

– as a by-product we studied some rare decays of $K^+$, measuring for the first time[3][4]

$$K^+ \rightarrow \pi^+e^+e^- \qquad (2.6 \pm 0.5)\ 10^{-7}$$

$$K^+ \rightarrow \mu^+\nu e^+e^- \qquad (10 \pm 3)\ 10^{-7}$$

$$K^+ \rightarrow e^+\nu e^+e^- \qquad (2^{+2}_{-1})\ 10^{-7}$$

and gave several limits

$$K^+ \rightarrow \pi^+\mu^+e^- \quad < \ 4.8\ 10^{-9}$$

$$K^+ \rightarrow \pi^\pm\mu^\mp e^+ \quad < \ 6.9\ 10^{-9}$$

$$K^+ \rightarrow \mu^-\nu\, e^+e^- \quad < \ 1.9\ 10^{-8}$$

$$K^+ \rightarrow \pi^-e^+e^- \quad < \ 9.2\ 10^{-9}$$

– we studied 225 000 $\tau^+$decays[5];

– and last the unexpected observation of $\pi^\circ \rightarrow e^+e^-$ decays $\left(2.2^{+2.4}_{-1.1}\right)\ 10^{-7}$ [6].

The Collaboration of two external laboratories (Geneva University and Saclay) at CERN was successful, pleasant and peaceful. It is a pleasure to see that twenty years later some results are still unique and resist time victoriously.

<u>1973-1984</u>

The possibilities offered by the SPS at CERN were very attractive and I wanted to continue to work on weak interactions, which at this time had just become the electroweak interaction. Neutrino physics had a good potential and I dared to join a new and young collaboration in spite of the frightful reputation of the group leader, Jack Steinberger. The beginning was hard and the new French group had to demonstrate its strength. I was very much helped by the strong support of the technical team which built large planar drift chambers.

The physics results on neutrino physics obtained by the CDHS Collaboration (<u>C</u>ern, <u>D</u>ortmund, <u>H</u>eidelberg, <u>S</u>aclay) during 10 years were the following :

– a "normal" $Y$ distribution was found[7] ;

– scaling violation, $\sigma_T$, structure functions ;

– multilepton physics :

- opposite sign dimuons, strange sea distributions[8] ;
- like sign dimuons with the anomalous cross section ;
- trimuons which appeared at the beginning like very exciting physics ;
- and even a quadrimuon event which we did not know what to do with.

– One of us (Frederik Dydak) had a long love story with neutral current physics and reached the ultimate precision on the $\sin^2\theta_W$ measurements[9]:

$$\sin^2\theta_W = 0.225 \pm 0.005(\text{exp}) \pm 0.003(\text{theor.}) \pm 0.013\left(m_c - 1.5\ \text{GeV/c}^2\right)$$

– Many other results on beam dump, neutrino oscillation, interaction of neutrinos on hydrogen were published.

The CDHS Collaboration has been for me the main stage of my physicist 's life. I have learnt to work rigorously, and have increased very much my knowledge of physics and had a much more complete view on physics. A very good group of physicists among them F. Dydak, F. Eisele, M. Holder, K. Kleinknecht, A. Para, K. Tittel, H. Wahl worked "sous la main de maître de Jack Steinberger". It was a very good school.

In parallel since 1978 I started a physics programme at FNAL. I worked again with V. Fitch on an experiment on charm meson production with a group from Torino University led by Rosana Cester-Regge. It was fun but the results fair.

In 1981, Jim Cronin invited me to join an experiment on $\varepsilon'/\varepsilon$ measurement and gave me the opportunity to be back in CP Physics. Bruce Winstein, spokesman of the E617 experiment, accepted the Saclay Collaboration which since is still working with him in experiment E731.

When in 1980 the construction of the LEP machine was decided, the CDHS Collaboration was a natural nucleus of a much larger group which became the ALEPH Collaboration. I stopped real work on ALEPH in 1984.

All this work was possible with the help of many people. First a succession of excellent students. I would like to give their name as they came along : A. Zylbersztejn, P. Basile, C. Bloch, A.M. Diamant-Berger, G. Marel, B. Devaux, N. Do-Duc, Ph. Bloch, S. Zylberajch, J.P. Maillard, B. Peyaud, S. Loucatos, C. Guyot, J.P. Schuller, P. Debu, B. Vallage, F. Perrier, P. Jarry, R. Daudin. An excellent group of technicians and engineers, J.C. Brisson, S. Brehin, M. Lemoine, H. Martin, P. Micolon, B. Paul, J. Pelle, G. Tarte and two senior physicists J. Rander and P. Perez.

In 1984 I was nominated head of the D.Ph.P.E. at Saclay (Département de Physique des Particules Elémentaires). No director of a laboratory can seriously claim that he is still deeply involved in an experiment or doing his own research, even if he wants to. At least I could not and the management (word which is taken very often in a pejorative meaning by many physicists) of a laboratory is an exhausting work. We can sometime nevertheless have the possibility to participate in physics. It was the opportunity I had when in 1986 Professor H. Shopper, Director of CERN, asked me to become chairman of the LEP experiment committee. It is the reason why I would like now to speak about the LEP experimental programme.

## THE LEP EXPERIMENTAL PROGRAMME

The $15^{th}$ of July of this year the LEP machine will be completed and ready to start commissioning.

Four experiments ALEPH, DELPHI, L3 and OPAL will be ready to take data.

### LEP machine

"In the week after the $15^{th}$ of July the SPS will work on the leptonic mode and the LEP ring will receive $e^+$ and $e^-$ beams" (Private communication, Emilio Picasso on the $17^{th}$ of May). All the elements of the machine are ready and positioned.

In the LEP committee of the $14^{th}$ of March, S. Myers[10] has presented plans for the LEP commissioning. A long list of tests will have to be done. The main steps are the following :

- 26 days are necessary to reach the first beam interactions at $Z^\circ$ energy. It will be a pilot run at $Z^\circ$ peak (SLC energy) without superconducting low $\beta$ quadrupoles. 3 or 5 days of running, at a luminosity of a few inverse nanobarn giving a hundred $Z^\circ$ to experiments to test their complete apparatus.

- Eleven more days are necessary to adjust the collision energy and then start a physics data-taking run.

The goal is to obtain 100 000 $Z^\circ$ per interaction region for Christmas 1989.

The number of days are very minimal and "everybody knows that many problems could arise, but we know also the excellence of the CERN team of the LEP machine".

**LEP detectors**

They are almost completed and the teams have started debugging the data acquisition systems. The apparati has been put on the beam line.

They are all on time, the largest difference between them is that, the most advanced one will take 2 or 3 weeks data on cosmic rays and the least will just be ready when the door will be closed on the $15^{th}$ of July.

Magnificent work has been done to build 4 detectors "able to find a needle in a haystack !"

**Physics programme at LEP I**

A very rich programme can be undertaken as soon as the design luminosity $\mathcal{L} = 1.7\ 10^{31} \mathrm{cm}^{-2} s^{-1}$ will be reached. Physicists will decide what they want to tackle first. In what follow I shall recall the conclusions of a survey of the physics which can be done at LEP I and presented in "Physics at LEP" [11], and in lectures given by R. Peccei[12].

*Remarks on top quark*

If the mass of the top quark is less than 55 GeV/c$^2$, toponium physics will belong to the LEP I programme. But at the last meeting held at CERN on the $18^{th}$ of April, results[13] were presented by the three groups UA1, UA2, CDF. They gave "no indication" or "top unlikely" at a mass less than 60 GeV/c$^2$.

This limit could be improved by these experiments with the 1989 data, or even the top can be found. If not :

- in 1993-1994 the LEP energy will reach 100 GeV per beam ;

- colliders can reach a mass top limit of 120 GeV in 1993 if adequate luminosity is delivered $\left(100\ \mathrm{pb}^{-1}\right)$.

It is interesting to underline that if the the LEP I programme is going well, the measurement of the precise value of $M_Z$ and $\sin^2\theta_W$ leading to good knowledge of radiative corrections can provide an indication at $\pm$ 20 GeV/c$^2$ of the top mass in 1992.

*Mass of the $Z^\circ$, measurement of $\sin^2\theta_W$, radiative corrections*

Description of interactions between fermions and gauge bosons in the Standard Model needs 3 parameters :

**– $g_1$, $g_2$ gauge coupling constant of U(1)×SU(2) group ; a Higgs field vacuum expectation value,**

**– or $\alpha$ electromagnetic fine structure constant $\alpha = 1/137.03604$ (11),**

**– $G_\mu$ the Fermi constant $G_\mu = 1.166344\ (11) \times 10^{-5}\ \mathrm{GeV}^{-2}$,**

**– $\sin^2\theta_W = 0.231 \pm 0.006$ or $M_Z = 91.3 \pm 1.7\ \mathrm{GeV/c^2}$ or $M_W = 80.9 \pm 1.3\ \mathrm{GeV/c^2}$.**

**We immediately see the big difference in precision of $\alpha$, $G_\mu$ and any of the 3 other possible constants $\sin^2\theta_W$, $M_Z$, $M_W$. On the other hand, $\sin^2\theta_W$ is defined in the Standard Model** in lowest order as follow :

1. for doublet Higgs breaking

$$\sin^2\theta_W = 1 - \frac{M_W^2}{M_Z^2}$$

2. relation between the Fermi constant and $M_W$

$$\frac{\pi\alpha}{\sqrt{\mu}\ G_\mu} = M_W^2\ \sin^2\theta_W$$

3. neutral current definition

$$J_{NC}^{\mu} = 2\left(J_3^{\mu} - \sin^2\theta_W\ J_{em}^{\mu}\right)$$

The present agreement between the values of $\sin^2\theta_W$ from collider results ($M_Z$ and $M_W$) and the low energy neutral current (neutrino physics) is not sufficiently precise to give indication on structure beyond the minimal model.

LEP and SLC, by providing precise values, in a first step of $M_Z$and $\sin^2\theta_W$, and in a second step LEP II with $M_W$, will provide a test of the theory at the ultimate level.

But these measurements will be so precise and will provide redundant cross checks, that one needs to go further than lowest order calculations. The effects on higher order terms necessitate the specifications of the parameters in which renormalization is performed. The calculation of radiative corrections has given rise to many articles. In the Marciano and Sirlin[14] scheme, these radiative corrections are summarized in $\Delta r$ term :

$$M_W^2 = \frac{\Pi\alpha}{\sqrt{2}\ G_\mu}\ \frac{1}{\sin^2\theta_W\,(1 - \Delta\mathrm{r})}$$

$\Delta$r is a function of $\alpha$, $M_W$, $M_Z$, $M_H$, $m_t$ and has a calculated value of 0.07.

Any effect of the Higgs or top masses or any effect due to new physics "beyond the Standard Model" model contribute to $\Delta r$ at a level of a few percent. To know more about the validity of the Standard Model will necessitate measurements of $M_Z$, $\sin^2\theta_W$ and $M_W$ at the corresponding accuracy.

*What can be done at LEP I*

- $Z°$ mass. A precision of $\Delta m = \pm$ 50 MeV/c$^2$ can be achieved at LEP I and can be decreased further at $\Delta m = \pm$ 20 MeV/c$^2$ with the use of polarization. This corresponds to a precision of $\delta\sin^2\theta_W = 0.004$.

- Forward-backward asymmetry :

$$A_{FB} \quad \text{in } e^+e^- \quad \rightarrow \quad \mu^+\mu^-$$

$$A_{\text{pol}} \quad \text{in } e^+e^- \quad \rightarrow \quad \tau^+\tau^-$$

will correspond respectively to :

$$\delta\sin^2\theta_W = 0.001 \quad (\mu^+\mu^-)$$

$$\delta\sin^2\theta_W = 0.002 \quad (\tau^+\tau^-)$$

the $\tau$ helicity being reconstructed via $\tau \rightarrow \pi\nu$.

These measurements are delicate and will entail considerable time to be obtained. We also see that they give a good precision on $\sin^2\theta_W$.

Better precision will be obtained

- either at LEP II, by measuring $M_W$ with $\Delta M_W = \pm$ 100 MeV/c$^2$ $\rightarrow$ $\delta\sin^2\theta_W = 0.0008$

- or with longitudinally polarized beam allowing the measurement of

$$A_{LR} = \frac{\sigma_L - \sigma_R}{\sigma_L + \sigma_R}$$

with corresponding error

$$\delta\sin^2\theta_W = 0.0004$$

*$Z°$ width and neutrino counting*

The relative rate for various $Z°$ decay channels can be calculated and the total width can be written as :

$$\Gamma_{Z^\circ \to \text{all}} = 0.17\, N_\nu + 0.086\, N_{\ell\pm} + 0.38\, N_{-1/3} + 0.30\, N_{2/3} \text{ GeV}$$

$$\begin{array}{lc} & = \left( \begin{array}{ccccccccc} 2645 & \pm & 20 & \pm & 30 & \pm & 10 & \pm & 4 \\ & & \downarrow & & \downarrow & & \downarrow & & \downarrow \\ & & \alpha_s & & m_t & & \sin^2\theta_W & & Z^\circ\text{mass} \end{array} \right) \text{MeV/c}^2 \\ \text{With theoretical errors due to} \rightarrow & \end{array}$$

At LEP I the precision on the width can be $\Delta\Gamma_Z = \pm\ 50$ MeV/c$^2$. Since any extra neutrino generation contributes 170 MeV/c$^2$, then "LEP experimental results should be able to determine any additional light neutrino or any light neutral object which weakly couples to $Z^\circ$ ".

Another possibility of neutrino counting is to look at the reaction $e^+e^- \rightarrow \gamma +$ nothing visible. The study of the differential cross section near the $Z^\circ$ pole versus the photon energy shows a peak in the photon energy distribution. This peak is proportional to the number of neutrino families. The determination of the absolute value of this peak and the background subtraction make this experiment more difficult than the determination of the total width.

*Neutral Higgs searches*

This research is carried out each time a new range in energy is available. Since Higgs couplings to fermions and bosons are proportional to their masses, we have ways to make Higgs searches at LEP I

$$Z^\circ \longrightarrow H^\circ\ \gamma$$

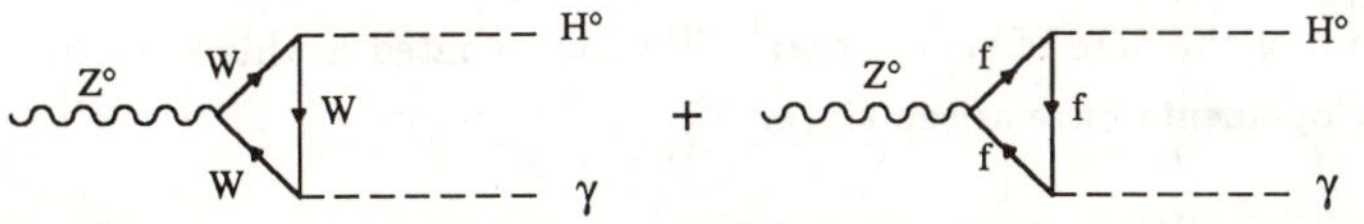


The branching ratio $\Gamma\left(Z^\circ \rightarrow H^\circ\gamma\right)/\Gamma\left(Z^\circ \rightarrow \mu^+\mu^-\right)$ is $10^{-6}$ for $m_H = 80$ GeV/c$^2$ and $6.10^{-5}$ for $m_H = 10$ GeV. $10^6$ $Z^\circ$ make possible the detection of Higgs up to a mass of 40 GeV/c$^2$

$Z^\circ \longrightarrow H^\circ\ \ell^+\ell^-$ following the graph :

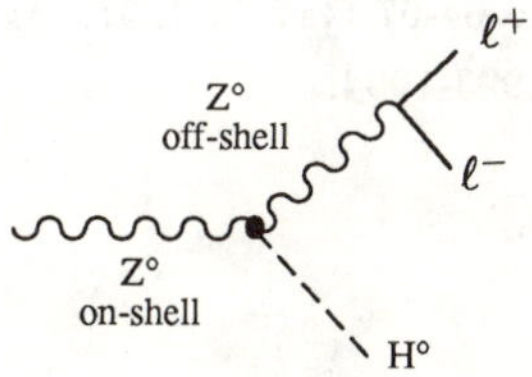

The branching ratio $\Gamma(Z^\circ \to H^\circ\mu^+\mu^-)/\Gamma(Z^\circ \to \mu^+\mu^-)$ is equal to :

$$\begin{array}{lll} 6.10^{-3} & m_H & = & 10\ \mathrm{GeV/c^2} \\ 10^{-4} & m_H & = & 40\ \mathrm{GeV/c^2} \end{array}$$

giving possibility to detect a Higgs particle up to a mass of 40 GeV/c$^2$.

*New particles at LEP I*

Searches for new particles such as heavy leptons, charged Higgs or supersymmetric partners have been intensively studied in [II]. It is clear that such searches will be undertaken as soon as the first data will be available.

## Upgrading LEP at 200 GeV

A strong push from physicists to get to the $W^+W^-$ production threshold as soon as possible was already undertaken and solidified by the ECFA workshop on LEP 200 held at Aachen in October 1986[15].

The physics goals are clear. The 3 gauge vertices behaviour is an independent test of the Standard Model and $W$ physics will extend our knowledge.

- $W$ mass measurement $\Delta M_W = \pm$ 100 MeV/c$^2$.
- $W$ decays.
- Higgs searches extended to 80 GeV/c$^2$.
- Non-standard particle searches can be pursued.

The question of the date of such a possibility was debated in this workshop and very interesting developements have arisen since.

Different scenarios[15] were presented. In the last LEP experimental committee of 14$^{th}$ of March, Emilio Picasso gave more precise details confirmed by the director of CERN, Professor Carlo Rubbia.

Four superconducting Nb cavities should be ready at the end of 1989 and four Nb Cu middle of 1990. This will allow an energy increase up to $\simeq$ 60 GeV in 1990. Ordering of 20 Nb s.c. and 4 Nb Cu s.c. has been done and they could be installed during 1990/1991 shutdown upgrading LEP energy to 64-67 GeV in 1991. As more become available, then $W$' Physics can be undertaken in 1993-1994.

## Polarization at LEP

A working group to study longitudinally polarized beams at LEP was created and reported at the November 1988 LEP Experimental Committee[16]. The main recommendations are the following :

1. Study of transverse polarization should start for early operation at LEP. The transverse polarimeter is already planned by the LEP Instrumental Group. Dedicated wigglers are essential to study transverse polarization. They reduce the polarization time to 36 minutes instead of 300, allowing thus an empirical correction of the machine imperfections in a reasonable time.

2. Longitudinal polarization could not be obtained then but a decision on which scheme to use for the spin rotator (Schwitters-Richter or Montague) needs further studies.

A very original running mode has been proposed by A. Blondel[16] which shows the big physics potential of polarized beams. The most striking contribution is the measurement of the asymmetry $A_{LR}$ with a remarkable precision, by far the best among all the measurements already proposed in asymmetry $(\delta \sin^2\theta_W = 0.0004)$. Such a precise measurement is a very powerful tool to determine the range of the Higgs mass and to reveal new physics beyond the Standard Model.

The planning for the implementation of the rotators is not so advanced as the energy upgrade of LEP. The first results on transverse polarization must be known to decide on construction and installation of the spin rotator which then can take place in 1992 at the earliest.

## Z° High luminosity

Recently a large increase (at least a factor 10) of the LEP design luminosity or "$Z^\circ$ factory" has begun to be discussed. This is based on the possibility of a "Bretzel" scheme already studied at CESR. Detector possibilities relative to such an increase of luminosity have still not been studied. To obtain $10^8$ $Z^\circ$ in 2 to 3 years of running time is very tempting. The $b$ quark physics can then be possible and the hope is to reach the possibility of CP violation in the $\bar{b}\bar{b}$ system.

To have more information, a working group has just been set up on the subject and should report in spring 1990.

## CONCLUSION

Over the next two years a straightforward programme is foreseen and work will be done to exploit LEP Phase I.

In parallel, thought and technical developments are being invested now to upgrade the machine to 200 GeV, to consider the possibility of running with longitudinally polarized electron and positron beams and to increase by a large factor the design luminosity. Physicists will have to decide among these possibilities considering the first results and their priorities.

Altogether this is a solid programme which will extend up to the end of the century.

A remarkable effort has been made by thousands of machine builders, technicians, engineers and physicists to build LEP and the 4 wonderful detectors .These large investments and the important technological advances over the past 10 years are ready and will bloom in a few months.

The quest will be done in a highly competitive atmosphere.... We are convinced that the results will be at the level of the efforts and the enthusiasm already displayed.

The Blois conference has been a wonderful week of physics and friendship. I said already how much we owe to Tran Thanh Van for this success. I would like to thank specially the organizers Anne-Marie Lutz and Bernard Peyaud who have constructed the scientific programme. To all my colleagues, thank you very much. "Une belle page est tournée".

I would like to thank Mesdames 0. Lebey and S. Frelat for their work in the preparation of this paper.

**References**


[1] P. Basile et al., Experimental study of the $K^{\circ}_{\mu 3}$ form factors. Phys. Rev. D2 (1970) 78.

[2] A. Zylbersztejn et al., Further results on $K_{e4}$ decay and energy dependence of low energy $\pi\pi$ phase shift, Phys. Lett. 38B (1972) 457.

[3] P. Bloch et al., Observation of the $K^+ \rightarrow \pi^+ e^+ e^-$ decay, Phys. Lett. 56B (1975) 201.

[4] A.M. Diamant-Berger et al., Study of some rare decays of the $K^+$mesons, Phys. Lett. 62B (1976) 485.

[5] B. Devaux et al., A study of 225 000 $\tau^+$ decays, Nucl. Physics B126 (1977) 11.

[6] J. Fischer et al., Observation of the $\pi^{\circ} \rightarrow e^+ e^-$ decay, Phys. Lett. 73B (1978) 364.

[7] M. Holder et al., Is there a high $y$ anomaly in antineutrino interactions ?, CDHS publication, Phys. Rev. Letters 39 (1977) 433.

[8] H. Abramowicz et al., Experimental Study of opposite-sign dimuons produced in neutrino and antineutrino interactions, Z. Physik C. Particles and Fields 15 (1982) 19.

[9] H. Abramowicz et al., Precision measurement of $\sin^2\theta_W$ from semileptonic neutrino scattering, Phys. Rev. Lett. 57 (1986) 298.

[10] S. Myers, LEP machine commissioning, Open Session of LEPC, 14$^{\text{th}}$ of March 1989.

[11] Physics at LEP by John Ellis and Roberto Peccei, CERN 86-02 - 1986 - vol. I and II.

[12] R. Peccei, Physics at LEP, Desy 86-163, December 1986.

[13] D. Denegri, private communication.

[14] W.J. Marciano and A. Sirlin, Phys. Rev. D22 (1980) 2695.

[15] ECFA Workshop on LEP 200, edited by A. Böhn and W. Hoogland, CERN 87-08 - vol. I and II, June 1987.

[16] G. Alexander et al., Polarization at LEP, CERN 88-06 - vol. I and II - September 1988.

*Daisy and Val Fitch, Annette Cronin, Kim Tran Thanh Van.*

# THEORETICAL INTRODUCTION AND RESULTS OF CURRENT EXPERIMENTS

*Jim Cronin, Victoria Christenson, Leon Lederman, Cecilia and Goran Jarlskog.*

# CP VIOLATION IN THE STANDARD MODEL

C. Jarlskog
Department of Physics
University of Stockholm
Stockholm, Sweden

ABSTRACT

I give an introduction to CP violation in the Standard Model. After a section on some history of the subject the talk is structured as follows:

Formalism
CP violation with three families
A parametrization without angles
Phenomenology
References

We are here at Blois to celebrate the 25th anniversary of the discovery of CP violation. At such a historical occasion one can not help wondering why is CP so fascinating (otherwise we would not be here)? Has it interested the physics community ever since it was discovered? Has there ever been a really attractive explanation of CP violation? Etc.,etc. The answer to the first question is probably not universal. Concerning the second and the third questions one can get a rough idea of the answers by consulting the proceedings of the biggest annual conference in particle physics and examining how much emphasize was put on CP violation and what was discussed. The result of such a "look" is the following. In the Proceedings of the Dubna Conference 1964 Fitch reported the discovery of the phenomenon and Treiman (who as you may know had been interested in CP for some time) made some comments about its origin. The proceedings of the following 10 conferences (1965-1974) show that CP violation quickly became a central theme and remained so during that period. Parallel as well as plenary sessions were devoted to it. Let me list the conferences and quote some of the relevant plenary talks:

Oxford (1965): Bell and Steinberger; Lee; Prentki

Berkeley (1966): Fitch; Lee

Heidelberg (1967): Okun and Rubbia

Vienna (1968): Cronin; Treiman

Lund (1969): Steinberger

Kiev (1970): Lobashev; Kobzarev

Amsterdam (1971): Winter

Chicago (1972): Rubbia

Aix-en-Provence (1973): Soergel

London (1974): Kleinknecht

During the above period (mostly in the first few years) the origin of CP violation was much debated. Was it due to strong interactions?, electromagnetic?, weak interactions?, superweak interactions? For example, one could argue that since $\alpha/\pi$ and $\epsilon$ were not so different the origin of CP violation is most probably electromagnetic. Some of the models proposed attributed CP violation to the existence of:

- a long-range (cosmic) force
- an octet of strongly interacting vector bosons
- new particles called the a-particles or chimerons
- mirror fermions
- leptoquarks
- gravitational dipole, etc.

That period was indeed a very exiting one for the theorists. On the experimental front, the proceedings of the conferences listed above witness that an enormous amount of work was done on improving the early results obtained in the neutral kaon system as well as on searches for CP violation in many other processes. During the above period it became gradually clear that there was no evidence for CP violation anywhere except in some decay modes of the $K_L$. Unfortunately, the situation has not changed since then and there is still no evidence for CP violation anywhere else. For a review of the theoretical as well the experimental situation of CP during those early years see, for example the articles in reference 1.

By 1974 the experiments had become quite accurate and within the experimental uncertainties one had $\eta_{+-} = \eta_{00}$. Indeed all results were in agreement with Wolfenstein's superweak model. A substantial improvement of the experimental results appeared to be so difficult that the subject was no longer much debated at our largest annual conference. In fact, as far as I could see, during the period 1975 to 1983, not a single parallel session or plenary talk was devoted to CP violation at those conferences (they were held in Palermo (1975), Tbilisi (1976), Budapest (1977), Tokyo (1978), Geneve (1979), Madison (1980), Lisbon (1981), Paris (1982) and Brighton (1983)). But a year later, at the Leipzig Conference there was a notable change in the CP atmosphere which was undoubtedly caused by the promise of new measurements of $\eta_{+-}$ and $\eta_{00}$ at CERN and at Fermilab. Since then the interest in CP violation has grown more and more as witnessed by the proceedings of the conferences at Bari (1985), Berkeley (1986), Uppsala (1987) and München (1988). Let me also add that in 1987 Lepton-Photon Conference the NA31 Collaboration from CERN presented evidence for nonequality of $\eta_{+-}$ and $\eta_{00}$, a result which would rule out the superweak model. By then **23 years** had passed since the original discovery of the CP violation. The message is very clear. We have to be very patient!

Fortunately, the dark ages of CP violation, with no new experimental results, no plenary or parallel sessions, did not stop the theorists from working on various aspects of CP violation. There were several reasons for this lively activity. With the discovery that the gauge theories made sense (i.e., their being renormalizable and unitary) and the discovery of the weak neutral currents the CP problem had to be reinvestigated within the new framework. As the Standard Electroweak Model [2] with only two families of quarks and leptons could not incorporate CP violation the extension to 3 families was proposed by Kobayashi and Maskawa [3] The possibility of spontaneous CP violation was also investigated. For an early review see Ref. 4. Models for CP violation within the Standard Model, through the extension of the Higgs sector, were also proposed [5]. Some authors proposed that the origin of CP violation be sought beyond the Standard Model, for example in the framework of left-right symmetric models. For a recent review see, for example Ref. 6. A second reason for interest in CP violation had to do with the discovery of new fundamental hadrons, charm and beauty, which called for an extension of the earlier works and promised new phenomena. Furthermore, the entrance of cosmology and the early universe into the lives of particle physicists triggered the study of the cosmological aspects of CP violation. For a recent review see, for example Ref. 7. To all this we should also add the discovery of the $\theta$ puzzle in QCD and its possible solutions on which a great deal of theoretical work was done during the "dark ages". For a very recent review see, for example, the articles listed in [8].

By far the most serious Model of CP violation in our time, i.e., 25 years after the discovery of the phenomenon, is the minimal version of the Standard Electroweak Model. Although the Standard Model is not expected to be the "final theory" it does seem to be a very good perhaps only "effective theory" and is highly successful in describing all relevant data.

**Formalism**

The (minimal version of the) Standard Model says that, in general there will be CP violation provided there are (at least) three families of quarks and leptons. Although CP violation resides only in the hadronic sector of the Model the leptons are necessary for the consistency of the Model (anomaly cancellation).

The Lagrangian density of the Electroweak Model is, symbolically of the form

$$\begin{aligned} L &= L(f,G) + L(f,H) + L(G,H) + L(G) - V(H) \\ f &= \text{fermions (quarks, leptons)} \\ G &= \text{gauge bosons } (\vec{W} \text{ and } B) \\ H &= \text{the Higgs doublet } \phi \end{aligned} \tag{1}$$

Leaving aside the so-called strong CP problem [8] all the terms in the above Lagrangian are, formaly, CP conserving (i.e., invariant under textbook CP transformations for fields) except the term $L(f,H)$ which describes the interactions of the massless-looking fermions with the Higgs doublet. The Higgs doublet interacts with quarks and leptons via scalar and pseudoscalar couplings. Due to presence of both such couplings CP is (formally) violated in the $L(f,H)$ sector. Under the spontaneous symmetry breaking, the field $\phi$, which represents four scalars, undergoes a shift viz., $\phi \to \phi + v/\sqrt{2}$, where $v$ is a real number called the vacuum expectation value of the field $\phi$. Three out of the four scalar fields in $\phi$ are "eaten" by the gauge bosons (which thereby become massive). One of them, called the physical Higgs, $\phi_0$ remains in the theory. The essential point is that what remains of $L(f,H)$ is

$$L(f,H) \stackrel{SSB}{\longrightarrow} - \sum_{j,k=1}^{N} \{m_{jk}\,\overline{q_{jL}}\,q_{kR} + m'_{jk}\overline{q'_{jL}}\,q'_{kR} \;+ h.c.\}(1+\frac{1}{v}\phi_0) + leptonic\ terms \quad (2)$$

where the quantities $m_{jk}$ and $m'_{jk}$ are the so-called quark mass matrices for the up-type quarks (denoted by $q_j$) and the down-type quarks (denoted by $q'_j$) respectively. The Standard Model does not tell us what the elements of the mass matrices should be. They are all free parameters which eventually end up as quark masses and mixings. In fact the bulk of the free parameters of the Model come just from these mass matrices. The sum over $j$ runs over the families, their number being N. The leptonic terms will in general have the same structure as the hadronic ones except that the first piece corresponding to the neutrino mass matrix will be absent because in the Minimal Electroweak Model there is no right-handed neutrino and thus the neutrino mass is automatically zero. The right-handed neutrino, if introduced into the theory, could be quite a remarkable object because it, unlike other fermions, would need no Higgs to "acquire" mass and thus the scale of its mass need not have anything to do with the scale of spontaneous symmetry breaking, $v$. The symmetry properties of the quark mass matrices are as follows. Consider, for example the up-kind sector. Suppressing the summation sign we have

$$\begin{aligned} m_{jk}\overline{q_{jL}}\;q_{kR} + \;h.c. &= \frac{1}{2}[\;m_{jk}\bar{q}_j(1+\gamma_5)q_k + m^*_{jk}\bar{q}_k(1-\gamma_5)q_j] \\ &= \frac{1}{2}\bar{q}\;[(m+m^\dagger)+(m-m^\dagger)\gamma_5]q \end{aligned}$$

Thus parity is conserved if the mass matrix is hermitian, because then the pseudoscalar terms, in the above equation, would be absent. Similarly, C-conjugation symmetry requires the mass matrix $m$ to be a symmetric. Finally, under the CP operation

$$m_{jk}\overline{q_{jL}}\;q_{kR} + \;h.c. \stackrel{CP}{\longrightarrow} \frac{1}{2}\bar{q}\;[(m+m^\dagger)^T - (m-m^\dagger)^T\gamma_5]q,$$

where, in addition, $\vec{x} \to -\vec{x}$. Thus CP invariance requires the mass matrices to be real

$$m \;=\; m^* \qquad m' \;=\; m'^* \qquad (3)$$

In order to find the physical fields we must diagonalize the quark mass matrices, $m$ and $m'$ which is done via

$$U_L\ m\ U_R^\dagger = D \equiv Diag.(m_u, m_c, m_t, ...) \qquad (4.a)$$

$$U_L'\ m'\ U_R'^\dagger = D' \equiv Diag.(m_d, m_s, m_b, ...) \qquad (4.b)$$

Here the $U_L$ and $U_R$ are unitary matrices and the daggers on the matrices, on the RHS, are introduced for the sake of notational convenience. Furthermore, $m_u$, $m_d$, etc. denote the eigenvalues, of the mass matrices, i.e., the quark masses. Thus

$$U_L m U_R^\dagger\ U_R m^\dagger U_L^\dagger\ = U_L m m^\dagger U_L^\dagger\ = D^2\ = Diag.(m_u^2, m_c^2, m_t^2, ...) \qquad (5)$$

Thus we see that the matrix $U_L$ diagonalizes the *hermitian* matrix $mm^\dagger$. Similarly, the matrix $U_R$ diagonalizes the hermitian matrix $m^\dagger m$. Furthermore, there are analogous relations for the down-kind sector. The physical fields are

$$q_L^{phys} = U_L q_L \qquad q_L'^{phys} = U_L' q_L'$$

and their analog for the R-handed quarks. Eq. (2) rewritten in terms of the physical fields reads

$$L^{phys}(f, H) = -(1 + \frac{\phi_0}{v})[m_u \bar{u}u + m_c \bar{c}c + ... + m_d \bar{d}d + m_s \bar{s}s + ...].$$

The essential feature of the above Lagrangian is that it conserves, seperately, P , C and thus also CP as well as T. This demonstrates that the apparent symmetry properties of a Lagrangian need not have anything to do with *physics*, if the Lagrangian in question is expressed in terms of unphysical fields. In rewriting the $L(f, H)$ in terms of the physical fields we find that there is no trace of the $U$ and $U'$ matrices left. The interactions of the Higgs with the fermions is automatically P and C conserving, as well as flavour conserving. In extensions of the Model these properties will, in general, not be valid.

What happens is that the original CP violation in the term $L(f, H)$ has now moved into a different sector of the theory, i.e., into the $L(f, G)$ sector which describes the interaction of the fermions with the gauge bosons. When one expresses the term $L(f, G)$ in terms of the physical fields several "miracles" happen. The neutral currents all conserve CP. In addition, there are no flavour changing neutral currents (the GIM mechanism [9]) because of two facts, (i) these terms are helicity conserving (i.e., they involve either the index L or R but not both) and (ii) the four matrices $U$ and $U'$ (with indices L or R) are unitary. The absence of flavour changing neutral currents means that there are no factors of the type $\bar{u}\gamma^\mu(a + b\gamma_5)c$, in the Lagrangian of the Standard Model. This absence of the flavour changing neutral currents is a very important feature of the Model. Note that the interactions of the physical vector bosons $A$ (the photon) and the $Z$ are then automatically flavour conserving because the photon and the Z are linear combinations of $W^3$ and $B$. As there are stringent limits, from experiment, on flavour changing neutral currents, the absence of such currents is one of the great successes of the Model. It should be noted that in the "extensions" of the Standard Model one generally encounters difficulties in this respect. However, in the charged current terms in $L(f, G)$ the up-type and the down-type quarks *do mix* simply because $W^\pm$ carry one unit of charge. Since the charged currents only involve left-handed quarks the matrices with the subscript R do not enter at all and the $U$ with the index L gets multiplied by $U'^\dagger$ with the same index. But $UU'^\dagger$ need not be equal to unity. All we know is that it is a unitary matrix. Dropping the numerical factors and the coupling constant, the charge current terms in Eq. (1) are given by

$$X_C \equiv [W^1_\mu - iW^2_\mu]\, \overline{q_L}\, \gamma^\mu q'_L \ + h.c. = [W^1_\mu - iW^2_\mu]\, \overline{q_L^{phys}}\, \gamma^\mu U_L U'^\dagger_L q'^{phys}_L + h.c.$$

$$= [W^1_\mu - iW^2_\mu]\, \overline{q_L^{phys}}\, \gamma^\mu V q'^{phys}_L \ + h.c.$$

where $V \equiv U_L U'^\dagger_L$ is the so-called quark mixing matrix. It is, of course a unitary N by N matrix. CP conservation now requires V to be real. A general unitary N by

N matrix has $N^2$ parameters $N(N-1)/2$ of these parameters may be taken as Euler angles which one introduces in dealing with rotations in $N$ dimensional (Euclidean) space. The remaining parameters are called phases. However, $2N-1$ of these phases are not measurables because we have $2N$ quark fields. The phases of these quark fields may be chosen at will. Rephasing of the quark fields rephases the elements of the quark mixing matrix. However, only the phase differences appear there and there are $2N-1$ such quantities. Therefore, $V$ has $N^2-(2N-1)=(N-1)^2$ parameters among which $N(N-1)/2$ are the rotation angles. Therefore, the number of phases in $V$ is $(N-1)(N-2)/2$. For the case of 2 families there is only 1 rotation angle and no phases. The quark mixing matrix is then of the form

$$\begin{pmatrix} cos\theta & sin\theta \\ -sin\theta & cos\theta \end{pmatrix} \tag{6}$$

The angle $\theta$ is the Cabibbo angle [10]. It is amusing to note that, for the leptonic sector, the matrix (6) was written down [11] as early as in 1962, in connection with neutrino oscillations.

**CP violation with 3 families**

Consider now the case of 3 families. The number of parameters in $V$ is $(3-1)^2=4$, i.e., 3 rotation angles and one phase. The first example of such a matrix was given by Kobayashi and Maskawa [3]. One can show that there are 36 different looking parametrizations of the "KM-matrix" [12]. Of course, they are all equivalent to each other. In the original KM parmetrization with the angles $\theta_j$, $j=1,2,3$ and the phase $\delta$ eight conditions on these angles must be satisfied in order to get CP violation. These are

$$\theta_j \neq 0,\ \frac{\pi}{2} \qquad \delta \neq 0,\ \pi, \qquad j=1,2,3 \tag{7}$$

Furthermore no two quarks with the same charge are allowed to be degenerate in mass because otherwise the phase can be removed and there will be no CP violation. Thus

$$m_u \neq m_c,\ m_c \neq m_t,\ m_t \neq m_u,\ m_d \neq m_s,\ m_s \neq m_b,\ m_b \neq m_d \tag{8}$$

It is interesting to note that the above 14 conditions, in Eqs. (7) and (8) are unified in a single relation, viz.,

$$det\ \mathbf{C} \neq 0 \tag{9}$$

where $\mathbf{C}$ is the commutator of the hermitian mass matrices (see Eq. (5))

$$\left[mm^{\dagger},\ m'm'^{\dagger}\right] \equiv i\mathbf{C} \tag{10}$$

Here the quantities $m$ and $m'$ are the 3 by 3 quark mass matrices which were introduced before (see after Eq. (2)). Again the theory does not tell us what these mass matrices are. In fact their "determination", from first principles, is one of the most outstanding problems in particle physics. Although we can not determine the mass matrices, we can measure their eigenvalues (the 6 quark masses) and the four independent measurables of the quark mixing matrix. What is highly remarkable about the above commutator is that its determinant is given by (see Ref. [12] and references therein)

$$\begin{aligned} det\ \mathbf{C} = -2J\ &(m_t^2 - m_c^2)(m_c^2 - m_u^2)(m_u^2 - m_t^2) \\ &\times(m_b^2 - m_s^2)(m_s^2 - m_d^2)(m_d^2 - m_b^2) \end{aligned} \tag{11}$$

where $J$ is an "invariant", related to CP violation. One can construct it quite simply for any parametrization. In the KM parametrization it is given by

$$J = sin^2\theta_1 sin\theta_2 sin\theta_3 cos\theta_1 cos\theta_2 cos\theta_3 sin\delta \tag{12}$$

Thus we see that the above 14 conditions are unified in the determinant. In other words **CP is violated if and only if** $det\ \mathbf{C} \neq 0$. This statement hardly needs a proof. If the determinant is nonzero then there is no way in which the phase can be removed since $sin\delta$ must be non-zero. For more details and references to the original papers see Ref. [12].

**A parametrization without angles**

The invariant measurables of the quark mixing matrix, in the case of 3 families, are the moduli of its elements i.e., the quantities $|V_{\alpha j}|$, and the quantity J. The angle and phase independent parametrization is based on the single fact [12, 13] that even the quantity J, up to a sign, is a function of the moduli of the quark mixing matrix. In fact it is given by

$$4J^2 = -\lambda(a^2, b^2, c^2) \tag{13}$$

where the $a, b, c$ are products of moduli of the elements of the quark mixing matrix. More precisely, they are the three sides of any of the 6 unitarity triangles [12, 13]. Let us first remind ourselves that the function $\lambda$ is the triangular function sometimes also refered to as the Källén $\lambda$ function. It is familiar from two body kinematics,

$$\lambda(x, y, z) = x^2 + y^2 + z^2 - 2xy - 2yz - 2zx \tag{14}$$

In fact all the six unitarity triangles have the same area and magnitude of J equals twice the area of any of these triangles (for a detailed discussion see Ref. [12]). If CP would be conserved all the six triangles collapse to lines and their area vanishes.

A very useful phenomenological parametrization of the quark mixing matrix was suggested by Wolfenstein [14] and is much used in the literature. See Wolfenstein's summary talk in these Preceeding.

**More than 3 families?**

A very essential point about the Standard Model is that, in its hadronic sector, going from one family to two leads to a new phenomenon, i.e., one gets flavour mixing. In other words, in general $\theta$ in Eq. (6) is non-zero. Going from two to three families again introduces a novel feature into the theory, i.e., in general CP is violated unless one of the 14 conditions mentioned before is not satisfied. However the transition from three to four is not known to lead to any new features. In fact, in the four family case there are six angles and three phases but these are of the same kind as before. Therefore, it is very hard to get excited about a possible fourth family although one can not exclude the existence of more standard families. For a recent review of the 4 family extension of the Standard Model see, for example, Ref. [15].

**Phenomenology**

In 1964 there were only 3 quark degrees of freedom (u,d,s) and $K^0$ and $\overline{K^0}$ were the only known conjugate weakly decaying neutral mesons. Nowadays the situation is very different as there is another such system in the charm sector, $D^0$ and $\overline{D^0}$, and two more in the beauty sector. Indeed a great deal of theoretical effort has gone into investigating the CP properties of these new systems. One can, in principle repeat the old experiments for the new systems and, for example look for CP violating charge asymmetries. One can compare the decay rates of CP conjugate decay modes, etc. It is not my task here to go through the theoretical possibilities as there are special talks schedueled for all such topics. However, it is clear that one does not expect to see, in the near future, any CP violation anywhere else than the $K$ system if the minimal version of the Standard Model is correct. The branching ratios to interesting final states are in general expected to be small. The analog of $\epsilon$ for these systems is expected to be small. Furthermore, the lifetimes of the new states are much shorter than those of the kaons. The seperation of the "long-lived" and the "short-lived" components appears very difficult. I shall not say any more as several speakers at this Conference will discuss the matter in detail. For a recent review, I refer you the recent book "CP Violation" (see the list of references).

A topic of great interest is the electric dipole moment of the neutron (EDM). Again, in the Standard Model the situation looks hopeless as the EDM of the neutron is expected to be smaller than $10^{-33} e - cm$, i.e., about 8 orders of magnitude below the present upper limit. For a recent review see Ref. [16]. However, the important point to keep in mind is that the neutron EDM is very sensitive to certain extensions of the Standard Model, specially where CP is due to Higgses. Then the EDM of the neutron tends to be large and often even larger than the present upper limit. Thus the best hope of learning something new on CP violation in the near future seems to be in the EDM of the neutron.

In the not so near future, with the construction of dedicated facilities the situation may change radically. Let's hope so!

**References**


[1] P. K. Kabir, The CP Puzzle (Academic Press, 1968)
J. S. Bell and J. Steinberger, Proc. Oxford Conf. on Elementary Particles (1965) p. 195

[2] S. L. Glashow, Nucl. Phys. 22, 579 (1961);
S. Weinberg, Phys. Rev. Lett. 19, 1264 (1967);
A. Salam, *in* Elementary Particle Theory, Ed. N. Svartholm (Almqvist and Wiksell, 1968)

[3] M. Kobayashi and T. Maskawa, Prog. Theor. Phys. 49, 652 (1973)

[4] T. D. Lee, Phys. Reports 9C, 143 (1973)

[5] S. Weinberg, Phys. Rev. Lett. 37, 675 (1976);
P. Sikivie Phys. Lett. 65B, 141 (1976)

[6] R. N. Mohapatra in "CP Violation", Ed. C. Jarlskog (World Scientific, 1989), p. 384

[7] P. Langacker in "CP Violation", loc. cit., p. 552

[8] R. D. Peccei in "CP Violation", loc. cit., p. 503
R. D. Peccei; P. Sikivie, these Proceedings

[9] S. L. Glashow, J. Iliopoulos and L. Maiani, Phys. Rev. D2, 1285 (1970)

[10] N. Cabibbo, Phys. Rev. Lett. 10, 531 (1963);
M. Gell-Mann and M. Levy, Nuovo Cimento 16, 705 (1960);
M. Gell-Mann, Phys. Lett. 8, 214 (1964)

[11] Z. Maki, M. Nakagawa and S. Sakata, Proc. of 1962 Inter. Conf. (CERN) p. 663;
Prog. Theor. Phys. 28, 870 (1962)

[12] C. Jarlskog in "CP Violation", loc. cit., p. 3

[13] See Ref. 12 and references thein. See also C. Jarlskog and R. Stora, Phys. Lett. B208, 268 (1988)

[14] L. Wolfenstein, Phys. Rev. Lett. 51, 1945 (1983) and these Proceedings

[15] A. Datta and E. A. Paschos in "CP Violation", loc. cit.

[16] S. M. Barr and W. J. Marciano in "CP Violation", loc. cit.

# THE NA31 $\epsilon'/\epsilon$ MEASUREMENTS

Ken Peach
University of Edinburgh

ABSTRACT

The CERN-Edinburgh-Mainz-Orsay-Pisa-Siegen (NA31) experiments on CP-violation in $K^0$ are described. The experimental arrangement is such that the acceptances largely cancel, through a "double ratio" technique. The result from the 1986 data is

$$R_\eta = |\eta_{00}/\eta_{+-}|^2 = 0.980 \pm 0.004 \pm 0.005$$

where the first error is statistical and the second systematic, from which $\mathrm{Re}(\epsilon'/\epsilon)$ is $0.0033 \pm 0.0011$, some three standard deviations from zero, and consistent with standard model expectations. The beam and detector have been upgraded and new data obtained in 1988 and 1989 will reduce substantially the statistical and systematic errors.

# 1 Introduction

The CERN-Edinburgh-Mainz-Orsay-Pisa-Siegen [1] (NA31) $K^0$ spectrometer [1] was designed and built to study in CP-violation in $K^0$ decays, in particular, to measure precisely the relative amounts of CP-violation in the $2\pi^0$ (neutral) and $\pi^+\pi^-$ (charged) amplitudes. The results [2] indicate for the first time an effect (at the three standard deviation level) outside the superweak model [3]. The beam and detector have been modified to reduce still further systematic uncertainties, and preliminary results from these studies are presented.

The next section discusses the principles of measurement, and the following section describes the beam and detector. The reconstruction and results are presented in the next two sections, and the future reviewed in the final section.

# 2 General Principles of Measurement

The theoretically interesting quantity is the amount of CP-violation in the $\Delta I = \frac{3}{2}$ amplitude relative to that in the $\Delta I = \frac{1}{2}$ amplitude, which is sensitive to the mechanism of CP-violation. The measured quantities are $\eta_{+-} \simeq \epsilon + \epsilon'$ and $\eta_{00} \simeq \epsilon - 2\epsilon'$ where $\epsilon$ is the CP-violation in the mass-matrix

$$K_S \simeq (K_1 + \epsilon K_2)/\sqrt{1+|\epsilon|^2} \quad \text{and} \quad K_L \simeq (K_2 + \epsilon K_1)/\sqrt{1+|\epsilon|^2}$$

and $\epsilon'$ is the amount of CP-violation in the $\Delta I = \frac{3}{2}$ amplitude (in the Wu-Yang phase convention) relative to the $\Delta I = \frac{1}{2}$ amplitude. Explicitly,

$$\epsilon' = (\imath Im(A_2))/(\sqrt{2}A_0)\exp[\imath(\delta_2 - \delta_0)]$$

Because of the phenomenological non-leptonic $\Delta I = \frac{1}{2}$ rule, $\epsilon'$ is expected to be small and therefore difficult to observe by measuring $\eta_{+-}$ and $\eta_{00}$ independently. Instead, recent experiments have concentrated on measurement of $R_\eta = |\eta_{00}/\eta_{+-}|^2$ related to $\text{Re}(\epsilon'/\epsilon)$ by

$$\text{Re}\left(\frac{\epsilon'}{\epsilon}\right) \simeq \frac{1}{6}(1 - R_\eta) = \frac{\Gamma(K_L \to 2\pi^0)}{\Gamma(K_S \to 2\pi^0)} \times \frac{\Gamma(K_S \to \pi^+\pi^-)}{\Gamma(K_L \to \pi^+\pi^-)} = \frac{B_L^{00}}{B_S^{00}} \times \frac{B_S^{+-}}{B_L^{+-}}$$

where the $\Gamma$'s are the partial decay rates and the $B$'s are the branching ratios. Because the phase of $\epsilon'$ is close to that of $\epsilon$, $|\epsilon'/\epsilon| \simeq \text{Re}(\epsilon'/\epsilon)$.

If we now consider an experiment, then $R_\eta$ is given by

$$R_\eta = \left\{\frac{N_L^{00}}{N_S^{00}} \times \frac{N_S^{+-}}{N_L^{+-}}\right\} \times \left\{\frac{C_L^{00}}{C_S^{00}} \times \frac{C_S^{+-}}{C_L^{+-}}\right\} \times \left\{\frac{A_S^{00}}{A_L^{00}} \times \frac{A_L^{+-}}{A_S^{+-}}\right\} \times \left\{\frac{F_S^{00}}{F_L^{00}} \times \frac{F_L^{+-}}{F_S^{+-}}\right\}$$

---

[1] E Augé, GD Barr, C Becker, L Bertanza, A Bigi, R Black, H Blümer, H Burkhardt, P Calafiura, M Calvetti, DJ Candlin, R Carosi, R Casali, C Cerri, RL Chase, P Clarke, M Corti, D Coward, DC Cundy, G Dietrich, N Doble, R Fantechi, L Fayard, D Fournier, G Gargani, L Gatignon, V Gibson, P Grafström, R Hagelberg, I Harrus, W Heinen, R Heinz, P Heusse, D Heyland, M Holder, M Kasemann, G Kessler, K Kleinknecht, A Kreutz, J van der Lans, A-M Lutz, I Mannelli, E Massa, P Mayer, T Miczaika, J Muir, A Nappi, HN Nelson, R Ossa, B Panzer, KJ Peach, GM Pierazzini, BJ Pijlgroms, G Quast, B Renk, H Röher, M Rost, HG Sander, AC Schaffer, IPJ Shipsey, P Steffen, J Steinberger, W Stephenson, H Taureg, H Wahl, W Weihs, R Werthenbach, C Youngman, G Zech

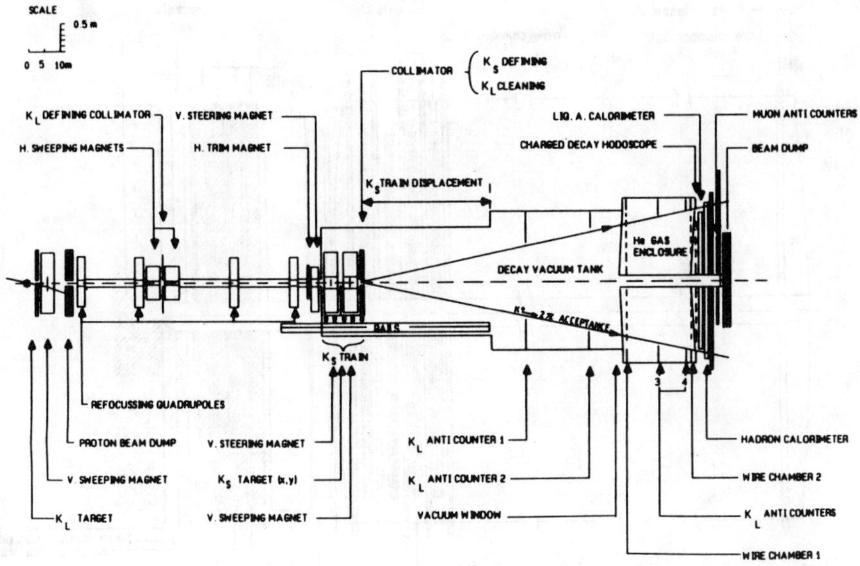


Figure 1: Beam Layout 1988

where $N$ is the number of events of each type, $C$ is the correction for backgrounds, scattering etc, $A$ is the (geometrical) acceptance of the apparatus including kinematic restrictions, and $F$ is the flux of kaons; all depend on the energy $E_K$ and vertex $Z_K$ of the decaying kaon.

The number of events is "given"; the only imperative is to obtain as many as possible, to minimise the statistical error. This means optimising the design for $K_L \to 2\pi^0$, as this has the smallest branching ratio and (usually) the smaller acceptance. There is a tendency, as the statistical error approaches the systematic error, for the two to be correlated, since the division of data into independent subsets is a powerful check of systematic effects.

The correction factors $C$ include backgrounds from other decay modes, scattered events leading to wrongly computed kinematic variables, effects from accidental hits in the detector, resolution etc. The design should keep these effects small, and where unavoidable, should arise from well-understood sources.

The acceptance is usually calculated by montecarlo simulation. An alternative is to arrange the experiment so that the acceptance cancels, at least to a good approximation. This is achieved by cunning design which allows the $K_S$ beam to "decay" with a distribution of vertices similar to that in $K_L$, with $K_S$ data taken at a series of target positions throughout the decay volume. There are residual differences in beam divergence, energy spectrum and vertex distribution to be calculated, but these are small ($< 0.3\%$) and thus the requirements on the montecarlo are modest.

Flux factors cancel identically by taking data on both modes concurrently.

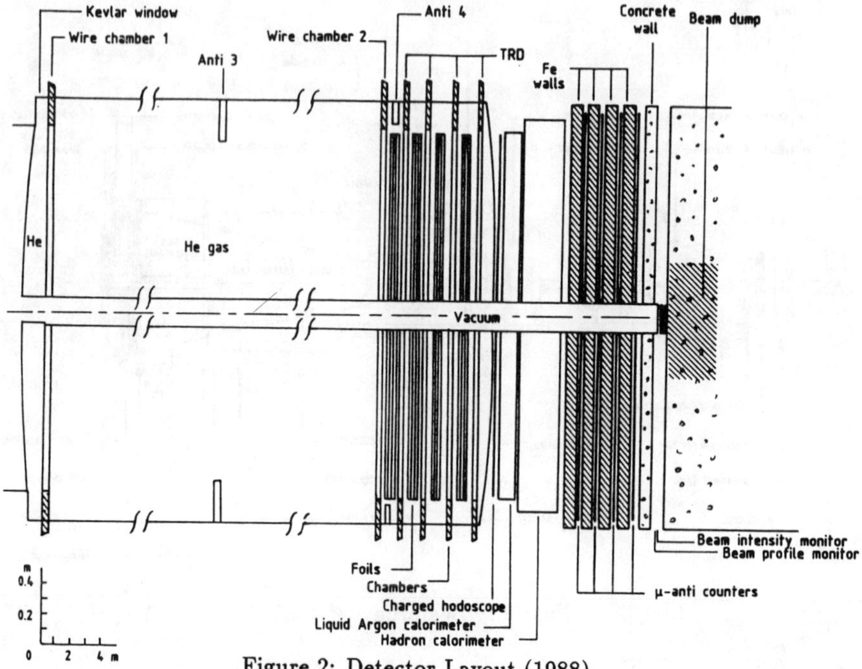


Figure 2: Detector Layout (1988)

# 3 The NA31 $K^0$ Spectrometer

An overview of the beam is shown in Figure 1. The $K_L$ and $K_S$ beams are transported in vacuum to reduce neutron interactions, leading to low counting rates in the detector; backgrounds arise mainly from $K^0$ decays. Good collimation reduces scattering, and there is no regenerator. Data taking alternates frequently between $K_L$ and $K_S$ (changing roughly every day) so that the result of the experiment is insensitive to changes in the detector. Short term drifts are monitored through continuous calibration. The beam has a mean energy of around 100 GeV, allowing good energy measurement with calorimetry.

The $K_L$ target is some 120m upstream from the decay volume. The $K_S$ target is mounted on a "train" which can be positioned at any one of 41 stations separated by 1.2m, so that a long $K_L$ fiducial volume is used. In 1986, the incident energy and angle of protons on the two targets were the same, giving the same production spectrum; the decay spectra have a difference of about 12 GeV in mean energy, giving a small systematic error in $R_\eta$. In 1988, the angle at the $K_L$ target was decreased, and the energy at the $K_S$ target decreased while the angle was increased, to achieve more equal decay spectra, which now differ by about 5 GeV, leading to a reduction in the systematic uncertainty due to the energy scale. The upstream limit of the decay volume in $K_S$ is defined by a lead sheet and an anti-counter at the final collimator, which is used to set the absolute energy and length scale.

The detector layout for 1988 is shown in Figure 2. The main features are:-

- **A large lead-liquid argon calorimeter, segmented in depth,** with good energy and position resolution and small non-linearities. The energy resolution is

$$\sigma(E)/GeV \quad = 0.075\sqrt{E} + 0.005E + 0.1$$

  where the three terms arise from sampling, uniformity and electronic noise; the spatial resolution is much better than 1mm

- An Iron-Scintillator calorimeter segmented in depth which, when combined with liquid argon calorimeter, gives an energy resolution of $65\%\sqrt{E}/GeV$

- Two 4-plane mini-drift chambers, separated by 25m in helium, with resolution $\sim 500\mu$

- Two trigger hodoscopes, one to trigger on $\pi^+\pi^-$ placed after the second wire chambers (charged), and one to trigger on $2\pi^0$ placed in the liquid argon calorimeter between the front and back halves of the stack (neutral)

- Ring-shaped anti-counters surrounding the decay volume to veto particles outside the acceptance of the detectors, and $\mu$-veto counters behind 4m iron-equivalent to veto $K_{\mu 3}$ decays, $\pi^\pm$ decays, and punch-through

These were unchanged for the 1987 phase difference run [4]. For 1988, the beam and detector were modified to reduce systematic errors identified in the 1986 analysis.

- New beam collimation to reduce scattering from the final collimator in $K_L$, which caused the first 10.5m to be cut out in the 1986 analysis. The distribution of the background [2] for the two data sets is shown in Figure 3 as a function of the vertex position. The collimation and shielding around the $K_S$ target were also improved.

- A transition radiation detector [5] between the second wire chamber and the charged trigger hodoscope, to check the $\pi^+\pi^-$ background.

- More $\mu$-veto planes, to study the $K_{\mu 3}$ background from muons radiating in the calorimeters with insufficient residual range to penetrate to the second $\mu$-veto plane.

- Additional photon counters in the beam and in front of the first wire chamber to identify some remaining $3\pi^0$ and $\pi^+\pi^-\pi^0$ background.

The trigger conditions (see Table 1) were loose. The stability and efficiency of the trigger were continously monitored by a special "unbiassed" trigger, which required only more than $\sim$ 25 GeV of energy in the calorimeters, and no anti-counter or $\mu$-veto counter. In addition, pulser and laser triggers to monitor calorimeter gains, and random triggers to determine pedestals and accidental activity, were taken each burst at a rate proportional to the beam intensity.

[2]The absolute level of background is apparently higher in 1988 because no tight mass cut has been applied.

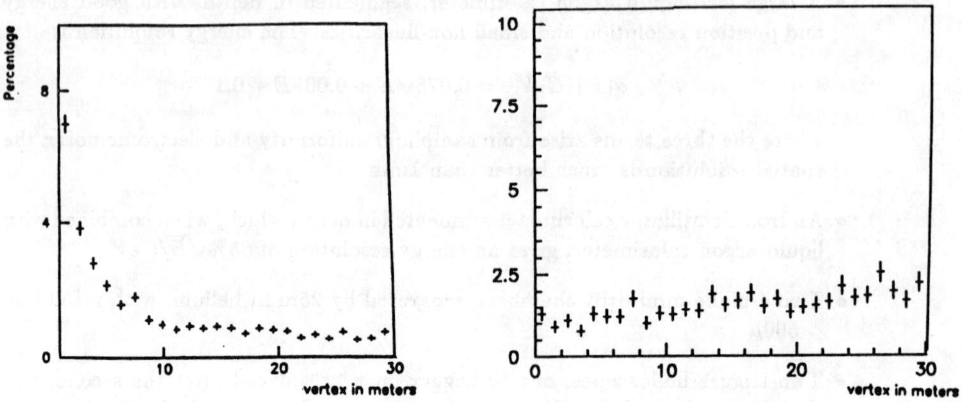


Figure 3: Distribution of high $d_T$ in $K_L$ as a function of vertex for 1986 and 1988 beam

| Trigger | "$2\pi^0$" | "$\pi^+\pi^-$" |
|---|---|---|
| Pretrigger (hodoscope) | left-right (neutral) | opposite quadrants (charged) |
| 1st Level | $> \sim 35$GeV (liquid argon) | $> \sim 35$Gev (total) electron rejection (longitudinal shower profile) |
| 2nd Level | 2-4 peaks (liquid argon) *and either* $< 7$ hits in 1$^{st}$ wire chamber *or* $< 12$GeV hadronic | $> 2$ hits in $> 2$ planes (1$^{st}$ wire chamber) |
| 3rd Level | Centre of Gravity Vertex refined energy cut | No photon peaks outside pion shower refined energy cut |

Table 1: Summary of the main trigger conditions

# 4 Reconstruction and Background Rejection

The separation of the triggers into charged and neutral candidates was done mainly from the information in the wire chambers. Neutral candidates were required to have no "space points" in the first wire chamber, and charged candidates were required to have exactly two; in this way, sensitivity to accidental hits in the first wire chamber was symmetric between the two decay modes. Charged triggers were required to have at least two "space points" in the second wire chamber; no condition on additional hits could be applied, because of hits associated with "backsplash" from the calorimeter; these are the same for both $K_L$ and $K_S$.

A "space point" was required to have hits in at least three of the four wire chamber planes; individual plane inefficiencies were less than 0.5%.

All restrictions to reduce the backgrounds were applied to both $K_L$ and $K_S$ data, so that any loss of genuine $2\pi$ events due to physical processes (such as shower fluctuations, $\delta$-rays etc) is the same for both beams, and so cancels in the ratio.

## 4.1 $2\pi^0$ Energy and Vertex

The reconstruction of $2\pi^0$ decays used the information from the liquid argon calorimeter.

Given $n$ observed photons from $K^0$ decay, with energies and positions $E_i, x_i, y_i$, the energy, vertex and two-photon invariant masses are given by

$$\text{Energy} \quad E_K = \sum_{i=1}^{n} E_i, \qquad \sigma_{E_K} \sim 1\%$$

$$\text{Distance from Calorimeter} \quad L_K = \frac{\sqrt{\sum_{i=1}^{n}\sum_{j=1}^{n} E_i E_j r_{ij}^2}}{m_K}, \qquad \sigma_{L_K} \sim 1\%$$

$$\text{Two-}\gamma \text{ invariant mass} \quad m_{\gamma_1\gamma_2} = \sqrt{\frac{E_1 E_2 r_{12}^2}{\sum_{i=1}^{n}\sum_{j=1}^{n} E_i E_j r_{ij}^2}}\, m_K, \qquad \sigma_{m_{\pi^0}} \sim 2\text{MeV}$$

where $r_{ij}^2 = (x_i - x_j)^2 + (y_i - y_j)^2$. Photons are associated into $\pi^0$'s by choosing the best of the three pairings to give $\pi^0$ masses. There are three remarks.

- The energy and length scales are completely correlated.
- Because length and energy scales are correlated, the two $m_{\pi^0}$'s are not independent - see Figure 4, which shows the two $m_{\pi^0}$ masses for $K_S$ and $K_L$.
- The background is from $3\pi^0$ decays in which two photons have been undetected. Missing $\gamma$'s reduce $L_K$ and so this background is a strong function of vertex.

The background is measured by defining ellipses of equal area

$$R_{ell} = [(m_{\gamma_1\gamma_2} + m_{\gamma_3\gamma_4})/(\sigma_+)]^2 + [(m_{\gamma_1\gamma_2} - m_{\gamma_3\gamma_4})/(\sigma_-)]^2$$

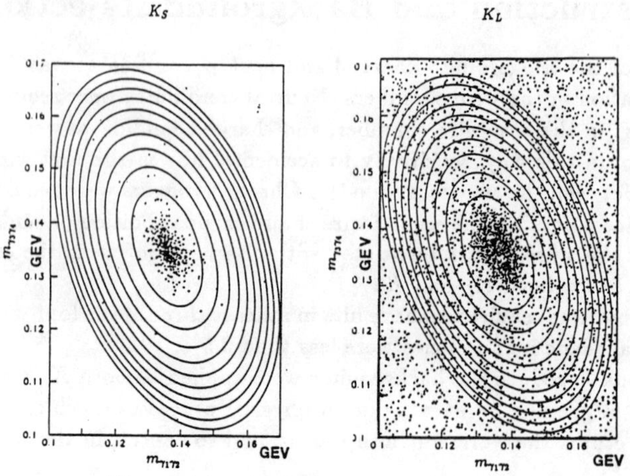


Figure 4: $m_{\pi^0}$ versus $m_{\pi^0}$ for $K_S$ and $K_L$

where $\sigma_+ \sim$ 2MeV ; $\sigma_- \sim$ 4MeV. $R_{ell} < 9$ represents a three standard deviation contour containing most of the signal (see Figure 4); $R_{ell} \gg 9$ is used to estimate the background. From montecarlo studies, and by observation, the background is almost flat in $R_{ell}$ (Figure 5); the distribution of the background as a function of decay vertex (Figure 6), shows the expected strong dependence.

## 4.2 $\pi^+\pi^-$ Energy and Vertex

Once all space points have been found in both wire chambers, those which lie in a plane and give a vertex within the beam are selected, with the vertex being defined as the mid-point of the "closest distance of approach"; the longitudinal uncertainty is $\sim 1m$.

Most of the $K_{e3}$ decays are rejected by cuts on the lateral and longitudinal shower profile in the calorimeters, the $K_{\mu 3}$ decays were efficiently rejected in the trigger, and the $\pi^+\pi^-\pi^0$ decays by the observation of photons in the liquid argon calorimeter outside the the pion showers. The small $\Lambda$ contamination was rejected by a requirement that the ratio of the pion energies should be less than 2.5.

It can be shown [3] that there is a simple relationship between the mass, total energy, opening angle and ratio of the energies of the pions. This is used in two ways. Firstly,

$$m_K \simeq \sqrt{m_{\pi^+}^2(2+R+R^{-1}) + E_K^2\theta_{\pi\pi}^2/(2+R+R^{-1})}, \quad R = E_+/E_-$$

is used to reject events inconsistent with the $K^0$ mass; this rejects most of the $K_{3\pi}$ background - see Figure 7. Then, assuming $\pi^+\pi^-$, the energy $E_K$ is found from

$$E_K \simeq \sqrt{m_K^2(2+R+R^{-1}) - m_{\pi^+}^2(2+R+R^{-1})^2/\theta_{\pi\pi}}, \qquad \sigma_{E_K} \sim 1\text{GeV}.$$

[3]This is an excellent exercise in relativistic kinematics for undergraduates

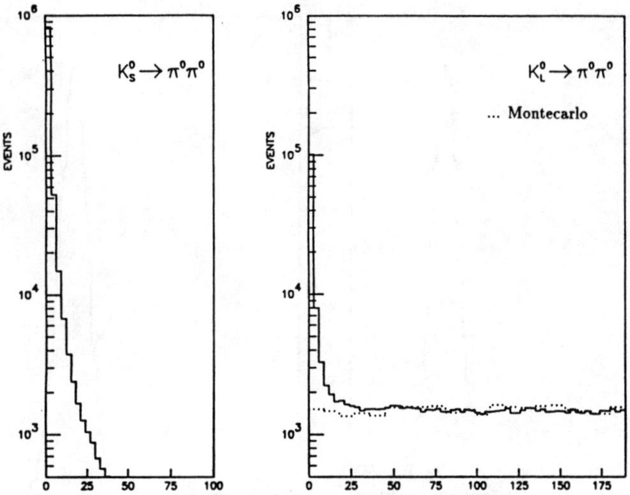


Figure 5: $R_{ell}$ distribution for $K_S$ and $K_L$

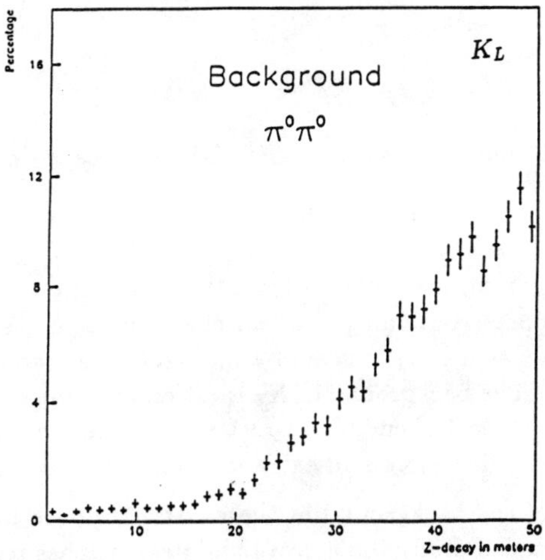


Figure 6: $3\pi^0$ background versus vertex position for $K_L$

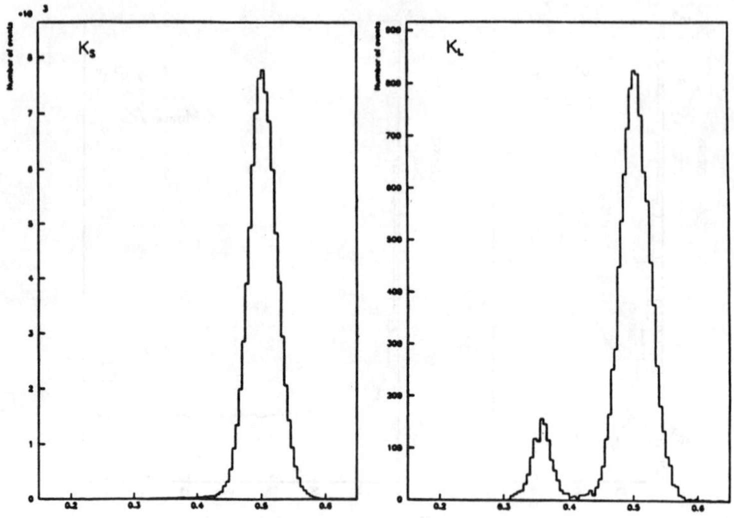


Figure 7: Mass distributions for $K_S$ and $K_L$

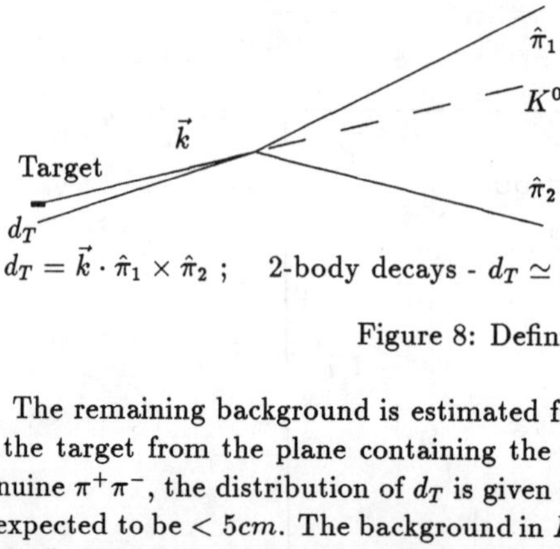


$d_T = \vec{k} \cdot \hat{\pi}_1 \times \hat{\pi}_2$ ;   2-body decays - $d_T \simeq 0$;   3-body decays - $d_T \neq 0$ in general

Figure 8: Definition of $d_T$

The remaining background is estimated from the distribution of $d_T$, the distance of the target from the plane containing the two charged tracks - see Figure 8. For genuine $\pi^+\pi^-$, the distribution of $d_T$ is given by the resolution, and (from $K_S$ decays) is expected to be $< 5cm$. The background in $K_L$ is estimated from events in a "control region" with $d_T$ between 7 and 12cm (Figure 9). There are two uncertainties with this (in contrast to the simple source and extrapolation in the $2\pi^0$ background).

- The composition of the background in the control region arises from $K_{e3}, K_{\mu 3}$ and $K_{3\pi}$. After detailed study the control and nearby mass regions, the contributions from the decay modes were estimated, and the results shown in Table 2.

- Once the composition in the control region is established, these must be extrapolated into the signal region. $K_{e3}$ and $K_{\mu 3}$ have similar kinematics, and so the extrapolations are similar, but the very different kinematics of $K_{3\pi}$ give an extrapolation much steeper.

With the TRD, the $K_{e3}$ background and extrapolation coefficient can be checked. Although this work is incomplete, preliminary estimates (also shown in Table 2)

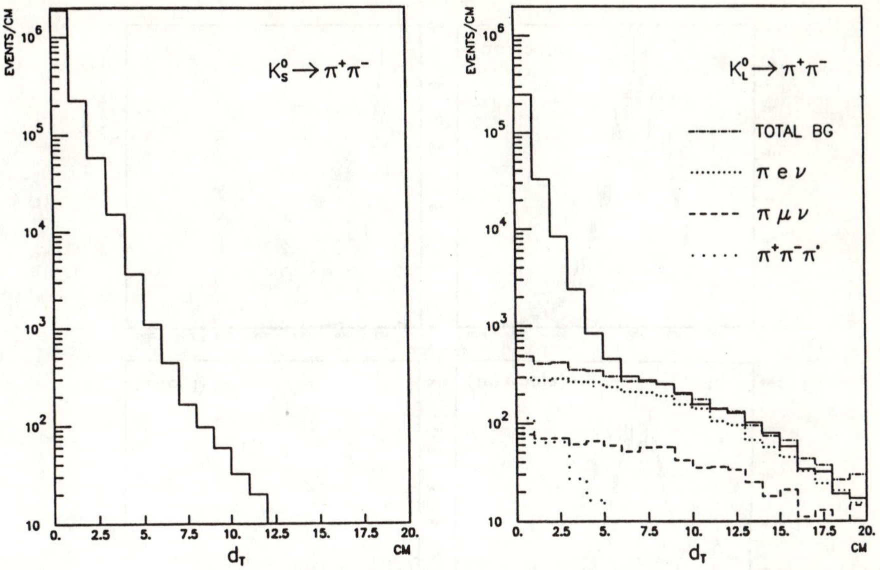


Figure 9: $d_T$ distributions for $K_S$ and $K_L$

| Decay | 1988 Check ($\times 10^3$) | 1986 estimate ($\times 10^3$) |
|---|---|---|
| $K_{3\pi}$ | $0.11 \pm 0.03$ | $0.1 \pm 0.1$ |
| $K_{e3}$ | $2.18 \pm 0.11$ | $2.8 \pm 0.2$ |
| $K_{\mu 3}$ | $1.11 \pm 0.15$ | $0.5 \pm 0.2$ |
| Total | $3.40 \pm 0.18$ | $3.40 \pm 0.3$ |
| Regenerated $K_S$ | $0.0 \pm 0.02$ | $0.2 \pm 0.1$ |

Table 2: Check of the 1986 charged mode background using the 1988 data

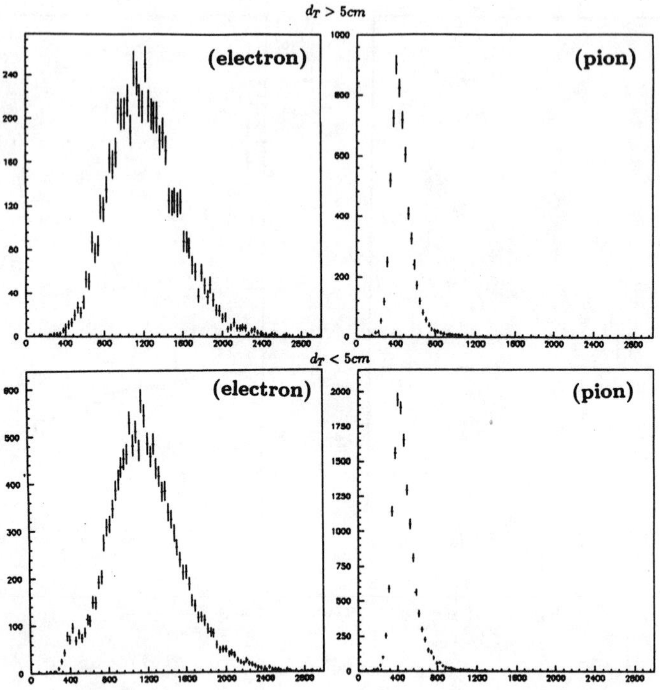


Figure 10: TRD Pulse Height Distributions for identified $K_{e3}$ events with $d_T < 5$cm and $d_T > 5$cm

broadly confirm the 1986 estimates - the proportion of $K_{e3}$ and $K_{\mu 3}$ decays is slightly different, but this has little effect on the total background because of the similarity of extrapolation coefficients. The TRD pulse-height spectra (mean pulse height of the three lowest pulse-heights) is shown for $K_{e3}$ events identified by calorimetry, for large and small $d_T$; a small $K_{2\pi}$ "contamination" in the $K_{e3}$ sample with $d_T < 5$cm is clearly seen.

## 4.3 Energy Scale

As noted above, the energy and length scales in the $2\pi^0$ decay are strongly correlated, and it is important that these be the same for both modes. The absolute energy scale is determined by fits to the anticounter position in $K_S$ about once each day. The distributions and fits are shown in Figure 11. The fit gives the anticounter position to much better than 10cm in 100m, so that the energy scales are known to better

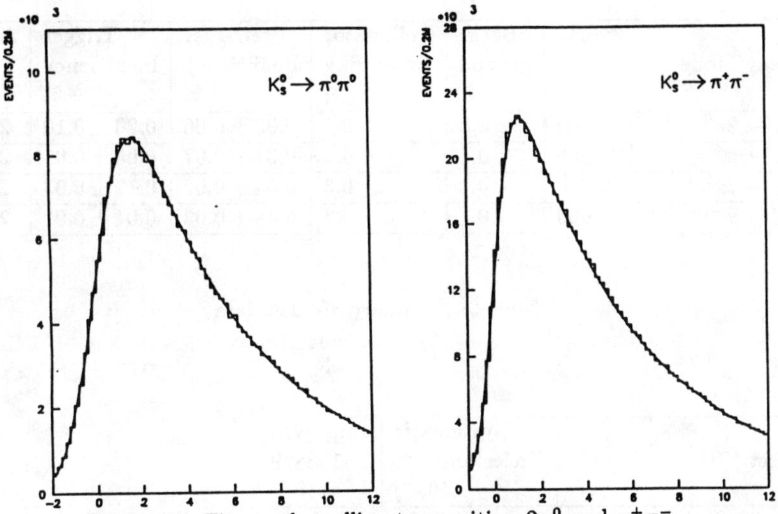


Figure 11: Fits to the collimator position $2\pi^0$ and $\pi^+\pi^-$

than 0.1%. This is crucial since a 0.1% uncertainty in the energy scale leads to an uncertainty of 0.3% in $R_\eta$. Between $K_S$ running, the energy scale is obtained by interpolation.

Finally, there is a small correction for the differential effect of accidental hits in the detector on the different modes in the two beams. This was estimated from the net loss of events when data were overlaid with the additional information from random triggers, taken at a rate proportional to beam intensity which were subsequently reconstructed and re-analyzed.

# 5 Results

The summary of the 1986 data with energy between 70 and 170 GeV, and with reconstructed vertex between 10.5 and 48.9m is shown in Table 3. The limits were determined (like all cuts) on criteria independent of $R_\eta$. The analysis is performed in "mini-periods" - self-contained and independent datasets of $K_L$ and $K_S$; thus time-dependent structure of the analysis can be identified, albeit with limited precision. Within each mini-period, the data is analyzed in bins of energy (10 GeV) and vertex (1.2m) so that the effect of the kinematic cuts and background rejection is harmonised. Care is required since many bins contain small numbers of events. Therafter, all remaining systematic effects are studied as a function, for example, of overall energy scale or background subtraction method.

The corrections are also shown in Table 3. The new beam and detector elements of the 1988 arrangement should reduce considerably the systematic error, as shown in Table 4. After corrections, the value for $R_\eta$ is found to be $0.980 \pm 0.004 \pm 0.005$, where the first error is statistical and the second is systematic. This is compared with previous measurements in Figure 12. The result as a function of energy, vertex position and mini-period (time) are shown in Figure 13.

| Decay Mode | Events | Back-ground % | Collimator Scattering % | Pretrigger Inefficiency % | Trigger Inefficiency % | Accidental Effects % |
|---|---|---|---|---|---|---|
| $K_L \to 2\pi^0$ | 109 000 | 4.0 | $< 0.1$ | $0.06 \pm 0.06$ | $0.20 \pm 0.10$ | $2.6 \pm 0.07$ |
| $K_L \to \pi^+\pi^-$ | 295 000 | 0.6 | $< 0.1$ | $0.37 \pm 0.07$ | $0.05 \pm 0.06$ | $2.6 \pm 0.05$ |
| $K_S \to 2\pi^0$ | 932 000 | <0.1 | 0.3 | $0.04 \pm 0.02$ | $0.12 \pm 0.03$ | $2.5 \pm 0.05$ |
| $K_S \to \pi^+\pi^-$ | 2 300 000 | <0.1 | 0.3 | $0.48 \pm 0.03$ | $0.01 \pm 0.01$ | $2.8 \pm 0.05$ |

Table 3: Summary of 1986 data

| Effect | Systematic Uncertainty % (1986) | Improved 1988/9 |
|---|---|---|
| Energy Scale | 0.3 | Yes |
| $2\pi^0$ background | 0.2 | Yes |
| $\pi^+\pi^-$ background | 0.2 | Yes |
| Accidental losses/gains | 0.2 | Yes |
| All other systematics | $<\sim 0.2$ | $<\sim 0.2$ |

Table 4: Main Systematic Effects on $R_\eta$

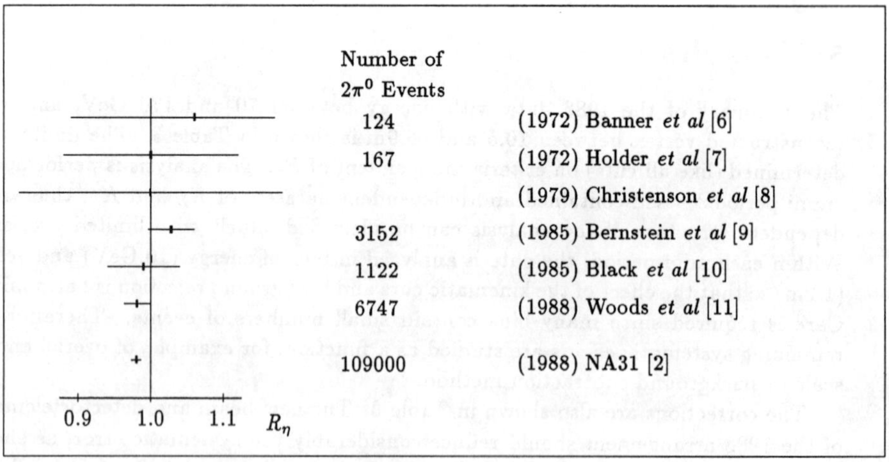


Figure 12: $R_\eta$ measurements

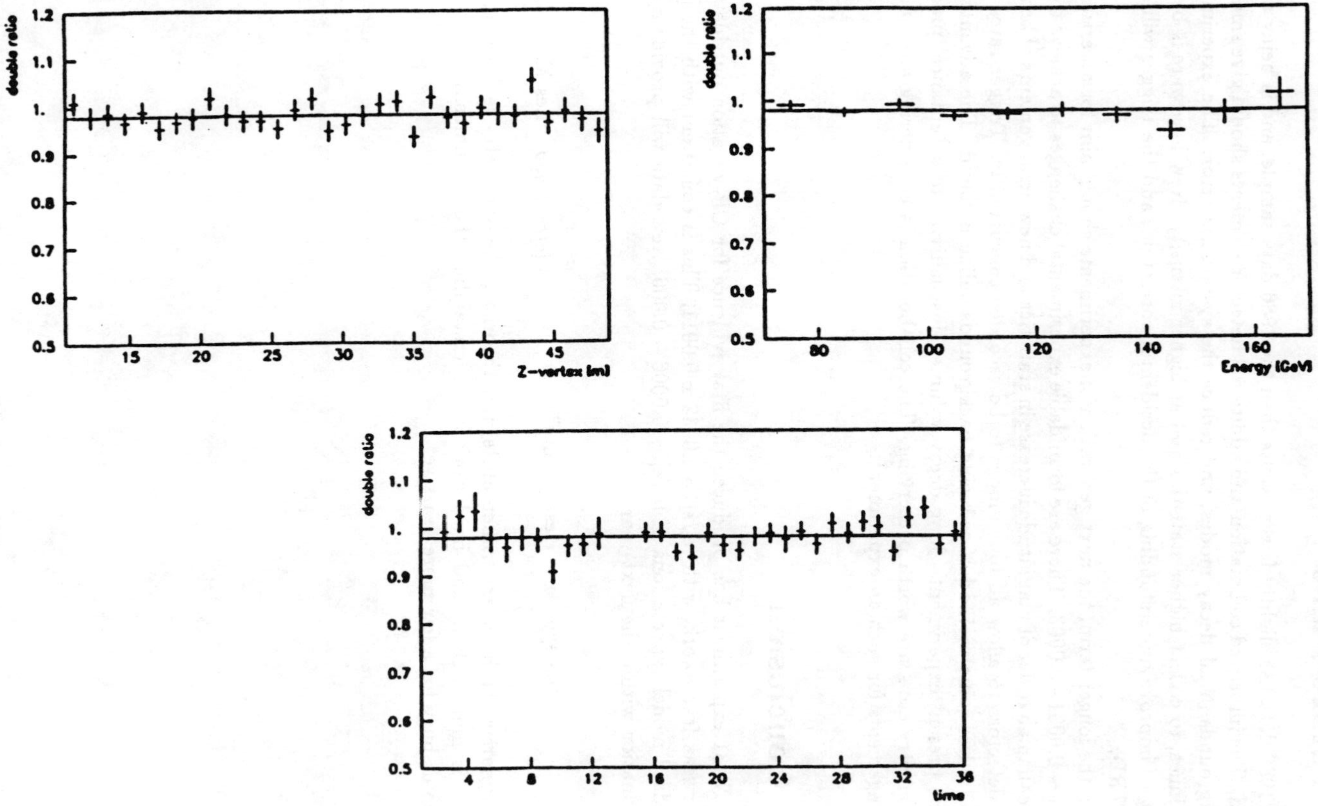


Figure 13: $R_\eta$ as a function of E,z,Time

# 6 Future Plans

The 1988 data has slightly fewer events than the 1986 data sample, and is being analyzed. The improved collimation and additional detector elements should give smaller backgrounds in all decay modes, and reduce the systematic error. The experiment continues, to collect higher statistics and at higher intensity. New hardware is being built to improve understanding of the accidental correction, and the trigger will use the TRD.

In the longer term, the next generation of experiments should aim for an error on $|\epsilon'/\epsilon| \sim 0.0001-0.0003$. There are formidable experimental challenges to achieve this - more than an order of magnitude increase in statistics and new arrangements of beams and detectors to allow all four modes to be recorded concurrently. Trigger rates are high, unless some method of reducing backgrounds online is found. The advantages of the present experiment - long decay volumes, insensitivity to acceptance, precise calorimetry etc - are worth preserving. The collaboration is considering alternative arrangements for such an experiment.

# 7 Conclusion

The NA31 experiment has produced the first evidence for CP-violation outside the superweak framework, with $|\epsilon'/\epsilon| = 0.0033 \pm 0.0011$. This is consistent with the [12] standard model expectations of $|\epsilon'/\epsilon| \sim 0.0006 - 0.006$. New data will provide more information within the next year or so.

"La symétrie, c'est l'ennui" *Victor Hugo* "Les Miserables"

Fortunately for both experimentalists and theorists, the symmetry is broken, and there is much delight and hard work in understanding the mechanism, which will keep us busy for a few more years yet.

# References


[1] H.Burkhardt *et al*, NIM A268 (1988) 116

[2] H. Burkhardt *et al*, Phys. Lett. B206 (1988) 169.

[3] L. Wolfenstein, Phys. Rev. Lett. 13 (1964) 569.

[4] V. Gibson, these proceedings.

[5] G.D. Barr *et al*, IEEE Trans. Nucl. Sci. 36 (1989) 66.

[6] M. Banner *et al*, Phys. Rev. Lett. 28 (1972) 1597.

[7] M. Holder *et al*, Phys. Lett. B40 (1972) 141.

[8] J.H. Christenson *et al*, Phys. Rev. Lett. 43 (1979) 1209.

[9] R.H. Bernstein *et al*, Phys. Rev. Lett. 54 (1985) 1631.

[10] J.K. Black *et al*, Phys. Rev. Lett. 54 (1985) 1628.

[11] M. Woods *et al*, Phys. Rev. Lett. 60 (1988) 1695.

[12] L. Wolfenstein, these proceedings.

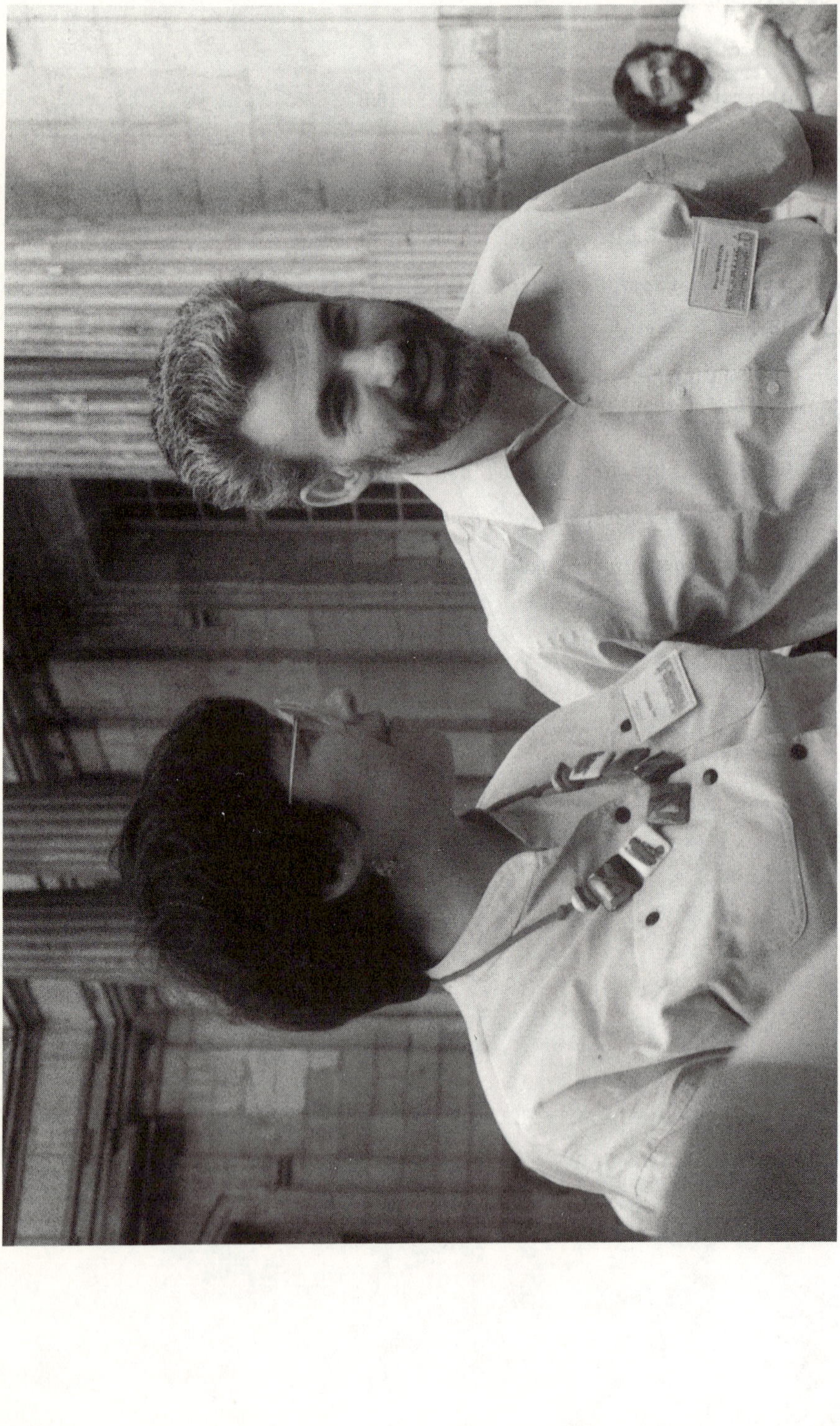

*Joan and Bruce Winstein.*

# A MEASUREMENT OF ε'/ε BY E731

B.Winstein
Enrico Fermi Institute and the Department of Physics
The University of Chicago
Chicago, Illinois 60637

## ABSTRACT

The technique used by Fermilab E731 for determining the CP violation parameter ratio ε'/ε is described. Systematic uncertainties are treated and a result based upon an analysis of a part of the data taken is presented.

## INTRODUCTION

This is a report from the E731 collaboration whose current members are: A. Barker, R. Briere, L. Gibbons, G. Makoff, V. Papadimitriou, J. R. Patterson, S. Somalwar, Y. Wah, B. Winstein, R. Winston, and H. Yamamoto, The University of Chicago; E. Swallow, Elmhurst College; G. Bock, R. Coleman, J. Enagonio, B. Hsiung, R. Stefanski, K. Stanfield, and T. Yamanaka, Fermilab; M. Karlsson, R. Tschirhart, and G. Gollin, Princeton University; P. Debu, B. Peyaud, R. Turlay, and B. Vallage, Saclay.

In order to observe direct CP violation in the decay of the neutral kaon to two pions, it is necessary to precisely determine the following four decay rates:

$$K_S \rightarrow \pi^0\pi^0, \qquad (1)$$

$$K_L \rightarrow \pi^0\pi^0, \qquad (2)$$

$$K_S \rightarrow \pi^+\pi^-, \qquad (3)$$

$$K_L \rightarrow \pi^+\pi^-. \qquad (4)$$

Then, by forming the double ratio:

$$R \equiv (1)/(2)/[(3)/(4)] \approx 1 + 6\,\mathrm{Re}(\varepsilon'/\varepsilon) \qquad (5)$$

$\varepsilon'/\varepsilon$ can be extracted. Given enough statistics, the major challenge is in designing the experiment and its subsequent analysis to minimize the resulting systematic uncertainty; this in practice involves the simultaneous collection of more than one of the above modes. At present, the CERN NA31 collaboration[1] has reported a 2.0% departure of R from unity, providing the first significant claim for direct CP violation; this was accomplished by separately measuring the $K_S$ and the $K_L$ modes. This paper will report upon a set of data in which all four modes were simultaneously collected.

## THE DETECTOR AND EXPERIMENTAL TECHNIQUE

The E731 detector[2] is shown schematically in Figure 1. Two nearly parallel $K_L$ beams enter the detector region, one passing through a thick $B_4C$ regenerator to provide $K_S$ decays. In this manner, $K_S$ and $K_L$ decays are viewed simultaneously by the same detector. The regenerator alternates between the beams on a pulse-by-pulse basis (roughly every minute) so that any small difference in beam intensity or detector acceptance is essentially eliminated. There are different triggers for charged and neutral decays; however, the trigger in no way distinguishes between $K_S$ and $K_L$. Thus the effect of the inevitable changes in chamber gas gain or drifts in calorimeter response over the course of a run will be reduced to high order. The same is the case for "noisy electronics" and similar effects. Moreover, changes in accelerator performance which affect the instantaneous rates in detector elements have little consequence with this technique.

With these advantages comes one principle disadvantage: the acceptance of the detector must be precisely known as a function of decay vertex because the lifetimes of the $K_S$ and $K_L$ are quite different.

The rates in the vacuum ($R_V$) and regenerated ($R_R$) beams are given (for example for charged decays) by:

$$R_V \propto | \eta_{+-} |^2 \tag{6}$$

and,

$$R_R \propto | \rho e^{-t/2\tau_s + i\Delta m t} + \eta_{+-} |^2 \tag{7}$$

where $\rho$ is the regeneration amplitude for the material. Since $|\rho| \approx 10|\eta|$, the ratio of the integrated numbers of regenerated to vacuum decays ( defined as $R_{+-}$) is given approximately by:

$$R_{+-} \propto | \rho/\eta_{+-} |^2. \tag{8}$$

Similarly,

$$R_{00} \propto | \rho/\eta_{00} |^2. \tag{9}$$

and, thus

$$R \approx R_{00}/R_{+-}. \tag{10}$$

The time dependence of the decay rate downstream of the regenerator, (7) above, can also be used to fit for $\tau_s$, the $K_S$ lifetime, $\Delta m$, the $K_L$-$K_S$ mass difference, and $\Phi_\eta$ , the phase of the CP violating decay amplitudes (relative to the phase of $\rho$).

The charged decays are reconstructed by means of magnetic analysis using the four precision drift chambers shown in Figure 1. A large lead-glass array serves to detect the neutral decays and permit off-line rejection of the $\pi e \nu$ decay mode; $\pi\mu\nu$ decays are eliminated with the muon filter. A background in the neutral mode arises from $K_L \rightarrow 3\pi^0$ decays where photons miss the lead-glass; these are largely eliminated by means of many planes of "photon vetoes" situated outside the solid angle of the lead-glass array.

## DATA COLLECTION AND TRIGGERS

The experiment took data from August of 1987 to February of 1988. About 75% of the data was taken with either the charged or neutral triggers in operation. For neutral mode running, a thin lead sheet was placed at the "trigger plane" in order to convert (usually) one photon. For the other 25%, all four modes were collected simultaneously (with no lead sheet) and it is from the bulk of this latter data set that we are now presenting a result.

The charged trigger required a two-track topology in the scintillator hodoscopes with no muon signal. In addition to the two pion decay, $\pi e \nu$ and $\pi^+\pi^-\pi^0$ events useful for calibration, alignment, and detector studies were collected.

The neutral trigger required[3] either four or six clusters and greater than about 30 GeV energy deposit in the lead glass; $3\pi^0$ decays were thus also collected.

The data set that we are reporting on consisted of approximately 20,000 spills (over an 18 day period) at the Fermilab Tevatron; each spill lasted for about 22 sec and the repetition rate was about one per minute. During each spill, a light flasher which illuminated each phototube of the lead-glass array was pulsed every second, determining the tube gains to about 1% per flash. Pedestal events were taken as frequently and an accidental trigger was taken about 70 times per spill to provide an unbiased sample of the underlying activity in the detector. Care was taken with the electronics so that both triggers were "loose" and unbiased with respect to $K_L$ vs $K_S$ decays.

For this analysis, an event was retained only if its reconstructed z position lay between 120m and 137m and its reconstructed momentum lay between 40 GeV and 150 GeV; this selection was made to reduce systematic uncertainties, arising from acceptance and lead-glass energy scale uncertainties.

## MONTE CARLO SIMULATION

The simulation of the detector response is important . The philosophy is to understand as much as is possible about the detector and the beam to insure confidence in the eventual result. The difference in detector acceptance for $K_L$ vs $K_S$ decays is needed and for this a monte carlo is used and is extensively checked with the very high statistics modes. The simulation:

a) incorporates an extensive photon cluster library using data and EGS-simulated photon showers,

b) treats the initial pair production process of a photon in the lead-glass,

c) includes (different) light attenuation lengths in the lead-glass blocks,

d) treats shower leakage (transverse and longitudinal),

e) includes dead-time in the drift chamber electronics,

f) includes the few dead or inefficient drift chamber wires,

g) can overlay accidental activity when necessary, and,

h) includes the small variations in energy and flux across the beam.

Elements that are not included from "first principles", those that need to be tuned to the data, are the beam shape (which is sensitive to collimator misalignments) and the "dilution factor", the ratio of $K^0$ and $\overline{K}^0$ production at the target. In addition, the momentum spectrum needs to be adjusted[4]. Tuning these elements gives better agreement in comparisons of data and monte-carlo distributions; however, the resulting value of ε'/ε does not vary significantly.

## RECONSTRUCTION OF $K_{L,S} \to \pi^+\pi^-$ CANDIDATES

In this section, we will treat the reconstruction of the charged decays. Each of the four drift chambers had two x and two y planes and the plane resolution was about 100μ. The transverse momentum of the analysis magnet was approximately 200 MeV/c and the resulting track momentum resolution was about 1%.

Pattern recognition and fiducial and kinematic cuts were all done without regard to whether an event was a candidate for a $K_L$ or a $K_S$ decay. The reconstruction efficiency, after all cuts, was about 90% for both modes. An event was classified as either $K_L$ or $K_S$ depending upon where its momentum vector extrapolated to the plane of the regenerator. Figure 2 shows the invariant mass distribution for candidate $K_S$ decays; on the low-mass side, the data is broader as a result of radiative $\pi\pi\gamma$ decays. Figure 3 shows the reconstructed transverse momentum-squared ($P_T^2$) distribution for the $K_S$ candidates (now with a mass selection between 484 and 512 MeV/c$^2$) showing a very sharp coherent peak above a background, consisting in diffractive and inelastic regeneration, of 0.13%. Figure 4 shows agreement between the data and monte carlo z-vertex distributions for these decays (with $P_T^2$ <250 (MeV/c)$^2$). The mass and z-vertex distributions for $K_L$ decays are shown in Figures 5 and 6 respectively; the $\pi\pi\gamma$ radiative tail is again present in addition to a residual background (0.31%) from semi-leptonic decays. The z-vertex distributions agree very well, and this coupled with the fact that the mean decay positions for $K_L$ and $K_S$ decays differ by only about 1.5m allows us to rule out a significant error in the determined ratio of $K_L$ to $K_S$ decays (see Figure 6b).

## RECONSTRUCTION OF $K_{L,S} \to \pi^0\pi^0$ CANDIDATES

We turn now to the neutral mode. The lead-glass calibration[5] was important; it was accomplished by means of frequent special runs using electron positron pairs; this gave a very clean "beam" with negligible background and the sample used for this data set consisted of over 1 million electrons, after all cuts. The non-linearity of the blocks was determined from this sample; the energy E in a block was approximately related to the ADC pulse-height P for that block by the following expression:

$$E = g(P)^{\alpha} \qquad (11)$$

where g is the tube gain and $\alpha$ was typically 0.977; the departure of $\alpha$ from unity results from light absorption in the lead-glass. This calibration procedure was extensively checked with several million electrons from $\pi e\nu$ decays and several hundred thousand $\pi^0$'s from $\pi^+\pi^-\pi^0$ decays.

The kaon vertex is determined by finding the best pairing for the $2\pi^0$ decay and using the constraint of the $\pi^0$ mass; then the $2\pi^0$ invariant mass is calculated.

Figure 7 shows the reconstructed center-of-energy distribution at the lead-glass of all 4 cluster events having invariant mass within 18 MeV/$c^2$ of the kaon mass. For this Figure, the event coordinates have been "flipped" according to the position of the regenerator. The two modes (coherently regenerated $K_S$ to $2\pi^0$ and vacuum $K_L$ to $2\pi^0$) are distinct, and one can see the presence of incoherent regeneration. Figure 8 displays the invariant mass distribution for the $K_L$ candidates; overlaid is the distribution, absolutely normalized, of $3\pi^0$ background (0.37%) as predicted by monte-carlo.

To subtract the background from incoherent regeneration, a plot of the event density in concentric rings away from each beam is made. This is shown for the vacuum beam in Figure 9; it can be seen that there is a flat distribution of "cross-over" events. The residual contribution under the peak can be estimated from the data alone, or it can be predicted absolutely by means of the $P_T^2$ distribution measured in the charged mode (Figure 3). This is our largest background (4.66%) but it is well understood. We remark that such an absolute prediction can be made since all four modes were simultaneously collected; otherwise, the rejection efficiency for inelastically regenerated events, which contain other particles in the final state, might depend upon the precise running conditions. However, since the predicted distribution in ring number can easily be normalized to the data, we believe that the systematic uncertainty coming from this subtraction will not be appreciably larger for the remainder of the data set where charged and neutral data were separately taken.

Figure 10 shows the z-vertex distribution for $K_L \rightarrow 2\pi^0$ events: the agreement with monte-carlo is excellent.

## EXTRACTION OF $\varepsilon'/\varepsilon$

Before fitting for the result, the backgrounds are subtracted and acceptance corrections are made.

The background summary and the estimated uncertainties are given in Table 1 below.

Table 1. Summary of Backgrounds

| Mode | Background | Correction [%] | Uncertainty [%] |
|---|---|---|---|
| $K_L \to \pi^+\pi^-$ | $\pi e \nu$ | 0.31 | 0.06 |
| $K_S \to \pi^+\pi^-$ | incoherent regeneration | 0.13 | 0.01 |
| $K_L \to 2\pi^0$ | "cross-over" | 4.66 | 0.14 |
| $K_L \to 2\pi^0$ | $3\pi^0$ | 0.37 | 0.07 |
| $K_S \to 2\pi^0$ | incoherent regeneration | 2.58 | 0.07 |

The magnitude of the acceptance and the event totals are given in the following Table.

Table 2. Event Totals and Corrections

| | Neutral | Charged | Double Ratio |
|---|---|---|---|
| Uncorrected Event Totals | | | |
| Vacuum | 52,226 | 43,357 | |
| Regenerated | 201,332 | 178,803 | 1.0698(78) |
| Background Fractions | | | |
| Vacuum | 0.0515 | 0.0042 | |
| Regenerated | 0.0257 | 0.0012 | 1.0445 |
| Acceptance | | | |
| Vacuum | 0.1885 | 0.5041 | |
| Regenerated | 0.1813 | 0.5064 | 1.0001 |

A few comments are in order. First, the indicated background fractions include a correction for the presence of residual $K_S$ decays in the $K_L$ beam; this is on the order of 0.1% and is largely common to the two modes. Second, the acceptances are seen to be about the same for $K_S$ and $K_L$ decays; this follows from the similarity in the decay distributions and the fiducial cuts. Third, we see that the raw double ratio is about 7% from unity and that after background and acceptance corrections, it is consistent with zero.

The extraction of $\varepsilon'/\varepsilon$ is then done by fitting (the integrated) decay distributions to the functional forms in (6) and (7). The regeneration amplitude $\rho$ is proportional to the difference in forward scattering amplitudes of $K^0$ and $\overline{K}^0$ from the regenerator material:

$$\rho \propto (f - \overline{f})/k \tag{12}$$

and this amplitude difference has a power-law dependence[6] upon momentum:

$$(f - \overline{f})/k \propto p^{-\alpha}. \tag{13}$$

Fits were done for each mode separately, assuming that $\varepsilon' = 0$ ; the amplitude difference at 70 GeV/c and the power $\alpha$ were fit. The results of the fits are given in the following Table and they are displayed in Figure 11.

Table 3. Results of fits for the regeneration amplitude

| | Amplitude Difference at 70 GeV [mb] | Power-law ($\alpha$) | $\chi^2$ / degrees of freedom |
|---|---|---|---|
| Charged | 5.972(18) | 0.602(10) | 11.5/9 |
| Neutral | 5.963(18) | 0.605(10) | 10.3/9 |

The power laws are seen to be consistent with each other; they are in addition consistent with previous determinations[6,7,8] using light elements. The charged and neutral extracted amplitude differences are also consistent which points to a small value of $\varepsilon'/\varepsilon$, as did the indication from the double ratio.

A grand fit is next done to both modes simultaneously for the value of $\varepsilon'/\varepsilon$, again allowing the regeneration parameters to float. The value obtained is

$$\varepsilon'/\varepsilon = -0.0005 \pm 0.0015 \tag{14}$$

where the error shown is statistical. We will now treat the sources of systematic uncertainty.

## SYSTEMATIC ERROR DETERMINATION

The possible sources of systematic error involve uncertainties in the backgrounds (which have already been given), the stability of the results as a function of time and intensity, the sensitivity to the energy scale and non-linearities in the lead-glass detector, sensitivity to accidental activity in the detector, and uncertainties in the detector acceptance.

Time Stability

In principle, because of drifts in the detector acceptance or in the beam position on the target, the yields could change with time. Indeed they do but with the method of recording long and short decays simultaneously, the effect of such drifts are greatly reduced. Alternating the regenerator every pulse eliminates effects which might change the relative acceptance between the two beams. In Figure 12 we show the ratios $R_{+-}$ and $R_{00}$ vs time into the run. Both ratios are time-independent even though the intensity varied.

Sensitivity to Energy Scale

The so called "energy scale" in the charged mode is easily determined: one simply adjusts the overall magnetic field so that the $K^0$ and $\Lambda$ masses agree with the accepted values. It is important that the non-uniformities in the field be well understood. The neutral energy scale must be accurate as well: the double ratio R is formed in momentum bins and the decay rates, which are inversely proportional to energy, must be compared at identical energies for charged and neutral decays. The lead-glass is calibrated using momentum-analyzed electrons (using the same magnetic field map as the charged mode analysis) so that in principle the energy scales will coincide. In practice, this is only valid to within a few tenths of one percent as a result of the difference in response of the lead-glass for photons and electrons. The overall neutral energy scale can be adjusted by aligning the observed distributions, but there remains a residual uncertainty on the order of 0.1%. The effects of this uncertainty of the ratio $R_{00}$ are largely eliminated by the choice of the fiducial region (in z-vertex): the region, which includes essentially all of the $K_S$ events, is chosen so that roughly the same number of $K_L$ events leave (enter) at the downstream boundary as enter (leave) at the upstream boundary when the gamma energies are decreased (increased). In Figure 13 we see that $R_{00}$ changes very little with lead-glass energy scale. However the energy resolution is not perfectly understood and as a result we estimate a possible 0.2% uncertainty in $R_{00}$.

Accidental Activity

In principle, the presence of accidental activity in the detector at the time of the event could result in a bias between $K_S$ and $K_L$ decays: the accidental activity tends to cluster near the vacuum beam and this could cause a systematic lowering of the efficiency for decays from that beam relative to decays from the other beam. Such a bias would obviously not cancel upon alternation of the regenerator position. The accidental activity, however, is at a low level: events have an extra cluster in the lead glass only 2.7% of the time (and only those which land on top of other clusters can cause a bias) and they have only about 8.5 extra chamber hits throughout the spectrometer (again only those which land on top of the true hits can cause a bias). We have searched for such biases, primarily using the high statistics modes; indeed, we have seen that distributions in track quality have a very slight broadening in the vacuum beam, but our cuts are sufficiently loose that this is inconsequential. We have used the technique of overlaying accidental events upon monte carlo events to investigate this possible bias. This procedure correctly reproduces a 3% loss in charged reconstruction efficiency (common to both modes) over the intensity range of the data (a factor of 4). There is, however, no asymmetry in $K_L$ vs $K_S$ for either mode within 0.07% so that the combined possible bias from accidental activity is taken to be 0.10%.

Acceptance

Possible uncertainties in the acceptance corrections are the most important source of systematic error for the measurement as we have chosen to perform it. The issue is of course the acceptance vs z-vertex; we have already seen that the agreement between data and monte-carlo is quite good and that, for there to be an effect at the percent level, the disagreements would be rather obvious (see Figure 6). In order to estimate the possible systematic error, we have relied upon the high statistics modes.

The z-vertex distribution for $K_L \to \pi e \nu$ decays is shown in Figure 14; the agreement over the fiducial decay volume is excellent, but with the high level of statistics (0.6 x $10^6$ monte-carlo events, 1.3 x $10^6$ data events) a small disagreement can be seen further upstream. The similar distribution for $3\pi^0$ decays is shown in Figure 15 for a portion of the decay region showing excellent agreement. The acceptance is easier to determine for the neutral mode: photons travel in straight lines and there is essentially only one aperture to model, the lead-glass array.

The monte-carlo includes the decay of the $K_L$ beam along the decay region and neglecting this gives poorer agreement. For our mean momentum of about 70 GeV/c, the decay rate should decrease by about 0.05% per meter; using a sample of $3\pi^0$ decays, we determine that $c\tau_L = 9.5\text{m} \pm 3.1\text{m}$ where the accepted[9] value is 15.5m. The error is statistical and the result gives confidence in the neutral mode acceptance determination.

For the charged mode where the z resolution is about 15cm, our result is consistent when fitting in 2m z-vertez bins, thereby nearly eliminating the dependence upon acceptance. (This is less reliable in the neutral mode where the z resolution is about 1m.) We have also systematically determined the effect that the acceptance disagreement observed in the $\pi e \nu$ decays would have on the result: we would have a –0.05% shift in $R_{+-}$. From this information and the stability of $R_{+-}$ and $R_{00}$ when apertures, beam shapes, cuts, and detector efficiencies are varied, we assign a systematic uncertainty of $\leq 0.18\%$ for each mode.

A summary of the systematic uncertainties is given in Table 4 below.

Table 4. Systematic Error Summary

| Source of Error | Uncertainty [%] |
|---|---|
| Accidental Effects | 0.10 |
| Backgrounds | 0.18 |
| Energy Scale/Resolution/ Non–linearity | 0.20 |
| Acceptance | 0.25 |
| TOTAL: | 0.38 |

## THE RESULT AND CONCLUSIONS

Before giving the result, we would like to present the outcome of additional fits to the data for other physical parameters. These may be regarded as internal consistency checks, and they provide interesting additional information. We have fit the data in 2m bins in z-vertex to determine the $K_S$ lifetime, $\tau_S$, and the $K_L$-$K_S$ mass difference, $\Delta m$, for each mode. We have determined the phase difference between $\eta_{+-}$ and $\eta_{00}$, $\Delta\Phi$. Finally, we give a value for the phase of $\eta_{+-}$ itself which can be determined with the assumption of analyticity[10] for the regeneration amplitude.

Table 5. Fits for other physical parameters

| | Charged Mode | Neutral Mode | PDG Value (ref.9) |
|---|---|---|---|
| $\tau_S$ [$10^{-10}$ sec] | 0.8909(63) | 0.8940(61) | 0.8923(22) |
| $\Delta m$ [$10^{10}$ $\hbar$ sec$^{-1}$] | 0.524(18) | 0.524(18) | 0.535(2) |
| $\Delta\Phi$ [degrees] | $2.3 \pm 3.0$ | | $\leq 0.1$ (by CPT) |
| $\Phi_{+-}$ [degrees] | $47.5 \pm 3.4$ | | $44.6 \pm 1.2$ |

The errors shown are statistical and these dominate. The results are obtained using the identical event selection and background subtractions as for the $\varepsilon'/\varepsilon$ analysis; ie, no optimization for the individual quantities has been made. The agreement of these results with the expected values gives confidence in the soundness of the data and its analysis.

Using (5) and (14) and Table 4, we then find:

$$Re(\varepsilon'/\varepsilon) = (-5 \pm 14 \pm 6) \times 10^{-4} \tag{15}$$

where the first error is statistical and the second is systematic. Combining the errors in quadrature then yields:

$$Re(\varepsilon'/\varepsilon) = (-5 \pm 15) \times 10^{-4}. \tag{16}$$

We show in Table 6 and in Figure 16 this result in relation to other recent measurements.

Table 6. Recent Measurements of Re(ε'/ε)

| Experiment | Result | Statistical Error | Systematic Error | Combined Error |
|---|---|---|---|---|
| BNL-Yale[11] | +17 | 72 | 43 | 83 |
| Chicago-Saclay[7] | −46 | 53 | 24 | 58 |
| E731 Test[8] | +32 | 28 | 12 | 30 |
| NA31[1] | +33 | 7 | 8 | 11 |
| This Result | −5 | 14 | 6 | 15 |

We can draw the following conclusions:

a) the result is consistent with zero and the systematic uncertainty has been significantly reduced;

b) we have simultaneously made precision determinations of an additional eight physical quantities ( $\tau_S$, $\Delta m$, and the regeneration slope for both modes; $\Delta\Phi$, and $\Phi_{+-}$ )

c) the result does not confirm the NA31 claim of the observation of direct CP violation- the discrepancy is at the two standard deviation level;

d) the result is not inconsistent with the "Standard Model"[12] although it would favor a rather high top quark mass; it is consistent with the superweak[13] model.

We believe that the situation will be better clarified when our full data set is processed and a result obtained. We have every reason to believe that the resulting error will be significantly reduced over what is here presented[14].

## ACKNOWLEDGEMENTS

I would like to express my appreciation for this very well run and stimulating conference commemorating the 25th anniversary of the discovery of CP nonconservation. J. Tran Thanh Van worked so hard and his efforts were felt by everyone present. Also, I would like to acknowledge a few individuals unable to attend who have over the years contributed to my appreciation for the subject, as well as to the field. G. Zweig gave me a first very enthusiastic introduction to the subject as I began graduate school in 1965. Val Telegdi, beginning in 1972, introduced me to the beauty of the subject of, and the experimental techniques in, neutral kaon physics. John Ellis, Mary K. Gaillard, and D. V. Nanopoulos, with their important paper (written in 1976), inspired me in 1978 to attempt to measure $\varepsilon'/\varepsilon$. And Andrzej Buras, whose writings on the subject in the 1980's are so remarkably clear.


## REFERENCES


[1] H. Burkhart *et. al.*, Phys. Lett. B **206**, 169 (1988).

[2] See M. Woods Ph.D. Thesis, The University of Chicago, 1988 (unpublished), for a full description of most of the detector elements.

[3] H. Sanders *et. al.*, IEEE Trans. Nucl. Sci **36**, 358 (1988).

[4] A. J. Malensek, Fermilab FN-341 (1981) gives an excellent starting spectrum.

[5] The calibration techniques will be described fully in the Ph.D. Thesis of J. R. Patterson, The University of Chicago, 1990.

[6] This follows from a single Regge exchange; in this case, the $\omega$ trajectory dominates. See J. Roehrig *et. al.*, Phys. Rev. Lett. **38**, 1116 (1977).

[7] R. H. Bernstein *et. al.*, Phys. Rev. Lett. **54**, 1631 (1985).

[8] M. Woods *et. al.*, Phys. Rev. Lett. **60**, 1695 (1988).

[9] Particle Data Group, G. P. Yost *et. al.*, Phys. Lett. B **204**, 1 (1988).

[10] The power $\alpha$ and the phase of $\rho$ are connected by a dispersion relation.

[11] J. K. Black *et. al.*, Phys. Rev. Lett. **54**, 1628 (1984).

[12] See, for example, F. Gilman, these proceedings.

[13] L. Wolfenstein, Phys. Rev. Lett. **13**, 562 (1964)

[14] At the time of the meeting in Blois, the E731 group was not ready to present a result but rather, because of the special nature of the conference, gave an indication of where the fits were coming out. It was said that the central value for Re $(\varepsilon'/\varepsilon)$ was between $-0.001$ and $+0.001$, with the statistical error ranging between 0.0013 and 0.0016 as cuts were changed.

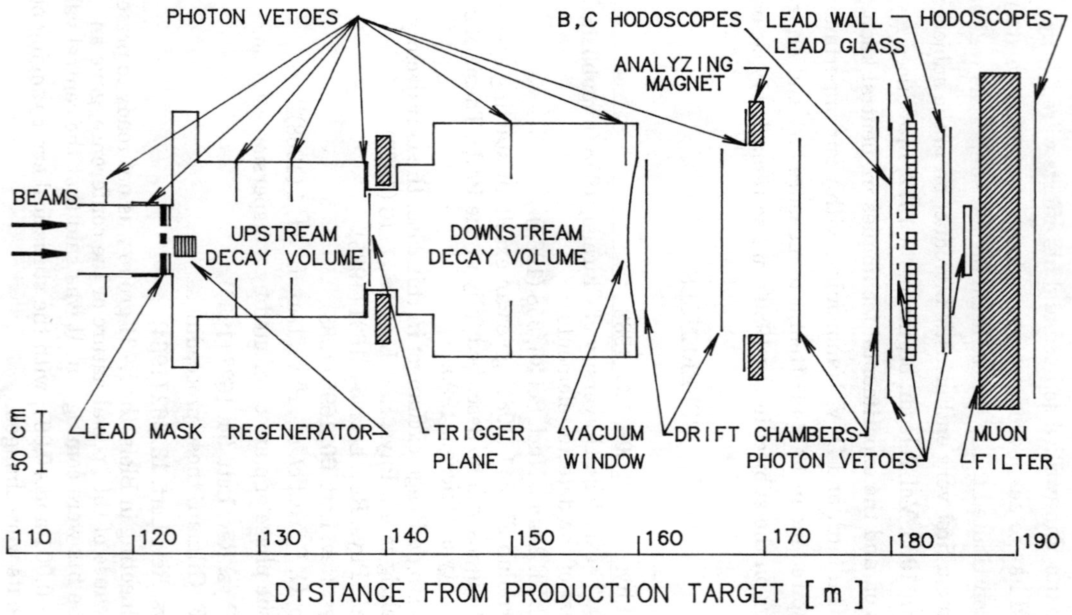


Fig. 1. Schematic of the E731 Detector.

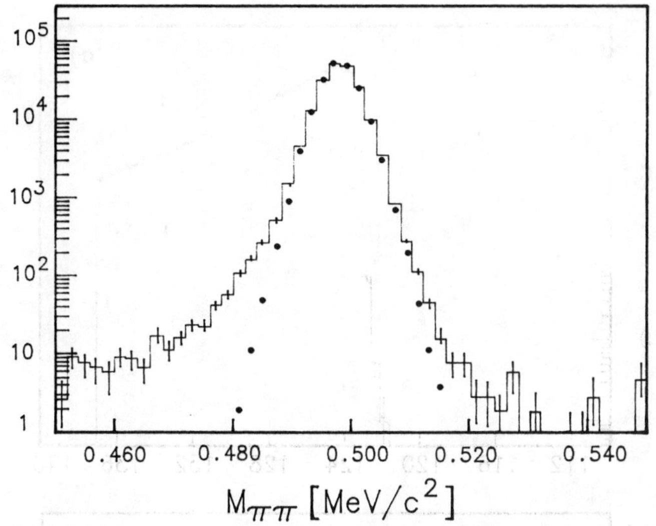


Fig. 2. Invariant mass distribution [events/2MeV/$c^2$] for $K_S \rightarrow \pi^+\pi^-$.The histogram is the data and the dots are the monte-carlo.

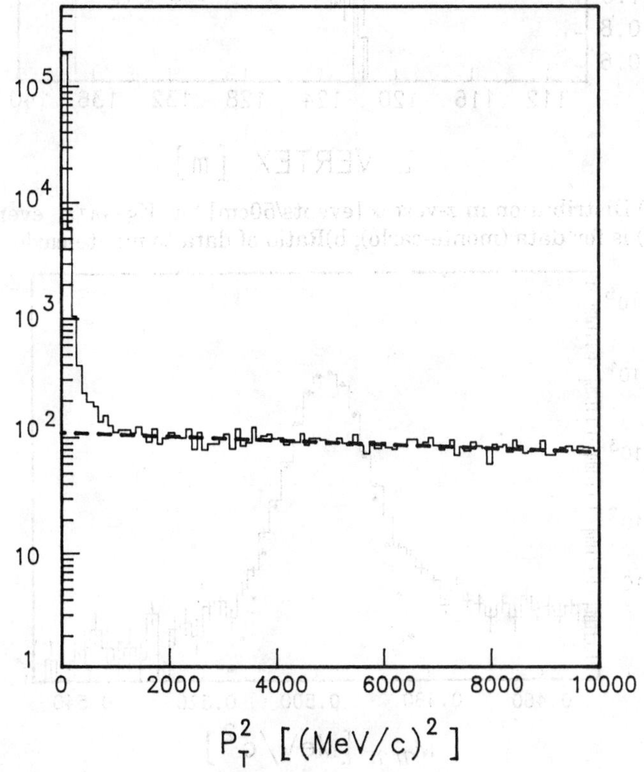


Fig. 3. $P_T^2$ distribution [events/(100MeV/c)$^2$] for $K_S \rightarrow \pi^+\pi^-$. The dashed line extrapolates the background under the coherent peak.

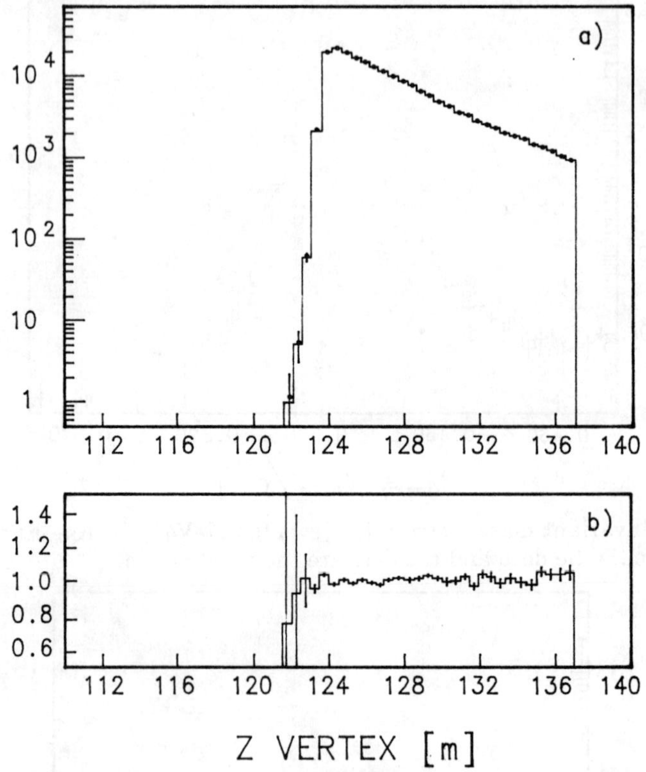


Fig. 4. a) Distribution in z-vertex [events/50cm] for $K_S \to \pi^+\pi^-$ events. The line (dot) is for data (monte-carlo); b)Ratio of data to monte-carlo.

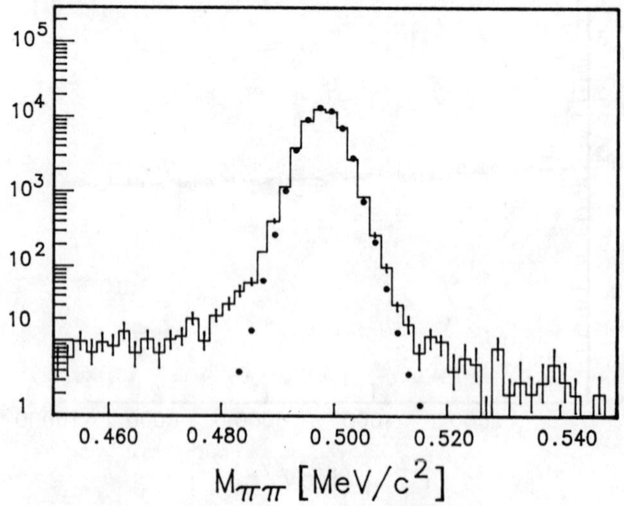


Fig. 5. Invariant mass distribution [events/2MeV/c²] for $K_L \to \pi^+\pi^-$.The histogram is the data and the dots are the monte-carlo.

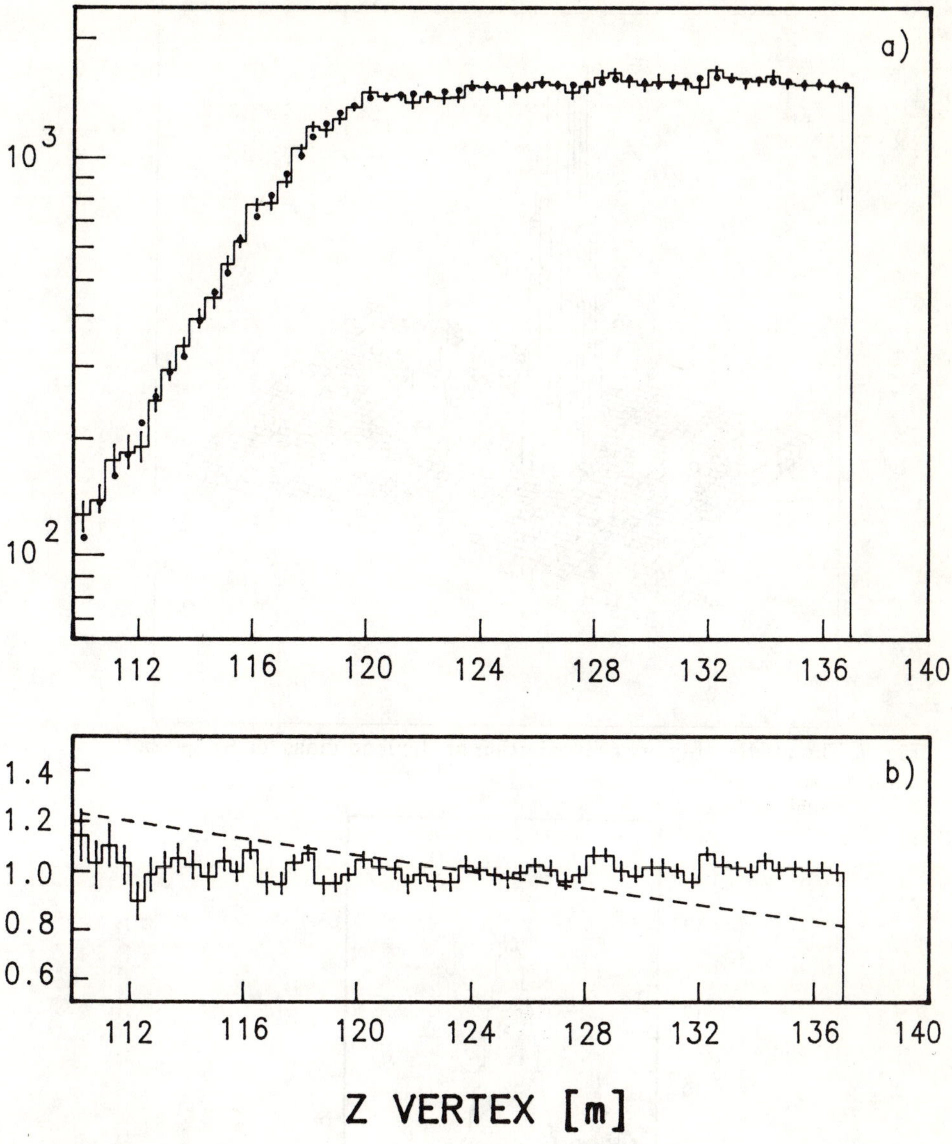


Fig. 6. a) Distribution in decay vertex [events/50cm] for $K_L \rightarrow \pi^+\pi^-$ events. The line (dot) is for data (monte-carlo); b) Ratio of data to monte-carlo. This ratio would follow the dashed line were there an acceptance error large enough to cause a 2% shift in the $K_S$ to $K_L$ ratio.

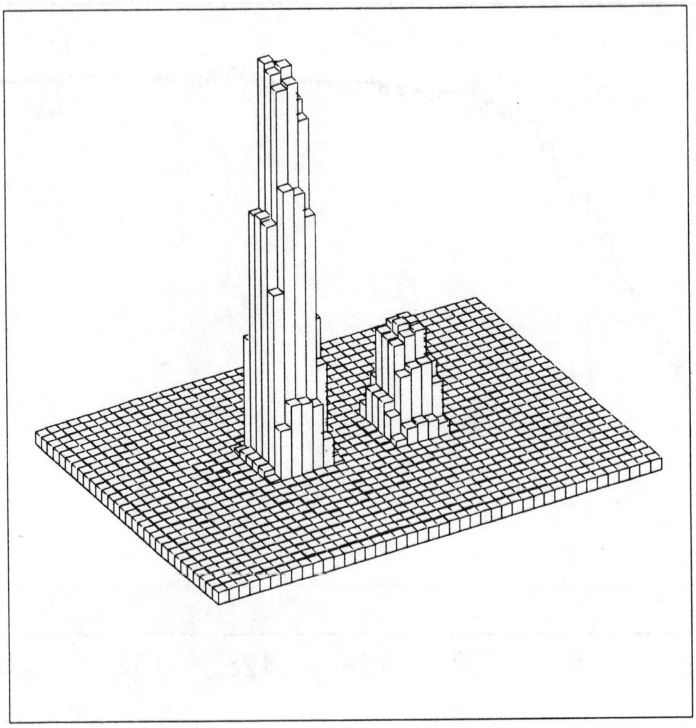

Fig. 7. The center-of-energy distribution at the lead-glass for $K_{L,S}\rightarrow 2\pi^0$ candidates.

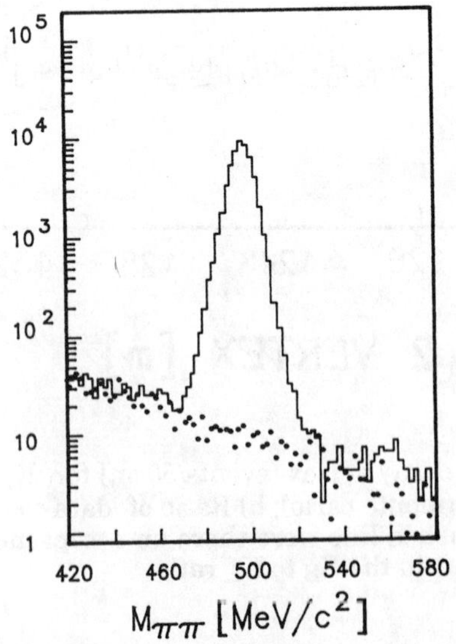


Fig. 8. Distribution in invariant mass [events/2MeV/c$^2$] for $K_L\rightarrow 2\pi^0$ events. The histogram is the data; the dots are the monte-carlo for the residual $K_L\rightarrow 3\pi^0$ background, absolutely normalized.

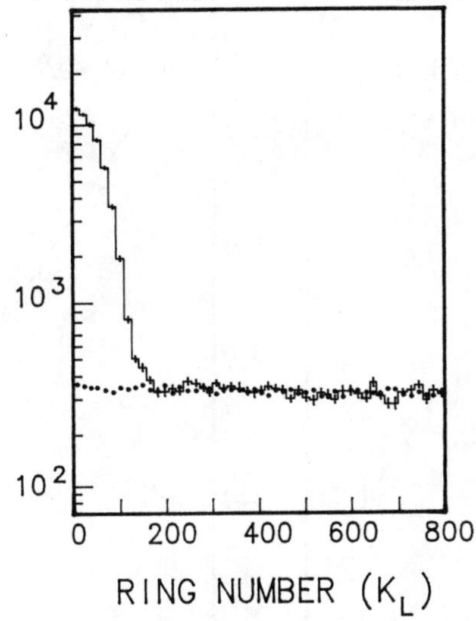


Fig. 9. The distribution in "ring number" around the vacuum beam for $2\pi^0$ events. The histogram is data; the dots are the (absolutely normalized) distribution for incoherent events as determined from $\pi^+\pi^-$.

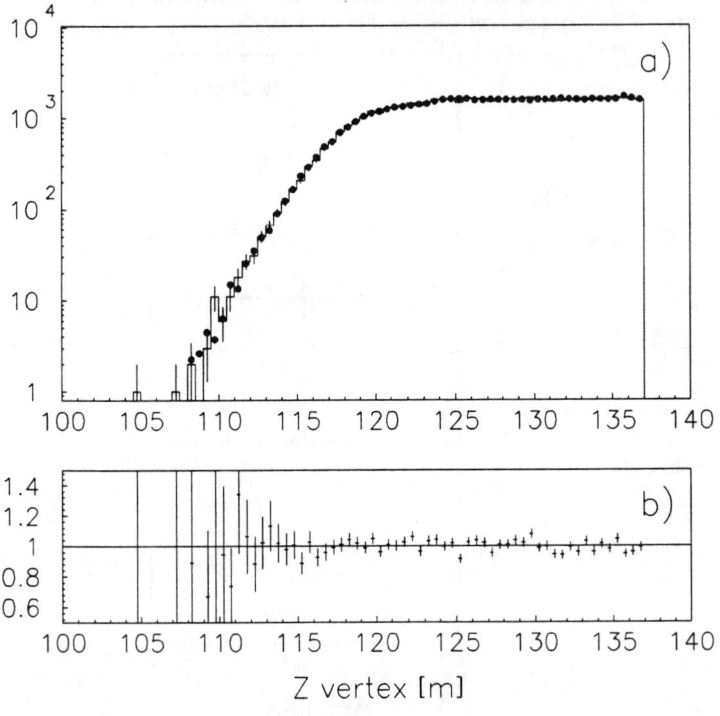


Fig. 10. a) Distribution in z-vertex [events/50cm] for $K_L \rightarrow \pi^0\pi^0$ . The line (dot) is for data (monte-carlo); b) The ratio of data to monte-carlo.

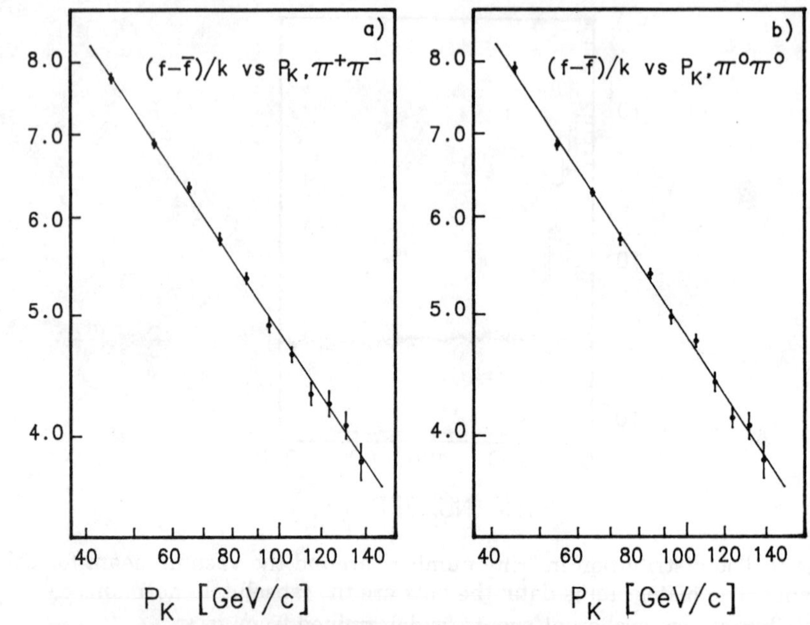


Fig. 11. The difference in forward scattering amplitudes [mb] vs. the kaon momentum. a) Charged mode; b) Neutral mode.

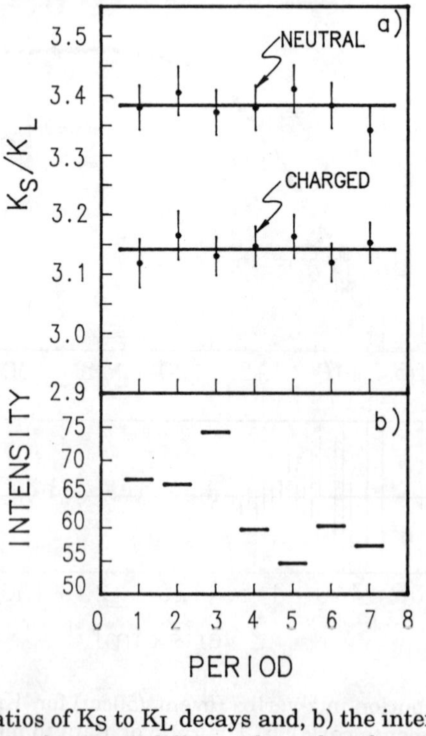


Fig. 12. a) The ratios of $K_S$ to $K_L$ decays and, b) the intensity, vs. the time into the run. (The selection is different from that described in the text.)

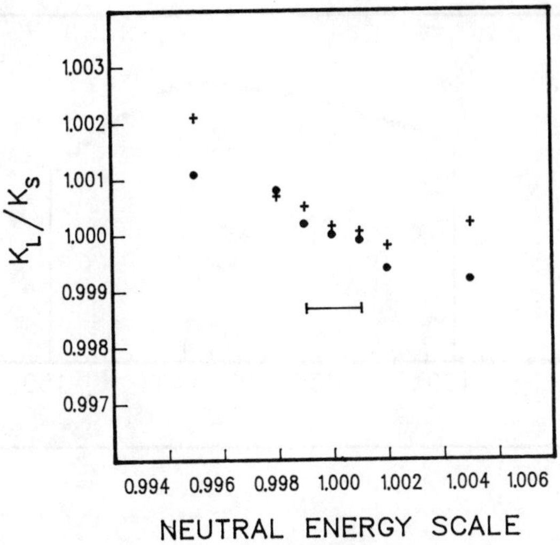


**Fig. 13. The factor by which the ratio of $K_L$ to $K_S$ for $\pi^0\pi^0$ events changes vs. the neutral energy scale. The pluses (dots) are the data (monte-carlo). Also shown is the estimated residual uncertainty in the energy scale.**

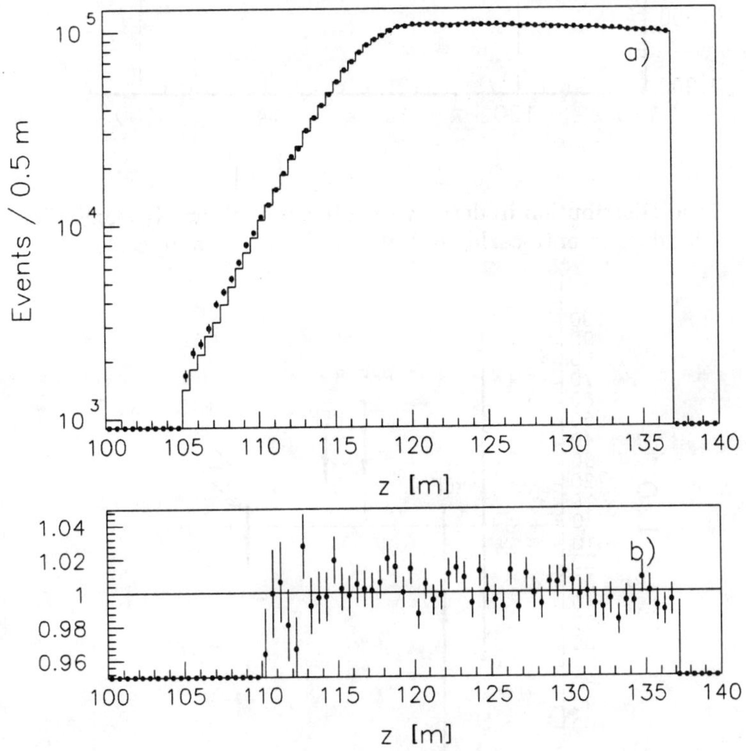


**Fig. 14. a) Distribution in z-vertex [events/50cm] for $K_L \rightarrow \pi e \nu$. The line (dot) is for data (monte-carlo); b) Ratio of data to monte-carlo.**

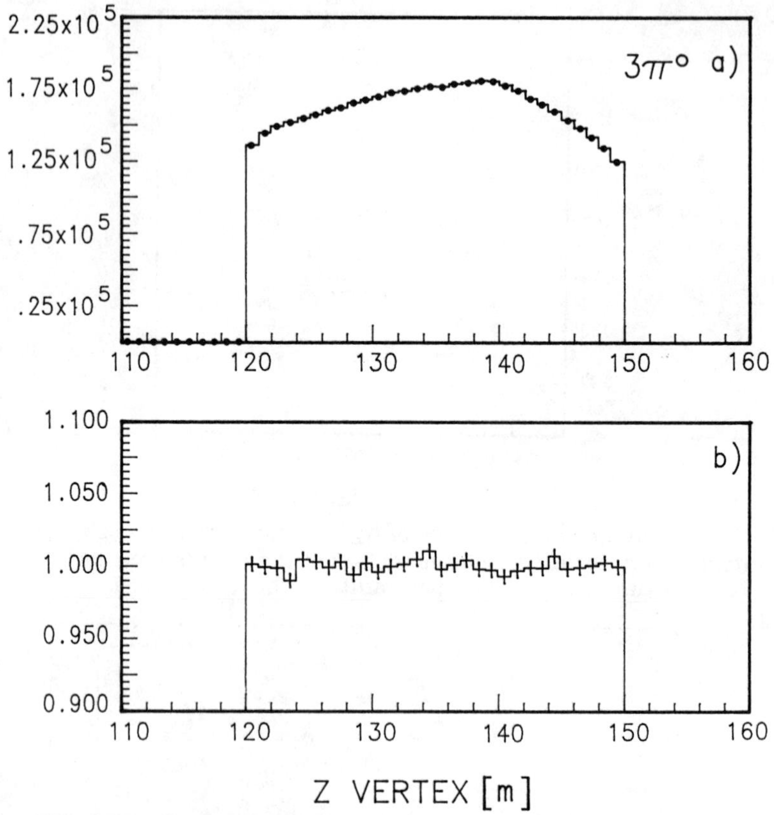


Fig. 15. a) Distribution in decay vertex [events/m] for $K_L \to 3\pi^0$. The line (dot) is for data (monte-carlo); b) Ratio of data to monte-carlo.

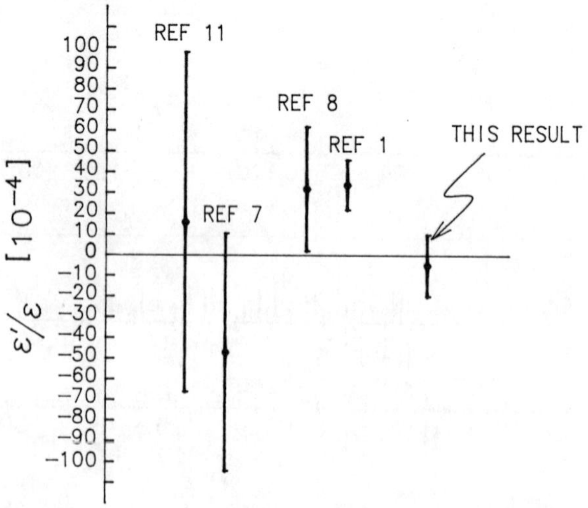


Fig. 16. This result with other recent determinations of Re(ε'/ε).

# CPT AND T INVARIANCE

*Robert Adair and Robert Sachs.*

# CP OR T VIOLATION?

Robert G. Sachs*

Enrico Fermi Institute and the Department of Physics
The University of Chicago
5640 South Ellis Avenue, Chicago, IL 60637

We are here to celebrate the beautiful experiment of Christenson, Cronin, Fitch and Turlay which led to the remarkable discovery of CP violation twenty-five years ago. Each of us who has had the privilege of participating in the tumult of physics during that period will want to take advantage of this opportunity to add some reminiscences to Pais' excellent historical summary of events surrounding the event itself.

My own inclination is to emphasize some even earlier events because they have had an important impact on my thinking about this subject. In fact it will be useful to step back in historical perspective in order to clarify the point of physics that I want to make here. In doing so I would like to suggest to those who are concerned with the history of

--------------------

*This work supported in part by the U.S. Department of Energy, Contract No. AC02 80 ER 10587.

physics that they should not only focus on the paths to success, which are of course of unlimited historical value, but should also give some attention to those errors which have offered useful guidance to the correct path.

Although every important event in physics is laid on a foundation of many other earlier events it is sometimes possible to identify a special connection with a single earlier event. In this sense I would say that the precursor that served as the beginning of the story of CP violation was the 1955 paper by Gell-Mann and Pais[1)] on what is now called $K^0,\bar{K}^0$ mixing. I remember being rather thick-headed about that paper when it appeared and that I finally understood it only after some rather emphatic discussion with George Snow who was visiting us at the University of Wisconsin during the summer of '55.

My interest in K mesons was stimulated by experiments of Fry and his collaborators that were being carried out at Wisconsin but the first theoretical work in which I was involved is reported in the 1956 paper[2)] with Sam Treiman on the time development of the K decay modes. As far as I know, this was the first explicit treatment of the $K^0,\bar{K}^0$ interference phenomena. We were led to this investigation by a stimulating discussion with our present chairman (R. Dalitz) at the 1956 Rochester Conference on the interpretation of $K^0$ lifetime measurements.

Experimental results[3)] of my colleagues at Wisconsin and their collaborators which suggested a violation of the $\Delta S = \Delta Q$ rule in $K^0$ decays initiated my involvement with the implications of CP violation. Sam Treiman and I pointed out[4)] that if the transition $\Delta S = -\Delta Q$ did indeed occur in the semileptonic decay of the $K^0$, interference phenomena in those decay processes would provide an excellent measure of CP violation. Of course the occurrence of $\Delta S = -\Delta Q$ transitions was not verified but that question was settled only some years later.

The discovery of CP violation that we are celebrating introduced a deep mystery into physics. CP violation appeared then, and appears now, to have no natural place in the structure of physical theory. This is in stark contrast to our first experience with a failure of a general symmetry principle, the discovery of P violation. In that case Lee and Yang[5)] and, independently, Salam[6)] saw how the chiral invariance of a massless neutrino could offer a natural origin for P violation. The consequent development of the role of chiral symmetry in physics has had profound implications for the structure of theoretical physics, as we all know.

A particular consequence of this theory of P violation was pointed out by Oehme,[7)] namely that C violation must occur to the same degree as P violation and in such a way as to maintain CP invariance. This observation led to the work of Lee, Oehme and Yang[8)] establishing the phenomenology of $K^0$, $\bar{K}^0$ mixing in terms of CP invariance and pointing out some of the ways in which the invariance could be tested.

Landau[9)] pointed out that CP invariance provided a form of "mirror" symmetry which would serve as a natural substitute for separate P and C invariance. But the discovery of CP violation destroyed this notion of a natural symmetry leaving us with a very uncomfortable situation. The only way we theorists seem to be able to account for CP violation is to introduce a model including a relative phase that manifests CP and T violation. Unfortunately the CP-violating phase is just another ad hoc parameter. Although some attempts have been made to fix the magnitude of this phase by using the notion of "maximal CP violation," an analogue of the maximal P and C violation characterized by chiral symmetry, it has been found that this notion is highly ambiguous and does not serve much purpose.

This situation was summarized nicely by Pais in his remark this morning that with the discovery of CP violation we were left with "nothing to replace the simple and elegant concept of CP symmetry" which had restored our confidence in symmetry principles after the fall of P invariance. Although I have shared this view of the CP situation for these 25 years, very recently I have come to the conclusion that the situation may not be as inelegant as we thought.

It is now my opinion that the very success of the theory of P violation, which led us to consider CP invariance to be the "natural" surrogate for mirror symmetry, has also led us to overlook another possible surrogate, namely, PT invariance. In fact in the context of the improper transformations associated with the Lorentz group, PT invariance might be considered the more natural surrogate since it is a symmetry of space-time. Of course, the requirement of PT invariance in the presence of P violation implies that there is compensating T violation (in place of the compensating C violation associated with CP invariance).

Although the theory of weak interactions is clearly dominated by nearly compensating P and C violations, we are familiar with another class of theories in which P violation is counterbalanced by T violation so as to maintain PT (and C) invariance. The example I have in mind is QCD for which the so-called "strong CP violation" is generated by a P-violating, T-violating, PT-invariant interaction of the form $\vec{E}_c \cdot \vec{B}_c$ where $\vec{E}_c$ and $\vec{B}_c$ are the color (gluon) electric and magnetic fields respectively. Another (and related) interaction is that generated by "imaginary" axial vector quark currents, that is spinor currents of the form $i\bar{\psi}_1 \gamma_5 \gamma_\mu \psi_2$. It is also a C-invariant, P- and T-violating interaction. I shall refer to such interactions as "Class II interactions." "Class I interactions" will denote P-violating, C-violating, CP-invariant interactions.

The two classes of theory based separately on Class I or Class II interactions lead to quite different experimental results. Neither one of them by itself leads to the results providing current evidence of CP violation. All existing evidence is based on some form of charge asymmetry which occurs only when there is C violation. In the examples with which we are familiar, both C violation <u>and</u> CP violation are needed. Although a Class II interaction is T violating and consistent with CPT invariance by itself it will <u>not</u> lead to the typical phenomena of CP violation.

The test of a Class II interaction is to be found in experiments measuring T violation (as well as P violation) directly. The most direct of these is the measurement of the electric dipole moment of a particle having non-zero spin. The existence of an electric dipole moment requires <u>both</u> P and T violation but is unrelated to the issue of C invariance (except that the CPT theorem guarantees that particle and antiparticle have equal and opposite electric dipole moments).

Because Class II interactions allow the possibility of T violation without the associated charge asymmetry phenomena characterizing CP violation, I believe it to be important to distinguish between the two concepts and prefer to use "CP violation" to refer only to those effects that can be associated with a charge (or particle-antiparticle) asymmetry. In this sense the term "strong CP violation" should be replaced by "strong T violation" because it is strictly a Class II effect.

Unless there is a violation of the CPT theorem, interactions that lead to CP violation (<u>i.e.</u>, cause a charge asymmetry) must be a linear combination of Class I and Class II interactions. Their relative strength determines the CP-violating phase but it is just the strength of the Class II interaction that determines the measure of T violation.

It may appear that the standard KM model stands in contradiction to the above conclusion because the weak quark currents are chiral currents carrying a common phase for each flavor combination. Therefore there is no phase difference between vector and axial vector currents as is required to separate Class II from Class I. However the effective weak interaction associated with a quark of a given flavor is generated by the second order effect of a W-boson loop. The sum over flavors of the virtual quarks in the loop includes terms with different phases and provides the required mix of Class I and II interactions.

I may summarize these remarks by noting the following distinctions between the experimental realizations of the classes of interactions:

Case I:

Experiments demonstrating maximal P and C violation.

Case II:

Electric dipole moment measurements

Motion reversal experiments such as the determination of spin and momentum triple correlations in decays. (Final state interaction effects must be determined and subtracted.)

Case I & Case II Combined:

Particle-antiparticle asymmetries. Includes all such measurements of CP violation.

Certain charge-blind correlations among jets produced in $e^+,e^-$ or $\bar{p},p$ collisions. (See talk by L. Golowich, this conference.)

From this characterization of the different experiments it is evident that, with sufficient effort, a separation of the weak interactions into terms belonging in each category might be possible. The importance of such a separation is that it may provide a clue to the fundamental origin of CP violation.[10)]

## REFERENCES AND FOOTNOTES


1. M. Gell-Mann and A. Pais, Phys. Rev. 97, 1387 (1955).
2. S.B. Treiman and R.G. Sachs, Phys. Rev. 103, 1545 (1956).
3. R.P. Ely, W.M. Powell, H. White, M. Baldo-Ceolin, E. Calimani, S. Ciampolillo, O. Gabbi, F. Farini, C. Filyppi, H. Huzita, G. Miari, U. Camerini, W.F. Fry, and S. Natali, Phys. Rev. Lett. 8, 132 (1962).
4. R.G. Sachs and S. Treiman, Phys. Rev. Lett. 8, 137 (1940).
5. T.D. Lee and C.N. Yang, Phys. Rev. 105, 1671 (1957).
6. A. Salam, Nuovo Cimento 5, 229 (1957).
7. Letter R. Oehme to C.N. Yang, see C.N. Yang, Selected Papers 1945-80, W.H. Freeman & Co., San Francisco (1983) p. 32.
8. T.D. Lee, R. Oehme, and C.N. Yang, Phys. Rev. 106, 340 (1957).
9. L. Landau, Soviet Physics JETP 5, 336 (1957); Nucl. Phys. 3, 127 (1957).
10. For a recent related remark see P. Castaldi, J.-M. Frére, and G.L. Kane, Phys. Rev. D39, 2633 (1989).

# $\phi_{00} - \phi_{+-}$

DEBU Pascal
CEN Saclay
DPhPE
91191 Gif sur Yvette Cedex
France

No written contribution received.

*from left : Jack Steinberger, Louis Jauneau, Heinrich Wahl, Yvonne and Pascale Jauneau.*

# AN EXPERIMENTAL TEST OF CPT INVARIANCE IN THE NEUTRAL KAON SYSTEM

CERN-Edinburgh-Mainz-Orsay-Pisa-Siegen Collaboration*)

Presented by
V. Gibson
CERN, Geneva, Switzerland

**ABSTRACT**

A measurement of the phases of the CP violating $K \rightarrow 2\pi$ amplitudes provides a unique tool with which to test the conservation of CPT. New results from NA31 for the phases, $\phi_{00}$ and $\phi_{+-}$, and in particular their difference, $\Delta\phi = \phi_{00} - \phi_{+-}$, are reported and the consequent improved limits on CPT violation are discussed.

*) see reference 1.

## 1. INTRODUCTION

CP violation is well established in the neutral kaon system and can be accommodated within quantized field theories such as the Standard Model[2,3]. The CPT theorem states that a local field theory which is Lorentz invariant is also invariant under the combined operation CPT[4,5]. A consequence of CPT symmetry is that the mass, lifetime and the magnitude of the magnetic moment for a particle and its antiparticle are equal. The neutral kaon system provides a sensitive test of CPT due to the small mass difference between the short and long lived $K^0$ states,

$$m_L - m_S = (3.521 \pm 0.014) \times 10^{-15}\,\text{GeV} .$$

The interference of CP violating $K \to 2\pi$ decays with CP conserving $K \to 2\pi$ decays provides a unique tool for the study of CPT violating effects in the neutral kaon system. It is convenient to define the CP eigenstates

$$K_1 = \frac{1}{\sqrt{2}}\,(K^0 + \overline{K^0}) \qquad \text{with} \qquad CP = +1$$

$$K_2 = \frac{1}{\sqrt{2}}\,(K^0 - \overline{K^0}) \qquad \text{with} \qquad CP = -1 .$$

The weak interaction was thought to be invariant under the operation CP until 25 years ago, when CP violation in the kaon system was discovered[6]. The physical particles are the mass eigenstates and can be written as

$$K_S \approx K_1 + \epsilon K_2 \quad \text{and} \quad K_L \approx K_2 + \epsilon K_1 ,$$

where $\epsilon$ is a measure of the CP violation induced by kaon state mixing. CP violation may also occur in the direct decay of the $K_2 \to \pi\pi$ with relative amplitude $\epsilon'$. The first evidence for direct CP violation has recently been reported by this experiment[7].

The CP violating parameters $\eta_{+-}$ and $\eta_{00}$ are defined as the ratios of the CP violating and allowed amplitudes for the $K_L$ and $K_S$ decays into two charged and neutral pions respectively. The relevant experimental measurements are the magnitude and phases of these two parameters.

$$\eta_{+-} = \frac{\langle \pi^+\pi^-|T|K_L\rangle}{\langle \pi^+\pi^-|T|K_S\rangle} = |\eta_{+-}|\, e^{i\phi_{+-}}$$

$$\eta_{00} = \frac{\langle \pi^0\pi^0|T|K_L\rangle}{\langle \pi^0\pi^0|T|K_S\rangle} = |\eta_{00}|\, e^{i\phi_{00}} .$$

These amplitude ratios consist of contributions from CP violation in the kaon state mixing ($\epsilon$) and from CP violation in the direct $K_2 \to 2\pi$ decay ($\epsilon'$) and can be written as

$$\eta_{+-} = \epsilon + \epsilon', \qquad \eta_{00} = \epsilon - 2\epsilon' .$$

The phase of $\epsilon$ is given by unitarity to be

$$\phi_\epsilon \approx \tan^{-1}\left(\frac{2(m_L - m_S)}{\Gamma_S}\right) = 43.7 \pm 0.2° .$$

The parameter $\epsilon'$ is proportional to the ratio of the amplitudes, $A_0$ and $A_2$, for the decay into isospin 0 and 2 two pion states:

$$\epsilon' = \frac{i}{\sqrt{2}} \, \mathrm{Im}\left(\frac{A_2}{A_0}\right) e^{i(\delta_2 - \delta_0)} .$$

As a consequence of CPT, the phase of $\epsilon'$ is entirely determined by the difference in the s-wave $\pi\pi$ scattering phase shifts, $\delta_0$ and $\delta_2$, for the isospin 0 and 2 states respectively:

$$\phi_{\epsilon'} = (\delta_2 - \delta_0) + \frac{\pi}{2} .$$

At the time of this conference, the experimental status of the CP violating parameters could be summarized as follows[8]:

$$|\eta_{+-}| = (2.275 \pm 0.021) \times 10^{-3}$$

$$|\eta_{00}/\eta_{+-}| = 0.9899 \pm 0.002 \pm 0.0025$$

$$\phi_{+-} = 44.6 \pm 1.2°$$

$$\phi_{00} = 54.5 \pm 5.3°$$

$$\mathrm{Re}(\epsilon) = (1.630 \pm 0.083) \times 10^{-3}$$

$$\mathrm{Re}(\epsilon'/\epsilon) = (3.3 \pm 1.1) \times 10^{-3}$$

$$\delta_0 - \delta_2 = 45 \pm 15°$$

A natural consequence of the magnitude of $\epsilon'$ is that the magnitude and phases of $\eta_{00}$ and $\eta_{+-}$ are close to that of $\epsilon$. Similarly, one would expect the phase difference, $\Delta\phi = \phi_{00} - \phi_{+-}$, to be no more than a fraction of a degree. The experimental situation is summarized in Figure 1. Both $\eta_{+-}$ and $\eta_{00}$

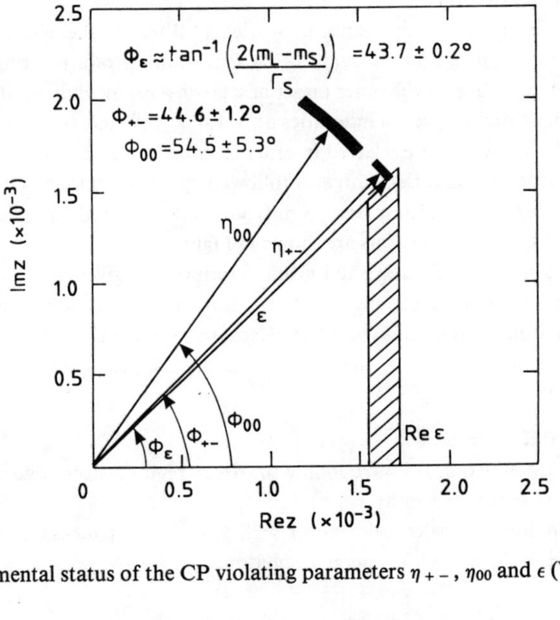


Figure 1 The experimental status of the CP violating parameters $\eta_{+-}$, $\eta_{00}$ and $\epsilon$ (Wu–Yang diagram).

have magnitudes approximately equal to the magnitude of $\epsilon$ and $\phi_{+-}$ is in good agreement with $\phi_\epsilon$. However, the most precise measurement of $\phi_{00}$ shows a two standard deviation discrepancy relative to $\phi_\epsilon$[9]. A direct comparison to $\phi_{+-}$ in the same experiment gives

$$\Delta\phi = 12.6 \pm 6.2° \ .$$

Confirmation of such a large phase difference would imply CPT violation in both the kaon state mixing and decay. Consequently, we have performed a dedicated experiment to measure the phases $\phi_{00}$ and $\phi_{+-}$ and in particular the phase difference $\Delta\phi$[10]. Detailed descriptions can be found in references 11-13.

## 2. PRINCIPLE OF THE MEASUREMENT

In this experiment, the phase $\phi$ of the CP violating parameter $\eta$ is determined from the time dependence of the rate of kaon decays into two pions:

$$I_{2\pi}(t) = S(p)\{e^{-t/\tau_S} + |\eta|^2 e^{-t/\tau_L} + 2\,D(p)|\eta| e^{-t/2(1/\tau_S + 1/\tau_L)} \cos(\Delta m t - \phi)\}$$

S(p) is the kaon production spectrum as a function of the kaon momentum. D(p), the so called dilution factor, is a measure of the initial admixture of $K^0$ and $\overline{K^0}$, $D(p) = (K^0 - \overline{K^0})/(K^0 + \overline{K^0})$. $\tau_S$ and $\tau_L$ are the lifetimes of $K_S$ and $K_L$ and $\Delta m = m_L - m_S$ their mass difference.

The sensitivity of the experiment is optimal in the region where the $K_S$ and $K_L$ decay rates are equal, at about 12 $K_S$ lifetimes. For this reason the beam and target arrangement were chosen for the phase experiment to obtain the maximum acceptance in the interference region. Two target stations were employed, one near to the detector ($K_{NEAR}$) and one far from the detector ($K_{FAR}$). The main advantage of this arrangement is that the acceptance correction is negligible if the ratio of the two target distributions is taken. At the average kaon momentum, ~ 100 GeV/c, the distance between the two targets, 14.4 m, corresponds to a phase shift of about $\pi/2$.

## 3. THE PHASE EXPERIMENT

The design of the interference beams is similar to that of the $K_L$ beam used in the $\epsilon'/\epsilon$ measurement[14]. Kaons of mean momentum ~ 100 GeV/c are produced from a 450 GeV/c proton beam focussed to intercept one or the other target at a downward inclination of 3.6 mrad with respect to the kaon beam line. Nominal proton intensities of $\sim 1.4 \times 10^{10}$ and $\sim 2.0 \times 10^{10}$ protons per pulse on the $K_{NEAR}$ and $K_{FAR}$ targets were selected in order to equalize the $K_L$ rates. The beams are selected by a two stage collimation system. Each target is followed by a first collimator 10 m downstream and a final collimator at a distance of 33.6 m and 48 m from the $K_{NEAR}$ and $K_{FAR}$ targets respectively. The momentum spectra of the accepted events are shown in Figure 2.

The apparatus is shown in Figure 3 and a full description is given in reference 15. Kaons which decay along a 50 m evacuated tank are detected a further 70 m downstream. At the kaon mean momentum, 1 $K_S$ lifetime corresponds to ~ 5 m. Decays into charged and neutral pions are detected concurrently. The principal features of the charged and neutral pion detection can be summarized as follows:

- Neutral Decays, $K \rightarrow 2\pi^0 \rightarrow 4\gamma$.
  - four ring-shaped anticounters surrounding the decay region detect large angle photons and thus reduce the $K \rightarrow 3\pi^0$ background;
  - a liquid argon/lead sandwich calorimeter with strip readout measures photons with < 1 mm position and $\pm 7.5\%/\sqrt{E}$ (GeV) energy resolution;

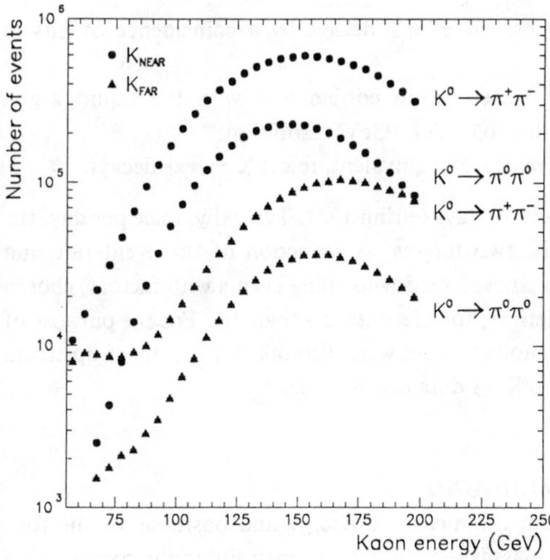


Figure 2  Decay spectra for $K \rightarrow 2\pi$ decays from the $K_{NEAR}$ and $K_{FAR}$ targets.

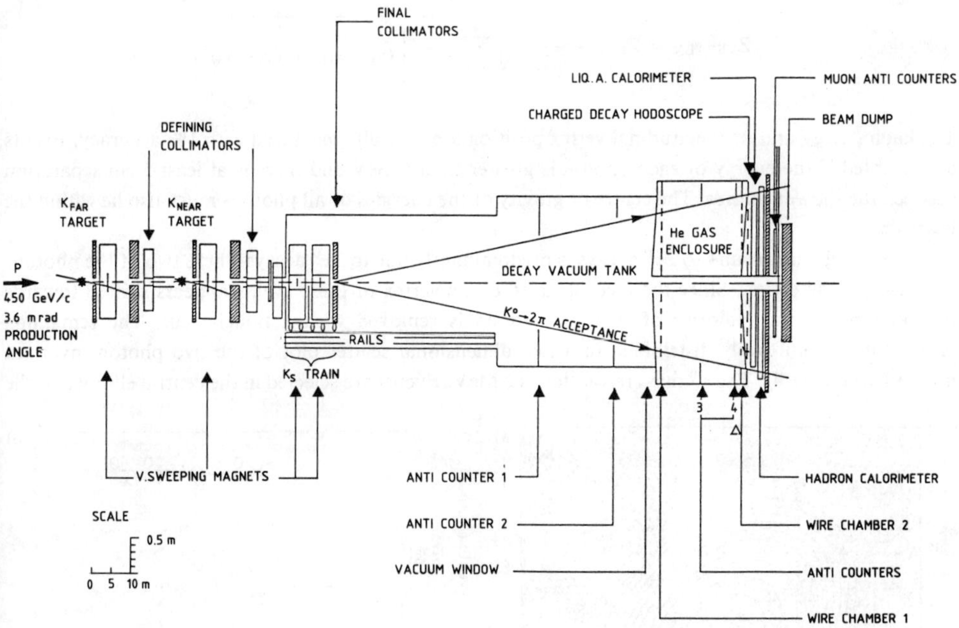


Figure 3  The Phase Experiment.

- a plane of scintillation counters, installed in the liquid argon after 13 radiation lengths of material, triggers on neutral decays.

Charged Decays, $K \rightarrow \pi^+\pi^-$.

- 2 wire chambers, with ± 0.5 mm resolution in each projection, spaced 25 m apart in a helium enclosure, track charged pions;

- a hodoscope of scintillation counters triggers on $\pi^+\pi^-$ decays by a coincidence of hits in opposite quadrants;
- an iron/scintillator sandwich calorimeter measures, in conjunction with the liquid argon calorimeter, the energy of charged pions with $\pm 65\%/\sqrt{E}$ (GeV) resolution;
- two planes of scintillators, after a total of 3 m of iron equivalent, reject $K \to \pi\mu\nu$ decays.

Data were taken over a period of approximately 70 days during 1987. Typically, once per day, the running conditions were interchanged between the two targets. A reduction of the event rate and increased statistics in the interference region were achieved by downscaling events with factors, chosen to give about the same number of events per $K_S$ lifetime, for lifetimes less than $7\tau_S$. For the purpose of the energy scale calibration, pure $K_S$ data from a mobile target were also taken at the most upstream end and centre of the decay region after each $K_{NEAR}/K_{FAR}$ data taking cycle.

## 4. NEUTRAL RECONSTRUCTION AND BACKGROUND

$K \to 2\pi^0 \to 4\gamma$ decays are reconstructed from the measured energy and positions of the four photons in the liquid argon calorimeter. The kaon mass is then used as a constraint in the computation of the distance of the decay vertex, $Z_{VERTEX}$, from the position of the calorimeter $Z_{CAL}$:

$$Z_{VERTEX} = Z_{CAL} - \frac{1}{M_K} \sqrt{\sum_{i,j<i}^{4} E_i E_j \left[ (x_i - x_j)^2 + (y_i - y_j)^2 \right]} \quad .$$

The kaon energy and its longitudinal vertex position are typically measured with 1% accuracy. Events are accepted if the energy of each photon is greater than 5 GeV and there is at least 5 cm separation between the shower centres. The centre of gravity of the energies of all photons must also lie within the beam region.

The small background to $2\pi^0$ decays is predominantly due to $3\pi^0$ decays where two of the photons have escaped detection. Since the reconstructed decay vertex of these events appears shifted towards the calorimeter, the majority of this background is removed with a fiducial cut. The remaining background is uniformly distributed in a two dimensional scatter plot of the two photon invariant masses (see Figure 4). The $\pi^0$ mass resolution is 2 MeV. Events are selected in the central ellipse and the

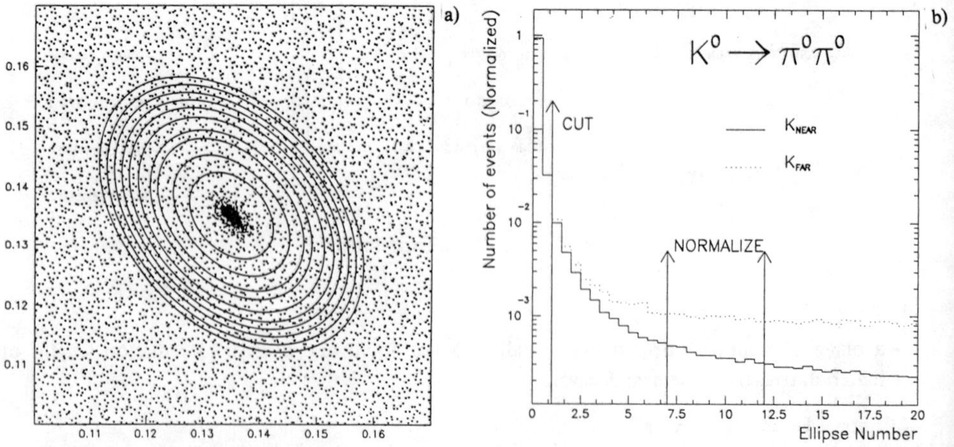


Figure 4 Two photon invariant mass combinations for $K_L$ decays and the number of accepted 4 $\gamma$ events as a function of ellipse number.

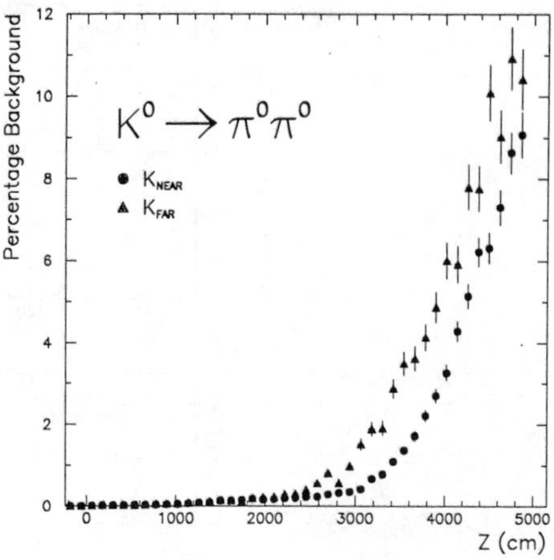


Figure 5 Percentage background as a function of decay vertex for neutral events.

background is subtracted by a flat extrapolation into the signal region from outer ellipses of equal area (see Figure 4). The amount of background subtracted, 0.1% for $K_{NEAR}$ and 0.3% for $K_{FAR}$, is different due to the relative number of $K_L$ and $K_S$ decays in each beam. Figure 5 shows the percentage background subtracted as a function of the position of the decay vertex. The total effect of the neutral background subtraction is to reduce $\phi_{00}$ by 0.6 degrees.

A change in the energy scale of the calorimeter has the effect of shifting the vertex position of the decay and in addition changing the calculated lifetime ($\tau_S \sim Z_{VERTEX}/p$). An uncertainty of $10^{-3}$ in the energy scale is equivalent to an error of 1° on the phase. The absolute energy scale is determined from a fit of the vertex distribution to the known position of the anticounter in the pure $K_S$ beam. The result of the anticounter fit is shown in Figure 6 for the two positions of the $K_S$ target, at the start and in the centre of the decay region. An accuracy of $10^{-3}$ in the energy scale is equivalent to a 10 cm position uncertainty at the centre of the decay volume. The energy scale is also checked with charged

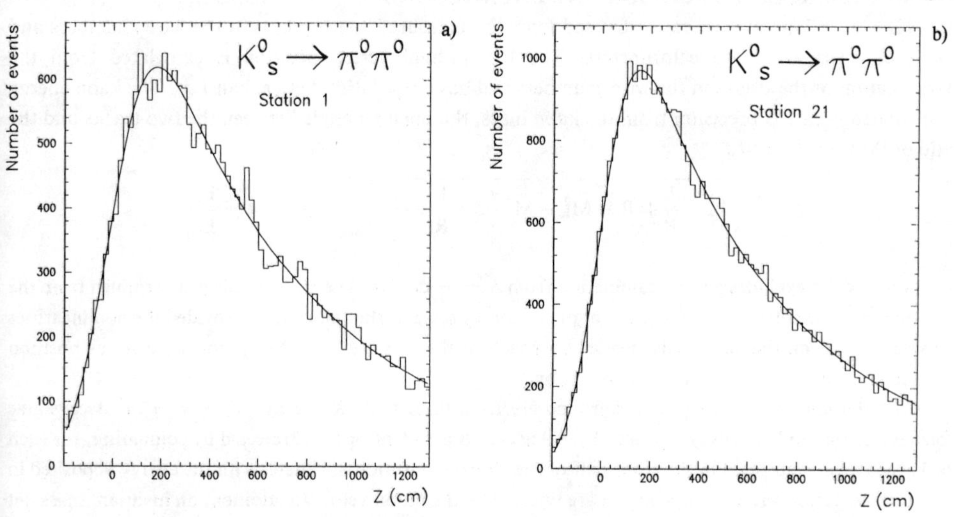


Figure 6 $K_S \rightarrow 2\pi^0$ event distributions, for all $K^0$ momenta, as a function of the distance from the collimator for $K_S$ data taken at the start (station 1) and centre (station 21) of the decay volume.

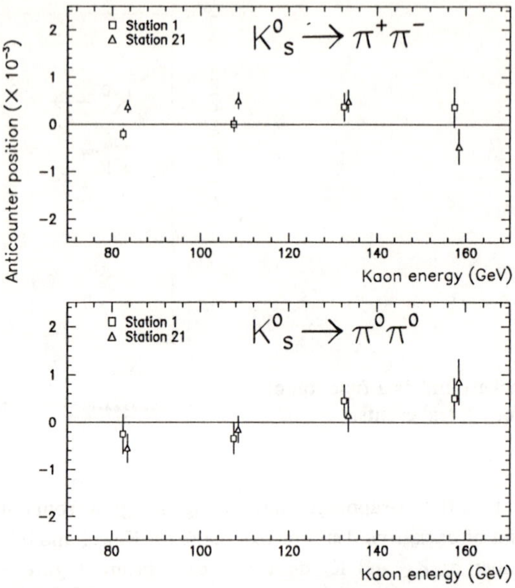


Figure 7 $K_S$ anticounter position as function of the kaon momentum for charged and neutral decays.

decays since the charged energy scale is relatively independent of the calorimeter energy scale (see next section). A comparison of the $K_S$ anticounter position as a function of kaon momentum for charged and neutral decays is shown in Figure 7.

## 5. CHARGED RECONSTRUCTION AND BACKGROUND

$K \rightarrow \pi^+\pi^-$ decays are reconstructed from the measured space points in the wire chambers and the track energies in the calorimeters. The longitudinal vertex position is calculated from the extrapolation of the tracks in the wire chambers and has a resolution better than 1 m. The kaon energy is calculated with 1% precision from the kaon mass, the opening angle between the two tracks and the ratio of the track energies:

$$E_K = \frac{1}{\theta}\sqrt{\left(2+\frac{1}{R}+R\right)\left[M_K^2 - M_\pi^2\left(2+\frac{1}{R}+R\right)\right]} \quad \text{where} \quad R = \frac{E_{\pi 1}}{E_{\pi 2}} .$$

A cut on $R < 2.5$ excludes any contamination from $\Lambda \rightarrow p\pi$ decays. The energy scale is determined from the wire chamber geometry and is independent of the energy scale of the calorimeter provided the nonlinearities are small. A cut on the closest distance of approach of the two tracks at the approximate vertex position guarantees a reasonable reconstruction precision.

Contributions to the charged background are from three body $K_L$ decays. $K \rightarrow \pi^+\pi^-\pi^0$ decays give isolated photons and are easily rejected. $K \rightarrow \pi e \nu$ decays are identified and rejected by comparing, for each track, the energy deposited in the front half of the electromagnetic calorimeter with the energy deposited in the hadron calorimeter. $K \rightarrow \pi\mu\nu$ decays are rejected by the muon vetos. In addition, an invariant mass cut around the kaon mass and a centre of gravity cut are applied. The residual background is subtracted from the shape of the $d_{target}$ distribution which is determined from a sample of $K \rightarrow \pi e \nu$ decays. $d_{target}$ is the perpendicular distance from the target to the decay plane. The control region for the background

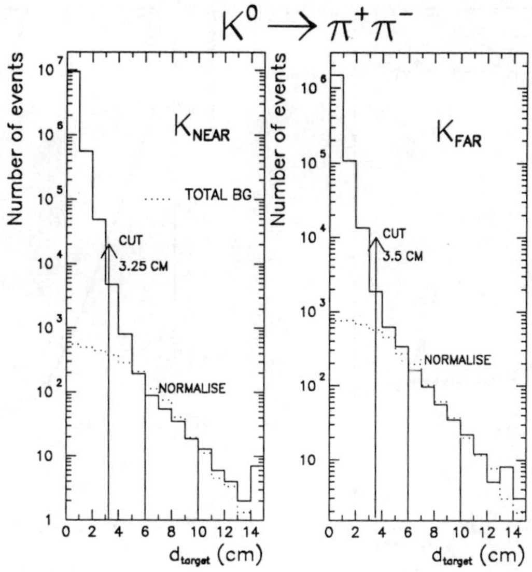


Figure 8 Event distributions for charged decays as a function of $d_{target}$, with signal and background normalization regions indicated.

extrapolation is selected to be 6 cm < $d_{target}$ < 10 cm and the signal region for $d_{target}$ < 3.25 cm and $d_{target}$ < 3.5 cm for the $K_{NEAR}$ and $K_{FAR}$ targets respectively. The $d_{target}$ distributions for $K_{NEAR}$ and $K_{FAR}$ are shown in Figure 8. Background events constitute $\leq 0.01\%$ of the total sample. The total charged background subtraction has the effect of reducing $\phi_{+-}$ by 0.2 degrees.

## 6. RESULTS AND DISCUSSION

The number of events after all cuts and in the range 1.2 m < Z < 49.2 m, 70 GeV < E < 170 GeV are given in Table 1. Statistically, the error on the phase difference is dominated by the number of $K \to \pi^0\pi^0$ decays. A total of $2.1 \times 10^6$ $2\pi^0$ events have been accepted; that is, about 3000 $K_L \to \pi^0\pi^0$ events per $K_S$ lifetime in the interference region, and is a factor 25 more than the previous experiment[9]. The quality of the data is demonstrated in Figure 9 and Figure 10. Figure 9 shows the rate of decay of kaons to charged and neutral pions as a function of $K_S$ lifetime. The combined target distributions have been summed over momentum and the lifetime distribution without interference superimposed. Interference results in an excess of events between 12 and 15 $K_S$ lifetimes and too few events in the region between 8 and 12 lifetimes. For demonstration purposes, it is possible to extract the interference term by subtracting the

**Table 1**

| | $K \to \pi^0\pi^0$ ($\times 10^6$) | $K \to \pi^+\pi^-$ ($\times 10^6$) |
|---|---|---|
| $K_{NEAR}$ | 1.81 | 2.24 |
| $K_{FAR}$ | 0.31 | 0.57 |

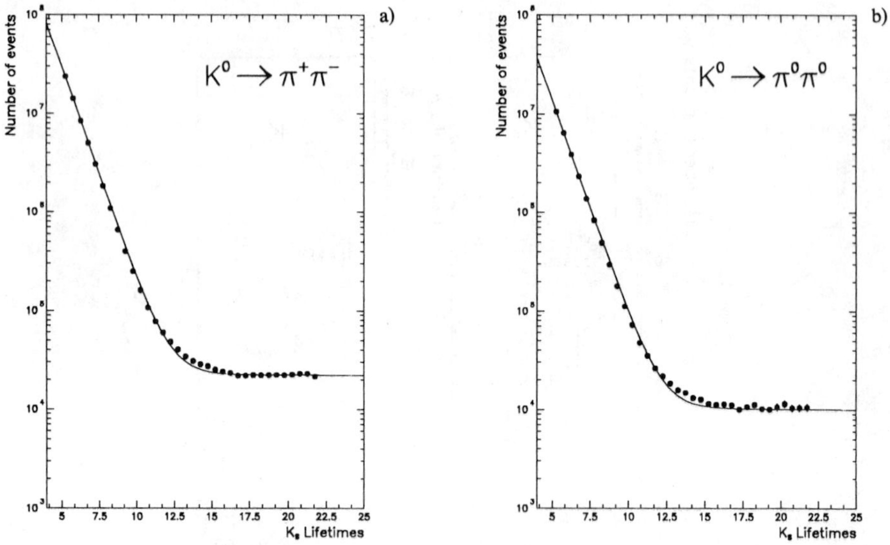


Figure 9 The rate of decay of kaons to charged and neutral pions as a function of the $K_S$ lifetime. Superimposed are the theoretical lifetime distributions without interference.

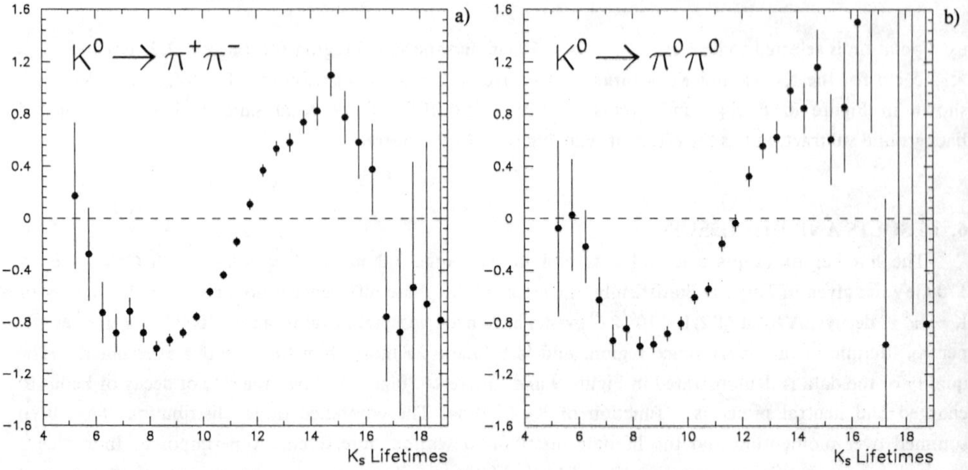


Figure 10 The extracted cosine interference terms.

theoretical distribution from the data. This is shown in Figure 10 and clearly shows the cosine term with frequency $\Delta$m.

The merit of the two target system is that the acceptance correction has a negligible effect on the ratio of intensity distributions. The ratio is calculated in 5 GeV momentum and $\tau_S/2$ lifetime bins, where the number of lifetimes is computed from the mid-point between the two targets. Lifetime limits are also imposed by the finite decay volume. A simultaneous fit to the charged and neutral ratio distributions is performed. The phases, and an unparameterized dilution factor in 10 GeV intervals are varied in the fit. In addition, a charged and a neutral normalization ($K_{NEAR}/K_{FAR}$) is allowed to vary as a function of momentum. The $K_S$ and $K_L$ lifetimes, the CP violating parameter $\eta$ and $\Delta m$ are assumed to be known and fixed to the values given below

$$\tau_S = 0.8923 \times 10^{-10}\,s$$

$$\tau_L = 5.183 \times 10^{-8}\,s$$

$$|\eta_{+-}| = 2.275 \times 10^{-3}$$

$$|\eta_{00}/\eta_{+-}| = 0.9899$$

$$\Delta m = 0.5349 \times 10^{10}\,\hbar\,s^{-1}$$

The result of the fit in a selection of momentum bins for charged and neutral decays is shown in Figure 11. The full effect of interference can be seen when all the momentum bins are combined and scaled to 100 GeV (see Figure 12). The distortion of the ratio distribution due to interference is clearly

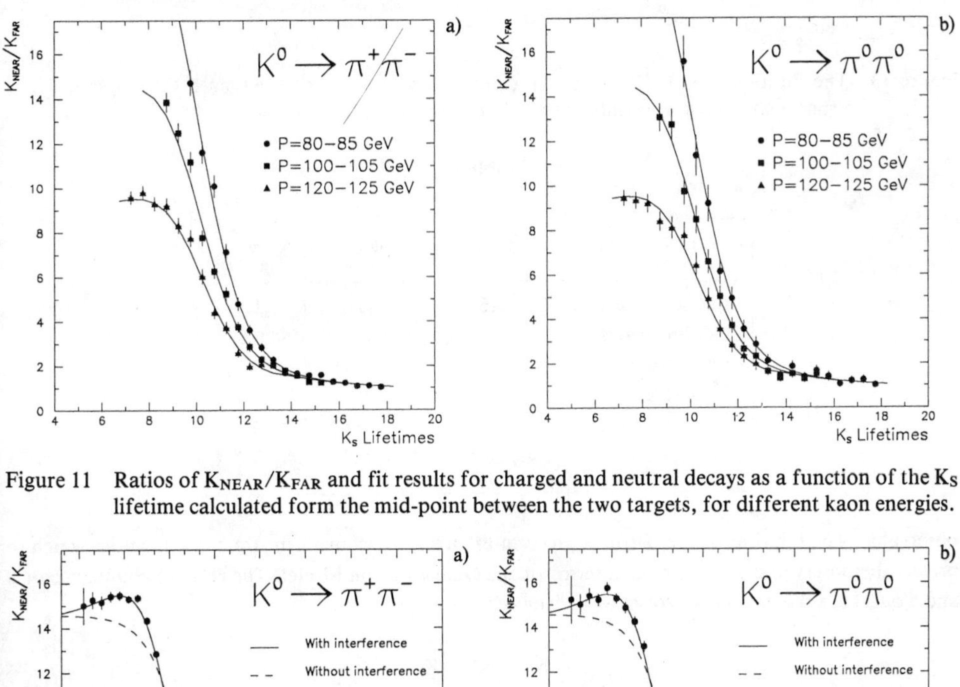


Figure 11 Ratios of $K_{NEAR}/K_{FAR}$ and fit results for charged and neutral decays as a function of the $K_S$ lifetime calculated form the mid-point between the two targets, for different kaon energies.

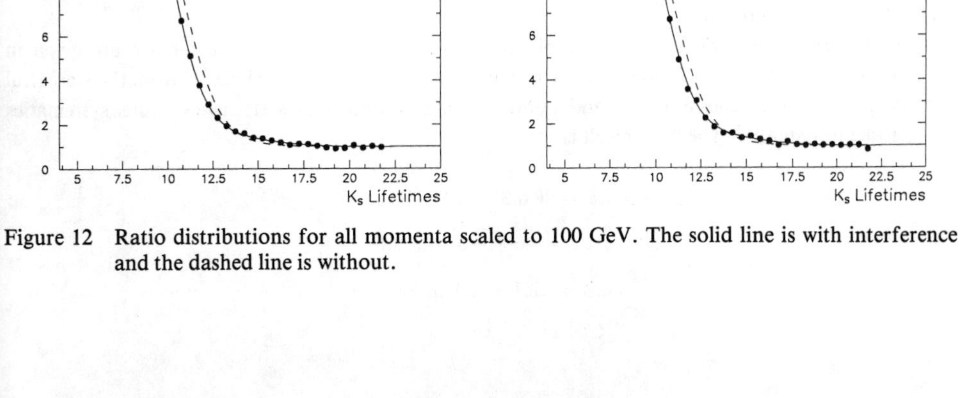


Figure 12 Ratio distributions for all momenta scaled to 100 GeV. The solid line is with interference and the dashed line is without.

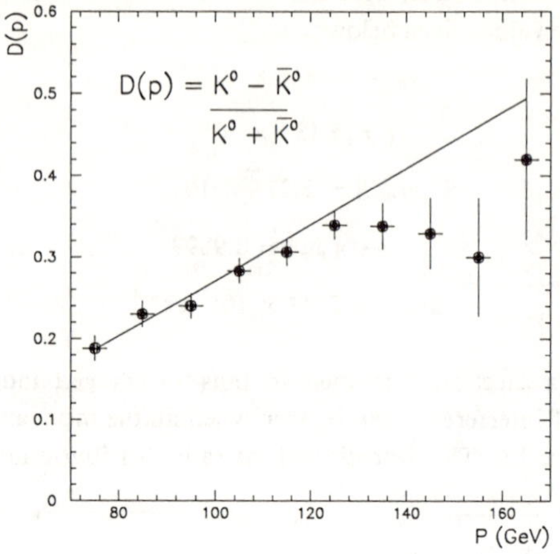


Figure 13 The dilution factor. The curve is a prediction in terms of the Quark Parton Model and experimental measurements of the $K^+/K^-$ production ratio.

**Table 2**

| Contributions | $\phi_{+-}$ | $\phi_{00}$ | $\Delta\phi$ |
|---|---|---|---|
| Energy scale | 0.6 | 0.8 | 1.0 |
| Nonlinearities | – | < 0.5 | < 0.5 |
| Background | 0.1 | 0.1 | 0.1 |
| Regeneration | 0.1 | 0.1 | – |
| MC acceptance | 0.1 | 0.1 | 0.1 |
| Resolution | negligible | 0.2 | 0.2 |

observable. The dilution factor, D(p), is given in Figure 13. A convenient framework within which to predict the observed distribution is in terms of the Quark Parton Model. The relative abundance of u and d quarks in the incoming proton would suggest that

$$D(p) = \frac{K^0 - \overline{K^0}}{K^0 + \overline{K^0}} = \frac{K^+ - K^-}{K^+ + 3K^-}$$

Using the experimental measurement of the $K^+/K^-$ production ratio[16]), this prediction has been superimposed on Figure 13.

The various systematic uncertainties on the absolute phases and their difference are given in Table 2. Of these, the dominant uncertainty is due to the energy scale, which arises from the statistical and systematic uncertainties of the $K_S$ anticounter fits and the anticounter efficiency. Other systematics are considerably smaller. The final result is

$$\Delta\phi = +0.3 \pm 2.6 \pm 1.1°$$

$$\phi_{+-} = 46.8 \pm 1.4 \pm 0.6°$$

$$\phi_{00} = 47.1 \pm 2.1 \pm 1.0° \quad ,$$

where the first error is statistical and the second is the quadratic sum of all the systematic uncertainties. The $\chi^2$ per degree of freedom is 796/768. In addition, there is an extra dependence of the absolute phases on the $K_S$ lifetime and $\Delta m$ due to their own experimental errors. Explicitly, the sensitivities of the absolute phases to these parameters can be written as

$$\phi_{+-} = \phi_{+-} + 0.8° \left(\frac{\tau_S \times 10^{10} - 0.8923}{0.0022}\right) + 1.2° \left(\frac{\Delta m \times 10^{-10} - 0.5349}{0.0022}\right)$$

$$\phi_{00} = \phi_{00} + 0.6° \left(\frac{\tau_S \times 10^{10} - 0.8923}{0.0022}\right) + 1.2° \left(\frac{\Delta m \times 10^{-10} - 0.5349}{0.0022}\right),$$

where the denominators are the one standard deviation errors of the parameters used in the fit.

Comparisons of the phases obtained from this experiment with previous experimental results of comparable precision are shown in Figure 14. The phase $\phi_{+-}$ is in good agreement with previous measurements; and $\phi_{00}$ and $\Delta\phi$ substantially improve the precision achieved by previous experiments.

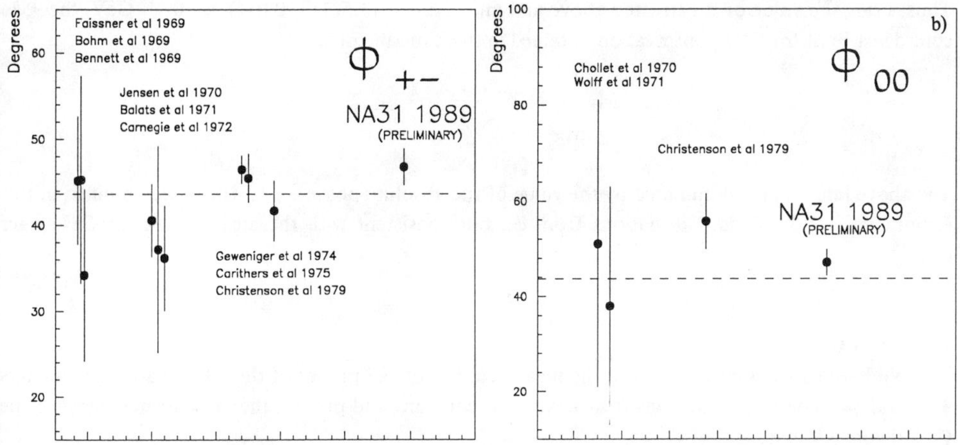


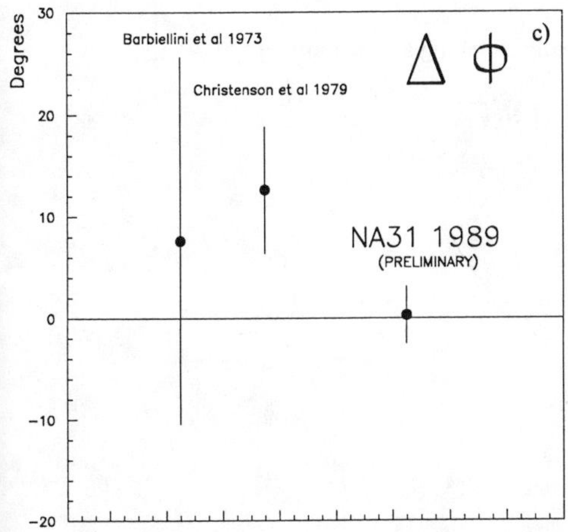


Figure 14 A comparison of $\phi_{+-}$, $\phi_{00}$ and $\Delta\phi$ with previous experimental results. The references are summarized in reference 8.

Since $\Delta\phi$ is consistent with zero, simultaneous CPT violation in the kaon state mixing and in the decay now seems unlikely. Limits on CPT violation can be estimated from the magnitude of the parameter $\Delta$, which can be introduced into the kaon state mixing to allow CP violation with CPT non conservation[17,18],

$$K_S \approx K_1 + (\epsilon + \Delta)K_2 \quad \text{and} \quad K_L \approx K_2 + (\epsilon - \Delta)K_1 \ .$$

In the phenomenology of kaon decay, $\Delta$ can be related to the phases of the CP violating parameters as follows

$$\Delta \approx |\eta| \left\{ \phi_{+-} - \phi_\epsilon + \frac{1}{3} (\phi_{00} - \phi_{+-}) \right\} \ .$$

Using the values of the phases measured in this experiment, $\Delta = (1.3 \pm 0.7) \times 10^{-4}$. Perturbation theory gives

$$\Delta \approx \frac{M_{\overline{K^0}} - M_{K^0}}{2(m_L - m_S)} \ .$$

Thus, using the value of $\Delta$ extracted above and $m_L - m_S = (3.521 \pm 0.014) \times 10^{-15}$ GeV, the 95% confidence limit for CPT conservation obtained in this experiment is

$$\frac{M_{K^0} - M_{\overline{K^0}}}{M_{K^0}} < 3.5 \times 10^{-18} \ .$$

The above limit is now dominated by the value of the absolute phase $\phi_{+-}$. The value obtained in this experiment is 1.5 standard deviations from $\phi_\epsilon$, but consistent with the measurement of Geweniger et al.[19].

## 7. CONCLUSIONS

We have substantially improved the measurements of the phases of the CP violating parameters $\eta_{+-}$ and $\eta_{00}$. The results are consistent with the predictions and provide the most stringent test of the CPT theorem to date.

My personal thanks to Dr. B. Panzer for systematically checking all the numbers.

## REFERENCES

1) G.D. Barr, R. Carosi, P. Clarke, D. Coward, D. Cundy, N. Doble, L. Gatignon, V. Gibson, P. Grafström, R. Hagelberg, G. Kesseler, J. van der Lans, H.N. Nelson, H. Wahl (CERN); R. Black, D.J. Candlin, J. Muir, K.J. Peach (Edinburgh); H. Blümer, R. Heinz, M. Kasemann, K. Kleinknecht, P. Mayer, B. Panzer, B. Renk, H. Rohrer (Mainz); E. Augé, R.L. Chase, D. Fournier, P. Heusse, L. Iconomidou-Fayard, I. Harrus, A.M. Lutz, H.G. Sander, A.C. Schaffer (Orsay); L. Bertanza, A. Bigi, P. Calafiura, M. Calvetti, R. Casali, C. Cerri, G. Gargani, I. Mannelli, A. Nappi, G.M. Pierazzini (Pisa); C. Becker, H. Burkhardt, M. Holder, A. Kreutz, G. Quast, M. Rost, W. Weihs, R. Wertenbach, G. Zech (Siegen).

2) M. Kobayashi and K. Maskawa, Prog. Theor. Phys. **49** (1973) 652.

3) J. Ellis, M.K. Gaillard and D.V. Nanopolous, Nuc. Phys. **B109** (1976) 213.

4) G. Lüders, Ann. Phys. **2** (1957) 1.

5) W. Pauli, 'Niels Bohr and the Development of Physics', Pergamon Press (1955).

6) J.H. Christenson et al., Phys. Rev. Lett. **13** (1964) 138.

7) H. Burkhardt et al., Phys. Lett. **B206** (1988) 169.

8) Particle Data Group, Phys. Lett. **B204** (1988) 1.

9) J.H. Christenson et al., Phys. Rev. Lett. **43** (1979) 1209.

0) P. Clarke et al., CERN/SPSC/86-6 (1986).

1) R. Black, University of Edinburgh thesis, RALT-074 (1988).

2) P. Calafiura, INFN Pisa thesis, (1989).

3) B. Panzer, University of Mainz thesis, (1989).

4) M. Clément et al., SPS/EBS/Note 87-1 (1987).

5) H. Burkhardt et al., Nuc. Inst. Methods **A268** (1988) 116.

6) H.W. Atherton et al., CERN Report, CERN 80-07.

7) T.D. Lee and C.S. Wu, Ann. Rev. of Nuc. Sci. **16** (1966) 511.

8) V.V. Barmin et al., Nuc. Phys. **B247** (1984) 293.

9) C. Geweniger et al., Phys. Lett. **48B** (1974) 487.

*from left : Robert Sachs, George and Cordelia Rose Gollin, Abraham Pais and Ronald Rau.*

# PHYSICS AT THE PLANCK SCALE? A TEST OF CPT INVARIANCE AT FERMILAB BY EXPERIMENT 773

G.D. Gollin
University of Illinois
Department of Physics
1110 West Green Street
Urbana, Illinois 61820 USA

Abstract

It is possible that CPT-violating amplitudes with sizes of order $m_K / m_{Planck}$ contribute to processes involving K mesons. We describe several tests of CPT invariance that could be carried out at Fermilab. One experiment in particular, E773, will perform a precision measurement of the phase difference between $\eta_{00}$ and $\eta_{+-}$ in 1990. Expected precision on $\mathrm{Arg}(\eta_{00}) - \mathrm{Arg}(\eta_{+-})$ is $0.5^{\circ}$.

## Introduction

Reasonable assumptions such as Lorentz invariance and the connection between spin and statistics result in the invariance of physical systems under the combined operations of charge conjugation, parity, and time reversal—CPT. The validity of CPT symmetry at current energy scales has not been seriously contradicted by experiment. Besides establishing relationships among decay amplitudes, CPT invariance guarantees the equality of masses and lifetimes between particles and antiparticles, even if charge conjugation itself is violated. It is here that the most stringent tests of CPT are found. The most powerful of these is $\delta m$, the $K^0$- $\overline{K}^0$ mass difference, inferred from the measured $K_L$-$K_S$ mass difference and other parameters of the neutral-kaon system to be $\delta m / m_K \leq 6 \times 10^{-19}$.[1)]

CPT violations may occur at the Planck mass scale[2), 3)] of $1.2 \times 10^{19}$ GeV/$c^2$ where conditions necessary for CPT symmetry (e.g. locality) might not hold. If these effects do occur in the K system at a scale of $m_K/m_{Planck} = 4 \times 10^{-20}$, current limits on $\delta m$ are greater by less than two orders of magnitude. The prospect of probing Planck- scale physics in kaon decay experiments is tantalizing!

## CP, T, and CPT in the Neutral K System

### Mixing and $\varepsilon$

CP violation has only been observed in the existence of a charge asymmetry in the semileptonic decays $K_L \to \pi\mu\nu$, $K_L \to \pi e\nu$ and in the decay $K_L \to \pi\pi$.[4)] The dominant source of CP violation is a T-violating asymmetry in the transition rates $K^0 \leftrightarrow \overline{K}^0$: $\Gamma(\overline{K}^0 \to K^0) > \Gamma(K^0 \to \overline{K}^0)$. This asymmetry causes both the (nonmixing) mass eigenstates, $K_L$ and $K_S$, to contain slightly more $K^0$ than $\overline{K}^0$. Neglecting normalizations,

$$|K_S> \sim (1+\varepsilon)|K^0> + (1-\varepsilon)|\overline{K}^0> \quad \text{and} \quad |K_L> \sim (1+\varepsilon)|K^0> - (1-\varepsilon)|\overline{K}^0>.$$

The $\Delta S = \Delta Q$ rule requires $K^0 \to \pi^- l^+ \nu$ and $\overline{K}^0 \to \pi^+ l^- \overline{\nu}$ so the observed charge asymmetry in a $K_L$ beam is proportional to Re(ε). Note that the same charge asymmetry would be observed in a pure $K_S$ beam. The magnitude and phase of ε are $(2.259 \pm 0.018) \times 10^{-3}$ and $(43.67 \pm 0.13)^\circ$ respectively. [5] Since $CP(|K^0\rangle) = |\bar{K}^0\rangle$ and $CP(|\bar{K}^0\rangle) = |K^0\rangle$, the mass eigenstates are not CP eigenstates. The CP eigenstates $K_1$ and $K_2$ are

$$|K_1\rangle \sim |K^0\rangle + |\bar{K}^0\rangle \text{ and } |K_2\rangle \sim |K^0\rangle - |\bar{K}^0\rangle.$$

In terms of $K_1$ and $K_2$ one may write

$$|K_S\rangle \sim |K_1\rangle + \varepsilon|K_2\rangle \text{ and } |K_L\rangle \sim |K_2\rangle + \varepsilon|K_1\rangle.$$

Because the $|\pi\pi\rangle$ final state from K decay must have CP = +1, a pure $K_L$ beam will yield ππ decays from the beam's small $K_1$ component.

Direct CP Violation and ε′

A second possible source for CP violation would be the existence of a non-zero amplitude for the decay $K_2 \to \pi\pi$. It is necessary that $\text{Amp}(K^0 \to \pi\pi) \neq \text{Amp}(\bar{K}^0 \to \pi\pi)$ if the $K_2$ is to decay into ππ. In particular, the quantity

$$\varepsilon' \equiv \frac{1}{\sqrt{2}} \frac{A_2 - \bar{A}_2}{A_0 + \bar{A}_0} e^{i(\delta_2 - \delta_0)}$$

must be nonzero where

$$A_0 \equiv \text{Amp}(K^0 \to \pi\pi|_{I=0}), \quad \bar{A}_0 \equiv \text{Amp}(\bar{K}^0 \to \pi\pi|_{I=0})$$
$$A_2 \equiv \text{Amp}(K^0 \to \pi\pi|_{I=2}), \quad \bar{A}_2 \equiv \text{Amp}(\bar{K}^0 \to \pi\pi|_{I=2}).$$

$\delta_2$ and $\delta_0$ are final state interaction phase shifts suffered by the pions moving away from the kaon decay vertex. After defining

$$\eta_{+-} \equiv \frac{\text{Amp}(K_L \to \pi^+\pi^-)}{\text{Amp}(K_S \to \pi^+\pi^-)} \text{ and } \eta_{00} \equiv \frac{\text{Amp}(K_L \to \pi^0\pi^0)}{\text{Amp}(K_S \to \pi^0\pi^0)},$$

some algebra reveals that $\eta_{+-} = \varepsilon + \varepsilon'$ and $\eta_{00} = \varepsilon - 2\varepsilon'$. A nonzero ε′ would split the values of the two η's by 3ε′. Searches for this "direct" CP violation through measurements of $Re(\varepsilon'/\varepsilon)$ are underway at Fermilab and CERN; results so far are inconclusive.[6] Since CPT invariance forces $\bar{A}_I = A_I^*$ and since the $\pi\pi$ phase shifts have been well measured, the phase of $\varepsilon'$ is known to be $(48 \pm 8)^o$. $Arg(\eta_{+-})$ has been determined to be $(44.6 \pm 1.2)^o$; recent results from Fermilab E731 and CERN NA31 indicate that $Arg(\eta_{00})$ is within three degrees of $Arg(\eta_{+-})$[5),6)].

CPT violation parameters

Two different avenues for CPT violation suggest themselves. The first is a CP-violating, but T-conserving mixing of $K^0$ and $\bar{K}^0$ to produce the mass eigenstates. This corresponds to a situation where the $K_S$ contains, for example, an excess of $K^0$ while the $K_L$ contains an excess of $\bar{K}^0$. The CPT violating parameter $\Delta$ enters into the description of $K_S$ and $K_L$ like this:

$$|K_S> \sim (1+\varepsilon+\Delta)|K^0> + (1-\varepsilon-\Delta)|\bar{K}^0> \quad \text{and} \quad |K_L> \sim (1+\varepsilon-\Delta)|K^0> - (1-\varepsilon+\Delta)|\bar{K}^0>.$$

Note the sign switch between $\varepsilon$ and $\Delta$ in the $K_L$ expression. The component of $\Delta$ which is perpendicular to $\varepsilon$ in the complex plane, $\Delta_\perp$, corresponds to a $K^0$ - $\bar{K}^0$ mass difference, while the component parallel to $\varepsilon$ corresponds to a lifetime difference.[1), 7)]

A second possible route to CPT violation is through a CPT-violating relationship among the various $K\rightarrow\pi\pi$ decay amplitudes. Because $A_I - A_I^* = 2i\, Im(A_I)$ and $A_I + A_I^* = 2Re(A_I)$, the phase of $\varepsilon'$ would be shifted from its value of $(48\pm8)^o$ by a CPT-disallowed relationship such as $\bar{A}_I \neq A_I^*$.

In terms of $\varepsilon$, $\varepsilon'$, and $\Delta$, $\eta_{+-} = \varepsilon + \varepsilon' - \Delta$ and $\eta_{00} = \varepsilon - 2\varepsilon' - \Delta$. A nonzero value of $\Delta$ will shift both $\eta$'s in the same direction in the complex plane and will split the $K_L$ and $K_S$ semileptonic charge asymmetries. An unusual $\varepsilon'$ phase will split the $\eta$'s apart without affecting the charge asymmetries.

## Experimental investigation of CPT violation

### Overview

A comparison ot the semileptonic charge asymmetries for $K_L$ and $K_S$ decays yields information about Re($\Delta$). Arg($\eta_{+-}$) - Arg($\eta_{00}$) ,when combined with a value of Re($\varepsilon'/\varepsilon$), determines the phase of $\varepsilon'$. Arg($\eta_{+-}$) - Arg($\varepsilon$) and Arg($\eta_{00}$) - Arg($\varepsilon$) provide determinations of $\varepsilon' - \Delta$ and $-2\varepsilon' - \Delta$, respectively. Because of the relationship between $\Delta_\perp$ and $\delta m$, the $K^0$ - $\bar{K}^0$ mass difference, the semileptonic rate study is the most interesting (and perhaps least feasible) of the possible measurements. Physics at the Planck scale (or string compactification scale) might give rise to processes which generate a nonzero $\delta m$, of size $\delta m / m_K \approx m_K / m_{Planck} = 4 \times 10^{-20}$. The current experimental limit is $\delta m / m_K < 6 \times 10^{-19}$, only a factor of 25 larger than $m_K / m_{Planck}$. Note that $\delta m$ will be zero if CPT is conserved while $\Delta m$, the $K_L$-$K_S$ mass difference, is not ($\Delta m = (3.521 \pm 0.014) \times 10^{-6}$ eV).[5] An experiment with this sensitivity would need to record about $10^{12}$ semileptonic decays and could be performed at the proposed Fermilab Main Injector[8]. We will discuss an experiment to measure Arg($\eta_{+-}$) - Arg($\eta_{00}$).

The interpretation of a precision measurement of Arg($\eta_{+-}$) - Arg($\eta_{00}$) as a CPT test is complicated by the fact that the $\eta_{+-}$ - $\eta_{00}$ phase difference can be small, but nonzero, in a CPT-invariant universe. Arg($\eta_{+-}$) - Arg($\eta_{00}$) is known to be within three degrees of zero; Fermilab E773 will measure it to an accuracy of 1/2 degree with first data in 1990. Interpretation of measurements considerably more precise than this would require improved determinations of the phase of $\varepsilon$ and the value of Re($\varepsilon'/\varepsilon$).

### Experimental determination of Arg($\eta_{00}$) - Arg($\eta_{+-}$)

The $\eta_{00}$ - $\eta_{+-}$ phase difference was measured in 1979 by an NYU group[9] to be (12.6±6.2) degrees, two standard deviations from the CPT-preferred small difference. Based on data from a dedicated run, CERN's NA31 has determined the phase difference to be within three degrees of zero; analysis of 20% of E731's 1987 run also shows that Arg($\eta_{00}$) - Arg($\eta_{+-}$) is less than three degrees[6]. Using a new experimental technique, E773 will measure the phase difference to an accuracy of 0.5 degrees. This level of precision requires careful attention to systematic uncertainties, so the new experiment has been designed to minimize sensitivity to various effects which have troubled earlier measurements. E773 is a collaboration of groups from the University of Chicago, Fermilab, Elmhurst College, and the University of Illinois.

Information about the phase of the η's comes from $K_L/K_S$ interference in the $K \to \pi\pi$ rate downstream of a regenerator. Where the regeneration amplitude ρ is the ratio of the $K_S$ and $K_L$ amplitudes after a regenerator, the resulting $\pi\pi$ rate is

$$\Gamma(K \to \pi\pi) = \Gamma(K_S \to \pi\pi) \times [|\rho|^2 e^{-\Gamma_S \tau} + |\eta|^2 e^{-\Gamma_L \tau} + 2|\rho\eta| e^{-(\Gamma_S + \Gamma_L)\tau/2} \cos(\Delta m \tau + \phi_\rho - \phi_\eta)].$$

Here $\Delta m$ is the $K_L$-$K_S$ mass difference, $\phi_\rho$ is the phase of ρ, and $\phi_\eta$ is the phase of η. E773 will run with a pair of beams separated vertically by 10cm; a regenerator is placed in each beam. The "upstream" beam's regenerator is 12m before the fiducial decay volume while the "downstream" beam's regenerator is at the beginning of the decay volume. $K \to \pi^0\pi^0$ and $K \to \pi^+\pi^-$ decays from both beams are recorded simultaneously. Figure 1 is a diagram of the E773 apparatus; it is similar to the E731 detector, described elsewhere in these proceedings.

The regenerator thicknesses and separation are chosen so that the ratios of the magnitudes of the $K_S$ and $K_L$ amplitudes inside the decay volume are comparable in the two beams. The thicker upstream regenerator produces more $K_S$ amplitude than the thinner downstream regenerator, but much of this amplitude has decayed by the time the beam enters the decay volume. As a result, the decay distributions are similar in each beam, reducing systematic uncertainties associated with imperfect modeling of experimental acceptence and resolution. Figure 2 shows a Monte Carlo prediction of the $\pi^0\pi^0$ decay spectra for both beams, including acceptance and resolution effects. After normalization, the two curves are always within ±10% of each other; the steep leading and trailing edges are caused by hardware-defined acceptance edges.

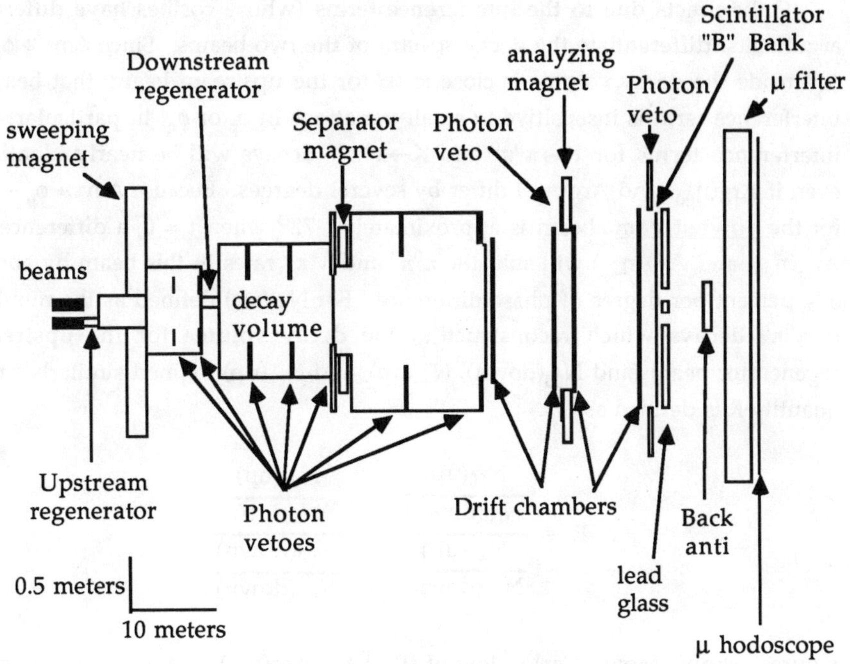


Figure 1. E773: $\phi_{+-} - \phi_{00}$ (elevation view)

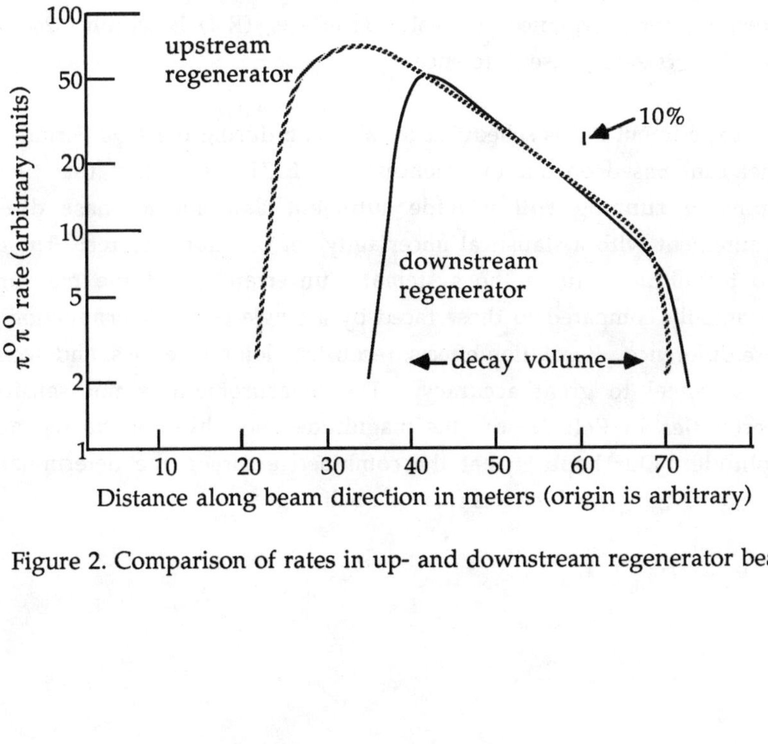


Figure 2. Comparison of rates in up- and downstream regenerator beams

Only effects due to the interference terms (whose cosines have different arguments) differentiate the decay spectra of the two beams. Since $\Delta m\tau + \phi_\rho - \phi_\eta$ inside the decay volume is close to $0^o$ for the upstream beam, that beam's interference term is insensitive to small variations in $\phi_\rho$ or $\phi_\eta$. In particular, the interference terms for $K\rightarrow\pi^0\pi^0$ and $K\rightarrow\pi^+\pi^-$ decays will be nearly identical even if Arg($\eta_{00}$) and Arg($\eta_{+-}$) differ by several degrees. Because $\Delta m\tau + \phi_\rho - \phi_\eta$ for the downstream beam is approximately $-72^o$ when $\tau = 0$, a difference in Arg($\eta_{00}$) and Arg($\eta_{+-}$) will split the $\pi^0\pi^0$ and $\pi^+\pi^-$ rates in this beam by about 3/4 percent per degree of phase difference. For $N_{00}$(up) defined as the number of $\pi^0\pi^0$ decays which reconstruct in the decay volume for the upstream regenerator beam (and $N_{00}$(down), $N_{+-}$(up), and $N_{+-}$(up) defined similarly), the quantity R is defined as

$$R \equiv \frac{\dfrac{N_{00}(\mathrm{up})}{N_{00}(\mathrm{down})}}{\dfrac{N_{+-}(\mathrm{up})}{N_{+-}(\mathrm{down})}} = \frac{\dfrac{N_{00}(\mathrm{up})}{N_{+-}(\mathrm{up})}}{\dfrac{N_{00}(\mathrm{down})}{N_{+-}(\mathrm{down})}}.$$

Figure 3 shows Monte Carlo plots of (R-1) for Arg($\eta_{00}$) - Arg($\eta_{+-}$) values of 0 and 1 degree *vs.* kaon energy. The effectiveness of proper choice of regenerator geometry in reducing systematic uncertainties is obvious: even without corrections for acceptance or resolution effects, (R-1) is within 0.2% of zero when there is zero phase difference.

Experiment 773 is scheduled to take data during the 1990 Fermilab fixed target run. Based on our experiences with E731, we believe that two good months of running will provide sufficient data for a phase difference measurement with a statistical uncertainty of one-third degree. The double ratio technique reduces the systematic uncertainties of the measurement substantially compared to those faced by a single-beam determination of the phase difference. Resolution effects, reconstruction efficiencies, and acceptance effects cancel to great accuracy. The measurement is not sensitive to uncertainties in Re($\varepsilon'/\varepsilon$) or the magnitude and phase of the regeneration amplitude. Our hope is that the combined error on the determination of Arg($\eta_{00}$) - Arg($\eta_{+-}$) will be less than one-half degree.

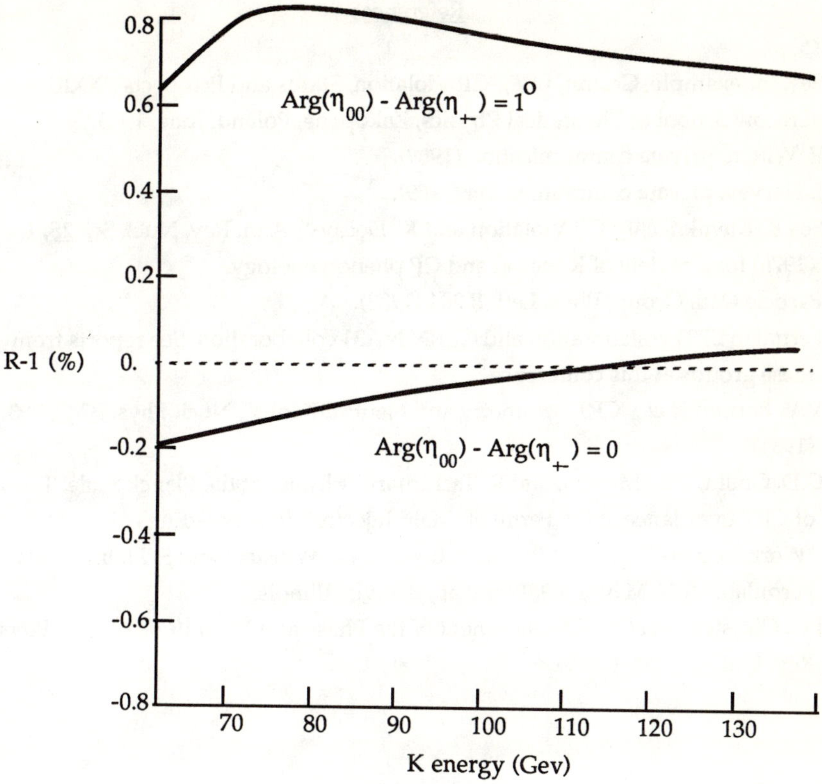


Figure 3. R-1 vs. kaon energy

## Conclusion

It is possible that CPT is not an absolute symmetry of the fundamental interactions. The level at which CPT invariance might be broken is less than two orders of magnitude smaller than the current experimental limit on the size of the $K^0$- $\overline{K}^0$ mass difference. Fermilab E773 will perform a different CPT test in 1990, making a new determination of the phase difference between $\eta_{00}$ and $\eta_{+-}$ to a precision of ±0.5 degree. The method to be employed by E773 is designed to minimize systematic uncertainties.

## References


1) See, for example, Cronin, J.W., "CP Violation, Status and Prospects", XXIII Cracow School of Theoretical Physics, Zakopane, Poland, June, 1983.

2) E. Witten, private communication (1989).

3) J. Harvey, private communication (1989).

4) See K. Kleinknecht, "CP Violation and $K^0$ Decays", Ann. Rev. Nucl. Sci. 25, 1 (1976) for a review of K meson and CP phenomenology.

5) Particle Data Group, Phys. Lett. B 204 (1988).

6) Fermilab E731 collaboration and CERN NA31 collaboration. See reports from these groups at this conference.

7) V.V. Barmin *et al.*, "CPT Symmetry and Neutral Kaons", Nucl. Phys. B247, 293 (1984).

8) G.D. Gollin, P.D. Meyers, and R. Tschirhart, "Physics at the Planck Scale: Tests of CPT Invariance at the Fermilab Main Injector", in proceedings of "Workshop on Physics at the Main Injector", B. Winstein and S. Holmes, eds., Fermilab, 16-18 May, 1989, Fermilab, Batavia, Illinois.

9) J.H. Christenson *et al*, "Measurement of the Phase and Magnitude of $\eta_{00}$", Phys. Rev. Lett. 43, 1209 (1979).

# THE MEASUREMENT OF THE NEUTRON ELECTRIC DIPOLE MOMENT

K.Smith,
School of Mathematical and Physical Sciences,
University of Sussex,
Falmer, Brighton, Sussex,
BN1 9QH, England.

ABSTRACT

The electric dipole moment (E.D.M.) of the neutron requires breakdown of parity conservation and time reversal invariance for a non-zero value. A brief account of the history of the measurement of the neutron E.D.M., including results obtained using ultra-cold neutrons by the group at Lenningrad, Russia, is followed by a more detailed description of the ultra-cold neutron apparatus at the Institute Laue Langevin, Grenoble, France. The analysis of most of the data taken there yields for the E.D.M. the tentative result $-(6\pm4\pm2)\times10^{-26}$ e.cm. where the 4 is an estimate of systematic error and the 2 represents the statistical error. Future plans to eliminate the systematic error are presented.

The relevence of the neutron EDM to CP invariance.

A permanent electric moment (EDM) observed for any fundamental particle would indicate violation of both time reversal symmetry (T) and parity symmetry (P). For practical reasons related to the absence of total charge and the simplicity of the interactions, the neutron is one of the best particles to use to test for T violation assuming P violation to be well established. Values of the dipole length D, which has to be multiplied by e to give the electric dipole moment, vary from $10^{-20}$ cm to $10^{-38}$ cm when P and T violation is included in the various elementary particle theories presented in other sections of this book.

The first experiment to measure the neutron electric dipole moment was set up on a neutron beam at Oak Ridge[1)] in the late 1940's in collaboration with Norman Ramsey of Harvard University, well before the fundamental significance of the measurement was appreciated, and it had yielded by 1951 the value $-(0.1\pm2.4)\times10^{-20}$ cm for the dipole length.

The EDM measurement technique.

Polarised neutrons, with spin + and magnetic moment μ, precess in a magnetic field B with frequency 2μB/h, at 30 Hz in a magnetic field of 10 m.gauss. The frequency change which occurs when an electric field E parallel to B is reversed, 4eD/h, is independent of the applied B field and its measurement, in principle, gives a value for D. A neutron EDM of $10^{-25}$ e.cm. and an E field of 10 kV/cm results in a frequency shift of $10^{-6}$ Hz.

The beam of neutrons at Oak Ridge was polarised by transmission through magnetised iron. The neutrons then passed through a long uniform E and B field region and into a second analysing iron plate which allowed neutrons which had not changed spin direction into a neutron detector. The precession frequency was deduced from the way the detector count rate varied with the frequency of the oscillating fields applied at the beginning and end of the uniform field region.

With increasing interest in symmetry violation several neutron beam experiments shown in Figure 1 were carried out in the USA and England until, finally, an optimised neutron beam experiment was set up at Oak Ridge in the mid 1960's. Data taking completed there by 1969 yielded the result $D < 5\times10^{-23}$ cm. The apparatus was then moved to ILL, Grenoble to take advantage of the more intense beam of slower neutrons (speed 150 m/s) and data taken in the period 1974/75 for what was the last and most precise measurement with a neutron beam, yielded the

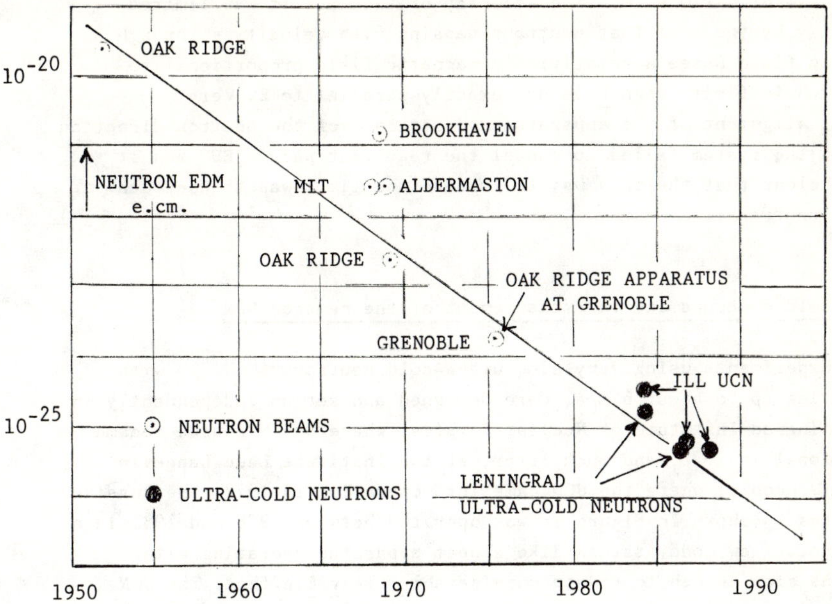


Figure 1. An indication of the way the experimental value for the neutron electric dipole moment has become smaller with time.

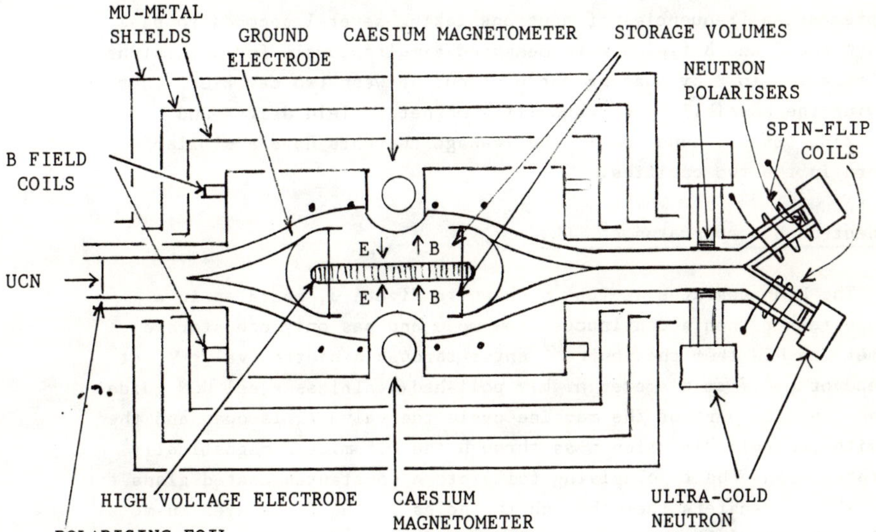


Figure 2. An outline drawing of the ultra-cold neutron EDM apparatus at Leningrad, USSR.

result $D=+(4\pm15)\times10^{-25}$ cm. The accuracy of this result was limited primarily by the fact that neutrons passing with velocity $\underline{v}$ through an electric field $\underline{E}$ see a relativistic magnetic field proportional to $\underline{v}\times\underline{E}$ which is finite when $\underline{v}$ is not exactly parallel to $\underline{E}$. Very careful alignment of the apparatus and reversal of the neutron direction through the system failed to cancel the resultant pseudo EDM and it became clear that the only way to improve the limit was to use much slower neutrons.

## Ultra-cold neutrons and the measurement of the neutron EDM.

Experiments using very slow ultra-cold neutrons (UCN)[2], with velocities up to about 6 m/s, were designed and set up independently in the Leningrad Institute of Nuclear Physics, where a UCN source became operational in 1972, and much later, at the Institute Laue Langevin (ILL), Grenoble, where the UCN came into operation in 1977. The Russian apparatus[3], shown in Figure 2, was operated between 1978 and 1982 in a continuous flow mode, rather like a beam apparatus operating with neutrons slow enough to reduce considerably the $\underline{v}\times\underline{E}$ effect. The UCN bounce elastically off the guides and container walls because of the average force exerted by the nuclei in the surface but, surprisingly, they do not gain energy from thermal motion of the nuclei or loose polarisation as a result of the spin dependence of the nuclear forces. The precession frequencies of neutrons taking several seconds to pass through the E and B fields were measured simultaneously for both signs of E relative to B by sharing the neutrons between two cavities, thereby allowing the possibility of cancelling magnetic field drifts and correcting for a pseudo EDM if the leakage currents have a similar pattern in the two cavities.

## The neutron EDM apparatus at ILL.

The EDM apparatus at ILL[5], shown in Figure 3, operates in a storage rather than a continuous flow mode and has only one storage volume. The UCN from the source[4] enter through a neutron valve $V_1$ at the end of a 6.7 cm diameter highly polished stainless steel UCN guide. During the fill part of the machine cycle the valve $V_2$ is open and the UCN with one spin direction pass through the 0.5 micron magnetically saturated iron-cobalt polarising foil into a constantan coated glass guide which transfers them through the holes in the five-layer mu-metal shield into the 5 litre storage volume. After a fill time of some 10 s the neutron density in the storage volume approaches its maximum value and the neutron valves are closed. The UCN then bounce around with very small probability of loss in the storage volume formed by a 10 cm high,

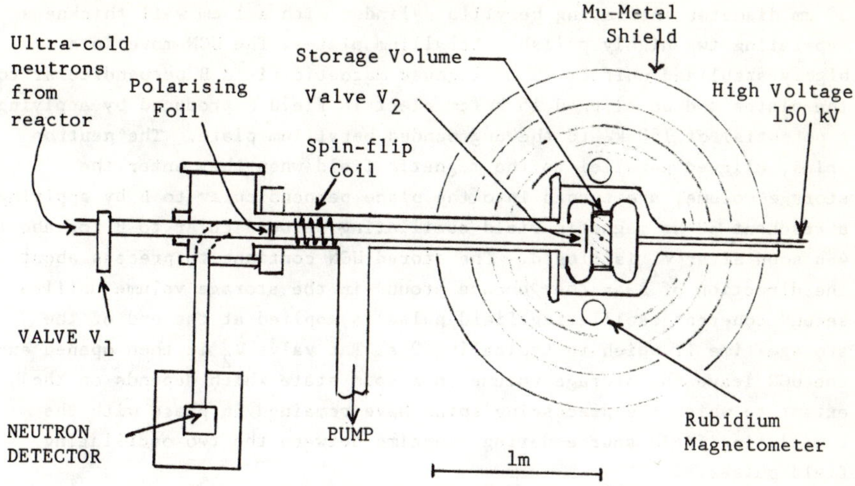


Figure 3. The ultra-cold neutron EDM apparatus at the Institute Laue Langevin, Grenoble, France.

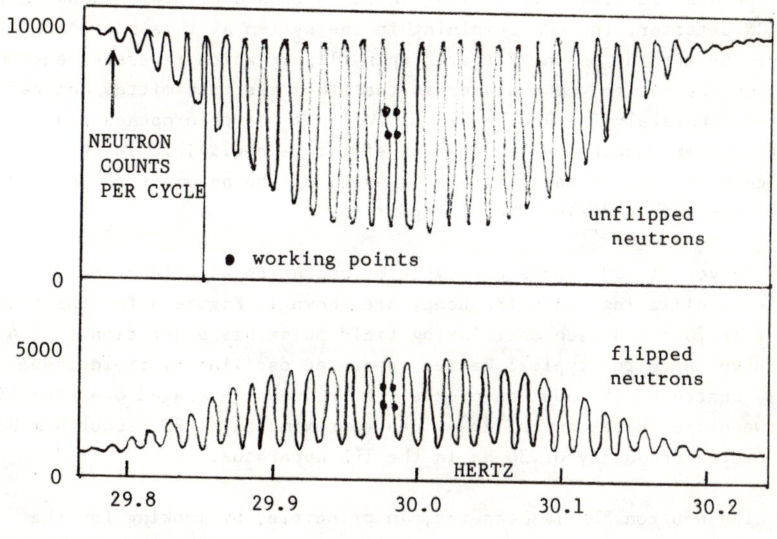


Figure 4. A plot of neutron counts per cycle for both signs of neutron spin as a function of the frequency of the applied oscillating field pulses separated by a storage time T of 70 seconds. The line-width is approximately 1/2T, about 0.006 Hz.

30 cm diameter insulating beryllia cylinder with a 1 cm wall thickness separating two highly polished beryllium plates. The UCN move in a highly stablised uniform 10 milligauss magnetic field B perpendicular to the plates and an aligned 15 kV/cm electric field E produced by applying a potential of 150 kV to the ungrounded beryllium plate. The neutron spins, aligned parallel to the magnetic field when they enter the storage volume, are turned into the plane perpendicular to B by applying a resonant 30 Hz magnetic field oscillating perpendicular to B for about 4 s soon after $V_2$ is closed. The stored UCN continue to precess about the direction of B as they bounce around in the storage volume until a second coherent oscillating field pulse is applied at the end of the storage time T, which is typically 70 s. The valve $V_2$ is then opened and the UCN leave the storage volume in a spin state which depends on the extent to which the precessing spins have remained in phase with the oscillating field source during the time between the two oscillating field pulses.

The UCN pass back along the guide to the iron-cobalt foil which now acts as an analyser by reflecting neutrons arriving with their spin direction reversed. The UCN transmitted follow the guide to a change-over system which directs them into a 50 cm vertical guide where they speed up sufficiently to pass though a thin aluminium window into a $^3$He UCN detector. The UCN remaining in the system at this stage are removed by turning on an adiabatic spin flipper which reverses neutron spins on the way to the analyser so that they are transmitted and can be counted separately for another 10 s. Valve $V_1$ is then opened again to start another machine cycle. The $\underline{v}\mathrm{x}\underline{E}$ effect is negligible in the storage mode because the centre of gravity of the assembly of neutrons moves very little during the storage time.

Curves of UCN counts per cycle of operation as a function of applied oscillating field frequency are shown in Figure 4 for the case when T is 70 s and each oscillating field pulse has a duration t of 4 s. The curves have the typical Ramsey separated oscillatory field shape with a centre at the neutron precession frequency averaged over the time T between the pulses and a line-width approximately 1/2T, about 6 m Hz at a centre frequency of 30 Hz in the ILL apparatus.

The neutron EDM is measured, in principle, by looking for the small change in precession frequency of the order $10^{-6}$ Hz caused by the interaction of the EDM with the applied electric field. In practice, the frequency is set to a working point about half way down the central valley and the counts per cycle are recorded on both sides of the valley for different E field directions. Observations are made at the four working points so that the computer can control the oscillator frequency to follow slow magnetic field drifts and compensate automatically for small UCN source intensity changes. The EDM is deduced by correlating the average count per cycle with the E field direction relative to the B field. The standard deviation in the estimate of D, if it is limited by counting statistics, is $\hbar/2e\alpha ET\sqrt{N}$ where $\alpha=(C_1-C_2)/(C_1+C_2)$, $C_1$ and $C_2$ are the neutron counts per cycle at the peak and trough of the resonance shown in Figure 4 and N is the total number of UCN counted in the period of measurement.

The high voltage is fed to the storage volume through a coaxial system fitted with remote indicators to record the current supplied by the generator and the leakage current into the grounded plate. This current is kept down to about 2 nA at the operating voltage of 150 kV by careful treatment of the plates and insulator, operating at a voltage lower than that applied during the start-up period and maintaining a pressure of $10^{-4}$ torr of helium in the storage volume. The coaxial arrangement minimises the stray pulsed magnetic fields due to occasional sparks, fields which would cause permanent changes in the mu-metal shield magnetisation and hence the B field.

The magnetic field is monitored by three optical pumping magnetometers placed as near as possible to the neutron storage volume and arranged to average the field readings over the storage cycle. The magnetometer data is analysed for correlation with electric field direction in the same way as the neutron counts in order to deduce a 'Magnetometer EDM', an indicator of any pseudo EDM associated with the magnetic fields produced by the leakage currents.

## Neutron EDM Results from Leningrad and ILL.

The neutron EDM results from Leningrad are summarised in Figure 5 so that they can be compared with those from ILL. The analysis of the results for both groups is complicated because magnetic field noise associated with other reactor users, the effects of sparks and the apparatus malfunctions likely to occur when data is taken for many months. Many cuts in the data runs are then necessary and these involve decisions by the analyst. The more recent Leningrad result were obtained after modifying the apparatus to work in a storage mode, still retaining the two cavities.

Since March 1986 nearly $4\times10^9$ neutrons have been detected in the EDM apparatus. All the data taken up to the end of 1988 has been fully analysed and grouped into 13 data sets, each of which covers

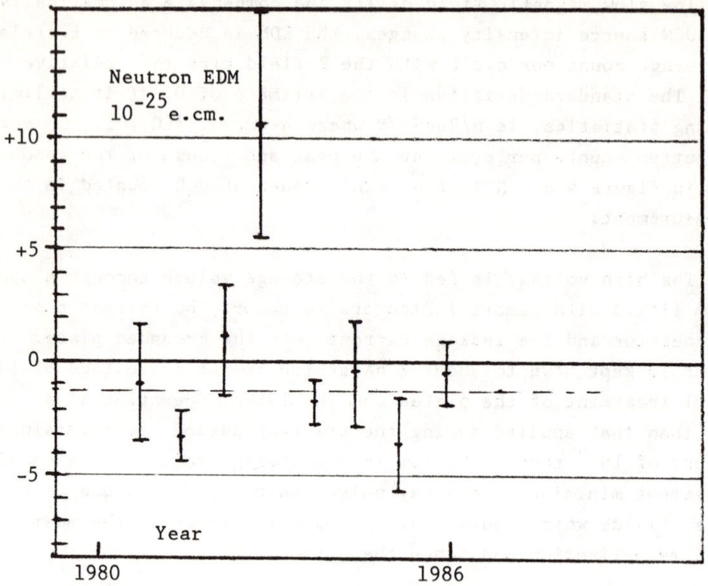


Figure 5. An indication of the results obtained up to 1986 using the ultra-cold neutron EDM apparatus at Leningrad, USSR.

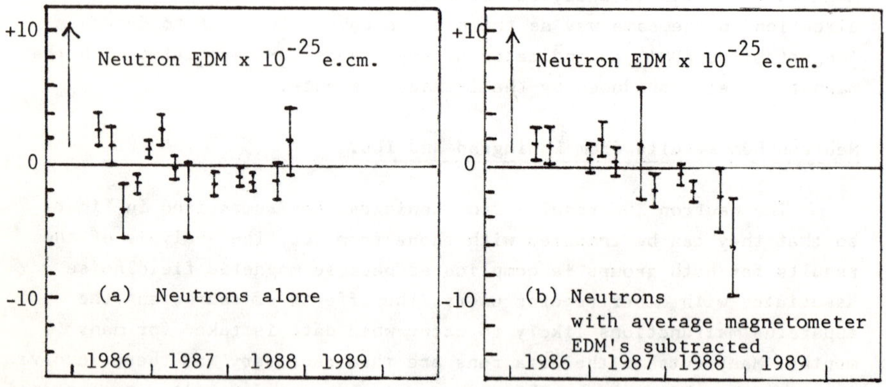


Figure 6. An indication of the results obtained with the ultra-cold neutron EDM apparatus at ILL, Grenoble, France up to the end of 1988.

approximately one reactor cycle. The derived EDM's for each set are shown in Figure 6 for (a) the raw neutron values and (b) the neutron values less the average magnetometer readings. The noise on the raw neutron data on a day to day basis is little above counting statistics but chi-squared for the 13 data sets is close to 4, indicating that the noise due to systematic effects on the time scale of a reactor cycle exists at about twice the amplitude due to counting statistics. Analysis indicates a weak but not very conclusive correlation between the neutron and average magnetometer EDM's, sufficient to use data in Figure 6b to estimate the mean neutron EDM. The statistical error quoted comes from the neutron counting statistics while the systematic error is related to the large value of chi-squared for the neutron data. We believe this systematic noise comes from (1) local magnetic fields due to leakage currents in the bottle and (2) any other E field correlated magnetic field changes.

## Changes proposed for the neutron EDM apparatus at ILL.

Starting in June 1989 we intend to increase the size of the storage volume at ILL to 50 litres to improve statistics by obtaining more neutron counts per cycle. We also intend to install a new magnetometer arranged so that polarised Hg atoms are stored in the same volume as the neutrons. This will allow direct measurement of the space and time averaged magnetic field seen by the neutrons and the possibility of cancelling the effects of any magnetic field changes correlated with the E field direction. Mercury is particularly favourable for this purpose because the electric dipole limit has already been measured by and the techniques for polarisation and detection have been studied in detail by Lamoreaux[6].

## Physicists concerned with the neutron EDM experiment at ILL.

| | | | |
|---|---|---|---|
| J.M.Pendlebury | Sussex, England | N.Ramsey | Harvard, USA |
| K.F.Smith | Sussex, England | B.Heckel | Seattle, USA |
| Y.Chibane | Sussex, England | S.Lamoreaux | Seattle, USA |
| K.Green | RAL, England | W.Mampe | ILL, France |
| A.I.Kilvington | RAL, England | | |
| D.Thomson | RAL, England | | |

References.

1. Dress, W. B. et al, 'Search for an Electric Dipole Moment of the Neutron', Phys. Rev. D15, 9-21, (1976).

2. Golub, R. and Pendlebury, J. M., Ultra-cold neutrons, Rep. Prog. Phys. 42, 439-501, (1979).

3. Altarev, I. S. et al, Search for an Electric Dipole Moment of the Neutron, JETP Lett, 44, 460-465, (1986).

4. Steyerl, A. et al, Phys. Lett. A116,347, (1986).

5. Pendlebury, J. M. et al, 'Search for a Neutron Electric Dipole Moment', Phys. Lett. B136, 327-331, (1984).

6. Lamoreaux, S. K. et al, 'Search for a Permanent Electric Dipole moment of Optically Pumped $^{199}$Hg', Phys. Rev. Lett. 59, 2275, (1987) and S K Lamoreaux, University of Washington thesis.

# CP VIOLATION AND THE NEUTRON ELECTRIC DIPOLE MOMENT

M.B. Gavela*

*Departamento de Física Teórica Universidad Autonoma de Madrid, Canto Blanco (Madrid), Spain*

A. Le Yaouanc, L. Oliver, O. Pène and J.-C. Raynal

*Laboratoire de Physique Théorique et Hautes Energies***
*Université de Paris-Sud, bâtiment 211, 91405 Orsay, France*

Contribution to the Blois Conference (May 1989) to celebrate the 25th anniversary of the discovery of CP violation (presented by O. Pène)

## ABSTRACT

After a brief review of the experimental status of the neutron electric dipole moment $D_n$, we examine the predictions of different theoretical schemes of CP violation: 1) the Standard Model with its possible sources of CP violation, the Kobayashi-Maskawa mixing matrix, and the $\theta$ parameter of QCD; 2) Higgs models of spontaneous breaking of CP; 3) left-right symmetric models; 4) N=1 Supergravity. In all these possibilities we examine the predictions for $D_n$ together with the constraints on these models from the parameters $\epsilon$ and $\epsilon'$ of the kaon system. We underline the general natural features of these theoretical schemes, leaving aside awkward particular models, or fine tuning of the parameters. If non zero values for $D_n$ and for $\epsilon'$ were established at the present level of experimental accuracy , the standard Kobayashi-Maskawa model would be unable to explain the data. One would need for example the Kobayashi-Maskawa phase for $\epsilon$ and $\epsilon'$ and a new flavor-blind CP violating phase that could come, for instance, from the pure supersymmetric sector in N=1 Supergravity. The LR symmetric model of CP violation with two generations could also marginally describe both data for $D_n$ and $\epsilon'$.

* On leave of absence of LPTHE, Université de Paris-Sud, Orsay, France.
** *Laboratoire associé au Centre National de la Recherche Scientifique*

# 1. Introduction

Taking into account the quark binding within the neutron and radiative corrections due to the Standard Model $SU(3) \otimes SU(2) \otimes U(1)$ (and possibly from other interactions beyond it), the electromagnetic coupling of the neutron is described in all generality by the form factors

$$\bar{u}_n(g_V\gamma_\mu - g_A\gamma_\mu\gamma_5 + ig_M\sigma_{\mu\nu}q^\nu - ig_T\sigma_{\mu\nu}q^\nu\gamma_5 + g_Sq_\mu - g_Pq_\mu\gamma_5)u_n \qquad (1)$$

The coupling we are interested with here is $g_T$, on which hermiticity tells us that $g_T = -g_T^*$, and T (or CP if we assume CPT) invariance imposes $g_T = g_T^*$, so that a non-vanishing $g_T$ is a signal of T violation, $g_T$ being purely imaginary, $g_T = iD_n$ with $D_n$ being the neutron electric dipole moment. We have therefore two couplings of the neutron spin to an external electromagnetic field

$$\begin{aligned} &g_M\bar{N}\sigma_{\mu\nu}q^\nu NA^\mu \quad \text{(magnetic moment, P even, CP even)} \\ &D_nN\sigma_{\mu\nu}q^\nu\gamma_5NA^\mu \quad \text{(electric dipole moment, P odd, CP odd)} \end{aligned} \qquad (2)$$

that give, in the static limit, respectively the couplings $\mu_n\sigma.\mathbf{B}$ and $D_n\sigma.\mathbf{E}$ , to the magnetic and electric fields respectively.

Since the pioneering proposal by Purcell and Ramsey in 1950[1)] there has been many attempts[2)] to measure $D_n$ particularly in Grenoble[3)] and Leningrad[4)] . The last reported values form these two experiments using ultracold neutrons are

$$\begin{aligned} D_n &= (-1.4 \pm 0.6)\, 10^{-25}\, e.cm \text{ (Leningrad, 1986)} \\ D_n &= (-0.6 \pm 0.6)\, 10^{-25}\, e.cm \text{ (Grenoble, 1988)} \end{aligned} \qquad (3)$$

To have an idea of the order of magnitude, from $\mu_n \simeq -2.10^{-14}$ e.cm, we see that $D_n/\mu_n \leq 10^{-11}$.

## 2. The Standard Model $SU(3) \otimes SU(2) \otimes U(1)$.

We have two sources of CP violation in the Standard Model: the Kobayashi-Maskawa phase in the flavor mixing matrix[5], and the term $\theta_{QCD}\epsilon_{\mu\nu\rho\sigma}Tr(F^{\mu\nu}F^{\rho\sigma})$ in the QCD Lagrangian[6]. Let us examine which are their consequences for $D_n$.

### 2.1. Kobayashi-Maskawa mixing matrix.

In the Kobayashi-Maskawa model of CP violation, we need to involve the three quark generations to obtain a non-vanishing CP violating effect like $D_n$. Typically, we need two W exchanges[7]. Early order of magnitude estimates including GIM suppressions, the product $s_1^2 s_2 s_3 s_\delta$ that enters in all quantities violating CP, and the light mass $m_d$ since a chiral suppression mass factor is needed for the helicity flipping dipole moment operator, resulted in $D_n/\mu_n \leq 10^{-15}$ , or $D_n \leq 10^{-30}\, e.cm$.

Fig. 1a

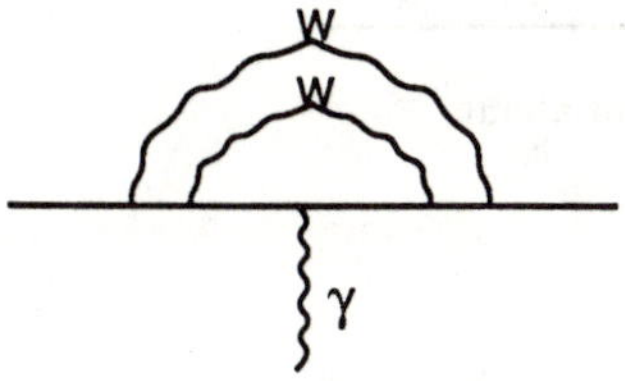


Fig. 1b

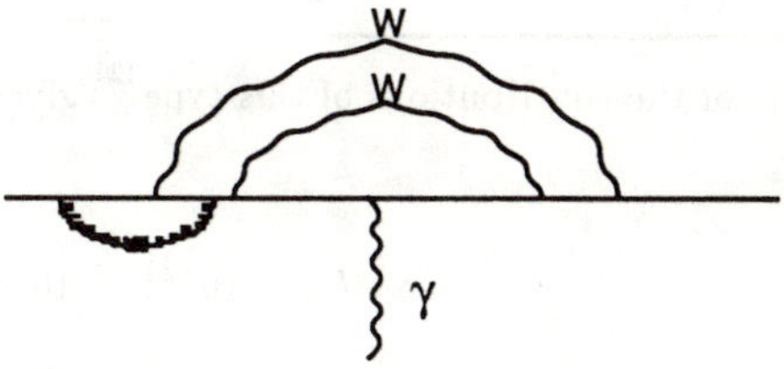


In more detail, we have to consider the **quark electric dipole moment** (Figs.1a, 1b) and contributions to $D_n$ that involve several quarks within the neutron (Figs.2a, 2b). Although the contribution of the pure electroweak two-loop graph (Fig.1a) is claimed to vanish in the static limit[8], gluon radiative corrections (Fig.1b) can give a non-vanishing result[9] giving an order of magnitude (here and in the following we take into account the present expectations for the top mass

$$D_d \quad \lesssim \quad \frac{\alpha_s \alpha_{ew}^2}{(2\pi)^2}\, \frac{m_t^2 - m_c^2}{m^2}\, \frac{m_t^2 - m_u^2}{m^2}\, \frac{m_c^2 - m_u^2}{M_W^2}\, \frac{m_b^2 - m_s^2}{M_W^2}\, \frac{m_d}{M_W^2} s_1^2 s_2 s_3 s_\delta \tag{4}$$

$m^2$ being of the order of $m_t^2$. In (4) we made explicit the minimal type of GIM cancellations needed, that is $D_d$ should vanish whenever two quarks of the same charge become degenerate. For $m_d = m_s$ ($m_d = m_b$) the cancellation would show up through the superposition of the $D_d$ and $D_s$ ($D_b$) contributions. For $D_u$ one gets an analagous formula exchanging up-type and down-type quark contributions. The result is very small: $D_n < 10^{-34}\, e.cm$ .

Among the diagrams that involve several quarks within the neutron, the ones of the type **quark transition electric dipole moment** of Fig.2a also give a very small contribution[10), of the order $D_n = 10^{-34} e.cm$. The probably dominant mechanism within the Standard Model is given by the diagrams of Fig.2b, with a Penguin interaction within the neutron followed by a radiative emission and an internal W exchange to restore the quark flavor[11)]. The intermediate states can be baryons of spin 1/2 with parities + or − and with different flavors. One obtains a ratio between the diagrams of type 2b to those of type 2a of the order

$$\frac{Diagrams(2b)}{Diagrams(2a)} \sim 10\pi\alpha_s \frac{log(m_t^2/m_c^2)}{(m_t^2 - m_c^2)/M_W^2} \sim 10^1 - 10^2 \tag{5}$$

Fig. 2a Fig. 2b

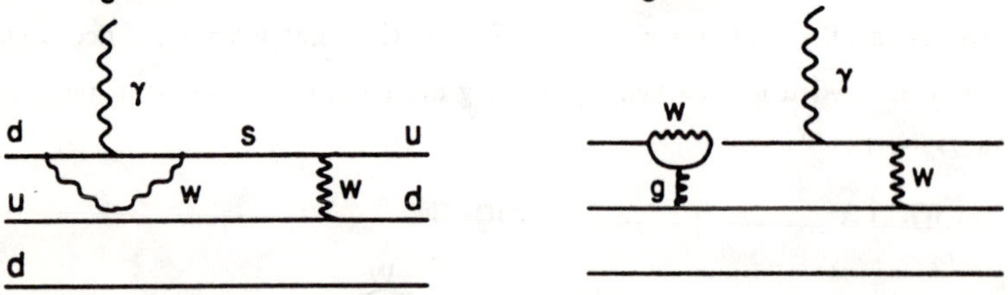


The sum of the contributions of this type[12)] gives a value of about

$$D_n \sim 10^{-31} - 10^{-32}\, e.cm. \tag{6}$$

There are other interesting contributions in the Standard Model. The so-called photopenguin graph[13)] (Fig.3a) turns out to be controlled by the axial anomaly AVV, where the axial vertex comes from the Fierz reordering of the W exchange, and the V vertices are coupled to a photon and a gluon; it is not anomaly within $SU(2) \otimes U(1)$ (that should cancel), but an external anomaly since it involves a gluon. Other contributions involve two W loops on the same quark and gluon exchange between different quarks[14)] (Fig.3b). Both these contributions give a value of about $10^{-33}$ $e.cm$.

Fig. 3a Fig. 3b

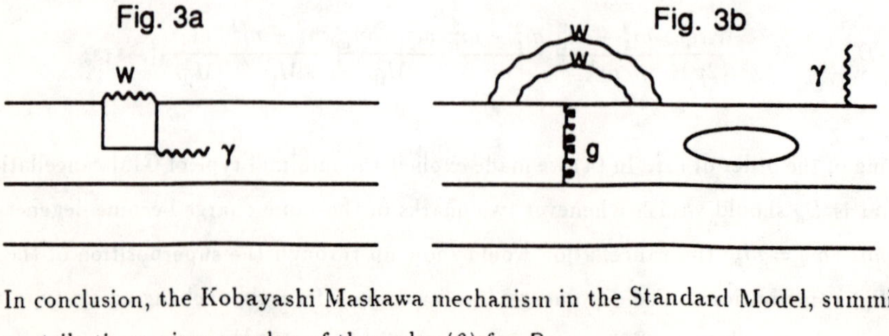


In conclusion, the Kobayashi Maskawa mechanism in the Standard Model, summing up all the contributions, gives a value of the order (6) for $D_n$.

### 2.2. $\theta_{QCD}$ AND STRONG CP VIOLATION.

Due to non-perturbative effects in non-abelian gauge theories, the Lagrangian of QCD has a CP violating term:

$$\mathcal{L} = -\frac{1}{4}Tr(F_{\mu\nu}F^{\mu\nu}) + i\bar{q}\gamma^{\mu}D_{\mu}q + \bar{q}_L\tilde{M}q_R + \bar{q}_R\tilde{M}^{\dagger}q_L - \theta_{QCD}\frac{\alpha_s}{16\pi}\epsilon_{\mu\nu\rho\sigma}Tr(F^{\mu\nu}F^{\rho\sigma}) \qquad (7)$$

where q is a vector in flavor space and M is a general complex mass matrix.

The term $\epsilon_{\mu\nu\rho\sigma}Tr(F^{\mu\nu}F^{\rho\sigma})$ violates CP since it is of the form $\sim Tr(\mathbf{E}.\mathbf{B})$ , C even and P odd. Although formally a total derivative, it modifies the action due to the non trivial topology of QCD through the presence of instantons. It is thus a long distance effect and can only be estimated through non perturbative techniques. Waiting for a direct lattice estimation of the effect of $F\tilde{F}$ on $D_n$, the best approach uses the anomalous Ward identity: Under a $U(1)$ chiral rotation $q \rightarrow (1 - i\delta\theta\gamma_5)q$ , and the Lagrangian (7) transforms like

$$\delta\mathcal{L} = \delta\theta\left(\frac{N_f\alpha_s}{32\pi}\epsilon_{\mu\nu\rho\sigma}Tr(F^{\mu\nu}F^{\rho\sigma}) + i\bar{q}_L\tilde{M}q_R + i\bar{q}_R\tilde{M}^{\dagger}q_L\right) \qquad (8)$$

where $N_f$ is the number of quark flavors. Since a $U(1)$ chiral rotation involves these two terms, one can see, performing a convenient $U(1)$ chiral rotation and diagonalizing the quark mass matrices, that the true observable parameter is not $\theta_{QCD}$ but [15)]

$$\bar{\theta} = \theta_{QCD} + Arg(det\tilde{M}) \qquad (9)$$

Moreover, the $U(1)$ chiral transformation (8) allows to express the pure gauge CP violating term in (7) as a fermion CP violating mass term [16)]

$$\mathcal{L}(CPodd) = i\bar{\theta}\frac{m_u m_d m_s}{m_u m_d + m_u m_s + m_d m_s}(\bar{u}\gamma_5 u + \bar{d}\gamma_5 d + \bar{s}\gamma_5 s) \qquad (10)$$

where three flavors have been considered.

The best experimental limit on the parameter $\bar{\theta}$ is provided by the data (3) on $D_n$. To have an order of magnitude estimate of the contribution that the interaction (10) gives for $D_n$ one considers [16)] pole diagrams of the type of Fig. 4, where the intermediate states are $\frac{1}{2}^-$ baryon resonances coupled to the neutron by the CP violating interaction (10) and having a E1 electromagnetic coupling to the neutron. One obtains an order of magnitude

$$\frac{D_n}{\mu_n} \sim \frac{m_q}{\Delta E}\bar{\theta}\, e \tag{11}$$

where $m_q$ is a quark mass and $\Delta E$ the energy difference between the neutron and the intermediate state . This gives $D_n \sim 2.7\, 10^{-16}\bar{\theta}\, e.cm.$, and from the data (3), one obtains

$$\bar{\theta} < 10^{-9} \tag{12}$$

One obtains a similar estimate from pion loops in chiral perturbation theory[17].

Fig. 4

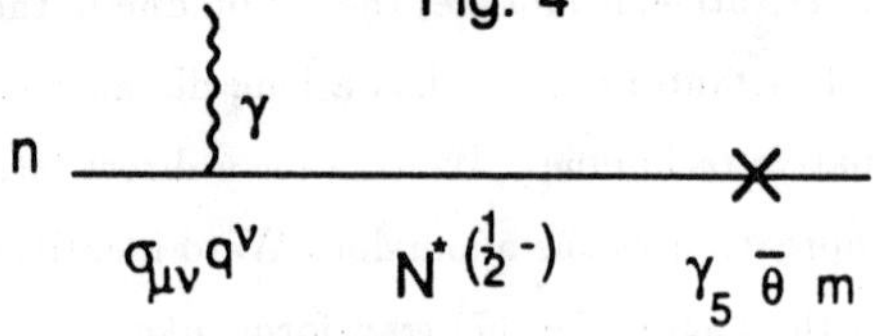


This strong limit indicates that $\bar{\theta}$ must be small to a great accuracy, an unnatural situation. The parameter $\bar{\theta} = \theta_{QCD} + Arg(det\bar{M})$ could be rotated away by a $U(1)$ chiral transformation if $det\bar{M}$ could be made real. A possibility realizing this situation would be if one quark mass, e.g. $m_u$ , would vanish. However, the analysis of pseudoscalar meson masses shows that this possibility is no realized in nature, since $(m_d - m_u)/(m_d + m_u) = 0.28 + 0.03$. Another possibility is the Peccei-Quinn symmetry[18] an extended $U(1)$ symmetry that is spontaneously broken, predicting a new pseudoscalar Goldstone boson, very weakly coupled to matter, the axion. However, this mechanism remains speculative, since for the moment there is no experimental evidence for the axion[19]. Besides some solutions in the context of spontaneous breaking of CP have been proposed[20].

Another recent speculation is a mechanism linked to a recent development of quantum gravity, the baby universe theory, that claims to explain why the cosmological constant is zero)[21], (another well known fine tuning problem). By a similar mechanism, $\bar{\theta}$ could vanish[22], although recent calculations[23] claim to find rather $\bar{\theta} = \pi$.

Finally we would like to make the following remark: in any model, the $\bar{\theta}$ generated by loop corrections (when they happen to be finite) leads to a contribution to $D_n$ of the same order of magnitude than the direct contribution to $D_n$ in the considered model.

## 3. General features of Non-Standard Models.

Before proceeding to specific theoretical schemes beyond the Standard Model, let us give a brief account of their general features. All these models have in common new contributions involving right-handed fermions besides the left-handed ones of the Standard Model. This is equivalent to consider new fermion flavors, and new CP violating phases. This has as a general consequence the fact that $D_n$ can occur at the one-loop level, through diagrams of the type of Fig.5a, where there is a mass insertion on the quark line allowing for helicity flip, and a relative phase between L and R fermions.

Fig. 5a

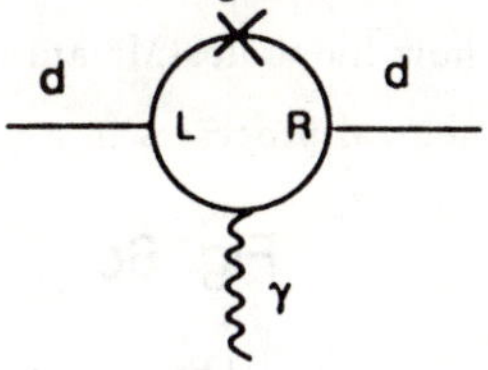


Fig. 5b

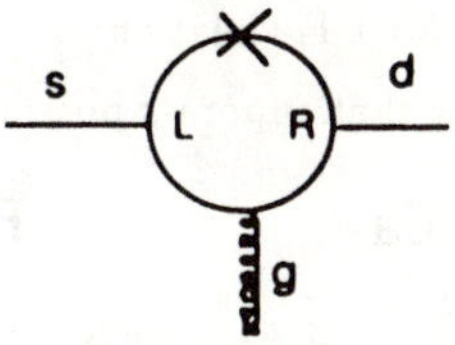


It is important to notice that in all these models there will be a qualitative relation between $\epsilon'$ and $D_n$, because a similar diagram, of the Penguin type (Fig.5b) will contribute to $\epsilon'$ . We will then expect an order of magnitude for $D_n$ ,

$$D_n/\epsilon \sim \mu_n G_F m^2 \epsilon' \sim 10^{-14} cm 10^{-5} 10^{-6} \sim 10^{-25}\, cm \qquad (13)$$

where $G_F$ is the Fermi constant, indicating a one-loop effect, and m is some hadronic mass. Therefore, in non-standard models, $D_n$ will be rather close to the present experimental limits, although more realistic estimations give results that are often smaller than (13) by one to two orders of magnitude, according to the particular model considered.

We see that there are three fundamental differences between these models and the Standard Model: In the Standard model one needs to involve three generations (i.e., **two W exchanges**) to get a non-vanishing CP violating phase. On the contrary, in these models, due to the presence of new phases in the right-handed fermion sector, one can obtain CP violation within two or even within one generation. Then, in these models, one can have CP violation at the level of **one loop**, at the cost of an insertion of the internal mass.

This explains why, beyond the Standard Model, the ratio $D_n/\epsilon'$ can be about $10^{-19}$ *e.cm.*, and is much smaller, of the order $10^{-26}$ *e.cm.* , in the Standard Model.

Let us now review in some detail a few non-standard models, each of them presenting some interesting theoretical features: the **Higgs model of spontaneous CP violation**, the **left-**

**right symmetric models** , where parity is spontaneously broken, and the "standard" N=1 **Supergravity model**. We will only underline the general natural features of these theoretical schemes, leaving aside awkward particular models, or fine tuning of the parameters.

## 4. Higgs model of spontaneous CP violation.

Among the models of spontaneous CP violation due to Higgs fields with complex vacuum expectation values[24] the scheme proposed by Weinberg[25] , with three Higgs doublets, has a number of interesting theoretical features: 1) CP violation can have as a unique origin the spontaneous breaking of the symmetry, and the Kobayashi-Maskawa matrix is real; 2) it gives a natural scale for CP violation; 3) the model has Glashow-Iliopoulos-Maiani cancellations in the Higgs sector that suppress possible flavor-changing neutral processes in a natural way.

Fig. 6a Fig. 6b Fig. 6c

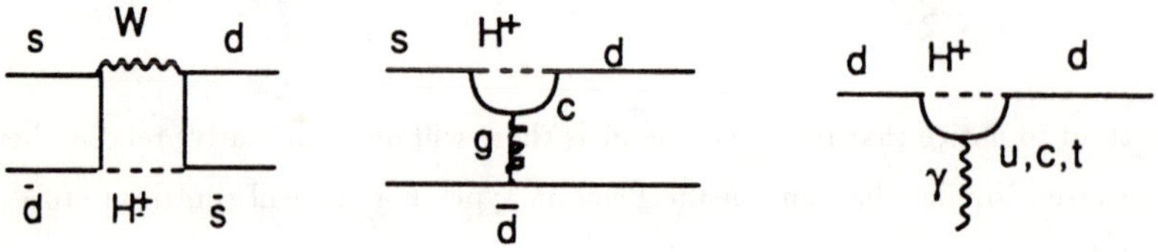


In this model, CP violation occurs only in the Higgs sector, and we have two types of contributions to CP violation: box diagrams with a W and a charged Higgs exchange (Fig.6a), that will contribute to the imaginary part of the mass matrix, i.e. to $\epsilon_m$ ,

$$\epsilon_m = \frac{Im\, M_{12}}{Re\, M_{12}} \quad (\Delta S = 2) \tag{14}$$

and Penguin-type diagrams due to Higgs loops (Fig.6b), that contribute to the CP violation in the decay amplitudes, i.e., to the parameter $\xi$ ,

$$\xi = \frac{Im\, A_0}{Re\, A_0} \quad (\Delta S = 2) \tag{15}$$

The order of magnitude of the physical parameters $\epsilon$ , $\epsilon'$ , is given by

$$\epsilon \sim \xi + \epsilon_m \qquad \epsilon' \sim -\xi/20 \tag{16}$$

where the suppression factor 20 comes from the $\Delta I = 1/2$ rule in kaon decays.

An important general feature of this model is that **Penguin-type diagrams largely dominate the box diagrams**. Therefore the $\xi$ contribution to $\epsilon$ will dominate over the $\epsilon_m$ contribution.

$$\epsilon_m \ll \xi \tag{17}$$

and, fitting $\xi$ to obtain the right magnitude for $\epsilon$ , $\epsilon'$ will, in general, be too large[26) .] Indeed, one finds, as an order of magnitude,

$$\frac{\epsilon'}{\epsilon} \sim -0.05 \tag{18}$$

This value is too large compared to the present experimental NA31 data, and moreover it has the wrong sign, opposite to the sign of the Standard Model.

On the other hand, as emphasized in the preceding chapter, the same type of diagram contributing to $\epsilon'$ will contribute to $D_n$ (Fig.6c), and we will have the order of magnitude,

$$D_n/\epsilon \sim 10^{-25} cm, \tag{19}$$

just as we have seen in (13).

These are the natural features of Higgs models of CP violation. There have been a number of proposals to save the model, correcting the order of magnitude prediction (18).

For instance long distance effects in the estimation of $\epsilon_m$ and $\xi$ were first introduced to solve the problem that the estimation (18) relies on a vacuum saturation approximation with an incorrect chiral limit[27)]. These effects lead to estimates[28,29)] for $\epsilon'/\epsilon$ from $-0.002$ to $-0.045$. Although these effects, hard to evaluate, go in the right direction to correct the large negative value (18), they do not go far enough, and the Higgs model seems to be now excluded by experiment. Note that the prediction (19) remains valid when long distance effects are considered[29,30)].

There is another proposal[31)] involving a definite hierarchy of vacuum expectation values of the three doublets:

$$v_1 \ll v_2 < v_3 \tag{20}$$

Then, the model gives an acceptable value

$$\frac{\epsilon'}{\epsilon} \sim \pm(1-2)\,10^{-3} \tag{21}$$

However, this scheme makes a definite prediction on the relative sign of $\epsilon'$ and $D_n$:

$$sign(\epsilon') = sign(D_n), \tag{22}$$

in contradiction with experiment if both non-null data, NA31 on $\epsilon'$ and Leningrad on $D_n$, turned out to be confirmed.

Finally one should mention a model[32)] with only two Higgs doublets and fined tuned flavor changing neutral currents suppression which leads to superweak-like results similar to what results generally from left-right symmetric model, see (26) and (32).

## 5. Left-right symmetric models.

In left-right symmetric models like $SU(2)_L \otimes SU(2)_R \otimes U(1)$, the simplest extension of the Standard Model, where parity is spontaneously broken, CP violation can have a two-fold origin: 1) Kobayashi-Maskawa-like phases in the mixing matrices of left-handed (L) and right-handed (R) quarks, and 2) spontaneous CP violation due to complex vacuum expectation values of Higgs fields, or both at the same time[33)]. We must again emphasize that, having at our disposal R fermions is equivalent to have new flavors, and therefore new freedom in the phases of the mixing matrices. An important consequence that follows is that one can have CP violation within the first two generations, or even within the first one. In the case of two generations we can have as much as three CP phases, that can be shifted to the R sector Cabibbo mixing matrix. One has the L and R mixing matrices for the first two generations[34)],

$$U_L = \begin{pmatrix} \cos\theta_L & \sin\theta_L \\ -\sin\theta_L & \cos\theta_L \end{pmatrix} \qquad U_R = e^{i\gamma} \begin{pmatrix} e^{-i\delta_2}\cos\theta_R & e^{-i\delta_1}\sin\theta_R \\ -e^{i\delta_1}\sin\theta_R & e^{i\delta_2}\cos\theta_R \end{pmatrix} \tag{23}$$

One has in general **three** CP phases for **two generations** if we consider spontaneous CP violation and also complex Yukawa couplings. We can read from the expression above that for one single generation we are still left with one phase $\gamma - \delta_2$.

An interesting particular case, from a theoretical point of view an its predictive power, is the case of only spontaneous CP violation[33,35)]. Even for the more economic way of breaking $SU(2)_L \otimes SU(2)_R \otimes U(1)$, one can have complex vacuum expectation values that lead to observable phases, because the Higgs structure is richer than in the Standard Model. The Higgs field $\phi$, responsible for the last breaking to $U(1)_{em}$, belonging to the $(1/2, 1/2)$ representation of $SU(2)_L \otimes SU(2)_R$, can acquire a complex vacuum expectation value

$$< \phi >= e^{i\alpha} \begin{pmatrix} \chi & 0 \\ 0 & \chi' \end{pmatrix} \tag{24}$$

and then the three phases of the general mixing matrices (22) can be expressed in terms of one single parameter $\alpha$.

Let us now give the general features of CP violation in L-R symmetric models. We will examine what is specific to LR models, namely the fact that, unlike the Standard Model, we can have CP violation at the two generation level.

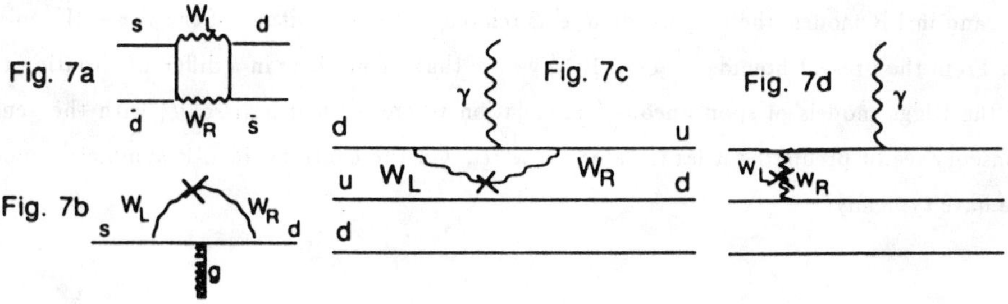


We will first remark that there are new kinds of box diagrams (Fig.7a) with the exchanges of one $W_L$ and one $W_R$. These diagrams contain large logarithms of the form $\log(M_{W_L}^2/m_c^2)$, very large LR hadronic matrix elements, and new small distance QCD corrections. All this results in an enhancement factor of the box diagram LR relative to the LL:

$$(\text{box diagram})_{LR} \sim 430(M_{W_L}^2/M_{W_R}^2)(\text{box diagram})_{LL} \tag{25}$$

The CP conserving piece of the box diagram contributes to $\Delta M_K$ , and since the LL diagram gives already the right order of magnitude for this mass difference, (25) gives a strong bound[36] : $M_{W_R} > 1.6TeV$.

Let us now consider the box diagram contribution to $\epsilon$, i.e., $\epsilon_m$. In the LR model with two generations described above we have the same type of diagrams that contributed to $\Delta M_K$ (Fig.7a). These LR box diagrams, due to the phases (23), will give CP violation, unlike in the LL ones with two generations. An immediate very interesting consequence is that $\epsilon_m$ will be proportional to the ratio

$$\epsilon_m \sim 430 \times \frac{M_{W_L}^2}{M_{W_R}^2} \sin(\delta_1 - \delta_2) \tag{26}$$

This formula tells us that in LR symmetric models the scale of CP violation is related to the spontaneous breaking of parity: $\epsilon_m$ is small because the energy scale at which parity is restored is large. Using the bound on $M_{W_R}^2$ from $\Delta M_K$ , one has the freedom of adjusting the angle $\delta_1 - \delta_2$ to fit $\epsilon$ .

In LR symmetric models there are tree diagram contributions to $\xi$ (Fig.7b) that come from the $W_R - W_L$ mixing $\zeta$ and the phases (23). One has the proportionality

$$|\epsilon'| \sim |\xi| \sim |\zeta| \, |\sin(\gamma - \delta_1) + \sin(\gamma - \delta_2)| \tag{27}$$

and in LR models the magnitude of $\epsilon'$ is related to the magnitude of the $W_R - W_L$ mixing $\zeta$ . From the present bounds[37] $|\zeta| \leq 10^{-3}$ we see that we are here in a different situation than in the Higgs models of spontaneous CP violation where we had $|\epsilon_m| \ll |\xi|$ with the general consequence of predicting a large value for $|\epsilon'/\epsilon|$. On the contrary, in L-R symmetric models we have typically

$$|\epsilon_m| \ll |\xi|. \tag{28}$$

and the model will predict in general a small value for $\epsilon'/\epsilon$ naturally of the order of the NA31 data or smaller,

$$\frac{\epsilon'}{\epsilon} \sim 10^{-3} - 10^{-4} \tag{29}$$

Note that the agreement with NA31 is only achieved at the price of having $|\delta_1 - \delta_2| \ll |\gamma - \delta_2|$ as we can see from (26), (27).

Moreover, the approximate proportionality[38] between $\epsilon'$ and $D_n$, characteristic of non-standard models, also applies here, because lowest order graphs contribute to $D_n$ through the $W_R - W_L$ mixing. There are two types of diagrams that in principle should dominate: the one loop contribution to the quark electric dipole moments[39] (Fig.7c) and the exchange diagram of Fig.7d with $N^{*(1/2-)}$ intermediate states, just as in the $\theta_{QCD}$ contribution considered above (Fig. 4). Both contributions are of the same order of magnitude (see G. Ecker et al.[39]), and we will only quote the value that results from the one-loop of Fig.7c,

$$\frac{D_n}{e} \sim \frac{g_L g_R}{72\pi^2 M_{W_L}^2} \zeta \sin^2\theta_C \times 5m_c \sin(\gamma+\delta_1) \tag{30}$$

with the consequence of predicting $D_n$ much larger than in the Standard Model but in general smaller than in the Higgs model, feature expected in view of the proportionality between $\epsilon'$ and $D_n$ (13), and of (29),

$$D_n/e \sim 10^{-21} cm \zeta \sin(\gamma+\delta_1) \tag{31}$$

Assuming the angles $\gamma+\delta_1$ in (31) and $\gamma-\delta_1$, $\gamma-\delta_2$ in (27) to be of the same order of magnitude, as it happens in the more economic model with spontaneous CP violation[35], then the comparison between (27) and (31) and the present limits on $\epsilon'$ give the bound[40]

$$D_n/e \quad \lesssim \quad 10^{-26} cm \tag{32}$$

To summarize the Standard Model, the Higgs model (HM) of spontaneous CP violation and the LR symmetric models, we have the following natural pattern. Once $\epsilon$ is fitted in the models, we have the hierarchy:

$$\begin{array}{ccccc} |\epsilon'/\epsilon|_{LR} & \lesssim & |\epsilon'/\epsilon|_{SM} & \ll & |\epsilon'/\epsilon|_{HM} \\ |D_n|_{SM} & \ll & |D_n|_{LR} & \lesssim & |D_n|_{HM} \end{array} \tag{33}$$

From this ordering of the predictions we see that the Standard Model and the Higgs model are in trouble to accommodate both data NA31 on $\epsilon'/\epsilon$ and Leningrad on $D_n$. The Standard Model could account for $\epsilon'/\epsilon$ but not for $D_n$, while the Higgs model could account for $D_n$ but not for $\epsilon'/\epsilon$ . The LR symmetric model could account for both although forcing a little on the parameters, since it tends to give values both for $\epsilon'/\epsilon$ and for $D_n$ that are naturally somewhat smaller than the NA31 and Leningrad data.

## 6. Supergravity N = 1.

Within the "Standard Model" of N = 1 Supergravity[41)] we have new sources[42)] of CP violation that we will classify as follows: 1) Supersymmetric Kobayashi-Maskawa model; 2) Pure Supersymmetric phases resulting from the superpotential phase, from the complex gaugino masses, and from the gaugino-higsino mixing whenever present.

The source 1) is of the same type as in the Standard Model, as it concerns the mixing matrices among matter fields. On the contrary, the source 2) concerns new phases stemming from supersymmetry which do not necessarily require changes of generation and are often flavor blind.

Let us now examine briefly each of these types of contributions:

### 6.1. Supersymmetric KM model.

In this scheme, the usual KM quark mixing matrix has to be completed by its effect on the scalar supersymmetric partners of the quarks, the squarks. An interesting result is that heavy quark-higgsino loops (Fig.8a) (recall that Higgs bosons and hence also their supersymmetric partners like to couple to heavy fermions) renormalize the down squark-left mass matrix[43)]

$$m^2_{\tilde{d}_L} = \mu^2 + M_d^2 - |c| V^{\dagger}_{KM} M_u^2 V_{KM} \tag{34}$$

where $M_d$ $M_u$ refer to the mass matrix of the down and up quark respectively.

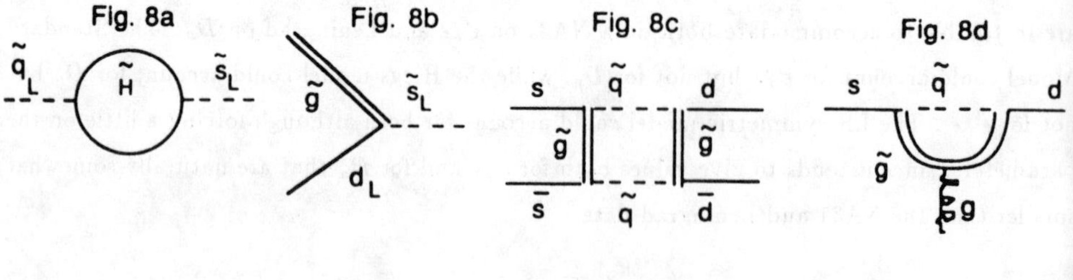


This mixing induced in the squark sector results in CP violating phases in the supersymmetric QCD sector that change flavor[44)]. It leads to complex couplings of the type: quark-squark-gluino (with negatively charged quarks and squarks), for example $d_L - s_L - gluino$ (Fig.8b). These couplings are affected by the KM matrix, resulting in a new box contribution to $\epsilon_m$ (Fig.8c) and other contributions to $\xi$, i.e., to $\epsilon'$ (Fig. 8d). Of course, the magnitude of these new contributions will depend on the actual gluon and squark masses. The recent CDF limits on these masses[45)],

$$m_{\tilde{g}}, m_{\tilde{q}} \geq 100GeV \tag{35}$$

exclude a significant contribution of this mechanism of CP violation to $\epsilon$ and to $\epsilon'$. Its contribution to $D_n$ is small: $D_n < 10^{-28} e.cm$.

Earlier two loop contributions using wino exchange give the same order of magnitude[46)]

Supplementary phases can appear in the squark-left squark-right mass matrix and it has been claimed[47)] that their effect can be large on $D_n$.

6.2. PURE SUPERSYMMETRIC PHASES

Superpotential phases.

These phases originate in vacuum expectation values that are at the origin of the spontaneous breaking of supersymmetry through the super-Higgs mechanism[48)]. These vacuum expectation values could be complex

$$A = e^{-2i\phi_A}|A| \tag{36}$$

and give rise to CP violation of the soft type, like the Higgs model of spontaneous CP violation. There is however a fundamental difference with the latter: here the CP violation is flavor-blind. This phase is not by itself observable, it has to be combined with the phases which we will describe next.

Complex phases in the gaugino and higgsino sector .

The dominant one seems to be the contribution due to **complex gluino masses**. The gluino fields (supersymmetric partners of the gluons of QCD) are complex, leading to complex masses

$$m_g = e^{-2i\phi_g}|m_g| \tag{37}$$

and new CP phases[42,48,49,50)]. This type of CP violation is also flavor blind . For example, this type of mechanism could give contributions to the quark electric dipole moment of the type of Fig.8e. Notice that this graph can be relatively large because the gluino couples through $\alpha_s$

There is however a mass suppression in the squark line because the gluino mass term (responsible for the CP violation) connects a quark L with the same flavor quark R. That is the mixing squark-left squark-right (mentioned at the end of the preceeding section) is needed . The latter is dominantly flavor diagonal. Contributions to $\epsilon'$ from this source need flavor-changing squark-left squark-right mixing.

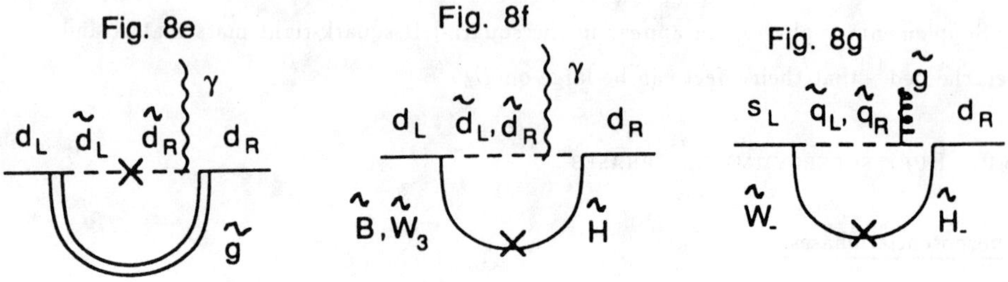


It should be pointed out that the physical observable phase in this type of CP violation is $\phi_A - \phi_g$.

Besides the global gaugino mass phases, other effects may arise in the **wino higgsino mixings**. Indeed, just as in the KM model for different quark flavors, the gauginos and Higgsinos will in general mix, leading to new CP violating phases in the mixing matrices [51,52) . In the neutral sector this mechanism is also flavor-blind .

For example, we can have contributions to $D_n$ of the type of Fig.8f. We do not need now a mass insertion on the squark line because the Higgsino field (coupling like a mass) performs the necessary helicity flip. The charged sector also provides a similar diagram (with a gluon instead of a photon) contributing to $\epsilon'$ since charged gauginos can change flavor (Fig.8g).

<u>Summary on supergravity:</u> Purely supersymmetric phases alone tend to give naturally a large value[52)] for $\epsilon'/\epsilon$ not compatible with NA31 ( $\epsilon'/\epsilon \sim -0.05$ ) or, if fitted to $\epsilon$, values too large[53)] for $D_n$ .

On the other side the super-Kobayashi-Maskawa mechanism alone seems not able to explain $\epsilon$ and $\epsilon'/\epsilon$, given the present bounds on supersymmetric masses.

Nevertheless it should be stressed that, in the presence of supersymmetry the dipole moment appears at one-loop level. In consequence , would the Leningrad result for $D_n$ be confirmed, essentially flavor blind pure supersymmetric phases remain a good candidate to explain $D_n$ while $\epsilon$ and $\epsilon'/\epsilon$ are most easily explained by the standard Kobayashi-Maskawa model.

As an aside comment we must emphasize that one can solve the strong CP problem in Supergravity N = 1 by the Peccei-Quinn mechanism with an elusive axion[42,49,55)]

## 7. Conclusion.

In Fig.9 we plot the predictions of these theoretical schemes for the ratios $|D_n/\epsilon'|$ and $|\epsilon'/\epsilon|$ together with the Leningrad and NA31 data. For simplicity we only represent absolute value of these quantities in the different models . We indicate only approximatively the region covered by each model, without fine tuning of the parameters of the model. We see that the Standard Model agrees very well with the NA31 value for $|\epsilon'/\epsilon|$, but is far away the Leningrad data on $D_n$. On the other hand, although forcing a little the parameters, the LR symmetric models could account for both $|\epsilon'/\epsilon|$, and $D_n$. Also, another possibility could be a combination of the Standard Model for $|\epsilon'/\epsilon|$ , with some flavor-blind interaction that could account for $D_n$, like a pure supersymmetric phase of the type considered in Section 6.2, without forcing the parameters of the supersymmetric model (although further work is necessary to confirm this point).

Fig. 9

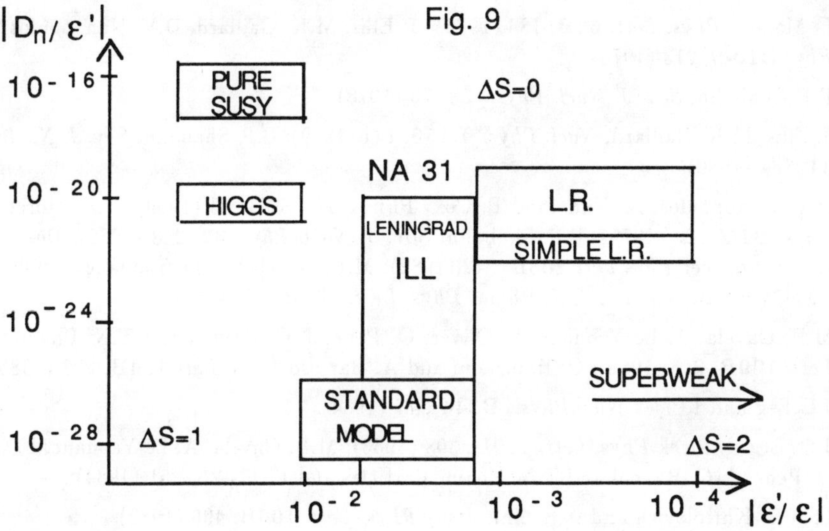


In conclusion, if the NA31 value for $|\epsilon'/\epsilon|$ and the Leningrad data on $D_n$ were confirmed (they are not however on the same degree of plausibility), both data could be accounted for by the LR symmetric model of CP violation. However, if one stands to the Standard KM model, a single source of CP violation is excluded: another flavor-blind mechanism, like those existing in Supersymmetric theories (in the pure gauge sector) has to be considered.

We want to make a final remark on the present Promethean search of new sources of CP violation. The neutron electric dipole moment is in an exceptional position to reveal possible new CP violation physics, since the Standard Model predicts such a small value for it. On the other hand, B-meson physics will also be in the future an important laboratory, $B_d$ mesons to confirm maybe the Standard KM Model, and $B_s$ mesons possibly to reveal new sources of CP violation. Finally, it is very important to measure the electron electric dipole moment, as here we can get evidence on flavor-blind CP violation not polluted by the strong CP problem, and indirect information on neutrino masses (for recent reviews on Dipole moments of neutron, electron, atoms and others, see[56]).

## REFERENCES


1. E.M.Purcell and N.F.Ramsey *Phys. Rev.* **78** 807 (1950).
2. For a review see N. Ramsey, *Ann. Rev. Nucl. Sci.* **32** , 211 (1982).
3. K. Smith these proceedings.
4. I.S. Altarev et al., *JETP Lett.* **44**, 461 (1986).
5. Kobayashi, Maskawa, *Progr. Theor. Phys.* **49**, 652 (1973).
6. R. Jackiw and C. Rebbi, *Phys. Rev. Lett.* **37**, 172 (1976).
7. L. Maiani, *Phys. Lett.* **62B**, 183 (1976); J. Ellis, M.K. Gaillard, D.V. Nanopoulos, *Nucl. Phys.***B10 9**, 213 (1976).
8. E.P. Shabalin, *Sov. J. Nucl. Phys.* **28**, 75 (1978).
9. J. Ellis, M.K. Gaillard, *Nucl. Phys.* **B150**, 141 (1979); E.P. Shabalin, *Sov. J. Nucl. Phys.* **31**, 864 (1980).
10. D.V. Nanopoulos, A. Yildz and P. Cox, *Phys. Lett.* **87B**, 53 (1979); B.F. Morel,*Nucl. Phys.* **B157**, 23 (1979); E.P. Shabalin, *Sov. J. Nucl. Phys.* **32**, 228 (1980); Deshpande, Eilam, Spencer *Phys.Lett* **108B** , 42 (1982); M.B. Gavela, A. Le Yaouanc, L. Oliver, O. Pène, J.-C. Raynal and T.N. Pham, *Phys. Lett.* **109B**, 83 (1982) .
11. M.B. Gavela, A. Le Yaouanc, L. Oliver, O. Pène, J.-C. Raynal and T.N. Pham, *Phys. Lett.* **109B**, 215 (1982); C. Hamzaoui and A. Barroso *Phys. Lett* **154B**, 202 (1985).
12. J.O.Eeg and I.Picek *Nucl.Phys.* **B244** , 77 (1984).
13. J.O. Eeg, I. Picek,*Phys. Lett.* **130B**, 308 (1983); M.B. Gavela, A. Le Yaouanc, L. Oliver, O. Pène, J.-C. Raynal and T.N. Pham, Particles and Fields **23**, 251 (1984).
14. 11. I.B. Khriplovich and A.R. Zhitnitsky,*Phys. Lett.* **109B**, 490 (1982).
15. for a recent review see R.D.Peccei in CP Violation, World Scientific Publishing Co., edited by C. Jarlskog (1988).
16. V. Baluni, *Phys. Rev.* **D19**, 2227 (1979).
17. R.J. Crewther, P. Di Vecchia, G. Veneziano and E. Witten, *Phys. Lett.* **88B**, 123 (1979).
18. R.D. Peccei, H.R. Quinn, *Phys. Rev. Lett.* **38**, 1440 (1977), *Phys. Rev.* **D16**, 1791 (1977).
19. P.Sikivie in these proceedings.
20. for a recent review see S.Barr these proceedings.
21. 15. S. Coleman, *Nucl. Phys.* **B310**, 643 (1988).
22. H. B. Nielsen and M. Ninomiya, preprint INS Rep.-721 (1988).
23. J. Preskill, S.P. Trivedi, M.B. Wise, *Phys. Lett.* **B223** , 26 (1989).

24. T.D. Lee, *Phys. Rev.* **D8**, 1226 (1976), *Phys. Rep.* **9**, 1037 (1974).

25. S. Weinberg, *Phys. Rev.* Lett. **37**, 657 (1976); see also G. Branco *Phys. Rev.* **D22**, 2901 (1980) .

26. A.I. Sanda, *Phys. Rev.* **D23**, 2647 (1981); N.G. Deshpande, *Phys. Rev.* **D23**, 2654 (1981).

27. Y. Dupont and T.N. Pham, *Phys. Rev.* **D28**, 2169 (1983).

28. J.F. Donoghue and B.R. Holstein, *Phy. Rev.* **D32** , 1152 (1985) ; J.-M. Frère, J. Hagelin and A.I. Sanda,*Phys. Lett.* **151B**, 161 (1985) .

29. H.Y.Cheng *Phys. Rev.* **D34** ,1397 (1986).

30. G. Beal and H. Deshpande *Phys. Lett.* **132B** , 427 (1983); I.I. Bigi and A.I. Sanda *Phys. Rev. Lett.* **58** , 1604 (1987); for a recent review see I.I. Bigi and A.I. Sanda in CP Violation, World Scientific Publishing Co., edited by C. Jarlskog (1988) .

31. G.C.Branco, A.J. Buras and J.-M. Gérard, *Nucl. Phys.* **B259**, 306 (1985).

32. J. Liu and L. Wolfenstein *Nucl. Phys.* **B289**, 1 (1987)

33. For a recent review, see R.N. Mohapatra, in CP Violation, World Scientific Publishing Co., edited by C. Jarlskog (1988).

34. R.N. Mohapatra and J.C. Pati, *Phys. Rev.* **D11** , 566 (1975); R.N. Mohapatra, R.E. Paige and D.P. Sidhu, *Phys. Rev.* **D17** , 2642 (1978); R.N. Mohapatra and D.P. Sidhu, *Phys. Rev.***D17** , 1876 (1978).

35. D. Chang, *Nucl. Phys.***B214** , 435 (1983).

36. G.Beall, M.Bender and A. Soni *Phys. Rev.* Lett. **48**, 848 (1982).

37. J. Bigi and J.-M. Frère *Phys. Lett.* **110B** , 255 (1982); J. Donoghue and B. Holstein *Phys. Lett.* **113B** , 382 (1982) .

38. G.C. Branco, J.-M. Frère and J.-M. Gérard *Nucl. Phys.* **B221**, 317 (1983).

39. G. Beall and A. Soni, *Phys. Rev.* Lett. **47**, 552 (1981); G. Ecker, W. Grimus and H. Neufeld, *Nucl. Phys.* **B229**, 421 (1983).

40. X.-G. He, B.H.J. Mc Kellar and S.Pakvasa *Phys. Rev. Lett.* **61**, 1267 (1988).

41. H.P. Nilles *Phys. Lett.* **115B**, 193 (1982), L. Ibanez *Phys. Lett.* **118B**, 73 (1982); A. Chamsedine, R. Arnowitt and P. Nath *Phys. Rev. Lett.* **49**, 970 (1982); L. Alvarez-Gaume, J. Polchinski and M. Wise *Nucl. Phys* **B221**, 495 (1983); H.P. Nilles Phys. Rept **110**, 1 (1984).

42. J. Ellis and D.V. Nanopoulos *Phys. Lett.* **110B**, 44 (1982); J. Ellis S. Ferrara and D.V. Nanopoulos *Phys. Lett.* **114B**, 230 (1982).

43. M.J. Duncan *Nucl. Phys.* **B221**, 285 (1983); J.F. Donoghue, H.P. Nilles and D. Wyler *Phys. Lett.* **128B**, 55 (1983); M. Dugan, B. Grinstein and L. Hall, *Nucl. Phys.* **B255**, 413 (1985); M.J. Duncan and J. Trampetic, *Phys. Lett.* **134B**, 439 (1984); A. Bouquet, J. Kaplan and C.A. Savoy,*Phys. Lett.* **148B**, 69 (1984); A. Langacker, and Sathiapalan *Phys. Lett.* **144B**, 395 (1984) .

44. M.J. Duncan *Nucl. Phys.* **B224**, 289 (1983); J.-M. Gérard, W. Grimus, A. Raychoudhury and G. Zoupanos,*Phys. Lett.* **140B**, 349 (1984); J.-M. Gérard, W. Grimus, A. Masiero, D.V. Nanopoulos and Raychoudhury, *Phys. Lett.* **141B**, 79 (1984).

45. M.J. Shochet in the Proceedings of the XXIV International Conference on High Energy Physics, Munich Aug. 4-10 1988.

46. S.P. Chia and S. Nandi *Phys. Lett.* **117B**, 45 (1982).

47. T. Kurimoto *Progr. Theor. Phys.* **76**, 654 (1986).

48. J. Polchinski and B. Wise *Phys. Lett.* **125B**, 393 (1983).

49. W. Büchmuller and D. Wyler *Phys. Lett.* **121B**, 321 (1982).

50. E. Franco and M. Mangano *Phys. Lett.* **135B**, 445 (1984); T. Kurimoto *Progr. Theor. Phys.* **73**, 209 (1985).

51. F. del Aguila, M.B. Gavela, J.A. Grifols and A. Mendez *Phys. Lett.* **126B**, 71 (1983).

52. J.M. Frère, M.B. Gavela *Phys. Lett.* **132B**, 107 (1983).

53. J.-M. Gérard, W. Grimus, A. Masiero, D.V. Nanopoulos and A. Raychaudhury *Nucl. Phys.* **B253**, (1985).

54. R. Akhoury, I.I.Bigi and H.E. Haber *Phys. Lett.* **135B**, 113 (1984).

55. D.V. Nanopoulos and M. Srednicki *Phys. Lett.* **128B**, 61 (1983).

56. S.M. Barr and W.J. Marciano in CP Violation, World Scientific Publishing Co., edited by C. Jarlskog (1988); X.-G. He, B.H.J. Mc Kellar and S.Pakvasa University of Hawaï preprint, UH-511-666-89.

# Generation Mixing

*Volker Soergel, Alexander Skrinsky and René Turlay, (salles des portraits, Château de Beauregard).*

# ARGUS Results on $B$ Mesons

Henning Schröder
DESY, Hamburg, Germany

Using the ARGUS detector at the $e^+e^-$ storage ring DORIS II at DESY new results on beauty physics have been obtained. With more data an update on $B^0\overline{B}^0$ mixing is given for the mixing parameter $r = (21 \pm 6)\%$. The lifetime ratio of charged and neutral $B$ mesons is measured to be $\tau_{B^+}/\tau_{B^0} = 1.00 \pm 0.23 \pm 0.14$. In semileptonic $B$ decays, as well as in exclusive $B$ decays, no significant signals are found for $b \to u$ transitions.

## 1 Introduction

The study of $B$ mesons allows one to determine the missing fundamental parameters of the standard model and to test it [1]. At ARGUS these investigations are performed by using as a source for $B$ mesons the $\Upsilon(4S)$ resonance which is formed in $e^+e^-$ annihilation at $\sqrt{s}$=10.58 GeV and decays with about equal probability into a $B^0\overline{B}^0$ or a $B^+B^-$ pair. From 1983 until 1988 a sample of 152 000 $\Upsilon(4S)$ decays has been collected with the ARGUS detector at the $e^+e^-$ storage ring DORIS II.

## 2 Update on $B^0\overline{B}^0$ Mixing

Large $B^0\overline{B}^0$ mixing is well established [2][3]. The aim of the present experiments is to improve the accuracy of the measurement and to remove uncertainties when connecting the experimental data with the theoretical predictions. Of importance here is the lifetime ratio between charged and neutral $B$ mesons, which will be discussed in section 3.

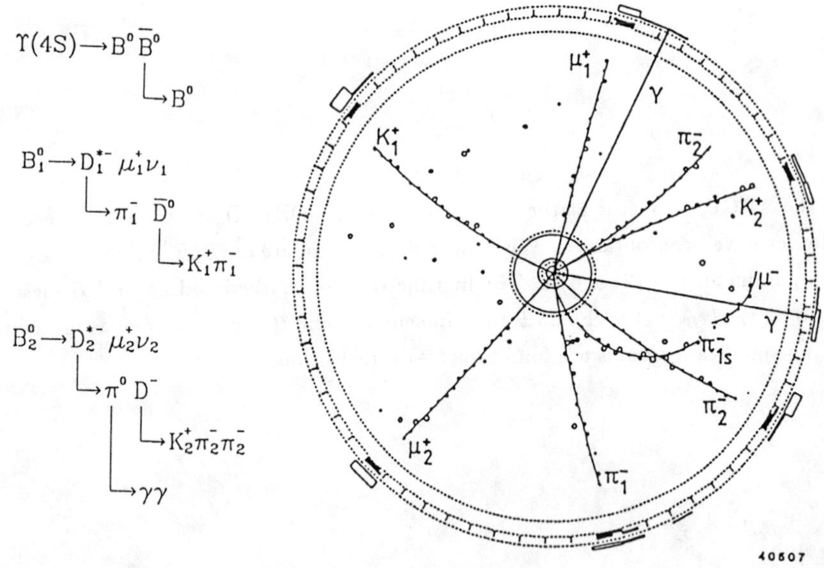


Figure 1: Completely reconstructed event consisting of the decay $\Upsilon(4S) \to B^0B^0$

Theoretically the mass difference $\Delta M$ between the CP eigenstates in the $B$ system is given by [4]

$$\Delta M = \frac{G_F^2}{6\pi^2} B_B f_B^2 m_b \left|V_{tb}^* V_{td}\right|^2 m_t^2 F(\frac{m_t^2}{M_W^2}) \eta_{QCD}$$

In this expression, several of the parameters are reasonably well known; these are, besides $G_F$, the mass of the $b$ quark, $m_b$, the QCD correction $\eta_{QCD}$, the functional dependence $F(\frac{m_t^2}{m_W^2})$ and, since 1983, the lifetime of $B$ mesons, $\tau_b$ [5]. Not known is the decay constant $f_B$ together with the bag factor $B$, the KM matrix element $|V_{td}|$ and, in particular the mass of the top quark, $m_t$. The measurable quantity is the mixing parameter $r$ which is connected to $\Delta M$ by

$$r = \frac{Prob(B^0 \to \overline{B}^0)}{Prob(B^0 \to B^0)} = \frac{(\Delta M \cdot \tau_b)^2}{2 + (\Delta M \cdot \tau_b)^2}$$

$B^0\overline{B}^0$ mixing in $\Upsilon(4S)$ decays is observed by tagging the flavor content of the $B^0$ meson. This is best done by the reconstruction of both $B$ mesons as demonstrated by the mixed event observed by ARGUS (Figure 1) where the the $\Upsilon(4S)$ apparently decays via $\Upsilon(4S) \to B^0 B^0$, and both $B^0$ mesons decay into $B^0 \to D^{*-}\mu^+\nu$. This event represents the first observation of $B^0\overline{B}^0$ mixing [2].

The discovery of this event was a consequence of the observation of the decay $B^0 \to D^{*-}\ell^+\nu$ [6]. For $B^0$ mesons produced in $e^+e^- \to \Upsilon(4S) \to B^0\overline{B}^0$ and decaying into $B^0 \to D^{*-}\ell^+\nu$ the quantity

$$M^2_{rec} = (E_{beam} - (E_{D^{*-}} + E_{l^+}))^2 - (\vec{p}_{D^{*-}} + \vec{p}_{l^+})^2$$

has to be $\approx 0$.

Figure 2a shows the $M^2_{rec}$ distribution for events taken at the $\Upsilon(4S)$ containing a $D^{*-}\ell^+$ combination. The $D^{*+}$ is reconstructed in the decay $D^{*+} \to D^0\pi^+$, followed by $D^0 \to K^-\pi^+$. The peak at $M^2_{rec} \approx 0$ is the signal for the decay $B^0 \to D^{*-}\ell^+\nu$, and has been analysed in detail [7]. For $D^{*+}\ell^+$ combinations the $M^2_{rec}$ distribution looks completely different (Figure 2b).

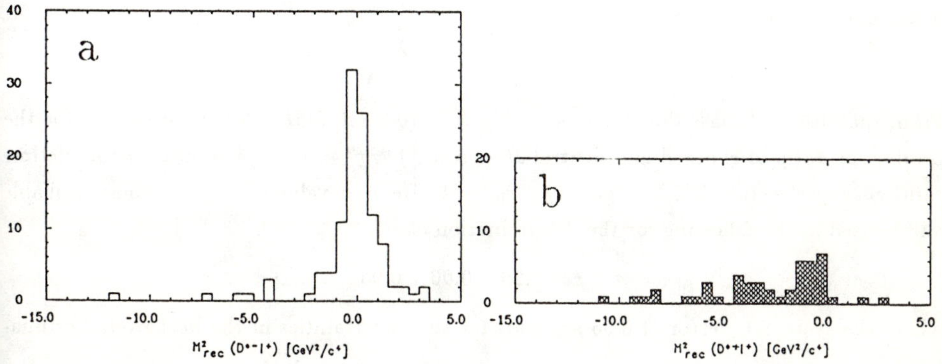


Figure 2: $M^2_{rec}$ distribution for (a) $D^{*-}\ell^+$ and (b) $D^{*+}\ell^+$ combinations

The traditional method to determine the amount of mixing in $\Upsilon(4S)$ decays is to search for $\ell^+\ell^+(\ell^-\ell^-)$ lepton pairs where both leptons tag one $B^0(\overline{B}^0)$. In table 1 the resulting rates are summarized from the ARGUS experiment, where new data have been taken in 1988. The cuts applied to obtain these numbers are similar to the ones in [2]. After subtraction of background a mixing signal of 34.6 events containing like-sign lepton pairs is obtained from 64 observed candidate events. For the events with unlike-sign lepton pairs one obtains from 414 candidate events a signal of 381.3 events with one lepton from one $B$ meson and the oppositely charged lepton from a $\overline{B}$ meson.

| | $e^\pm e^\pm$ | $\mu^\pm\mu^\pm$ | $e^\pm\mu^\pm$ | $e^+e^-$ | $\mu^+\mu^-$ | $e^\pm\mu^\mp$ |
|---|---|---|---|---|---|---|
| $\Upsilon(4S)$ + Continuum | 15 | 21 | 34 | 102 | 97 | 204 |
| Continuum | 0 | 1 | 1 | 2 | 3 | 2 |
| $\Upsilon(4S)$ direct | 15.0 | 18.1 | 31.1 | 96.1 | 88.2 | 198.1 |
| Corrected for $J/\psi$ cut | 15.0 | 18.1 | 31.1 | 112.3 | 103.2 | 198.1 |
| | ±4.9 | ±5.2 | ±6.3 | ±10.9 | ±11.1 | ±14.9 |
| **Background** | | | | | | |
| Fakes | 1.8 | 4.0 | 6.7 | 3.6 | 7.0 | 13.2 |
| Conversion | 0.3 | - | 0.3 | 0.3 | - | 0.3 |
| Secondary decays | 4.6 | 2.9 | 7.2 | 1.9 | 1.2 | 3.0 |
| $J/\psi$ decays | 0.6 | 0.3 | 0.8 | 0.6 | 0.3 | 0.8 |
| **Signal** | 7.7 | 10.9 | 16.2 | 105.9 | 94.7 | 180.7 |
| | ±4.9 | ±5.2 | ±6.3 | ±10.9 | ±11.1 | ±14.9 |
| **Average mixing parameter r** = (**20** ± **6**)% | | | | | | |

Table 1. Dilepton rates

In calculating $r$ one has to account for the decay $\Upsilon(4S) \to B^+B^-$ by a factor:

$$\lambda = \frac{f^+}{f^0}\left(\frac{\tau_{B^+}}{\tau_{B^0}}\right)^2$$

which enters the formula for $r$:

$$r = \frac{N_{\ell^\pm\ell^\pm}(1+\lambda)}{N_{\ell^+\ell^-} - N_{\ell^\pm\ell^\pm}\cdot\lambda}$$

From the observed mass difference $m_{B^-} - m_{B^0} = (0.2 \pm 0.7) MeV/c^2$ [8] one infers for the production ratio of charged and neutral $B$ mesons, $f^+/f^0 \approx 1$. New results on the lifetime ratio $\tau_{B^+}/\tau_{B^0}$ (section 3,[8]) suggest $\tau_{B^+}/\tau_{B^0} \approx 1$. Hence a value of $\lambda = 1$ is most probable and is used in the following for the determination of $r$:

$$r = 0.20 \pm 0.06 \pm 0.05$$

where the systematic error of 0.05 accounts for the uncertainties in the background estimations.

The complication concerning production rates and lifetime differences is avoided by searching for events containing $D^{*+}\ell^-\ell^-$ combinations where the $D^{*+}$ [1] together with one $\ell^-$ originates from the decay $\overline{B}^0 \to D^{*+}\ell^-\nu$. Comparing this rate with the number of unmixed events containing $D^{*+}\ell^-\ell^+$ combinations one gets

$$r = \frac{N(D^{*+}\ell^-\ell^-)}{N(D^{*+}\ell^-\ell^+)}$$

After background subtraction ARGUS gets the numbers in table 2 from which one obtains

$$r = 0.24 \pm 0.12$$

[1]The $D^{*+}$ is reconstructed in the decay $D^{*+} \to D^0\pi^+$, followed by $D^0 \to K^-\pi^+, K^-\pi^+\pi^0$ and $K^-\pi^+\pi^+\pi^-$

| | $N(D^{*+}\ell^-\ell^-)$ | $N(D^{*+}\ell^-\ell^+)$ |
|---|---|---|
| At $\Upsilon(4S)$ | 8 | 27 |
| Background | $1.9 \pm 0.5$ | $1.7 \pm 0.5$ |
| Signal | $6.1 \pm 2.8$ | $25.3 \pm 5.9$ |

Table 2. Dilepton events with $D^{*+}$ mesons from ARGUS

Another method, which is independent of the first two, assumes that $D^{*+}$ mesons are only produced in the decay of $\overline{B}^0$ mesons. The search for events containing a $D^{*+}$ together with a $\ell^\pm$ is performed by using the $M^2_{rec}$ distributions in Figure 2a and 2b. That the $D^{*+}$ and the $\ell^\pm$ are coming from different $B$ mesons is guaranteed by requiring

$$M^2_{rec} < -2.5GeV^2/c^4$$

From this analysis ARGUS obtains the numbers in table 3.

| | $N(D^{*+}\ell^-)$ | $N(D^{*+}\ell^+)$ |
|---|---|---|
| At $\Upsilon(4S)$ | 7 | 23 |
| Background | $2.0 \pm 0.5$ | $2.3 \pm 0.5$ |
| Signal | $5.0 \pm 3.1$ | $20.7 \pm 5.5$ |

Table 3. $D^{*+}\ell^\pm$ events from ARGUS

For the mixing parameter $r$, ARGUS then obtains, under the above assumption,

$$r = \frac{N_{D^{*+}\ell^-}}{N_{D^{*+}\ell^+}} = 0.24 \pm 0.16$$

This value is, within the large errors, close to the measured value obtained by the other methods, indicating that indeed most of the $D^{*+}$ mesons are decay products of $\overline{B}^0$ mesons. Possible decays of charged $B$ mesons into $D^{*+}$ mesons would increase the value for $r$ obtained with this method.

Combining all results on $r$ ARGUS obtains

$$r = 0.21 \pm 0.06$$

in perfect agreement with the original observation.

The consequences for the determination of the standard model parameters remain essentially unchanged [9]. Using the above result together with the estimate $B_B^{\frac{1}{2}} f_B = (140 \pm 40) MeV$ [10], $\mid V_{td} \mid < 0.018$ from the unitarity of the KM-matrix [11] and $m_t < 190 GeV/c^2$ from the measurements of the electroweak radiative corrections [12] one obtains

$$60 \leq m_t \leq 190 \ GeV/c^2$$

$$0.007 \leq \mid V_{td} \mid \leq 0.018$$

in the standard model with three families.

Improvements on the determination of the standard model parameters will mainly come from solving the theoretical uncertainties in $B_B^{\frac{1}{2}} f_B$, a quantity which is very difficult to measure.

## 3 Measurement of $\tau_{B^+}$ / $\tau_{B^0}$

The lifetime ratio $\tau_{B^+}$ / $\tau_{B^0}$ is given by

$$\tau_{B^+}/\tau_{B^0} = \frac{BR(B^+ \to \ell x)}{BR(B^0 \to \ell x)}$$

assuming equal semileptonic decay rates for charged and neutral $B$ mesons. The above ratio can be approximated by

$$\frac{BR(B^+ \to \ell x)}{BR(B^0 \to \ell x)} = \frac{BR(B^+ \to D^{*0}\ell\nu, D^0\ell\nu)}{BR(B^0 \to D^{*+}\ell^-\nu, D^+\ell\nu)}$$

This would imply that in semileptonic decays of charged $B$ mesons we observe only $D^0$ mesons, whereas for neutral $B$ mesons we observe mainly $D^+$ mesons and $D^0$ mesons originating from the well measured decay $\overline{B}^0 \to D^{*+}\ell^-\nu, D^{*+} \to \pi^+ D^0$ [6,7]. This picture of semileptonic $B$ decays is only slightly altered by considering the possible decay $B \to D^{**}\ell^+\nu$, which can account only for a small fraction of semileptonic $B$ decays [14],[8]. Within the present experimental accuracy this effect can be negelected and the lifetime ratio is then given by

$$\frac{\tau_{B^+}}{\tau_{B^0}} = \frac{N'(D^0\ell^-) - N'(D^{*+}\ell^-, D^{*+} \to D^0\pi^+)}{N'(D^+\ell) + N'(D^{*+}\ell^-, D^{*+} \to D^0\pi^+)}$$

for a production rate $f^+/f^0 = 1$. $N'$ are the acceptance and efficiency corrected numbers of events containing the $D\ell$ combination. Figure 3a shows the $K^-\pi^+$ mass spectrum for events taken at $\Upsilon(4S)$ energies containing a negative lepton $\ell^-(p_{\ell^-} > 1.0GeV/c)$, where the $K^-\pi^+$ momentum $p$ is restricted to $x_p < 0.5 (x_p = p/p_{max})$. Similarly, in Figure 3b the corresponding $K^-\pi^+\pi^+$ mass spectrum is shown and in Figure 3c the $K^-\pi^+\pi^+$ mass spectrum, where the $K^-\pi^+$ mass is constrained to the $D^0$ mass. Prominant $D^0$ , $D^+$ and $D^{*+}$ signals are observed.

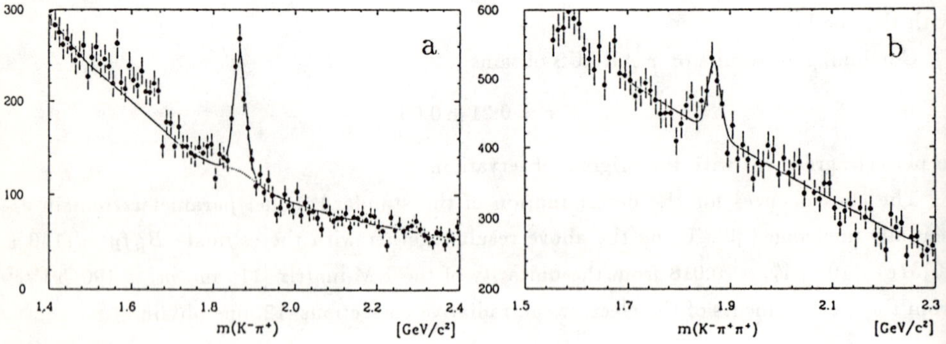


Figure 3: (a) $M_{K^-\pi^+}$ and (b) $M_{K^-\pi^+\pi^+}$ distribution in events containing a lepton $\ell^-(p_{\ell^-} > 1.0GeV/c, x_p(K^-\pi^+(\pi^+)) < 0.5)$.

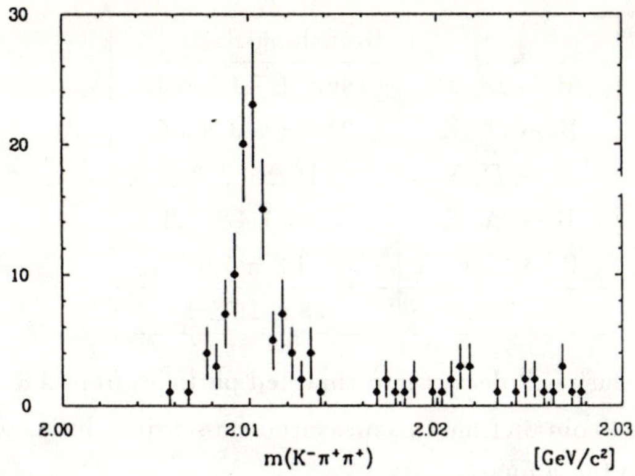


Figure 3c: $M_{K^-\pi^+\pi^+}$ distribution in events containing a lepton $\ell^-(p_{\ell^-} > 1.0 GeV/c$ , $x_p(K^-\pi^+\pi^+) < 0.5)$. The $K^-\pi^+$ mass is constrained to the $D^0$ mass.

From the spectra the number of $D\ell$ combinations is evaluated (Table 4).

| | $D^0\ell^-$ | $D^+\ell^-$ | $D^{*+}\ell^-$ |
|---|---|---|---|
| $N_{D\ell}$ | $373 \pm 27$ | $243 \pm 36$ | $66 \pm 9$ |
| Background : | | | |
| $c\bar{c}$ continuum | $26 \pm 5$ | $34 \pm 7$ | $3 \pm 2$ |
| Cascade leptons | $6 \pm 3$ | $5 \pm 3$ | $1.1 \pm 0.6$ |
| $B^0\overline{B}^0$ Mixing | $6 \pm 3$ | $10 \pm 5$ | $2.9 \pm 1.3$ |
| Fake leptons | $11 \pm 2$ | $11 \pm 2$ | $1.4 \pm 0.3$ |
| Signal | $324 \pm 28$ | $183 \pm 37$ | $58 \pm 9$ |

Table 4. $D\ell^-$ rates from ARGUS at the $\Upsilon(4S)$

Using these measurements one obtains, after correcting for branching ratios [15] and acceptances

$$\frac{\tau_{B^+}}{\tau_{B^0}} = 1.00 \pm 0.23 \pm 0.14$$

This ratio is consistent with unity, as expected in the spectator model.

## 4 Inclusive $B$ decays

The inclusive $B$ decays into the lowest lying charmed particles $D^0, D^+, D_s, \Lambda_c$ and $J/\psi$ show that $b \to c$ transitions are dominant (Table 5).

| | Branching Ratio [%] |
|---|---|
| B $\rightarrow D^0$ X | $46.6 \pm 7.1 \pm 6.3$ |
| B $\rightarrow D^+$ X | $23.2 \pm 5.3 \pm 3.5$ |
| B $\rightarrow$. $D_s$ X | $16 \pm 4 \pm 3$ |
| B $\rightarrow$' $\Lambda_c'$ X | $7.6 \pm 1.4 \pm 1.8$ |
| 2·B $\rightarrow$ '$J/\psi$' X | $4.2 \pm 1.0$ |
| $\Sigma$ | $98 \pm 10 \pm 8$ |

Table 5. Inclusive $B$ decays into charmed particles from ARGUS

However, it is not possible to obtain from this measurement sensitive limits on the coupling strength of the $b$ quark to the $u$ quark.

As a by product of the study of the inclusive $B$ deacy rates one finds [13]

$$BR(B \rightarrow D^{*+} + X)BR(D^{*+} \rightarrow \pi^+ D^0)BR(D^0 \rightarrow K^-\pi^+) = 0.0071 \pm 0.0006 \pm 0.0012$$

Using the measured $D$ branching ratios [15] one derives

$$BR(B \rightarrow D^{*+} + X) = (29.7 \pm 5.1 \pm 5.0)\%$$

where the second error is due to the uncertainty of the $D$ branching ratios. Comparing this value to the ARGUS numbers on $BR(B \rightarrow D^0 X)$ and $BR(B \rightarrow D^+ X)$ (Table 5), and assuming that $BR(B \rightarrow D^{*}X) = BR(B \rightarrow D^{*0} X)$, one concludes that $D$ mesons in $B$ decays are mainly decay products of $D^*$ mesons.

The best way to search for a signal for $b \rightarrow u$ transitions is through study of the inclusive lepton spectrum (Figure 4) in $B$ decays.

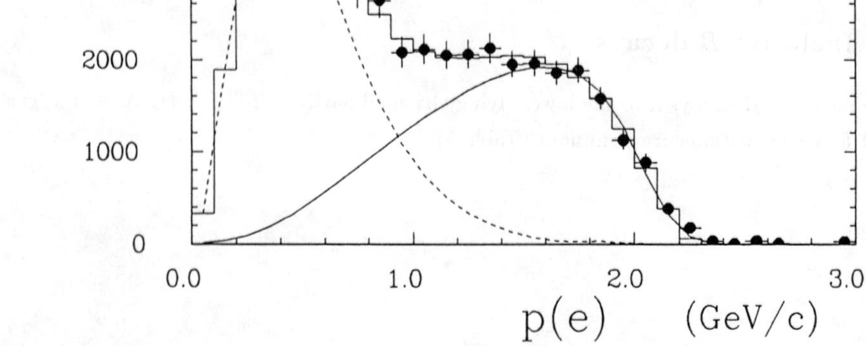


Figure 4: Inclusive electron spectrum in $\Upsilon(4S)$ decays from ARGUS

The shown electron spectrum is well described by assuming only $b \to c\ell\nu$ transitions (solid curve) and cascade decays $b \to cX, c \to s\ell\nu$ (dashed curve). The branching ratio for $B \to \ell X$ is measured to be

$$BR(B \to \ell^+ X) = (10.0 \pm 0.8)\%$$

using the model of Altarelli et al. [16]. This number is too small to be understood in a straightforward way in the spectator model where branching ratios of 12 – 15% are predicted.

From the absence of a significant signal beyond the kinematical limit for $b \to c\ell\nu$ transitions beyond $p = 2.3 GeV/c$ a limit is obtained on the ratio of the Kobayashi-Maskawa matrix elements $|V_{ub}/V_{cb}| < 0.16$.

## 5 Rare $B$ decays

In the search for $b \to u$ transitions ARGUS tried to reconstruct $B$ mesons in the channels $B \to n\pi (n = 1, 2, 3)$ and $B \to p\bar{p}\pi(\pi)$. The search for "penguin"-type transitions were performed through the reconstruction of $B$ meson in the decays $B \to' K'\pi^+,' K'\rho^0,' K'\phi$ and $K^*\gamma$. $'K'$ represents $K^0_s, K^+, K^{*0}$ and $K^{*+}$.

The mass of the $B$ mesons in determined from its momentum by an energy constraint fit which uses the fact that the energy of a $B$ meson has to coincide with the beam energy, e.g.:

$$M^2_{n\pi} = E^2_{Beam} - p^2_{n\pi}$$

No signal is found in the channels $B \to n\pi$. The corresponding 90 % C.L. upper limits are shown in Table 6.

| Decay mode | BR (90% C.L.) |
|---|---|
| $B^+ \to \pi^+\pi^0$ | $< 5.8 \cdot 10^{-4}$ |
| $B^0 \to \pi^+\pi^-$ | $< 1.7 \cdot 10^{-4}$ |
| $B^+ \to \pi^+\pi^+\pi^-$ | $< 7.4 \cdot 10^{-4}$ |
| $B^+ \to \rho^0\pi^+$ | $< 3.2 \cdot 10^{-4}$ |
| $B^0 \to \pi^+\pi^-\pi^0$ | $< 1.5 \cdot 10^{-3}$ |
| $B^0 \to \rho^0\pi^0$ | $< 4.8 \cdot 10^{-4}$ |
| $B^0 \to \rho^+\pi^-$ | $< 1.4 \cdot 10^{-3}$ |
| $B^+ \to \rho^+\pi^0$ | $< 1.8 \cdot 10^{-3}$ |
| $B^+ \to \pi^+\pi^0\pi^0$ | $< 2.3 \cdot 10^{-3}$ |

Table 6: Limits on charmless $B$ decays from ARGUS

In principle, the observed upper limits can be converted into measurements of upper limits for the KM-matrix element $V_{ub}$. However, the limits are less sensitive than the ones derived from the endpoint of the lepton spectrum in $B$ decays. For example, the obtained upper limit for the branching ratio $BR(B^0 \to \pi^+\pi^-) < 1.7 \cdot 10^{-4}$ corresponds to an upper limit of $|V_{ub}/V_{cb}| < 0.26$ [19].

In the decays $B \to p\bar{p}\pi^+(\pi^-)$ ARGUS observed in 1987 a signal of $25 \pm 8$ events[20]. A preliminary analysis of the new data, corresponding to 65 % of the previous sample, does not show the expected signal. New data, which ARGUS is taking in 1989, will help to clarify the situation.

For the "penguin"-type $b \to s$ transition no signals are seen, leading to the limits shown in Table 7 [21]. Some of these limits are close to the theoretical predictions [22].

| Decay mode | $N_{events}$ (90% CL) | BR (90% CL) |
|---|---|---|
| $B^0 \to K^+\pi^-$ | $< 9.4$ | $< 1.8 \cdot 10^{-4}$ |
| $B^+ \to K^0_S\pi^+$ | $< 3.2$ | $< 1.0 \cdot 10^{-4}$ |
| $B^0 \to K^{*+}(892)\pi^-$ | $< 2.3$ | $< 6.2 \cdot 10^{-4}$ |
| $B^+ \to K^{*0}(892)\pi^+$ | $< 3.6$ | $< 1.7 \cdot 10^{-4}$ |
| $B^0 \to K^0_S\rho^0$ | $< 2.3$ | $< 1.6 \cdot 10^{-4}$ |
| $B^+ \to K^+\rho^0$ | $< 6.4$ | $< 1.8 \cdot 10^{-4}$ |
| $B^0 \to K^{*0}(892)\rho^0$ | $< 7.7$ | $< 4.6 \cdot 10^{-4}$ |
| $B^+ \to K^{*+}(892)\rho^0$ | $< 3.4$ | $< 9.0 \cdot 10^{-4}$ |
| $B^0 \to K^0_S\phi$ | $< 2.3$ | $< 3.6 \cdot 10^{-4}$ |
| $B^+ \to K^+\phi$ | $< 2.9$ | $< 1.8 \cdot 10^{-4}$ |
| $B^0 \to K^{*0}(892)\phi$ | $< 2.3$ | $< 3.2 \cdot 10^{-4}$ |
| $B^+ \to K^{*+}(892)\phi$ | $< 2.3$ | $< 1.3 \cdot 10^{-3}$ |
| $B^0 \to K^{*0}(892)\gamma$ | $< 7.9$ | $< 4.2 \cdot 10^{-4}$ |
| $B^+ \to K^{*+}(892)\gamma$ | $< 2.3$ | $< 5.2 \cdot 10^{-4}$ |

Table 7: Limits on "penguin-type" $B$ decays from ARGUS

## References


[1] See for example H.Schröder, Physics of $B$ Mesons, Rep.Progr.Phys. **52** (1989) 765 and references herein.

[2] H.Albrecht et al (ARGUS collaboration), Phys.Lett. **192B** 245 (1987)

[3] M.Artuso et al (CLEO collaboration), Cornell preprint CLNS 89/889 February 1989

[4] M.K.Gaillard and B.W.Lee, Phys.Rev. **D10** (1974) 897, J.S.Hagelin, Phys.Rev. **D20** 2893 (1979)

[5] B.Fernandez et al. (MAC collaboration), Phys.Rev.Lett. **51** (1983) 1022, N.Lockyer et al. (MARKII collaboration), Phys.Rev.Lett. **51** (1983) 1316.

[6] H.Albrecht et al (ARGUS collaboration), Phys.Lett. **197B** 452 (1987)

[7] H.Albrecht et al (ARGUS collaboration), Phys.Lett. **219B** 121 (1989)

[8] S.Stone (CLEO collaboration), in the Proceedings of the 12th International Workshop on Weak Interactions and Neutrinos, April 9-14, 1989, Ginosar, Israel.

[9] H. Schröder, (ARGUS Collaboration), Proc. of the XXIV International Conference on High Energy Physics, München (1988), p.73.

[10] G.Altarelli, in Proceedings of the International Europhysics Conference on High Energy Physics, Uppsala (1987), p.1002.

[11] K.Kleinknecht and B.Renk, Z.Phys. **C34** (1987) 209.

[12] G.Costa et al., Nucl. Phys. **B297** 244 (1988).

[13] G.Harder (ARGUS collaboration), Ph.D. Thesis, Universität Hamburg, Hamburg, Germany, 1989

[14] H. Schröder, (ARGUS Collaboration), Proc. of the 23rd Rencontre de Moriond, 1988, p.311

[15] J.Adler et al, (MARK III collaboration), Phys.Rev.Lett. **60**, 89 (1988), Phys.Lett. **B208**, 152 (1988)

[16] G.Altarelli et al Nucl.Phys. **B208** 365 (1982)

[17] J.G.Körner and G.A.Schuler, Z.Phys. **C38** 511 (1988)

[18] T.Altomari and L.Wolfenstein, CMU-HEP87-20 (1987)

[19] M.Bauer et al, Z.Phys. **C34** 103 (1987)

[20] H.Albrecht et al (ARGUS collaboration), Phys.Lett. **209B** 119 (1988)

[21] J.Swain (ARGUS collaboration), Ph.D. thesis, University of Toronto, Toronto, Canada, in Preparation.

[22] M.B.Gavela et al, Phys.Lett. **154B** 425 (1985), L.-L.Chau and H.Y.Cheng, Phys.Rev.Lett. **59** 958 (1987), N.G.Deshpande et al, Phys.Rev.Lett. **59** 183 (1987)

# $B^0\bar{B}^0$ MIXING and CHARMLESS B * DECAYS: RECENT CLEO RESULTS

PERSIS S. DRELL

*Floyd R. Newman Laboratory of Nuclear Studies,*

*Cornell University, Ithaca, NY 14853, USA*

## ABSTRACT

Recent CLEO results on $B^0\bar{B}^0$ mixing and charmless B decays are presented.

* Work supported by the National Science Foundation.

# 1. INTRODUCTION

Twenty-five years ago, CP violation was observed in the neutral kaon system. There is still no fundamental theoretical understanding of this stunning experimental discovery, but there is a phenomenological model whereby CP violation is a natural byproduct of quark mixing. In the standard SU(2) x U(1) electroweak model with 3 generations of quarks and leptons, the quark eigenstates of the weak interactions are related to the quark mass or flavor eigenstates by a 3 x 3 unitary matrix.[1] All 9 elements of this CKM mixing matrix can be described in terms of 3 real angles, and one complex phase which is responsible for the CP violation in the neutral K system. This phenomenological model must be tested, and one place this can be done is in the B meson system.

There are several ways that the B system can be used to probe the CKM phenomenology of weak decays.

1. we can measure $B^0\bar{B}^0$ mixing: the neutral B mesons should mix, and they do.
2. we can try to measure $V_{ub}$, the matrix element that describes the coupling of the b quark to the u quark. If the CKM description of CP violation in the K system is correct with 3 generations, then we must have $V_{ub} \neq 0$.
3. ultimately, we want to measure CP violation in the B meson system. If the CKM picture is correct, CP violation should exist in the B system, and while seeing it will not be the surprise that the observation of CP violation in the neutral kaon system was, it is in many ways the holy grail of B physics.

One warning is that if SLC and LEP tell us that there are more than three generations of quarks and leptons, then testing an "extended" CKM model with mixing between all 4 generations will be much more difficult.

The CLEO collaboration is a group of 90 physicists from 12 universities. The results presented in this paper are based on data from the 1987 run of the CLEO detector at the Cornell Electron Storage Ring (CESR).

CLEO is a general purpose magnetic detector which is described in detail elsewhere.[2] The momentum resolution of the CLEO tracking system for charged particles is

$$(\frac{\Delta p}{p})^2 = (0.0023p)^2 + (0.007)^2 \tag{1}$$

with $p$ and $\Delta p$ in units of GeV/c. Particle identification is done through energy loss measurements made in the 51 layer drift chamber with a dE/dX resolution of 6.5%. The integrated luminosity of the 1987 data set is $212pb^{-1}$ in the $\Upsilon(4S)$ and $101pb^{-1}$ in the continium just below $B\bar{B}$ threshold. This corresponds to a data set of approximately 242,000 $B\bar{B}$ pairs.

## 2. $B^0\bar{B}^0$ Mixing

Using the 1987 data sample, CLEO has confirmed the ARGUS report of large mixing in the neutral B meson system.[3] At the $\Upsilon(4S)$ we produce charged or neutral B mesons in pairs. The neutral mesons can mix via box diagrams and this mixing is observable if a $B^0$ mixes into a $\bar{B}^0$ at a rate comparable to the decay rate of the $B^0$.

The cleanest measure of mixing is full event reconstruction of $B^0B^0$ or $\bar{B}^0\bar{B}^0$ events. We have not been able to do that and hence must rely on tagging the flavor of events in inclusive B decays. Specifically, we use the lepton sign in semi-leptonic B decays to tag the charge of the W emitted when the b quark decays. Leptons are particularly useful because they have a large ($\sim$ 10%) branching ratio, they are easy to detect, and there is low background at high momentum.

In order to measure mixing, we want to measure the ratio of the probability that a particle created as a $B^0$ will decay as a $\bar{B}^0$, to the probability that it will decay as a $B^0$. On the $\Upsilon(4S)$, in the absence of background, this can be expressed in terms of the ratio of same sign to opposite sign dilepton events from neutral B decays.

$$r = \frac{\Gamma(B^0 \to \bar{B}^0 \to \bar{X}^0)}{\Gamma(B^0 \to X^0)} = \frac{N(\ell^+\ell^+) + N(\ell^-\ell^-)}{N(\ell^+\ell^- from B^0 decay)} \tag{2}$$

Whereas counting the number of same sign and opposite sign dilepton events is simple in practice, there are several experimental difficulties:

1. fakes: hadrons misidentified as leptons can contribute to the dilepton signal
2. cascade decays from charm: one B can decay semileptonically and the other B can decay to a charmed meson which then decays semileptonically, contributing like sign dilepton events even in the absence of mixing.
3. charged B's produced at the $\Upsilon(4S)$ can decay semileptonically and contribute to the denominator of the mixing ratio r.

For the mixing analysis we select hadronic events with 2 leptons where both leptons are required to have a momentum of greater than 1.4 GeV/c. Below 1.4 GeV, the lepton spectrum becomes deminated by secondary leptons from the cascade decay $\bar{B} \rightarrow X'D \rightarrow \ell^+\nu Y$. The cut at 1.4 GeV/c supresses events where one lepton is from a B semileptonic decay and the second lepton in the event is from a D semileptonic decay. Hadrons misidentified as leptons are suppressed by requiring the lepton momentum to be less than 2.4 GeV/c which is close to the kinematic upper limit for B decaying to charm, and by requiring the angle between the two leptons, $\theta_{\ell_1\ell_2}$ be

$$-0.8 \leq cos\theta_{\ell_1\ell_2} < 0.9 \tag{3}$$

The upper limit on $cos\theta_{\ell_1\ell_2}$ supresses photon conversions.

These cuts leave 470 unlike sign and 71 like sign dilepton events. Of these 71 like sign events, cascade decays from charm and misidentified hadrons account for almost 50% of the events and must be subtracted. Table 1[4] gives the numbers of like sign and opposite sign dilepton events in the CLEO data sample.

In order to calculate the mixing ratio, r, from the numbers in table 1, we must subtract the contribution to the opposite sign dilepton sample from the semileptonic decays of charged B mesons. This is difficult to do since the fraction of the total number of dileptons coming from charged B decays depends on quantities that are not well known.

**Table 1. Numbers of Dilepton Events**

| | $N(\ell^+\ell^-)$ | | | $N(\ell^\pm\ell^\pm)$ | | |
|---|---|---|---|---|---|---|
| | $e^+e^-$ | $\mu^+\mu^-$ | $e^\pm\mu^\mp$ | $e^\pm e^\pm$ | $\mu^\pm\mu^\pm$ | $e^\pm\mu^\pm$ |
| on $\Upsilon(4S)$ | 186 | 66 | 218 | 26 | 6 | 39 |
| Fakes + | 19 | 9 | 16 | 15.2 | 5.1 | 18.2 |
| Continuum + | ±5.1 | ±3.7 | ±5.4 | ±3.4 | ±1.1 | ±3.9 |
| Cascades + $\psi$ | | | | | | |
| Signal | 167 | 57 | 202 | 10.8 | 0.9 | 20.8 |
| (statistical) | ±14 | ±8 | ±15 | ±5.1 | ±2.4 | ±6.2 |
| (systematic) | ±2.8 | ±2.1 | ±5.3 | ±3.4 | ±1.1 | ±3.9 |

The number of dileptons from charged B decays depends on:

(a) The ratio of the number of charged B mesons produced at the $\Upsilon(4S)$ to the number of neutral B mesons produced at the $\Upsilon(4S)$: $f_{+-}/f_{00}$.

(b) The ratio of the semileptonic branching ratios of charged to neutral B mesons, which is equal to the lifetime ratio of charged to neutral B mesons: $b^+/b_0$.

If we choose what we think are reasonable values for these numbers:[3]

$$\frac{f^{+-}}{f_{00}} = \frac{.55}{.45}; \quad \frac{b^+}{b_0} = 1, \tag{4}$$

we get a value for the mixing ratio $r = .188 \pm .055 \pm .056$ where the first error is statistical and the second error is systematic. The errors in r do not reflect the uncertainty in our knowledge of $f^{+-}/f_{00}$ and $b_+/b_0$ since they do not affect the statistical significance of our result. However, we want to know r absolutely.

In order to do this we estimate that

$$.897 \leq \frac{f_{+-}}{f_{00}} \leq 1.233 \tag{5}$$

based on the particle data book[5] value for the $B^0 - B^+$ mass difference and a new CLEO result

$$PDG : m_{B^0} - m_{B^+} = 1.9 \pm 1.1 MeV \tag{6}$$

$$CLEO : m_{B^0} - m_{B^+} = -.2 \pm .7 MeV\,(preliminary) \tag{7}$$

and assuming that the production ratio is proportional to $\beta^3$ as determined by phase space. We use a new CLEO measurement[6] for the ratio of the semileptonic branching ratio of charged to neutral B mesons from exclusive reconstruction of $\bar{B}^0 \to D^{*+}\ell^-\bar{\nu}$ and $B^- \to D^{*0}\ell^-\bar{\nu}$ which gives

$$\frac{b_+}{b_0} = .848 \pm .19^{+.16}_{-.22}. \tag{8}$$

Combining these results gives us that the central value for the mixing ratio r must be bounded:

$$.09 < r < .33 \tag{9}$$

at the 90% confidence level.

## 3. Charmless B Decays

CLEO has searched for evidence that the coupling of the b quark to the u quark is different from zero in a variety of ways. The most sensitive technique is to examine the inclusive single lepton spectrum for leptons from the decay $\bar{B} \to X\ell^-\bar{\nu}$ which are incompatible with coming from a charm decay of the B meson. This technique relies on the fact that the minimal hadronic mass in charm decay is m(D) while in a charmless decay it can be as light as $m(\pi)$. This is reflected in the momentum of the lepton from the decaying B. Leptons from the decay $\bar{B} \to D\ell^-\bar{\nu}$ have $p_\ell < 2.46$ GeV/c in the lab while in $\bar{B} \to \pi\ell^-\bar{\nu}$, the lepton momentum can extend to 2.7 GeV. This analysis used the full $212pb^{-1}$

of the 1987 data set. The events are continuum supressed by a cut on event shape and the remaining continuum contribution is fitted and subtracted using continuum data. A known source of leptons in the endpoint region is the tail of the $\bar{B} \to D(D^*)\ell^-\bar{\nu}$ decay spectrum. This contribution is estimated from Monte Carlo using the model of Isgur et al.[7] to be $36 \pm 6$ leptons in the 2.3 to 2.4 GeV region, and $2 \pm 3$ leptons in the 2.4 to 2.6 GeV region. We are left with an excess of $4 \pm 17$ events in 2.3 - 2.4 GeV/c and $56 \pm 21$ events in the 2.4 - 2.6 GeV/c for a total of $60 \pm 27$ events. We cannot account for these leptons by any known source and assume that they result from semileptonic B decays proceeding through a $b \to u$ transition.[8] The cross section integrated over 2.3 to 2.6 GeV/c is

$$\sigma(2.3 < p_\ell < 2.6) = 0.33 \pm 0.15 pb \tag{10}$$

In order to relate the observed cross section to the Kobayashi-Mashawa matrix element, $V_{ub}$, requires theoretical models. There are a variety of models: inclusive or spectator models start at the quark level and predict the full lepton momentum spectrum; exclusive or form factor models start with a few low multiplicity channels ($\bar{B} \to \pi\ell^-\bar{\nu}, \rho\ell^-\bar{\nu}$) and predict only the high end of the momentum spectrum where those channels dominate.[14] Results on $|\frac{V_{ub}}{V_{cb}}|$ are listed in Table 2 for a variety of models.

**Table 2. Preliminary Values for $|V_{ub}/V_{cb}|$ Using Various Models.**

| | $\|V_{ub}/V_{cb}\|$ |
|---|---|
| ISGW[7] | $0.12 \pm 0.03$ |
| KS[12] | $0.06 \pm 0.02$ |
| WBS[10] | $0.08 \pm 0.02$ |
| ACM[13] | $0.07 \pm 0.02$ |

A second technique to search for charmless B decays is to attempt to exclusively reconstruct a charmless final state in B decay. A fully reconstructed decay would be striking evidence that $|V_{ub}| \neq 0$, although the theoretical interpretation in terms of a measurement of $V_{ub}$ is difficult. CLEO has attempted to reconstruct B decays to a variety of low multiplicity charged particle final states not containing a charm or strange particle.[9] For each decay mode listed in Table 3, the mass of a candidate hadronic resonance was required to be within $\frac{3}{4}\Gamma$ of its known value. The total energy of the tracks of a B candidate had to be within $2\sigma$ of the beam energy where $25 < \sigma < 37 MeV$ for all the modes investigated.

**Table 3. Upper Limits (90% C.L.) for Exclusive $b \to u$ Decays.**

The ranges given for $p\bar{p}\pi^-(\pi^+)$ correspond to searches including different regions of final state phase space.

| Mode | Upper Limits | |
|---|---|---|
| | B.R. (units of $10^{-4}$) | $\lvert V_{ub}/V_{cb}\rvert$ |
| $\bar{B}^0 \to \pi^+\pi^-$ | 0.9 | 0.21 |
| $\bar{B}^0 \to a_1^{\pm}\pi^{\mp}$ | 5.7 | 0.30 |
| $\bar{B}^0 \to a_2^{\pm}\pi^{\mp}$ | 3.5 | - |
| $\bar{B}^0 \to a_1^+ a_1^-$ | 32.2 | 0.66 |
| $\bar{B}^0 \to \rho^0\rho^0$ | 3.4 | - |
| $B^- \to f_0\pi^-$ | 1.2 | 0.57 |
| $B^- \to f_2\pi^-$ | 2.1 | - |
| $B^- \to \rho^0 a_1^-$ | 5.4 | 0.39 |
| $B^- \to \rho^0 a_2^-$ | 6.3 | - |
| $B^- \to \pi^-\pi^+\pi^-$ | 1.7 | - |
| $B^- \to \rho^0\pi^-$ | 1.5 | - |
| $\bar{B}^0 \to p\bar{p}$ | 0.4 | - |
| $\bar{B}^0 \to p\bar{p}\pi^+\pi^-$ | 1.2-11.9 | - |
| $B^- \to \Delta^0\bar{p}$ | 3.3 | - |
| $B^- \to \bar{\Delta}^{--}p$ | 1.3 | - |
| $B^- \to p\bar{p}\pi^-$ | 1.1-3.0 | - |

The dominant background was from continuum events. This was supressed by cutting on the event shape, exploiting the fact that the decay products of 2 B's are uncorrellated while continuum events are jet like. The upper limits on the branching ratios to various modes was determined by counting events within 5 MeV of the B mass and estimating the efficiency for detecting each mode from the Monte Carlo. Table 3 gives the 90% confidence level upper limits for a variety of decay modes. These branching ratio upper limits can be converted into upper limits on $V_{ub}/V_{cb}$ using the form factor model of Wirbel, Bauer, and Stech,[10] and these are listed in Table 3.

A final search for charmless B decays was motivated by the ARGUS report of a relatively large branching ratio for $B \to p\bar{p}\pi$, $B \to p\bar{p}\pi\pi$. CLEO searched for this signal in a variety of ways. We did an unbiassed search very similar to the analysis just described where no region of final state phase space was favored, and then searches were done to explore regions of phase space which might be favored in B decay or disfavored in background. In particular CLEO looked at decays where the $p$ and $\bar{p}$ were almost back to back where ARGUS saw a signal. The upper limits on the branching ratio for these decays is given in Table 3.

The CLEO detector is turning on in the summer of 1989 after a 15 month shutdown, during which the entire detector (with the exception of the drift chambers) has been replaced. A central feature of the new CLEO detector is a CsI shower detector which will enable us to detect photons. There is a planned luminosity upgrade of the CESR storage ring to reach a peak luminosity of $5 * 10^{32} cm^{-2} sec^{-1}$ from the current $1 * 10^{32}$ luminosity, which will correspond to $5 * 10^6 B\bar{B}/$ year in a typical operating year.

The current machine will not allow us to measure CP violation in the B system unless we are very lucky.[11] There is an R&D program underway at Cornell to address the technical issues in accelerator physics so that we can propose a design for a machine that will have a luminosity high enough to get within range of measurements of CP violation in the near future.

## REFERENCES

1) M. Kobayashi and T. Maskawa, Prog. Theor. Phys. 49, 652 (1973).

2) D. Andrews et al., Nucl. Instr. and Methods 211, 47 (1983). D. Cassel et al., Nucl. Instr. and Methods A252, 325 (1986).

3) H. Albrecht et al., Phys. Lett. 192B, 245 (1987).

4) M. Artuso et al., Phys. Rev. Lett. 62, 2233, (1989).

5) Particle Data Group: G.P. Yost et al., Phys. Lett. 204B, 1, 1988.

6) Sheldon Stone, CLNS 89/924

7) Isgur et al., Phys. Rev. D39, 799 (1989).

8) Michael Procario, "Charmless B Decays", in Proceedings of 1989 Heavy Quark Symposium, ed. P. Drell and D. Rubin American Institute of Physics, NY (to be published).

9) D. Borteletto et al., CLNS 89/887

10) M. Wirbel et al., Z. Phys. C29, 637 (1985).

11) G. Feldman et al., In High Energy Physics in the 1990's, DPF 1988 Summer Study, Snowmass, Colorado (to be published).

12) J. G. Körner and G. A. Schuler, Z. Phys. C38, 511, (1988).

13) G. Altarelli et al., Nucl. Phys. B208, 365, 1982.

14) K. Berkelman, "B Meson Decays" in Proceedings of 1988 SLAC Summer Institute, (to be published).

# Generation Mixing and the Parameters of the Standard Model

Harari Haim
Weizmann Institut of Science
Dept of Nuclear Physics
76100 REHOVOT
Israel

No written contribution received.

*Harry Lipkin and Haim Harari.*

# ORIGIN OF CP VIOLATION AND NEUTRAL K-SYSTEM

Yoshio YAMAGUCHI
Tokai University

ABSTRACT

Existence of a CP violating but C conserving interaction $H_X^C$ is proposed, effective strength of which is ~$10^{-3}$ as compared to that of strong interactions. Implications to particle physics, in particular to neutral kaons, are briefly discussed.

## §1 Introduction

The $2\pi$-decay of the long-lived neutral kaon $K_L$ is unique and clear-cut proof of CP-violation. Its branching ratio may be used to derive an amplitude $\sim 10^{-3}$ of $K_1$, CP even state, mixed in the $K_L$ state dominated by CP odd state $K_2$. This fact clearly shows the existence[1)] of a CP violating interaction $H_X$, whose effective strength (namely, second order effects in $H_X$, symbolically denoted as $(H_X^2)$ hereafter) is of the order of $10^{-3}$ as compared to that of the strong interaction $H_S$. Such a straightforward possibility has long been ignored during the past quarter of century since the discovery of $K_L \to 2\pi$ in 1964.

A CP violating interaction $H_X$ is shown to be equal to a sum of two parts, the one which is $H_X^C$ is CP violating but C conserving and the other $H_X^P$ which is CP violating but P conserving, under a general assumption of CPT invariance.

A CP violating $H_X$ is effectively weaker by $10^{-3}$ than the strong interaction $H_S$ and stronger by at least $10^3$ than the weak interaction $H_W$ a la Weinberg-Salam. Accordingly, $H_X$ must conserve all flavours of quarks.[1)] Otherwise, $H_X$ will completely spoil the success of the Standard Model (SM) explaining all hadronic weak processes. Moreover, it is simple to consider that $H_X$ will not act in lepton sectors in order not to disturb beautiful validity of QED and $H_W$ in all leptonic processes.

These requirements suffice to exclude $H_X^P$, since $H_X^P$ must accompany flavour change (i.e., $\Delta S = \pm 2$) is order to produce mixing between $K_2(0^-)$ and $K_1(0^-)$. Therefore, $H_X$ must be of the type $H_X^C$ (for details, see §2).

We assume that the SM Hamiltonian, i.e., strong, electromagnetic and weak interactions:

$$H_{SM} = H_S + H_{EM} + H_W$$

to be CP conserving, whereby the Kobayashi-Maskawa quark mixing matrix $U_{KM}$ to be real and orthogonal (in which the CP-phase is taken to be zero). All CP violating effects are supposed to be induced dynamically by $H_X$ (more specifically, $H_X^C$) in this paper.

## §2 CP violating and C conserving interaction $H_X^C$

Firstly, let us discuss a case where $H_X$ is CP violating but C conserving one $H_X^C$. An example of such interaction is given by

$$\sum_j f_j \bar{q}_j (1+i\gamma_5) q_j Y \qquad \left(\text{or} \quad \sum_j g_j \bar{q}_j (1-i\gamma_5) q_j Y'\right), \tag{2.1}$$

where $f_j$ is a real coupling constant which may depend on quark-flavour, $q_j$ is a quark field of a type j and Y is a spinless neutral boson field. We have chosen here (2.1) to be flavour conserving interaction. We may consider the coupling constants $f_j$ to be similar to the gauge coupling constant $g_{WS}$ of the Weinberg-Salam theory:

$$f_j^2 \sim 0.1 \tag{2.2}$$

from the spirit of GUTs and/or Kaluza-Klein theory. The mass $m_Y$ of CP violating force-carrier Y can be estimated by considering a ratio ~$10^3$ of effective strengths of ($H_X^2$):

$$f_j^2/m_Y^2 \tag{2.3}$$

and of ($H_W^2$):

$$g_{WS}^2/m_Z^2 \, . \tag{2.4}$$

We find in this way

$$m_Y \simeq \{10^{-3}(f_j^2/g_{WS}^2)\}^{1/2} m_Z \sim 3 \text{ GeV}/c^2 \tag{2.5}$$

For illustrative purpose, let us take the non-relativistic quark model of kaons. The $K^0$ meson is the $^1S_0$ state $|^1S_0,\bar{s}d\rangle$ of a $\bar{s}$ quark and d quark. The CP violating force in momentum representation between $\bar{s}$ and d can be derived from (2.1):

$$\frac{if_sf_d}{m_Y^2+q^2}\{1 - \frac{1}{2m_d}(\sigma_d q) + \frac{1}{2m_s}(\sigma_s q) + O(q^2)\} , \qquad (2.6)$$

where q is the momentum transfer: q = p(inicial) - p(final) , $m_s$ and $m_d$ are (constituent) masses of s and d quarks, respectively. The CP violating force (2.6) connects the pseudoscalar kaon state $|^1S_0,\bar{s}d\rangle$ to the scalar kaon state $|^3P_0,\bar{s}d\rangle$. Hence, the physical kaon will be represented by

$$|K^0\rangle = |^1S_0,\bar{s}d\rangle + \frac{\langle ^3P_0|(2.5)|^1S_0\rangle}{m(^1S_0)-m(^3P_0)}|^3P_0,\bar{s}d\rangle . \qquad (2.7)$$

Adopting the harmonic oscillator model for the $\bar{s}$-d system (and ignoring spin dependence), the coefficient of $|^3P_0,\bar{s}d\rangle$ in R.H.S. of (2.7) may approximate to

$$\frac{\langle 2p|(2.5)|1s\rangle}{m(1s)-m(1p)} = \frac{\langle 2p|(2.5)|1s\rangle}{-\omega} , \qquad (2.8)$$

where the value of $\omega$ may be fixed by fitting the electromagnetic m.s.r. radius of the 1s-wave function of the $K^+$ to the experimentally measured value:

$$\omega \simeq 0.4 \text{ GeV}. \qquad (2.9)$$

Then, we find in the p-representation

$$(\sin\phi)e^{i\Delta} = \frac{\langle 2p|(2.5)|1s\rangle}{-\omega} \simeq -(2.5\times10^{-3})e^{i(90°)}, \qquad (2.10)$$

namely $\sin\phi = -2.5\times10^{-3}$ and $\Delta = 90°$, where we have used

$$f_s f_d \simeq 0.1, \quad m_Y \simeq 3\ \mathrm{GeV}/c^2. \tag{2.11}$$

CP-mixing in $|K^0\rangle$ expressed in (2.7) implies that $|K_L\rangle$ should be written as

$$|K_L\rangle = \frac{1}{\sqrt{2}}\{|{}^1S_0,\bar{s}d\rangle + |{}^1S_0,\bar{d}s\rangle\}$$
$$+ \frac{\langle 2p|(2.5)|1s\rangle}{-\omega} \times \frac{1}{\sqrt{2}}\{|{}^3P_0,\bar{s}d\rangle + |{}^3P_0,\bar{d}s\rangle\},$$

or more generally

$$\begin{aligned} |K_L\rangle &= (\cos\phi)|K_2(0^-)\rangle + (\sin\phi)e^{i\Delta}|K_1(0^+)\rangle \\ |K_S\rangle &= (\cos\phi)|K_1(0^-)\rangle + (\sin\phi)e^{i\Delta}|K_2(0^+)\rangle \end{aligned} \tag{2.12}$$

The interaction $H_X^C$ given in (2.1) guarantees

$$\Delta = 90°, \tag{2.13}$$

while experimental data suggest $|\tan\phi| \sim 10^{-3}$.

It should be notice[1] that our $|K_S\rangle$ and $|K_L\rangle$ are orthogonal

$$\langle K_S|K_L\rangle = 0.$$

Moreover, the CP violating interaction (2.1) is "imaginary" (i times "real") in the sense of Wigner.[2]

Another possibility $H_X^P$ of CP violating interaction will predict a CP mixing which is distinctly different from (2.16):

$$\begin{aligned} |K_S\rangle &= (\cos\zeta)|K_1(0^-) - (\sin\zeta)e^{i\Delta}|K_2(0^-)\rangle \\ |K_L\rangle &= (\cos\zeta)|K_2(0^-)\rangle + (\sin\zeta)e^{i\Delta}|K_1(0^-)\rangle \end{aligned} \quad \text{for } H_X^P, \tag{2.14}$$

where $|\tan\zeta| \sim 10^{-3}$. To induce the type of mixing in eq.(2.14), $H_X^P$ must contain the coupling between $K_1(0^-)$ and $K_2(0^-)$, and hence between $K(0^-)$ and $\bar{K}(0^-)$, which requires $\Delta S = \pm 2$ transition of the effective strength $\sim 10^{-3}$ as compared to the strength of strong interactions. This should clearly be ruled out. So that the CP violating interactions $H_X$ must be restricted to those of C conserving type, $H_X^C$.

The expressions (2.16) [and (2.17)] are accurate up to $10^{-4}$ (or $10^{-5}$) at most. To be accurate up to $10^{-6}$ - $10^{-7}$, we have to take into account P and C violating mixing of 4 states, $K^0(0^\pm)$ and $\overline{K^0}(0^\pm)$, due also to $(H_W^2)$. Whereas, the expressions (2.16) [and (2.17)] can be used analyze $K_L \to 2\pi$ and the CP asymmetry of $(K_L)_{3\ell}$ up to the accuracy of $10^{-4}$ (or even to $10^{-5}$).

## §3 Minimal CP violation

It is convenient to introduce a concept of the minimal CP violating interactions. As stated earlier, the interaction, shown in eq.(2.1), is "imaginary" in the sense of Wigner. The total Hamiltonian consists of

strong interaction, $H_S$ (C, P, T separately
electromagnetic interaction $H_{EM}$ conserving)

and CP violating interaction $H_X$ in eq.(2.1). We shall define minimal CP violation to be CP violating interactions which are "imaginary" in the sense of Wigner. Clearly, (2.1) is of minimal CP violating. Also, an interaction

$$f\bar{u}(1+i\gamma_5)dY + f^*\bar{d}(1+i\gamma_5)uY^+, \qquad (3.1)$$

where f is complex and Y is a charged spinless boson field, is shown to be of minimal CP violating.

The S-matrix S of systems described by a minimal CP violating Hamiltonian

$$H = H_S + H_{EM} + H_X^C \tag{3.2}$$

can be expressed in the form

$$(S)_{mn} = e^{-i\ell_m \pi/2} (\tilde{S})_{mn} e^{i\ell_n \pi/2} \tag{3.3}$$

where $(\tilde{S}_{mn})$ is unitary and symmetric, and base vectors, $|n\rangle$, are labelled by total angular momentum, parity and other quantum, numbers including relevant orbital angular momentum $\ell_n$. S matrices of the time reversal invariant systems are symmetric and satisfy reciprocity. In the minimal CP violating case, S-matrices have the structure shown in (3.3), which is often sufficient to guarantee for many reactions to satisfy reciprocity. In this sense, previous tests of time reversal invariance are not necessarily the check of TRI and are often consistent with (3.3), namely the minimal CP violation. This is very important to realize. Moreover, it is remarkable that flavour conserving CP violating and C conserving interactions are normally of minimal CP violating. Non minimal CP violating local interactions requires flavour change (or introduction of new particles).

Let us consider a simple but remarkable consequence of minimal CP violation. Take two identical nucleon systems at low energies, where only elastic scattering should be considered. Then, the scattering of longitudinally polarized nucleons by unpolarized (target) nucleons will exhibit in general a longitudinal asymmetry

$$A_L = \frac{\sigma_R - \sigma_L}{\sigma_R + \sigma_L} \tag{3.4}$$

where $\sigma_R$ ($\sigma_L$) is a total cross-section for incident right(left)-handed helicity. $A_L$ is proportional to

$$A_L \propto (\sin 2\varepsilon)(\cos\Delta)$$

where $\tan\varepsilon$ and $\Delta$ are the relative amplitude and relative CP phase of two parity states, ${}^1S_0$ and ${}^3P_0$, of the same total angular momentum, 0. A minimal CP violating Hamiltonian predicts $\Delta$ to be $\pi/2$ (or odd integer times $\pi/2$). So that, the longitudinal asymmetry vanishes:

$$A_L = 0 \quad \text{(for NN scattering at low energies including minimal CP violating interactions),} \tag{3.5}$$

pretending the parity conserving situation. In practise, the weak interaction $H_W$ of SM must operate in two nucleon systems, so that $A_L$ would not vanish but $|A_L| \sim 10^{-6}$ to $10^{-7}$ at low energies ($\lesssim$ 100 MeV). This is a new possibility which has not been emphasized up to now.

## §4 $K_L \to 2\pi$

The $2\pi$ decays of $K_L$ consist of two contributions, the one is the $2\pi$ decays of the $K_1$ component in the $K_L$-state and the other is due to CP violating final state interactions (CP mixing between $2\pi$ and $3\pi$ states) of the pure $K_2$ state in the $K_L$-meson. First, we shall discuss the case, where these CP violations are induced by C conserving $H_X^C$, which is of minimal CP violation. Then, the relevant $2\pi$-decay amplitudes are given by

$$A(K_1(0^-) \to \pi^+\pi^-) = i(\cos\phi_\pi)(A_0+A_2F)e^{i\delta_0},$$

$$A(K_1(0^+) \to \pi^+\pi^-) = (\cos\phi_\pi)(B_0+B_2F)e^{i\delta_0},$$

$$A(K_2(0^-) \to \pi^+\pi^-) = (\sin\phi_\pi)(A_0+A_2F)e^{i\delta_0},$$
$$A(K_2(0^+) \to \pi^+\pi^-) = (\sin\phi_\pi)(B_0+B_2F)e^{i\delta_0}, \quad (4.1)$$
$$A(K_1(0^-) \to \pi^0\pi^0) = -i(\cos\phi_\pi)(A_0-2A_2F)e^{i\delta_0},$$
$$A(K_1(0^+) \to \pi^0\pi^0) = -(\cos\phi_\pi)(B_0-2B_2F)e^{i\delta_0},$$

and
$$F = e^{i(\delta_2-\delta_0)}, \quad (4.2)$$

where $A_0$ is chosen to be real and positive (by convention), and $\delta_I$ (I = 0 or 2) is the s-wave $\pi\pi$ elastic scattering phase shift with the total iso-spin I. $K_L$-states can be expressed as

$$|K_L\rangle = (\cos\phi)|K_2(0^-)\rangle + i(\sin\phi)|K_1(0^+)\rangle \quad (4.3)$$

up to the accuracy of $10^{-4}$ (or $10^{-5}$). Using decay amplitudes in (4.1), we find

$$A(K_L \to \pi^+\pi^-) = (\cos\phi\sin\phi_\pi)(A_0+A_2F) + i(\sin\phi\cos\phi_\pi)(B_0+B_2F),$$
$$A(K_S \to \pi^+\pi^-) = (\cos\phi\cos\phi_\pi)(A_0+A_2F) + i(\sin\phi\sin\phi_\pi)(B_0+B_2F), \quad (4.4)$$

The $2\pi^0$ amplitudes $A(K_j \to \pi^0\pi^0)$ (j = K and S) are give by (4.4) in which $A_2$ and $B_2$ are replaced by $-2A_2$ and $-2B_2$, respectively, $|\phi|$ and $|\phi_\pi|$ are dimensionless quantities of the order of $10^{-3}$. Branching ratios, $BR(K_S \to \pi^+\pi^-)$ and $BR(K_S \to \pi^0\pi^0)$, show that the $\Delta I = 1/2$ rule holds in $K_S \to 2\pi$ decays and give

$$-A_2/A_0 \simeq 0.02 \ll 1. \quad (4.5)$$

The usual ratios

$$\eta_{ab} \overset{\text{def}}{=} \frac{A(K_L \to \pi^a \pi^b)}{A(K_S \to \pi^a \pi^b)}, \qquad (ab) = +- \text{ and } 00 \qquad (4.6)$$

are easily derived from (4.4):

$$\begin{aligned} i\eta_{+-} &= \tan\phi_\pi + i\tan\phi \left\{\frac{B_0}{A_0} + \frac{B_2}{A_0}F\right\}, \\ i\eta_{00} &= \tan\phi_\pi + i\tan\phi \left\{\frac{B_0}{A_0} - \frac{2B_2}{A_0}F\right\}, \end{aligned} \qquad (4.7)$$

which are correct up to $10^{-4}$ (or $10^{-5}$).

Fitting (4.7) to measured magnitudes

$$\begin{aligned} |\eta_{+-}| &= (2.27 \pm 0.02)\times 10^{-3}, \quad |\eta_{00}/\eta_{+-}| = 1.00 \pm 0.01, \\ \arg(\eta_{+-}) &= 44.6° \pm 1.2° \simeq \arg(\eta_{00}) \end{aligned} \qquad (4.8)$$

we find

$$\tan\phi_\pi = -\frac{B_0}{A_0}\tan\phi \simeq +1.5\times 10^{-4}, \quad |B_2| \ll |B_0|, \qquad (4.9)$$

$$\arg(\eta_{+-}) \simeq \arg(\eta_{00}) \simeq 45°, \qquad (4.10)$$

Next, we shall consider the case where a CP violating interaction is P conserving, $H_X^P$. The $K_L$ state is, then, given by (2.17) with $|\tan\zeta| \sim 10^{-3}$, which leads to

$$\eta_{+-} = \eta_{00} = \tan\zeta \, e^{i\Delta}.$$

However, this case $H_X^P$ is unacceptable because $H_X^P$ induces too large $\Delta S = \pm 2$ transitions as stated earlier.

Another CP violating phenomenon,

$$\delta_\ell = \frac{\Gamma(K_L\to\pi^-\ell^+\nu)-\Gamma(K_K\to\pi^+\ell^-\bar{\nu})}{\Gamma(K_L\to\pi^-\ell^+\nu)-\Gamma(K_K\to\pi^+\ell^-\bar{\nu})}$$

can easily be explained (see, ref.1)).

## §5 Implication of $H_X^C$ to Particle Physics

Implication of an existence of new CP violating interaction $H_X^C$ to nuclear physics has been given.[1] Also the nature of the CP violating force carrier, a new boson, has been described.[1]

We shall give a few comments on implication of $H_X^C$ in particle physics. $\pi^0$ and $K_L^0$ are dominantly pseudoscalar states, but they contain non-strange scalar components (in amplitudes of $\sim 10^{-3}$ and $\sim 10^{-9}$, respectively). The non-strange scalar "mesons" are essentially nuclei of mass number zero and can decay into vacuum ($0^+$) by emitting a charged lepton pair, $e^+e^-$ or $\mu^+\mu^-$ ($0^+$-$0^+$ transition of nuclei). Taking the off-matrix element of (electro-magnetic radius)$^2$, of the order of electro-magnetic radii of $\pi^+$ or $K^+$, we can fit the branching ratio $BR(\pi^0\to e^+e^-)$ and $BR(K_L\to\mu^+\mu^-)$ (which were unable to explain theoretically so far).

Moreover, $\eta$ can decay into $2\pi$ through its tiny scalar component induced by $H_X^C$.

## Reference

1) A preliminary version of this new possibility has been published, Y. Yamaguchi, INS-Rep.-736, 5 Feb. 1989.

2) E. P. Wigner, Göttinger Nachrichten 31, (1932) 546.

*from left : Jacques Haïssinski, André Martin, Fred Gilman, Yoshio Yamaguchi and Martine Rebière.*

# RECONSTRUCTION OF THE KOBAYASHI-MASKAWA MATRIX FROM THE MODULI OF ITS ELEMENTS

André Martin
CERN - Geneva

## ABSTRACT

We show that if there are more than three generations of leptons and quarks, it is not necessarily true that one can reconstruct the Kobayashi-Maskawa matrix from the knowledge of the moduli of the matrix. In other words, while, for three generations, the knowledge of the decay rates of all "up" quarks into "down" quarks fixes the phase of the matrix and hence the CP violating effects in principle, the situation for four generations and above is more complicated. There may be continuous or discrete ambiguities.

I am presenting here a work done in collaboration with G. Auberson and G. Mennessier from Montpellier [1]. The problem, which some of you, in this audience composed mostly of experimentalists, could consider as rather academic, is the following.

In the Standard Model with $N$ generations of quarks and leptons, charged current weak interactions are described by a $N \times N$ unitary Kobayashi-Maskawa matrix. Assume we know with good accuracy the probabilities of decay of all charge $\frac{2}{3}$ quarks into charge -$\frac{1}{3}$ quarks, i.e., we know all **moduli** of the elements of the Kobayashi-Maskawa matrix; do we know the phases of the Kobayashi-Maskawa matrix?

In more physical terms, if we know all decay widths involving charged currents of superficially CP conserving processes, do we know automatically the phases necessary to calculate CP violating effects?

For the case of three generations of leptons and quarks, the problem has been solved by many authors. I shall quote only Cecilia, who is attending this Conference [2]. The answer is **YES**. The knowledge of the moduli fixes the unique non-trivial phase. Explicit construction of this phase can be given [3]. This is of course for the time being rather academic since the moduli are not very well known.

However, there may be more than three generations. This was true at the time of the conference and remains true at the time of writing this report, since the preliminary SLAC and CDF data [4] indicate that within one standard deviation, a fourth neutrino is not yet excluded. Furthermore, if, in the future, a fourth light neutrino were to be excluded by a measurement of the $Z$ width at LEP, for instance, there would still be the possibility of a fourth family with a superheavy neutrino with a mass larger than half the $Z_0$ mass, or some other possibility that we do not imagine at the present time.

I must admit that our interest in this problem is partly due to the mathematical challenge but I feel that this problem is nevertheless much closer to experimental physics than what the string people do.

Let me state our provisional conclusions in advance for those who do not want to read any further.

1. Either there are continuous ambiguities for given moduli. Explicit examples have been given some time ago by Auberson [5] and we now know all of them for four generations [1].

2. Or, if the Kobayashi-Maskawa matrix has a structure "à la Wolfenstein" , i.e., with matrix elements decreasing rapidly when one goes away from the main diagonal, we believe that there are discrete ambiguities - as many as $2^{\frac{N^2-3N}{2}}$ for $N$ generations. This is based on what is happening in the case of **unitary symmetric** matrices and on an approximate construction invented by Bjorken [6].

Now what do we mean by determining the phases? If we take a $N \times N$ matrix we can multiply the lines by $\exp i\alpha_1$, $\exp i\alpha_2$,..., $\exp i\alpha_N$ and the columns by $\exp i\beta_1$, $\exp i\beta_2$,..., $\exp i\beta_N$ and the physics will not be changed. What counts is the phases of the "plaquettes":

$$\phi_{i,j} - \phi_{i+1,j} + \phi_{i+1,j+1} - \phi_{i,j+1}$$

Furthermore the physics is unchanged if the matrix $M$ is replaced by its complex conjugate.

We shall now demonstrate the existence of discrete ambiguities in a special case.

## THE CASE OF SYMMETRIC MATRICES

Naturally, Kobayashi-Maskawa matrices have no reason to be symmetric, but they could be, by accident. So if ambiguities are present for symmetric matrices we expect ambiguities in general.

The case $N = 4$ was treated long ago by Babelon, Basdevant, Caillerie and Mennessier [7] (they were looking at four-channel scattering amplitudes), and they found that for a certain range of parameters they had four solutions, given the moduli.

In the case of general $N$, one removes the arbitrariness of the phase by deciding that the elements of the main diagonal are real positive. Then, under a certain condition which will come soon, one finds that there are

$$2^{\frac{N^2-3N}{2}}$$

different solutions. To make this plausible, assume that the matrix

$$U = \exp iH,$$

$H$, Hermitian operator, is very close to unity, then

$$U \simeq 1 + iH,$$

and $H$, being Hermitian symmetric is **real** symmetric and the moduli of $U$ are approximately $| h_{ij} |$. In this approximation you can change the signs of the $h_{ij}$ without changing the moduli. Counting the number of ways to change the sign and remembering that two complex conjugate matrices are equivalent gives the desired result.

To make this more rigorous we write $U = 1 + 2iT$, and we write the unitary condition as

$$\frac{1}{i}(T - T^{+}) = TT^{+} + T^{+}T \tag{1}$$

and

$$TT^{+} = T^{+}T \tag{2}$$

We have been able to show that if $T$ is small enough in norm,

$$||T|| < \frac{1}{\sqrt{2}},$$

the second equation (2) is automatically satisfied. We write the matrix elements as

$$T_{ij} = t_{ij} e^{i\delta_{ij}} \quad \text{with} \quad | \delta_{ij} | < \frac{M}{2}$$

$t_{ij}$ real but not necessarily positive. However, we can take $t_{i1} > 0$. Then Eq. (1) becomes

$$t_{ij} \sin \delta_{ij} = \frac{\sum_k' t_{ik} t_{jk} \cos(\delta_{ik} - \delta_{jk})}{1 - t_{ij} - t_{jj}},$$

where $\sum_k'$ is a sum over $k$ excluding $i$ and $j$.

Once the signs of the $t_{ij}$ have been chosen, this equation can be considered as defining a mapping: if we put a set of $\delta_{ij}$'s on the right-hand side, we get another set of $\delta_{ij}$'s on the left-hand side. It is not too difficult to prove

**THEOREM**

If

$$\sup \frac{\sum' | t_{ik} || t_{jk} |}{| t_{ij} | [1 - t_{ij} - t_{jj}]} < \frac{1}{2} \tag{3}$$

the iteration of the mapping converges towards the solution. For any choice of the signs of the $t_{ij}$ there is one and only one solution. Counting all possible signs and removing complex conjugation we find $2^{\frac{N^2-3N}{2}}$ solutions. This means one solution for $N = 3$, four solutions for $N = 4$, as expected.

In fact, if (3) holds, one can prove that one can forget the restriction on the norm of $T$. Condition (3) says that in a certain way the matrix $U$ is close to unity. However, it will be violated if one particular element is exceptionally small.

**THE CASE OF NON-SYMMETRIC MATRICES**

The most striking fact was the discovery of Auberson that there are continuous families of unitary matrices with given moduli of the elements [4]. Only very recently, cleaning my office, I realized that long ago I had written, at the request of Peter van Nieuwenhuizen, the most general unitary **antisymmetric** $4 \times 4$ matrix

$$U = \begin{pmatrix} 0 & a & b & c \\ -a & 0 & c^* e^{i\phi} & -b^* e^{i\phi} \\ -b & -c^* e^{i\phi} & 0 & a^* e^{i\phi} \\ -c & b^* e^{i\phi} & -a^* e^{i\phi} & 0 \end{pmatrix}$$

and this has the continuous parameter $\phi$ for given moduli.

We (i.e., in fact. G. Auberson) have been able to find **all** possible continuous ambiguities of $4 \times 4$ unitary matrices. I shall content myself with one class. I give only the moduli

$$| U_{ij} |= \begin{pmatrix} a & b & c & d \\ b & a & d & c \\ i & j & k & l \\ j & i & l & k \end{pmatrix}$$

with, naturally the sum of the squares of the lines and of the columns equal to unity (this means we have five independent moduli). Then there is **one** phase arbitrary.

There are two other classes of matrices with continuous ambiguities, but I shall not give them here. They can be found in Ref. [1]. Auberson succeeded in proving that these are the **only** possible continuous ambiguities

What happens outside these classes? For $4 \times 4$ matrices, there are at most 12 solutions, but most likely some of these are inacceptable (complex or corresponding to a cosine larger than one, etc.). Lavoura [5] has shown that the number of solutions for $4 \times 4$ matrices is even and has worked out explicit examples and exhibited four solutions. We give here one of the examples.

For a given matrix of the **squares** of the moduli,

$$(| U_{jk} |^2) = \begin{pmatrix} .95 & .0484 & .0001 & .0015 \\ .0489 & .9460 & .0024 & .0027 \\ .0005 & .0040 & .5000 & .4955 \\ .0006 & .0016 & .4975 & .5003 \end{pmatrix}$$

he gets, for the phases in degrees:

$$\begin{pmatrix} 0 & 0 & 0 & 0 \\ 0 & 179.89 & -25.04 & -173.86 \\ 0 & -74.73 & -137.18 & 138.04 \\ 0 & 73.54 & 117.45 & -147.55 \end{pmatrix}$$

$$\begin{pmatrix} 0 & 0 & 0 & 0 \\ 0 & 179.59 & 160.04 & -123.19 \\ 0 & -103.22 & 23.84 & 156.92 \\ 0 & -8.69 & -144.28 & 168.53 \end{pmatrix}$$

$$\begin{pmatrix} 0 & 0 & 0 & 0 \\ 0 & 179.58 & 165.58 & -122.72 \\ 0 & -32.74 & 168.09 & 167.15 \\ 0 & -123.78 & -20.18 & 159.16 \end{pmatrix}$$

$$\begin{pmatrix} 0 & 0 & 0 & 0 \\ 0 & 179.51 & -160.69 & -123.21 \\ 0 & -42.71 & -169.52 & 156.83 \\ 0 & -113.88 & 21.95 & 168.61 \end{pmatrix}$$

The fact that he has not more than four solutions, and the fact that in the symmetric case there are four solutions makes us suspect that outside the continuous ambiguities, and if the matrix is sufficiently close to unity, there are never more than four solutions in the $4 \times 4$ case.

Now comes the last indication which makes us believe that there are at most $2^{\frac{N^2-3N}{2}}$ solutions if one stays away from continuous ambiguities. Bjorken [6] assumes that the Kobayashi-Maskawa matrix has a structure "à la Wolfenstein", i.e., the matrix elements decrease very rapidly when you go away from the main diagonal of the matrix:

$$| U_{i,j} | \gg | U_{i,j+1} | \quad if \quad i < j, etc.$$

He makes the phase convention

$$U_{ii} \quad real > 0 \qquad U_{i,i+1} \quad real > 0$$

(notice that this convention would not be acceptable for an exactly symmetric matrix!). The exact unitarity condition can be written

$$U_{ii}U_{ji} + U_{ij}^*U_{jj} = -\sum_k{}' U_{ik}^*U_{jk}$$

where $\sum_k'$ represents a summation over $k$ excluding $i$ and $j$.

However, if the "Wolfenstein" property holds, this becomes, approximately, for $i < j$

$$U_{ji} + U_{ij}^* = -\sum_{k=i+1}^{k=j-1} U_{ik}^*U_{jk}$$

This equation is used by Bjorken to make an approximate, iterative construction of the Kobayashi-Maskawa matrix, starting from the main diagonal of the matrix to get the elements of the next-to-main diagonal, next-to-next-to-main, etc. In the first step, the "Bjorken" equation reduces to

$$U_{i+1,i} = -U_{i,i+1}^* = -U_{i,i+1}$$

In the next step, we get

$$U_{i+2,i} + U_{i,i+2}^* = -U_{i,i+1}^*U_{i+2,i+1}$$

The right-hand side is already known, and on the other hand, the moduli of $U_{i+2,i}$ and $U_{i,i+2}$ are known. So we have to construct a triangle for which we know one side, with its orientation in the complex plane, and the lengths of the two other sides. There are two (or zero) solutions to this problem, and so there is a twofold ambiguity, which was first overlooked by Bjorken and which I pointed out to him (I am very grateful to B.J. for publicizing my remark). At each following step the ambiguity appears so that, if there is no obstacle to the construction, i.e., if there is nowhere a violation of triangular inequalities, one gets $2^{\frac{N^2-3N}{2}}$ approximate solutions.

One can show that if the matrix is such that

$$| U_{ij} | + | U_{ji} | > \sum{}' | U_{ik} || U_{jk} |$$

and has moduli sufficiently close to being **symmetric**, these $2^{\frac{N^2-3N}{2}}$ approximate solutions can be constructed.

Now, does this mean that exact solutions also exist, and that they are isolated? We believe this but for the time being we have no idea on how to prove it. Let us remark that for the $4 \times 4$ case all matrices leading to continuous ambiguities violate the structure "à la Wolfenstein".

## REFERENCES


1. G. Auberson, G. Mennessier and A. Martin, CERN Preprint TH. 5404 (1989), to appear in Léon Van Hove's Festschrift, A. Giovannini and W. Kittel editors.
2. C. Jarlskog, *Z.Phys.* **C29** (1985) 491;, *Phys.Rev.Letters* **55** (1985) 1039.
3. C. Hamzaoui, *Phys.Rev.Letters* **61** (1988) 35.
4. Preliminary results presented at the Lepton-Photon Conference, SLAC (August 1989).
5. G. Auberson, *Phys.Lett.* **B216** (1988) 167.
6. J.D. Bjorken, *Phys.Rev.* **D39** (1989) 1396.
7. O. Babelon, J.L. Basdevant, P. Caillerie and G. Mennessier, *Nucl.Phys.* **B113** (1976) 445.

# CP VIOLATION BEYOND THE STANDARD MODEL

*Cynthia and Pierre Sikivie.*

# THE STRONG CP PROBLEM AND THE AXION

P. Sikivie
Department of Physics
University of Florida
Gainesville, FL 32611 USA

I. The Strong CP Problem

The strong CP problem raised its head when it was discovered[1] that QCD, the theory of the strong interactions, depends not only on a gauge coupling $g_s$ and quark masses $m_j$ (j=1...n) but also on an angle $\theta$:

$$\mathcal{L}_{QCD} = -\frac{1}{4} G^a_{\mu\nu} G^{a\mu\nu} + \sum_{j=1}^{n} [\bar{q}_j i\gamma^\mu D_\mu q_j - m_j q_{Lj}^\dagger q_{Rj} - m_j^* q_{Rj}^\dagger q_{Lj}]$$

$$+ \frac{\theta g_s^2}{32\pi^2} G^a_{\mu\nu} \tilde{G}^{a\mu\nu}. \qquad (1.1)$$

The $q_j$ are the quark fields, the $G^a_{\mu\nu}$ are the $SU^c(3)$ field strengths (a=1...8) and $\tilde{G}^a_{\mu\nu} \equiv \frac{1}{2} \epsilon_{\mu\nu\alpha\beta} G^{a\alpha\beta}$. The last term in (1.1) is actually a 4-divergence and hence contributes only a surface term to the QCD action. Surface terms do not contribute to the classical equations of motion and they do not contribute to the perturbative expansion of the quantum theory. However, they can produce non-perturbative quantum effects and these effects can be large in a strong coupling regime. This is what happens in the case of QCD in which there are non-perturbative effects associated with the existence of color instantons[2]. The latter are Yang-Mills field configurations in Euclidean space for which the surface term $\frac{g_s^2}{32\pi^2} \int d^4x_E\, G^a_{\mu\nu} \tilde{G}^{a\mu\nu}$ is non-zero even though $\int d^4x_E\, -\frac{1}{4} G^a_{\mu\nu} G^{a\mu\nu}$ is finite (the field configuration is a pure gauge as $|x| \to \infty$) and which describe the quantum-mechanical tunneling between inequivalent vacua of QCD. For an excellent account of the properties and significance of color instantons, we refer the reader to S. Coleman's 1977 Erice lectures[3]. For the topic of concern here, the two following points are the crucial ones:

1. Instanton effects exist and hence the physics of QCD is $\theta$ dependent. How can one be so sure of that? The main reason for our confidence is that the non-perturbative instanton effects of QCD are used to solve[1] the $U_A(1)$ problem. This is a major success of QCD. The $U_A(1)$ problem[4] is the problem why there is no quasi-Nambu-Goldstone boson associated with the spontaneous breaking of the $U_A(1)$ quasi-symmetry

under which $q_j \to e^{i\alpha\gamma_5} q_j$ (j=1...n) whereas the $\pi$, K and $\eta$ mesons can be successfully identified as the quasi-Nambu- Goldstone bosons associated with the spontaneous breaking of $SU_L(3)xSU_R(3) \to SU_{L+R}(3)$. The non-perturbative effects associated with QCD instantons produce a large explicit breaking of $U_A(1)$ and thus solve the $U_A(1)$ problem. But they also make the physics of QCD depend upon the parameter $\theta$.

2. The physics of QCD does not change if one applies the transformations

$$q_j \to e^{i\alpha_j\gamma_5} q_j, \quad m_j \to e^{-2i\alpha_j} m_j \qquad \text{for } j=1\ldots n$$

$$\theta \to \theta - 2 \sum_{j=1}^{n} \alpha_j . \tag{1.2}$$

This result follows from the statement of the anomaly in the chiral $U_A(1)$ current:

$$\partial_\mu(\bar{q}_j\gamma^\mu\gamma_5 q_j) = 2m_j i\, \bar{q}_j\gamma_5 q_j - \frac{g_s^2}{16\pi^2} G^a_{\mu\nu}\, \tilde{G}^{a\mu\nu} . \tag{1.3}$$

The invariance (1.2) implies that QCD is periodic in $\theta$ with period $2\pi$. It also implies that QCD depends upon $\theta$ only through the combination

$$\bar{\theta} = \theta - \arg \det m_q \tag{1.4}$$

and that the dependence of QCD upon $\theta$ disappears in the limit where one of the quark masses goes to zero.

If $\theta \neq 0$ and/or the quark masses are complex, P and CP are in general violated, with C remaining conserved, since $G^a_{\mu\nu}\tilde{G}^{a\mu\nu} = -4\, \vec{E}^a\cdot\vec{B}^a$ and $i\bar{q}\gamma_5 q$ are both P odd, CP odd and C even. Using (1.2), one can shuffle the P and CP violation back and forth between the $\theta$-term and the complex quark masses. However, if $\bar{\theta} \neq 0$ (mod $\pi$) one cannot simultaneously set $\theta$=0 and have all the quark masses real. $\bar{\theta}$ is the parameter that sets the amount of P and CP violation in QCD. From the experimental standpoint, no P or CP violation has been seen in the strong interactions. Hence $\bar{\theta}$ must be small (mod $\pi$). The best constraint on $\bar{\theta}$ is provided by the present experimental upper limit[5] on the neutron electric dipole moment[6]:

$$d_n \approx 5.10^{-16}\,\bar{\theta}\text{ e cm} \lesssim 6.10^{-25}\text{ e cm} \qquad (1.5)$$

which implies $\bar{\theta} \lesssim 10^{-9}$. The strong CP problem is the problem of explaining why $\bar{\theta}$ is so small. The quark masses originate in the electroweak sector of the theory which violates P and CP. There is no reason why arg det $m_q$ should exactly cancel $\theta$ in Eq. (1.4). In fact, in the standard model with CP violation à la Kobayashi-Maskawa[7], the Yukawa couplings are arbitrary complex numbers and hence arg det $m_q$ and $\bar{\theta}$ have no reason to take on any special value at all.

Peccei and Quinn[8] proposed a simple and elegant solution to the strong CP problem. The Peccei-Quinn solution and the concomitant axion will be described in the next section. Before getting there it may be good, however, to briefly mention alternative points of view:

- Ellis and Gaillard[9] pointed out that in the Kobayashi-Maskawa model of CP violation, and in most other models as well, the infinite renormalization of $\bar{\theta}$ occurs only in very high order in perturbation theory. This means that $\bar{\theta}$ varies only very slowly with energy scale. Hence, if the theory of everything (TOE) sets $\bar{\theta} = 0$ at the Planck scale, then $\bar{\theta} \ll 10^{-9}$ at the QCD scale. The question how TOE set $\bar{\theta} = 0$ is left unanswered; the point, however, is that the strong CP problem does not necessarily have to be solved in our present low energy effective theory.

- QCD is $\bar{\theta}$-independent and hence CP is conserved by the strong interactions if there is a massless colored fermion. One may consider the possibility that $m_u = 0$. This disagrees however with the lowest order current algebra relations between the pseudo-scalar meson masses[10]. One may also consider the possibility that the massless colored fermion is a sextet, octet or other large n-plet of $SU^c(3)$. This second alternative (as well as variations on this idea in technicolour models) is equivalent to the Peccei-Quinn mechanism. It predicts an axion with a decay constant v of order the binding energy of $\bar{Q}Q$ pairs in the vacuum where Q is the massless colored fermion.

- It is possible that CP is violated spontaneously. In that case $\bar{\theta}$ is calculable. It is possible to construct models of spontaneous CP violation in which $\bar{\theta} << 10^{-9}$. This alternative to the Peccei-Quinn mechanism is discussed in detail in the next talk by S. Barr.

- Recently, it has been shown that wormhole physics[11] predicts[12] the value of $\bar{\theta}$ in terms of low energy physics, i.e. physics below the weak interaction scale. It appears unfortunately that the predicted value of $\bar{\theta}$ is always such that poor current algebra relations among the pseudo-scalar meson masses are implied.

II. The Peccei-Quinn Mechanism and the Axion

Peccei and Quinn solved the strong CP problem by postulating a global $U_{PQ}(1)$ quasi-symmetry. It must have the following properties:

1. it is a symmetry of the classical theory, i.e. a symmetry of the theory at the Lagrangian level,
2. it is broken explicitly by those non-perturbative QCD effects (instantons and the like) which makes the physics of QCD depend upon the paramter $\theta$,
3. it is broken spontaneously by the vacuum expectation value of some scalar field.

To see how the existence of a $U_{PQ}(1)$ quasi-symmetry yields $\bar{\theta} = 0$ (up to tiny corrections due to the CP violating interactions responsible for $K_L \to 2\pi$) consider the theory defined by

$$\mathcal{L}_{PQ} = -\frac{1}{4} G^a_{\mu\nu} G^{a\mu\nu} + \frac{1}{2} \partial_\mu \varphi^\dagger \partial^\mu \varphi + \sum_{i=1}^{n} \left[ \bar{q}_i \gamma^\mu D_\mu q_i - (K_i\, q^\dagger_{Li}\, q_{Ri}\, \varphi + h.c.) \right]$$
$$+ \frac{\theta g_s^2}{32\pi^2} G^a_{\mu\nu} \tilde{G}^{a\mu\nu} - V(\varphi^\dagger \varphi). \qquad (2.1)$$

$\mathcal{L}_{PQ}$ has a classical $U_{PQ}(1)$ symmetry under which $\varphi \to e^{i\alpha}\varphi$ and $q_j \to e^{-i/2\alpha\gamma_5} q_j$ for j=1...n. Assuming the potential V has the shape of a "Mexican hat", the $U_{PQ}(1)$ symmetry is spontaneously broken by the vacuum expectation value of the scalar field $\varphi$:

$$\langle\varphi(x)\rangle = v\, e^{i\alpha(x)} . \qquad (2.2)$$

The quarks acquire masses

$$m_i = K_i\, v\, e^{i\alpha} \qquad (2.3)$$

and hence

$$\bar\theta = \theta - \arg\,(m_1 \ldots m_n)$$

$$= \theta - \arg\,(K_1 \ldots K_n) - n\alpha . \qquad (2.4)$$

The important difference between the theory defined by Eq. (1.1) and the theory defined by Eq. (2.1) is that in the former $\bar\theta$ is a function solely of the parameters in the theory, whereas in the latter $\bar\theta$ is a function also of the dynamical field $\alpha$. As a result, those non-perturbative effects which make the physics of QCD depend upon the parameter $\bar\theta$ will, in the theory defined by Eq. (2.1), produce an effective potential $V_{eff}(\alpha)$ for the dynamical field $\alpha$. It can be shown that the minimum of $V_{eff}(\alpha)$ occurs at a value of $\alpha$ such that

$$\bar\theta = 0 \ (\mathrm{mod}\ \pi) , \qquad (2.5)$$

i.e. such that the theory conserves P and CP. The value of $\bar\theta$ is always such as to yield the successful Gell-Mann-Oakes-Renner[13)] current algebra relations amongst the pseudoscalar meson masses. For pedagogical reasons, we did not include the electroweak interactions in our example of Eq. (2.1). Actually, the electroweak interactions only add a few inessential complications to the implementation of the Peccei-Quinn mechanism. The main difference from the description given above is that the CP violating interactions responsible for $K_L \to 2\pi$ will induce a small value for $\bar\theta$. This induced value of $\bar\theta$ is however always much smaller than $10^{-9}$. For example, Georgi and Randall[14)] found $\bar\theta \approx 10^{-17}$ for the Kobayashi-Maskawa model of CP violation.

Weinberg and Wilczek[15)] pointed out that the Peccei-Quinn solution to the strong CP problem implies the existence of a light pseudo-scalar particle, called the axion. The axion is the quasi-Nambu-Goldstone boson associated with the spontaneous breaking of the Peccei-Quinn quasi-symmetry. The axion field is

$$a(x) = v\,\alpha(x) \tag{2.6}$$

where v and $\alpha(x)$ are defined by Eq. (2.2). One can calculate the properties of the axion using current algebra or chiral Lagrangian techniques[16]. The axion mass is given by (in the limit $m_s \gg m_u, m_d$)

$$m_a = |N| \frac{f_\pi m_\pi}{v} \frac{\sqrt{m_u m_d}}{m_u + m_d} \tag{2.7}$$

where

$$N = 2 \sum_f t_f\, Q_f^{PQ} . \tag{2.8}$$

In Eq. (2.8), the sum is over all colored left-handed Weyl fermions f, the $Q_f^{PQ}$ are the Peccei-Quinn charges of these fermions and the $t_f$ are given by $Tr(T_f^\alpha T_f^\beta) = t_f\, \delta^{\alpha\beta}$ where the $T_f^\alpha$ ($\alpha = 1...8$) are the $SU^c(3)$ generators for the color representation to which f belongs ($t_3 = t_{\bar{3}} = 1/2$, $t_6 = t_{\bar{6}} = 5/2$, etc.). For example, N = n in the model of Eq. (2.1). I will sometimes use $f_a \equiv \frac{v}{N}$. The coupling of the axion to two photons is given by

$$\mathcal{L}_{a\gamma\gamma} = \frac{\alpha}{8\pi} \frac{a}{f_a} \left[\frac{N_e}{N} - \left(\frac{5}{3} + \frac{m_d - m_u}{m_d + m_u}\right)\right] F_{\mu\nu} \tilde{F}^{\mu\nu} . \tag{2.9}$$

Here $\alpha$ is the fine structure constant and

$$N_e = 2 \sum_f (Q_f^\gamma)^2\, Q_f^{PQ} , \tag{2.10}$$

where $Q_f^\gamma$ is the electric charge of fermion f in units of e. In many axion models $\frac{N_e}{N} = \frac{8}{3}$. In particular, this is true in all grand unified axion models which implement the successful Georgi-Quinn-Weinberg[17] prediction of $\sin^2\theta_w$. Comparing Eq. (2.7) with (2.9), one finds that in these models $\mathcal{L}_{a\gamma\gamma}$ is given uniquely in terms of the axion mass by

$$\mathcal{L}_{a\gamma\gamma} = - g_\gamma \frac{\alpha}{\pi} \frac{m_a}{0.6\ 10^{16}\ eV^2}\ a\ \vec{E}\cdot\vec{B} \qquad (2.11)$$

with $g_\gamma = \frac{m_u}{m_d+m_u} \approx 0.36$. In the limit where CP is conserved, the coupling of the axion to fermion f has the form

$$\mathcal{L}_{a\bar{f}f} = i\ g_f \frac{m_f}{v}\ a\ \bar{f}\gamma_5 f. \qquad (2.13)$$

Formulas which give the $g_f$ in terms of the parameters that define an axion model are given in Ref. (16), including those for the case where f is a proton or neutron.

## III. The Experimental and Observational Constraints on the Axion

When the axion was first proposed, it was thought that the breaking of $U_{PQ}(1)$ occurs at the electroweak scale, i.e. v ~ 250 GeV. The corresponding axion would have had mass in the 100 kev -- few MeV range and be relatively strongly coupled. It was searched for in various laboratory experiments but was not found. Soon, however, it was discovered how to construct axion models with arbitrarily large values of v[18]. As far as the solution to the strong CP problem is concerned, the value of v is arbitrary. So, we must a-priori consider an infinite range of possible axion masses. For v >> 250 GeV, the axion is so weakly coupled that the event rates in the axion search experiments mentioned above are hopelessly small. Such axions were called "invisible" because of that. However it has turned out that such axions are still constrained by astrophysical and cosmological considerations, and that they may even be searched for in new types of experiments or observations.

The high energy physics type laboratory searches rule out all axions with mass larger than a few tens of kev. These searches include electron and proton beam dumps, nuclear reactions and rare particle decays such as $K \to \pi a$, $J/\psi \to \gamma a$, $\Upsilon \to \gamma a$ and $\pi^{\dagger} \to a\ \nu e^{\dagger}$ where the axion escapes from the apparatus or decays into $\gamma\gamma$ or $e^+e^-$. From a historical viewpoint, it is interesting to note that axions with mass larger than one MeV which decay promptly into $e^+e^-$ ($\tau_a < 10^{-11}$ sec) were not ruled out until relatively recently. It was the discovery of a peak (at 1.8 MeV) in the energy spectrum of $e^+e^-$ pairs produced in heavy ion collisions which prompted physicists to take another look at the possibility of a heavy axion and then soon rule out this possibility altogether. For an excellent recent summary of the constraints on the "visible" axion, see R. Peccei in Ref. (19).

Next, let us consider the constraints from stellar evolution. Light weakly coupled bosons, such as the axion, are severely constrained by stellar evolution because stars emit such particles from their whole volume whereas they emit the more strongly coupled photons only from their surface. This idea, originally due to H. Sato and K. Sato[20)], has been very extensively applied to the axion[21,22)] using a variety of stellar objects (the sun, red giants, neutron stars, while dwarfs, and more recently the supernova SN 1987a) and a variety of axion producing thermal processes in those objects (the Primakoff process, Compton-like scattering, axion bremstrahlung in electron-nucleon and nucleon-nucleon scattering, etc). The main constraints come from the evolution of red giants and from the duration of the neutrino pulse in SN 1987a. In regard to the constraint from red giants one must distinguish between axions which have generic couplings to electrons (DFSZ-type axions) and axions whose coupling to the electron is much suppressed ("hadronic" or KSVZ-type axions). Incidentally, let us note that there is nothing artificial about an axion having a suppressed coupling to the electron. It is perfectly straightforward to build an axion model with that property and such a model can even be grand-unified. On the other hand, it is difficult to suppress the coupling of the axion to photons or to the nucleons. The evolution of red giants rules out the axion mass range 200 kev - $10^{-2}$ eV in the case of DFSZ-type axions, and the range 200 kev - 2 eV in the case of KSVZ-type axions. If the axion is heavier than 200 kev, it is too heavy to be thermally produced in red giants. Whereas, if the axions are sufficiently light, then they are sufficiently weakly coupled and hence their production is sufficiently suppressed that they do not appreciably alter the evolution of these stars. The constraint from SN 1987a rules out the axion mass range 2eV - $10^{-3}$ eV. Production of axions in the supernova core depletes the energy available for neutrino production and hence would have shortened the neutrino signal observed by the Kamiokande and IMB collaborations. If $m_a \lesssim 10^{-3}$ eV, too few axions are produced to affect the neutrino signal. If $m_a \gtrsim 2$ eV, the axions produced get reabsorbed in the supernova core resulting in no net loss of energy. As far as particle physics is concerned, the limits from SN 1987a are rather model independent because the process by which axions are emitted from the supernova core is axion bremstrahlung in nucleon-nucleon scattering. The relevant couplings are those of Eq. (2.13) with f = proton and neutron. The corresponding $g_p$ and $g_n$ are somewhat model-dependent but it is very difficult, as well as artificial, to construct axion models in which $g_p$ and $g_n$ are suppressed simultaneously. Note that in the case of KSVZ-type axions, there is possibly a window left open near $m_a$ = O(eV) between the upper limit from the evolution of red giants and the lower limit from SN 1987a. An excellent, up-to-date discussion[23] of the astrophysical constraints on the axion by M. Turner has recently appeared.

The cosmological bound on the axion mass is of order $m_a \gtrsim 10^{-5}$ eV. It will be discussed in detail in the next section. There is an additional constraint due to the presence of domain walls[24] in axion models. The domain walls must be gotten rid of before they dominate the energy density of the universe. This can, in fact, be achieved through inflation or by constructing the axion model in such a way that it has a unique vacuum.

## IV. The Cosmological Axion Energy Density

There are two main contributions to the present cosmological axion energy density. The first is due to axion production caused by the realignment of the vacuum during the QCD phase transition. The second is due to axions radiated by cosmic axion strings. We will discuss the two contributions in succession.

### a. The contribution from initial vacuum misalignment[25]

At very high temperatures the $U_{PQ}(1)$ symmetry is restored. It becomes spontaneously broken when the temperature drops below $T_{PQ} \approx v$ and the $\varphi$-field acquires vacuum expectation value:

$$\langle\varphi(x)\rangle = v\, e^{i\alpha(x)}. \qquad (4.1)$$

At these high temperatures, however, the non-perturbative QCD effects which will give the axion its mass and thereby lift the degeneracy at the bottom of the "Mexican hat" potential $V(\varphi^\dagger\varphi)$ are highly suppressed. When the axion mass turns on near the QCD phase transition, the axion field $a(x) = v\,\alpha(x)$ will start to oscillate about one of the CP conserving minima of the effective potential $V_{eff}(\alpha)$. The oscillation will start approximately at cosmological time $t_1$ such that

$$t_1 m_a[T(t_1)] = O(2\pi) \qquad (4.2)$$

where $m_a(T)$ is the temperature dependent axion mass. After time $t_1$, the axion mass changes sufficiently slowly that the total number of axions in the oscillation of $\alpha(x)$ about the CP conserving minimum is approximately an adiabatic invariant. $T_1 = T(t_1)$ has been estimated to equal about one GeV[25]. The number density of axions at time $t_1$ due to the initial vacuum misalignment is of order

$$n_a^{vac}(t_1) \approx \rho_a^{vac}(t_1)\,\frac{1}{m_a(t_1)} \approx \frac{1}{2}\,v^2\, m_a(t_1)\,\overline{\alpha^2(t_1)} \approx f_a^2\,\frac{1}{t_1} \qquad (4.3)$$

where $f_a = \frac{v}{N}$. In Eq. (4.3) we have used the fact that the field $a(x) = v\alpha(x)$ is approximately homogeneous on the horizon scale $t_1$. Wiggles in $\alpha(x)$ which entered the horizon long before $t_1$ have been red-shifted away. In Eq. (4.3) we have also used Eq. (4.2) and the fact that the initial departure $\alpha(t_1)$ of $\alpha$ from the nearest local minimum is of order $\frac{1}{N}$ because N is the number of CP conserving minima at the bottom of the "Mexican hat" potential. The axions of Eq. (4.3) are decoupled and non-relativistic. Assuming that the ratio of the axion number density to the entropy density is contant from time $t_1$ till today, one finds that the present energy density in axions is[25)]

$$\rho_a^{vac}(t_o) \approx \rho_{crit}(t_o)\left(\frac{0.6\ 10^{-5}\ \mathrm{eV}}{m_a}\right)^{7/6} \cdot \left(\frac{200\ \mathrm{MeV}}{\Lambda_{QCD}}\right)^{3/4} \left(\frac{75\ \mathrm{km/s\ Mpc}}{H_o}\right)^{2} \quad (4.4)$$

where $\rho_{crit}(t_o)$ is the present critical energy density for closing the universe, $H_o$ is the present value of the Hubble constant and $\Lambda_{QCD}$ is the QCD scale factor. Eq. (4.4) implies the bound $m_a \gtrsim 0.6\ 10^{-5}$ eV.

It should be emphasized however that there are many sources of uncertainty in the estimate of the cosmological axion energy density and that Eq. (4.4) only provides us with a very rough estimate. For example, the axion energy density can be diluted by the entropy release due to the decay of heavy particles which decouple before the QCD epoch but decay afterwards[26)]. It can also be diluted by the entropy release associated with a first order QCD phase transition. On the other hand, if the QCD phase transition[27)] is first order, the sudden change in the axion mass at the critical temperature may increase $\rho_a(t_o)$ as compared to Eq. (4.4). Finally, if inflation occurs after the Peccei-Quinn phase transition at which $U_{PQ}(1)$ gets spontaneously broken, then the right-hand side of Eq. (4.4) must be multiplied by $(\alpha(t_1)N)^2$ where $\alpha(t_1)N$ is a randomly chosen number between zero and O(1).

The axions produced during the QCD phase transition when the axion mass turns on are "cold dark matter" because these axions are non-relativistic even from the moment of their first appearance at 1 GeV temperature[28)]. Cold dark matter with a flat (Zel'dovich-Harrison) spectrum of primordial density perturbations and some "biasing" in the extent to which light traces matter yields at present a popular and apparently viable scenario of galaxy formation. It is thus conceivable that the dark matter which appears clustered in halos around galaxies is made of axions with mass in the $10^{-5}$ eV range.

b. The contribution from cosmic axion strings[29,30)]

Cosmic axion strings make another sizable contribution to the axion cosmological energy density if inflation does not occur between the Peccei-Quinn phase transition where $U_{PQ}(1)$ becomes spontaneously broken and the QCD epoch. An axion string is a topological knot in the vacuum expectation value (4.1). Its energy per unit length is $\tau \approx \pi v^2 \ell n\, vL$ where L is an infra-red cut-off which in the early universe is the average distance between strings. Cosmic axion strings dissipate their energy by radiating axions. This is a very efficient process. As a result the number density of axion strings in the early universe is expected to be of order one string per horizon, i.e. the lowest possible consistent with causality. The energy density in axion strings at cosmic time t is thus of order

$$\rho_s(t) \approx \frac{\tau}{t^2} = \frac{\pi v^2}{t^2} \ell n\, \mu t. \tag{4.5}$$

Using energy conservation, one obtains from (4.5) the number density of axions radiated by cosmic axion strings:

$$n_a^{str}(t) \approx \frac{1}{t^{3/2}} \int_{t_{PQ}}^{t} \frac{dt'}{t'^{3/2}} \frac{\tau(t')}{\omega(t')} , \tag{4.6}$$

where

$$\frac{1}{\omega(t)} = \int \frac{dE}{d\omega}(t) \frac{d\omega}{\omega} \Big/ \int \frac{dE}{d\omega}(t) d\omega \tag{4.7}$$

Here, $\frac{dE}{d\omega}(t)$ is the energy spectrum of axions radiated at time t.

The axions that are radiated at time t are radiated by cosmic axion strings which are bent over a distance scale of order t and which are relaxing to lower energy configurations. There is at present disagreement in the literature about the behavior and the energy spectrum of such a string. One view[29)] is that an axion string which is initially bent over a distance scale t oscillates 10 to 20 times before it has reached its minimum energy configuration and that the spectrum of radiated axions is peaked near $\frac{2\pi}{t}$. In that case, using Eqs. (4.6) and (4.7), one finds that the number density of radiated axions at time $t_1$ when the axion mass turns on [Eq. (4.2)] is

$$n_a^{str}(t_1) \approx \frac{v^2}{t_1} \ln vt_1 \qquad (4.8)$$

Comparing Eqs. (4.3) and (4.8), one would conclude that the cosmological bound on the axion mass from axions radiated by cosmic axion strings is roughly a factor $N^2 \ln vt_1 \approx 70\ N^2$ times more severe than the bound from initial vacuum misalignment. In that case there may be little or no room left between the cosmological bound ($m_a \gtrsim 4.10^{-4}\ N^2$ eV) and the bound from the supernova ($m_a \lesssim 10^{-3}$ eV). Diego Harari and myself[30], on the other hand, have argued that, in the limit of large $\ln \mu t$, an axion string which is initially bent over a distance scale t will straighten itself out at once (without any large scale oscillations) and that the energy spectrum of the radiated axions is 1/k with a high energy cutoff of order v and a low energy cutoff of order $\pi/t$. In that case, one has $\omega(t) \approx \frac{2\pi}{t} \ln vt$ and

$$n_a^{str}(t_1) \approx \frac{v^2}{t_1}\ . \qquad (4.9)$$

One would then conclude that the cosmological bound on the axion mass is, at least in the case N=1, roughly of the same order of magnitude as the bound from initial vacuum misalignment.

During the past year, the question of what are the qualitative characteristics of global string decay (in particular, the number of oscillations before the string has decayed away and the spectrum of radiated Nambu-Goldstone bosons) has received increased attention[31]. Hopefully, we can look forward in the near future to a resolution of the issues discussed in this subsection.

## V. The Cavity Detector of Galactic Halo Axions

As we saw in the preceding section, it is conceivable that the earth is bathed in a sea of galactic halo axions. If the galactic halo is made up exclusively of axions, their density in the solar neighborhood[32] would be of order $\frac{1}{2}\ 10^{-24}$ gr/cm$^3$ and their velocity dispersion would be of order $10^{-3}$ times the speed of light. A most likely mass range for such axions would include

$$2\pi(242\ \text{MH}_z) = 10^{-6}\ \text{eV} \lesssim m_a \lesssim 10^{-4}\ \text{eV} = 2\pi(24.2\ \text{GH}_z). \tag{5.1}$$

The possibility of detecting these dark matter axions constitutes an exciting prospect and one which, as we will see, is not entirely unrealistic.

a. The general principle[33)]

Axions can be searched for by stimulating their conversion to photons in a strong magnetic field. The relevant coupling is given in Eq. (2.11). In particular, an electromagnetic cavity permeated by a strong magnetic field can be used to detect galactic halo axions. The latter have velocities $\beta$ of order $10^{-3}$ and hence their energies

$$\epsilon_a = m_a + \frac{1}{2} m_a \beta^2 \tag{5.2}$$

have a spread of order $10^{-6}$ above the axion mass. Consider a cylindrical electromagnetic cavity of arbitrary cross-sectional shape, permeated by a large static approximately homogeneous longitudinal magnetic field $\vec{B} = B_o\hat{z}$. When the frequency $\omega = 2\pi f$ of an appropriate cavity mode equals $m_a[1+O(10^{-6})]$, galactic halo axions can convert to quanta of excitation (photons) of that mode. Only the $TM_{n\ell o}$ modes couple in the limit where the cavity is much smaller than the de Broglie wavelength $\lambda_a = 2\pi(\beta m_a)^{-1} \approx 2\pi\ 10^3\ m_a^{-1}$ of the galactic halo axions. The power on resonance ($\omega_{n\ell} = m_a[1 + O(10^{-6})]$) from axion → photon conversion into the $TM_{n\ell o}$ mode is

$$P_{n\ell} = \left(\frac{\alpha}{\pi} g_\gamma \frac{N}{v}\right)^2 V B_o^2 \rho_a C_{n\ell} \frac{1}{m_a} \min(Q_L, Q_a)$$

$$= 0.2\ 10^{-19}\ \text{Watt} \left(\frac{V}{500\ \text{liter}}\right) \left(\frac{B_o}{8\ \text{Tesla}}\right)^2 C_{n\ell}\ \cdot$$

$$\cdot \left(\frac{\rho_a}{\frac{1}{2}\ 10^{-24}\ \text{gr/cm}^3}\right) \left[\frac{m_a}{2\pi(3\text{GH}_z)}\right] \text{Min}\left(\frac{Q_L}{Q_a}, 1\right) \tag{5.3}$$

where V is the volume of the cavity, $\rho_a$ is the density of galactic halo axions on earth, $Q_L$ is the loaded quality factor of the cavity and $Q_a \equiv 10^6$ is the "quality factor" of the galactic halo axion signal, i.e. the ratio of their energy to their energy spread. $C_{n\ell}$ is a model dependent form factor defined by

$$C_{n\ell} = \frac{\left(\int_V \vec{B}_o \cdot \vec{E}_{n\ell}\, d^3x\right)^2}{V B_o^2 \int_V \epsilon\, \vec{E}_{n\ell} \cdot \vec{E}_{n\ell}\, d^3x} \qquad (5.4)$$

where $\vec{E}_{n\ell}(\vec{x})$ is the electric field for the cavity mode.

Because the axion mass is only known in order of magnitude at best, the cavity must be tunable and a large range of frequencies must be explored seeking a signal. The cavity can be tuned by inserting a dielectric rod into it[34], or by moving sideways inside the cavity a dielectric rod or a metal post[35]. Using Eq. (5.3), one finds that to obtain a given signal to noise ratio s/n, the required maximum search rate is

$$\frac{df}{dt} = \frac{2.6\ \text{GHz}}{\text{year}} \left(\frac{3n}{s}\right)^2 \left(\frac{V}{500\ \text{liter}}\right)^2 \left(\frac{B_o}{9\ \text{Tesla}}\right)^2 C^2 \cdot$$

$$\cdot \frac{\rho_a}{\frac{1}{2}\, 10^{-24}\ \text{gr/cm}^3} \left(\frac{20K}{T_n}\right)^2 \left(\frac{f}{3\text{GHz}}\right)^2 \frac{Q_W}{Q_a} \qquad (5.5)$$

where $T_n$ is the sum of the physical temperature of the cavity plus the noise temperature of the microwave receiver, and $1/Q_W$ is the contribution to $1/Q_L$ due to losses in the cavity walls. It was assumed in deriving Eq. (5.5) that $Q_L < Q_a$ in which case the search rate is optimized by having $Q_L = \frac{1}{3} Q_W$.

In fact, the best possible quality factors attainable at present, using oxygen free copper, are only of order $10^5$ in the GHz range. The factor $Q_W/Q_a$ on the RHS of Eq. (5.5) is therefore of order $10^{-1}$. On the other hand, we have measured[36] noise temperatures as low as 3K (in the 1.2 - 1.6 GHz range) in commercially available microwave receivers. Provided C = O(1), Eq. (5.5) suggests then that a galactic halo axion search is feasible with presently available technology. One of the problems such a search faces however is that a large volume empty cylindrical cavity has a low resonance frequency in its lowest TM mode: $f = 0.29\ \text{GHz}\ (\frac{40\text{cm}}{R})$ where R is the radius of the cavity. A large cylindrical cavity is convenient for exploring the low frequency end of the range (5.1). However, one wishes to extend the search to frequencies well above 1 GHz. How is this to be done? Use of the higher $TM_{n\ell o}$ modes entails an unaffordable loss in sensitivity because their form factors fall off like the square of the mode number. One can however increase the frequency of the lowest TM mode by inserting one or several metal posts into the cavity. In this way one can in fact obtain simultaneously large volume, high frequency and form factors C of order one. Alternatively one can adopt the brute force approach of filling up the available volume with many small cavities all resonant at the same frequency. These issues are discussed extensively in Ref. (35) which presents the results of computer simulations which optimize the cavity design to obtain the highest possible cosmic axion search rate for a given magnet size.

b. The status of ongoing experiments

There are at present two cavity detectors of galactic halo axions taking data, one at Brookhaven National Laboratory[34], the other at the University of Florida[36]. In addition, a group at KEK is working on the construction of a third detector[37].

The Brookhaven experiment has been running for about three years. It uses a 11 liter cavity placed in a 6 Tesla superconducting solenoid when searching near 1.2 GHz. To search at higher frequencies, an insert is placed inside the 6 Tesla coil which increases its magnetic field to 9 Tesla at some expense in available volume. The Brookhaven experiment system noise temperature is of order 15K. They have searched over a relatively large frequency band ranging from about 1.1 GHz to about 4 GHz. The upper limit they place on the local axion density over this range is of order $10^3$ times larger than the local halo density.

The assembly of the Florida detector has been completed and we have started to take data. Our detector has a 7 liter cavity placed inside a 9 Tesla superconducting solenoid. During the next few months, we are planning to search in the 1.3 to 1.7 GHz range. In this band- width, our

HEMT microwave receiver/amplifier has noise temperature as low as 3K. When the detector is cooled to 2.2K by pumping on the $H_e$ bath, the system noise temperature is 3K + 2.2K = 5.2K. In the 1.3 to 1.7 GHz range, we hope to place a limit on the local axion density which is somewhat more stringent than that obtained by the Brookhaven experiment.

## VI. Searching for Axions of Mass $m_a = O(eV)$

As was emphasized in Section III, there is an additional window, at $m_a = O(eV)$, where the axion may lurk provided its coupling to the electron is suppressed. Two ways to search for such axions have been proposed: 1) to search for axions emitted by the sun via their conversion to photons in a magnetic field, and 2) to search for monochromatic photons from the $2\gamma$ decay of relic thermal axions. Let us briefly describe these two opportunities.

### a. A solar axion detector or axion helioscope[33)]

The flux of solar axions on earth is[38)]

$$f_{\Theta,a} \approx \frac{7.4\ 10^{11}}{\text{sec. cm}^2} \left(\frac{m_a}{1\ \text{eV}}\right)^2 \tag{6,1}$$

in the case where the axion does not couple to the electron. Solar axions have a broad spectrum of energies of order the temperature in the solar interior, i.e. energies ranging from one to a few keV.

The cross-section for $a \to \gamma$ conversion in a region of volume V permeated by a static magnetic field $\vec{B}_o(\vec{x})$ is

$$\sigma = \frac{1}{16\pi^2 \beta_a} \left(\frac{\alpha}{\pi} \frac{g_\gamma}{f_a}\right)^2 \int d^3k_\gamma\, \delta(E_a - \omega) \left| \int_V d^3x\, e^{i(\vec{k}_\gamma - \vec{k}_a)\cdot\vec{x}}\, \vec{n}\times\vec{B}_o(x) \right|^2 \tag{6.2}$$

where $(E_a, \vec{k}_a) = E_a(1, \vec{\beta}_a)$ is the 4-momentum of the axion and $(\omega, \vec{k}_\gamma) = \omega(1, \vec{n})$ is the 4-momentum of the photon. Because the magnetic field is static, $E_a = \omega$ and therefore extra momentum $\vec{q} = \vec{k}_\gamma - \vec{k}_a$ is in general necessary for the transition to occur. Indeed the cross-section (6.2) is proportional to the power in the Fourier component of $\vec{B}_o(\vec{x})$ corresponding to the wave-vector $\vec{q} = \vec{k}_\gamma - \vec{k}_a$.

Consider a solar axion detector of volume V in which there is a magnetic field $B_o$ transverse to the direction of the sun. Multiplying the solar axion flux (6.1) by the cross-section (6.2), one finds the event rate

$$\frac{\#\text{events}}{\text{time}} \approx \frac{200}{\text{day}} \frac{V\ell}{(\text{meter})^4} \left(\frac{B_o}{8\ \text{Tesla}}\right)^2 \left(\frac{m_a}{1\ \text{eV}}\right)^4 \quad (6.3)$$

The final state photons are soft x-rays. In Eq. (6.3), $\ell$ is the coherence length for the $a \to \gamma$ conversion. If the magnetic field is homogeneous and the volume V is empty, the coherence length is[33)]

$$\ell \approx \frac{1}{q} = \frac{1}{E_a - \sqrt{E_a^2 - m_a^2}} \approx \frac{2E_a}{m_a^2} = 0.04\ \text{cm}\ \left(\frac{E_a}{\text{keV}}\right) \left(\frac{1\ \text{eV}}{m_a}\right)^2 . \quad (6.4)$$

However, one can increase $\ell$ up to a value equal to the depth of the volume V by filling this volume by a buffer gas[38)] whose pressure is adjusted in such a way that the plasma frequency, which acts as an effective mass for the photon, equals the axion mass inside V. A proposal[39)] has been put forth for the construction of an axion heliscope that could detect solar axions in the $0.1 \lesssim m_a \lesssim 5$ eV range.

b) Monochromatic photons from the $a \to 2\gamma$ decay of relic thermal axions[40)]

It so happens that if the axion mass is in the eV range, axions have the following interesting properties:

1) the present energy density in thermal axions is a sizable fraction of the total energy density

$$\Omega_a^{\text{thermal}} \approx 2.10^{-2} \left(\frac{m_a}{\text{eV}}\right) \quad (6.5)$$

2) their decay rate into $2\gamma$

$$\Gamma(a \to 2\gamma) = \frac{1}{0.67\ 10^{25}\ \text{sec}} \left(\frac{m_a}{\text{eV}}\right)^5 \quad (6.6)$$

is sufficiently long so that not all the relic axions have decayed and sufficiently short so that a substantial fraction are decaying at present. As a result, one can search for axions of mass in the multi-eV range by looking for photons from the $2\gamma$ decay of relic thermal axions. Such axions would have clustered in the same places as other forms of dark matter, i.e. the halos of galaxies and clusters of galaxies. The photon signal from $a \rightarrow 2\gamma$ decay in these objects would be highly monochromatic. This makes it of course easier to detect. It appears that a search for a monochromatic axion line from a cluster of galaxies will be sensitive to an axion as light as about 2 eV and an observational program is underway to search for such a line. See Refs. (40) for more details.

## VII. Conclusions

The axion has had a venerable and relatively long life as a theoretical hypothesis and, as such, it is not dead yet. However, by the standards celebrated at this conference -- i.e. of physical fact as opposed to attractive theoretical hypotesis -- we must pursue the question: does the axion exist in reality? It appears there are still two possible windows for the axion mass. The $10^{-3}$ eV $\lesssim m_a \lesssim 10^{-6}$ eV window can be searched for using an electromagnetic cavity in a magnetic field. Such searches are now under way but the present detectors are still a factor 100 to 300 too small to detect galactic halo axions. The $m_a = O(\mathrm{eV})$ window can be searched for by looking for solar axions, via their conversion to photons in a strong magnetic field, or by looking for monochromatic photons from the $2\gamma$ decay of relic thermal axions.

Acknowledgement:

This work was supported in part by the U. S. Department of Energy under contract No. DE-FG05-86ER-40272.

References:

1. G. 't Hooft, Phys. Rev. Lett. 37 (1976) 8, and Phys. Rev. D14 (1976) 3432; R. Jackiw and C. Rebbi, Phys. Rev. Lett. 37 (1976) 172; C. G. Callan, R. F. Dashen and D. J. Gross, Phys. Lett. 63B (1976) 334.

2. A. A. Belavin, A. M. Polyakov, A. S. Schwartz and Y. S. Tyuplin, Phys. Lett. 59B (1975) 85.
3. S. Coleman, "The many uses of instantons" in the Proceedings of the 1977 International School of Subnuclear Physics, Ettore Majorana [Erice, July 23 - August 10].
4. S. Weinberg, Phys. Rev. D11 (1975) 3583.
5. I. S. Altev et al., Phys. Lett. B136 (1984) 327.
6. V. Baluni, Phys. Rev. D19 (1979) 2227; R. J. Crewther, P. Di Vecchia, G. Veneziano and E. Witten, Phys. Lett. 88B (1979) 123.
7. M. Kobayashi and K. Maskawa, Progr. Theor. Phys. 49 (1973) 652.
8. R. D. Peccei and H. Quinn, Phys. Rev. Lett. 38 (1977) 1440, and Phys. Rev. D16 (1977) 1791.
9. J. Ellis and M. K. Gaillard, Phys. Lett. 74B (1978) 374.
10. It has been argued however that higher order corrections are important and that these allow $m_u = 0$. See D. B. Kaplan and A. V. Manohar, Phys. Rev. Lett. 56 (1986) 2004.
11. S. W. Hawking, Phys. Lett. 195B (1987) 337; S. Coleman, Nucl. Phys. B307 (1988) 867; S. Giddings and A. Strominger, Nucl. Phys. B307 (1988) 854; S. Coleman, Nucl. Phys. B310 (1988) 643; T. Banks, Nucl. Phys. B309 (1988) 493.
12. H. B. Nielsen and M. Ninomiya, U. of Tokyo preprint INS-Rep.-727 (Dec. 1988); K. Choi and R. Holman, Phys. Rev. Lett. 62 (1989) 2575; J. Preskill, S. P. Trivedi and M. B. Wise, Phys. Lett. B223 (1989) 26.
13. M. Gell-Mann, R. J. Oakes and B. Renner, Phys. Rev. 175 (1968) 2195.
14. H. Georgi and L. Randall, Nucl. Phys. B276 (1986) 241.
15. S. Weinberg, Phys. Rev. Lett. 40 (1978) 223; F. Wilczek, Phys. Rev. Lett. 40 (1978) 279.
16. W. A. Bardeen and S.-H. H. Tye, Phys. Lett. 74B (1978) 229; J. Ellis and M. K. Gaillard, in Ref. (9); T. W. Donnelly et al., Phys. Rev. D18 (1978) 1607. More recent treatments include: M. Srednicki, Nucl. Phys. B260 (1985) 689; P. Sikivie, in "Cosmology and Particle Physics", Ed. by E. Alvarez et al., pp. 143-169, World Scientific, 1987.
17. H. Georgi, H. R. Quinn and S. Weinberg, Phys. Rev. Lett. 33 (1974) 451.

18. J. Kim, Phys. Rev. Lett. 43 (1979) 103; M. A. Shifman, A. I. Vainshtein and V. I. Zakharov, Nucl. Phys. B166 (1980) 493; M. Dine, W. Fischler and M. Srednicki, Phys. Lett. 104B (1981) 199; A. P. Zhitnitskii, Sov. J. Nucl. Phys. 31 (1980) 260.
19. R. D. Peccei, "The Strong CP Problem", DESY preprint 88-109 (Aug. 1988).
20. K. Sato and H. Sato, Progr. Theor. Phys. 54 (1975) 1564.
21. D. Dicus, E. Kolb, V. Teplitz and R. Wagoner, Phys. Rev. D18 (1978) 1829 and Phys. Rev. D22 (1980) 839; M. Fukugita, S. Watamura and M. Yoshimura, Phys. Rev. Lett. 48 (1982) 1522 and Phys. Rev. D26 (1982) 1840; N. Iwamoto, Phys. Rev. Lett. 53 (1984) 1198; L. M. Krauss, J. E. Moody, and F. Wilczek, Phys. Lett. 144B (1984) 391; D. Dearborn, D. N. Schramm and G. Steigman, Phys. Rev. Lett. 56 (1986) 26; G. G. Raffelt, Phys. Lett. 166B (1986) 402; G. G. Raffelt and D. Dearborn, Phys. Rev. D36 (1987) 2201.
22. G. Raffelt and D. Seckel, Phys. Rev. Lett. 60 (1988) 1793; M. S. Turner, Phys. Rev. Lett. 60 (1988) 1797; R. Mayle et al., Phys. Lett. 203B (1988) 188; T. Hatsuda and M. Yoshimura, Phys. Lett. 203B (1988) 469; A. Burrows, M. S. Turner and R. P. Brinkman, Phys. Rev. D39 (1989) 1020; N. Iwamoto, Univ. of Toledo preprint (1988).
23. M. Turner, "Windows on the Axion", Fermilab preprint 89/104-A (April 1989).
24. P. Sikivie, Phys. Rev. Lett. 48 (1982) 1156; A. Vilenkin and A. E. Everett, Phys. Rev. Lett. 48 (1982) 1867; T. W. Kibble, G. Lazarides and Q. Shafi, Phys. Rev. D26 (1982) 435; P. Sikivie, "Axions in Cosmology" in Where are the Elementary Particles, Proc. 14th Summer School of Particle Physics (Gif-sur-Yvette, 1982) eds. P. Fayet et al.
25. L. Abbott and P. Sikivie, Phys. Lett. 120B (1983) 133; J. Preskill, M. Wise and F. Wilczek, Phys. Lett. 120B (1983) 127; M. Dine and W. Fischler, Phys. Lett. 120B (1983) 137.
26. P. J. Steinhardt and M. S. Turner, Phys. Lett. B129 (1983) 51.
27. For discussions of the uncertainties in the cosmological axion energy density due to the first order nature of the QCD phase transition see: W. G. Unruh and R. M. Wald, Phys. Rev. D32 (1985) 831; T. De Grand, T. W. Kephart and T. J. Weiler, Phys. Rev. D33 (1986) 910.

28. J. Ipser and P. Sikivie, Phys. Rev. Lett. 50 (1983) 925.

29. R. Davis, Phys. Lett. 180B (1986) 225. See also: A. Vilenkin and T. Vachaspati, Phys. Rev. D35 (1987) 1138.

30. D. Harari and P. Sikivie, Phys. Lett. 195B (1987) 361. See also: T. Vachaspati, "Global String Dynamics", Cambridge preprint (Oct. 1988).

31. R. L. Davis and E. P. S. Shellard, ITP preprint 88-195 (Jan. 1989); C. Rosenzweig and A. M. Srivastava, Syracuse preprint SU-4228-401 (Feb. 1989); B.-A. Gradwohl et al., MIT preprint (March 1989); R. L. Davis and E. P. S. Shellard, MIT preprint, MIT-CTP-1728 (March 1989); A. Dabholkar and J. M. Quashnock, Princeton preprint PUPT-1129 (May 1989); C. Hagmann and P. Sikivie, in preparation.

32. M. S. Turner, Phys. Rev. D33 (1986) 889.

33. P. Sikivie, Phys. Rev. Lett. 51 (1983) 1415; P. Sikivie, Phys. Rev. D32 (1985) 2988; L. Krauss, J. Moody, F. Wilczek and D. Morris, Phys. Rev. Lett. 55 (1985) 1797.

34. S. DePanfilis et al., Phys. Rev. Lett. 59 (1987) 839.

35. C. Hagmann et al., "Cavity design for a cosmic axion detector", Univ. of Florida preprint (Dec. 1988).

36. S.-I. Cho et al., Japanese J. of Appl. Phys. 26 (1987) Suppl. 26-3, 1705.

37. S. Inagaki et al., "Study of Galactic Axions by a Cavity Haloscope", KEK preprint (1988).

38. K. van Bibber, P. M. McIntyre, D. E. Morris and G. G. Raffelt, LBL preprint 25908 (Sept. 1988).

39. K. van Bibber et al., Fermilab letter of intent, P-794 (March 1988).

40. T. Kephart and T. Weiler, Phys. Rev. Lett. 58 (1987) 171; M. Turner, Phys. Rev. Lett. 59 (1987) 2489.

# THE STRONG CP PROBLEM: SOLUTIONS WITHOUT AXIONS

S.M. Barr
Bartol Research Institute
University of Delaware
Newark, Delaware 19716

ABSTRACT

Models which solve the strong CP problem without axions are reviewed. These are based on spontaneous breaking of CP at large scales. The current status of such models is discussed including the constraints imposed by an observation of $\epsilon'/\epsilon$ in the range $(3.3 + 1.1) \times 10^{-3}$.

1. **The Strong CP Problem.** The strong CP problem is the problem of explaining why strong interaction CP violation is so weak, when weak interaction CP violation is so strong. The link between the two is through the masses of the quarks, which feel both the strong and weak forces. The well-known strong CP parameter, $\overline{\Theta}$, can be written as $\overline{\Theta} = \theta_{QCD} + \theta_{QFD}$, where $\theta_{QFD}$ = arg det $M_Q$. $\theta_{QCD}$ is the vacuum angle of QCD and $M_Q$ is the quark mass matrix.

In the KM model weak CP violation is the consequence of $M_Q$ being complex. By a chiral redefinition of the quark fields we can always go to a basis where $\Theta_{QCD} = 0$ and $\overline{\Theta}$ = arg det $M_Q$. Since $\delta_{KM}$ is not very small, and $|\overline{\Theta}| \lesssim 10^{-9}$ we have the problem of explaining why the complex matrix $M_Q$ has a determinant that is real to at least nine decimal places. Some people shrug this off and seem to think that zero is a natural value for $\overline{\Theta}$. But it is not. If CP were exactly conserved then of course $\overline{\Theta}$ would vanish. The whole problem is precisely that CP is not conserved and so setting $\overline{\Theta}$ to zero does not increase the symmetry of the theory in any way. $\overline{\Theta} = 0$ is not a symmetry point, and therefore it is 'unnatural' for $\overline{\Theta}$ to be small.

2. **Two approaches.** There are two presently viable approaches to the strong CP problem, the dynamical approach and the symmetry approach. The dynamical approach is the Peccei-Quinn mechanism[1)] which leads to axions.[2)] We have heard about the status of this very beautiful idea from Pierre Sikivie. Since it is such a very beautiful idea, why, it may be asked, look for something else? There are two reasons. First, axions may not exist! And, second, as we shall, the alternative approach leads to models which are interesting in their own right.[3)]

The symmetry approach is based on the following principle. Departures from a symmetry relation which is exact at tree level due to spontaneous (or soft) breaking of that symmetry are 'finite' and 'calculable' (and typically small). So the smallness of $\overline{\Theta}$ would be because $\overline{\Theta} = 0$ is a symmetry relation that holds at tree level. What symmetry gives the relation $\overline{\Theta} = 0$? Only two that I know of: P and T (equivalently CP). Since P is badly and explicitly broken in the weak interactions, the logical candidate is CP. Thus we are led to consider theories in which CP is a fundamental symmetry at the lagrangian level which is spontaneously broken. This should gladden the hearts of those who were shocked or disturbed by the discovery of 25 years ago which we are honoring at this conference. This calls for some comments. (1) Why should CP be a good symmetry of the lagrangian? Now that we have gotten used to the idea of its being broken CP no longer seems so fundamental. My best answer to this is that in some theories, especially simple ones, CP is automatically conserved. An example is QED. Indeed the fewer parameters a theory has and the simpler its structure the more likely it is that any complex phases can be 'rotated away'. That is why three generations are needed in the KM model. I speculate that the ultimate theory may be so simple as to automatically conserve CP. CP might be spontaneously broken by the compactification of extra dimensions, or in a more conventional way. (2) Could not $\Theta_{QCD} \neq 0$ be thought of as a spontaneous symmetry breaking since $\Theta_{QCD}$ is a vacuum angle? And, if so, once we allow spontaneous CP breaking is not any value of $\Theta_{QCD}$ allowed? This is to some extent an aesthetic question, but I think it makes more sense here to think of $\Theta_{QCD}$ as a coupling constant that appears in the 'hard' interaction term $F\tilde{F}$. Then CP invariance at the lagrangian level would imply $\Theta_{QCD} = 0$ in the basis in which all the Yukawa couplings are real. (3) CP is easy to break spontaneously. One way is to have a non-trivial relative phase between the VEVs of two Higgs fields. For example, consider two complex scalar fields $\phi_1$ and $\phi_2$ which have a higgs potential (with, by CP, real parameters) containing terms of the form $(\phi_1^\dagger\phi_2)^2$+h.c., and $\phi_1^\dagger\phi_2$+h.c. Then the relative phase of $\langle\phi_1\rangle$ and $\langle\phi_2\rangle$, call it $\alpha$, will have a potential of the form $V(\alpha) = A\cos 2\alpha + B\cos\alpha$. This gives $\alpha = \cos^{-1}(B/4A)$ which in general is nontrivial even though A and B are real. (4) CP must be broken at superlarge scales so that the cosmic domain walls that arise from the spontaneous breaking of CP (or any discrete symmetry) may be inflated away. (Also to have baryogenesis possible at high temperatures.) (5) A final crucial comment is that CP spontaneously broken is not enough to ensure that $\bar{\Theta} = 0$ at tree level. It is also necessary that the VEVs that break CP spontaneously do not at tree level contribute to det $M_Q$. There are two possibilities. The first is that the CP-violating VEVs do not at tree level contribute at all to $M_Q$ so that $M_Q$ is approximately real.[4] In this case one must envision another source of weak-interaction CP violation than the KM mechanism, since $\delta_{KM}$ will be a higher order effect and thus very small (presumably $|\delta_{KM}| \sim |\bar{\Theta}|$). The second possibility is that the CP-violating VEVs do contribute at tree level to $M_Q$ but that $M_Q^{tree}$ has some special form so that det $M_Q^{tree}$ is nonetheless real. The first non-axion models[5] based on spontaneous CP violation were proposed in 1978 and were based on the latter possibility. The 'special form' of $M_Q$ was achieved through the use of ad hoc family or discrete symmetries. These models generally suffered from serious difficulties including flavor-changing neutral currents, $\bar{\Theta}$ arising at one-loop and consequently being too big, incompatibility with grand unification, and artificiality. Many of these problems arose from the way in which CP violation was introduced into $M_Q$ at tree level. It was assumed that there were two (or more) higgs doublets whose VEVs broke SU(2)xU(1) and had a relative phase that broke CP. This phase appeared in $M_Q$ as a consequence of both higgs contributing to the quark and lepton masses. This of course gave rise to flavor-changing couplings of the Higgs fields and it was these, in turn, that also gave one-loop contributions to $M_Q$ and $\Theta_{QFD}$.

3. **Newer models**. A model with none of the defects of the older models was invented by A. Nelson[6] in 1984 and quickly generalized by this author.[7] The beauty of these models is that they reduce at low energy to the KM model. The new structure that is introduced to give $M_Q$ a 'special form' appears at large (presumably grand unification) mass scales. The one key ingredient is the existence of super-heavy quarks in addition to the usual light quark families.

Suppose we have a model whose full symmetry group is G (ie., not just its gauge symmetry). Let the fermions come in three sets of representations of G: F, C, and $\bar{C}$. The F (for 'family') are the representations that have to be there to give the three families of quarks and leptons. C + $\bar{C}$ is an extra 'mirror set' of fermions. C is complex under G and $\bar{C}$ is its conjugate. So C + $\bar{C}$ is a (reducible) real representation. Such extra real fermion representations should have no effect on the number or type of light fermions. This is because of the Georgi Survival Hypothesis[8] according to which C + $\bar{C}$ should become heavy and decouple since there is no symmetry to prevent them from doing so. Since $\bar{C}$ is conjugate to C, C x $\bar{C}$ contains a singlet (and remember G is the whole symmetry group of the model) and a mass term $M\psi_{\bar{C}}\psi_C$ is allowed. One would expect M to be very large, of order the Planck or GUT scale. Thus at low energy one sees only F (or that part of F that is complex under SU(3)xSU(2)xU(1)) no matter what C + $\bar{C}$ is. So that the fact we only observe F says nothing about the likelihood that such real fermion representations exist. And that likelihood would seem to be great since we see such mirror pairs emerging naturally in a number of contexts such as superstring theory and Kaluza-Klein theory.

Though the $C + \bar{C}$ fermions decouple at low energy they nevertheless can perform a useful service through mixing with the F fermions. We are supposing CP is spontaneously broken at large (eg. GUT) scales. Call the Higgs multiplet which does that $\Omega$. The second key ingredient of these models is that $\Omega$ couples the F fermions to the $C + \bar{C}$; specifically that there is a Yukawa coupling of the form $\lambda(\psi_F \psi_C)\ \Omega$. (or the form $\psi_F \psi_{\bar{C}}\ \Omega$; we don't want both. Since F, C and $\bar{C}$ are all complex rrepresentations it is very unlikely that $\Omega$ could couple to both FC and $F\bar{C}$.) Now let us look at the kind of quark mass matrix that results. There are quarks in F, in C, and in $\bar{C}$. So the quark mass matrix has a 3x3 block form. Schematically:

$$\bar{Q}\, M_Q^{(tree)} Q = \begin{pmatrix} \bar{Q}_F & \bar{Q}_C & \bar{Q}_{\bar{C}} \end{pmatrix} \left(\begin{array}{c|c|c} \lambda' v & \lambda\langle\Omega\rangle & 0 \\ \hline \lambda\langle\Omega\rangle & 0 & M \\ \hline 0 & M & 0 \end{array}\right) \begin{pmatrix} Q_F \\ Q_C \\ Q_{\bar{C}} \end{pmatrix} \qquad (1)$$

Here v represents the usual SU(2)xU(1)-breaking VEV and $\lambda'$ the usual standard model Yukawa couplings. M represents the range masses $M\, \psi_{\bar{C}} \psi_C$ that we argued had to be there since no symmetry prevents them. The CP violation comes into $M_Q$ through the phases in $\langle\Omega\rangle$. When one diagonalizes $M_Q$ one will find, as one must, three light families. These will be mixtures of F and $C + \bar{C}$ with mixing angles of order $\lambda\langle\Omega\rangle/M$. These mixtures will have CP-violating phases from $\langle\Omega\rangle$ and so the light quark mass matrix will be complex. The low energy theory is just the KM-standard model - with one difference. det $M_Q$ is real at tree level. This is trivial to see. Since there are as many quarks in C as antiquarks in $\bar{C}$, M is a square matrix. From this it follows that det $M_Q$ = det $(\lambda' v)$ det M det M -- it does not depend on $\langle\Omega\rangle$. Thus $\bar{\Theta} = 0$ at tree level. The picture we have described requires that certain rules be observed. It is not allowed to have certain kinds of higgs fields in the model. One may not have Higgs fields that could couple F to $\bar{C}$ or C to C or $\bar{C}$ to $\bar{C}$. If there were, the zeroes in our eq. (1) would be non-zero and det $M_Q^{(tree)}$ would depend on $\langle\Omega\rangle$. But there is no reason necessarily that Higgs in such representations of G would have to be present. A second condition is that CP-violating phases are not permitted to appear in either M or v in eq. (1) (at tree level). This condition, also, is not onerous. If M is an explicit mass term it is real by the CP invariance of L. For M to be complex at tree level would require the existence of Higgs fields (more than one) which were in the product $C \times \overline{C}$ and whose VEVs had a CP violating relative phase. Again there is no reason why such fields have to be introduced. If there is only a single higgs which breaks SU(2)xU(1) then its VEV (v) cannot have a CP-violating phase since its phase is not physically meaningful (it can be changed by doing a global weak hypercharge transformation.)

How big is $\overline{\Theta}$ in such models? There are several contributions to $\overline{\Theta}$.[9] There are one loop contributions that involve virtual superheavy particles. These go as $\frac{1}{16\pi^2} f^2$ where f is some Yukawa coupling of superheavy particles. In order for $|\overline{\Theta}| \lesssim 10^{-9}$ it must be that $f \lesssim 10^{-3}$. This is neither technically unnatural nor particularly implausible considering that most of the Yukawa couplings in the standard model are that small or smaller. ($f_e \cong 2 \times 10^{-6}$). There are also two-loop contributions to $\overline{\Theta}$ which are model dependent. In Nelson's original model these come out to be of order $10^{-11}$ which is probably typical.

There are two relatively minor criticisms that can be leveled against these models. The first is the requirement just alluded to that certain couplings be $\lesssim 10^{-3}$. The other is that it would seem difficult to make such models supersymmetric. In the first place, SUSY models have two higgs that break SU(2)xU(1) whose relative phase would violate the rule that v in eq. (1) be real. In the second place, SUSY introduces many possible new CP-violating parameters and contributions to $\overline{\Theta}$. For example there are the phases of the $\mu$ parameter, the gaugino masses, the higgsino masses, and the parameters A and B. In ref. 10 the question of implementing this idea in supersymmetric models was investigated, and it was found that it is not hard provided a few further rules are observed relating to how CP and SUSY are broken and the form of the superpotential. An added bonus is that the one-loop contributions to $\overline{\Theta}$ that required some coupling(s) to be less than $10^{-3}$ are suppressed by $(M_{3/2}/M_{GUT})$. The leading contributions to $\overline{\Theta}$ arise only at three loops.

There are few predictions of such models. First, $\overline{\Theta}$ is small but not hopelessly so. Probably $\overline{\Theta}$ is in the range $10^{-11}$ to $10^{-9}$, compared to $\overline{\Theta} \sim 10^{-15}$ for invisible axion models. Second, of course, there is no axion. And, third, at low energy the model looks like the KM model or like the supersymmetric KM model with all 'new' phases (arg $\mu$, arg A, arg B, and arg $m_g$) extremely small ($\lesssim \bar{\theta}$).

4. **Real quark mass matrix**. The other possibility for having $\bar{\theta} = 0$ at tree level is simply for $M_Q$ to be real at tree level.[4] This means giving up the KM mechanism, and invoking a new superweak or milliweak force which must be mediated, it turns out, by new colored scalar fields.[11,12] The reason is that color-singlet scalars would develop CP-violating VEVs and contribute a tree-level phase to $M_Q$. There are three kinds of models: those with a superweak force mediated by 'diquark' scalars,[11] or a milliweak force mediated by 'leptoquark' scalars,[13] or a milliweak force mediated by diquark scalars.[12] The first example of such models was in ref. 11. The superweak force in that model is mediated by a diquark which is a (6, 1, -2/3) of SU(3)xSU(2)xU(1). As such it couples to ($d_{iR}$ C $d_{jR}$) where $d_i$ = d,s,b. It can give a tree-level $\Delta S = 2$ interaction of the form $\lambda(s_R\ C\ s_R)^+ (d_R\ C\ d_R)$ + h.c. It is easy to see that there are no one-loop contributions to $\bar{\theta}$ (ie. to $M_Q$) since such a diagram would require the diquark scalar to couple to a left-handed quark which is forbidden by SU(3)xSU(2)xU(1). In fact it is typical of these models that $\bar{\theta}$ arises first at two loops.

The recent measurement[14] of $\epsilon'/\epsilon$ if it is confirmed would rule out the first of the three types of real-quark-mass model, the superweak diquark models. It would also virtually rule out the second type, the milliweak lepto-quark models. In the leptoquark models the same combination of Yukawa couplings enters the Penguin-type diagrams that give rise to $\epsilon'/\epsilon$ and the box diagrams that give rise to $\epsilon$. This allows one to compute[15] the mass of the leptoquark (or the lightest if there is more than one) in terms only of an unknown CP violating phase angle.

$$M_{LQ} \simeq 1.7\ \text{GeV} \times \sqrt{\tan\phi} \quad \left( \frac{3.3\times10^{-3}}{|\epsilon'/\epsilon|} \right)$$

There are large uncertainties, probably at least a factor of 3, in $M_{LQ}$ due to gluonic corrections to the Penguin diagram, and uncertainties in the values of $\alpha_s$, B, and hadronic matrix elements. The $\tan\phi$ can be bounded from above if one can bound the standard model contribution to $ReM_{12}$. $\tan\phi \sim (ReM_{12})^{SM}/(ReM_{12})^{expt}$ (for $\tan\phi$ large). Given the theoretical and experimental uncertainties one cannot absolutely rule out these models by the current limit of $M_{LQ} > 20.5$ GeV, but they are in trouble. This is especially so if we consider muon-number violating processes $\mu \to e\gamma$ and $\mu N \to eN$. Hall and Randall[13)] estimated that $M_{LQ} \gtrsim 10^5$ GeV to satisfy the limits on $\mu Ti \to eTi$. So these models are already ruled out unless one imposes (rather artifically) muon-number conservation.

The diquark milliweak models are not in trouble from $\epsilon'/\epsilon$ since that arises at tree level (by a $\Delta S = 1$ diquark exchange) rather than from a Penguin. In different versions of such models $M_{DQ}$ can vary over a large range (easily from 10 GeV to several TeV) and is quite model-dependent.

If $\epsilon'/\epsilon$ turns out to be consistent with zero then all of these models are still healthy. Since they predict the existence of new particles there should be a very rich phenomenology associated with them, such as rare decays of $B^o$, $D^o$ in channels suppressed in the standard model, CP violating effects in the $B^o$ - $\overline{B^o}$ system that do not fit the KM model, discovery of new colored scalars, and (in the leptoquark models) large polarization asymmetries in $K \to \mu\mu$ and $K_{\ell 3}$ decays. These are under investigation presently.[15)]

**References**


1. R. Peccei, and H.R. Quinn, Phys. Rev. Lett. 38, 1440 (1977); Phys. Rev. D16, 1791 (1977).

2. S. Weinberg, Phys. Rev. Lett. 40, 223 (1978); F. Wilczek, Phys. Rev. Lett. 40, 271 (1978).

3. For good reviews see J. Kim, Phys. Rep. 150, 1 (1987), and H.-Y. Cheng, Phys. Rep. 158, 1 (1988).

4. S.M. Barr and D. Seckel, Nucl. Phys. B233, 116 (1984).

5. M.A.B. Beg and H.S. Tsao, Phys. Rev. Lett. 41, 278 (1978); H. Georgi, Hadronic Journal 1, 155 (1978); R.N. Mohapatra and G. Senjanovic, Phys. Lett. 126B, 283 (1978); G. Segre and A. Weldon, Phys. Rev. Lett. 42, 1191 (1979); S.M. Barr and P. Langacker, Phys. Rev. Lett. 42, 1654 (1979).

6. A. Nelson, Phys. Lett. 136B, 387 (1983).

7. S.M. Barr, Phys. Rev. Lett. 53, 329 (1984); Phys. Rev. D30 1805 (1984). (The second paper has improved rules for constructing such models compared to the first.)

8. H. Georgi, Nucl. Phys. B156, 126 (1979).

9. A. Nelson, Phys. Lett. 143B, 165 (1984).

10. S.M. Barr and A. Masiero, Phys. Rev. D38, 366 (1988).

11. S.M. Barr and A. Zee, Phys. Rev. Lett. 55, 2253 (1985).

12. S.M. Barr, Phys. Rev. D34, 1567 (1986).

13. L. Hall and L. Randall, Nucl. Phys. B274, 157 (1986).

14. H. Burkhardt, et al., Phys. Lett. 206B, 169 (1988).

15. E.M. Freire and S.M. Barr, in preparation.

# Strong CP conservation from quantum gravity wormholes

H. Bech Nielsen

Niels Bohr Institute
Blegdamsvej 17
DK-2100 COPENHAGEN
Danmark

No written contribution received

*Listening to the guide at the "Clos Lucé".*

# RARE DECAYS AND CP VIOLATION IN SUPERSYMMETRY

John S. Hagelin
Maharishi International University
Fairfield, IA 52556

ABSTRACT

We present a detailed discussion of rare decays and CP violation in the framework of supersymmetry. Special attention is paid to "direct" flavor-changing neutral currents induced by renormalization-group scaling both in and beyond the minimal supersymmetric Standard Model.

## I. INTRODUCTION

Flavor-changing neutral currents (FCNC's) provide an important test of the radiative structure of the Standard Model and a sensitive probe of physics beyond the Standard Model. Indeed, many theoretical frameworks beyond the Standard Model predict large flavor-changing neutral currents compared to the Standard Model predictions, even for new particle masses beyond the reach of present and proposed accelerators. In this talk, I will present the results of a recent, comprehensive analysis[1)] of flavor-changing neutral currents, CP violation and rare decays in the framework of supersymmetry.

With the gauge structure and fermion content of the Standard $SU(2)_L xU(1)_Y$ Electroweak Model firmly in place, the primary challenge is to understand the mechanics of electroweak symmetry breaking through which the $W^{\pm}$ and Z, as well as all quarks and leptons, receive their masses. For this purpose, the Higgs boson was historically introduced. However the Higgs, having spin-0, is badly behaved at the quantum-mechanical level, and must either be rendered meaningful by supersymmetry[2)] or else replaced by fermion condensates in a dynamical symmetry breaking ("technicolor") scheme.[3)*] The key point is that, in either case, quantum-mechanical consistency *requires* physics beyond the Standard Model. This new physics associated with electroweak symmetry breaking represents the most important and exciting frontier of physics in the next decade.

If this new physics is supersymmetry, then the masses of the new superpartners must be $\lesssim$ TeV (a few TeV for scalar quarks and gluinos) if the desired improvement in the quantum-mechanical behavior of the theory is to be preserved. This offers hope that these particles will be discovered in the relatively near future, either directly in particle accelerators or through their virtual contributions to flavor-changing neutral currents and rare decays, which is the primary subject of this talk. We will show that if the sparticle masses are $\lesssim$ few hundred GeV, then rare decays provide a viable test of supersymmetry, as well as a sensitive probe of any new physics beyond the minimal supersymmetric Standard Model extending to scales as high as $M_{GUT}$ or $M_{Pl}$.

## II. $K^+ \to \pi^+ \nu\bar{\nu}$

The rare decay $K^+ \to \pi^+ \nu\bar{\nu}$ has long been advocated[4)] as an important test of the induced flavor-changing neutral current (FCNC) structure of the Standard Model and a sensitive probe of new physics beyond the Standard Model.[5-11)]

*Alternatively, one must fine tune the Higgs mass$^2$ order-by-order in perturbation theory to an accuracy better than one part in $10^{34}$ in order to maintain $m_H \ll M_{Pl}$.

We will use this interesting and important decay as an opportunity to introduce much of our notation and formalism, which we will subsequently use in our analysis of other decays.

IIa: The Standard Model

$K^+ \to \pi^+\nu\bar{\nu}$ receives Standard Model contributions at the one-loop level from the $W^\pm$-exchange box diagram and induced sdZ vertex shown in Figure 1. These "short-distance" processes give rise to an effective interaction lagrangian whose coefficient depends on the masses of the heavy quarks (j=c,t) exchanged in the loops and the Cabibbo-Kobayashi-Maskawa (KM) angles $V_{js}, V_{jd}$ coupling these heavy quarks to the s and d quarks. From the analysis of ref. 12, we have

$$\mathcal{L} = \frac{G_F}{\sqrt{2}} \frac{\alpha}{\pi \sin^2\theta_W} \sum_{j\neq u} V^*_{js} V_{jd}\, D(x_j)\, \bar{s}_L \gamma^\mu d_L\, \bar{\nu}_L \gamma_\mu \nu_L \tag{1a}$$

where $x_j = m_j^2/M_W^2$ and

$$D(x) = \frac{1}{8}\left[1 + \frac{3}{(1-x)^2} - \frac{(4-x)^2}{(1-x)^2}\right] x\ln x + \frac{x}{4} - \frac{3}{4}\frac{x}{1-x} = \begin{cases} 11/8; & m_q = M_W \\ -\frac{3}{2}x\ln x - \frac{x}{2}; & m_q \ll M_W \end{cases} \tag{1b}$$

In writing (1) we have assumed, for simplicity, that the mass of the charged lepton exchanged in the loop (Figure 1) is small compared to $M_W$, which is a good approximation for the three known lepton generations. The matrix element of the induced weak current $\bar{s}_L \gamma_\mu d_L$ between the initial $K^+$ and final state $\pi^+$ is related by isospin to the known matrix element of the charged-current operator $\bar{s}_L \gamma_\mu u_L$ measured in $K^+_{e3}$ decay. The resulting branching ratio for $K^+ \to \pi^+\nu\bar{\nu}$ per neutrino generation is given by

$$\frac{B(K^+ \to \pi^+\nu_i\bar{\nu}_i)}{B(K^+ \to \pi^o e^+\nu)} = \frac{\alpha^2}{8\pi^2 \sin^4\theta_W} \frac{1}{V^2_{us}} \left| \sum_{j\neq u} V^*_{js} V_{jd}\, D(x_j)\right|^2 \tag{2}$$

or

$$B(K^+ \to \pi^+\nu_i\bar{\nu}_i) = 0.70 \times 10^{-6} \left| D(x_c) + s_2(c_1 s_2 c_3 + c_2 s_3 e^{i\delta}) D(x_t)\right|^2 \tag{3}$$

where the numerical coefficient is obtained using $\sin^2\theta_W = 0.228$, $\alpha(\mu = M_W) \simeq 1/128$, and by incorporating the electromagnetic enhancement $\left[1 + (\alpha/\pi)\ln[M_W/(\mu = 1\ \mathrm{GeV})]\right] \sim 1.01$ of the $K^+ \to \pi^o e^+\nu$ amplitude. There are also short-distance QCD corrections[5)] to the branching ratio (3). The primary effect of these QCD corrections is to suppress the charm-quark contribution to the amplitude relative to (1) by a factor ~ (0.70 - 0.62) depending on $\Lambda_{QCD}$.[13)*] These QCD corrections have been included in our numerical estimates (Figure 2).

In addition to the short-distance contributions to the $K^+ \to \pi^+\nu\bar{\nu}$ amplitude shown in Figure 1, in general one expects "soft" contributions from processes involving distance scales > $1/m_{c,t,W,Z}$. These long-distance contributions have been analyzed in ref. 13 and represent a negligible uncertainty in the case of $K^+ \to \pi^+\nu\bar{\nu}$, which is a major virtue of this rare decay.

The short-distance charm-quark contribution $D(x_c)$ in (3) gives rise to a branching ratio ~ $2 \times 10^{-11}$ for three neutrino generations. From the measured magnitude[14)] of $B^o$-$\bar{B}^o$ mixing, the top-quark contribution is expected to be at least as large.[15)] One can show[16)] that the interference between the c- and t-quark contributions is essentially constructive ($ReV^*_{ts} > 0$) for experimentally favored values of $\bar{R} \equiv B(b \to u)/B(b \to c) < 12\%$. However, the presence of CP violation in $K^o$-$\bar{K}^o$ mixing requires $\sin\delta \neq 0$ at some level, so this constructive interference cannot be maximal.[17)] A proper quantitative analysis of $K^+ \to \pi^+\nu\bar{\nu}$ must therefore consider constraints on the KM angles $s_2, s_3, s_\delta$ arising from $\tau_b$, $\bar{R}$, $B^o$-$\bar{B}^o$ mixing, and $\epsilon_K$ simultaneously.[13)]

The allowed range of $K^+ \to \pi^+\nu\bar{\nu}$ in the Standard Model is shown in Figure 2 as a function of $m_t$. The lower bound on $K^+ \to \pi^+\nu\bar{\nu}$ results from a lower bound on $V^*_{ts}V_{td}$ that follows from $\tau_b$ and $B^o$-$\bar{B}^o$ mixing. For $m_t \lesssim 150$ GeV, the most stringent upper bound on $V_{td}$ and hence on $K^+ \to \pi^+\nu\bar{\nu}$ follows directly from $\tau_b$ and $\bar{R} \equiv B(b \to u)/B(b \to c)$, and is therefore free of the hadronic uncertainties associated with $B^o$-$\bar{B}^o$mixing.[13)] This relatively clean theoretical upper bound on the branching ratio over much of $m_t$ makes $K^+ \to \pi^+\nu\bar{\nu}$ an especially critical test of the Standard Model and a sensitive probe of physics beyond the Standard Model. In addition to testing the Standard Model, a precision measurement of $K^+ \to \pi^+\nu\bar{\nu}$ would also provide the most accurate determination of the KM angle $s_2$ (or $V_{td}$) as a function of $m_t$ of any foreseeable experiment. This measurement of $V_{td}$ does not suffer from the type of hadronic uncertainties that accompany $B^o$-$\bar{B}^o$ mixing (5), which currently limit our ability to predict the $K^+ \to \pi^+\nu\bar{\nu}$ branching ratio based on present data. There are, however, remaining uncertainties associated with $m_c$ and $|V_{ts}|$ which would render a measurement of $B(K^+ \to \pi^+\nu\bar{\nu})$ to an accuracy *better* than ±30% of limited theoretical value.

IIb: "Trivial" Supersymmetry

In the framework of low-energy supersymmetry, the first question one might ask is whether there are any new, supersymmetric decay modes that could

* See also F.J. Gilman, this conference.

bolster the $K^+ \to \pi^+$ + "unobserved" branching ratio. In this context, it is logical to consider $K^+ \to \pi^+\chi\chi$, where $\chi$ represents the lightest supersymmetric particle (LSP).[5-7] The most likely candidates for the LSP are a light spin-1/2 "neutralino" (e.g., a photino $\tilde{\gamma}$ or a Higgsino $\tilde{H}$)[18] or possibly a scalar neutrino $\tilde{\nu}$.[19] In ref. 13 we show that for reasons of phase space and/or calculated branching ratio, none of these supersymmetric decay channels are likely to bolster the $K^+ \to \pi^+$ + "unobserved" branching ratio above its Standard Model prediction, and we will therefore move on to potentially greener pastures.

We now turn to the more interesting subject of whether supersymmetry can significantly modify the Standard Model amplitude for $K^+ \to \pi^+\nu\bar{\nu}$. The answer to this question is certainly yes.[1] In the limit of exact supersymmetry, the supersymmetric contributions to many FCNC processes are precisely equal to the Standard Model ones. For example, the $\tilde{W}^{\pm}$ box contribution to $K^o$-$\bar{K}^o$ mixing equals the Standard Model box both in magnitude and in sign. Similarly, the $\tilde{W}^{\pm}$ box contribution to $K_L \to \mu\bar{\mu}$ is equal to the standard $W^{\pm}$ box contribution. With the introduction of global soft supersymmetry breaking, these contributions ultimately fall off as $(M_W/m_{\tilde{q}})^4$. It is interesting to note, however, that even for sparticle masses $\lesssim M_W$, these large supersymmetric contributions to $K^o$-$\bar{K}^o$ mixing and $K_L \to \mu\bar{\mu}$ would probably go unnoticed. The decay $K_L \to \mu\bar{\mu}$ is principally due to long-distance physics, which is dominated by the $2\gamma$ intermediate state. The short-distance box contribution to $K_L \to \mu\bar{\mu}$ is small for $m_t \lesssim 200$ GeV.[15] If this short-distance contribution were doubled through the addition of the supersymmetric $\tilde{W}^{\pm}$ box, such an effect would probably go unnoticed.

Similarly, $K^o$-$\bar{K}^o$ mixing receives large contri utions from long distance physics which are uncertain in both their magnitu and their sign.[20] If the short-distance charm-quark contribution (which accounts for about half the experimental $K_L$-$K_S$ mass difference) were doubled by adding the supersymmetric box, this effect would be lost in these long-distance uncertainties. Hence, there exist no useful constraints on these "trivial" supersymmetric contributions to FCNC's, which opens the possibility of large [O(1)] corrections to the $K^+ \to \pi^+\nu\bar{\nu}$ amplitude from sparticle masses $\lesssim M_W$ (see Figure 3). Since $K^+ \to \pi^+\nu\bar{\nu}$ is free of the long-distance uncertainties associated with these other processes, it is an excellent place to look for these potentially striking supersymmetric effects.[1]

All such supersymmetric contributions, however, fall rapidly as $(M_W/m_{\tilde{q}})^4$, and for sparticle masses $\gtrsim 150$ GeV, are totally unobservable. In our subsequent discussion, we will examine a new class of supersymmetric effects which can lead to experimentally interesting signatures over a much wider range of supersymmetric particle masses—so-called "direct" FCNC effects.

IIc: "Direct" FCNC's in Supersymmetry

In addition to the obvious supersymmetric contributions to FCNC processes in direct analogy with the Standard Model ones, there are also "direct" tree-level FCNC's that result from soft supersymmetry breaking. Even if the soft supersymmetry-breaking masses start out purely flavor-diagonal, renormalization-group scaling inevitably induces nontrivial mixing among squark and slepton generations.[21] This appears at low energies as off-generational couplings for the neutral gauginos $\tilde{g},\tilde{\gamma},\tilde{Z}$, along with departures from the naive expectation that the supersymmetric KM angles linking $\tilde{W}^{+}\tilde{q}q$ be equal to the ordinary KM angles. We will choose to work in a "super-KM" basis in which all such couplings are assumed to be the canonical ones, and where these direct FCNC's appear as off-generational entries in the $\tilde{u}$-$\tilde{c}$-$\tilde{t}$ and $\tilde{d}$-$\tilde{s}$-$\tilde{b}$ squark mass matrices. It is then straightforward[1] to show that renormalization-group (RG) scaling between $M_{Pl}$ and $M_W$ leads to off-generational squark mass terms $\delta\tilde{m}^2_{ij}$ of the form

$$\left(\delta\tilde{m}^2_{d_L d_L}\right)_{ij} = -\frac{1}{8\pi^2}\left(V^{\dagger}\lambda_u^2 V\right)_{ij} \ln\frac{M_{Pl}}{M_W}\left[3m_o^2 + A^2\right]$$
$$\left(\delta\tilde{m}^2_{u_L u_L}\right)_{ij} = -\frac{1}{8\pi^2}\left(V\lambda_d^2 V^{\dagger}\right)_{ij} \ln\frac{M_{Pl}}{M_W}\left[3m_o^2 + A^2\right] \qquad (8)$$

where V is the KM matrix, $\lambda_{u,d}$ are the diagonalized u-c-t and d-s-b quark Yukawa couplings, and $m_o$,A are global soft supersymmetry-breaking scalar masses and trilinear scalar couplings, respectively.* If soft supersymmetry breaking takes the form of a pure gaugino mass term $m_{1/2}$, then $\delta\tilde{m}^2_{d_L d_L}$, $\delta\tilde{m}^2_{u_L u_L}$ get generated at the two-loop level.) In minimal supersymmetry, the right-handed squark mass matrices remain diagonal at one loop, while the effects of left-right mass terms $\delta\tilde{m}^2_{LR}$ are suppressed in physical processes by additional powers of external light quark masses and momenta. Hence $\delta\tilde{m}^2_{LL}$ (8) generally represents the dominant contribution to induced FCNC processes.

Since the KM matrix V contains off-diagonal elements, $\delta\tilde{m}^2_{d_L d_L}$ and $\delta\tilde{m}^2_{u_L u_L}$

*Note that the form of (8) differs markedly from the expression $\delta\tilde{m}^2_{LL} \propto cm_t^2$ ($c \simeq 0.5$) often found in the literature.[22] This expression hides the obvious fact that off-generational $\delta\tilde{m}^2$'s grow proportionally to the scale of soft supersymmetry breaking, and falsely implies that direct FCNC contributions to most physical processes fall off as the eighth power of the sparticle mass, instead of the fourth power. This has resulted in many incorrect conclusions in the literature. For a complete discussion, see ref. 1.

(8) will have off-generational entries. These $\delta\tilde{m}^2$'s give rise to off-generational mass insertions in squark propagators in Feynman graphs, leading to "direct" FCNC effects resulting (for example) in $K^+ \to \pi^+\tilde{\gamma}\tilde{\gamma}$ at the tree level. Since we have argued that this particular process is probably forbidden by phase space, we will now examine the effects of these direct FCNC's on the $K^+ \to \pi^+\nu\bar{\nu}$ amplitude.

The dominant contribution to the $K^+ \to \pi^+\nu\bar{\nu}$ amplitude in the presence of direct FCNC's results from $\tilde{Z}$ exchange (Figure 4). These diagrams give rise to an effective interaction lagrangian[1])

$$\mathcal{L} = \frac{g^4}{768\pi^2\cos^4\theta_W}\left[-\frac{1}{2} + \frac{1}{3}\sin^2\theta_W\right]^2 \frac{\delta\tilde{m}^2_{s_L d_L}}{m^4_{\tilde{q}}}\ \bar{s}_L\gamma^\mu d_L\ \bar{\nu}_L\gamma_\mu\nu_L \cdot F(m^2_{\tilde{\nu},\tilde{Z}}/m^2_{\tilde{q}}) \tag{9a}$$

where

$$F(x) = \frac{6(7 + 2\ell n x + 4x + 12x\ell n x - 11x^2 + 4x^2\ell n x)}{(1-x)^4} = \begin{cases} 1; & m_{\tilde{\nu},\tilde{Z}} = m_{\tilde{q}} \\ 42+12\ell n x; & m_{\tilde{\nu},\tilde{Z}} \ll m_{\tilde{q}} \end{cases} \tag{9b}$$

By direct comparison with Eqs. (1-3), we observe that a $K^+ \to \pi^+\nu\bar{\nu}$ branching ratio at the $10^{-10}$ level would require an off-diagonal $\delta\tilde{m}^2_{s_L d_L}$ mass term (in units of $m^2_{\tilde{q}}$) of order

$$\frac{\delta\tilde{m}^2_{s_L d_L}}{m^2_{\tilde{q}}} = \frac{0.17}{F}\left(\frac{m_{\tilde{q}}}{100\ \text{GeV}}\right)^2 \tag{10}$$

To see whether a $\delta\tilde{m}^2_{s_L d_L}$ of this magnitude is consistent with known data, we will examine a closely related constraint from the $K^o$-$\bar{K}^o$ system. For $K^o$-$\bar{K}^o$ mixing, gluino exchange[22]) provides the dominant direct FCNC contribution (Figure 5). By summing these graphs, we obtain the effective $\Delta S$=2 interaction lagrangian[1])*

$$\mathcal{L} = -\frac{\alpha_s^2}{216 m^2_{\tilde{q}}}\left(\frac{\delta\tilde{m}^2_{s_L d_L}}{m^2_{\tilde{q}}}\right)^2 (\bar{s}_L\gamma^\mu d_L)^2 \cdot G(m^2_{\tilde{g}}/m^2_{\tilde{q}}) \tag{11a}$$

*Note that the color structure of the effective $\Delta S$=2 operator in (11) is of the usual Singlet•Singlet form. This differs from ref. 22, in which Octet•Octet operator structures were also found to arise. Whereas these additional color structures do appear in the Feynman amplitude, they also arise from a Singlet•Singlet operator and hence do not belong in the effective lagrangian.

where

$$G(m_{\tilde{g}}^2/m_{\tilde{q}}^2) = \begin{cases} 1; & m_{\tilde{g}} = m_{\tilde{q}} \\ 22; & m_{\tilde{g}} \ll m_{\tilde{q}} \end{cases} \tag{11b}$$

By requiring that the contribution (11) to the $K_L$-$K_S$ mass difference not exceed the experimentally measured value $\Delta M_K = 3.52 \times 10^{-15}$ GeV, one obtains

$$\frac{\delta\tilde{m}^2_{s_L d_L}}{m^2_{\tilde{q}}} < \frac{1 \times 10^{-2}}{G^{1/2}} \sqrt{\frac{0.6}{B_K}} \left(\frac{m_{\tilde{q}}}{100 \text{ GeV}}\right) \tag{12}$$

where $B_K$ is the usual parametrization of the $\Delta S=2$ matrix element in terms of vacuum insertion. Comparing (12) with (10), we find that $K^o$-$\bar{K}^o$ mixing appears to rule out any significant contribution to $K^+ \to \pi^+\nu\bar{\nu}$ from direct FCNC's for typical values of $F(x), G(x) \sim 1$. The gluino couplings appearing in $K^o$-$\bar{K}^o$ (Figure 5) are simply too large, giving too strong a constraint on $\delta\tilde{m}^2_{s_L d_L}$. Thus it is only the trivial supersymmetric contribution to $K^+ \to \pi^+\nu\bar{\nu}$ (Figure 3) that makes an important contribution, and then only for sparticle masses $\lesssim$ 100-150 GeV. For these trivial contributions, $K^o$-$\bar{K}^o$ mixing has no strong-coupling advantage, and indeed suffers from long-distance uncertainties whereas $K^+ \to \pi^+\nu\bar{\nu}$ does not, as we have discussed already.

Before leaving the subject of direct FCNC's in supersymmetry, we note that there is an even stronger constraint on these effects which results from CP violation in the $K^o$-$\bar{K}^o$ system. Using Eq. (8) and the fact that $\lambda_t \gg \lambda_{c,u}$, we observe that

$$\frac{\text{Im } M_{12}^{susy}}{\text{Re } M_{12}^{susy}} \simeq 2 \frac{\text{Im } \delta\tilde{m}^2_{s_L d_L}}{\text{Re } \delta\tilde{m}^2_{s_L d_L}} \simeq 2 \frac{\text{Im } V^*_{31}V_{32}}{\text{Re } V^*_{31}V_{32}} \simeq \frac{2s_1 s_2 s_3 s_\delta}{s_1 s_2(s_2 + s_3 c_\delta)} \tag{13}$$

which is *larger* than the corresponding Standard Model ratio. This is because, in the Standard Model, the short-distance contribution to $M_{12}$ is dominated by the charm-quark contribution, which is real. If we require that the supersymmetric contribution to $\epsilon_K$ be no larger than the Standard Model contribution, we obtain the following very strong bound:

$$\frac{\delta\tilde{m}^2_{s_L d_L}}{m^2_{\tilde{q}}} < \frac{2 \times 10^{-3}}{G^{1/2}} \left(\frac{m_{\tilde{q}}}{100 \text{ GeV}}\right) \tag{14}$$

A comparison of Eqs. (14), (10) and (12) reveals that CP violation $\epsilon_K$ is the *first* place to look for direct FCNC's in the kaon sector. As a bound, however, Eq. (14) is somewhat specious, for if one *were* to double the Standard Model contribution to $\epsilon_K$ via supersymmetry, one could simply compensate for this by reducing the CP-violating phase $\delta$ (13). For this reason, it is critical to have a second, precision measurement of CP violation in the Standard Model. In principle, $\epsilon'/\epsilon$ affords such a measurement, but at the present time, theoretical and experimental uncertainties are too large to place meaningful constraints on these direct FCNC contributions. Other possibilities include the rare decays $K_L \to \pi^o ee$ and $K_L \to \pi^o \nu\bar{\nu}$, which also measure direct CP violation, but without the same hadronic uncertainties. (This is especially true of $K_L \to \pi^o \nu\bar{\nu}$.) These decays will form the subject of the next two sections.

## III. $K_L \to \pi^o \nu\bar{\nu}$

The decay $K_L \to \pi^o \nu\bar{\nu}$ is closely related to $K^+ \to \pi^+ \nu\bar{\nu}$. Both decays result from the same Feynman diagrams (Figure 1) and interaction lagrangian (1). The $K_L \to \pi^o \nu\bar{\nu}$ amplitude, however, vanishes in the limit of exact CP conservation. Using the isospin relation $\langle \pi^o | \bar{s}_L \gamma_\mu d_L | K^o \rangle = (1/\sqrt{2}) \langle \pi^+ | s_L \gamma_\mu d_L | K^+ \rangle$, we have[23)]

$$A(K_L \to \pi^o \nu\bar{\nu}) = \frac{1}{2}\Big[(1+\epsilon_K)A(K^+ \to \pi^+\nu\bar{\nu}) - (1-\epsilon_K)A(K^- \to \pi^-\nu\bar{\nu})\Big] \quad (15a)$$

$$\simeq \text{Im}\, A(K^+ \to \pi^+\nu\bar{\nu}) \quad (15b)$$

where we have dropped in (15b) a contribution due to $\epsilon_K$ (which gives a branching ratio $< 5 \times 10^{-15}$) and used the fact that $A(K^- \to \pi^- \nu\bar{\nu}) = A^*(K^+ \to \pi^+ \nu\bar{\nu})$ for the short-distance charm- and top-quark contributions. Since the charm-quark contribution to $K^+ \to \pi^+ \nu\bar{\nu}$ (1) is CP-conserving, the $K_L \to \pi^o \nu\bar{\nu}$ branching ratio into three generations is just

$$\begin{aligned} B(K_L \to \pi^o \nu\bar{\nu}) &= 3 \times \frac{\tau_{K_L}}{\tau_{K^+}} \times \frac{0.70 \times 10^{-6}}{V_{us}^2} \left| \text{Im}(V_{ts}^* V_{td}) D(x_t) \right|^2 \\ &\simeq 0.88 \times 10^{-5} \left[ s_2 s_3 s_\delta D(x_t) \right]^2 \end{aligned} \quad (16)$$

The short-distance QCD corrections to $K_L \to \pi^o \nu\bar{\nu}$ are the same as for $K^+ \to \pi^+ \nu\bar{\nu}$,[4)] but now only involve the top-quark and are thus very small. Taking $m_t \sim M_W$ and a typical value of $s_2 s_3 s_\delta \sim 6 \times 10^{-4}$ derived from $\epsilon_K$,[16)] we find a $B(K_L \to \pi^o \nu\bar{\nu}) \simeq 5 \times 10^{-12}$. Estimates will vary by factors of 2-4 depending on $m_t$ and on the CP-violating product $s_2 s_3 s_\delta$, which is quite sensitive to $B_K$ used in evaluating $\epsilon_K$. Because such hadronic uncertainties are absent in (16), $K_L \to \pi^o \nu\bar{\nu}$ in principle affords the most precise measurement of these angles and of direct CP violation in the Standard Model. With the development of improved detectors and kaon sources, such a measurement might well become feasible,[13,23)] providing us with the cleanest available test of the CP-violating structure of the Standard Model.

In a supersymmetric world, the "trivial" supersymmetric contribution to $K_L \to \pi^o \nu\bar{\nu}$ (Figure 3) is expected to be as small or smaller than the Standard Model contribution. Moreover, these trivial supersymmetric contributions rapidly fall off as $(M_W/m_{\tilde{q}})^4$ for sparticle masses $\gtrsim M_W$, and are not very interesting in the present context. While "direct" FCNC's of the type discussed in the previous section might appear to hold greater promise, these turn out to be small in $K_L \to \pi^o \nu\bar{\nu}$ for the same reason they were small in $K^+ \to \pi^+ \nu\bar{\nu}$—they are constrained by the $K_L$-$K_S$ mass difference.

However, these direct FCNC's do affect $K_L \to \pi^o \nu\bar{\nu}$ indirectly.[1] For $\delta\tilde{m}^2_{s_L d_L}/m^2_{\tilde{q}} > 2 \times 10^{-3}$ (14), the Feynman graphs in Figure 5 start to dominate the CP-impurity parameter $\epsilon_K$. This added contribution to $\epsilon_K$ necessitates a readjustment of the CP-violating phase $\sin\delta$ down from its Standard Model value. This in turn causes a decrease in the magnitude of CP violation elsewhere in the Standard Model—in $\epsilon'/\epsilon$, in $K_L \to \pi^o ee$, and in $K_L \to \pi^o \nu\bar{\nu}$.*

To estimate the magnitude of this proposed effect, we use the renormalization-group analysis of $\delta\tilde{m}^2_{s_L d_L}$ in ref. 1, where results are plotted for various possible assumptions about the ratios of soft supersymmetry-breaking parameters $m_o$, A and $m_{1/2}$. We observe that

$$\frac{\delta\tilde{m}^2_{s_L d_L}}{m^2_{\tilde{q}}} \simeq 0.5 \times 10^{-4}\ (1 + r^2) \left(\frac{m_t}{80\ \text{GeV}}\right)^2 f\left(\frac{m_o}{m_{1/2}}, \frac{A}{m_{1/2}}\right) \tag{17}$$

where r is the ratio of Higgs vacuum expectation values $H^o_d/\bar{H}^o_u$ (generally < 1) and f is a smooth function of $m_o/m_{1/2}$ and $A/m_{1/2}$ which is equal to 1 for pure gaugino-type breaking and otherwise ranges from ~ 0.5 to 5. In light of (17), it is unlikely that direct FCNC effects in the minimal supersymmetric model would contribute noticeably to physics in the kaon sector, including $\epsilon_K$ (14).

However, the minimal supersymmetric Standard Model is certainly not the Theory of Everything, and it is reasonable to expect that minimal supersymmetry will be embedded in a supersymmetric GUT, or possibly the superstring, at some high mass scale. Since direct FCNC's are induced by off-diagonal Yukawa couplings (and thus necessarily small in the supersymmetric Standard Model), once supersymmetry is embedded in a larger structure containing additional Yukawa couplings, much larger $\delta\tilde{m}^2$'s are generally expected.[24]

*It is important to note that this argument could be reversed in a non-minimal supersymmetric model, where the interference in $\epsilon_K$ with the Standard Model contribution could be destructive rather than constructive, depending upon the sign of any new CP-violating phases. We will return to the subject of non-minimal supersymmetry shortly.

The question of direct FCNC's has recently been explored[1)] in a supersymmetric Flipped SU(5)xU(1) GUT[25)] derived from the superstring.[26)] This theory has been extolled[27)] as a simpler and more successful alternative to conventional grand unified theories. For our present purposes, the theory provides a simple and theoretically well-motivated example of a non-minimal supersymmetric model in which direct FCNC's can be large. In ref. 1 it is shown that Yukawa couplings $\lambda_6^{ij} F_i \bar{H} \phi_j$ giving mass to right-handed neutrinos induce large off-generational squark masses of the form

$$\begin{aligned} \left(\delta\tilde{m}^2_{d_L d_L}\right)_{ij} &= \left(\delta\tilde{m}^2_{u_L u_L}\right)_{ij} = -\frac{1}{8\pi^2}\left(V^\dagger \lambda_6^\dagger \lambda_6 V\right)_{ij} \ln\frac{M_{Pl}}{M_{GUT}}\left[3m_o^2 + A^2\right] \\ \left(\delta\tilde{m}^2_{d_R d_R}\right)_{ij} &= \Phi_{ik}\left(\delta\tilde{m}^2_{d_L d_L}\right)_{kl}\Phi^*_{lj} \end{aligned} \qquad (18)$$

where $\Phi$ is a diagonal matrix of phases. Assuming arbitrary $|\lambda_6^{ij}| \lesssim 1$, this yields[1)]

$$\frac{\delta\tilde{m}^2_{sd}}{m_{\tilde{q}}^2} \lesssim \begin{cases} 0.2 & (m_o) \\ 0.01\ A^2/m_{1/2}^2 & (A) \\ 0.004 & (m_{1/2}) \end{cases} \qquad (19)$$

for initial soft supersymmetry breaking in the form of $m_o$, A or $m_{1/2}$, respectively. These $\delta\tilde{m}^2$'s are easily large enough to induce observable FCNC effects in $\epsilon_K$ (14) and thereby to affect, indirectly, the Standard Model predictions for $\epsilon'/\epsilon$, $K_L \to \pi^o ee$ and especially $K_L \to \pi^o \nu\bar{\nu}$. Also note that since the Yukawa couplings $\lambda_6^{ij}$ responsible for this effect are uncertain with respect to their magnitude and their phase, the sign of the supersymmetric contribution to $\epsilon_K$ is unknown. Thus the effect on the CP-violating phase $\delta$ could be either to decrease it or to increase it, which would lead to either a suppression or an enhancement of $\epsilon'/\epsilon$, $K_L \to \pi^o ee$ and $K_L \to \pi^o \nu\bar{\nu}$.

In this section, we have seen that $K_L \to \pi^o \nu\bar{\nu}$ in principle affords the most precise measurement of direct CP violation in the Standard Model. As such, it also provides the most sensitive probe within the kaon sector of "direct" FCNC effects, which would result from embedding minimal supersymmetry into a larger structure like a GUT. It is important to emphasize that these direct FCNC's provide a window on the structure of physics at extremely high scales ~ $M_{GUT}$ - $M_{Pl}$, where these effects are actually generated (18). Such an experimental opportunity is unique apart from proton decay, and should clearly be exploited to the fullest extent possible.

IV. $K_L \to \pi^o ee$

From a theoretical standpoint, $K_L \to \pi^o ee$ is similar to $K_L \to \pi^o \nu\bar{\nu}$. In $K_L \to \pi^o ee$, however, a somewhat larger branching ratio is obtained at the expense of added theoretical complexity.[13,28] There are now three potentially significant Standard Model contributions to $K_L \to \pi^o ee$. There are contributions due to indirect CP violation $\epsilon_K$ and due to direct CP violation (both resulting from one photon exchange), and there is a CP-conserving contribution that results from a two-photon intermediate state. The direct CP-violating contribution is believed to be somewhat larger than the indirect contribution—a few $\times 10^{-11}$ compared with $6 \times 10^{-12}$ in the branching ratio.[29] The hope is that one can separate these two CP-violating contributions and use this decay as a precise measurement of direct CP violation in the Standard Model.

This would be essential from a supersymmetric standpoint, since we have argued previously that direct FCNC's would reveal themselves (indirectly) in the magnitude of direct CP violation [even though they contribute directly only to indirect CP violation!].

Unfortunately, there is still some controversy regarding the magnitude of the CP-conserving "soft" contribution from $2\gamma$ exchange. Chiral perturbation theory gives a branching ratio from this effect of $O(10^{-14})$,[30] while a recent calculation by Sehgal[31] based on a vector dominance model suggests that this $2\gamma$ contribution could dominate the branching ratio. If Sehgal is right, there should be striking interference effects in the Dalitz plot that would help to sort things out (given enough events). All of which makes this decay an interesting laboratory for testing chiral perturbation theory, if somewhat less desirable as a precision measurement of direct CP violation.

V. $K_L \to \mu e$; $K^+ \to \pi^+ \mu e$

These lepton number-violating strange decays have long been advocated as a sensitive laboratory for dynamical symmetry breaking and composite models. Dynamical symmetry breaking (or Technicolor) schemes are typically accompanied by horizontal gauge interactions and other particles, like technipions and leptoquarks, with flavor-changing interactions. $K_L \to \mu e$ and $K^+ \to \pi^+ \mu e$ have been used to place stringent bounds on the masses of these particles (especially horizontal gauge bosons) at the ~ 100 TeV level.[32] (Potentially more stringent bounds actually follow from $K^o$-$\bar{K}^o$ mixing, but it is possible to worm out of these constraints by invoking a "conserved generation number" that accompanies heavy particle exchange.)

In the minimal supersymmetric Standard Model, there are no lepton number violations resulting either from "trivial" supersymmetric exchanges or from direct FCNC's induced by RG scaling. In non-minimal supersymmetry, however, this is typically not the case.[24)] In supersymmetric Flipped SU(5)xU(1),[25-27)] for example, off-generational elements $\delta\tilde{m}^2_{\mu_L e_L}$ in the slepton mass matrices are generated by RG scaling between $M_{Pl}$ and $M_{GUT}$:[1,33)]

$$\left(\delta\tilde{m}^2_{e_L e_L}\right)_{ij} = -\frac{1}{2\pi^2}\left(U^\dagger \lambda_u^2 U\right)_{ij} \ln\frac{M_{Pl}}{M_{GUT}}\left[3m_o^2 + A^2\right] \tag{20}$$

where $\lambda_u$ are the diagonalized u-c-t quark Yukawa couplings and U is an unknown unitary matrix. This same expression (20) also applies to the conventional GUT's SU(5) and O(10), except that in these theories U gets replaced by the KM matrix V. This amounts to a considerable suppression, since the product of angles $V^*_{ts}V_{td}$ that accompanies the dominant top-quark contribution to $\delta\tilde{m}^2_{\mu_L e_L}$ is of order $10^{-3}$. If we instead take arbitrary $|U_{ij}| \le 1$ relevant to Flipped SU(5)xU(1), then (20) yields[1)]

$$\frac{\delta\tilde{m}^2_{\mu_L e_L}}{m^2_{\tilde{\ell}}} \lesssim \begin{cases} 0.03 & (m_o) \\ 0.002\ A^2/m^2_{1/2} & (A) \\ 0.001 & (m_{1/2}) \end{cases} \tag{21}$$

for initial soft supersymmetry breaking in the form of $m_o$, A and $m_{1/2}$, respectively. These off-generational mass terms give rise to off-generational mass insertions in slepton propagators in Feynman graphs, resulting in a $K_L \to \mu e$ branching ratio[1)]

$$B(K_L \to \mu e) \simeq 1.5 \times 10^{-2}\ \mathrm{GeV}^4\ \ H^2(m^2_{\tilde{\ell},\tilde{Z}}/m^2_{\tilde{q}})\ \frac{m^4_{\tilde{\ell}}}{m^8_{\tilde{q}}}\left(\frac{\delta\tilde{m}^2_{s_L d_L}}{m^2_{\tilde{q}}}\right)^2\left(\frac{\delta\tilde{m}^2_{\mu_L e_L}}{m^2_{\tilde{\ell}}}\right)^2 \tag{22a}$$

where

$$H(x) = -\frac{30(3 + \ell n x + 4x\ell n x - 3x^2 + x^2\ell n x)}{(1-x)^5} = \begin{cases} 1; & m_{\tilde{\ell},\tilde{Z}} = m_{\tilde{q}} \\ -90-30\ell n x; & m_{\tilde{\ell},\tilde{Z}} \ll m_{\tilde{q}} \end{cases} \tag{22b}$$

For a branching ratio of $10^{-11}$ (as proposed by BNL E791[34)]) we would need

$$\left(\frac{\delta\tilde{m}^2_{s_L d_L}}{m^2_{\tilde{q}}}\right)\left(\frac{\delta\tilde{m}^2_{\mu_L e_L}}{m^2_{\tilde{\ell}}}\right) \simeq \frac{0.25}{H}\frac{m^2_{\tilde{q}}}{m^2_{\tilde{\ell}}}\left(\frac{m_{\tilde{q}}}{100\ \mathrm{GeV}}\right)^2 \tag{23}$$

Recalling that the $K_L$-$K_S$ mass difference (12) sets an upper limit on the magnitude of $\delta\tilde{m}^2_{s_L d_L}/m^2_{\tilde{q}} < 1 \times 10^{-2}$, we observe that a very large $\delta\tilde{m}^2_{\mu_L e_L}/m^2_{\tilde{\ell}} > 1$ would be needed to satisfy (23). The function $H(m^2_{\tilde{\ell},\tilde{Z}}/m^2_{\tilde{q}})$ may be increased by A factor of ~ 10-20 by taking $m_{\tilde{q}}/m_{\tilde{\ell},\tilde{Z}} \sim 3$-$4$, but the increased sensitivity to $\delta\tilde{m}^2_{\mu_L e_L}$ this would afford is offset by a comparable factor $(m^2_{\tilde{q}}/m^2_{\tilde{\ell}})$ which appears in (23).

We conclude that $K_L \to \mu e$ is not a viable place to look for direct FCNC's in supersymmetric Flipped SU(5)xU(1) or in any other simple extension of the minimal supersymmetric model. [Recall that in supersymmetric SU(5) and O(10), the branching ratio (22) is a million times smaller.] These same conclusions also apply to $K^+ \to \pi^+\mu e$, whose analysis is similar, but where the branching ratio is further suppressed by phase space.

VI. $\mu \to e\gamma$

The rare decay $\mu \to e\gamma$ provides another sensitive and independent test of lepton number violation. While not quite competitive with $K_L \to \mu e$ and $K^+ \to \pi^+\mu e$ in the framework of dynamical symmetry breaking (since $K_L \to \mu e$ requires a loop, while the latter occur at the tree level), it can be very competitive in other frameworks, like supersymmetry, in which *all* low-energy FCNC's involve loops. The theoretical motivation for this decay in supersymmetry is similar to the previous section. The principal difference is that whereas $K_L \to \mu e$ and $K^+ \to \pi^+\mu e$ were tightly constrained by the $K_L$-$K_S$ mass difference, $\mu \to e\gamma$ is not.

The analysis of $\mu \to e\gamma$ in supersymmetry is somewhat complicated and can be found in ref. 1. The current experimental upper limit[35)] $B(\mu \to e\gamma) < 4.9 \times 10^{-11}$ places an upper bound on the off-generational slepton masses

$$\frac{\delta\tilde{m}^2_{\mu_L e_L}}{m^2_{\tilde{\ell}}} \lesssim \frac{1 \times 10^{-3}}{K}\left(\frac{m_{\tilde{\ell}}}{60\ \text{GeV}}\right)^2 \tag{24a}$$

whereas a signal at the $10^{-13}$ level sought by MEGA would require

$$\frac{\delta\tilde{m}^2_{\mu_L e_L}}{m^2_{\tilde{\ell}}} \sim \frac{5 \times 10^{-5}}{K}\left(\frac{m_{\tilde{\ell}}}{60\ \text{GeV}}\right)^2 \tag{24b}$$

These values are well within the possible range (21) expected from supersymmetric Flipped SU(5)xU(1), even for slepton masses as heavy as 200-1000 GeV. Indeed, (24a) already provides a useful constraint on the matrix U and the pattern of soft supersymmetry breaking. Information concerning the latter could help to sort out one of the major theoretical puzzles of supersymmetric theories generally—namely, how supersymmetry is broken and how this breaking enters the observable sector.

A comparison of (24) and (14) reveals that $\mu \to e\gamma$ provides the most sensitive test of direct supersymmetric FCNC's of any foreseeable experiment. It should be emphasized, however, that (24) and (14) are logically independent, since the Yukawa couplings $\lambda_u$ responsible for lepton number violations and those contributing to CP violation ($\lambda_6$) are unrelated, and may be present or absent in various models. Either discovery would offer powerful evidence for supersymmetric structure beyond the minimal supersymmetric Standard Model. However, their interpretation as such will depend on the discovery of supersymmetric particles at LEP, the Tevatron, or the SSC.

The analysis of $\mu \to 3e$ and muon conversion is similar, and their details can be found in ref. 1. Although very interesting in other contexts,[32] these lepton number-violating processes are not competitive with $\mu \to e\gamma$ in supersymmetry.

## VII. Conclusion

In this talk, we have considered rare decays and CP violation in the context of supersymmetry. We have seen that the rare decay $K^+ \to \pi^+\nu\bar{\nu}$ provides a useful test of supersymmetry if the sparticle masses $m_{\tilde{q}}, m_{\tilde{g}} \lesssim$ 100-150 GeV. The CP-violating decays $K_L \to \pi^o\nu\bar{\nu}$ and $K_L \to \pi^o ee$ are indirectly affected by "direct" FCNC contributions to $\epsilon_K$, and are therefore sensitive to much heavier sparticle masses *if* minimal supersymmetry is embedded in a larger theoretical structure (like a GUT) at high energies. Whereas the lepton number-violating strange decays $K_L \to \mu e$ and $K^+ \to \pi^+\mu e$ are tightly constrained in supersymmetry by the $K_L$-$K_S$ mass difference, the purely leptonic process $\mu \to e\gamma$ provides another, very sensitive probe of direct FCNC's which is logically independent from $K^+ \to \pi^o\nu\bar{\nu}$ and $K_L \to \pi^o ee$.

In the framework of low-energy supersymmetry, these CP-violating and lepton number-violating processes provide an experimental window on the structure of physics at extremely high scales ~ $M_{GUT}$ or $M_{Pl}$ which is unique apart from proton decay, and which should be pursued with comparable experimental zeal.

Acknowlegements

I would like to thank S. Kelley, whose collaboration led to ref. 1 and to many aspects of the present talk. I also wish to thank L. Littenberg for valuable discussions.

REFERENCES


1. Hagelin, J.S. and Kelley, S., "Flavor-Changing Neutral Currents and Rare Decays in Supersymmetry", MIU preprint no. MIU-THP-88/34 (1988) submitted to Nucl. Phys. B.
2. Fayet, P. and Ferrara, S., Phys. Rep. 32C, 249 (1977); Witten, E., Nucl. Phys. B185, 513 (1981).
3. Weinberg, S., Phys. Rev. D13, 974 (1976); Phys. Rev. D19, 1277 (1979); Susskind, L., Phys. Rev. D20, 2619 (1979).
4. Gaillard, M.K. and Lee, B.W., Phys. Rev. D10, 897 (1974).
5. Ellis, J. and Hagelin, J.S., Nucl. Phys. B217, 189 (1983).
6. Gaillard, M.K., Kao, W.-C., Lee, I.-H. and Suzuki, M., Phys. Lett. 123B, 241 (1983).
7. Suzuki, M., U.C. Berkeley preprint no. UCB-Pth 82/8 (1982); Shrock, R.E., Stony Brook preprint no. ITP-SB-82-38 (1982).
8. Peccei, R.D. and Quinn, H., Phys. Rev. Lett. 38, 1440 (1977); Phys. Rev. D16, 1791 (1977); Weinberg, S., Phys. Rev. Lett. 40, 223 (1978); Wilczek, F., Phys. Rev. Lett. 40, 279 (1978); Frere, J.-M., Gavela, M.B. and Vermaseren, J.A.M., Phys. Lett. 103B, 129 (1981).
9. Wilczek, F., Phys. Rev. Lett. 49, 1549 (1982).
10. Bertolini, S. and Santamaria, A., Carnegie Mellon preprint no. CMU-HEP88-2 (1988).
11. Fischbach, E., Sudarsky, D., Szafer, A., Talmadge, C. and Aronson, S.H., Phys. Rev. Lett. 56, 3 (1986); Aronson, S.H., Cheng, H.-Y., Fischbach, E. and Haxton, W., Phys. Rev. Lett. 56, 1342 (1986).
12. Inami, T. and Lim, C.S., Prog. Theor. Phys. 65, 297 (1981).
13. Hagelin, J.S. and Littenberg, L.S., Prog. Part. and Nucl. Phys., to appear.
14. Albrecht, H. et al., Phys. Lett. 192B, 245 (1987).
15. Ellis, J., Hagelin, J.S. and Rudaz, S., Phys. Lett. 192B, 201 (1987); Ellis, J., Hagelin, J.S., Rudaz, S. and Wu, D.-D., Nucl. Phys. B304, 205 (1988).
16. Gilman, F.J. and Hagelin, J.S., Phys. Lett. 126B, 111 (1983); Phys. Lett. 133B, 443 (1983).
17. Nir, Y., Nucl. Phys. B306, 14 (1988).
18. Ellis, J., Hagelin, J.S., Nanopoulos, D.V., Olive, K.A. and Srednicki, M., Nucl. Phys. B238, 453 (1984).
19. Hagelin, J.S., Kane, G.L. and Raby, S., Nucl. Phys. B241, 638 (1984).
20. See Wolfenstein, L., Nucl. Phys. B160, 501 (1979) and references therein.
21. Duncan, M.J., Nucl. Phys. B221, 285 (1983); Donoghue, J.F., Nilles, H.P. and Wyler, D., Phys. Lett. 128B, 55 (1983); Bouquet, A., Kaplan, J. and Savoy, C.A., Phys. Lett. 148B, 69 (1984); Gerard, J.M., Grimus, W., Masiero, A., Nanopoulos, D.V. and Raychaudhuri, A., Nucl. Phys. B253, 93 (1985); Dugan, M., Grinstein, B. and Hall, L., Nucl. Phys. B255, 413 (1985).
22. See, e.g., Bertolini, S., Borzumati, F. and Masiero, A., Phys. Lett. 194B, 545 (1987) and *Erratum*, Phys. Lett. 198B, 590 (1987).
23. Littenberg, L., BNL preprint (1988), submitted to Phys. Rev. D.
24. Hall, L.J., Kostelecki, V.A. and Raby, S., Nucl. Phys. B267, 415 (1986).
25. Antoniadis, I., Ellis, J., Hagelin, J.S. and Nanopoulos, D.V., Phys. Lett. 194B, 231 (1987).
26. Antoniadis, I., Ellis, J., Hagelin, J.S. and Nanopoulos, D.V., Phys. Lett. 205B, 459 (1988); Phys. Lett. 208B, 209 (1988).
27. Ellis, J., Hagelin, J.S., Kelley, S. and Nanopoulos, D.V., Nucl. Phys. B311, 1 (1988).

28. Atiya, M. et al., Snowmass 88 Proceedings; Littenberg, L., in *Proceedings of the TRIUMF CP-Violation Workshop*, Vancouver, B.C., December, 1988.
29. Dib, C.O., Dunietz, I. and Gilman, F.J., SLAC-PUB-4762 (1988).
30. Donoghue, J.F., Holstein, B.R. and Valencia, G., Phys. Rev. D35, 2769 (1987);
Ecker, G., Pich, A. and de Rafael, E., Nucl. Phys. B303, 665 (1988).
31. Sehgal, L.M., Phys. Rev. D38, 808 (1988).
32. See, e.g., Cahn, R.N. and Harari, H., Nucl. Phys. B176, 135 (1980).
33. Gabbiani, F. and Masiero, A., INFN preprint no. DFPD 16/88 (1988); Farragi, A.E., Lopez, J.L., Nanopoulos, D.V. and Yuan, K., Univ. of Wisconsin preprint no. MAD/TH/88-26 (1988).
34. Molzon, W.R. et al., BNL report no. KL-164 (1988).
35. Bolten, R.D. et al., Phys. Rev. Lett. 56, 2461 (1985).

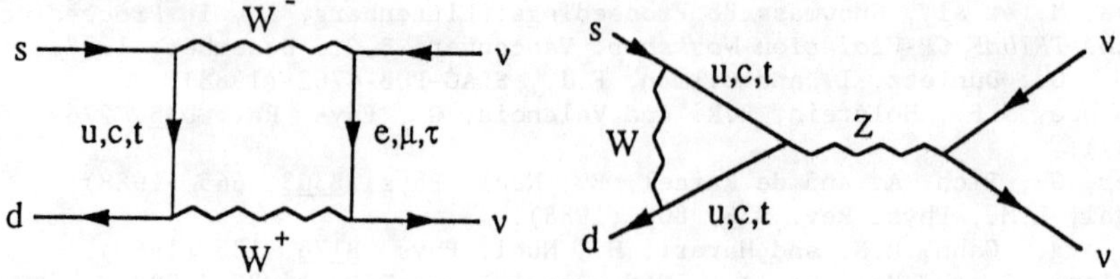


Figure 1. One-loop diagrams responsible for $K^+ \to \pi^+ \nu\bar{\nu}$ in the Standard Model.

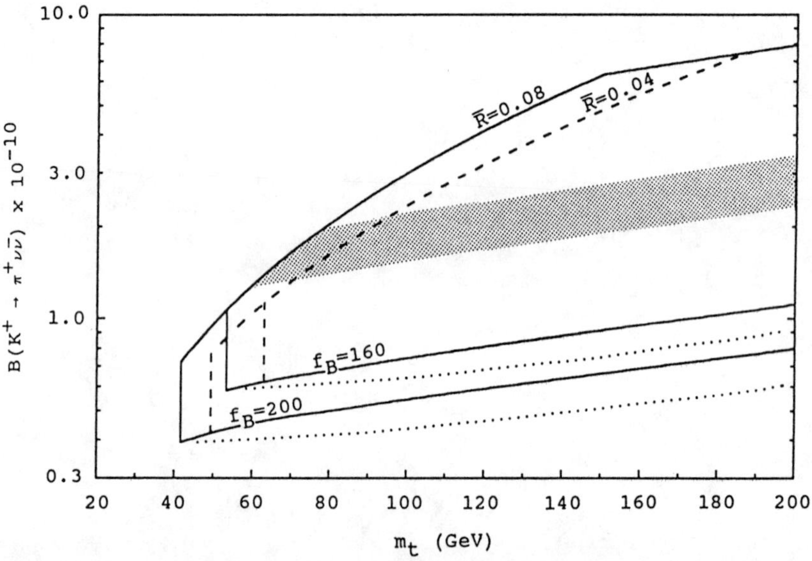


Figure 2. The $K^+ \to \pi^+ \nu\bar{\nu}$ branching ratio as a function of $m_t$ in the Standard Model with three generations. Upper bounds for $m_t > 150$ GeV are derived from $B_d^o$-$\bar{B}_d^o$ mixing, assuming $f_{B_d} B_{B_d}^{1/2} = 80$ MeV, $r_d = 20\%$, $m_c = 1.4$ GeV, and $\tau_b = 0.97 \times 10^{-12}$ s. For $m_t < 150$ GeV, upper bounds follow directly from $\tau_b$ and $\bar{R} \equiv B(b \to u)/B(b \to c) \leq 0.08$. Lower bounds are displayed for both $f_{B_d} B_{B_d}^{1/2} = 160$ MeV and $f_{B_d} B_{B_d}^{1/2} = 200$ MeV, with $r_d = 10\%$, $m_c = 1.2$ GeV, and $\tau_b = 1.25 \times 10^{-12}$ s. Dotted lines indicate the effects of less-than-maximal interference between the charm- and top-quark contributions. The shaded region represents a favored scenario in which $r_d = 18\%$, $m_c = 1.35$ GeV, $\tau_b = 1.11 \times 10^{-12}$ s, and $f_{B_d} B_{B_d}^{1/2} = 110 - 140$ MeV.

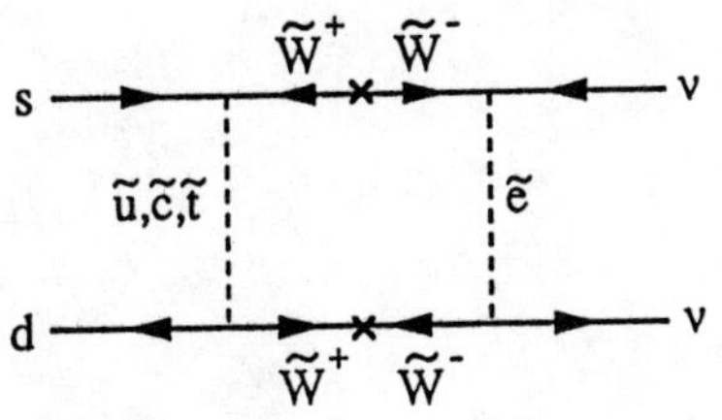


<u>Figure 3.</u> A "trivial" supersymmetric contribution to $K^+ \to \pi^+ \nu\bar{\nu}$.

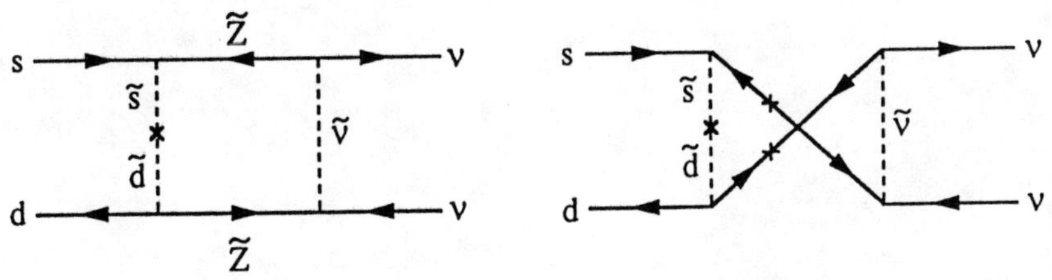


<u>Figure 4.</u> "Direct" FCNC contributions to $K^+ \to \pi^+ \nu\bar{\nu}$ resulting from soft supersymmetry breaking.

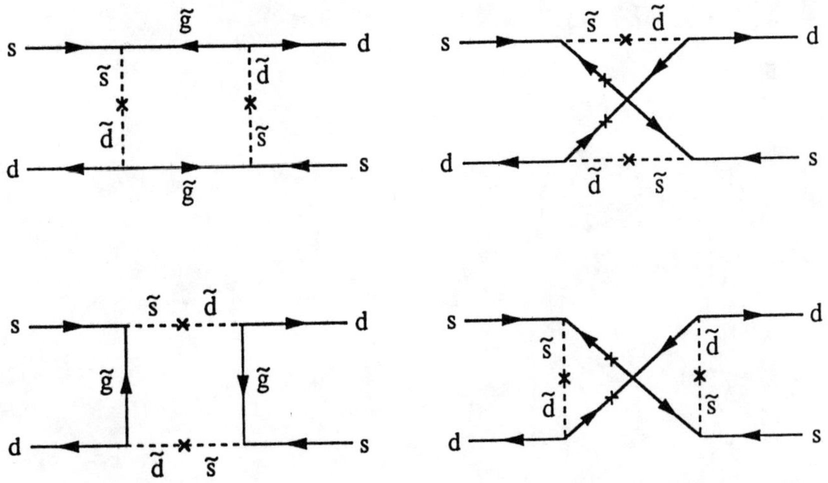


<u>Figure 5.</u> The dominant supersymmetric contributions to $K^o$-$\bar{K}^o$ mixing in the presence of "direct" flavor-changing neutral currents.

# Heavy and Light Meson Decays in Lattice QCD

Martinelli Guido
CERN
Division TH
1211 GENEVE 23
Switzerland

No written contribution received.

# CP VIOLATION : A MATTER OF GRAVITY ?

G. Chardin
Département de Physique des Particules Elémentaires
CEN Saclay, Gif/Yvette, France

Abstract – The idea that CP violation, only observed in the $K^\circ - \overline{K}^\circ$ system, may be explained by antigravity —a different acceleration of matter and antimatter in a gravitational field— is proposed. A critical review of the arguments which led, thirty years ago, to the rejection of antigravity leads, by reformulating one of the arguments of impossibility, to a simple expression for ε, the CP violating parameter, as a function of the intensity of the gravitational field. Some predictions of this model are presented. Finally, several experimental tests are proposed, which can be realised in a near future by two experiments in preparation at CERN.

## 1. INTRODUCTION

Twenty five years ago, CP violation was discovered experimentally [1]. This seemed to define an absolute distinction between matter and antimatter [2]. Whereas Lee and Yang [3] had noted in 1956 that parity symmetry had no reason to be respected in weak interaction processes, Landau [4] soon proposed that symmetry would be restored under the combined CP transformation. The demonstration, under apparently extremely general assumptions, of the invariance of physical laws under the combined CPT transformation [5] thereby linked any manifestation of CP asymmetry to a failure of time-reversal invariance. Any macroscopic phenomenon in the universe exhibits such a temporal asymmetry which is, however, usually related to boundary conditions [6], whereas fundamental processes are usually considered to be of little relevance to the cosmological "arrow of time". In this context, the announcement by Christenson et al. [1] of the observation of CP asymmetry in the study of the $K^\circ - \overline{K}^\circ$ system was received with considerable surprise. Significantly, Bell and Perring [7], together with Bernstein, Cabibbo and Lee [8] proposed independently to restore CP symmetry by invoking a long-range coupling to a vector field, which would reflect the predominance in our environment of matter over antimatter. Gravitation itself appeared to be excluded as a candidate from the arguments of Schiff [9], based on the constraints imposed by the Eötvös-Dicke-Panov experiments, and of Good [10], based on the existence of a long-lived $K_L$ component in the $K^\circ - \overline{K}^\circ$ system and the non-observation, at this time, of any CP violation in this system. These arguments, which will be briefly discussed in the following, appeared to demonstrate that the action of gravity should be identical, to a considerable precision, on matter and antimatter. In any case, the linear dependence of the CP violating parameter $\varepsilon$ as a function of the kaon beam energy, predicted by these simple models, is not observed experimentally. Today, although the Standard Model provides the possibility to accommodate CP violation, no model of such a violation seems particularly compelling [11].

Recently, however, the hypothesis that the principle of equivalence is an exact law has been questioned : Donoghue, Holstein and Robinett [12] have built an explicit example of violation of the principle of equivalence in the context of field theory at finite temperature. Similarly, Goldman and Nieto [13] have recalled that supergravity theories incorporate a violation of the principle of equivalence in its classical sense, and lead to reconsider the impossibility arguments of Schiff [9] and Good [10] already mentioned. Furthermore, black hole radiation [14], which was introduced for questions of consistency with the second law of thermodynamics and, more recently, wormholes [15] suggest a relation between time asymmetry and gravitation. Without pretending to scientific rigour, and in the spirit of Morrison's paper thirty years ago [16], it seems opportune to reconsider the possibility that CP violation in the $K^\circ - \overline{K}^\circ$ system may be due to a violation of the principle of equivalence.

## 2. ARGUMENTS OF IMPOSSIBILITY

Proposed by Morrison [16] in a celebrated paper, antigravity appears to violate the sacrosanct CPT symmetry, exclude the existence of the long-lived component in the neutral kaon system in the presence of the Earth gravitational field, violate energy conservation, imply vacuum instability and contradict the results of the Eötvös-Dicke-Panov experiments. These impossibility arguments have recently been reviewed by Bell [17], and criticized by Goldman and Nieto [13] and by Beverini et al. [18].

### 2.1 CPT violation

Considering more closely these impossibility arguments, it appears, however, that the demonstration of the CPT theorem presupposes a flat space-time, a condition which is obviously not met in the context of gravity. Indeed, Wald [19] and Penrose [20] have suggested that any consistent quantum gravity theory should incorporate a violation of time reversal symmetry and, in a limited sense, of CPT symmetry [19]. Therefore, within our present understanding, an (apparent) CPT violation on curved space-time cannot be used as an argument against antigravity.

### 2.2 Vacuum instability and violation of energy conservation

Facing the problems of vacuum instability and violation of energy conservation, it is interesting to consider, with the aid of a gedanken experiment invented by Wigner [21], the typical energy of the photons created in virtual processes in the presence of a gravitational field of intensity $g$. The principle of the gedanken experiment can be summarized in the following way: let us consider an $e^+ e^-$ pair at rest in a constant gravitational field of intensity $g$ and in a null electric field, at an altitude $z=z_0$. We would like to transport this $e^+ e^-$ pair at a lower altitude $z=z_1$. In order to do this, we may choose to transport separately both the electron and the positron in the gravitational field. But we have also the possibility to transport the pair of particles in a different way: we may first annihilate the $e^+ e^-$ pair into two photons, propagate downwards the two photons to the altitude $z=z_1$, and there recreate the $e^+ e^-$ pair (Fig.1).

Initially, each annihilation photon has an energy $mc^2$, where $m$ is the electron mass and $c$ the velocity of light. After the fall in the gravitational field, however, it is known, from the experiments of Pound and Rebka [22] and Vessot and Levine [23], that the two photons have gained an energy $mg(z_0 - z_1)$. This means that recreating the $e^+ e^-$ pair at the lower altitude $z=z_1$ would result in an excited state and, for small values of $mg(z_0 - z_1)$, in releasing a deexcitation photon of this energy.

For the purpose of simplicity, let us imagine that electrons and positrons have strictly "opposite" gravitational charges: i.e. that in a gravitational field, an electron and a positron are submitted to opposite forces; we define this as "pure antigravity". Considered as a composite system, the $e^+ e^-$ pair can then be carried (adiabatically) anywhere in the gravitational field at no cost, because the resultant force acting on the $e^+ e^-$ pair is zero. We have then the possibility to realize a cyclic process resulting in the extraction of a photon from the gravitational field (Fig.1). Namely, an $e^+ e^-$ pair is annihilated at the altitude $z=z_0$, the annihilation photons transported to the lower altitude $z=z_1$ where a pair is recreated in an excited state (emission of a photon) and the pair is carried back, at no cost, at the altitude $z=z_0$.

This global argument (the construction of a perpetual engine of the second kind) has a local counterpart which results in the instability of the vacuum in the presence of a gravitational field, even if there are no "real" particles present. Let us find the typical energy of photons created in virtual creation-annihilation

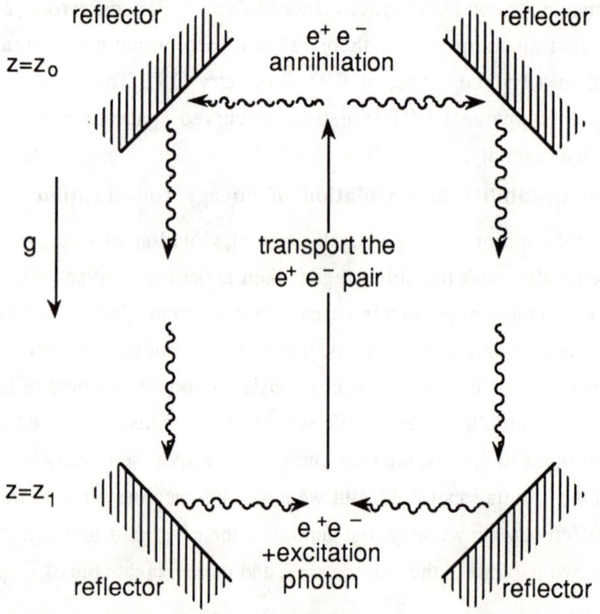


Fig.1: Schematic representation of a cyclic process in which a photon is extracted from the gravitational field.

processes in the vacuum. From the Heisenberg inequalities, a virtual creation-annihilation process will take place over a length scale $\Delta z \approx \hbar/mc$, where $m$ is the mass of the propagated particle. Due to antigravity, the typical energy of a photon which can be extracted from the gravitational field is then: $m\, g\, \Delta z \approx \hbar g/c$ (note that this quantity is independent of the mass of the propagated particle). The associated wavelength is $\lambda \approx c^2/g$, approximately one light-year at the surface of the Earth. Whereas such an instability cannot be used to reject the hypothesis of antigravity since it leads to unobservable effects in testable conditions, it should be noted that a similar expression of a "vacuum temperature" enters the Unruh [24] and Hawking [14] effects. A physicist disturbed by this aspect of antigravity should face the fact that the Hawking radiation of black holes provides a similar example of "vacuum instability". It should also be noted that the apparent violation of energy conservation here is not unexpected. From Noether's theorem, a conserved quantity named energy can be defined whenever invariance under time translations is respected. Any process defining an "arrow of time" will then forbid the construction of energy as a conserved quantity (instead, *entropy* becomes the relevant quantity).

### 2.3 Schiff's argument

Schiff [8] claimed that the results of the Eötvös-Dicke-Panov experiments, which verify the validity of the equivalence principle for ordinary matter to a precision of $10^{-11}$, implied that the principle of equivalence should also apply to antimatter. The argument is based on the fact that, due to vacuum polarization, virtual electron-positron pairs contribute differently to the energy momentum tensor of different materials. Estimating this contribution, Schiff concluded that the null results of the Eötvös-Dicke-Panov experiments constrain the gravitational mass of the electron and the positron to be equal to within one percent. However, Goldman and Nieto [13] have pointed out that Schiff's argument is based on an incorrect renormalization procedure. Without providing an exact calculation, these authors argue that the relative contribution of the polarization term calculated by Schiff would be $\approx 10^{-16}$, well beyond the existing limits.

### 2.4 Good's argument

Good's argument [9] seems to constrain very strongly (at the $10^{-7}$ level) a possible violation of the equivalence principle when antimatter is concerned. It uses the very small mass difference between the two propagation eigenstates of the $K^\circ - \overline{K}^\circ$ system, approximately 3.6 $10^{-6}$ eV. Noting that the potential energy $mV$ of a neutral kaon in the gravitational field of the Earth is $\approx$ 0.4 eV, the seemingly natural frequency prescription $E/\hbar \rightarrow (E \pm mV)/\hbar$ in the presence of a potential appears to imply that, in the gravitational potential of the Earth, a $K_L$ would be instantly regenerated in $K_S$. The almost complete absence of such regeneration (CP violation had not been discovered at this time) enabled Good to conclude that the gravitational masses of the $K^\circ$ and of the $\overline{K}^\circ$ cannot differ by more than one part in $10^7$. However, this argument refers to an absolute potential, unknown from gauge theories. Indeed, the reference to the Earth potential is mysterious to Good himself; why not use the Sun or the Galactic potential ? Goldman and Nieto [13] find a counter-example to this argument in the vacuum polarization by an electric field intense enough to create electron-positron pairs. Obviously, the frequency difference between the newly created electron and positron starts from zero rather than being instantaneously created. Applying this natural idea to the $K^\circ - \overline{K}^\circ$ system leads, surprisingly, to a very simple expression for $\varepsilon$ which lies close to the experimental value and leads to several experimental predictions.

## 3. EXPRESSION OF THE CP VIOLATING PARAMETER. CONSEQUENCES

Needless to say, the existence of antigravity would imply major revisions in the present theoretical framework. Antigravity obviously violates the equivalence principle, the basic principle of general relativity, and defines an "arrow of time" (through entropy production): any attempt to build a Lagrangian field theory then appears doomed. Nevertheless, it is interesting to derive an estimate of the (apparently) CP violating parameter which would result from antigravity — in the static limit. Relativistic corrections to this expression will then be examined.

The relevant quantities entering the expression of ε are the characteristic mixing time introduced by weak interactions, $\tau \approx \hbar/\Delta mc^2$, and the frequency difference introduced by pure antigravity between a K° and a $\overline{\mathrm{K}}$° after a time $\Delta t$: $\Delta\nu \approx 2m\, g\, c\, \Delta t/\hbar$ (strictly speaking, the mass entering this expression is not equal to the kaon mass since a neutral kaon is made of a quark and an antiquark). In order to see this, assume that there exists a static limit and estimate the frequency difference for a upward-going kaon beam with $\gamma = E_K/m_K$ of order unity. The ε parameter is then simply defined as the phase difference developed during time τ [1]:

$$\varepsilon \approx 2\hbar m\, g/\Delta m^2 c^3 \approx 1.7\ 10^{-3}.$$

A simple phenomenological model describing a fermion-antifermion coupled pair in a potential leads to a similar expression for ε. The COW experiment [26], which measures the perturbation induced by the Earth gravitational field on neutron interferometry, is the only existing (non relativistic) quantum gravity experiment. Relativistic corrections to the phase shift induced by gravity have been calculated by Anandan [27] and involve a $(1 + \beta^2)$ dependence, where β is the neutron velocity. The measurement of the ε parameter using semi-relativistic kaons appears therefore crucial. In addition, the dependence on $g$ in the expression of ε leads to consider the influence of the zenith angle θ of the kaon beam. The relevant quantity here is the ratio between the variation of potential energy during a $K_S$ lifetime and the quantity $\Delta mc^2$, leading to expected correction factors of order $\gamma \cos\theta$.

In addition, using the scaling expression in $m/\Delta m^2$, the value of the CP violating parameter in the B°– $\overline{\mathrm{B}}$° system is predicted to be equal to 2-3 $10^{-6}$, differing from the Standard Model (SM) prediction, although such a precision is hardly attainable in practice. Besides, the value of the ε' parameter describing direct CP violation is zero, again differing from the SM prediction, and the phases of the $\eta_{+-}$ and $\eta_{oo}$ parameters are equal. The value taken by these parameters is also a prediction. It is well known that a long-range coupling with a potential entering as a *real constant term* in the mass matrix will lead to a phase for ε equal to – Arctan $(2\Delta m/\Gamma)$ (modulo 180°) [6], [7], which is at ≈ 87° from the "natural phase" Arctan $(2\Delta m/\Gamma)$. Here, the parametric

---

[1]That the quantity $\hbar m\, g/\Delta m^2 c^3$ is of the order of the experimental value of ε has been noted by Fischbach [25]. The first force, not the fifth.

oscillation introduced by antigravity leads a priori to a phase in quadrature with –Arctan ($2\Delta m/\Gamma$), i.e. 90° – Arctan ($2\Delta m/\Gamma$) (modulo 180°). This phase (≈46.5°) is only ≈ 3° from the natural phase (≈43.5°).

## 4. EXPERIMENTAL TESTS

Even in the absence of a consistent theoretical formalism, the expression that we have proposed for $\varepsilon$ suggests the measurement of this parameter in new experimental conditions. In particular, the PS-195 experiment [28], at the Low-Energy Antiproton Ring (LEAR) at CERN, currently in the stage of installation, will, for the first time, determine with sufficient precision the $\varepsilon$ parameter in conditions where the $(1 + \beta^2)$ factor, expected from the relativistic corrections to the COW experiment, is significantly different from its ultrarelativistic limit. Using kaons in the momentum range [400, 700] MeV/c, generated from the annihilation at rest of a proton with an antiproton, this experiment will also be able to test the existence, as a function of the zenith angle $\theta$, of corrections terms of order $\gamma \cos\theta$. Such corrections, for an isotropic emission, should be easily measurable by the PS-195 experiment, designed to achieve a precision of a few parts in a thousand in the determination of the CP violating parameters. In addition, this experiment will also determine, within ≈0.2°, the value of the $\eta_{+-}$ parameter and will thus be able to test without ambiguity the antigravity hypothesis. It should be noted that the only experiment for which the $\gamma \cos\theta$ correction is expected to be significant [29] has led to a determination of the $\eta_{+-}$ parameter equal to $(2.09 \pm 0.02)\ 10^{-3}$, in disagreement by 9 standard deviations from the usually accepted value (see Ref.[30] for a review on the inconsistencies in the determination of the $\eta_{+-}$ parameter). In this last experiment, the neutral beam used made an angle of 6.25 mrad with the horizontal, at a typical energy of 70 GeV.

Another experiment at LEAR, the PS-200 experiment [18], proposes to measure, with a precision of the order of 1%, the difference between the gravitational masses of proton and antiproton. However, since the quarks inside the proton carry only a small fraction of its mass, the most promising experiment allowing the observation of an explicit violation of the equivalence principle consists in measuring the gravitational mass of the positron, an experiment advocated by Fairbank and Witteborn [31, 32].

## 5. CONCLUSIONS

We have proposed the idea that antigravity may be responsible for the (apparent) CP violation in the $K^\circ - \overline{K}^\circ$ system. This suggests an expression of the CP violating parameter $\varepsilon$ which depends a priori on the intensity of the gravitational field, and on the energy and the angle relative to the horizontal of the kaon beam. All precision measurements of the $\varepsilon$ parameter have involved nearly horizontal and ultrarelativistic kaon beams and have therefore been unable to test crucial aspects of our hypothesis. These will be tested in a near future by the PS-195 and PS-200 experiments at CERN. Any absolute distinction between matter and antimatter would vanish, and any explicit CP violating effect, notably in the charge asymmetry in the $K_L$ decay products, disappears. Whereas CP violation allows to find whether a remote galaxy is made of matter or antimatter [2], this distinction becomes impossible if antigravity is the source of apparent CP violation in the $K°-\bar{K}°$ system: in an antiworld, antiphysicists performing a charge asymmetry measurement will find the same result as physicists in a world of ordinary matter (assuming the experiments are performed in gravitational fields of the same intensity). On the contrary, the violation of the equivalence principle implied by antigravity appears to be linked to a violation of time-reversal symmetry, in a context where the CPT theorem has not been demonstrated. Whereas general relativity is invariant under time-reversal, antigravity may provide the "arrow of time" which seems to be an integral part of any quantum theory incorporating gravity.

## ACKNOWLEDGMENTS

It is a pleasure to thank J. Tran Thanh Van, the organiser of this conference, for his invitation and all the physicists, especially from Saclay, who have shared my enthusiasm for this provocative idea.

REFERENCES

[1] J. H. Christenson, J. W. Cronin, V. L. Fitch and R. Turlay, Phys. Rev. Lett. **13**, 138 (1964)
[2] J. J. Sakurai and A. Wattenberg, Phys. Rev. **161**, 1449 (1967)
[3] T. D. Lee and C. N. Yang, Phys. Rev. **104**, 254 (1956)
[4] L. D. Landau, JETP **32**, 405 (1957)
[5] G. Lüders, Ann. Phys. **2**, 1 (1957); W. Pauli in *"Niels Bohr and the development of physics"* (McGraw Hill, New York, 1955)
[6] R. G. Sachs, in *"The Physics of Time Reversal"*, (The University of Chicago Press, Chicago, 1987)
[7] J. S. Bell and J. K. Perring, Phys. Rev. Lett. **13**, 348 (1964)
[8] J. Bernstein, N. Cabibbo and T. D. Lee, Phys. Lett. **12**, 146 (1964)
[9] L. I. Schiff, Proc. Nat. Acad. Sci. **45**, 69 (1959); Phys. Rev. Lett. **1**, 254 (1958)
[10] M. L. Good, Phys. Rev. **121**, 311 (1961)
[11] L. Wolfenstein, Ann. Rev. Nucl. Part. Sci., **36**, 137 (1986)
[12] J. F. Donohue, B. R. Holstein and R. W. Robinett, Phys. Rev. **D30**, 2561 (1984)
[13] T. Goldman and M. M. Nieto, Phys. Lett. **112B**, 437 (1982); *"Physics with antiprotons at LEAR in the ACOL era"*, eds U. Gastaldi, R. Klapisch, J. M. Richard and J. Tran Thanh Van (Ed. Frontières, Paris, 1985), p. 639.
[14] S.W. Hawking, Nature **248**, 30 (1974), Comm. Math. Phys. **43**, 199 (1975)
[15] M. S. Morris, K. S. Thorne and U. Yurtsever, Phys. Rev. Lett. **61**, 1446 (1988); J. Ellis, S. Mohanty and D. V. Nanopoulos, Phys. Lett. B**221**, 113 (1989)
[16] P. Morrison, Am. J. Phys. **26**, 358 (1958)
[17] J. S. Bell, in *"Fundamental Symmetries"*, eds P. Bloch, P. Pavlopoulos and R. Klapisch, (Plenum, New York, 1987)
[18] N. Beverini et al., CERN/PSCC/86-2 (1986) *"A measurement of the gravitational acceleration of the antiproton"*
[19] R. M. Wald, Phys. Rev. D**21**, 2742 (1980)
[20] R. Penrose, in *"Highlights in gravitation and cosmology"*, eds. B. R. Iyer, A. Kembhari, J. V. Narlikar and C. V. Vishveshvara, (Cambridge Univ. Press, 1988)
[21] E. P. Wigner, Proc. Am. Phil. Soc. **93**, 521 (1949)
[22] R. V. Pound and G. A. Rebka, Jr., Phys. Rev. Lett. **4**, 337 (1960); R. V. Pound and J. L. Snider, Phys. Rev. **140**, B788 (1965)
[24] W. G. Unruh, Phys. Rev. D**14**, 870 (1976)
[23] R.F.C. Vessot and M.W. Levine, Proceedings of the 2nd Frequency Standards and Metrology Symposium, ed. H. Hellwig, p.659 (1976)
[25] E. Fischbach, in *"Cosmology and Gravitation"*, eds P. G. Bergmann and V. de Sabbata (Plenum, New York, 1980), p. 359
[26] R. Colella, A. W. Overhauser and S. A. Werner, Phys. Rev. Lett. **34**, 1472 (1975)
[27] J. Anandan, Phys. Rev. D**15**, 1448 (1977)
[28] L. Adiels et al., *"Tests of CP violation with $K^\circ$ and $\bar{K}^\circ$ at LEAR"*, CERN/PSCC/85-6 (1985)
[29] S. H. Aronson, G. J. Bock, H. Y. Cheng and E. Fishbach, Phys. Rev **D28**, 476, 495 (1983)
[30] Particle Data Group, Phys. Lett. B**75**, 1 (1978)
[31] F. C. Witteborn and W. M. Fairbank, Phys. Rev. Lett. 19, 1049 (1967); Nature 220, 436 (1968)
[32] W. M. Fairbank, in *"Fifth Force, Neutrino Physics"*, eds O. Fackler and J. Tran Thanh Van, (Ed. Frontières, Paris, 1988)

# PHI FACTORIES

*José Bernabeu and Lalit Seghal.*

# COHERENT DECAY OF $(K^0 \bar{K}^0)$: CP-EFFECTS AND $K_s$- DECAYS *)

J. Bernabéu, F.J. Botella, and J. Roldán.

*Departament de Física Teòrica, Universitat de València, e IFIC, Centre Mixt Univ. Valéncia-CSIC, 46100 Burjassot, Spain.*

## ABSTRACT

This paper discusses the observable effects for CP-violation studies resulting from the decay of the (C=-) state of the $(K^0-\bar{K}^0)$ system, such as it is produced in an $e^+e^-$ $\Phi$-factory. After some comments on the characteristics of the initial state, a coherent superposition of $(K_L\ K_s)$ states, we study the decay amplitude to the CP-violating $\{2\pi, 2\pi\}$ channels. The time dependent intensity asymmetry for the $\pi^+\pi^-\pi^0\pi^0$ channel measures $Re(\epsilon'/\epsilon)$ in the non-interference region, whereas the relative phase between $\epsilon$ and $\epsilon'$ is accessible in the interference region. The CP-parameter can also be obtained from the relative time-integrated rates for different decay channels. Furthermore, these experiments can be used, with less demand of statistics, to do systematic studies of $K_s$-decays.

## 1. INTRODUCTION

CP violation is well established in the neutral kaon system and our presence in this magnificent Château de Blois to cellebrate the 25th Discovery of CP violation[1] is a proof of it. The physical particles, with definite mass and lifetime, correspond to

$$K_S \simeq K_1 + \tilde{\varepsilon}\, K_2 \qquad \text{and} \qquad K_L \simeq K_2 + \tilde{\varepsilon}\, K_1 \tag{1}$$

where $K_{1,2}$ are the CP eigenstates of the $K^o - \overline{K}^o$system with eigenvalues ±1, respectively. $\tilde{\varepsilon}$ is the (phase convention dependent) mixing amplitude responsible of the CP violation induced by the kaon mixing matrix[2,3]. CP violation may also occur in the direct decay transition amplitude. We consider the $2\pi$ decay modes of $K^o$, $\overline{K}^o$. In terms of the amplitudes to definite isospin states

$$\frac{1}{\sqrt{2}}\left\{|\pi^+\pi^-> + |\pi^-\pi^+>\right\} = \frac{1}{\sqrt{3}}\left[\sqrt{2}\;|\,I=0> + |\,I=2>\right]$$

$$|\pi^o\,\pi^o> = \frac{1}{\sqrt{3}}\left[|I=0> - \sqrt{2}\;|\,I=2>\right] \tag{2}$$

we can describe the observables in terms of four amplitudes $<0|S>$, $<2|S>$, $<0|L>$, $<2|L>$. As the state of $2\pi$' s is eigenstate of CP with eigenvalue +1, a non-vanishing value of the last two amplitudes is a signal of CP-violation. In the limit of the $\Delta I = 1/2$ rule for the transition, the second and fourth amplitudes would be suppressed. Thus the natural parameters for the phenomenology of the $K^o$, $\overline{K}^o$system decaying to $2\pi$ states are

$$\varepsilon = \frac{<0|L>}{<0|S>}\,, \qquad \omega = \frac{<2|S>}{<0|S>}\,, \qquad \varepsilon_2 = \frac{<2|L>}{<0|S>} \tag{3}$$

where the first is CP-violating, the second the piece violating the $\Delta I = 1/2$ rule and the third violating both CP and $\Delta I = 1/2$.

The CP-violating parameters $\eta_{+-}$ and $\eta_{oo}$ are defined as the ratios of the CP violating and allowed amplitudes for the $K_L$ and $K_S$ decays into two charged and neutral pions, respectively. These are given by

$$\eta_{+-} \equiv \frac{<+-|L>}{<+-|S>} = \varepsilon + \frac{\varepsilon'}{1+\frac{1}{\sqrt{2}}\omega}$$

$$\eta_{oo} \equiv \frac{<00|L>}{<00|S>} = \varepsilon - \frac{2\varepsilon'}{1-\sqrt{2}\,\omega} \qquad (4)$$

where $\varepsilon' = \frac{1}{\sqrt{2}}(\varepsilon_2 - \omega\varepsilon)$

If $\tilde{\varepsilon}$ in Eq. (1) were the only source of CP-violation, so that $<2\pi|K_2> = 0$, then

$$\varepsilon = \tilde{\varepsilon} \quad , \quad \omega = \frac{<2|K_1>}{<0|K_1>} \quad , \quad \varepsilon_2 = \omega\tilde{\varepsilon}$$

$$\longrightarrow \varepsilon' = 0 \quad , \quad \eta_{+-} = \eta_{oo} \qquad (5)$$

Experimentally we know $\varepsilon' << \varepsilon$ [4] and, furthermore, $\omega$ is suppressed by the $\Delta I = 1/2$ rule, so, to a good approximation, we can write

$$\eta_{+-} = \varepsilon + \varepsilon' \quad , \quad \eta_{oo} = \varepsilon - 2\varepsilon' \qquad (6)$$

The comparison with Eq.(5) shows that $\varepsilon' \neq 0$ measures CP-violation in the decay amplitude $<2\pi|K_2> \neq 0$. It is worth to emphasize that $\varepsilon' \neq 0$ requires the presence of both $\Delta I = 1/2$ and $\Delta I = 3/2$ components in the decay amplitude.

Instead of considering single Kaon states, our interest is focussed on the decay of coherent states of the two particle ($\bar{K}^o - K^o$) system, such as it is produced from the decay of the $\Phi(1020)$ resonance, with a branching ratio of 34.4±0.9 % .

## 2. THE INITIAL $K^o - \bar{K}^o$ STATE

The ($K^o - \bar{K}^o$) state with well defined angular momentum $J = L$ , and well defined charge conjugation $C = (-1)^L$, has to be symmetric, due to Bose statistics, under the combined operation of particle-antiparticle exchange C and permutation $\mathcal{P}$ of the space coordinates. In other words, the physical state $|\psi>$ has to be an eigenstate of $C\,\mathcal{P}$ with eigenvalue +1, as a consequence of quantum mechanics indistinguishibility. In terms of the direct product state $|LM;\ K^o \bar{K}^o>$ of a $K^o$ and a $\bar{K}^o$, we can write explicitely

$$|\psi> = \frac{1}{\sqrt{2}} (\mathbb{1} + C\mathcal{P}) \mid L M ;\ K^o \bar{K}^o > \\ = \frac{1}{\sqrt{2}} \left\{ |K^o(\vec{k})\bar{K}^o(-\vec{k})> + (-1)^L |\bar{K}^o(\vec{k}) K^o(-\vec{k})> \right\} \tag{7}$$

In general, without any use of CPT, CP or T symmetries, the state (7) for L=1, (C=−) can be expressed in terms of the definite mass and lifetime states of the neutral kaon as

$$|K^o \bar{K}^o (C=-)> = \frac{1+|\alpha|^2}{2\sqrt{2}\,\alpha} \left\{ |K_L(\vec{k}) K_S(-\vec{k})> - |K_S(\vec{k}) K_L(-\vec{k})> \right\} \tag{8}$$

with no presence of $K_S K_S$ or $K_L K_L$ states: Bose statistics says that, for L=1 , the state of two identical bosons is forbidden. This statement is applied to any time in the evolution of the system. So, coming back for L=1 to definite flavour states (7) for arbitrary time, we find that there cannot be simultaneous $K^oK^o$ or $\bar{K}^o\bar{K}^o$ states in this case. The parameter $\alpha$ is given by $\alpha = (1+\tilde{\varepsilon})/(1-\tilde{\varepsilon})$.

We realize that the $(K^o - \bar{K}^o)$ state, such as is produced in an $e^+ e^-$ $\Phi$-factory, is a coherent superposition of $(K_S\ K_L)$ states of the single neutral kaon. Thus this method is a simple and natural way of producing simultaneously (correlated) beams of $K_S$ and $K_L$.

### 3. THE DECAY OF THE (C=−) STATE.

The decay amplitude corresponding to the appearance of $X_1$ at time $t_1$ and $X_2$ at time $t_2$, as illustrated in the Figure,

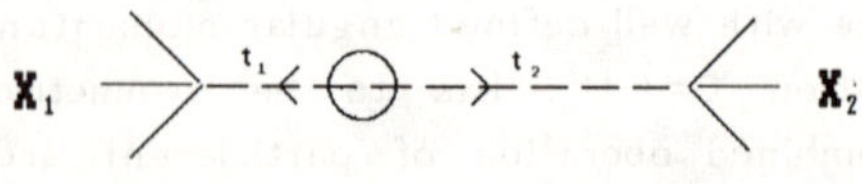


is easily obtained from the time evolution of Eq. (8). We get

$$A(X_1, X_2) = \frac{1+|\alpha|^2}{2\sqrt{2}\,\alpha} <X_1|K_S><X_2|K_S> \exp\left[-i(\lambda_L+\lambda_S)t/2\right]$$

$$\cdot \left[\eta_1 \exp(-i\Delta\lambda\, \Delta t/2) - \eta_2 \exp(i\Delta\lambda\, \Delta t/2)\right] \qquad (9)$$

where $\eta_K \equiv \frac{<X_K|K_L>}{<X_K|K_S>}$, k=1, 2; and the parameters and variables

$$\lambda_{S,L} = M_{S,L} - \frac{i}{2}\Gamma_{S,L} \; ; \quad \Delta\lambda = \lambda_S - \lambda_L$$

$$t \equiv t_1 + t_2 \quad ; \quad \Delta t \equiv t_2 - t_1 \qquad (10)$$

In fact, the use of the variables t and $\Delta t$, instead of $t_1$ and $t_2$, is very useful to show correlations. Suppose that we detect the two decay products $X_1$ and $X_2$ at equal times $t_1=t_2$. For $\Delta t=0$, Eq. (9) gives

$$A(X_1, X_2) \propto (\eta_1 - \eta_2)$$

In particular, for $X_1=X_2$ one finds

$$A\left[X_1(t_1), X_1(t_1)\right] = 0 \qquad (11)$$

which is a quantum mechanical correlation of the Einstein–Podolsky–Rosen type implied by Bose statistics. This correlation was noticed by H.J.Lipkin [5] in 1968.

If we select the two different $2\pi$ decay channels of the neutral kaon, i.e. $X_1 \equiv \pi^+\pi^-$ and $X_2 \equiv \pi^o\pi^o$, we find at equal times

$$A\left[\pi^+\pi^-(t_1), \pi^o\pi^o(t_1)\right] \propto (\eta_{+-} - \eta_{oo}) \propto \varepsilon' \qquad (12)$$

This result, first found in Ref. [6], has been discussed by Dunietz, Hauser and Rosner [7] in the context of Φ–factories, trying to maximize the ε' effect of Eq. (12) by comparing the observables at different time slices. One interesting question is to see how the time smearing affects the dip in the amplitude for the $\pi^+\pi^-\ \pi^o\pi^o$ channel implied by the result (12), in a situation for which the "natural" value is of order ε.

## 4. INTENSITY ASYMMETRY

Studying the events corresponding to the $\pi^+\pi^- \ \pi^o\pi^o$ channel, integrate the decay probability over all "accessible" times t, keeping $\Delta t$ a constant

$$I(\Delta t) \equiv \frac{1}{2}\int_{|\Delta t|}^{\infty} dt\ |A[X_1, X_2]|^2 ; \qquad \begin{array}{l} X_1 \equiv \pi^+\pi^- \\ X_2 \equiv \pi^o\pi^o \end{array} \tag{13}$$

An intensity asymmetry arises under the change $\Delta t \rightarrow -\Delta t$ when $\epsilon'/\epsilon \neq 0$. One gets [7]

$$A_I(\Delta t) \equiv \frac{I(\Delta t) - I(-\Delta t)}{I(\Delta t) + I(-\Delta t)} = \tag{14}$$

$$= \frac{2\left[\eta_r \sinh \frac{\Delta\Gamma \Delta t}{2} - \eta_i \sin \Delta M\, \Delta t\right]}{(\eta_+ + \eta_-) \cosh \frac{\Delta\Gamma\,\Delta t}{2} + (\eta_- - \eta_+) \cos \Delta M\ \Delta t}$$

where the $\eta$ parameters are given in terms of $\epsilon$ and $\epsilon'$ by

$$\eta_+ = |\varepsilon|^2 \left[4 - 4\ \mathrm{Re}\left(\frac{\epsilon'}{\epsilon}\right) + \left|\frac{\epsilon'}{\epsilon}\right|^2\right]$$

$$\eta_- = |\varepsilon|^2\, 9 \left|\frac{\varepsilon'}{\varepsilon}\right|^2 \tag{15}$$

$$\eta_r = |\varepsilon|^2 \left[6\,\mathrm{Re}\left(\frac{\epsilon'}{\epsilon}\right) - 3\left|\frac{\varepsilon'}{\varepsilon}\right|^2\right]$$

$$\eta_i = |\varepsilon|^2 \left[-6\ \mathrm{Im}\left(\frac{\epsilon'}{\epsilon}\right)\right]$$

The last quantity is sensitive to $(\Phi_{+-} - \Phi_{oo})$, the phases of the two parameters (4) for the CP–violating decays. The phase of $\epsilon$ is determined [3] by unitarity to be $43.7 \pm 0.2°$. As a consequence of CPT, the phase of $\epsilon'$ is entirely determined by the difference in the s–wave $\pi\pi$ scattering phase shifts, with a value of $45 \pm 15°$. A natural consequence of the small magnitude of $\varepsilon'$

is that the phases of $\eta_{+-}$ and $\eta_{oo}$ are close to that of $\varepsilon$ and one expects the phase difference $(\Phi_{+-} - \Phi_{oo})$ to be no more than a fraction of a degree.

A typical intensity asymmetry[7] , as function of $\Delta t$, is given in the Figure

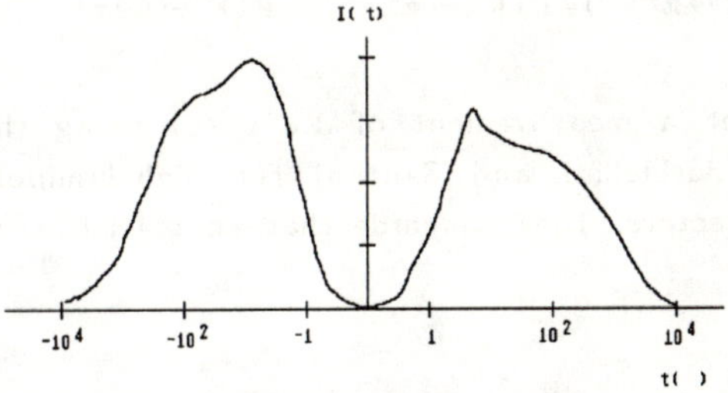


where $\Delta t$ is given in units of the $K_S$ lifetime $\tau_S$ . The interference region, with $|\Delta t| \sim \tau_S$, is the one sensitive to the phase difference. F. Erné has estimated[8] that smearing each vertex distance with $\sigma$ of the order of $\sigma \sim 2\,mm \sim 1/3\ d_S$, where $d_S$ is the $K_S$ mean decay path, keeps enough signal in $I(\Delta t)$ to extract the phase difference. If one considers as observable the rate asymmetry

$$A_\Gamma(\tau, 0) = \frac{\Gamma(\tau, 0) - \Gamma(0, -\tau)}{\Gamma(\tau, 0) + \Gamma(0, -\tau)}$$

$$\Gamma(\tau_2, -\tau_1) \equiv \int_{-\tau_1}^{\tau_2} d(\Delta t)\ I(\Delta t) \tag{16}$$

its value for $\tau = 4\tau_S$, for an integrated luminosity of $\int L dt = 1000\ pb^{-1}$, allows to get a precision in the phase difference $(\Phi_{+-} - \Phi_{oo})$ of one degree.

For the events in the non-interference region, $\Delta t \gg \tau_S$, the intensity asymmetry limit from Eq. (14) gives

$$A_I(\Delta t) \underset{\Delta t \gg \tau_S}{\longrightarrow} 3\,\mathrm{Re}\left(\frac{\varepsilon'}{\varepsilon}\right) \tag{17}$$

where terms of order $(\varepsilon'/\varepsilon)^2$ have been neglected. This limit corresponds to the more conventional physics discussion of the value of $(\varepsilon'/\varepsilon)$ from the ratio of ratios of $K_{L,S}$ decay probabilities to $\pi^+\pi^-$, $\pi^o\pi^o$. From the definition of the intensity asymmetry, one has in this limit

$$A_I(\Delta t >> \tau_S) \longrightarrow \frac{P(K_S\to\pi^o\pi^o)\,P(K_L\to\pi^+\pi^-) - P(K_S\to\pi^+\pi^-)\,P(K_L\to\pi^o\pi^o)}{P(K_S\to\pi^o\pi^o)\,P(K_L\to\pi^+\pi^-) + P(K_S\to\pi^+\pi^-)\,P(K_L\to\pi^o\pi^o)} \tag{18}$$

$$= \frac{P(K_L\to\pi^+\pi^-)/P(K_S\to\pi^+\pi^-) - P(K_L\to\pi^o\pi^o)/P(K_S\to\pi^o\pi^o)}{P(K_L\to\pi^+\pi^-)/P(K_S\to\pi^+\pi^-) + P(K_L\to\pi^o\pi^o)/P(K_S\to\pi^o\pi^o)}$$

The feasibility of a measurement of $Re(\varepsilon'/\varepsilon)$ using this method has been discussed[9)] by Barbiellini and Santoni for high-luminosity $e^+e^-$ rings and sophisticated detectors. They conclude that an statistical error of few $\times 10^{-4}$ is reachable.

## 5. TIME INTEGRATED RATES

Instead of comparing the rates at different time intervals, for a given decay channel $\pi^+\pi^-\pi^o\pi^o$ of the $(K^o - \overline{K}^o)$ system, we can adopt a different strategy: the comparison of different decay channels in time integrated rates.[10)]

In the time interval $(\tau_1, \tau_2)$ the integrated rate is given by

$$N(X_1, X_2) = \int_{\tau_1}^{\tau_2} dt_1 \int_{\tau_1}^{\tau_2} dt_2 \,|A[X_1, X_2]|^2$$

$$= \frac{(1+|\alpha|^2)}{8\,|\alpha|^2}\,\frac{1}{\Gamma_L\,\Gamma_S}\,|\langle X_1|K_S\rangle|^2\,|\langle X_2|K_S\rangle|^2\,. \tag{19}$$

$$\cdot\ \{C_1(|\eta_1|^2 + |\eta_2|^2) - 2C_2\,Re\,(\eta_1\,\eta_2{}^*)\}$$

where $C_1$ and $C_2$ are functions of $(\tau_1, \tau_2)$. For an infinite detector, $C_1(0,\infty) = 1$, $C_2\,(0, \infty) \simeq 3\times10^{-3}$. For the interval $(\tau_S, \tau_L)$, the corresponding values are $C_1(\tau_S, \tau_L) = 0.23$, $C_2(\tau_S, \tau_L) \simeq 10^{-3}$. If these events come from $\Phi$'s, the mean decay paths $d_S$ and $d_L$ for $K_S$ and $K_L$ are $d_S \sim 5$ cm and $d_L \sim 3$ m. The parameter $C_1$ controls the number of decay events. In general, for $\tau_2 \geq 15\tau_S$ the interference term $C_2$ satisfies $C_2 << C_1$.

The total decay rate of this system is dominated by $X_1 \in X_L$, $X_2 \in X_S$, and viceversa, where $X_L \equiv (\pi\pi\pi,\ \pi l \nu)$ are the leading decay modes of $K_L$ and

$X_s \equiv (\pi\pi)$ are the leading decay modes of $K_s$. Then this rate is

$$R \equiv \sum_{X_1, X_2} N(X_1, X_2) \simeq C_1 \; ; \qquad C_1(0,\infty) = 1 \tag{20}$$

With this result, the branching ratio to the arbitrary $(X_1, X_2)$ channel is given by

$$Br(X_1, X_2)\Big|_{\tau_1}^{\tau_2} = C_1 Br(K_s \to X_1)\, Br(K_s \to X_2) \frac{\Gamma_s}{\Gamma_L} \left\{ (|\eta_1|^2 + |\eta_2|^2) - 2\frac{C_2}{C_1} Re(\eta_1 \eta_2^*) \right\} \tag{21}$$

Let us choose again the CP-violating channels $\{2\pi, 2\pi\}$. Their expected branching ratios, following Eq. (21), are of order,

$$Br \simeq |\varepsilon|^2 \frac{\Gamma_s}{\Gamma_L} C_1 \sim 10^{-3} \tag{22}$$

If we compare these time-integrated rates for the different decay channels we get

$$\frac{Br(\pi^+\pi^-, \pi^+\pi^-)}{Br(\pi^o\pi^o, \pi^o\pi^o)} = \left[\frac{Br(K_s \to \pi^+\pi^-)}{Br(K_s \to \pi^o\pi^o)}\right]^2 \left\{1 + 6\, Re\left(\frac{\varepsilon'}{\varepsilon}\right)\right\} \tag{23}$$

and similarly [10] for the other possible ratio. To extract the $(\varepsilon'/\varepsilon)$ effect at the $10^{-3}$ level, one would need about $10^6$ events for these channels, which means at least $10^9$ $\Phi$' s.

At present, the $K_s \to 2\pi$ branching ratios are known to a precision of $2.6 \times 10^{-3}$. It is interesting to decrease the error.

***i)*** to extract $\varepsilon'/\varepsilon$

***ii)*** by itself, because the comparison

$\left\{ \frac{Br(K_s \to \pi^+\pi^-)}{Br(K_s \to \pi^o\pi^o)} - 2 \right\}$ is sensitive to deviations from the $\Delta I = 1/2$ rule.

## 6. THE LEADING $K_s \to \pi\pi$ CHANNELS

Is it possible to extract the ratio of $K_S$ branching ratios, present in Eq. (23), from the same experiment? If, instead of choosing $(2\pi, 2\pi)$ channels, we choose one of the leading $(2\pi, X_L)$ decay modes of the $(K_S, K_L)$ system, we get

$$\frac{Br(K_S \to \pi^+\pi^-)}{Br(K_S \to \pi^o\pi^o)} = \frac{Br(\pi^+\pi^-, X_L)}{Br(\pi^o\pi^o, X_L)} \left\{1 + 6\frac{C_1}{C_2}\, Re\left(\frac{\varepsilon'}{\eta_L}\right)\right\} \tag{24}$$

where $\eta_L \geq 1$, so the bracket correction {....} would be $1+(<10^{-6})$. For a precision at the $10^{-3}$ level, one would need about $10^6$ $\Phi$' s.

As said before, a good determination of the $K_S \to 2\pi$ branching ratios would allow the extraction of the isospin amplitudes

$$A(K_S \to \pi^+\pi^-) = \sqrt{2}\left[A_o + \frac{1}{\sqrt{2}}(A_2 + A'_2)\, e^{i(\delta_2 - \delta_o)}\right]$$

$$A(K_S \to \pi^o\pi^o) = \left[A_o - \sqrt{2}\,(A_2 + A'_2)\, e^{i(\delta_2 - \delta_o)}\right] \tag{25}$$

$$A(K^+ \to \pi^+\pi^o) = \frac{3}{2}A_2 - A'_2$$

where $A_o$, $A_2$ and $A'_2$ are the irreducible isospin amplitudes. One has the possibility to disentagle the $\Delta I = 3/2$ components of the weak hamiltonian, present in $A_2$, from the electromagnetic corrections $A'_2$ able to induce $\Delta I = 5/2$ terms. A. Pich and E. de Rafael have emphasized [11] the interest in the extraction and the understanding of these amplitudes, which should be analyzed together with the $K_S \to \pi\pi\gamma$ channels. Present results indicate [11] that $A'_2 \neq 0$ at the $2\sigma$ level.

## 7. OTHER $K_s$ CHANNELS

From the time-integrated rate given in Eq. (21), we can choose a final state $(X_L, X_L)$ from our coherent $(K_L, K_S)$ system. It is immediate to show that this choice allows to extract the branching ratios of $K_S$ to the $X_L$ channels, where $X_L$ are leading decay modes of $K_L$. One gets

$$Br(X_L, X_L) = C_1\left(1 - \frac{C_2}{C_1}\right) Br(K_L \to X_L)\; Br(K_S \to X_L) \qquad (26)$$

where the last factor is the output. For the semileptonic channel $X_L = (\pi l \nu)$, the theoretically expected branching ratio is of order $10^{-3}$. For $X_L = (\pi^o \pi^o \pi^o)$, the decay is CP-forbidden and, in the superweak ansatz, one expects $Br(K_S \to \pi^o \pi^o \pi^o) \simeq 2\times 10^{-9}$. The same strategy can be used, in a similar fashion, for other decay channels of $K_S$.

## 8. CONCLUSIONS

In this contribution we have discussed several ways by which the decay observables of coherent $(K^o - \overline{K}^o)$ states, such as they are produced in $e^+ e^-$ $\Phi$-factories, can provide CP-violating effects.

The intensity asymmetry, under $\Delta t \to -\Delta t$, for the $\{\pi^+\pi^-, \pi^o\pi^o\}$ channel is able to measure both Re $(\varepsilon'/\varepsilon)$ and Im $(\varepsilon'/\varepsilon)$. If the time difference of decay events for the two K's is outside the interference region, $\Delta t >> \tau_S$, one gets Re$(\varepsilon'/\varepsilon)$. If the decay events are chosen for $\Delta t \leq$ few $\tau_S$, then one has effects from Im $(\varepsilon'/\varepsilon)$.

The comparison of time-integrated rates for the different $\{2\pi, 2\pi\}$ decay channels $\{\pi^+\pi^-, \pi^+\pi^-\} \to \{\pi^+\pi^-, \pi^o\pi^o\} \to \{\pi^o\pi^o, \pi^o\pi^o\}$ provides also a measurement of Re $(\varepsilon'/\varepsilon)$.

Finally, we wish to emphasize that a high-luminosity $e^+e^-$ machine of the kind that is being discussed by the scientific community, able to give $10^{10}$ $\Phi$'s or more at one's disposal, is not only a beautiful device to measure $\varepsilon'/\varepsilon$ but also a unique facility for the study of $K_S$ decay modes, such as $\{\pi\pi\}$, $\{\pi\pi\gamma\}$, $\{\pi l \nu\}$, $\{\pi\pi\pi\}$, with great precision.

## ACKNOWLEDGEMENTS

The CP-Violation Conference in the Château de Blois was a great event. It is a pleasure to thank Tran Thanh Van for the magnificant organization of the Conference and the stimulating atmosphere during these days. This work has been supported by CICYT under Grant AEN 88-0021. One of us (J. R.) is indebted to Generalitat Valenciana for a fellowship.

**REFERENCES**


1. J.H. Christenson, J.W. Cronin, V.L. Fitch and R. Turlay, Phys. Rev. Lett. 13 (1964) 138.
2. L.L. Chau, Phys. Rep. 95 (1983) 1.
3. K. Kleinknecht in "CP VIOLATION", Ed. C. Jarlskog, World Scientific, (1989).
4. W. Burkhardt et al., Phys. Letters B206 (1988) 169.
5. H.J. Lipkin, Phys. Rev. 176 (1968) 1715.
6. T. Kamae, T. Kifume and K. Tsumemoto, Prog. Theor. Phys. 41 (1967) 1267.
7. I. Dunietz, J. Hauser and J.L. Rosner, Phys. Rev. D35 (1987) 2166.
8. F. Erné, Proc. of NIKHEF Workshop, April 1989.
9. G. Barbiollini and C. Santoni, Proc. of NIKNEF Workshop, April 1989.
10. J. Bernabéu, F.J. Botella and J. Roldán, Phys. Lett. B211 (1988) 226.
11. E. de Rafael, in RINGBERG Workshop on Hadronic matrix Elements and Weak Decays., April 1988.
A. Pich, private comunication.

# THE BACKGROUND PROBLEM IN Φ FACTORY

Tran N. TRUONG
Centre de Physique Théorique
Ecole Polytechnique
91128 Palaiseau, France.

ABSTRACT

The background problem in the future ϕ factory is discussed.There are considerable uncertainties in theoretical estimates.

Part of this work is done with the collaboration of S. Nussinov[1]. The process $e^+e^- \rightarrow "\gamma" \rightarrow K^0 \overline{K}^0$ is an attractive way to obtain correlated $K^0 \overline{K}^0$ pairs. Assuming that charge conjugation C is conserved in the strong and electromagnetic interations, the $K^0 \overline{K}^0$ state produced via the single photon exchange (C=odd) is purely $K_L K_S$ which can provide a valuable way to study time-dependent correlations when both $K_L$ and $K_S$ decay into the same channel. The cross section for this $K \overline{K}^0$ production is strongly enhanced when the $e^+e^-$ energy is at the $\phi$ meson mass : this is the $\phi$ factory[2].

For obvious reasons, it is important to study the background problem in the $\phi$ factory before this facility can be used to study the CP violation in the $K \overline{K}$ system. There are two types of C=even background which leads to the final states of $K_L K_L$ and $K_S K_S$. The first is the intrinsic one where the $K \overline{K}$ pairs are produced in the even partial wave via the 2 photon exchanged mechanism. The second one is the allowed decay $\phi \rightarrow K^0 \overline{K}^0 \gamma$ where the $K \overline{K}^0$ pair are in the relative S state with the emission of a 20 Mev. soft photon which may be undetected by experimental apparatus unless special care is taken. We discuss first the 2 $\gamma$ exchanged mechanism.

**a) S wave $K \overline{K}$ production by 2 $\gamma$ exchange :**

The cross section for this process is previously calculated and found to be very small[2]:

$$r_s = \frac{\sigma\,(e^+e^- \rightarrow \gamma\gamma \rightarrow K\overline{K}\text{ (s wave)})}{\sigma\,(e^+e^- \rightarrow \gamma \rightarrow \phi \rightarrow K\overline{K})} \approx 10^{-10} \qquad (1)$$

This is due to the fact that the S-wave cross section is suppressed by a factor of $m_e^2$ due to the conservation of the helicity.

**b) D wave $K \overline{K}$ production by 2$\gamma$ exchange.** This amplitude is not proportional to $m_e$ because helicity conservation is possible, but it is suppressed by the centrifrigual barrier. This amplitude is estimated by two different methods. In the first method, we assume the $f^0$ (I = 0 and J = 2) dominance in the intermediate state connecting the two photon and the $K \overline{K}$ state. The unitarity can be used to calculate the absorptive part of $e^+e^- \rightarrow 2\gamma \rightarrow K^0 \overline{K}^0$ (D state) in terms of the experimental $f_0 \rightarrow 2\gamma$ and $f_0 \rightarrow K\overline{K}$ widths. Assuming the real part of the amplitude is of the same order of magnitude as the imaginary part, a straightforward calculation yields :

$$r_D = \frac{\sigma\,(e^+ e^- \rightarrow 2\,\gamma \rightarrow K\overline{K}\text{ (D state)})}{\sigma\,(e^+ e^- \rightarrow \gamma \rightarrow \phi \rightarrow K\overline{K})} \approx 10^{-10} \qquad (2)$$

In the second method the D wave $K\overline{K} \rightarrow 2\gamma$ amplitude is generated by the vector meson $K^*$ exchange. Using the experimental width $K^* \rightarrow K\gamma$, one finds

$$r_D = 2.10^{-10} \qquad (3)$$

Because the two different estimates yield essentially the same result, we can conclude that the D wave C = even background is of the same order of magnitude as the S-wave and hence is negligible.

**Soft Photon Emission in $\phi \rightarrow K\overline{K}\gamma$ Decay** : $\phi$ can decay into a pair $K\overline{K}$ in S-state and a low energy 20 MeV photon. This is the E1 transition. If care is not taken to detect this soft photon ( or to use the colinearity test ), this process could simulate the C = even $K\overline{K}$ background. There is however considerable uncertainty in estimating the $\phi \rightarrow K\overline{K}\gamma$ decay rate.

In the following we use two different methods to estimate this amplitude. In the first method, the Gell-Mann Goldberger and Thirring (G.G.T.) sum rule is used; in the second method a Feynman graph calculation is done.

In both methods, because of the small Q value of the $\phi \rightarrow K^0\overline{K^0}\gamma$ decay, it is reasonable to assume the scalar mesons dominance, the I = 0 $S^*$ (975) and the I = 1 $\delta$(980) which are both quite near to the $K\overline{K}$ threshold. Because these scalar mesons are likely to have a 4 - quark structure, while the $\phi$ meson is a pure $\overline{s}s$ state, the overlap of the 4-quark and 2-quark wave function is difficult to calculate, hence there are considerable uncertainties in these two methods which reflect in their widely different results.

**a)G.G.T. Sum Rule method**. We want to calculate first the decay of $\phi \rightarrow S^*\gamma$.The G.G.T.sum rule method is a generalisation of the classic Thomas Reich Kuhn method; it is fully relativistic and can be applied to any type of interacting potentials. For this purpose let us follow G.G.T. by considering the spin average foward compton scattering on the $S^*$ target and subtracting from it the foward compton scattering on its constituents which are supposed to be the 4-quark $K\overline{K}$ state. Notice that the wavefunction of the $S^*$ is $\frac{1}{\sqrt{2}}(K^+ K^- + K^0 \overline{K}^0)$, we are led to examine the foward Compton amplitude $f = f_{\gamma S} - \frac{1}{2} f_{\gamma K^+} + f_{\gamma K^-} + f_{\gamma K^0} + f_{\gamma \overline{K}^0})$ and use the low energy theorem (Thompson limit) in combination with the unsubtracted dispersion relation and the optical theorems for these amplitudes. (The unsubtracted dispersion relation should be interpreted as a finite energy sum rule). We denote $f_{\gamma S}$ is the foward Compton on the $S^*$ target etc. Because the spin-averaged foward Compton amplitude vanishes at zero energy for a neutral system, assuming C-invariance,we have the following sum rule:

$$\frac{1}{4\pi^2} \int (\sigma_{\gamma S} - \sigma_{\gamma K^+} - \sigma_{\gamma K^0})\, dE + R = \frac{\alpha}{m_K} \qquad (4)$$

where R is the high energy contribution due to the closed circle at E = N which is needed to apply the Cauchy theorem and to cancel out the divergence of the integral if any; this term will be neglected in the following ( $\alpha$ = 1/137 and $m_K$ is the Kaon mass). Saturating $\sigma_{\gamma i}$ where i = S, $K^+$... with the intermediate state n

$$\sigma^n_{\gamma i} = \frac{\pi}{q^2} \frac{2J_n + 1}{2J_i + 1} \frac{\Gamma^n_{\gamma i} \Gamma^n}{(E - E_n)^2 + \Gamma^2_n / 4} \qquad (5)$$

where $\Gamma^n$ is the total width of the n intermediate state, E and q are, respectively, the total energy and photon momentum in the c.o.m. system. Saturating $\sigma_{\gamma S}$ with $\phi(1020)$ and $\phi'(1680)$ intermediate states and $\sigma_{\gamma K}$ with the $K^*(892)$ intermediate states, the contribution of $\sigma_{\gamma K}$ is negligible and we therefore have :

$$\frac{\Gamma(\phi \rightarrow S^* \gamma)}{(m_\phi - m_{S^*})^2} + \frac{\Gamma(\phi' \rightarrow S^* \gamma)}{(m_{\phi'} - m_{S^*})^2} = \frac{2}{3}\alpha \qquad (6)$$

The reason that we have to take into account of the $\phi'(1680)$ is because the first term on the L.H.S of Eq.(6) being proportional to $| M |^2 (m_\phi - m_{S^*})$ is less important than the 2nd term which is proportional to $| M' |^2 (m_{\phi'} - m_{S^*})$ where M's are the matrix elements. In order to get an order of magnitude estimate, let us take M = M' in Eq.(6) which yields

$$\Gamma(\phi \rightarrow S^* \gamma) = 1.5 \times 10^{-3} \text{ Mev.} \qquad (7)$$

We can similarly calculate $\Gamma(\phi \rightarrow S^* \gamma)$ width with $S^*$ as a bound state using the GGT sum rule. The result is the value of Eq.(7) reduced by a factor $\frac{2}{9}\left(\frac{m_K}{m_S}\right)$. Taking the value of the constituent quark mass $m_S$ = 400 Mev, one gets $\Gamma(\phi \to S^*\gamma) \approx 4 \times 10^{-4}$ Mev.

One can also estimate $\phi \to S^* \gamma$ width by using the standard $E_1$ transition:

$$\Gamma(\phi \to S^* \gamma) \approx \frac{4}{9} \alpha\, (q_i^2)\, |\langle r \rangle|^2\, (E_i - E_f)^3$$

where $q_i^2 = 1/9$ and $|\langle r \rangle|^2$ is the radius of the P state splitting from S-state. From the analysis of the $q\bar{q}$ system[5], $|\langle r \rangle|^2 \approx 8\text{Gev}^{-2}$, one has $\Gamma(\phi \to S^*\gamma) = 2.6 \times 10^{-4}$ Mev., bearing in mind that this result could be wrong due to the 4-quark structure of the $S^*$.

In conclusion our estimate based on the GGT sum rule or quark model yields $\Gamma(\phi \to S^*\gamma) \approx 3.10^{-4} \sim 10^{-3}$ Mev.

**b) Feynman Graph calculation.**

Let us consider the Feynman triangle graph where the $\phi$, $S^*$ and $\gamma$ are attached to each vertex. The $\phi \to S^*\gamma$ decay amplitude is given by

$$M(\phi \to S^* \gamma) = \hat{\varepsilon}_\mu(p)\, \varepsilon_\nu(q)\, M_{\mu\nu}(p, p') \qquad (7)$$

and

$$M_{\mu\nu}(p^2, p'^2) = (p_\nu q_\mu - p.q\, g_{\mu\nu})\, H(p^2, p'^2)$$

where p, p' q the momenta of $\phi$, $S^*$ and $\gamma$, and $\hat{\varepsilon}$, $\varepsilon$ the $\phi$ and $\gamma$ polarisation vertors. Let m, M M' be respectively K $\phi$ and $S^*$ masses.

A straightforward calculation yields :

$$H(M^2, M'^2) \approx (0.04 - 0.03i)\ \text{Gev}^{-1}$$

where we have used $g_\phi^2 / 4\pi = 1$ and $g^2 / 4\pi = 0.6\ \text{Gev}^2$ from the previous analysis of $S^* \to 2\gamma$ decay[5], with $g_\phi$ is the $\phi$ $K\bar{K}$ coupling and g the $S^*$ $K^+ K^-$ coupling. The $\phi \to S^*\gamma$ width is

$$\Gamma(\phi \to S^*\gamma) = 5.7 \times 10^{-6}\ \text{Mev}. \qquad (8)$$

which is a factor of 300 smaller than the estimate of Eq.(7). Where could this calculation go wrong?[8]

To see this, let us calculate the decay of $\phi$ into a light Higgs boson and a photon rate $\Gamma(\phi \to H\gamma)$, using the reliable Wilczek mechanism[6] and compare it with the above triangle graph calculation. For this purpose, we can use the Higgs $K\bar{K}$ coupling $g_{HK^+K^-}$

$= m_K^2 / v$, where v = 250 Gev to replace $g_{S^*K^+K} = 2.7$ Gev. Let us choose $m_H = 900$ Mev which is outside the $S^*$ resonance so that the $g_{HK^+K^-}$ dependence on Higgs mass can be neglected as a first approximation. Using $m_S = 150$ Mev and neglect the relativistic effect for an order of magnitude estimate, the triangle graph calculation yields a rate for $\phi \to H\gamma$ which is smaller by a factor of 150 compared with the standard Wilczek mechanism.

The conclusion is that the triangle graph calculation, treating $\phi$ as a bound state of the $K\overline{K}$ system, underestimated $\Gamma(\phi \to S^*\gamma)$ by a large factor. This is due to the probability of finding a pair of $K^+ K^-$ in P state over a region near the origine is much less than the corresponding probability of finding a pair of $\overline{s}s$ quark in S-state which is the correct description of the $\phi$ meson..

From the above analysis, it is clear that, no reliable calculation can be made for the rate of $\phi \to S^*\gamma$. It is my opinion that Eq.(7) gives a more realistic estimate than that given by Eq.(8)

For an orientation, let us use Eq.(7) to estimate $\phi^0 \to K^0\overline{K}^0\,\gamma$ due to $S^*$ :

$$\Gamma(\phi \to \gamma K^0 \overline{K}^0)_{S^*} \sim 3 \times 10^{-6}\ \text{Mev} \qquad (9)$$

A similar figure is also obtained for the same decay via the isospin 1 scalar meson $\delta$. Within the frame work of the use of the G.G.T. sum rule, we cannot determine the relative sign of the S* and the $\delta$ dominance amplitudes.

This magnitude of the background as given by Eq(9) is of no consequences in the measurement of $\varepsilon'/\varepsilon$ [7] but could be important in experiments designed for detecting, for example, the CP violation effect in $K_S \to 3\pi$ decay.

In order to **eliminate** this kind of the background it is simplest to use a 20 Mev photon detector to veto the events coming from $\phi \to K^0\overline{K}^0\,\gamma$ decays. Last but not least, because there are two orders of magnitude of difference between the sum rule approach and the Feynman triangle graph calculation, it is important to detect the decay mode $\phi \to \pi\pi\gamma$ which has now an experimental upper limit branching ratio $B(\phi \to \pi^+\pi^-\gamma) \leq 0.7$ %. Assuming all the $\pi^+\pi^-$ comes from $S^* \to \pi^+\pi^-$ decay which is an overestimate, one gets $\Gamma(\phi \to S^*\gamma) < 3.10^{-2}$ Mev which is one order of magnitude too high compared with the sum rule estimate of Eq.(7).

CONCLUSION :

$\phi$ factory is an important project for future experiments to test CP and CPT. As a by-product, but with much less importance, the detection of the rare $\phi$ decays such as those discussed here, could provide more insight to the quark models. With the present difficulty in obtaining funding for new accelerators, it is important to construct a $\phi$ linear collider with high luminosity $\mathbb{L} = 10^{33}$ $cm^2$/sec to demonstrate the principle of the linear collider is perfectly sound. Upgrading this facility to $\tau$-charm factory and B $\overline{B}$ factory and even to CLIC energy is certainly the dream of many experimentalists!.

It is a pleasure to thank T.N. Pham for useful discussions and to the organisers of the Blois Conference for their invitation to present this talk.

**References.**

1 S. Nussinov and T.N.Truong U.C.L.A. preprint (1989)

2 I. Dunietz, J. Hauser and J. Rosner Phys. Rev. D35, 2166 (1987)

3 M.Gell-Mann, M. L. Goldberger and W. Thirring, Phys. Rev.95,1612 (1954)

For the application of the G.G.T. sum rule to particle physics see T.N.Pham and T.N.Truong Phys. Lett. 64B,51 (1976)

4 F. E.Close in " An Introduction to Quarks and Partons"Academic Press, N.Y.(1979)

5 T.N.Truong Ecole Polytechnique Preprint A857.1088 (unpublished)

6 F. Wilczek , Phys. Rev. Lett 39, 1304 (1977)

7 J. Bernabeu (private communication)

8 After completion of this work, we are informed a similar analysis of the triangle graph, as discussed in the text, was previously given by N.N. Achasov and V.N.Ivanchenko, Nucl. Phys. B315, 465 (1989).

# PHI-FACTORY PROJECT IN NOVOSIBIRSK

Barkov L.M., Dikansky N.S., Ivanov P.M., Koop I.A., Perevedentsev E.A.
Shatunov Yu.M., Skrinsky A.N., Tumaikin G.M., Vasserman I.B.
Institute of Nuclear Physics, 630090 Novosibirsk, USSR

## ABSTRACT

The project of a dedicated 510 MeV electron-positron storage ring with the luminosity above $10^{33}\,cm^{-2}s^{-1}$ is proposed. Its energy corresponds to a maximum of the phi-meson resonance production ( Phi-factory). An essential feature of the project is the solenoidal focusing used to obtain round beams at the interaction point.

## 1. Introduction

At the INP (Novosibirsk) a project for a set of facilities with colliding electron-positron beams is in progress [1].

It includes:

— a $2\times6.5$ GeV B-meson factory with the luminosity above $5\cdot10^{33}\,\mathrm{cm}^{-2}\,\mathrm{s}^{-1}$;
— a $2\times510$ MeV Phi-meson factory with the luminosity above $1\cdot10^{33}\,\mathrm{cm}^{-2}\,\mathrm{s}^{-1}$;
— an injector comprising the linacs and the damping ring to provide the factories with intense and perfect electron and positron beams.

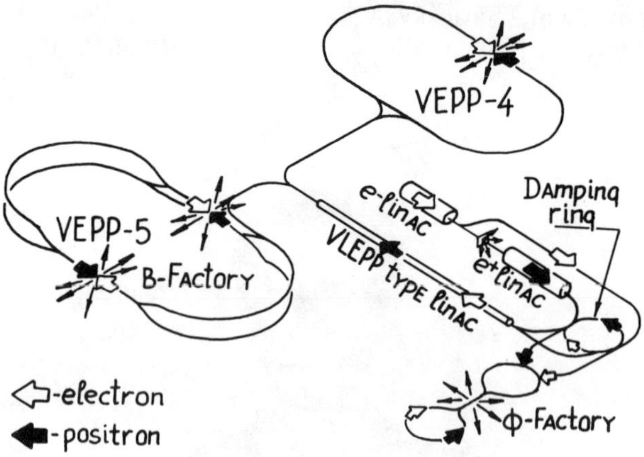


*Fig. 1.* Layout of VEPP-5, the facility with colliding electron-positron beams.

The Phi-meson factory is a new generation facility with colliding $e^+e^-$ beams in the Phi-meson resonance energy range (1020 MeV). In this project a single ring scheme is used to reach the very high luminosity. This seems to be preferable as compared with the idea of employing separate electron and positron storage rings with the electrostatic convergence of the two beams in the common interaction points [2].

The first stage envisages the Phi-factory operation in the single bunch mode with attainable luminosities in excess of $1\cdot10^{33}\,\mathrm{cm}^{-2}\,\mathrm{s}^{-1}$. With a view to attain higher luminosity, the lattice and the construction of the storage ring are capable to operate in the three-bunch mode using an electrostatic beam separation at the side interaction points. With the even number of bunches in each beam $e^{\mp}e^{\pm}$ colliding beams can be arranged with the simultaneous collision of four bunches. In case of exact coincidence of orbits in the interaction point the space charge compensation mode is realized. The possibility to enhance the luminosity in this mode can be proved experimentally.

The Phi-factory project is based on the idea of round colliding beams, with the operating point dwelling on the main coupling resonance near the integer one. The high-luminosity mode of operation is achieved with equal beta-functions at the interaction point taken as low as possible, and at equal transverse emittances of the beams.

## 2. Physical Experiments on the Phi-Meson Factory

The Phi-factory having a luminosity of $(1\div3)\cdot10^{33}\,cm^{-2}\,s^{-1}$ offers the unique possibility to study the CP-parity violating interactions. Such a high luminosity makes it possible to attain the luminosity integrals of $3\cdot10^{40}-3\cdot10^{41}\,cm^{-2}$ in two to five years. Assuming that the efficiency of luminosity utilization is about 50 per cent, we will be able to obtain $5\cdot10^{10}$ to $5\cdot10^{11}$ monochromatic $K_S$ and $K_L$ mesons in the $e^+e^-\to\Phi\to K_SK_L$ reaction.

When using rather perfect universal detector, high monochromaticity permits an effective separation of various decay channels for both long- and short-lived kaons as well as a considerable advance in investigation of the CP-parity violating processes.

In addition to the higher measurement accuracy for the already known processes where the CP-parity violation was discovered: $K_L\to\pi^+\pi^-$, $K_L\to\pi^0\pi^0$, and where the charge assymetry in semi-leptonic decays of the $K_L$-meson was clearly demonstrated

$$\delta=\frac{\Gamma(K_L\to\pi^-e^+\nu)-\Gamma(K_L\to\pi^+e^-\nu)}{\Gamma(K_L\to\pi^-e^+\nu)+\Gamma(K_L\to\pi^+e^-\nu)}$$

there appears a real possibility to measure the parameters of the CP-parity violating interaction in the decays of short-lived kaons: $K_S\to\pi^+\pi^-\pi^0$, $K_S\to\pi^0\pi^0\pi^0$.

The experimental limitations on the widths of the corresponding decays are 4—5 orders of magnitude higher than those theoretically expected. Experiments at the Phi-factory open the possibility to observe these decays. The possibility to reach a 5 orders of magnitude higher accuracy at this factory is due to its high luminosity and to production of pure, perfectly marked, monochromatic $K_S$-mesons. Note that on hadron accelerators all the studies on CP-violation can practically have been performed using the extracted beams of $K_L$-mesons, and therefore the $K_S$-mesons decay channels remain insufficiently studied.

In experiments at the Phi-factory the measurement accuracy for the relative amplitudes of the $K_L$ decay into $\pi^+\pi^-$ and $\pi^0\pi^0$

$$\eta_{+-}=\frac{A(K_L\to\pi^+\pi^-)}{A(K_S\to\pi^+\pi^-)}=|\eta_{+-}|\,e^{i\Phi_{+-}},$$

$$\eta_{00}=\frac{A(K_L\to\pi^0\pi^0)}{A(K_S\to\pi^0\pi^0)}=|\eta_{00}|\,e^{i\Phi_{00}}$$

and for the phase differences $\Delta\Phi=\Phi_{+-}-\Phi_{00}$ is expected to be improved by about one order of magnitude. This will make it possible to check the Bell—Steinberger relation and the validity of the CPT-theorem at quite a new level of accuracy.

Thus, the possibility to improve nearly by an order of magnitude the measurement accuracy of such known parameters for the CP-violating interaction in decays of $K_L$-mesons as $|\eta_{+-}|$, $|\eta_{00}|$, $\varepsilon'/\varepsilon=1/6\,[1-|\eta_{00}/\eta_{+-}|^2]$, $\Delta\Phi$, and by five orders for $|\eta_{+-0}|^2$, $|\eta_{000}|^2$ in $K_S$-meson decay channels will considerably contribute to our understanding of the fundamental laws of nature, and current theoretical ideas will be tested at a qualitatively new level.

Besides measurement of the constants of the CP-violating interaction, the high luminosity Phi-factory will also enable a considerable advance in the search for the exotic decay modes of $\rho$-, $\omega$- and $\varphi$-mesons. In addition, the monochromatic $K^0$ and $K^{\pm}$ meson intensity ($\approx 10^3$ per sec, $\Delta W = 300$ keV, at kinetic energy of $W = 3-30$ MeV) may be of great interest for investigation of their interaction with nuclei.

In experiments with colliding $e^+e^-$ beams on the Phi-factory the background must be extremely low. For events identification the detector should possess a high detection efficiency in all $e^+e^-$ interaction channels, high measurement accuracy for the characteristics of charged and neutral particles. These properties are inherent in a CMD-type detector with a liquid xenon calorimeter [3]. The possibility is considered of using localized-charge counters which provide high spatial accuracy in detection of charged and neutral particles and their flighttime characteristics measurement at a level of tens of picoseconds.

### 3. Phi-Factory Lattice

In the lattice with 8-shaped orbit suggested for the Phi-factory, two opposite points of beam-beam collisions are geometrically superimpcsed due to reverse bend magnets, and hence the capacity of the facility is doubled (Fig. 2). Taking

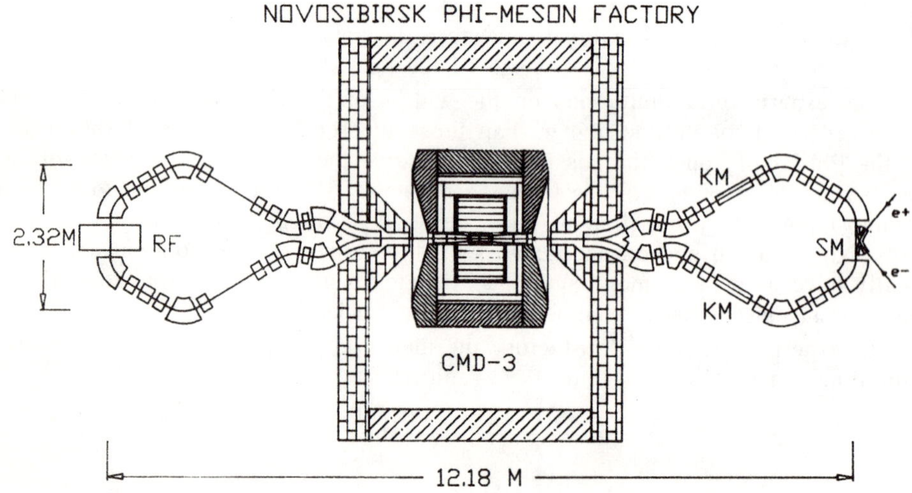


*Fig. 2.* A schematic view of the Phi-meson factory.

into account the complexity of the detector and its sizes comparable with those of the storage ring, it is expedient to have a single interaction point for only one detector.

To obtain low beta-functions in two planes simultaneously, the magnetic focusing of solenoidal type has been chosen. The optical scheme consists of two pairs of superconducting solenoids with a maximum field of 11.0 T, which are placed symmetrically with respect to the centre of straight section. As construction components, the solenoids are incorporated in the detector housing. In contrast to the quadrupole focusing, the use of solenoids in the common straight section provides quite symmetric focusing properties during forward and reverse passages of beams. The azimuthal dependence of the longitudinal magnetic field throughout the experimental straight section is given in Fig. 3. In each pair, the solenoids are connected in opposition. This enables one to keep the longitudinal

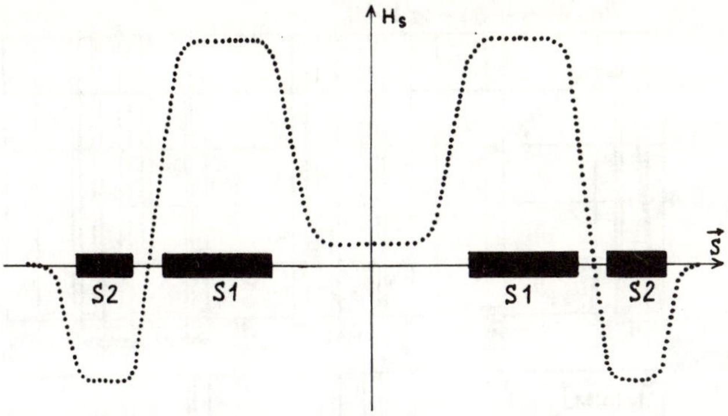


*Fig. 3.* Azimuthal distribution of the longitudinal magnetic field in the experimental straight section of the Phi-factory.

field integral over the straight equal to $\pi \cdot HR$, regardless of variation of the required focusing parameters. As a result, the betatron oscillation planes are rotated by $+\pi/2$ or $-\pi/2$ angle during each passage of the beam. It is easy to see that the eigenvectors of the transition matrix for the complete turn are «flat» in the magnetic arcs and dwell in the $X$ or $Z$ planes, and they are inclined at $\varphi = |\pi/4|$ at the interaction point. The betatron oscillation normal mode is thus vertical in one arc and horizontal in the other. The projections of the envelopes

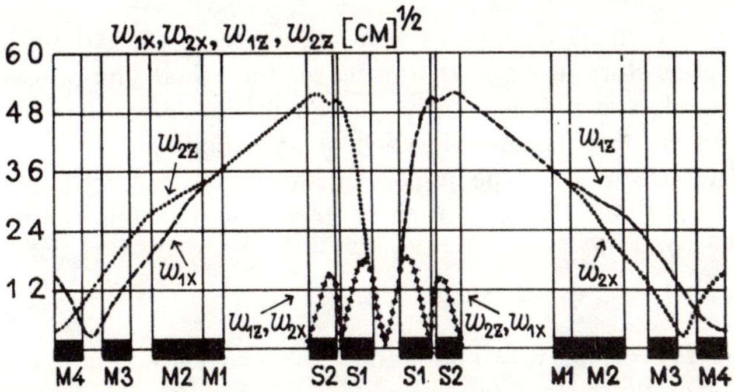


*Fig. 4.* The Phi-factory lattice (central section).

of normal modes onto the $X$ and $Z$ planes are depicted in Figs 4, 5. The beam and the lattice parameters were calculated in terms of the formalism usually applied to the systems with strongly-coupled linear betatron oscillations [4].

In the planes natural for the storage rings, i. e. in the $X$ and $Z$ planes, the r.m.s. sizes of the beams are:

$$\sigma_z^2(s) = w_{1z}^2(s)\, \mathscr{E}_1 + w_{2z}^2(s)\, \mathscr{E}_2,$$

$$\sigma_x^2(s) = w_{1x}^2(s)\, \mathscr{E}_1 + w_{2x}^2(s)\, \mathscr{E}_2 + \eta_x^2(s)\, (\sigma_{\Delta E/E})^2,$$

where $w_{1,z,x}$ and $w_{2,z,x}$ are the projections of the envelopes of normal modes onto the $X$, $Z$ planes; $\mathscr{E}_1$ and $\mathscr{E}_2$ are the emittances of the normal betatron modes.

The energy dispersion function $\eta_x(s)$ always lies in the median plane of magnetic arcs and is zero in the experimental straight section. There is no

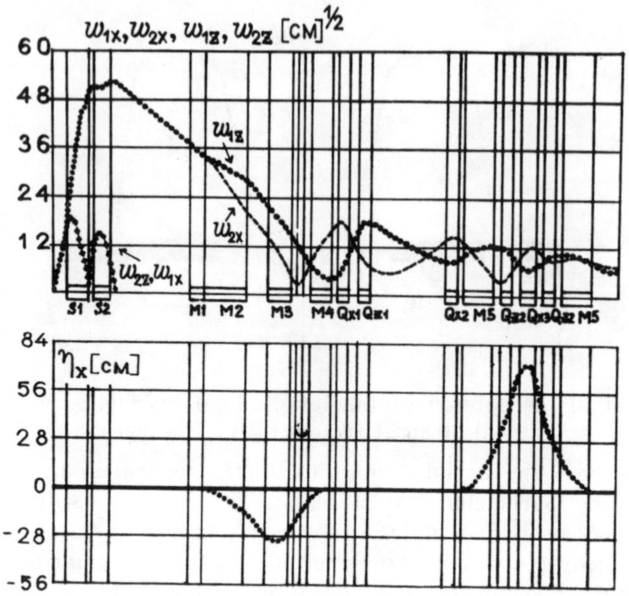


*Fig. 5.* One-fourth of the Phi-factory lattice period, mirror-symmetry with respect to the interaction point.

excitation of the vertical dispersion, and therefore, $\eta_z(s) \equiv 0$ for the whole circumference of the storage ring. In this case, the equal emittances of each normal betatron mode are independently formed in the corresponding magnetic arc due to the quantum fluctuations of synchrotron radiation.

In the structure of the type under consideration, the frequencies of normal betatron modes are half the sum of the betatron oscillations frequencies calculated without the rotation:

$$\nu_{1,2} = 1/2(\nu_x + \nu_z).$$

In this project, $\nu_x = 7.05$, $\nu_z = 5.05$ and $\nu_{1,2} = 6.05$. A distinctive feature of the suggested type of the solenoidal focusing lattice is the absence of the coupling between the transverse modes of betatron oscillations. The working point is placed exactly on the basic coupling resonance $\nu_1 - \nu_2 = 0$ since there is no splitting of the normal modes' frequencies.

To substantially increase the thresholds of different kinds of instabilities, including the coherent instability on the colliding beam, it is necessary to have large enough decrements of radiation damping. For this purpose, superconducting bending magnets with 6.0 T field are envisaged in magnetic arcs. A pair of dipole magnets with a quadrupole triplet in between form a 120″ achromatic bend. An RF resonator is positioned in one of the straight sections between the achromats. The septum-magnets designed for the injection of electron and positron beams in opposite directions, are placed in the opposite straight section in another arc. There are two symmetric kicker-magnets in the straight sections on either side of the septum straight.

Table 1

**Basic Parameters of the Phi-Factory and Beams at 510 MeV**

| Parameters | | Units | Phi-factory | VEPP-2M with wiggler |
|---|---|---|---|---|
| Energy | $E$ | MeV | 510 | 510 |
| Circumference | $c$ | m | 28.0 | 17.88 |
| Accelerating voltage frequency | $f_0$ | MHz | 700 | 200 |
| Momentum compaction factor | $\alpha$ | — | 0.003 | 0.167 |
| Emittances | $\mathscr{E}_{x0}$<br>$\mathscr{E}_{z0}$ | cm·rad<br>cm·rad | $2.8\cdot10^{-5}$<br>$2.8\cdot10^{-5}$ | $4.6\cdot10^{-5}$<br>$5.5\cdot10^{-7}$ |
| Radiative energy loss per turn | $\Delta E_0$ | keV | 40 | 9.1 |
| Dimensionless damping decrements between interaction points | $\delta_z$<br>$\delta_x$<br>$\delta_s$ | —<br>—<br>— | $2.3\cdot10^{-5}$<br>$2.3\cdot10^{-5}$<br>$4.7\cdot10^{-5}$ | $0.44\cdot10^{-5}$<br>$0.38\cdot10^{-5}$<br>$0.94\cdot10^{-5}$ |
| R.m.s. energy spread in the beam | $\sigma_{\Delta E/E}$ | — | $5\cdot10^{-4}$ | $6\cdot10^{-4}$ |
| Beta-functions at the I.P. | $\beta_z^*$<br>$\beta_x^*$ | cm<br>cm | 0.5<br>0.5 | 4.5<br>48 |
| R.m.s. longitudinal bunch size | $\sigma_s$ | cm | 0.4 | 3.5 |
| Betatron tunes | $\nu_z$<br>$\nu_x$ | —<br>— | 6.05<br>6.05 | 3.09<br>3.06 |
| Number of particles per bunch | $N$ | $e^+, e^-$ | $8.9\cdot10^{10}$ | $3.7\cdot10^{10}$ |
| Space charge parameters | $\xi_z$<br>$\xi_x$ | —<br>— | $\gtrsim0.07$<br>$\gtrsim0.07$ | 0.05<br>0.02 |
| Luminosity in a single-bunch mode | $\mathscr{L}_{max}$ | $cm^{-2}\cdot s^{-1}$ | $\gtrsim1\cdot10^{33}$ | $\sim1\cdot10^{31}$ |
| Luminosity in a three-bunch mode | $\mathscr{L}_{max}$ | $cm^{-2}\cdot s^{-1}$ | $\sim3\cdot10^{33}$ | — |

The basic parameters of the lattice and beams are given in Table 1. According to calculations the beam lifetime with ultimate parameters is limited by the effect of single intrabunch scattering and is 1000 s within the design aperture. At very high luminosity the lifetime is expected to degrade down to 300 s because of the losses, caused by the single bremsstrahlung on colliding beam. The injection frequency of the next portions of the beams is assumed to be ranged from 0.1 to 0.2 Hz, so that the average luminosity achieves as high value as possible.

#### 4. Luminosity and Beam-Beam Effects

In the approximation $\sigma_s \ll \beta^*_{z,x}$, the luminosity of a storage ring with colliding beams is defined by the well known expression:

$$\mathscr{L} = \frac{\pi \gamma^2 \xi_x \xi_z f_0 (1+\sigma^*_z/\sigma^*_x)^2}{r_e^2 \beta^*_z} \frac{\mathscr{E}_{x0}}{(1+\varkappa^2)},$$

where $\xi_z$, $\xi_x$ are the space charge parameters whose maximum value, $(\xi_z)_{max}$ and $(\xi_x)_{max}$, are limited by the beam-beam effects; $\sigma^*_z$, $\sigma^*_x$ are the r.m.s. sizes of the beams at the interaction point; $\beta^*_z$ is the vertical beta-function at the interaction point; $\mathscr{E}_{x0}$ is the horizontal emittance in the absence of the coupling between the transverse oscillations modes; $\varkappa = \sqrt{\mathscr{E}_z/\mathscr{E}_x}$ is the coupling coefficient; $f_0$ is the collision frequency; $r_e$ is the classical radius of electron; and $\gamma$ is a relativistic factor.

In the Phi-factory lattice, an independent formation of the equal transverse emittances ($\mathscr{E}_{x0} = \mathscr{E}_{z0} = \mathscr{E}_0$) implies the absence of the coupling for betatron oscillations modes ($\varkappa = 0$). In this case, the maximum luminosity for round colliding beams ($\sigma^*_z = \sigma^*_x = \sigma^*_0$) can be higher by a factor of 4 as compared with the flat beams ($\sigma^*_z \ll \sigma^*_x$). The ultimate emittances are limited by a permissible dynamic aperture in focusing solenoids used to create the low beta-functions at the interaction point.

In general case the maximum luminosity is limited by the electromagnetic beam-beam interaction effects. The interaction force is characterized by the space charge parameters defined the round beams as follows:

$$\xi_{z,x} = \xi_0 = \frac{N r_e \beta^*_0}{4\pi\gamma\sigma_0^{*2}}.$$

In the maximum luminosity regime, the threshold of the space charge parameter is likely to be determined by the coherent colliding-beams instability conditions. This effect is theoretically treated in Refs 6—10.

The coherent shift of betatron tunes, which is caused by the colliding beam field, results in the appearance of the bands where coherent oscillations are unstable, near the machine resonance lines. The band width depends on the achieved value of $\xi_{z,x}$ and on the order of the effective resonance. As the multipole number of the oscillation modes grows, the widths of the coherent betatron resonances and the instability increments die out according to the power law. The enhancement of radiation damping can decrease the order of the effective resonance and thus allows to attain the larger values of the space charge parameters.

In our project the operating point was chosen in close proximity to the integer resonance with eqeal frequencies of normal betatron oscillations modes $\nu_1 = \nu_2$. In terms of the coherent beam-beam instability the operating point location on the main coupling resonance offers some advantages [9]. The tune shift for round beams, which is caused by the colliding beam, takes place along the line $\nu_1 - \nu_2 = 0$; note that the coherent instability bands formed by powerful two-dimensional coupling satellite resonances are not trespassed. Exactly on the resonance $\nu_1 - \nu_2 = 0$, no instability of coherent betatron and synchrobetatron oscillations appears provided that $\sigma^*_x = \sigma^*_z$ and $\xi_{x,z} = \xi_0$. On the other hand, sufficiently large radiation damping decrements are capable of suppressing the coherent instability near one-dimensional high-order betatron resonances which intersect the coupling resonance. On the existing electron-positron colliders the attainable $(\xi_{zx})_{max}$ usually is at a level of 0.03—0.06. The lattice properties and the choice of the operating point favour the attaining of the highest values of the space charge parameters. This can contribute efficiently to the realization of the very high luminosity regime due to the quadratic dependence: $\mathscr{L}_{max} \sim (\xi_0)^2_{max}$.

The behaviour of incoherent beam-beam effects on the main coupling resonance has been numerically simulated in the «strong-weak» approximation in Ref. 11.

The lattice of a storage ring with very low beta-functions at the interaction point necessitates an obtaining of short intense bunches. A possible bunch lengthening can lead to a considerable degradation of luminosity because of an increased geometrical factor at the interaction point. For $\beta_0^*/\sigma_s \leqslant 1$ modulation of the beta-functions throughout the interaction region results in enlarging the effective interaction area. At the interaction point, these functions are characterized by the azimuthal dependence:

$$\beta_0(s) = \beta_0^* + s^2/\beta_0^*,$$

$$\beta_0^* = w_{1z}^{*2} + w_{2z}^{*2} = w_{1x}^{*2} + w_{2x}^{*2}.$$

Here $s$ is a longitudinal coordinate counted off from the interaction point. The luminosity of round colliding beams with a charge density of the form:

$$\rho_\pm(x, z, s, t) = \frac{N^\pm}{(2\pi)^{3/2}\sigma_0^{*2}\sigma_s} \exp\left(-\frac{x^2}{2\sigma_0^{*2}}\right) \exp\left(-\frac{z^2}{2\sigma_0^{*2}}\right) \exp\left[-\frac{(s \pm ct)^2}{2\sigma_s^2}\right]$$

can be determined by

$$\mathscr{L} = \mathscr{L}_0 \sqrt{\pi} \left(\frac{\beta_0^*}{\sigma_s}\right) \exp\left(\frac{\beta_0^{*2}}{\sigma_s^2}\right) \left[1 - \mathrm{erf}\left(\frac{\beta_0^*}{\sigma_s}\right)\right],$$

where $\mathscr{L}_0 = \frac{N^+ N^- f_0}{4\pi\sigma_0^{*2}}$, erf $(\beta_0^*/\sigma_s)$ is the probability integral of the parameter $\beta^*{}_0/\sigma_s$. The geometric factor is equal to 0.76 at $\beta_0^*/\sigma_s = 1$.

One should also notice one more important advantage of round colliding beams. In the case $\mathscr{E}_{x0} = \mathscr{E}_{z0}$ and $\beta_z^* = \beta^*{}_x$ the space charge parameters are no more dependent on the colliding bunch length. For flat beams, the modulation effect establishes the threshold value of $\beta^*{}_z/\sigma_s \approx 1.5$ starting with which the maximum attainable value of $(\xi_z)_{max}$ begins falling off, and further decrease of $\beta_z^*$ does not lead to better luminosity.

### 5. Radiative Polarization on the Phi-Factory

The lattice of this factory has some peculiarities which result in the electron and positron radiation polarization effects. Application of superconducting bending magnets with high magnetic fields leads to a characteristic polarization time [12]:

$$\tau_p^0 = \left[\frac{5\sqrt{3}}{8} \frac{me^2c}{\hbar^2} \gamma^2 \frac{\langle H_z^3 \rangle}{H_0^3}\right]^{-1},$$

$$H_0 = \frac{m^2c^3}{e\hbar^3} = 4.41 \cdot 10^{13}\,\mathrm{Oe}; \quad \langle H_z^3 \rangle = \frac{1}{2\pi R} \int_0^{2\pi R} |H_z|^3\, ds$$

equal to 9.5 mins for $E = 510$ MeV, and this can occur to be comparable with «luminosity life time». The attainable polarization degree outside the depolarizing resonances approaches to $S_\infty \sim 70\%$. It is evident that the radiative polarization should be taken into account since the angular deformations of the elastic $e^+e^-$ scattering and cross-sections of secondary particles production are possible.

On passing the experimental straight with the longitudinal field integral $|\pi HR|$ the particle spin rotates around the velocity direction by $\varphi = \pi$. In this scheme, the equilibrium precession axis $\vec{n}_s(\theta)$ in the storage ring arcs is vertical and oppositely directed, while it is horizontal at the interaction point. If the directions of the guide fields are chosen to be symmetrically opposite in the opposite arcs, then the radiative polarization with $\tau_p^0$ and $S_\infty$ indicated above will occur. For uni-directed fields in the opposite arcs, the self-polarization degree is identically equal to zero. This can be useful in some physical experiments in order to eliminate the angular deformations of cross-sections which are associated with the beam polarization.

The spin precession frequency in the Phi-factory lattice is proportional, just as in the usual «round» storage rings, to the particle energy, and the resonance depolarization method [13] is applicable to the absolute calibration of the mean energy in the beams.

The Phi-factory can also be employed for longitudinal polarization experiments. For these purposes, it is proposed to inject accelerated in the linac polarized $e^-$. After injection the direction of polarization can be rotated by the angle $\pi/2$ with a high frequency magnetic field whose frequency is resonant with the spin motion [14]. In this case, the polarization direction at the interaction point changes from turn to turn, according to the known law. Thus, phase of coherent spin precession can be measured for each event.

It is probable that the luminosity of the Phi-factory with polarized beams will be somewhat lower as compared with the conventional regime because of a depolarizing influence of the colliding bunches. This problem and the problem of machine depolarizing resonances need further investigation.

**Acknowledgements**

The authors are indebted to D.V. Pestrikov for helpful discussions on the coherent beam-beam effects' problems, E.P. Solodov for his significant contribution to realizing the problems of physical experiments at the Phi-factory. We would like to thank A.P. Lysenko for his help with the computer simulation.

**References**

1. *Tumaikin G.M.* Proposals on B- and Phi-meson Factories in Novosibirsk. To be published in Proc. of IEEE Part. Accel. Conf. Chicago, March 19—22, 1989.
2. *Eichler R.A.* ETHZ-IMP-P 85/4, Zurich (1986).
3. *Aksenov G.A. et al.* Proposal of CMD-2 detector for VEPP-2M. INP Preprint 85-118. Novosibirsk (1985).
4. *Litvinenko V.N. and Perevedentsev E.A.* Proc. of VIth All-Union Conf. on Charged Part. Accelerators. Dubna (1979), v.2, p.254.
5. *Anashin V.V. et al.* INP Preprint 84-123. Novosibirsk (1984).
6. *Balbekov V.I. and Kolomensky A.A.* ZhTF (1969), v.37, N 11, p.2021.
7. *Derbenev Ya.S.* Proc. of IIIrd All-Union Conf. on Charged Part. Accelerators. Moscow, 1972, Nauka (1973), v.2, p.382.
8. *Chao A. and Keil E.* CERN-ISR-TH/79-31, Geneva (1979).

9. *Dikansky N.S. and Pestrikov D.V.* Part. Accel. (1981), v.12, p.27.
10. *Chao A. and Ruth R.* Part. Accel. (1985), v.16, p.201.
11. *Vasserman I.B. et al.* INP Preprint 79-74, Novosibirsk (1979).
12. *Sokolov A.A. and Ternov I.M.* USSR Ac. Sc. Doklady (1963), p.1052.
13. *Derbenev Ya.S. et al.* Part. Accel. (1980), v.10, p.177.
14. *Vasserman I.B. et al.* Phys. Lett. (1987), B198, p.302.

# A LINEAR COLLIDER $\phi$ FACTORY AND BEAM DYNAMICS TEST MACHINE

**[1,3]W.A. Barletta, [1]R.C. Berg, [1]A. Boden, [1]C. Buchanan**
**[1]D.B. Cline, [1,4]A. Fridman, [1]C.H. Ho, [2]C. Pellegrini**

[1] Physics Department, University of California, Los Angeles
405 Hilgard Avenue, Los Angeles, California 90024-1547

[2] Brookhaven National Laboratory, 75 Brookhaven Avenue
Building 725, Upton, Long Island, New York 11973

[3] Lawrence Livermore National Laboratory, P.O. Box 808
Livermore, California 94550

[4] IN2P3, France

**ABSTRACT**

We present the design of a high luminosity Linear Collider $\phi$ Factory. This machine can be used to search for violation of CPT invariance and the study of CP violation ($\epsilon'$).It can also serve as a test collider to study the feasibility of $\mathcal{L} > 10^{34}$ cm$^{-2}$sec$^{-1}$ $\bar{\text{B}}$B factory and the beam dynamics of large disruption systems. The machine will be asymmetric in beam energy, thus giving the $\phi$'s a boost.

## 1. Introduction

The development of linear electron-positron colliders is extremely important for the future of elementary particle physics. The recent success in observing Z° particles at the SLC linear collider at SLAC is an enormous step in this direction. We believe linear colliders will be of equal importance as $\bar{B}B$ and $\phi$ factories. For example, a recent linear collider $\bar{B}B$ factory design would reach a luminosity of $> 10^{34}$ $cm^{-2}sec^{-1}$ and an energy spread of $10^{-3} \times E_{beam}$ (1). The study of the beam-beam collision dynamics of linear colliders is just starting (2). A low energy test collider would allow a systematic study of the beam-beam interaction dynamics. In this brief report we show that a unique high luminosity $\phi$ factory could result from such a test collider.

## 2. Tests of CPT and Study of CP Violation Using a $\phi$ Factory

The $\phi$ particle is a state of $J^{PC} = 1^{--}$ and decays partially into

$$\phi \rightarrow K_S K_L$$

This decay mode is prescribed by Quantum Mechanics and is independent of CP or CPT violation. The subsequent decays of the $K_S$ and $K_L$ into various final states can be used to provide new measurements of the CP parameters for the K° system and a new extremely sensitive test for CPT violation. While there have been several discussions of the use of $\phi$ mesons for this study, we follow here the recent work of Dunietz, Hauser and Rosner for illustration [3]. The basic concept is to observe the time development of the $K_S^\circ$ and $K_L^\circ$ system with the $K_L^\circ$ decay mode a CP violating mode such as $\pi^+\pi^-$ or $\pi^\circ\pi^\bullet$. Consider $K_S \to \pi^+\pi^-$ at $t_1$ and $K_L^\circ \to \pi^\circ\pi^\circ$ at $t_2$ and form

$$\Delta = t_2 - t_1$$

and observed the decay intensity of $I(\Delta t)$ as a function of $\Delta$. This is qualitatively similar to the technique to measure the $K_S^\circ - K_L^\circ$ mass difference starting with $K^\circ$ beams. The direct CP violation process can cause an asymmetry in the rates as a function of $\Delta$ as shown in Fig. 1. The test for CPT violation depends on a non-zero value of $e'$.

In Fig. 2 we show the number of $\phi$'s required to observe various values of Re $\epsilon'/\epsilon$ and Im $\epsilon'/\epsilon^4$. It is clear that $\mathcal{O}$ $(10^9)$ and greater are required to test measure $\epsilon'/\epsilon$ and to test CPT at a sensitive level. This requires an $e^+e^-$ machine with a luminosity in excess of $10^{33}$ $cm^{-2}sec^{-1}$. We will describe later some concepts to design such a machine.

The search for CPT violation at a sensitive level is a very important test of our current understanding of elementary particles and fields. Unfortunately, there is no known way to break CPT although it is sometimes suggested that quantum gravity may cause such effects.

### 3. Conceptual Design of a Linear Collider $\phi$ Factory

The basic concept of the high luminosity $\bar{B}B$ factory is to collide the beams with very small radius and at high enough repetition rate to achieve a luminosity of greater than $10^{34}$ $cm^{-2}sec^{-1}$. The most serious problem is the high intensity of positrons required. In the scheme of reference[(1)] this is taken care of by recirculating the positrons with continuous damping. This requires a single damping ring for the machine. In Fig. 3 we show a schematic of a $\bar{B}B$ factory using this concept.

A one tenth scale model of this $\bar{B}B$ factory can be used to test this principle and would provide the basis for the study of beam-beam interaction dynamics and could serve as well as a dedicated $\phi$ factory. A schematic of this machine is shown in Fig. 4 and Fig. 5, which is a second variant of the scheme.

The parameters of a $\phi$ factory machine is given in Tables 1a and 1b. Note that the design luminosity reaches $10^{33}$ $cm^{-2}sec^{-1}$. This would give a rate of greater than $10^{10}$ $\phi$'s per year.

We have designed an asymmetric $\phi$ factory for two reasons:

A. The $K^{\circ}_{L}$, $K^{\circ}_{S}$ are folded forward, thus reducing the size of the experimental detector required. Fig. 7 shows the angular distribution for a machine with the beam configuration which maximizes the $K^{\circ}$ acceptance for a detector spanning the azimuth range from 5° to 40°. This choice corresponds to a Lorentz boost of 0.35c and beam energies of 354 MeV on 735 MeV.

B. To simulate the asymmetric beam-beam interaction of a high luminoisty $\bar{B}B$ factory[(1)].

We have carried out a preliminary study of the wake fields for the collider using a code developed by Chernin and Mondelli[(2)]. The results of this simulation indicates a feasible solution was found. The optimal parameters are shown in Fig. 6. In Table 2 we give a preliminary set of parameters for a $\phi$ factory constructed using a high gradient linear accelerator as shown in Fig. 5.

These preliminary studies indicate the feasiblity of constructing a test collider and $\phi$ factory that will reach a luminosity of $10^{33}$ $cm^{-2}sec^{-1}$. The construction of this machine could lead to the design of a $\bar{B}B$ factory that might operate in the luminosity range of $10^{34} - 10^{35}$ $cm^{-2}sec^{-1}$. A possible site for the $\phi$ factory is the new science center on the UCLA west campus shown in Fig. 8.

The electron and positron beam densities will be several orders of magnitude greater than SLC and the disruption parameters can be varied over a wide range. Thus, this collider will allow a study of beam-beam interaction dynamics as a totally new regime, thus advancing the state of the art in this field as well.

We wish to thank D. Chernin and A. Mondelli for help with the wakefield calculations for the $\phi$ factory.

## REFERENCES

1. D. Cline and C. Pellegrini, "A High Luminosity $\bar{B}B$ Factory Using Positron Recirculation", published in the 1988 Snowmass proceedings and submitted to Nuclear Inst. and Methods (1989).

2. Chernin and Mondelli, Design Constraints for $e^+e^-$ Linear Super Colliders, PTN 88-002, Plasma Technology Division Science Applications, Inc., Goodridge Dr., McLean, VA (1988).

3. I. Dunietz, J. Hauser and J. Rosner, Phys. Rev. D. 35, 2166 (1987).

**Table 1a**

| Test Collider and $\phi$ Factory<br>General Interaction Parameters | |
|---|---|
| Positrons | Electrons |
| $E_1 = 1$ GeV | $E_2 = 0.3$ GeV |
| $P_1 = 2.03$ MW | $P_2 = 17.3$ KW |
| $N_1 = 1.79 \times 10^{11}$ | $N_2 = 5.06 \times 10^{9}$ |
| $\nu = 71.1$ KHz | $\nu = 71.1$ KHz |
| $\delta_{BS_1} = 4.52 \times 10^{-6}$ | $\delta_{BS_1} = 1.48 \times 10^{-4}$ |
| $D_1 = 1.5$ | $D_1 = 2000$ |
| $\sigma_{x_1} = \sigma_{y_1} = 7 \times 10^{-6}$ | $\sigma_{x_2} = \sigma_{y_2} = 7 \times 10^{-6}$ |
| $\sigma_{Z_1} = 1.2 \times 10^{-3}$ m | $\sigma_{Z_2} = 1.1 \times 10^{-4}$ m |

**Table 1b**

$\phi$ Factory

Values of Interaction Point Quantities

$$\gamma_1\ mc^2 = 1\ GeV$$

$$\gamma_2\ mc^2 = 0.3\ GeV$$

$$R = 1$$

$$L = 10^{33}$$

$$P_{b1} = 3.86\ MW$$

$$P_{b2} = 0.034\ MW$$

$$N_1 = 1.786 \times 10^{11}$$

$$N_2 = 5.319 \times 10^9$$

$$\nu = 135\ KHz$$

$$D_1 = 1.5$$

$$D_2 = 2000$$

$$\delta_{BS1} = 4.923 \times 10^{-4}$$

$$\delta_{BS2} = 7.49 \times 10^{-6}$$

$$\sigma_{y1} = 7.15 \times 10^{-7}\ m$$

$$\sigma_{y2} = 7.15 \times 10^{-7}\ m$$

$$\sigma_{z1} = 1.192 \times 10^{-3}\ m$$

$$\sigma_{z2} = 1 \times 10^{-4}\ m$$

Table 2a.

## Asymmetric PHI Factory with By-pass

| | | | |
|---|---|---|---|
| **Italics denote inputs* | | Accelerator | |
| Beam parameters | | rf | |
| * *Energy (GeV)* | 0.27 | RF (GHz) | 5.74 |
| * *Peak current (kA)* | 1.00 | * *rf-efficiency* | 0.65 |
| * *Pulse length (ps)* | 4.5 | * *Iris a (mm)* | 7.84 |
| * *Norm Emit (mm-rad)* | 0.010 | * *Fill T/atten . T* | 0.10 |
| * *rep rate (Hz)* | 5000 | * *Bunch (full phase)* | 9.3 |
| Pulse length (mm) | 1.4 | * *Fill fraction* | 0.82 |
| N-part | 2.7e+10 | wavelength(mm) | 52.3 |
| n bunches | 9 | iris/lambda | 0.15 |
| min. bunch space (ns) | 1.05 | cavity/iris | 2.7 |
| in wavelengths | 6 | Cavity size (mm) | 21.1 |
| | | uncor elastance | 2.5e+14 |
| Derived values | | v-group/c | 0.036 |
| Gamma | 5.3e+02 | atten. time (s) | 5.5e-07 |
| Emittance (m-rad) | 1.89e-08 | Q | 9901 |
| Pulse energy (J) | 1.17 | Energy/m | 9.0e+00 |
| Macro-pulse (J) | 10.5 | E-max (MV/m) | 111 |
| E-beam power (kW) | 52.5 | E-accel (MeV/m) | 49.9 |
| Brightness (a/m2-r2) | 2.0e+13 | Active length (m) | 6.6 |
| | | Gradient (MeV/m) | 40.9 |
| Size & Costs | | Fill length (m) | 0.59 |
| * *I-induction (A)* | 1500 | Number of sections | 11 |
| * *r-inner (cm)* | 4.5 | cavities/section | 34 |
| * *Grad-induction* | 0.9 | rf P (MW/m) | 171 |
| V-induction (MeV) | 1.2 | rf P (MW/section) | 122 |
| Ind. length (m) | 1 | struc. efficiency | 0.91 |
| "Tube" voltage (MV) | 1.2 | loading % | 2.3 |
| Number of tubes | 1 | t-refill (ns) | 0.9 |
| rf-strc k$/m | $52.8 | rf-supply (ns) | 53 |
| rf-power k$/m | $460.0 | Beam efficiency % | 16.6 |
| rf-cost (M$) | $3.6 | Total rf (TW) | 1.2e-03 |
| rf-strc cost (M$) | $0.4 | | |
| Cost of ring (M$) | $7.1 | Tolerances | |
| Positron source (M$) | $1.0 | Alignment (m) | 1263.1 |
| Total hardware M$ | $12.0 | Vibration tol. (m) | 17.4 |
| Facility cost (M$) | $5.3 | sigma-p to focus(%) | 0.30 |
| Total cost (M$) | $17.3 | Multi-bunch deltaE(%) | 0.06 |
| Accel. length (m) | 7.4e+00 | N-max in sigma-p | 14 |
| Wall power (MW) | 1.0 | Wall load (W/cm2) | 3.2 |

* *All 's & 's) are standard deviations* *program revision - 5/5/89*

Table 2b.

**Asymmetric PHI Factory with By-pass**

| | | | |
|---|---|---|---|
| Wakes | | **Beam 2** | |
| * *Dipole de-Q* | 100 | * *Energy (GeV)* | 0.97 |
| single bunch | | * *Pulse T (ps)* | 3.00 |
| wt (v/C/m^2) | 6.3e+15 | E-n (mm-rad) | 0.128 |
| Landau p/p (%) | 0.81 | Peak I (kA) | 3.65 |
| L to correct (m) | 0.2 | N-part | 6.6e+10 |
| Sigma-y (m) | 123.4 | gamma | 1898 |
| multi-bunch wake | | sigma-z2 (m) | 4.5e-04 |
| Q*(1/f) (s) | 1.7 | sigma-y2 (m) | 2.4e-06 |
| Wmb (V/C/m^2) | 1.5e+16 | | |
| Landau-m sigma-p % | 1.96 | Focus 2 | |
| L to correct (m) | 0.4 | * *Focus B (T)* | 1 |
| Longitudinal wake | | * *radius a (m)* | 0.04 |
| Wake at 0 V/C/m | 4.1e+14 | * *p/p (%) to focus* | 0.35 |
| non-lin sigma-p accel | 0.34 | * *Chromatic correction* | yes |
| Ph to corr. Landau | -4.7 | e-z | 0.21 |
| | | B-rho (T-m) | 3.23 |
| Linac focussing | | Theta-accept | 3.0e-02 |
| Focus system | FODO | focal scale | 0.36 |
| * *Linac focus (T)* | 0.5 | Lens L (m) | 1.87 |
| * *Quad. fraction* | 0.05 | Length (m) | 2.88 |
| # of quads | 8 | beta* (corr.) | 3.3e-04 |
| | | | |
| Final focus triplet-1 | | Beamstrahlung | |
| * *Focus stength (T)* | 1.2 | upsilon1 | 1.5e-05 |
| * *radius a (m)* | 0.025 | upsilson2 | 5.4e-05 |
| * *Ideal FDDF (m)* | 5.2 | H-upsilon1 | 1.00 |
| * *Chromatic correction* | yes | H-upsilon2 | 1.00 |
| B-rho (T-m) | 0.90 | deltaE-1 | 6.3e-06 |
| Theta-accept | 4.9e-02 | deltaE-2 | 1.6e-05 |
| focal scale | 0.1 | | |
| Lens L (m) | 0.7 | Disruption | |
| Length (m) | 1.1 | * *Flatness-R (1)* | 14 |
| beta* | 9.3e-5 | sigma-z1(m) | 6.8e-04 |
| Focus rad. (m) | 6.6e-01 | sigma-y 1 (m) | 6.6e-07 |
| | | D-y1 | 29.7 |
| | | D-y2 | 1.0 |
| Interaction point | | H-D1 | 2.4 |
| E-cm (GeV) | 1.024 | H-D2 | 2.0 |
| Total E spread (MeV) | 5.07 | oscillations1 | 1.2 |
| Luminosity | 1.0e+33 | oscillations2 | 0.2 |

Table 2c.

**Asymmetric PHI Factory with By-pass**

| | | | |
|---|---|---|---|
| Positron source | | e+ accelerator (post-ring) | |
| * *e- bunch/e+ bunch* | 180 | * *Accelerate (GeV)* | 0.3 |
| Beam energy (GeV) | 0.27 | Cost (M$) | 2.84 |
| psuedo-efficiency | 2.43 | Buncher 2 | |
| beam power (kW) | 52.49 | 4 dipole chicane | |
| e+ / bunch | 6.6e+10 | * *Bend angle -* | 1.00 |
| e+ / sec | 3.0e+15 | * *Dipole field (T)* | 1.50 |
| Cost (M$) | 1.15 | rho (m) | 2.15 |
| | | M34 (ps/%p/p) | 405.7 |
| | | Dipole sep (m) | 1.24 |
| | | Chicane length (m) | 14.75 |

| | | | |
|---|---|---|---|
| Damping ring | | | |
| * *DBA or FODO* | fodo | * *RF freq (GHZ)* | 0.5 |
| * *Dipole field (T)* | 2.5 | * *phase advance/cell* | 1.50 |
| * *Vert/horiz mixing* | 0.05 | * *Jx=Jy* | 1.05 |
| * *Dipole fraction* | 0.20 | * *aperture (m)* | 0.1 |
| * *beta-x/beta-y* | 0.5 | * *Quad fraction* | 0.23 |
| * *B-wiggler (T)* | 0.001 | * *Quad field (T)* | 1.4 |
| * *Wiggler fraction* | 0 | bunch spacing (m) | 3.1 |
| * *Wiggle period (cm)* | 5 | | |
| Energy (GeV) | 0.67 | Jz | 1.9 |
| gamma | 1311 | Jx+mix*Jy | 1.05 |
| N-ring | 5.9e+11 | B-local/B-ave | 1.07 |
| B-rho | 2.23 | Rev. freq | 1.1E+07 |
| Bend radius (m) | 0.9 | harmonic # - h | 4.7E+01 |
| Ring Radius (m) | 4.5 | E loss/turn | 2.0E+04 |
| Circumference (m) | 28.0 | Rf kVolts/turn *h | 9.3E+02 |
| Dipole length (m) | 5.6 | Momentum compactior | 0.059 |
| Drift lengths (m) | 16.0 | nu-synch | 3.6E-03 |
| L-dipole (m) | 0.39 | p in ring | 6.6E-04 |
| Dipole number | 14 | sigma-z in ring (m) | 4.8E-02 |
| K-wiggle | 0.0 | emit-z | 0.04 |
| T-wig / T-dipole | 1.00 | I (A) | 1.01 |
| Damping t (ms) | 6.26 | P-synch (kW) | 20.2 |
| emit | 1.3e-04 | U-crit (keV) | 0.83 |
| beta-x | 1.30 | U-wig (keV) | 0.08 |
| beta-y | 2.59 | Z/n (Ohms) | 6.02 |
| Tune Qx | 3.44 | n | 92.2 |
| Tune Qy | 1.7 | Z (Ohms) | 555.3 |
| Dispersion | 3.8e-01 | Fill time (s) | 5.64 |

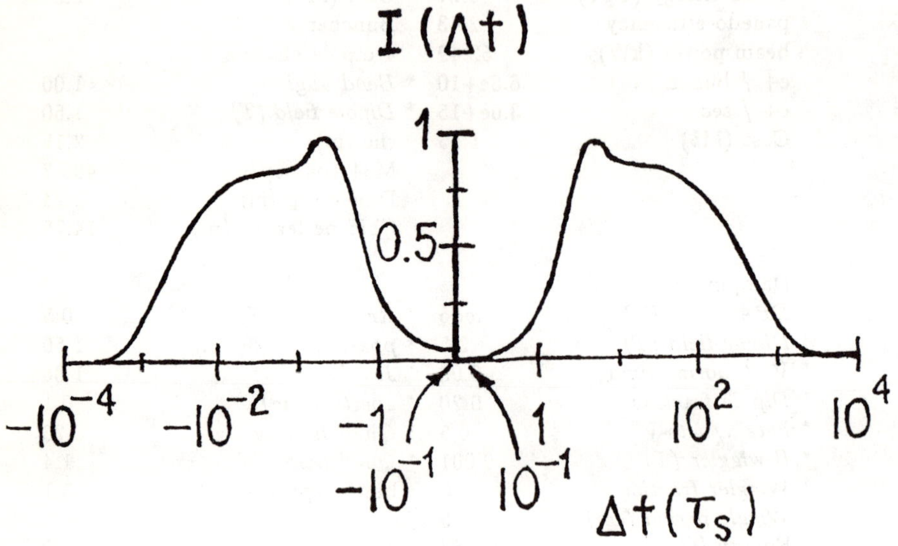


Figure 1. The time dependent intensity plot where I(Δt) refers to the difference of $K^{\circ}_{S/L_{(1)}} \to \pi^{\circ}\pi^{\circ}$ and $K^{\circ}_{S/L_{(2)}} \to \pi^{+}\pi^{-}$ rates as a function of $\Delta t = \tau_1 - \tau_2$ for $\epsilon'/\epsilon = 0.005e^{i(270^{\circ})}$.

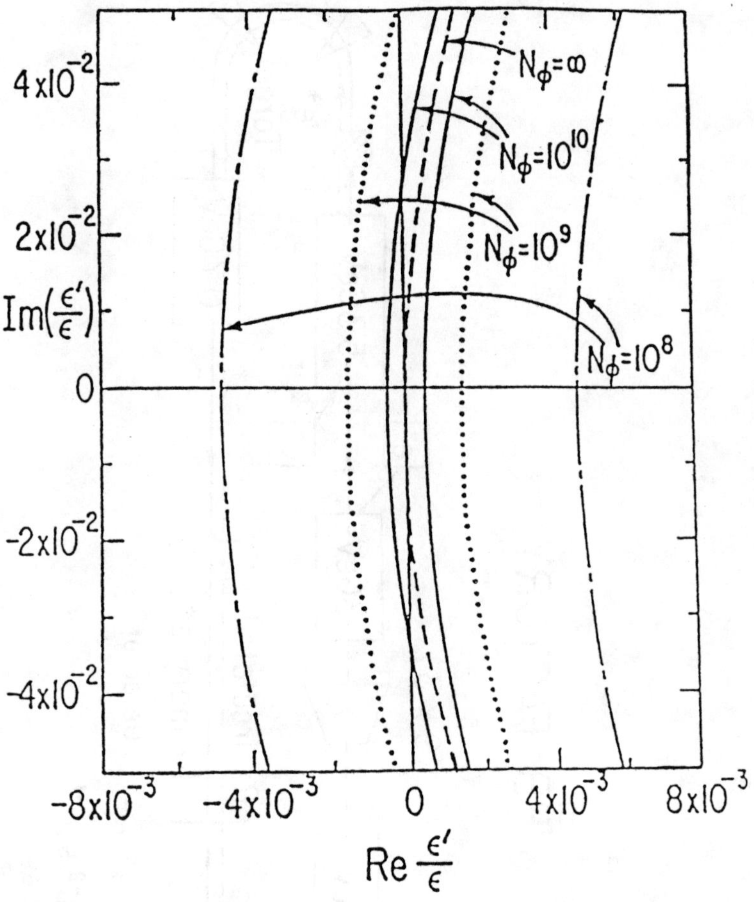


Figure 2. Contours of constant $N_\phi(\tau = \infty)$, and the corresponding asymmetry $A_\Gamma[\tau = \infty, 0]$. Dashed curve, $N_\phi = \infty$, $A_\Gamma = 0$; solid curve, $N_\phi = 10^{10}$, $A_\Gamma = \pm 1.4 \times 10^{-3}$; dotted curve, $N_\phi = 10^8$, $A_\Gamma = \pm 4.5 \times 10^{-2}$. Here the three right hand most contours have negative asymmetries; all others are positive.

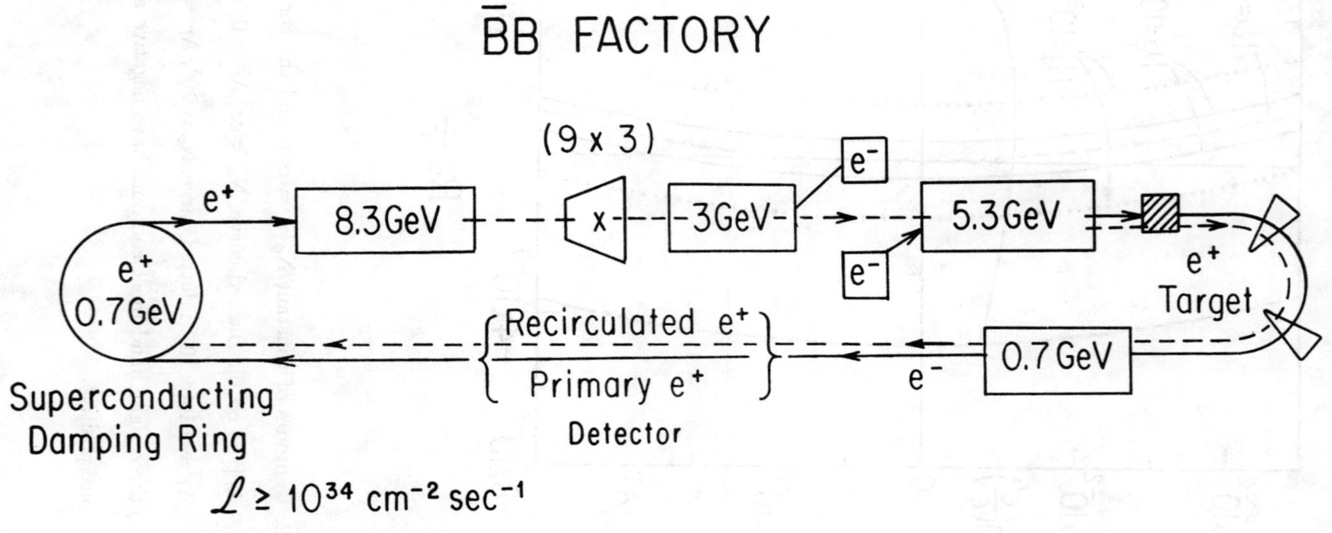
B̄B FACTORY
(9 x 3)
e+
8.3GeV
x
3GeV
e−
e−
5.3GeV
e+
Target
e+
0.7GeV
0.7 GeV
e−
Recirculated e+
Primary e+
Detector
Superconducting
Damping Ring
$\mathcal{L} \geq 10^{34}$ cm$^{-2}$ sec$^{-1}$

Figure 3.

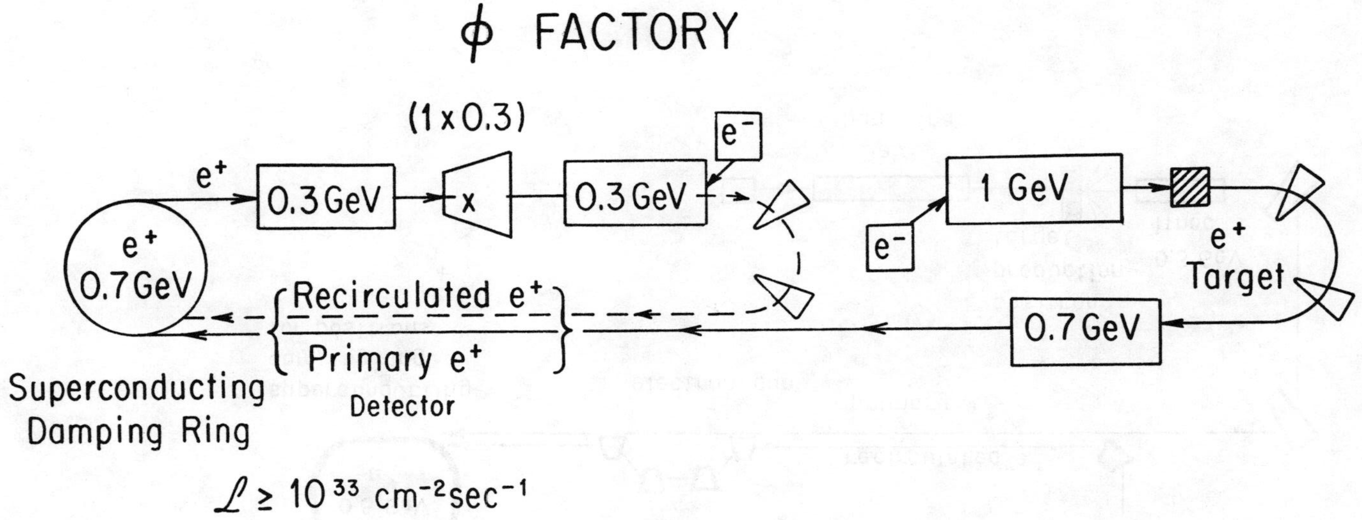
ϕ FACTORY
(1 x 0.3)
e+
0.3 GeV
x
0.3 GeV
e−
1 GeV
e−
e+
Target
e+
0.7 GeV
Recirculated e+
Primary e+
0.7 GeV
Superconducting
Damping Ring
Detector
ℒ ≥ 10 33 cm-2sec-1

Figure 4.

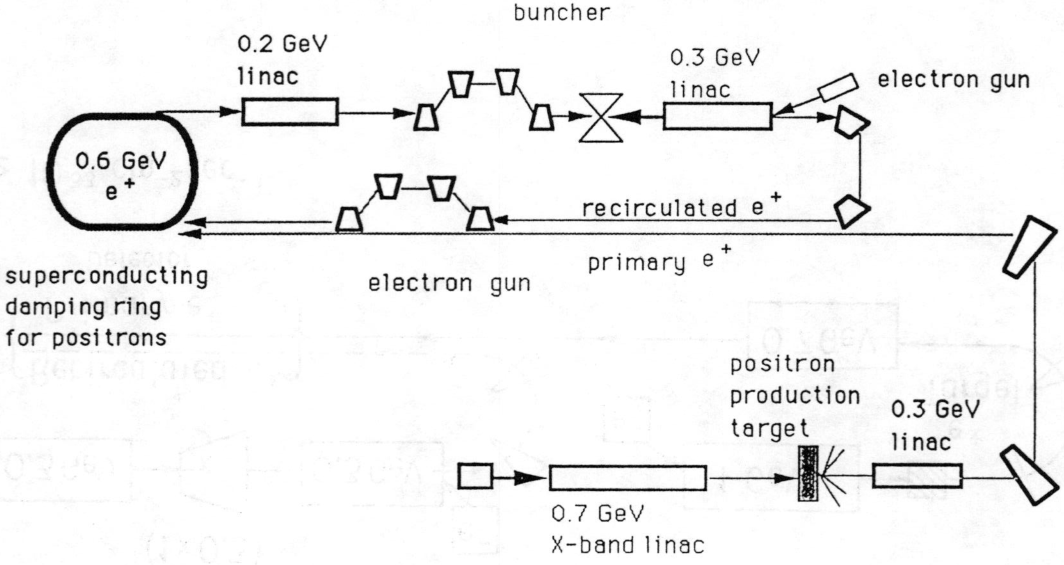


Figure 5. Schematic of asymmetric $\phi$ factory with by-pass as described in Table 2.

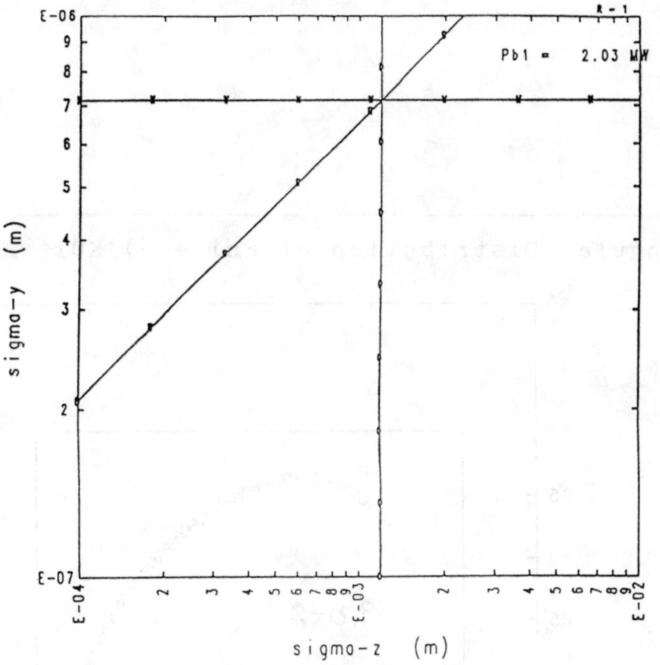


Figure 6. Positron beam interation point bunch profile constraints (after reference (2)). Shown in the figure are the three curves of constraint on the space of transverse ($\sigma_y$) and longitudinal ($\sigma_z$) bunch size. The horizontal line represents the constraint of the number of particles per bunch, and the operational point must lie on or below this line. The diagonal line represents a maximum beamstrahlung radiative loss constraint, and again the operational point must lie on or below this line. Finally, the vertical line represents the disruption parameter constraint, and our operational point must lie on or to the left of this line. For the parameters of Table 1a the three constraint curves meet at a single point which fix the positron bunch site given in the table.

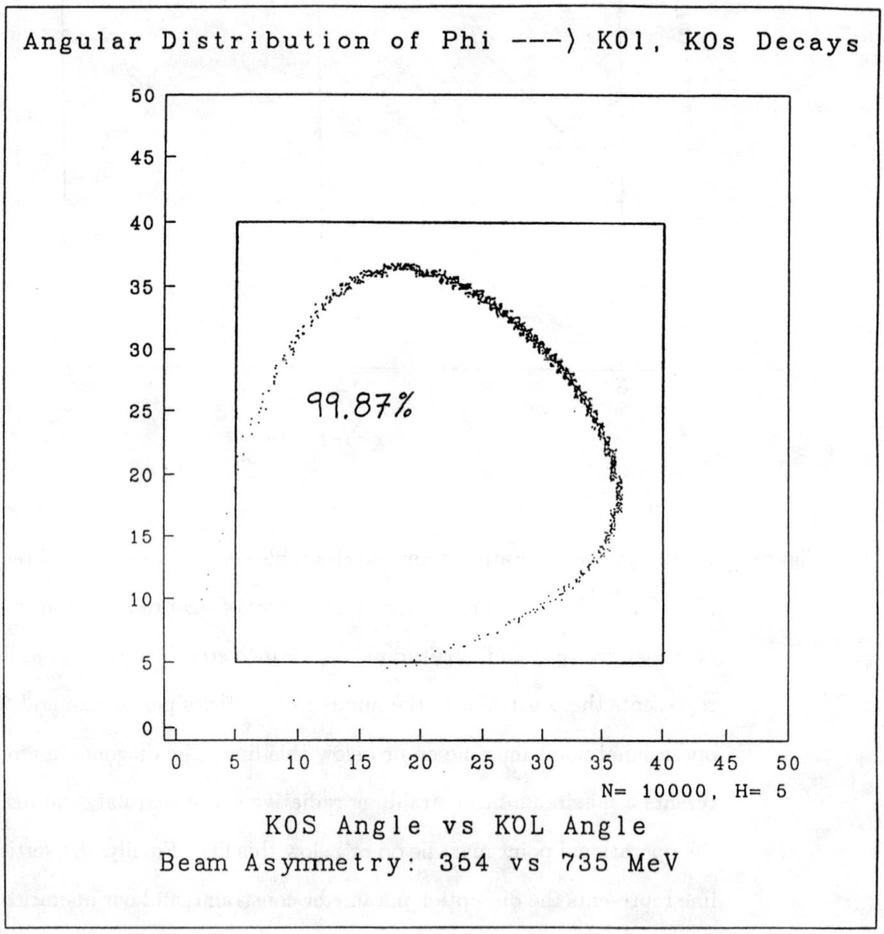
Angular Distribution of Phi ---⟩ K0l, K0s Decays
50
45
40
35
30
25
20
15
10
5
0
0 5 10 15 20 25 30 35 40 45 50
99.87%
N= 10000, H= 5
K0S Angle vs K0L Angle
Beam Asymmetry: 354 vs 735 MeV

Figure 7.

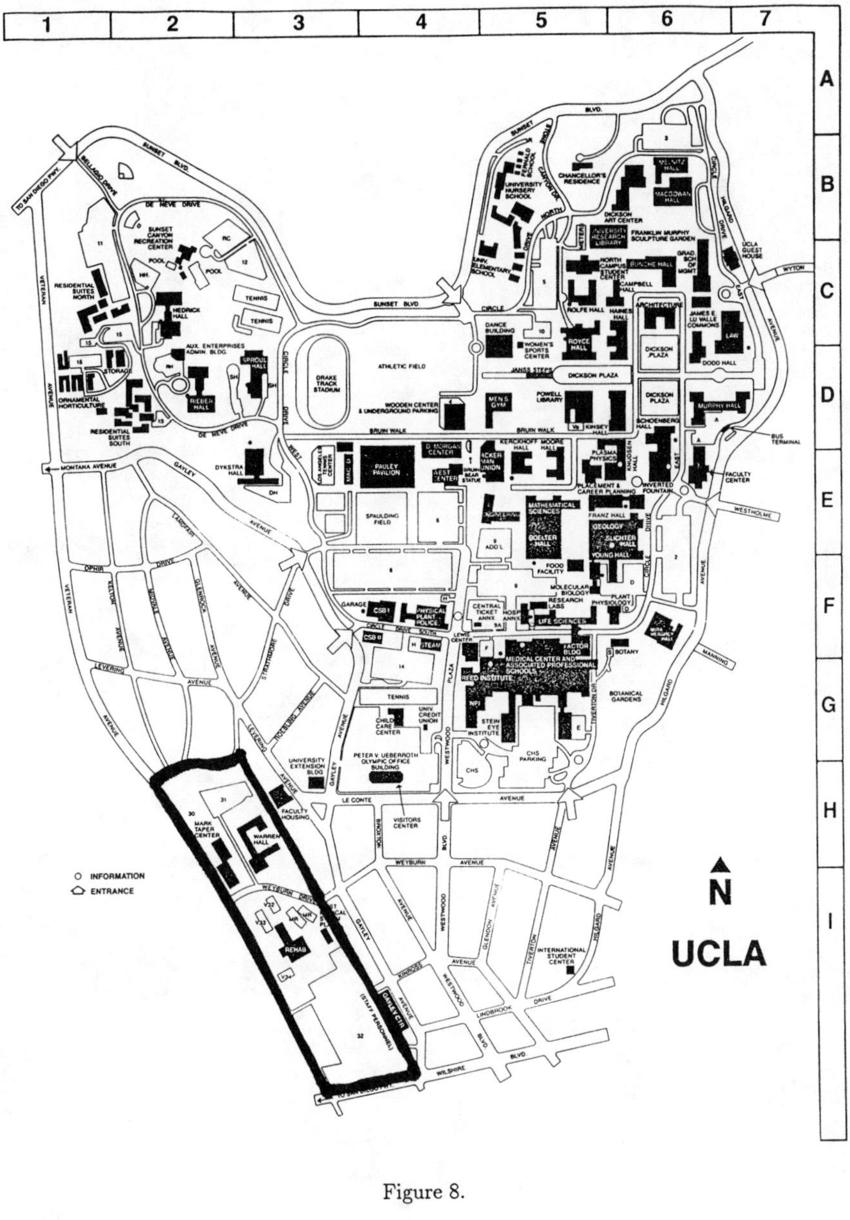
1
2
3
4
5
6
7
A
B
C
D
E
F
G
H
I
SUNSET BLVD.
BELLAGIO DRIVE
TO SAN DIEGO FWY.
DE NEVE DRIVE
SUNSET CANYON RECREATION CENTER
POOL
TENNIS
RESIDENTIAL SUITES NORTH
HEDRICK HALL
AUX. ENTERPRISES ADMIN. BLDG.
SPROUL HALL
RIEBER HALL
ORNAMENTAL HORTICULTURE
RESIDENTIAL SUITES SOUTH
STORAGE
VETERAN AVENUE
MONTANA AVENUE
DYKSTRA HALL
GAYLEY AVENUE
LANDFAIR
OPHIR DRIVE
KELTON AVENUE
MIDVALE AVENUE
GLENROCK AVENUE
LEVERING AVENUE
STRATHMORE DRIVE
DRAKE TRACK STADIUM
ATHLETIC FIELD
CIRCLE DRIVE WEST
WOODEN CENTER & UNDERGROUND PARKING
BRUIN WALK
PAULEY PAVILION
SPAULDING FIELD
GARAGE
CSB 1
PHYSICAL PLANT POLICE
CIRCLE DRIVE SOUTH
STEAM
TENNIS
CHILD CARE CENTER
UNIV. CREDIT UNION
UNIVERSITY EXTENSION BLDG.
PETER V. UEBERROTH OLYMPIC OFFICE BUILDING
VISITORS CENTER
LE CONTE AVENUE
FACULTY HOUSING
MARK TAPER CENTER
WARREN HALL
WEYBURN DRIVE
REHAB
WEYBURN AVENUE
KINROSS
WESTWOOD BLVD.
LINDBROOK DRIVE
WILSHIRE BLVD.
TO SAN DIEGO FWY.
(STAFF PERSONNEL)
INTERNATIONAL STUDENT CENTER
UNIVERSITY HURSERY SCHOOL
UNIV. ELEMENTARY SCHOOL
CHANCELLOR'S RESIDENCE
MELNITZ HALL
MACGOWAN HALL
DICKSON ART CENTER
UNIVERSITY RESEARCH LIBRARY
FRANKLIN MURPHY SCULPTURE GARDEN
BUNCHE HALL
GRAD SCH OF MGMT
UCLA GUEST HOUSE
WYTON
NORTH CAMPUS STUDENT CENTER
CAMPBELL HALL
ROLFE HALL
HAINES HALL
ARCHITECTURE
JAMES E. LU VALLE COMMONS
LAW
DANCE BUILDING
WOMEN'S SPORTS CENTER
ROYCE HALL
JANSS STEPS
DICKSON PLAZA
DODD HALL
MEN'S GYM
POWELL LIBRARY
MURPHY HALL
SCHOENBERG HALL
KINSEY HALL
KERCKHOFF HALL
MOORE HALL
BUS TERMINAL
FACULTY CENTER
ACKERMAN UNION
WEST CENTER
PLASMA PHYSICS
KNUDSEN HALL
PLACEMENT & CAREER PLANNING
INVERTED FOUNTAIN
WESTHOLME
MATHEMATICAL SCIENCES
FRANZ HALL
GEOLOGY
SLICHTER HALL
BOELTER HALL
YOUNG HALL
FOOD FACILITY
MOLECULAR BIOLOGY
PLANT PHYSIOLOGY
CENTRAL TICKET ANNX
RESEARCH LABS
LIFE SCIENCES
LEWIS CENTER
FACTOR BLDG
MEDICAL CENTER AND ASSOCIATED PROFESSIONAL SCHOOLS
REED INSTITUTE
BOTANY
MANNING
BOTANICAL GARDENS
HILGARD AVENUE
TIVERTON DR.
STEIN EYE INSTITUTE
CHS PARKING
CHS
INFORMATION
ENTRANCE
N
UCLA

Figure 8.

# CP VIOLATION AT LEAR

*Lalit Seghal and Eugene Golowich.*

# TRIPLE PRODUCT CORRELATIONS

Eugene Golowich
Department of Physics and Astronomy
University of Massachusetts, Amherst MA 01003

## ABSTRACT

Triple product correlations as experimental signals for CP violation are defined and surveyed. A short checklist is given for anyone who would contemplate such an experiment. The issue of fake signals arising from unitarity phases is then discussed, followed by consideration of several possibilities (hyperon decay, jet variables, $Z^0$ decay) for collider experiments. Finally, an example from semileptonic meson decay is given wherein an observed triple product correlation would indicate new physics.

Progress in the study of CP violations comes slowly. In the 25 years since the discovery of CP violations in the $K^0 - \overline{K}^0$ system, only five real numbers relating to the effect have been detected. These are the two complex-valued amplitude ratios $\eta_{\pm}$, $\eta_{00}$ and the real-valued charge asymmetry $\delta$.

Here we consider a distinct class of signals for CP violations, the so-called triple product correlations. Quite often in a scattering or decay process, it is possible to measure at least some of the momentum $(\mathbf{p})$ and spin $(\mathbf{s})$ variables of the final state particles. Under the operation of time reversal, both the momentum and spin become reversed. Thus a scalar-valued triple product of momentum and/or spin is likewise reversed. Examples of such products are $\mathbf{p}_1 \cdot \mathbf{p}_2 \times \mathbf{p}_3$, $\mathbf{s}_1 \cdot \mathbf{p}_2 \times \mathbf{p}_3$, etc. If, as we assume here, CPT is an invariance of the theory, then such products are potential signals for violations of CP invariance as well as T invariance.

Needless to say, triple-product correlations are not a panacea for the field of CP violation studies. The fact that not a single example has yet been detected speaks for itself, so *caveat emptor*! In fact searches of this type were carried out long ago[1] in a hyperon decay experiment, more recently in nuclear physics,[2],[3] and also in a kaon decay experiment[4].

However, a conference like the one we are attending forces us not only to think back over the past 25 years, but also to look with fresh thinking to the next 25 years. Just as there will be renewed efforts to detect tiny electric dipole moments both in the neutron and in a variety of atomic systems, the effect of CP violations in kaon rare decay modes, and partial rate asymmetries in heavy meson decays, so should we not leave the topic of decay correlations unconsidered. Certainly there is a broad range of possibilities. These include nonleptonic and semileptonic modes in heavy quark systems, semileptonic kaon decay, and various collider studies which we shall survey shortly. For the sake of argument, suppose that we were discussing plans for doing one of these. Then, at least to begin, I would emphasize the following three points: (i) decide what *kind* of correlation to detect, (ii) if such a correlation is observed, determine whether it is *really* a manifestation of CP violation, and (iii) if CP violation is indeed found, try to decide if it is necessarily a consequence of the Standard Model. In the following we shall address aspects of these three points.

Point (i) is concerned with a potentially fatal flaw with the triple product correlation method, the possibility of a fake signal associated with a *unitarity phase*. It is convenient to consider this topic from the viewpoint of a specific system, which we take to be the decay of a heavy pseudoscalar meson into a pair of vector mesons,[5] e.g., $B(\mathbf{p}) \to V_1(\mathbf{k}, \mathbf{s}_1) + V_2(\mathbf{q}, \mathbf{s}_2)$. The invariant amplitude for this process

$$\mathcal{M} = a\epsilon_1 \cdot \epsilon_2 + \frac{b}{m_1 m_2} p \cdot \epsilon_1 p \cdot \epsilon_2 + i\frac{c}{m_1 m_2} \epsilon^{\alpha\beta\gamma\delta} \epsilon_{1\alpha} \epsilon_{2\beta} k_\gamma p_\delta \tag{1}$$

contains three kinematically independent subamplitudes

$$\begin{aligned} a &= |a| e^{i(\delta_s + \phi_s)} \\ b &= |b| e^{i(\delta_d + \phi_d)} \\ c &= |c| e^{i(\delta_p + \phi_p)} \ . \end{aligned} \tag{2}$$

There are seen to be two kinds of phases. The $\{\delta_i\}$ are called unitarity phases, and they exist as a consequence of the interaction between the final state vector mesons. The other phases $\{\phi_i\}$ occur if and only if CP violation is present somewhere in the decay process. Incidentally, for notational simplicity we have suppressed the sum over isospin channels in the above. The object now is to select triple product observables which are sensitive to the $\{\phi_i\}$. The most obvious choice, $\mathbf{p} \cdot \mathbf{k} \times \mathbf{q}$ will not do because the momentum vectors $\mathbf{p}, \mathbf{k}, \mathbf{q}$ are not independent. In fact, one needs four or more particles in the final state in order to construct a nontrivial triple product composed entirely of momentum vectors. Of course there are no stable vector mesons, so after each vector particle decays there will be at least four final state particles. This situation is treated in Ref. 6. Let us instead turn to another possible quantity, $\mathbf{k} \cdot \epsilon_1 \times \epsilon_2$. The experimental asymmetry would then be

$$A = \frac{N(\mathbf{k} \cdot \epsilon_1 \times \epsilon_2 > 0) - N(\mathbf{k} \cdot \epsilon_1 \times \epsilon_2 < 0)}{N_{tot}} , \tag{3}$$

defined over the phase space distribution of events. For this particular asymmetry, we can compute that

$$\begin{aligned} A &\sim Im\ ac^* \\ &\sim |ac| \sin(\delta + \phi) \ , \end{aligned} \tag{4}$$

where $\delta \equiv \delta_s - \delta_p$. There is a problem with this particular kind of signal and it is self-evident. Even if there is no CP violation, there is a nonzero triple product correlation due to unitarity phase. This is what is meant by a fake signal. With no further information, we could not conclude from detection of a nonzero asymmetry that CP is violated - it may or may not be. Now suppose that by a quirk of fate, someone else has decided to study the antiprocess $\overline{B}(\mathbf{p}) \to \overline{V}_1(\mathbf{k}, \mathbf{s}_1) + \overline{V}_2(\mathbf{q}, \mathbf{s}_2)$ with

$$\overline{\mathcal{M}} = \overline{a}\epsilon_1 \cdot \epsilon_2 + \frac{\overline{b}}{m_1 m_2} p \cdot \epsilon_1 p \cdot \epsilon_2 - i\frac{\overline{c}}{m_1 m_2}\epsilon^{\alpha\beta\gamma\delta}\epsilon_{1\alpha}\epsilon_{2\beta}k_\gamma p_\delta \tag{5}$$

and

$$\begin{aligned} \overline{a} &= |a|e^{i(\delta_s - \phi_s)} \\ \overline{b} &= |b|e^{i(\delta_d - \phi_d)} \\ \overline{c} &= |c|e^{i(\delta_p - \phi_p)} \ . \end{aligned} \tag{6}$$

Observe that the unitarity phases are unaffected, but the CP-violating phases change sign. Now assume that by some miracle the two experiments are performed and analyzed in exactly the *same* manner. Then we can freely combine both sets of data. The sum

$$A + \overline{A} \sim |ac| \cos\delta \sin\phi \tag{7}$$

is nonzero only if CP violation is present. Complications of isospin notwithstanding, this is as clean a test for the presence of CP violation in the decay as one could get. Moreover, the difference

$$A - \overline{A} \sim |ac| \sin\delta \cos\phi \tag{8}$$

probes the unitarity phases in like manner.

The only problem with the above thought experiment is that it could never happen in practice. Small differences will always creep into experiments which are performed at different times and places. The tiny CP violating effect being sought would almost certainly be lost amid much larger systematic effects. However it is not necessary to brood over this point. Instead, use a collider! The point is that the initial state is an eigenstate of CP. Consider for example a proton-antiproton momentum state in the centre-of-mass,

$$\begin{aligned} \mathcal{P}C|P(\mathbf{k})\overline{P}(-\mathbf{k})> &= e^{i\eta_c}\mathcal{P}|\overline{P}(\mathbf{k})P(-\mathbf{k})> \\ &= -e^{i\eta_c}\mathcal{P}|\overline{P}(-\mathbf{k})P(\mathbf{k})> \\ &= e^{i\eta_c}|P(\mathbf{k})\overline{P}(-\mathbf{k})> \ . \end{aligned} \tag{9}$$

Thus any triple product correlation in the final state must be induced by a CP impurity somewhere in the scattering/production process. There are no fake signals, and all observables involving final state particles and antiparticles have a common systematics and background. The overall situation is as good as it could be for the undertaking of a search for a CP violating asymmetry.

One can search for CP violations in collider experiments at either low energies or high energies. An example of the former involves production and subsequent decay of hyperon-antihyperon pairs at a facility such as LEAR.[7],[8] The most obvious such reaction is $P + \overline{P} \to \Lambda + \overline{\Lambda}$. Any observation of CP violation would arise from the weak decay $\Lambda \to P + \pi^-$. We recall that the amplitude for $\Lambda$ weak decay can be expressed as

$$\mathcal{M} = S + P\sigma \cdot \hat{\mathbf{p}} \; , \tag{10}$$

where $\hat{\mathbf{p}}$ is the decay proton momentum. The lambda is assumed to be characterized by a polarization vector $\mathbf{s}$. We learn about the amplitudes $S$, $P$ by performing different kinds of measurements. There is the total transition rate

$$\Gamma \sim |S|^2 + |P|^2 \; , \tag{11a}$$

the proton angular distribution

$$W(\theta) = \frac{1}{4\pi}(1 + \alpha \mathbf{s} \cdot \hat{\mathbf{p}}) \; , \tag{11b}$$

and the proton polarization

$$< \vec{\sigma} >_{pr} = \frac{(\alpha + \mathbf{s} \cdot \hat{\mathbf{p}})\hat{\mathbf{p}} + \beta \; \mathbf{s} \times \hat{\mathbf{p}} + \gamma \; \hat{\mathbf{p}} \times (\mathbf{s} \times \hat{\mathbf{p}})}{1 + \alpha \mathbf{s} \cdot \hat{\mathbf{p}}} \; , \tag{11c}$$

where

$$\begin{aligned} \alpha &= \frac{2 \; Re(S^* P)}{|S|^2 + |P|^2} \\ \beta &= \frac{2 \; Im(S^* P)}{|S|^2 + |P|^2} \\ \gamma &= \frac{|S|^2 - |P|^2}{|S|^2 + |P|^2} \; . \end{aligned} \tag{12}$$

Analogous quantities $\overline{\Gamma}$, $\overline{\alpha}$, $\overline{\beta}$, $\overline{\gamma}$ describe the decay of an antilambda. Since under the operation of $\mathcal{CP}$, we have $\alpha \leftrightarrow -\overline{\alpha}$, $\beta \leftrightarrow -\overline{\beta}$, the parameters

$$\begin{aligned} B &= \frac{\beta+\overline{\beta}}{\beta-\overline{\beta}} \\ C &= \frac{\alpha+\overline{\alpha}}{\alpha-\overline{\alpha}} \\ \Delta &= \frac{\Gamma-\overline{\Gamma}}{\Gamma+\overline{\Gamma}} \end{aligned} \tag{13}$$

can serve as signals for CP violation. It is clear that of the three, $B$ is the most difficult to measure, $C$ is less difficult, and $\Delta$ is the easiest of all. Each of these can be cast in terms of unitarity phases $\delta_I^\ell$ and CP-violating phases $\phi_I^\ell$, where $I$ is the isospin and $\ell$ is the orbital angular momentum. Calculation reveals that

$$\begin{aligned} B &= \tan\left(\phi_{1/2}^s - \phi_{1/2}^p\right) \\ C &= \tan\left(\phi_{1/2}^s - \phi_{1/2}^p\right)\tan\left(\delta_{1/2}^p - \delta_{1/2}^s\right) \\ &\sim B/10 \\ \Delta &= \frac{2S_{3/2}}{S_{1/2}}\tan\left(\phi_{3/2}^s - \phi_{1/2}^s\right)\tan\left(\delta_{3/2}^s - \delta_{1/2}^s\right) \\ &\sim B/100\ , \end{aligned} \tag{14}$$

where the numerical factors are respectively associated with the small size of the strong interaction phase shifts $\delta_I^\ell \sim 0.1$ at low energy and with the $\Delta I = 1/2$ rule suppression $S_{3/2}/S_{1/2} \sim 0.05$. Thus we are faced with the ironical situation that the difficulty of measurement goes inversely with the size of the effect!

Consider the reaction

$$\begin{aligned} P(\mathbf{k}) + \overline{P} &\to \Lambda(\mathbf{q}) + \overline{\Lambda} \\ &\qquad \to P(\mathbf{p}) + \pi^- + \overline{P}(\overline{\mathbf{p}}) + \pi^+\ , \end{aligned} \tag{15}$$

for which it is possible to define the correlation

$$\begin{aligned} \tilde{A} &= \mathbf{k}\times\mathbf{q}\cdot(\mathbf{p}+\overline{\mathbf{p}}) \\ &= \frac{N_{P+\overline{P}}(up) - N_{P+\overline{P}}(down)}{N_{P+\overline{P}}(up) + N_{P+\overline{P}}(down)} \\ &= \frac{|\mathbf{s}|}{2}(\alpha+\overline{\alpha})_{\Lambda\to P+\pi}\ , \end{aligned} \tag{16}$$

where the designations *up*, *down* refer to the production plane defined by the vector $\mathbf{k}\times\mathbf{q}$. Theoretical estimates of this are clouded with the usual uncertainties related to quark wavefunctions, but the estimate

$$(\alpha+\overline{\alpha})_{\Lambda\to P+\pi}\simeq -1\times 10^{-4} \tag{17}$$

indicates that the signal will be quite small.

A potentially larger CP-violating effect is associated with the chain of reactions

$$\begin{aligned} &P+\overline{P}\to \Xi^{-}+\overline{\Xi}^{+}\,,\\ &\Xi^{-}+\overline{\Xi}^{+}\to \Lambda^{-}(\mathbf{q})+\pi^{-}+\overline{\Lambda}(\overline{\mathbf{q}})+\pi^{+}\\ &\qquad\to P(\mathbf{p})+\pi^{-}+\pi^{-}+\overline{P}(\overline{\mathbf{p}})+\pi^{+}+\pi^{+}\,. \end{aligned} \tag{18}$$

Denoting the $\Xi$ polarization by the vector $P_{\Xi}\hat{\mathbf{n}}$, we define the triple product correlation

$$\begin{aligned} \tilde{B}&=\hat{\mathbf{n}}\cdot(\mathbf{p}\times\mathbf{q}-\overline{\mathbf{p}}\times\overline{\mathbf{q}})\\ &=\frac{N_P(up)-N_{\overline{P}}(up)-N_P(down)+N_{\overline{P}}(down)}{N_{tot}}\\ &\sim\frac{\pi}{8}P_{\Xi}\alpha_{\Lambda}(\beta+\overline{\beta})_{\Xi}\,. \end{aligned} \tag{19}$$

This appears to be the simplest experimental configuration which involves the theoretically favorable CP violating hyperon parameter $\beta$. The experiment itself is not so simple.

Now we move to the other limit in collider experiments, that of high energy. It is a novel but attractive idea to contemplate employing *jet* degrees of freedom to investigate the presence of CP violations.[9] As the maximum available collider energy has been increased in recent years, the contribution of jet events to the total cross section has become dramatically larger. Already, multi-jet events are common. By a 'jet' we mean a substantial flow of energy and momentum which can be interpreted physically in terms of hadronic constituents like quarks and gluons. Of course, each experiment must establish its own set of criteria for deciding which events contain single-jet or multi-jet excitations. At any rate, such measurements occur within the context of inclusive or semi-inclusive reactions. To keep the discussion within reasonable bounds, we shall focus on triple product correlations which can be constructed *entirely* in terms of jet variables. For the first such example, suppose that the only information available for a set of events containing three or more jets is the individual jet energy and momentum. One can choose event-by-event the jets with the three largest energies, denoted as $E_1 > E_2 > E_3$. Then given the triple product $K_1 = \mathbf{p}_1 \cdot \mathbf{p}_2 \times \mathbf{p}_3$, one can in principle measure the asymmetry

$$A = \frac{N(K_1 > 0) - N(K_1 < 0)}{N_{tot}} . \tag{20}$$

Is this really a signal for CP violations? If the experiment is carried out so that the sum over particles and antiparticles is $\mathcal{C}$-blind then the parity operation reverses all momenta, and the above asymmetry is indeed an appropriate an experimental signature for CP violation. Another example is provided by assuming that associated with the initial state is a spin variable $\mathbf{s}$ which is perpendicular to the beam momentum $\mathbf{p}$. A CP-odd triple product then exists for the set of *one*-jet events. Letting the chosen jet (e.g., the one with maximal energy per event) have momentum $\mathbf{p}'$, one can define the triple product $K_2 = \mathbf{p}' \cdot \mathbf{s} \times \mathbf{p}$ and construct an appropriate asymmetry observable as above.

If the set of observables is broadened to include both jet and final state hadron variables, then there are a variety of additional CP-odd triple products which can be constructed. However rather than continue this discussion of jet variables, we turn instead to yet another application, in which the collider is tuned to the mass of the $Z^0$ vector boson and CP violation is sought in the pattern of decay products. If no polarizations of the final states are observed, then one must consider at least three-body modes in order to encounter CP-odd correlations. Unfortunately, studies of the Standard Model reveal in this case that triple product correlations are predicted to be quite tiny. However, one should not give up, because attempts at such measurements can put meaningful limits on new physics. But how is one to proceed without knowing what the new physics is like? The answer is to construct an effective Lagrangian which has the desired transformation properties. As an example, for the Yukawa coupling of the $Z^0$ to a fermion-antifermion pair $\overline{f}f$, an effective Lagrangian which is CP-odd, has dimension five (i.e., contains as few derivatives as possible), and which is $SU(2)_L \times U(1) \times SU(3)_c$ invariant is given by

$$\mathcal{L}_{eff} = \frac{-i}{2} g \, \overline{\psi} \sigma^{\mu\nu} \gamma_5 \psi (\partial_\mu Z_\nu - \partial_\nu Z_\mu) . \tag{21}$$

This effective Lagrangian represents the effect of CP violations induced by new physics defined by some unknown but presumably large energy scale $\Lambda$. Because we have no idea of the form of this new physics, the coupling constant $g$ is not prescribed. However because the effective Lagrangian has dimension five, the coupling constant has units of inverse energy, and we anticipate $g \sim \Lambda^{-1}$. Thus any detection of CP violations in situations where the Standard Model is predicted to have little or no effect can provide a valuable clue about interactions which are ordinarily inaccessible to direct accelerator study. For the interaction of Eq. (21), a more sensitive, although less model-specific, probe than asymmetry measurements is the shift of the $Z^0$ width,[10]

$$\Delta\Gamma(Z \to f\bar{f}) \simeq \frac{\eta}{6}\left(\frac{gM_z \sin 2\theta_w}{e}\right)^2 , \tag{22}$$

where $\eta$ is three (one) if $f$ is a quark (lepton), and $\Delta\Gamma$ is in GeV. Thus a measurement of the $Z^0$ width to an accuracy of 50 MeV. would yield interesting limits on an energy scale like $\Lambda$.

Another example relating to the issue of new physics comes from meson semileptonic decay.[6] The lowest order semileptonic transition at the quark level is given by a $W$-exchange tree-level amplitude. Triple product signals arise from the interference of two or more amplitudes having a relative CP-violating phase. The question we wish to pose is whether radiative corrections generated entirely within the Standard Model can induce such a CP-violating phase. The most obvious corrections are a quark-antiquark vacuum polarization contribution to the $W$ propagator and a correction to the $W$-quark vertex from $Z^0$ and photon loops. It turns out that neither interferes with the lowest order amplitude to give rise to a CP-violating effect. The reason is that for both these processes each new vertex (which generally *does* contribute a CP-violating phase) is accompanied by the conjugate vertex which cancels the phase. One is left with the overall phase of the tree-level amplitude. Consequently there is no CP violating phase in the squared sum of amplitudes. However not all radiative corrections share this property. There is a process whose phase differs from the tree amplitude, and thus will generate a CP violating signal.[6] It consists of a penguin-amplitude transition which converts an incoming quark to some other quark of the same electric charge, which then creates a lepton-neutrino pair via $W$-emission, and thereby is itself converted to a final state quark. We can actually estimate the magnitude of this effect relative to that of the CP invariant amplitude, and thereby obtain an estimate of expected correlation effect. We imagine this process as a long-range amplitude in which an incoming meson $P_1$ undergoes a penguin-induced transition to an intermediate meson $P'$ which in turn emits the lepton-neutrino pair to convert to a final-state meson $P_2$. This CP-violating (cpv) amplitude is

$$\mathcal{M}_{cpv} =< P_2\ \ell\ \overline{\nu}_\ell | H_w | P' > (m'^2 - m_1{}^2)^{-1} Im < P' | H_w | P_1 > , \tag{23}$$

and the ratio to the CP-conserving (cp) amplitude is

$$\mathcal{M}_{cpv} / \mathcal{M}_{cp} = R\ (m'^2 - m_1{}^2)^{-1} Im < P' | H_w | P_1 > , \tag{24}$$

where $R$ is some ratio of quark-mixing angles dependent on the specific process. For the transition $K \to \pi\ \ell\ \overline{\nu}_\ell$ all the quantities in Eq. (24) are known, and we find

$$\begin{aligned} \mathcal{M}_{cpv} / \mathcal{M}_{cp} &= \cot\theta_c (m_K^2 - m_\pi{}^2)^{-1} F_\pi \epsilon' Re A_0 \frac{\sqrt{2}}{\omega} \\ &\simeq \epsilon \cdot 10^{-7} . \end{aligned} \tag{24}$$

This effect is so tiny that any observed correlation will either be a sign of new physics, or depending on the magnitude, perhaps a spurious unitarity phase.

In conclusion we wish to stress several points made earlier. The key decisions to be made in detecting triple product correlations are to first choose the apppropriate correlation from among the set of possibilities, next to make sure that any observed effect is not a fake signal, and then to estimate the magnitude predicted by the Standard Model. It is also necessary to be careful that the experiment (detection efficiencies, cuts, etc.) is carried out in a CP-blind manner, or else the analysis will produce its own spurious signals. Finally, it is worthwhile to keep triple product correlations in mind as new experimental tools like colliders and meson factories become available. Even in cases where the Standard

Model predicts undetectably small signals, it should be possible to constrain new physics, especially if energy dependent CP-violating interactions are present.

## REFERENCES


1. O. Overseth and R. Roth, Phys. Rev. Lett. **19**(1967)391.
2. R. Baltrusaites and F. Calaprice, Phys. Rev. Lett. **38**(1977)464.
3. A. Hallin et al., Phys. Rev. Lett. **52**(1984)337.
4. M. Campbell et al., Phys. Rev. Lett. **47**(1981)1032.
5. G. Valencia, Phys. Rev. **39**(1989)3339.
6. E. Golowich and G. Valencia, "Triple-product Correlations in Semileptonic Decays", Phys. Rev. D (to appear).
7. J. Donoghue and S. Pakvasa, Phys. Rev. Lett. **55**(1985)162; J. Donoghue, X. He, and S. Pakvasa, Phys. Rev. **D34**(1986)833.
8. J. Donoghue, B. Holstein, and G. Valencia, Phys Lett. **178B**(1986)319.
9. J. Donoghue and G. Valencia, Phys. Rev. Lett. **58**(1987)451.
10. W. Bernreuther, U. Low, J. Ma, and O. Nachtmann, "CP Violation and $Z$ Boson Decays", Heidelberg preprint HD-THEP-88-26 (1988).

[illegible] predicts undetectably small signals it should be possible to constrain [illegible] [illegible] usually frequency [illegible] QCD [illegible] interaction [illegible]

## REFERENCES


1. [illegible] Phys. Rev. Lett. [illegible]
2. [illegible] Phys. Rev. [illegible]
3. [illegible]
4. [illegible] Rev. [illegible]
5. [illegible] Rev. [illegible]
6. [illegible]
7. [illegible]
8. [illegible] Phys. Rev. [illegible] Dorsaneo, [illegible]
[illegible]
9. [illegible]
10. [illegible]
11. [illegible]
12. [illegible]

# CP VIOLATION STUDIES AT LEAR

Adiels L.[15],Angelopoulos A.[1],Apostolakis A.[1],Aslanides E.[12],Backenstoss G.[2],
Bee C.P.[10],Bennet J.[10],Bloch P.[4],Bula Ch.[13],Burgun G.[14],Carlson P.[15],
Carvalho J.[5],Chardalas M.[16],Charalambous S.[16],Dedoussis S.[16],Dejardin M.[14],
Derre J.[14],Dodgson M.[10],Dousse J-C.[8],Ealet A.[4],Faravel L.[8],Fassnacht P.[12],
Faure J.L.[14],Felder Ch.[2],Ferreira-Marques R.[5],Fetscher W.[17],
Fidecaro M.[4],Francis D.[15],Fry J.[10],Fuglesang C.[4],Gabathuler E.[10],
Gamet R.[10],Garreta D.[14],Geralis T.[12],Gerber H.J.[17],
Gumplinger P.[17],Guyot C.[14],Harrison P.F.[10],Hayman P.J.[10],Heyes W.G.[4],
Hollander R.W [6],Johner H.U.[8],Jon-And K.[15],Jansson K.[15],Kerek A.[15],
Kern J.[8],Kettle P-R.[13],Kokkas P.[7,9],Lawry T.[3],Maley P.[10]Mall U.[2]
Marel G.[14],Mikuz M.[11],Miller J.[3],Montanet F.[12],Nakada T.[13],Onofre A.[5],
Pagels B.[2],Pavlopoulos P.[4,7],Pinto da Cunha[5],Policarpo A.[5],Postma H.[6],
Rickenbach R.[2],Roberts L.[3],Sacks L.[10],Sanders P.[10],Santoni C.[2],Sarigianis K.[1,7],
Schaller L.[8], Schopper A.[4],Szilagyi S.[15],Tauscher L.[2],Touramanis C.[7,16],
Triantis F.[9],Troester D.A.[2],Van den Putte M.[6],Van Eijk C.W.E.[6],Vlachos S.[7,16],
Witzig C[17],Zavratnik D.[11]

Athens[1] - Basel[2] - Boston[3] - CERN[4] - Coimbra[5] - Delft[6] - Democritos IPN[7] -
Fribourg[8] - Ioannina[9] - Liverpool[10] - Ljubljana[11] - Marseille[12] - PSI--Villigen[13] -
Saclay[14] - Stockholm MSI[15]- Thessaloniki[16] - Zurich ETH[17]

presented by P. Bloch,CERN

Abstract

The CPLEAR experiment will study CP violation effects in the decay of neutral kaons using low energy antiprotons as a source of tagged $K^0$ and $\overline{K}^0$ .This new approach will allow a precise measurement of $\varepsilon'/\varepsilon$, tests of the CPT symmetry and for the first time the evidence of CP violation in other decay modes of the neutral kaon such as $K^0 \rightarrow 3\pi$ or $2\gamma$. In this paper, we review the physics motivations for the CPLEAR experiment, discuss the expected results and describe shortly the experimental layout.

**Introduction**

Some time ago, it was realized that low energy antiproton interactions could become a source of tagged $K^0$ and $\bar{K}^0$ due to strangeness conservation in exclusive annihilations containing one charged and one neutral kaon. However, early experiments suffered from lack of statistics and therefore could not contribute much to the physics of CP violation.

This idea was reconsidered in 1982[1)] following the construction of an improved source of antiprotons and of the Low Energy Antiproton Ring (LEAR) at CERN. The possibility of observing directly $K_sK_l$ interference with tagged $K^0$ and $\bar{K}^0$ beams offers several advantages compared to the classical approach which uses measurements of $K_s$ and $K_l$ branching ratios:

1) The observation of an interference between the amplitudes when the $K_s$ and $K_l$ decay into a non-leptonic final state is clear evidence for CP violation. As Seghal and Wolfenstein pointed out [2)], this statement holds true even when the final state is not a state of definite CP, as is the case for $\pi^+\pi^-\pi^0$ or $\gamma\gamma$.

2) As will be shown in the following, the interference term is only first order in the CP violation parameter $\eta$, whereas the rate of CP violating decays is proportional to its modulus squared. Therefore, the measurement of the interference as a function of the proper time allows the determination of both the phase and the modulus of the CP violation parameter. For the same reason, the interference term becomes in some cases (e.g. $\pi^0\pi^0\pi^0$) accessible to experiments, whereas the CP violating branching ratios are too small.

3) Unlike $K_s$ and $K_l$, $K^0$ and $\bar{K}^0$ have identical decays . Therefore, systematic errors should be minimized in comparing $K^0$ and $\bar{K}^0$ decay rates. This is of particular interest for the measurement of $\varepsilon'/\varepsilon$.

For these reasons, a group of about 80 physicists decided to build a new detector around LEAR devoted to CP violation studies. This experiment, called CPLEAR, will start data taking at the end of the current year.

We recall the basic formalism of $K_s$ $K_l$ interferometry in section I, explain the source of tagged neutral kaons in Section II, list the expected physics parameters and measurement accuracies in section III (with a particular emphasis on $\varepsilon'/\varepsilon$ ) and finally describe the CPLEAR detector and its current status.

**I) $K_s$ $K_l$ interferometry**

The $K^0$ ( $\bar{K}^0$ ) decay rate to any final state is given to first order as:

$$R(\overset{(-)}{K}{}^0 \to f) = \left[\frac{1}{2} \pm Re(\varepsilon)\right]\Big\{ |<f\,|\,T\,|\,K_s>|^2\, e^{-\gamma_s t} + |<f\,|\,T\,|\,K_l\,|>|^2\; e^{-\gamma_l t}$$

$$\mp\; 2\,|<f\,|\,T\,|\,K_s><f\,|\,T\,|\,K_l>|\; e^{-(\gamma_s+\gamma_l)t/2}\cos(\Delta mt - \varphi_f)\Big\} \qquad (1)$$

where $\gamma_s$ and $\gamma_l$ are the decay width of the $K_s$ and $K_l$ , $\Delta m$ is their mass difference and $\varphi_f$ is the phase of the amplitude ratio

$$\eta_f = \frac{<f\,|\,T\,|\,K_l>}{<f\,|\,T\,|\,K_s>} \qquad (2)$$

As $\Delta m$ is approximately equal to $\gamma_s/2$, the interference term can be observed before the $K_s$ component has died away. The interference term changes its sign when going from $K^0$ to $\bar{K}^0$ and the measurement of the asymmetry as a function of the proper time t

$$A_f(t) = \frac{R(\bar{K}^0 \to f)(t) - R(K^0 \to f)(t)}{R(\bar{K}^0 \to f)(t) + R(K^0 \to f)(t)}$$

$$= \frac{2[Re(\varepsilon) - |\eta_f|\, e^{-(\gamma_s-\gamma_l)t}\cos(\Delta mt - \phi_f)]}{1 + |\eta_f|^2\, e^{(\gamma_s-\gamma_l)t}} \qquad (3)$$

allows a determination of both the modulus and the phase of the complex parameter $\eta_f$. Fig. 1 and 2 show this asymmetry for $\pi^+\pi^-$ and for $\pi^0\pi^0\pi^0$ final state, respectively. In the first case, the interference term shows up between 5 and 15 $K_s$ lifetimes, whereas the interference is concentrated in the first 5 lifetimes for the 3 $\pi$ final state. Note that in the latter case, the interference term produces an asymmetry of the order of $10^{-3}$ requiring a few $10^6$ events to be measured, whereas the branching ratio of the CP violating mode $K_s \to \pi^0\pi^0\pi^0$ is expected to be $O(10^{-9})$.

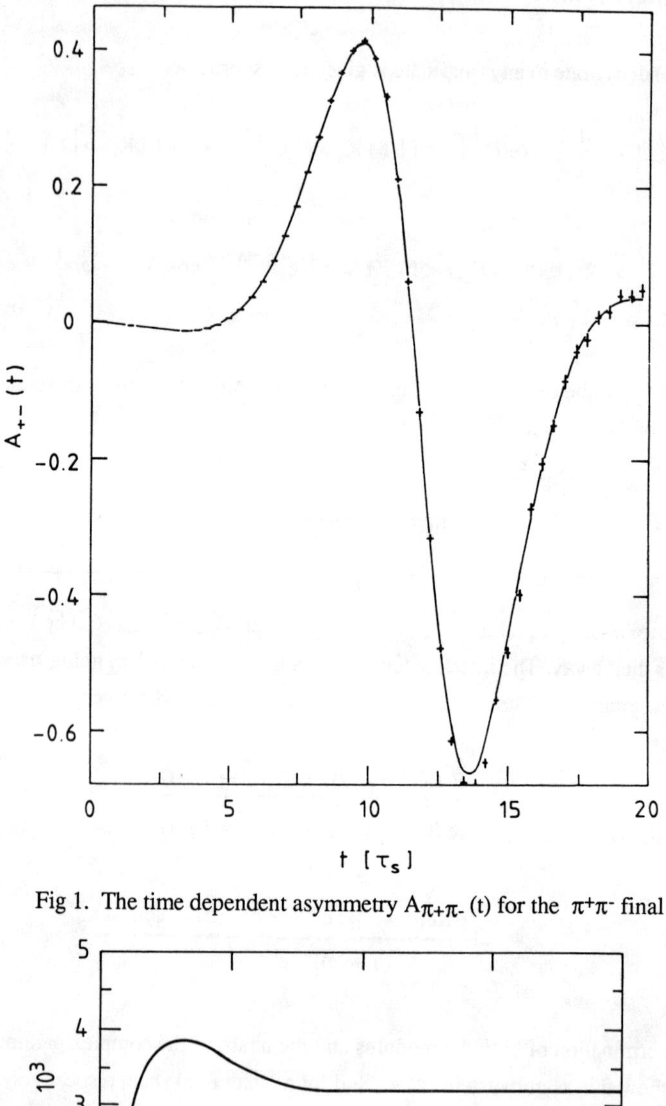


Fig 1. The time dependent asymmetry $A_{\pi^+\pi^-}$ (t) for the $\pi^+\pi^-$ final state.

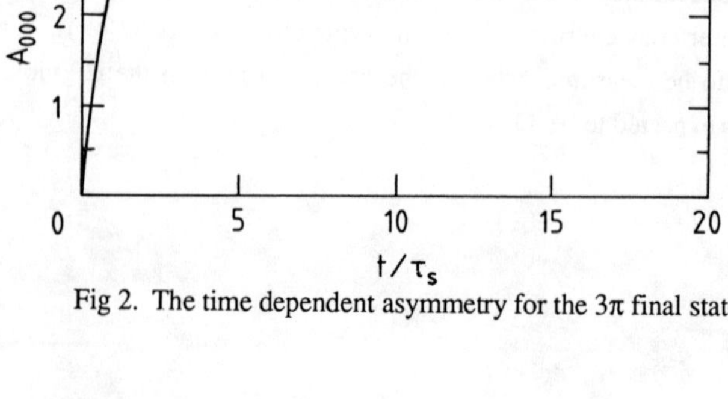


Fig 2. The time dependent asymmetry for the $3\pi$ final state

As already mentioned, this interference technique can be applied to any non leptonic final state, even if it is not a CP eigenstate. In this experiment, we hope to measure for the first time the CP violation effects in $\pi^+\pi^-\pi^0$ and $\gamma\gamma$ final states.

The determination of $\varepsilon'/\varepsilon$ could, in principle, be performed using the two independent measurements of the CP parameters $\eta_{00}$ and $\eta_{+-}$ obtained by measuring the two time dependent asymmetries $A_{00}(t)$ and $A_{+-}(t)$ (corresponding respectively to $\pi^0\pi^0$ and $\pi^+\pi^-$ final states) and by applying the well known formula:

$$\mathrm{Re}\,(\varepsilon'/\varepsilon) = \frac{1}{6}\left[1 - \frac{|\eta_{00}|^2}{|\eta_{+-}|^2}\right] \tag{4}$$

However, the measurement of the time dependent asymmetry requires a determination of the proper time of the decay, and hence of the $K^0$ path length. The finite resolution in the path length determination distorts the measured asymmetry . The necessary corrections are very different in the case of the charged final state (where the $K^0$ decay vertex is directly determined from the intersection of the two emerging charged tracks) to the case of the neutral final state (where the vertex is determined via a kinematical fit using the measurements of all the charged tracks and all the photons in the event). It is therefore not easy to control all these corrections in such a way that the systematic biases in the calculation of $\varepsilon'/\varepsilon$ are comparable with those of ongoing experiments.

For this reason, the determination of $\varepsilon'/\varepsilon$ in our experiment will instead be obtained from the integrated asymmetries:

$$I_{2\pi}\,(t_{max}) = \frac{\int_0^{t_{max}} R(\bar{K}^0 \to 2\pi)\,dt - \int_0^{t_{max}} R\,(K^0 \to 2\pi)\,dt}{\int_0^{t_{max}} R(\bar{K}^0 \to 2\pi)\,dt + \int_0^{t_{max}} R\,(K^0 \to 2\pi)\,dt} \tag{5}$$

As shown on Fig. 3, the value of $I_{2\pi}(t_{max})$ is insensitive to the upper integration limit as long as $t_{max} > 15\,\tau_s$ and $t_{max} << \tau_l$ (where $\tau_s$ and $\tau_l$ are the $K_s$ and $K_l$ life times).

The value of $\varepsilon'/\varepsilon$ can be inferred from the measurements of $I_{+-}$ and $I_{00}$, according to the equation :

$$\frac{\mathrm{Re}\,(\varepsilon')}{\mathrm{Re}\,(\varepsilon)} = \frac{1}{6}\left[1 - \frac{I_{00}}{I_{+-}}\right] \tag{6}$$

which gives a determination of $\varepsilon'/\varepsilon$ completely independent of the $K^0$ vertex resolution

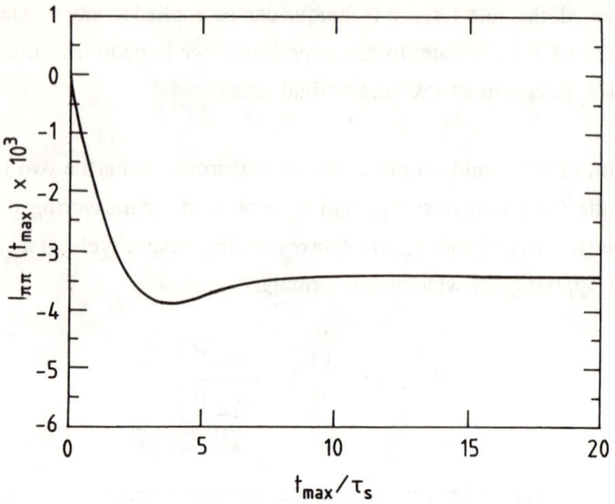


Fig 3. The integrated asymmetry $I_{2\pi}$ as function of the upper limit of integration $t_{max}$

**II) The tagged $K^0$ beam at LEAR.**

A high flux of tagged $K^0$ or $\bar{K}^0$ can be obtained at LEAR by stopping low energy antiprotons in a low density gaseous hydrogen target and selecting the 3 body-annihilations

$$\bar{p}p \rightarrow K^+ \pi^- \bar{K}^0$$

$$\bar{p}p \rightarrow K^- \pi^+ K^0 \qquad (7)$$

which each amounts to 2 $10^{-3}$ of the total annihilation rate The charge, i.e. the strangeness of the charged kaon, allows a unique identification of the strangeness of its neutral companion. Since the annihilation takes place at rest, the four-momentum of the neutral $K^0$ is known on an event by event basis.

In principle, C invariance ensures an equal branching ratio of the two reactions. However, the tagging efficiency in the detector may differ, mainly because of the different nuclear cross sections in matter for $K^+$ and $K^-$, as well as for $\pi^+$ and $\pi^-$. Therefore, an independent normalisation of the two channels will be performed.

The low density of the gaseous target minimizes the regeneration effects. Due to the small kinetic energy of the incoming antiprotons, the annihilation takes place in a limited spatial region (5 cm long, 2 cm diameter) and the study of the $K^0$ decay can be extended to zero life time.

This approach has a very attractive experimental advantage: Since $K^0$ and $\bar{K}^0$ have identical decays and are produced in the detector at the same time, many experimental effects such as solid angle, efficiencies, dead times, etc. cancel in the asymmetry. We can therefore minimize our systematic errors, in particular for the $\varepsilon'/\varepsilon$ determination. However, there are also large difficulties to overcome: one has to reject the large background of the other annihilation channels, and this rejection must be already efficient at the triggering level. Furthermore, the momentum of the produced $K^0$s and of their decay products are rather low:

$<p_t^{K^0}> = 550$ Mev/c $\quad <E_\gamma> = 160$ Mev for $K^0 \to 2\pi^0 \to 4\gamma$

making particle identification and photon detection more difficult at these energies than in the several GeV energy range.

The CPLEAR detector has been designed to accept a high (2 MHz) rate of stopped antiprotons. In June 1989, the CERN Antiproton Source reached its design value of more than $10^{12}$ stacked antiprotons per day. Taking into account the usual efficiencies in transferring, decelerating and extracting the antiprotons from LEAR, we will stop more than $10^{11}$ antiprotons per day in the CPLEAR experiment and therefore produce more than $10^8$ tagged $K^0$'s during the same time interval. In the following, we will assume a total of $10^{13}$ stopped antiprotons, which corresponds to approx. 100 days of running time.

### III) The expected measurements

#### a) The measurement of $\varepsilon'/\varepsilon$

The determination of $\varepsilon'/\varepsilon$ will be obtained following equation (6) from the ratio of the integrated asymmetries $I_{00} / I_{+-}$. The expected statistical accuracy is $\delta(\varepsilon'/\varepsilon) = 1.5\ 10^{-3}$. As already mentioned, most of the systematic errors will cancel in the ratio of the two asymmetries. There is no need to control the background due to $K_l$ decays with high precision, since only 3% of the $K_l$ decay in the detector volume.

A large effort has been invested in the study of the systematic errors which may arise, for example, in the background subtraction or may be due to slight variation of the reconstruction efficiency as function of the proper time . We have also investigated the effect of possible differences between the tagged $K^0$ and $\bar{K}^0$ momentum spectrum. In all the cases, the systematic error was found to be substantially smaller than the statistical error.

We therefore think that the accuracy of our experiment will be similar to that of the present high energy experiments, but the systematic effects will be completely different, allowing a valuable contribution to the determination of $\varepsilon'/\varepsilon$.

**b)CP violation in the 3$\pi$ decay**

The observation of the interference in the $\pi^+\pi^-\pi^0$ channel is a clear indication of CP violation. The expected accuracy on the parameter $\eta_{+-0}$ is $\delta(\eta_{+-0}) = 6\ 10^{-4}$. Thus ,if $\eta_{+-0}$ is of order $\varepsilon$, the experiment should show for the first time evidence of CP violation in the 3$\pi$ channel.

The detection efficiency of the $3\pi^0$ channel is lower than for the $\pi^+\pi^-\pi^0$ channel, mainly because the CPLEAR photon detector covers only 80% of the 4$\pi$ solid angle, but also because the average energy of the decay photons is very low (100 MeV). Hence, it will probably be impossible to detect the CP violation in this channel. However we expect to improve the actual limit $|\eta_{000}|^2<0.1$ [3] by at least a factor 50. The importance of improving this limit has been stressed many times, in particular at this conference[4].

**c)Measurement of the phase difference $\varphi_{00} - \varphi_{+-}$**

It can be shown [5] that, if one assumes CPT invariance, the expected phase $\varphi_{\varepsilon'}$ of $\varepsilon'$ must be in the range of 40 ° to 65 ° [6,7]. From the measured values of $\varphi_{+-} = 44.6° \pm 1.2°$ [3] and the upper limit on the value of $\varepsilon'$, we can conclude that the value of $\varphi_{00}$ must be equal to the value of $\varphi_{+-}$ within 1 °, or else $\varphi_{\varepsilon'}$ can not even be in the range from 0 to 90 °. Data published more than 10 years ago[8] seem to be inconsistent with CPT invariance, but only at the 2 standard deviation level. At this conference, the two high energy experiments presented preliminary results on the phase difference, which is now compatible with 0°:

$\varphi_{00} - \varphi_{+-} = 0.3° \pm 2.8° \pm 1°$ [9]

$\varphi_{00} - \varphi_{+-} = 2.8° \pm 2°$ [10]

In the CPLEAR experiment, the phase difference can be measured through the time dependent asymmetries $A_{00}(t)$ and $A_{+-}(t)$. The expected statistical error on $\varphi_{00} - \varphi_{+-}$ is of the order of 0.2 °. Many uncertainties (in the $K_S$-$K_l$ mass difference and in the absolute $K_S$ life time, for example) cancel in the difference. However systematic errors can arise from ignorance of the shape, sigma and mean value of the vertex resolution, in particular for the case of the neutral decay channel. This resolution will be monitored for instance with events where one of the $\pi^0$ undergoes a Dalitz decay and we expect to keep the systematic error below 1 °.

**d) CP violation in 2$\gamma$ decay**

This decay channel is of particular interest since the CP violation in the decay amplitude is not suppressed by the $\Delta I=1/2$ rule and could therefore contribute to as much as 10% of the CP violation effects.

Following the recent measurements of the $K_s$ and $K_l$ branching ratios into 2 $\gamma$ [11], we expect ≈9 $10^4$ $K^0 \to 2\gamma$ decays to be detected in our experiment. This will be enough to detect mass matrix CP violation, but unfortunately two orders of magnitude too small to show evidence of direct CP violation.

**e)T violation**

If we assume that the $\Delta S=\Delta Q$ rule is exactly true, then $K^0$ and $\bar{K}^0$ can be unambiguously identified by their semileptonic decay:

$$K^0 \to \pi^- e^+ \nu \quad \text{and} \quad \bar{K}^0 \to \pi^+ e^- \bar{\nu}$$

Any asymmetry such as :

$$A_T(t) = \frac{R(\bar{K}^0 \to e^+) - R(K^0 \to e^-)}{R(\bar{K}^0 \to e^+) + R(K^0 \to e^-)} \qquad (8)$$

exhibits a difference in the probabilities $P(K^0 \to \bar{K}^0)$ and $P(\bar{K}^0 \to K^0)$ and therefore a violation of time reversal invariance. If CPT holds, T violation must be observed and $A_T$ is expected to be ≈4Re($\varepsilon$). We expect to measure $A_T$ with a statistical accuracy of $\delta(A_T)/A_T$ =6%. This measurement requires e/$\pi$ separation on an event by event basis, which will be achieved with the photon detector and by using time of flight measurements.

In conclusion, the CPLEAR experiment will not only be an experiment dedicated to the measurement of $\varepsilon'/\varepsilon$ but it will also allow a more general survey of CP violation in the neutral kaon system.

**IV) The CPLEAR detector.**

A detailed description of the CPLEAR detector can be found elsewhere [12]. We will recall here the main features of the detector, shown on Fig 4 .

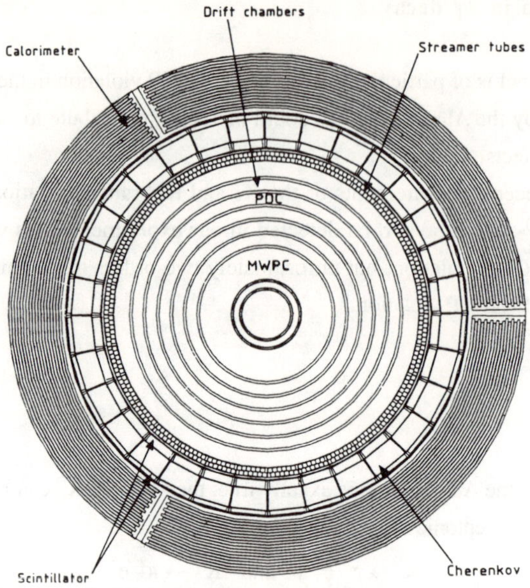


Fig 4 The layout of the CPLEAR detector

The design of the detector has to reflect the requirements imposed by the $K^0$ tagging and the detection of the decay products. Since the annihilation takes place at rest, the detector must cover the largest possible solid angle. A magnetic field is needed to measure the charge and momenta of the emitted particles. For these reasons, the detectors are assembled inside a solenoidal magnet previously built for the DM2 detector installed on the $e^+e^-$ DCI ring at Orsay. It provides a uniform 0.44 T field in a volume of 3.6 m length and 2m diameter.

A central tracking detector composed of 2 Multi-Wire Proportional Chambers and 6 layers of drift chambers allows the detection of charged tracks in the radius interval from 9 to 50 cm (the latter corresponding to 17 $K_S$ life times at our average kaon energy) . The MWPC's (with 1mm wire spacing) and the drift chambers (with only 5 mm drift distance and doublets of wires to resolve the left-right ambiguity) provide within a few hundred nanoseconds the tracking information which is used in our trigger system. Particular care has been taken to design thin chambers (they amount in total to $8.10^{-3}$ radiation lengths) so as to optimize the momentum resolution and to avoid large regeneration effects. All the chambers are equipped with cathode strips running at various angles between 25 and 45 degrees to measure offline the track coordinates along the beam.

The central detector is completed with two layers of large cross-section streamer tubes which provide online within 500 ns information on the coordinate of the tracks along the beam by measuring the time difference between signals from both ends of the tubes. The achieved resolution has a standard deviation of less than 2 cm.

One essential feature of the CPLEAR detector is the separation of kaons from pions so as to identify the charged kaon in reaction (7) and reject quickly the background due to multi-pion annihilations. This is achieved with the Particle Identification Detector, which consists of 32 thick threshold Cerenkov counters (filled with liquid freon of refractive index n=1.26) placed between two layers of scintillators. The scintillators are used to make sure that the kaon candidate has crossed the Cerenkov counter. In addition they also provide information about the multiplicity of the charged particles in the event . Measurements with a test beam have shown that the efficiency of the Cerenkov counter is greater than 99.9% for particles having a velocity $\beta$=0.9. The first layer of scintillators also provides time of flight (TOF) information. We expect to achieve a resolution of 200 ps with the TOF, giving another factor 5 improvement in the K/$\pi$ separation.

The photons are detected in an electromagnetic calorimeter located inside the coil. Monte Carlo simulations have shown that the spatial resolution of the photon detector is more important than its energy resolution for the vertex reconstruction of the $K^0$ decays into neutral states Thus we have built a gas sampling calorimeter which consists of 18 layers of lead interspaced with small (4x4.5 mm2) streamer tubes, specially developed for this experiment. The coordinates along the beam line are obtained with a system of strips , positioned above and below each streamer tube layer and running at $\pm$ 30° from the wire axis. All the strips and wires (a total of ≈65000 channels) are read digitally. A prototype was exposed to a tagged photon beam at PSI, showing that the energy resolution is on average $\sigma(E) / E = 13\% / \sqrt{E}$ and that the spatial resolution of the photon impact point is better than 5 mm.

The trigger in the CPLEAR experiment is a crucial item: it is based on several hardwired processors. A fast decision requires the existence of a kaon candidate, signed by the absence of signal in a Cerenkov counter when the two corresponding scintillators have fired. In order to reject slow pions which may fake a kaon, one further checks within ≈300 ns, using the chamber wires, that the kaon candidate has a transverse momentum in excess of 250 Mev/c. If this condition is satisfied, a processor reconstructs the tracks, computes their momenta and calculates the missing mass to ensure the presence of a "missing" $K^0$. The whole process will take less than 2 $\mu$s . Further trigger stages will then refine the K/$\pi$ separation using the amplitude of the Cerenkov signals, the TOF information and finally compute the photon mutiplicity in the calorimeter.

The acceptance of the selected channel is expected to be ≈15%, including solid angle, trigger efficiency and deadtime.

The CPLEAR detector is now completed and closed. Most of the electronics is in place and the missing items will gradually be installed this fall. At present, we have a first technical run to set up our equipment and data acquisition.

During the remaining part of this year we intend to improve our understanding of the detector and start full data taking early next year.

## Conclusion

The CPLEAR experiment, using tagged $K^0$ and $\overline{K}^0$ beam allows an independent check of $\varepsilon'/\varepsilon$ with systematic errors which are completely different from the higher energy $K_S K_l$ experiments. Moreover it will permit a complete survey of CP violation in other decay modes of the neutral kaons such as the $3\pi$, semileptonic and $2\,\gamma$ decay channels.

## References

1) Gabathuler E. and Pavlopoulos P., Strong and weak CP violation at LEAR, Proceedings of the Workshop on physics at LEAR with low energy cooled antiprotons ed. by U. Gastaldi and R. Klapish, Plenum publisher, New York, 747 (1982)

2) Sehgal L.M. and Wolfenstein L., Phys. Rev. 162,1362 (1967)

3) Particle Data Group, Review of Particle Properties, Phys. Lett. 204B (1988)

4) Robert Sachs, this conference.

5) Barmin V.V. et al.,Nucl. Phys. B247, 293 (1984)

6) Devlin Y.J. and Dickey J.O., Rev. Mod. Phys.51, 237, (1979)

7) Biswas N.N. et al., Phys. Rev. Lett. 47, 1378 (1981)

8) Christenson J.H. et al., Phys. Rev. Lett. 43, 1209 (1979)

9) Preliminary results of NA31 experiment, presented by V. Gibson at this conference.

10) Preliminary results of E731 experiment, presented by P.Debu at this conference

11) Burkhardt H. et al., Phys Lett 199B, 139 (1987)

12) Adiels L. et al., Proposal CERN PSCC/85-6 P82 (1985); PSCC/85-30 P82 (1985); PSCC/86-34 M263 (1986); PSCC/87-14 M272 (1987)

# TESTING CP INVARIANCE IN HYPERON DECAYS.

G. Bassompierre
LAPP, Annecy-le-Vieux, France.

**Abstract**

In order to study CP violation outside the $K^0$ system, we investigate the experimental possibilities of measuring signals in hyperon decays. Asymmetries at the level of $10^{-3}-10^{-4}$ can be measured in the reactions $\bar{p}p\to\bar{\Lambda}\Lambda$ and $\bar{p}p\to\bar{\Xi}\Xi$. Such a study can be done at FERMILAB $\bar{p}$ accumulator, or at SUPERLEAR with an internal gas-jet target. Some considerations on the errors and on the design criteria for the detector are given.

## 1. INTRODUCTION

Twenty five years after 1964, when Christenson et al. (ref.1) discovered evidence that CP is violated in $K^0$ decay, no system other than $K^0$ has exhibited CP or T violation. The origin of the observed violation of CP is still not known, but it is one of the fundamental questions in particle physics. It is therefore crucial to search for CP violation in systems other than $K^0$ meson decays, and hyperons seem to be one of the interesting candidates (ref.2,3,4).

## 2. PHYSICS

The CP violating $\Delta S = 1$ processes could be observed in the non-leptonic decays of hyperons. The CP violation arises from an interference between different amplitudes (S or P wave or/and $I = 1/2$, $I = 3/2$ final states).

For $\Lambda$ decay we have :

$$S = S_{1/2}\, e^{i(\delta^S{}_{1/2} + \Phi^S{}_{1/2})} + S_{3/2}\, e^{i(\delta^S{}_{3/2} + \Phi^S{}_{3/2})}$$

$$P = P_{1/2}\, e^{i(\delta^P{}_{1/2} + \Phi^P{}_{1/2})} + P_{3/2}\, e^{i(\delta^P{}_{3/2} + \Phi^P{}_{3/2})}$$

but for $\bar{\Lambda}$ decay :

$$\bar{S} = - S_{1/2}\, e^{i(\delta^S{}_{1/2} - \Phi^S{}_{1/2})} - S_{3/2}\, e^{i(\delta^S{}_{3/2} - \Phi^S{}_{3/2})}$$

$$\bar{P} = P_{1/2}\, e^{i(\delta^P{}_{1/2} - \Phi^P{}_{1/2})} + P_{3/2}\, e^{i(\delta^P{}_{3/2} - \Phi^P{}_{3/2})}$$

where S(P) is the S wave (P wave) parity-violating (-conserving) amplitude, and where $\Phi'_S$ are the CP violating phases and $\delta'_S$ the $\pi$-nucleon phase-shifts. The simple measurement of CP violation is the partial rate difference $\Delta$ :

$$\Delta = \frac{\Gamma(\Lambda \to p\pi^-) - \Gamma(\bar{\Lambda} \to \bar{p}\pi^+)}{\Gamma(\Lambda \to p\pi^-) + \Gamma(\bar{\Lambda} \to \bar{p}\pi^+)} = -2\, \frac{S_{3/2}}{S_{1/2}}\, \sin(\delta^S{}_{3/2} - \delta^S{}_{1/2}) \,.\, \sin(\Phi^S{}_{3/2} - \Phi^S{}_{1/2})$$

in the limit of S wave dominance, and to leading order. Using experimental values, $\Delta \approx 10^{-2}\sin(\Phi^S_{3/2}-\Phi^S_{1/2})$. This rate difference is thus suppressed both by the $\Delta I = 1/2$ rule and the smallness of the strong phase-shifts $\delta$. An alternative is to compare the decay asymmetry parameter $\alpha$ for the two decays. Neglecting the $I = 3/2$ final state, the CP violating effect is :

$$A = \frac{\alpha + \bar{\alpha}}{\alpha - \bar{\alpha}} = \sin(\Phi^S_{1/2}-\Phi^P_{1/2})\sin(\delta^P_{1/2}-\delta^S_{1/2}) \approx 0.1\sin(\Phi^S_{1/2}-\Phi^P_{1/2})$$

A still more difficult parameter to measure is $\beta$, which is related to the polarization of the final baryon. The CP violating effect can be given by

$$B' = \frac{\beta + \bar{\beta}}{\alpha - \bar{\alpha}} = \sin(\Phi^S_{1/2}-\Phi^P_{1/2}) \quad \text{or} \quad B = \frac{\beta + \bar{\beta}}{\beta - \bar{\beta}} = \frac{B'}{\tan(\delta^P_{1/2}-\delta^S_{1/2})}$$

where the $I = 3/2$ final state is still neglected. If we use $\chi$ to denote the strength of $\Delta S = 1$ CP violation, we find :

$$\begin{aligned} B' &= \chi \\ A &= \chi\sin(\delta^P_{1/2}-\delta^S_{1/2}) && \approx 1/10\,\chi \\ \Delta &= \chi\sin(\delta^S_{3/2}-\delta^S_{1/2})\, S_{3/2}/S_{1/2} && \approx 1/100\,\chi \end{aligned}$$

Theoretical estimates of these quantities were given in (ref.3,4,5). As an illustration, we give in table 1 the predictions of Donoghue et al. (ref.4) for asymmetries in the $\Lambda \to p\pi^-$ decay.

TABLE 1

| ASYM. $\chi$ | SUPERWEAK $\chi=0$ | STAND.MODEL $\chi\approx 20\varepsilon'$ | WEINBERG–HIGGS $\chi\approx\varepsilon$ | LEFT–RIGHT SYM. $\chi\approx\varepsilon/10$ |
|---|---|---|---|---|
| $\Delta$ | 0 | $-5.4\ 10^{-7}$ | $-7.8\ 10^{-6}$ | 0 |
| A | 0 | $-0.5\ 10^{-4}$ | $-2.5\ 10^{-5}$ | $-1.1\ 10^{-5}$ |
| B | 0 | $3.0\ 10^{-3}$ | $1.6\ 10^{-3}$ | $7.0\ 10^{-4}$ |
| B' | 0 | $-3.7\ 10^{-4}$ | $2.0\ 10^{-4}$ | $8.0\ 10^{-5}$ |

A new set of values for $A_\Lambda$ was given by J.F. Donoghue in the framework of the Standard Model. (ref.6)

$$A_\Lambda = \begin{array}{ll} -(0.3-1.0)\ 10^{-4} & \text{Bag model} \\ -(1.2-4.0)\ 10^{-4} & \text{Non-relativistic model} \end{array}$$

The present experimental result is $A_\Lambda = -0.023 \pm 0.057$. (ref.7)

### 3. ASYMMETRIES

The CP violating asymmetries correspond to triple products which are odd under CP (ref.8). It is to be noticed that in the concerned reactions, the final state interactions give no asymmetries. For the reaction $\bar{p}p \rightarrow \bar{\Lambda}\Lambda$ we can consider the asymmetry $\tilde{A}_\Lambda$ which just counts in the charged mode of the $\Lambda$ decay, the number of protons and antiprotons above and below the production plane.

$$\tilde{A}_\Lambda = \frac{Np(up) + N\bar{p}(up) - Np(down) - N\bar{p}(down)}{Nevts} = 1/2\ P_\Lambda\ (\alpha + \bar{\alpha})_\Lambda$$

where $P_\Lambda$ is the $\Lambda$ ($\bar{\Lambda}$) polarization. It is important to note that background processes cannot generate this asymmetry. The reaction $\bar{p}p \rightarrow \bar{\Xi}\Xi \rightarrow \bar{\Lambda}\pi\Lambda\pi$ is particularly interesting because $\Lambda(\bar{\Lambda})$ analyses its own spin, and a counting asymmetry corresponding to the spin observable $\beta$ could be derived :

$$\tilde{B}_\Xi = \frac{Np(up) - N\bar{p}(up) - Np(down) + N\bar{p}(down)}{Nevts} \approx \pi/8\ P_\Xi\ \alpha_\Lambda\ (\beta + \bar{\beta})_\Xi$$

where up(down) refers to the sign of $\hat{y} \cdot \vec{P}_p x \vec{P}_\Lambda$ in the $\Xi$ frame. ($\hat{y}$ : normal to the production plane, $\vec{P}_p$ : final proton momentum, $\vec{P}_\Lambda$ : $\Lambda$ momentum). However, this last relation was derived under the assumption that $(\alpha+\bar{\alpha})_\Lambda$ is an order of magnitude smaller than $(\beta+\bar{\beta})_\Xi$. In principle, we can also obtain $(\alpha + \bar{\alpha})_\Lambda$ from the cascade decay of the $\Xi$, but it is better to measure this asymmetry from the direct reaction $\bar{p}p \rightarrow \bar{\Lambda}\Lambda$. The measurements of the $\Sigma$ decay asymmetries $(\alpha+\bar{\alpha})_\Sigma$, even if they are at the level of $10^{-3}$, are difficult, since the final baryon (antibaryon) is a neutron (antineutron). In this case, it would be better to make separate measurements of $\alpha_\Sigma$ and $\bar{\alpha}_\Sigma$ by methods like those of weighted sums (ref.9). In the next sections, we will consider only the $\bar{\Lambda}\Lambda$ and $\bar{\Xi}\Xi$ channels.[1]

## 4. EXPERIMENTAL APPARATUS

As first noticed by the PS185 coll. (ref.10), the use of a thin internal target in LEAR is the best way to study CP violating asymmetries in the $\bar{\Lambda}\Lambda$ system. This is still true at higher energies, and also for the other channels. The working energy which should be determined by an optimization process, is such that the baryons (antibaryons) are located in the forward cone ($\lesssim 45°$).

### 4.1 H2 jet target :

The use of an internal target will provide very high luminosity, which is needed in such experiments, and $L = 10^{32}$ cm$^{-2}$s$^{-1}$ is expected in the SUPERLEAR project (ref.11). Systematic errors related to the secondary interactions ($Y, \bar{Y}$ and decay products) in the target will be reduced. The jet-target will allow a good vertex definition, which is very useful in the geometrical and kinematical reconstruction of the event. The target could be similar to the gas-jet target used by the R-704 experiment at the CERN ISR or to that of the JETSET experiment (ref.12) at LEAR, which is shown in fig. 1. With a density of $8.10^{13}$ atoms/cm$^2$ and with 4 $10^{10}$ $\bar{p}$ in the LEAR ring, the luminosity will be of the order of $10^{31}$ cm$^{-2}$s$^{-1}$.

---

[1] More details about the experimental study of the $\bar{\Lambda}\Lambda$ channel can be found in G. Bassompierre, Nuovo Cim. 101 (1989) 309.

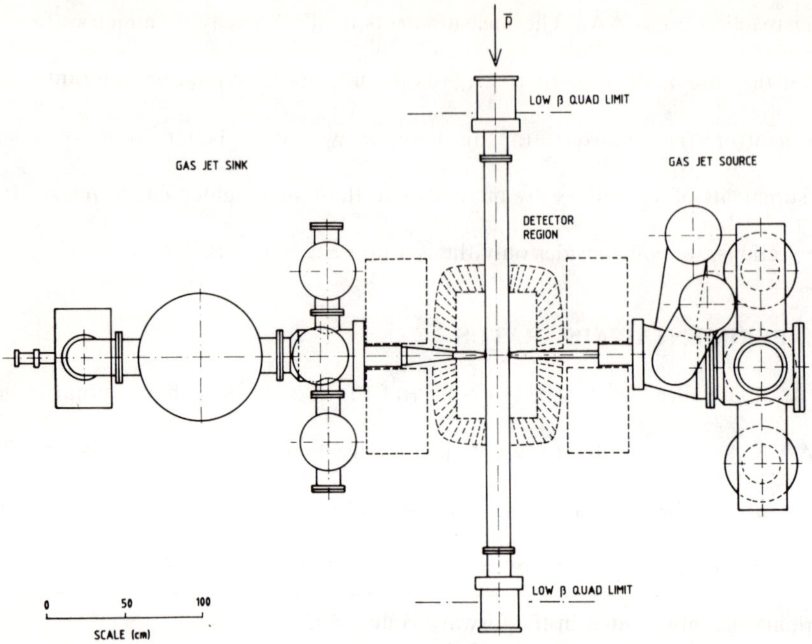


Fig.1. Top view of the Jet-Target to be installed at LEAR by the JETSET-collaboration.

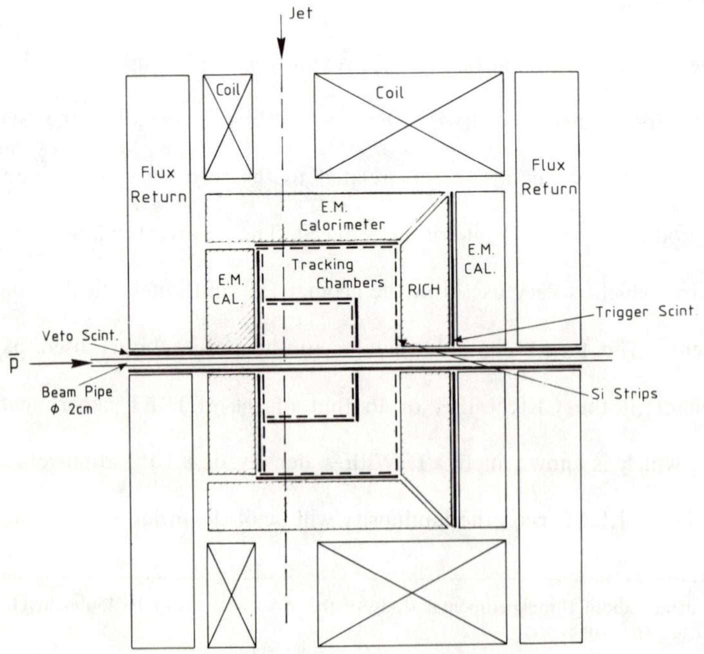


Fig.2. Schematic view of the detector.

**4.2 Detector :**

The detection set-up should accommodate the very high rates corresponding to a high luminosity (total rate: $10^6$ to $10^7$ event/s; useful rate : $\geq 10^7$ events/day). Therefore, to be able to handle such high trigger and acquisition rates, the different detectors, including their read-out, should be fast. In order to minimize the errors on the location of the baryons (antibaryons), an effort should be made to achieve a very precise tracking throughout the whole acceptance region, and especially in the forward region. This reconstruction implies a magnetic analysis. The precision is limited by the multiple scattering, in particular in the vacuum pipe. The apparatus should identify the final $p(\bar{p})$. Cherenkov counters (RICH) can fulfil this function. TOF or/and dE/dx devices may help. All these requirements could be fulfilled by the detector shown in fig.2. It is a compact detector with almost $4\pi$ geometry and high-rate devices :

- a fast ring-imaging Cherenkov counter (RICH) (identification and trigger)
- silicon microstrips and straw chambers or multi drift modules (tracking)
- scintillators or scintillating fibres (trigger)
- an electromagnetic calorimeter (BG0)

A solenoid gives the magnetic field along the beam direction.

**4.3 Trigger and acquisition :**

Because of high rates and the relative complexity of the events, several trigger levels are needed. Fast processors for the acquisition and the kinematical reconstruction can speed up the on-line analysis. The data acquisition and data reduction are not trivial, and require advanced techniques. In the case of $\bar{\Lambda}\Lambda$ and $\bar{\Xi}^0\Xi^0$, a neutral trigger is the best way to select interesting events. For the $\bar{\Xi}^+\Xi^-$ channel, an impact parameter trigger could be used. The neutral trigger requires a small beam pipe, and maybe the possibility of bypassing the apparatus at the initial cooling. Multiplicity criteria are obviously important ingredients in the trigger. Finally, the p($\bar{p}$) signature must be incorporated in the trigger. Systematic errors coming from the trigger lead to asymmetries which are similar to those due to the uncertainties in the efficiencies (see 5.2).

**4.4 Rates :**

The number of events is related to the asymmetry. For the reaction $\bar{p}p \to \bar{\Lambda}\Lambda$, if we want an asymmetry $A_\Lambda$ up to n st.dev., we need Nevts $= 2n^2/A_\Lambda{}^2\, \alpha_\Lambda\, P_\Lambda$. For example, if $A_\Lambda = 10^{-4}$ and $n=3$, we need at least 2.8 $10^9$ events. With a luminosity $L=10^{31}$ cm$^{-2}$ s$^{-1}$, a cross section $\sigma=40$ $\mu$b, and a global acceptance $\varepsilon$ = 50 % ,we have to run about 160 days. In these conditions, we get a rate of 200 evts/s for a total $\bar{p}p$ rate of $\approx 10^6$/s. In the case of the reaction $\bar{p}p \to \bar{\Xi}\Xi$ and for $B_\Xi$, we have the following expression : Nevts $\approx 8n^2/\pi\; B_\Xi{}^2\, \alpha_\Lambda\, \beta_\Xi\, P_\Xi$. Taking $B_\Xi = 10^{-3}$ up to $3\sigma$, we need at least $10^8$ events ($\Xi^0$) or 6 $10^8$ events ($\Xi^-$). With $L = 10^{32}$ cm$^{-2}$ s$^{-1}$, $\sigma = 1\mu$b and $\varepsilon$ = 50 %, we have to run 24 days ($\Xi^0$) or 140 days ($\Xi^-$). **Statistics is one of the fundamental limitations of the experiments.**

**4.5 Background :**

Background reactions without hyperons are not dangerous as far as the asymmetry is concerned. They cannot fake a spurious asymmetry. But some hyperon production reactions such as $\bar{p}p \to \bar{\Lambda}\Sigma^0, \Lambda\bar{\Sigma}^0$, which can simulate the $\bar{p}p \to \bar{\Lambda}\Lambda$ reaction if the $\gamma$ from $\Sigma^0(\bar{\Sigma}^0)$ decay escapes detection, can give unwanted signals. A very efficient electromagnetic calorimeter (BGO), a good identification, and a precise kinematical reconstruction should reduce this background.

## 5. EXPERIMENTAL ASYMMETRIES :

We have first the intrinsic asymmetries of the experiment. The $\bar{Y}$ are forward produced in the CM, and therefore more forward-peaked in the Lab than Y. In addition, the interactions of particle and antiparticle with matter (vacuum pipe, detectors) are also not symmetric. Apart from these intrinsic asymmetries, we also have asymmetries coming from the different parts of the apparatus (trigger, efficiencies, position errrors, magnetic field,...). Note that we have no problems with the normalization, because we measure simultaneously the Y and $\bar{Y}$ decays.

### 5.1 Asymmetries coming from the beam or the target :

The beam cooling process leads to a beam divergence of a few mrads at the target position. This and the target spatial extension ($\leq$ 1 $cm^3$) give uncertainties in the angular measurements.

### 5.2 Asymmetries coming from the detector :

The secondary interactions, and the multiple scattering of Y($\bar{Y}$) or decay products in the jet target, vacuum pipe and counters, give an important limitation on the measurement precision. The Y($\bar{Y}$) secondary interactions in the target are minimized by the use of a gas-jet target. But the $\Xi^-$ and the charged decay products give multiple scattering in the vacuum pipe and the detectors. This leads to distorsions in the spectra of the kinematical variables. The large-angle scattering of Y($\bar{Y}$) or decay products can be filtered out by the reconstruction program or even by multiplicity criteria at the trigger level. To first order, the asymmetry is not affected by counter inefficiency, because we work on the basis of a well-reconstructed event. To higher order, the inefficiencies can lead to more p($\bar{p}$) up than down, or vice-versa. This effect can be strongly reduced if one of the essential design criteria is the azimuthal symmetry condition. With this condition, we have only to know some integrated efficiencies. This should be investigated in more detailed Monte-Carlo simulations. Kinematical reconstruction should be good, but is not critical. This reconstruction should provide an additional rejection of the wrong hyperon (like $\Sigma^0$ in $\bar{\Lambda}\Sigma^0$) and radiative decay events.

**5.3 Radiative and leptonic decays :**

In the $\Lambda$ decay, the radiative decay $p\,\pi^-\gamma$ occurs with a BR of about $10^{-3}$. The $\gamma$ can be detected by the electromagnetic calorimeter, and then the event is rejected. Only a small fraction of the $\gamma$ escapes detection. As for the detection of the $\gamma$ from $\bar{\Lambda}(\Sigma^0)$, high detection efficiency of the low energy $\gamma$ is very important. Otherwise, the electromagnetic calorimeter is useful for the rejection of the electron coming from the leptonic decay $\Lambda \rightarrow pe^-\nu$, which occurs also with a BR of about $10^{-3}$, and for the rejection of electromagnetic signals associated with the wrong hyperon decays background in the $\bar{\Xi}\Xi$ channels.

**6. CONCLUSIONS**

In order to make a comparative study of the different hyperon systems, we need additional information which is presently missing. In particular, the polarization of the hyperon ($P_\Lambda$,$P_\Xi$,...) in the production reaction is an important parameter. Nevertheless, we can make some comments.

1. The most interesting system to study is $\bar{\Xi}\Xi$, for which we have to measure the asymmetry $B_\Xi$. Remember that a meaningful measurement of $B_\Xi$ needs the additional knowledge of $A_\Lambda$ at least at the same level of precision. This could be done by an independant measurement in the reaction $\bar{p}p \rightarrow \bar{\Lambda}\Lambda$.

2. The choice of the final state (neutral or charged) is more delicate. Concerning the rates, the neutral channel is clearly the best candidate. Also, the trigger is simpler if the related problems can be overcome. On the other hand, we can have better kinematical reconstruction in the charged channel, and also better background and efficiency control.

3. We have not studied the case of the $\Sigma'_S$. The $\Sigma^+ \rightarrow n\pi^+$ decay could be interesting in despite of the difficult neutron detection, if the asymmetry A is high ($\geq 10^{-3}$). (ref.5,13).

In conclusion, a CP violation study seems feasible in the $\bar{Y}Y$ systems. The precision of measurements could be at, or not far from, the level needed to test models. In addition, detectors and high intensity hyperon sources exist. If experiments were starting now, the time scale for such studies can be very competitive compared to the time scales of experiments studying the $\bar{B}B$ system.

## REFERENCES.


1) J.H. Christenson et al.,Phys.Rev.Lett.13(1964)138
2) L.L. Chau, Phys.Rep.C95(1983)62; T.Brown et al., Phys.Rev.Lett.51 (1983)1823; O.E. Overseth and S. Pakvasa, Phys.Rev.184(1969)1663
3) J.F. Donoghue and S. Pakvasa, Phys.Rev.Lett.55(1985)162;
4) J.F. Donoghue et al., Phys.Rev.D34(1986)833
5) L.L. Chau and H.Y. Cheng, Phys.Lett.131B(1983)202
6) J.F. Donoghue, Third Conf. on the Intersections between Particle and Nuclear Physics, Rockport, U.S.A. 14 – 19 May 1988 and UMHEP-301
7) N. Hamann, Third Conf. on the Intersections between Particle and Nuclear Physics, Rockport, U.S.A. 14 – 19 May 1988
8) J.F. Donoghue et al.,Phys.Lett.178B(1986)319
9) D. Besset et al., NIM 166 (1979)515
10) K. Kilian et al.,(PS185 Coll.) in CERN/PSCC/83 – 6, PSCC/M152
11) P. Dalpiaz et al., CERN/EP 87 – 27
12) G. Bassompierre et al., CERN/PSCC 86 – 23, PSCC/P97, 25 March 1986.
13) L.L. Chau, private communication

# RARE DECAYS

*Fred Gilman and Yoshio Yamaguchi.*

# CP VIOLATION IN RARE K DECAYS*

Frederick J. Gilman

*Stanford Linear Accelerator Center*
*Stanford University, Stanford, CA 94309*

## ABSTRACT

Recent theoretical and experimental progress on CP violation, particularly for rare $K$ decays, is summarized.

* Work supported by Department of Energy contract DE–AC03–76SF00515.

**Introduction**

As is the case for much of contemporary research in high energy physics, the area of CP violation in $K$ decays is to be seen in the context of a much broader effort of looking for physics beyond the Standard Model. There are two principal avenues:

(1) The high energy route involves the direct observation of new quarks, new leptons, heavy Higgs bosons.... Of necessity, this involves accelerators which are at the high energy frontier. That frontier, in the past few years, has begun to yield quark-quark (from proton-antiproton colliders) collisions at total energies of order 100 GeV. We now have lepton-lepton (from electron-positron colliders) collisions in this range and quark-quark collisions probing physics up to several hundred GeV. Experiments at the SSC will allow us to explore physics at the 1000 GeV scale and above. This is the natural continuation of the field of high energy physics to higher and higher mass scales.

(2) The "low energy" route also can involve the direct observation of new particles such as additional light neutrinos. The confirmation of nonzero neutrino mass and mixing would indicate physics beyond the Standard Model as well. However, much of the work at low energy aims to be sensitive to new physics through the indirect effects of virtual, heavy particles. These, through precision measurements, give us a window on the high energy world which others attack directly.

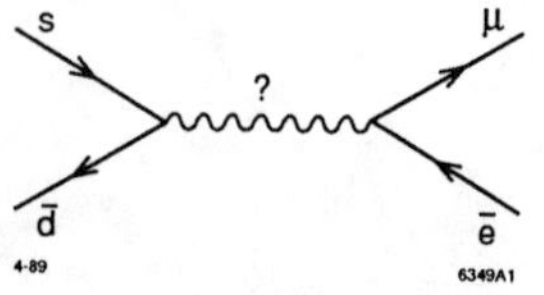


*Figure 1: Tree-level diagram involving a flavor-changing gauge boson.*

In examining the possibilities for new physics in rare decays, one needs to be acquainted with relatively few generic Feynman diagrams. There are some processes which are forbidden in the Standard Model to any order. An example is leptonic flavor-changing neutral-currents. They might occur at "tree-level," as shown in the diagram in Fig. 1, which could represent the exchange of a flavor-changing "horizontal" gauge boson, for example. There are also processes, which while forbidden at tree-level in the Standard Model, can occur at "one-loop," as indicated by the penguin and box diagrams shown in Fig. 2. Thus, we search for such new physics through:

($a$) Processes forbidden in the Standard Model, such as would be induced by lepton-flavor-changing neutral currents.

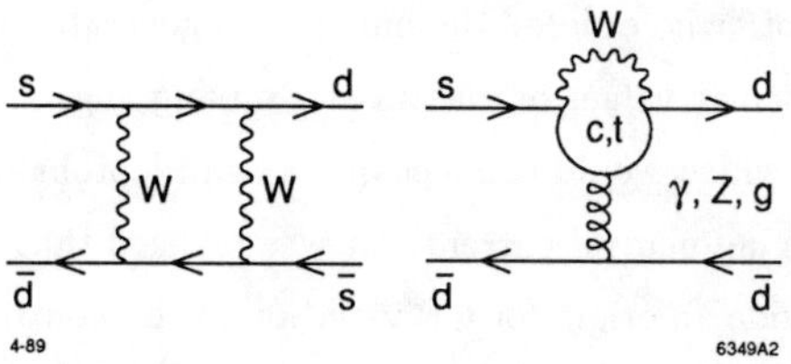


*Figure 2: One-loop diagrams giving rise to flavor-changing processes.*

(*b*) Indications that CP-violating phenomena have an origin other than from the non-trivial phase in the quark flavor-mixing matrix of the Standard Model.

(*c*) Deviations from expected rates, especially for rare processes which are sensitive to heavy virtual particles (from a fourth-generation, supersymmetry, left-right electroweak gauge symmetry, etc.) This is especially true of CP-violating processes, which, in some cases, are especially sensitive to the top quark and possible other high mass particles.

As we pin down and measure the parameters associated with each of the particles in the Standard Model, we use these numbers, together with our improved calculational skills, to obtain updated predictions. Then we can return to the former perspective of looking for new physics by comparing these predictions with all previous data and by pointing to further experiments which are yet more sensitive to new physics.

In this context, we will take a closer look at (b), and particularly at how rare $K$ decays can help determine whether CP violation is a phenomenon whose origin lies inside or outside the Standard Model.

### The "Rebirth" of $K$ Physics

The late 1960s and early 1970s marked a peak in experiments on $K$ decays, sparked by the discovery of CP violation.[1)] This effort tailed off as many important measurements were completed and new areas of physics opened up in the 1970s at electron-positron and hadron machines.

Then, in the late 1970s and early 1980s, both theoretical and experimental developments led to a rebirth of $K$ physics. On the theoretical side, the establishment of gauge theories for the strong and electroweak interactions provided a well-defined basis for calculations. The three-generation Standard Model could be used to make predictions of what, by definition, was inside, and by its complement, outside the Standard

Model. The question of "who ordered the muon" was generalized to "who ordered three generations with particular values of masses and mixing angles," and attention was directed at interactions which would connect quarks and leptons of different generations, producing flavor-changing neutral currents. It was realized that not only did the three-generation model provide an origin for CP violation in the nontrivial phase in the quark mixing matrix, but that CP violation should affect the $K^0$ decay amplitude as well as the $K^0 - \overline{K}^0$ mass matrix, resulting in values of $\epsilon'/\epsilon$ in the $10^{-3}$ to $10^{-2}$ range.[2] There were also predictions for short-distance contributions to a number of other rare $K$ decay amplitudes induced at one-loop, both CP conserving and CP violating.[3]

On the experimental side, great strides were made: to create high flux beams, handle high data rates, incorporate "smart triggers," improve detectors (especially for photons), and to be able to analyze enormous data samples. These matched, at least to some degree, the requirements in precision and rarity being demanded by the theory for incisive tests of the Standard Model. The last few years have seen the beginning of a parade of results which are the culmination of a decade of work in perfecting and performing the needed experiments. Much more is yet to come.

**The Rise of the Top Quark**

Over the past decade, the "typical" or "best" value of the top quark mass used in theoretical papers has risen monotonically, somehow always remaining one step, or maybe one-and-a-half steps, ahead of the experimental then-current lower bound. Values of 15, 25, 30, 45... GeV have been used in various papers (some of them mine), but all of which have fallen by the wayside as experiments have been able to search at higher and higher masses. The present lower limit is around 60 GeV, below which a top quark is said[4] to be "unlikely." It seems that lower limits even higher than this will be quoted at high confidence within a few months, as the analysis of the present round of collider data (which is still being taken as I speak) is completed. An upper limit of around 200 GeV follows from analysis of neutral and charged current data and the measured $W$ and $Z$ masses (*i.e.*, consistency of the $\rho$ parameter with unity).[5] Here again, we will know much more in a few months when we have a much more accurate $Z$ mass from electron-positron colliders. I suspect that we are headed for a lower limit (or a top mass value?!) in the neighborhood of 100 GeV later this year.

The rise of the top quark mass has important consequences when we go to calculate one-loop contributions. For the penguin diagrams in Fig. 2 involving a top and charm quark and a virtual photon (the "electromagnetic penguin"), the conserved nature of the current demands that a factor of $q^2$, the square of the four-momentum carried by the virtual photon, be present in the numerator of the amplitude. This cancels the $1/q^2$ from the photon propagator; the leading term for small (compared to $M_W^2$) top mass in the coefficient of the appropriate operator behaves as $\ln(m_t^2/m_c^2)$. By contrast, the "$Z$ penguin" or "$W$ box" involve nonconserved currents: the factor $q^2$ in the numerator is replaced by the square of the quark mass in the loop and the propagator by $1/(q^2+M_Z^2) \approx 1/M_Z^2$ or $1/M_W^2$. The corresponding coefficient behaves like $[(m_t^2/M_W^2)\ln(m_t^2/M_W^2) - (m_c^2/M_W^2)\ln(m_c^2/M_W^2)]$ when the top mass is small. In days when $m_t^2 \ll M_W^2$, it was completely justified to throw away the $Z$ penguin and $W$ box contributions to such amplitudes in comparison to that of the electromagnetic penguin. Not so any more. The various graphs give comparable contributions, as we will see later in specific examples. Moreover, the contributions from the top quark become the dominant ones to various rare $K$ decays when $m_t^2 \gg M_W^2$. In the three-generation Standard Model, as $m_t$ rises farther and farther above $M_W$, more and more of one-loop $K$ physics is top physics, and we are in the interesting situation where those working at the highest energy hadron colliders are pursuing another aspect of the same physics as those working on the rarest of $K$ decays at low energies.

## CP Violation in the Three-Generation Standard Model

The matrix[6)] that describes the mixing of three generations of quarks has three real angles and one nontrivial phase. Any difference of rates between a given process and its CP conjugate process (or of a CP-violating amplitude) always has the form:

$$\Gamma - \overline{\Gamma} \propto s_1^2\, s_2\, s_3\, c_1\, c_2\, c_3\, \sin\delta_{KM} = s_{12}\, s_{23}\, s_{13}\, c_{12}\, c_{23}\, c_{13}^2\, \sin\delta_{13} \quad , \tag{1}$$

where we express things first in the original parametrization of the quark mixing matrix[6)] and then in the "preferred" parametrization adopted by the Particle Data Group,[7)] using the shorthand that $s_i = \sin\theta_i$ and $c_i = \cos\theta_i$. Our present experimental knowledge assures us that the approximation of setting the cosines to unity, which we often adopt in the following, induces errors of at most a few percent. In that case, the combination of angle-dependent factors in Eq. (1), involving the invariant measure of CP violation,[8)] becomes the approximate combination:

$$s_1^2\, s_2\, s_3\, \sin\delta_{KM} = s_{12}\, s_{23}\, s_{13}\, \sin\delta_{13} \quad , \tag{2}$$

which was recognized earlier as characteristic of CP-violating effects in the three-generation

Standard Model.[9] Equation (1) shows us immediately that all three generations of quarks are necessary for CP violation; in particular, none of the angles can be zero, nor can any of the Kobayashi–Maskawa (K–M) matrix elements.

The K–M factors in Eq. (1) define the "price of CP violation" in the Standard Model. This price must be paid somewhere. It could be paid in a specific process by having many of these factors in both $\Gamma$ and $\overline{\Gamma}$, corresponding to a very small branching ratio for that process; then, when we form the asymmetry,

$$A_{CP\ violation} \ = \ \frac{\Gamma - \overline{\Gamma}}{\Gamma + \overline{\Gamma}} \ , \tag{3}$$

the smallness of the denominator results in a large asymmetry. On the other hand, the price could be paid by having few of these factors in $\Gamma$ and $\overline{\Gamma}$ separately (and hence in their sum), but only in their difference; the asymmetry is correspondingly small. There is, therefore, a very rough correspondence between rarer decays and bigger asymmetries. This rule of thumb is only that; it can be mitigated or exacerbated by other factors: hadronic matrix elements, dependence of one-loop amplitudes upon internal quark masses, and the possible presence of K–M factors in addition to those demanded by Eq. (1). A prime example of luck in this regard is provided by CP-violating effects which depend on $B - \overline{B}$ mixing, where the large top quark mass allows fairly big asymmetries between $B$ and $\overline{B}$ decays to occur in modes which are themselves not suppressed in rate by K–M factors.

Given this price of CP violation, we can "naturally" understand why

$$|\epsilon| \ \approx \ 2.28 \times 10^{-3} \tag{4}$$

is so small, *i.e.*, why there is a "near miss," and CP comes so close to being a symmetry in $K$ decays.[10] When all the factors are put in, the size of $|\epsilon|$ is roughly governed by that of $s_2 s_3 s_\delta$. This is naturally of the right size in the technical sense that to have $s_2 s_3 s_\delta$ of order $10^{-3}$ does not require any angle to be fine-tuned to be either especially small or especially large.

This same factor of $s_2 s_3 s_\delta$ pervades all CP violation observables in the $K$ system, so it is then not so surprising that after 25 years the total evidence for CP violation in Nature consists of a nonzero value of $\epsilon$, and one statistically significant measurement[11] of a nonzero value of the parameter $\epsilon'/\epsilon = 3.3 \pm 1.1 \ \times 10^{-3}$, representing CP violation in the $K \to \pi\pi$ decay amplitude itself. Experiments at Fermilab[12)] and at CERN[11)] are continuing with the aim of reducing the statistical and systematic errors. The value of $\epsilon'$ from Ref. 11 is consistent[13–15)] with the three-generation Standard Model. Unfortunately, this is not a very strong statement. Other values of $\epsilon'$ would be consistent as well because of our lack of knowledge both on the experimental and theoretical fronts:

- The hadronic matrix elements of the penguin operators, upon which the prediction of $\epsilon'$ depends, are fairly uncertain. Definitive results will presumably come from lattice QCD calculations which still seem several years away.
- The predictions depend on the value of $s_2 s_3 s_\delta$, which, in turn, depends (aside from another hadronic matrix element) on $m_t$ through imposing the constraint of obtaining the experimental value of $\epsilon$. Very roughly, as $m_t$ goes up, the range allowed for $s_2 s_3 s_\delta$ goes down, and so does the prediction for $\epsilon'$.
- Also as $m_t$ rises, the contributions from $Z$ penguin and $W$ box diagrams begin to be significant. For sufficiently large $m_t$, a recent calculation[16)] contends that most of the usual (strong) penguin contribution to $\epsilon'$ can be cancelled in this way.

Experimental and theoretical progress over the next few years should clarify these points. But even if the situation becomes that the value of $\epsilon'$ is in significant accord with the three-generation Standard Model, this single number is unlikely to be regarded as conclusively establishing that the origin of CP violation lies in the K–M matrix. We would demand additional evidence: A single set of K–M angles (including the phase) must be able to fit several different processes which exhibit CP-violating effects, providing a redundant check on the theory.

Where can we look to get this additional evidence? One place is the $B$ meson system. Here, there isn't a near miss, as CP-violating asymmetries potentially can be very large: of order $10^{-1}$ or more.[17)] Another place is the $K$ system, where we turn to other $K$ decays in which CP-violating effects, although very small, may occur with a different weighting (from that in $K \to \pi\pi$) between effects originating in the mass matrix and in the decay amplitude. Possible $K$ decays which come to mind include $K \to 3\pi$, $K \to \gamma\gamma$, and $K \to \pi\pi\gamma$,[18–20)] and especially $K_L \to \pi^0 \ell^+ \ell^-$ and $K_L \to \pi^0 \nu\bar{\nu}$. We follow $K$ decays in the rest of this talk, briefly discussing some rare CP-conserving processes to set the stage for CP-violating decays.

**Strangeness-Changing Kaon Amplitudes at One-Loop**

$K^0 - \overline{K}^0$ Mixing

The grandfather of all the calculations of amplitudes which are forbidden in lowest order of the electroweak theory is that of the off-diagonal elements of the $K^0 - \overline{K}^0$ mass matrix which generate the $K_L - K_S$ mass difference and $\epsilon$. This still provides the tightest constraint on quark flavor-changing neutral currents (provided, of course, that they contribute to this process). The one-loop, short-distance contribution to $\epsilon$ has been already alluded to in our discussion of CP violation.

$K^+ \rightarrow \pi^+ \ell^+ \ell^-$ and $K_S \rightarrow \pi^0 \ell^+ \ell^-$

Both of these processes receive short-distance contributions from the electromagnetic penguin with a charm quark in the loop. However, there are very large QCD corrections,[21] (so big as to change the sign of the amplitude) and the result is very untrustworthy. Not surprisingly, for the real (CP-conserving) part of the amplitude which enters both these processes, it is necessary to understand significant long-distance contributions. These may be best calculable in chiral perturbation theory.[22]

The measured branching ratio[7] for $K^+ \rightarrow \pi^+ e^+ e^-$ is $2.7 \pm 0.5 \times 10^{-7}$. We may expect hundreds, if not thousands, of events from ongoing experiments, as well as some events of $K^+ \rightarrow \pi^+ \mu^+ \mu^-$. The predicted branching ratio for $K_S \rightarrow \pi^0 \ell^+ \ell^-$ is in the neighborhood of several times $10^{-9}$, and will be of importance both for a check on the chiral perturbation theory calculations[22] and for CP violation in the decay $K_L \rightarrow \pi^0 \ell^+ \ell^-$, to be discussed later.

$K^+ \rightarrow \pi^+ \nu \overline{\nu}$

Here, the short-distance contribution from charm and especially top quarks in $Z$ penguin and $W$ box graphs provides the dominant contribution to the amplitude: all the estimates of long-distance effects show them to be negligible.[23] The QCD corrections are moderate in magnitude. They particularly need to be applied to the contribution of the charm quark. The original QCD corrections,[24] have been recently updated to the case where the top mass is comparable to $M_W$.[25] The resulting branching ratio for $K^+ \rightarrow \pi^+ \nu_e \overline{\nu}_e$ is shown in Fig. 3, with the dashed lines representing upper and lower bounds (given our present freedom in choosing K–M parameters, particularly $V_{td}$) without QCD corrections and the solid lines giving the corresponding bounds with those corrections.[25] The branching ratio ranges between about 0.2 and $2 \times 10^{-10}$ per neutrino flavor.

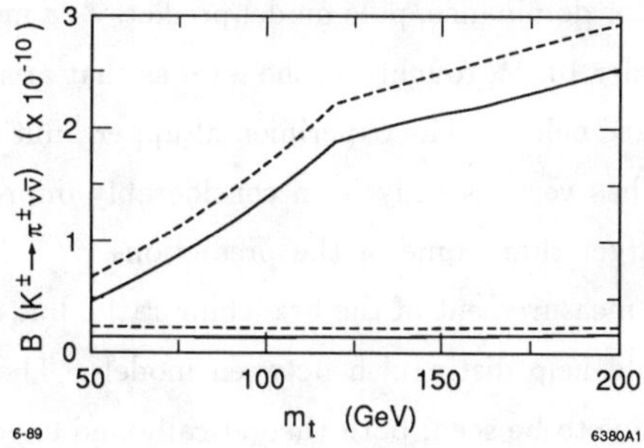


*Figure 3: The maximum and minimum of the branching ratio (per neutrino flavor) for $K^{\pm} \to \pi^{\pm}\nu\bar{\nu}$ without (dashed curve) and with (solid curve) QCD corrections ($\Lambda_{QCD} =$ 150 MeV). From Ref. 25.*

The upper limit on this process has recently been considerably improved to $3 \times 10^{-8}$ by a dedicated Brookhaven experiment.[26)] There are prospects of getting to the $10^{-9}$ level in the next year, and eventually reaching a sensitivity where there should be a few events if the Standard Model gives the correct rate. In the meantime, there is a large window still left open for new physics between where we are now and the Standard Model prediction.

## $K_L \to \pi^0 \ell^+ \ell^-$

If we define $K_1$ and $K_2$ to be the even and odd CP eigenstates, respectively, of the neutral $K$ system, then $K_L \to \pi^0 e^+ e^-$ has three contributions:

(1) Through a two-photon intermediate state:

$$K_2 \to \pi^0 \, \gamma\gamma \to \pi^0 e^+ e^- \quad .$$

This is higher order in $\alpha$, but is CP conserving. With two real photons, there are two possible Lorentz invariant amplitudes for $K_L \to \pi^0 \, \gamma\gamma$. One is the coefficient of $F^{(1)}_{\mu\nu} \, F^{(2)}_{\mu\nu}$ , which corresponds to the two photons being in a state with total angular momentum zero. Consequently, it picks up a factor of $m_e$ when contracted with the QED amplitude for $\gamma\gamma \to e^+e^-$, as the interactions are all chirality conserving, and its contribution to the $K_L \to \pi^0 e^+ e^-$ decay rate is totally negligible.[27)] The other invariant amplitude is the coefficient of a tensor which contains two more powers of momentum, and one might hope for its contribution to be suppressed by angular momentum barrier factors. In chiral perturbation theory, an order-of-magnitude estimate[28)] for the resulting branching ratio of $K_2 \to \pi^0 \; e^+e^-$ is

$10^{-14}$. However, a vector dominance, pole model predicts[29] a much bigger result: a branching ratio of order $10^{-11}$, roughly at the level as that arising from the CP-violating amplitudes (see below). The experimental upper limit on the branching ratio for $K_L \to \pi^0\gamma\gamma$ has very recently been considerably improved,[30] and now is only a few times larger than some of the predictions.[22,29] In the future, we might have not only a measurement of the branching ratio, but also a Dalitz plot distribution which could help distinguish between models. The final answer for this contribution remains to be seen, both theoretically and experimentally.

2) Through the small (proportional to $\epsilon$) part of the $K_L$ (*i.e.*, $K_1$), due to CP violation in the mass matrix:

$$K_L \approx K_2 + \epsilon K_1$$
$$K_1 \to \pi^0 \ \gamma_{virtual} \to \pi^0 e^+ e^- \quad .$$

We call this "indirect" CP violation and may calculate its contribution to the decay rate once we know the width for the CP-conserving process $K_1 \to \pi^0 e^+ e^-$. Eventually, there will presumably be an experimental measurement of $\Gamma(K_S \to \pi^0 e^+ e^-)$, which will take all the present theoretical model dependence away. For now, equating this width to the measured one for $K^+ \to \pi^+ e^+ e^-$ gives the estimate:

$$B(K_L \to \pi^0 e^+ e^-)_{\text{indirect}} = 0.58 \times 10^{-11} \quad . \tag{5}$$

3) Through the large part of the $K_L$ (*i.e.*, $K_2$), due to CP violation in the decay amplitude:

$$K_2 \to \pi^0 \ \gamma_{virtual} \to \pi^0 e^+ e^- \quad .$$

We call this "direct" CP violation, and the amplitude for it arises from the diagrams shown in Fig. 4. For values of $m_t \ll M_W$, it is the electromagnetic penguin that gives the dominant short-distance contribution to the amplitude, which is summarized in the Wilson coefficient, $C_{7V}$, of the appropriate operator,

$$Q_{7V} = \alpha \left(\bar{s}\gamma_\mu(1-\gamma_5)d\right)\left(\bar{e}\gamma^\mu e\right) \quad . \tag{6}$$

Values of $m_t \sim M_W$ allow the $Z$ penguin and W box contributions to become comparable to that of the electromagnetic penguin, and bring in another operator,

$$Q_{7A} = \alpha \left(\bar{s}\gamma_\mu(1-\gamma_5)d\right)\left(\bar{e}\gamma^\mu \gamma_5 e\right) \quad . \tag{7}$$

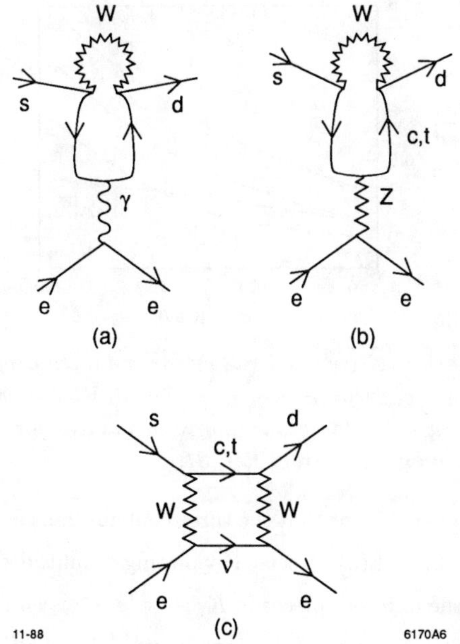


*Figure 4: Three diagrams giving a short-distance contribution to the process $K \to \pi\ell^+\ell^-$: (a) the electromagnetic penguin; (b) the Z penguin; (c) the W box.*

The QCD corrections are substantial for the electromagnetic penguin contribution and have been redone for the case[31,32] when $m_t \sim M_W$. In contrast, the top quark contributions from the $Z$ penguin and $W$ box live up at the weak scale and get only small QCD corrections. Still, the coefficient $C_{7V}$ comes largely from the electromagnetic penguin, even after its reduction from QCD corrections. On the other hand, the electromagnetic penguin cannot contribute to $C_{7A}$, and here it is the $Z$ penguin which gives the dominant contribution. The overall decay rate due to the direct CP-violating amplitude can be obtained by relating the hadronic matrix elements of the operators $Q_{7V}$ and $Q_{7A}$ to that which occurs in $K_{e3}$ decay. Then we find that

$$B(K_L \to \pi^0 e^+ e^-)_{direct} \approx 1 \times 10^{-5} \, (s_2 s_3 s_\delta)^2 \, [|\widetilde{C}_7|^2 + |\widetilde{C}_{7A}|^2] \quad . \tag{8}$$

The last factor, shown in Fig. 5, ranges[31] between about 0.1 and 1.0. As $s_2 s_3 s_\delta$ is typically of order $10^{-3}$, the corresponding branching ratio induced by this amplitude alone for $K_L \to \pi^0 e^+ e^-$ is around $10^{-11}$. Note that when $m_t \gtrsim 150$ GeV, the contribution from $C_{7A}$ overtakes that from $C_{7V}$, and it is the $Z$ penguin and $W$ box, coming from the top quark with small QCD corrections, which dominate the decay rate.

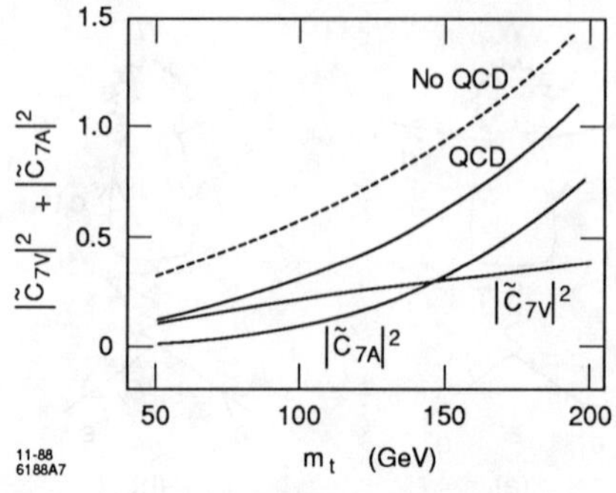


*Figure 5: The quantities $(\widetilde{C}_{7V})^2$ and $(\widetilde{C}_{7A})^2$ as a function of $m_t$, and their sum, $(\widetilde{C}_{7V})^2 + (\widetilde{C}_{7A})^2$, with (solid curve, $\Lambda_{QCD} = 150$ MeV) and without (dashed curve) QCD corrections, which enter the branching ratio induced for $K_L \to \pi^0\ell^+\ell^-$ by CP violation in the decay amplitude. From Ref. 31.*

Thus, it appears at this point that the three contributions from: (1) CP-conserving, (2) indirect CP-violating, and (3) direct CP-violating amplitudes could all be comparable. The weighting of the different pieces in $K_L \to \pi^0 e^+e^-$ is entirely different from that in $K \to \pi\pi$. The present experimental upper limit[33,34] is $4 \times 10^{-8}$, with prospects of getting to the Standard Model level of around $10^{-11}$ in the next several years.[35] Hopefully, the CP-conserving and indirect CP-violating amplitudes will be pinned down much better by then, permitting an experimental measurement of this decay to be interpreted in terms of the magnitude of the direct CP-violating amplitude.

## $K_L \to \pi^0\nu\overline{\nu}$

Having descended to miniscule branching ratios, we now add the impossible in detection: the decay $K_L^0 \to \pi^0\nu_\ell\overline{\nu}_\ell$ is an even more striking example of a process in which the relative sizes of various contributions to the decay rate are totally different[36] than in $K \to \pi\pi$. There is, of course, neither an electromagnetic penguin nor a two-photon, CP-conserving, contribution to the amplitude. Furthermore, the indirect CP violation arising from the neutral $K$ mass matrix gives a negligible contribution to the decay rate. That leaves us with just the $Z$ penguin and $W$ box, and the V–A character of the gauge boson couplings to neutrinos allows only the operator:

$$Q_\nu = \frac{e^2}{4\pi}(\overline{s}_\alpha\gamma_\mu(1-\gamma_5)d_\alpha)(\overline{\nu}_\ell\gamma^\mu(1-\gamma_5)\nu_\ell) \quad . \tag{9}$$

Being CP violating, it is the imaginary part of $C_\nu$ that is required:

$$Im\ C_\nu = \left(s_2 s_3 s_\delta\right)\left(\widetilde{C}_{\nu,t} - \widetilde{C}_{\nu,c}\right) \quad , \tag{10}$$

which is totally dominated by the top quark contribution. The branching ratio (per neutrino flavor) is

$$B(K_L^0 \to \pi^0 \nu_\ell \overline{\nu}_\ell) \approx 2.1 \times 10^{-5} \left(s_2 s_3 s_\delta\right)^2 |\widetilde{C}_{\nu,t} - \widetilde{C}_{\nu,c}|^2 \quad , \tag{11}$$

with the latter quantity shown in Fig. 6. Again, as $s_2 s_3 s_\delta$ is of order $10^{-3}$, the branching ratio with three generations of neutrinos is of order $10^{-11}$. The QCD corrections to the $t$ quark contribution should be small, making this theoretically an ideal decay in which to study CP violation in the decay amplitude. Experimentally, the problems are perhaps best represented by the statement that nobody has yet shown that a measurement of this decay is absolutely impossible.

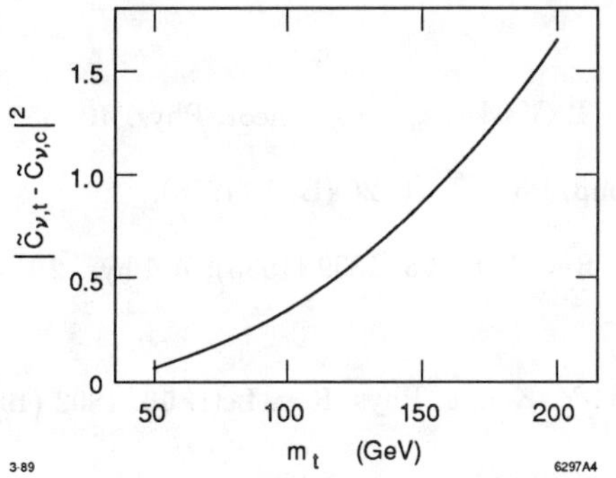


*Figure 6: The quantity $|\widetilde{C}_{\nu,t} - \widetilde{C}_{\nu,c}|^2$, which enters the branching ratio for the CP-violating decay $K_L \to \pi^0 \nu_\ell \overline{\nu}_\ell$, as a function of $m_t$. From Ref. 25.*

## Conclusion

After 25 years, we are still faced with answering the question of the origin of CP violation: is it a first, tiny bit of physics from beyond the Standard Model, or does it originate from inside the Standard Model, where it is the first evidence that there are three or more generations, all quark masses unequal, and all weak mixing angles nonzero? Indeed, the issue is somewhat more muddled now than it seemed to be a couple of years ago. It has taken longer to gain understanding than we had hoped, but with time, we will sort out the parameters of the Standard Model (including the top quark mass), do the theoretical work that will sharpen the predictions, and carry out the experiments to see CP-violating effects beyond those in the neutral $K$ mass matrix. Nature has been performing an elegant striptease; we just have to be patient and enjoy it.

## REFERENCES


1. J. H. Christenson, J. W. Cronin, V. L. Fitch, and R. Turlay, Phys. Rev. Lett. **13**, 138 (1964).
2. F. J. Gilman and M. B. Wise, Phys. Lett. **83B**, 83 (1979) and Phys. Rev. **D20**, 2392 (1979).
3. See, for example, the recent review of J. S. Hagelin and L. S. Littenberg, MIU–THP–89/039 (1989), to be published in Prog. Part. Nucl. Phys.
4. UA1, UA2, and CDF, any talk in early 1989.
5. U. Amaldi *et al.,* Phys. Rev. **D36**, 1385 (1987); G. Costa *et al.,* Nucl. Phys. **B297**, 244 (1988).
6. M. Kobayashi and T. Maskawa, Prog. Theor. Phys. **49**, 652 (1973).
7. Particle Data Group, Phys. Lett. **204B**, 1 (1988).
8. C. Jarlskog, Phys. Rev. Lett. **55**, 1839 (1985); Z. Phys. **29**, 491 (1985); and these proceedings.
9. L.-L. Chau and W.-Y. Keung, Phys. Rev. Lett. **53**, 1802 (1984).
10. A. Pais, these proceedings.
11. H. Burkhardt *et al.,* Phys. Lett. **206B**, 169 (1988); K. Peach, these proceedings.
12. M. Woods *et al.,* Phys. Rev. Lett. **60**, 1695 (1988); B. Winstein, these proceedings, reports a central value consistent with zero within the statistical error bars of $\pm 1.4 \times 10^{-3}$, based on 20% of the data from Fermilab experiment E731.
13. M. A. Shifman, *Proceedings of the 1987 International Symposium on Lepton and Photon Interactions at High Energies,* Hamburg, July 27–31, edited by W. Bartel and R. Ruckl (North Holland, Amsterdam, 1988), p. 289.
14. F. J. Gilman, *International Symposium on the Production and Decay of Heavy Flavors,* Stanford, September 1–5, 1987, edited by E. Bloom and A. Fridman (New York Academy of Sciences, New York, 1988), vol. 535, p. 211.
15. G. Altarelli and P. J. Franzini, CERN preprint CERN–TH–4914/87 (1987), unpublished.
16. J. M. Flynn and L. Randall, Phys. Lett. **224B**, 221 (1989).

17. K. J. Foley *et al., Proceedings of the Workshop on Experiments, Detectors, and Experimental Areas for the Supercollider*, Berkeley, July 7–17, 1987, edited by R. Donaldson and M. G. D. Gilchriese (World Scientific, Singapore, 1988), p. 701, review CP violation in $B$ decay and give references to previous work; S. Sanda, these proceedings.

18. L.-F. Li and L. Wolfenstein, Phys. Rev. **D21**, 178 (1980).

19. L.-L. Chau and H.-Y. Cheng, Phys. Rev. Lett. **54**, 1768 (1985) and Phys. Lett. **195B**, 275 (1987); J. O. Eeg and I. Picek, Phys. Lett. **196B**, 391 (1987).

20. G. Ecker, A. Pich, and E. de Rafael, Nucl. Phys. **B303**, 665 (1988).

21. F. J. Gilman and M. B. Wise, Phys. Rev. **D21**, 3150 (1980).

22. G. Ecker, A. Pich, and E. de Rafael, Phys. Lett. **189B**, 363 (1987); Nucl. Phys. **B291**, 691 (1987) and Ref. 20.

23. See, for example, D. Rein and L. M. Sehgal, Phys. Rev. **D39**, 3325 (1989), and references therein.

24. J. Ellis and J. Hagelin, Nucl. Phys. **B217**, 189 (1983).

25. C. O. Dib, I. Dunietz, and F. J. Gilman, SLAC preprint SLAC–PUB–4840 (1989), unpublished.

26. D. Marlow, talk at the Twelfth International Workshop on Weak Interactions and Neutrinos, Ginosar, Israel, April 6–14, 1989; A. J. S. Smith, these proceedings.

27. J. F. Donoghue, B. R. Holstein, and G. Valencia, Phys. Rev. **D35**, 2769 (1987).

28. Ref. 22 and E. de Rafael, these proceedings.

29. L. M. Sehgal, Phys. Rev. **D38**, 808 (1988); T. Morozumi and H. Iwasaki, KEK preprint KEK–TH–206 (1988), unpublished; J. Flynn and L. Randall, Phys. Lett. **216B**, 221 (1989).

30. V. Papadimitriou *et al.,* Phys. Rev. Lett. **63**, 28 (1989).

31. C. Dib, I. Dunietz, and F. J. Gilman, Phys. Lett. **218B**, 487 (1989) and Phys. Rev. **D39**, 2639 (1989).

32. Other recent work on the subject is found in J. Flynn and L. Randall, LBL preprint LBL–26310 (1988), unpublished.

33. L. K. Gibbons *et al.,* Phys. Rev. Lett. **61**, 2661 (1988).

34. G. D. Barr *et al.,* Phys. Lett. **B214**, 1303 (1988).

35. R. Patterson, these proceedings; M. Schmidt, these proceedings; G. Bock, these proceedings.

36. L. Littenberg, Phys. Rev. **D39**, 3322 (1989).

# LONG DISTANCE EFFECTS IN RARE K DECAYS

Eduardo de Rafael
Centre de Physique Théorique* , Section II
CNRS - Luminy, F-13288 MARSEILLE, France

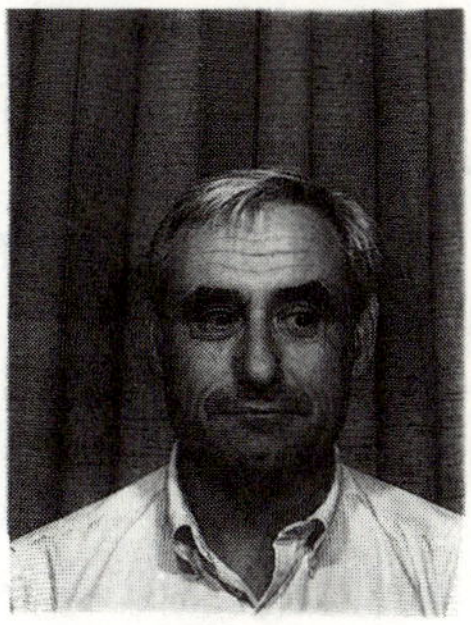

ABSTRACT

We give an overview of the chiral effective Lagrangian approach to the calculation of radiative K-decays. Results for various decay modes are presented. We discuss in particular the $K_2 \rightarrow \pi^0 \gamma\gamma \rightarrow \pi^0 e^+e^-$ transition and the status of its theoretical understanding.

---

* Laboratoire Propre du Centre National de la Recherche Scientifique.

## Introduction

We shall be concerned with the evaluation of matrix elements of the effective weak Hamiltonian of the standard model for radiative K-decays. There is a framework to organize these calculations, which has the merit of being consistent with the symmetries of the standard model. It is the chiral effective Lagrangian approach. I shall begin this talk with an overview of the basic ideas of this approach. Then, I shall present the results of calculations done within this framework. I shall concentrate on the mode $K_L \to \pi^0 e^+e^-$ which is a good candidate for CP violation physics ; as well as the related mode $K_L \to \pi^0 \gamma\gamma$ which has recently attracted much attention. This will bring me to another topic, which is the rôle of vector (axial) mesons in chiral Lagrangians. I shall summarize results of recent work on this subject. Finally I shall comment on the decay $K^+ \to \pi^+ e^+e^-$, a radiative mode with known branching ratio which was measured many years ago by René Turlay and collaborators[1].

## Basics of the Chiral Perturbation Theory Approach

In the limit where the masses of the up (u), down (d) and strange (s) quarks are set to zero, the QCD Lagrangian is invariant under rotations $(V_L, V_R)$ of the left, right handed quark triplets $\begin{pmatrix} u_L \\ d_L \\ s_L \end{pmatrix}$ and $\begin{pmatrix} u_R \\ d_R \\ s_R \end{pmatrix}$. These rotations generate the so called chiral $SU(3)_L \otimes SU(3)_R$ group. At the level of the physical states, this symmetry of the QCD-Lagrangian is spontaneously broken down to the $SU(3)_V$ generated by the sum of the $SU(3)_L$ and $SU(3)_R$ generators. This reduced invariance is the famous SU(3)-symmetry of the Eightfold Way[2].

The pattern of spontaneously broken symmetry implies specific constraints on the dynamics of the strong interactions between the low lying pseudoscalar states $(\pi, K, \eta)$, which are the Golstone bosons associated to the "broken" generators. The best way to spell out these constraints is by means of an effective Lagrangian[3], and the most general Lagrangian compatible with the symmetry pattern described above is a non-linear Lagrangian[4]. This is why in this approach the octet of pseudoscalar fields $\vec{\varphi}(x)$ appears in an exponential form ($f_\pi$ = 93.3 MeV) :

$$U(x) = \exp\left(\frac{i}{f_\pi}\, \vec{\lambda}.\vec{\varphi}(x)\right) \; ; \qquad (1a)$$

$$\frac{\vec{\lambda}}{\sqrt{2}}\cdot\vec{\varphi}(x) = \begin{pmatrix} \pi^0/\sqrt{2} + \eta/\sqrt{6} & \pi^+ & K^+ \\ \pi^- & -\pi^0/\sqrt{2} + \eta/\sqrt{6} & K^0 \\ K^- & \bar{K}^0 & -2\eta/\sqrt{6} \end{pmatrix} \equiv \phi(x) \ , \qquad (1b)$$

where $\vec{\lambda}$ are the Gell-Mann 3x3 matrices. The matrix U is unitary $UU^\dagger = 1$, unimodular detU=1, and it has been fixed so that under chiral rotations $U \to V_L U V_R^\dagger$. The most general chiral effective Lagrangian is a sum of terms with increasing number of derivatives of U and powers of quark masses - which break explicitly the chiral symmetry -. For example, to lowest order in derivatives, only one term can be constructed which is invariant under $(V_L, V_R)$ transformations :

$$\mathcal{L}_{\text{eff}} = \frac{1}{4} f_\pi^2 \, \text{tr}(\partial_\mu U \partial^\mu U^\dagger) \ , \qquad (2)$$

where the normalization is fixed in such a way that the axial-current deduced from this Lagrangian induces the experimentally observed $\pi \to \mu\nu$ transition. Because of the non-linearity in $\varphi$, processes with different number of pseudoscalars are then related. These are the successful current algebra relations[5] of the 60's which this Lagrangian formulation incorporates in a compact way.

Higher order terms in derivatives and masses are modulated by coupling constants which, in the absence of a detailed knowledge of the dynamics of confinement in QCD, have to be extracted from low energy hadron phenomenology. This requires the calculation of loop terms generated by the lowest order Lagrangian term (2), which appear at the same level of accuracy. Loops have to be taken into account to guarantee S-matrix unitarity at the level of approximation one is working. This program has been successfully carried out in the strong interaction sector up to order $O(p^4)$ in a series of beautiful papers by Gasser and Leutwyler[6].

The non-leptonic weak interactions appear as a perturbation to the strong effective chiral Lagrangian with transformation properties dictated by the chiral structure of the four quark operators of the effective weak Hamiltonian which results after the short-distance effects have been factorized[7].

In the chiral perturbation theory (CHPT) approach, physical amplitudes are calculated as a power expansion in dot products of external momenta normalized to the scale of spontaneous chiral symmetry breaking $\Lambda_\chi$. Typically, $\Lambda_\chi \approx 4\pi f_\pi$=1.2 GeV[8]. The calculations are made up to a few unknown coupling constants which have to be fixed from physical processes. Once this is done, predictions for other processes follow. Kaon decays are particularly well adapted to this approach because the characteristic momenta involved is small compared to $\Lambda_\chi$ : $M_k^2/16\pi^2 f_\pi^2 \approx 20\%$. Various radiative K-decay modes have been recently discussed within the CHPT-approach. The results are presented in the next section.

**Radiative Kaon Decays in Chiral Perturbation Theory**

As discussed in a series of papers[9], nonleptonic radiative K-decays with at most one pion in the final state share the common feature that their amplitudes vanish to lowest order $O(p^2)$ in CHPT. These decay modes are then particularly interesting because they are directly sensitive to chiral loops and, in most cases, to the coupling constants of the higher order effective chiral Lagrangian.

**Table 1. - Radiative Kaon Decays calculated in CHPT**

| | Mode | BR(CHPT) | References | BR(EXP) |
|---|---|---|---|---|
| I. | $K_S \to \gamma\gamma$ | $2.0 * 10^{-6}$ | 10,9b | $(2.4\pm1.2)*10^{-6}$ NA31 |
| | $K_L \to \pi^0\gamma\gamma$ | $6.8 * 10^{-7}$ | 9b,11 | $<2.7 * 10^{-6}$ E731 |
| | $K_S \to \pi^0\gamma\gamma$ | $3.8 * 10^{-8}$ | 9b | |
| | $K_S \to \gamma e^+e^-$ | $3.2 * 10^{-8}$ | 9c | |
| | $K_S \to \gamma\mu^+\mu^-$ | $7.5 * 10^{-10}$ | 9c | |
| II. | $K^\pm \to \pi^\pm\gamma\gamma$ | $\geq 4.10^{-7}$ | 9c | $<8 * 10^{-6}$ PDG |
| III. | $K^+ \to \pi^+e^+e^-$ | fit | 9a | $(2.7\pm 0.5) *10^{-7}$ PDG |
| | $K^+ \to \pi^+\mu^+\mu^-$ | $\begin{cases} 6.1 *10^{-8} \\ 4.5 *10^{-8} \end{cases}$ | 9a | $<2.1 * 10^{-7}$ BNL 787 |
| | $K_S \to \pi^0e^+e^-$ | $\begin{cases} 4.8 *10^{-10} \\ 4.9 *10^{-9} \end{cases}$ | 9a | $<4.5 * 10^{-5}$ E731 |
| | $K_S \to \pi^0\mu^+\mu^-$ | $\begin{cases} 1.0 *10^{-10} \\ 1.0 *10^{-9} \end{cases}$ | 9a | |
| | $K_L \to \pi^0e^+e^-$ | $\begin{cases} 1.5 *10^{-12} \\ 1.5 *10^{-11} \end{cases}$ | 9a,12,13,14 | $<4. * 10^{-8}$ NA31<br>$<4.2 * 10^{-8}$ E731 |
| | $K_L \to \pi^0\mu^+\mu^-$ | $\begin{cases} 5 *10^{-12} \\ 10^{-11} \end{cases}$ | 9a | $<1.2 *10^{-6}$ PDG |

Table 1 gives a list of the radiative kaon decays which so far have been calculated within this approach. The second column gives the predicted total branching ratios ; and the fourth column the present experimental values. The predictions in this table are those to lowest non trivial order i.e., $0(p^4)$ in this case. At that order, the calculation of the first class of processes I) involves a one loop integral which is finite ; and they are free from unknown constants. The calculation of $K^{\pm} \to \pi^{\pm} \gamma\gamma$ also involves a chiral loop which is finite but here there are $0(p^4)$ tree level terms which contribute as well to this process. The combination of the coupling constants which appears in the $K^{\pm} \to \pi^{\pm} \gamma\gamma$ amplitude is unknown, and this is why at present we can only predict a lower bound for this decay rate. (See however ref.9b for a discussion on predictions for the $\gamma\gamma$ invariant mass spectrum.) The third class of processes involves divergent loops. They are renormalized by some of the $0(p^4)$ tree level counterterms. One of the coupling constants which appears here has both long and short distance contribution (from the so called electromagnetic Penguin diagrams discussed by Gilman in his talk[13].) We use the fact that the rate for $K^+ \to \pi^+ e^+e^-$ is known to fix a combination of the unknown coupling constants. There is then a twofold solution, which explains the two predictions made for the other decays in the same class.

**The $K_2 \to \pi^0(\gamma\gamma) \to \pi^0 e^+e^-$ transition.**

Of the decays in Table 1, $K_L \to \pi^0 e^+e^-$ is perhaps the most interesting one because the amplitude $K_2 \to \pi^0 e^+e^-$ via one-photon exchange is forbidden by CP-invariance. ($K_1$ and $K_2$ denote CP eigenstates.) The question of the relative size of the CP-allowed two-photon exchange $K_2 \to \pi^0(\gamma\gamma) \to \pi^0 e^+e^-$ amplitude arises then naturally. To lowest non trivial order in CHPT, i.e., $0(p^4)$ this amplitude is suppressed by an helicity conservation $m_e/M_K$-factor[12]. The corresponding branching ratio, from the absorptive $\gamma\gamma$-amplitude alone, has been calculated[9c], with the result

$$\frac{\Gamma(K_2 \to \pi^0\gamma\gamma \to \pi^0 e^+e^-)}{\Gamma(K_L \to \text{all})}) = 8 * 10^{-15} \ , \qquad (3)$$

indeed a very small contribution. At the same $0(p^4)$ level, there is also the contribution from the real part of the $\gamma\gamma$-exchange amplitude to be considered. The dispersive integral, which is suppressed by the same $m_e/M_K$ factor, is not expected to add significantly since it is convergent ; and there are no further contributions to the real part because to $0(p^4)$ it is not possible to construct a local CP-conserving effective coupling which can contribute to $K_2 \to \pi^0 e^+ e^-$ at tree level.

The suppression $m_e/M_K$ factor is not however a general property of the $\gamma\gamma$-exchange amplitude[9c]. The amplitude structure for $K_2(p) \to \pi^0(p') + \gamma(q_1) + \gamma(q_2)$ with all particles on-shell is $\epsilon_\mu(q_1)\epsilon_\nu(q_2)M^{\mu\nu}(p,q_1,q_2)$ where gauge invariance and CP-conservation restrict $M^{\mu\nu}$ to depend on two invariant amplitudes $(x_i = \frac{p \cdot q_i}{M_K}, i = 1,2)$ ,

$$M^{\mu\nu}(p,q_1,q_2) = A(x_1,x_2)\left[q_2^\mu q_1^\nu - q_1 \cdot q_2 g^{\mu\nu}\right] +$$

$$+ B(x_1,x_2)\frac{1}{M_K^2}\left[-p \cdot q_1 p \cdot p_2 g^{\mu\nu} - q_1 \cdot q_2 p^\mu p^\nu + p \cdot q_1 q_2^\mu p^\nu + p \cdot q_2 p^\mu q_1^\nu\right] \quad (4)$$

The term proportional to A leads to configurations of the final $\gamma\gamma$-system with total angular momentum zero. This is the origin of the $m_e/M_K$ factor when this amplitude combines with the $\gamma\gamma \to e^+e^-$ QED amplitude. However, the term proportional to B contributes to $K_L \to \pi^0\gamma\gamma \to \pi^0 e^+ e^-$ even in the limit $m_e \to 0$. The B term has more powers of momenta and this is why the B amplitude, from the point of view of CHPT, originates in $O(p^6)$ contributions at least. Naïve chiral power counting, which consists in pulling out the extra $16\pi^2 f_\pi^2$ chiral suppression factor and letting the overall dimensionless coupling constant to be of order one, leads to an estimate

$$B \approx G_8 \frac{\alpha}{\pi} \frac{M_K^2}{16\pi^2 f_\pi^2} \bar{\omega} \text{ ; with } \bar{\omega} = O(1) , \quad (5)$$

where $G_8 = \frac{G_F}{\sqrt{2}} s_1 c_1 c_3\, g_8$ is the overall Fermi constant ($G_F$) modulated by the flavour mixing angles and the octet enhancement factor $|g_8| \approx 5.1$. This results in a branching ratio for $K_2(p) \to \pi^0 e^+ e^-$ at the $10^{-14}$ level [9c]. (See also ref. 14 for details of this calculation.)

This estimate however is not supported by a vector meson dominance (VMD) tree level calculation made by Sehgal[15] which results in a branching ratio comparable to the one expected from the CP-violating amplitude i.e., at the $10^{-11}$ level. A recent reanalyse of Sehgal's mechanism by Flynn and Randall[14] which includes $\eta$-$\eta'$ mixing and does not rely on the experimentally observed rate for $\eta \to \pi^0\gamma\gamma$, results in a smaller branching ratio ; but still their conclusion is that the VMD model prediction for the effective coupling $\bar{\omega}$ above is about 20 times larger than the naïve chiral power counting estimate. What is the reason for such a large discrepancy ?

Let me first emphasize that a large VMD-amplitude would manifest rather characteristically in the Dalitz plot distribution of the $K_L \to \pi^0 \gamma\gamma$ events.[16] Figure 1 shows the spectrum in the $\gamma\gamma$ invariant mass squared, $q^2=2M_K^2(x_1+x_2)+m_\pi^2-M_K^2$ , predicted by the lowest order CHPT calculation of ref. 9b. (The dotted line is the phase space prediction.)

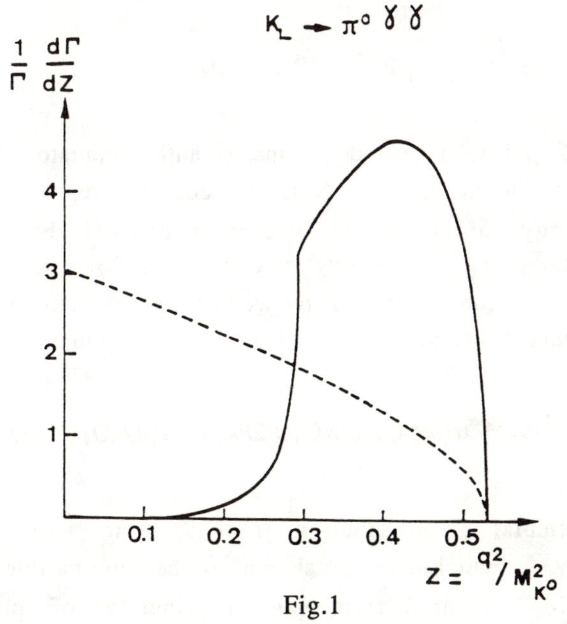


Fig.1

To $O(p^4)$ in CHPT, and at fixed Z, the predicted distribution in the $x_1$ - $x_2$ variable is flat. A large VMD-amplitude leads to sizable deviations from these predictions. (See also refs. 17 and 18 which have appeared after this conference.)

**Vector meson dominance and chiral symmetry**

There are a priori different ways of including massive spin-1 particles in a chiral invariant effective Lagrangian. These ambiguities have been recently analyzed[19,20] within the context of CHPT to $O(p^4)$. In ref. 20

been recently analyzed[19,20] within the context of CHPT to $O(p^4)$. In ref. 20 it has been shown that, provided consistency with QCD asymptotic behaviour is incorporated, the structure of the effective couplings induced by vector and axial-vector exchange is model independent. In this way, chiral symmetry and the phenomenologically successful notion of vector meson dominance[21] become perfectly consistent : when the heavy vector (axial vector) mesons are integrated out, the coupling constants of the resulting local $O(p^4)$ terms practically saturate the phenomenological values found in ref.6.

The vector field Lagrangian has $O(p^3)$ couplings linear in the octet vector field $\hat{V}_\mu$ [20], which when expanded up to terms linear in $\phi$ and $\gamma$ reproduce the basic $V \to \varphi\gamma$ coupling of the Sehgal mechanism[15]

$$h_2 \, 2e \frac{\sqrt{2}}{f_\pi} \varepsilon_{\mu\nu\rho\sigma} F^{\rho\sigma} \, \mathrm{tr} \, \hat{V}^\mu \{\partial^\nu \phi, Q\} + \ldots , \qquad (6)$$

where $Q = \mathrm{diag}\,(2/3, 1/3, -1/3)$ and $\{,\}$ means anticommutator. $F_{\mu\nu} = \partial_\mu A_\nu - \partial_\nu A_\mu$ with $A_\mu$ the photon field. In fact, this coupling reproduces very well the phenomenology of radiative vector decays[22]. From the latest measurement of the $\rho^- \to \pi^- \gamma$ branching ratio one gets $|h_2| = (3.6 \pm 0.2) * 10^{-2}$.

The effective action generated by vector meson exchange between two $h_2$-like couplings leads to local $O(p^6)$ effective couplings :

$$\mathcal{L}^{(6)} = \frac{h_2^2}{M_V^2} e^2 \frac{2}{f_\pi^2} \left[ F_{\rho\sigma} F^{\rho\sigma} \mathrm{tr}\{\partial_\nu \phi, Q\}\{\partial^\nu \phi, Q\} + 2 F_{\rho\sigma} F^{\sigma\nu} \mathrm{tr}\{\partial_\nu \phi, Q\}\{\partial^\rho \phi, Q\} \right] \qquad (7a)$$

There is in particular a contribution from $\mathcal{L}^{(6)}$ to forward Compton scattering $\gamma\pi^0 \to \gamma\pi^0$, which can be shown to be compatible[23] with the asymptotic behaviour required from general principles of quantum field theory.

In the presence of the lowest $O(p^2)$ chiral weak Lagrangian with $\Delta S=1$, there are couplings in $\mathcal{L}^{(6)}$ which contribute to both the A and B amplitudes of $K_L \to \pi^0 \gamma\gamma$ in eq. (4). It appears then that the VMD-calculations of refs. 15 and 14 can be consistently accommodated within the chiral Lagrangian approach. (There are some differences however, which will be clarified in a forthcoming publication[23].)

The point I wish to emphasize is that this is not by any means the full $O(p^6)$ contribution to $K_L \to \pi^0 \gamma\gamma$. As already mentioned in ref. 9c, there are other direct $O(p^6)$ weak couplings, like e.g. the term ($L_\mu = i \, f_\pi^2 U \partial_\mu U^\dagger$) :

$$G_8 e^2 \frac{\omega}{16\pi^2 f_\pi^4} F^{\mu\rho} F^{\nu}{}_{\rho} \left(Q\{L_\mu, L_\nu\} U Q U^\dagger\right)_{23} + \text{h.c.} \ , \tag{7b}$$

which contribute to the B-amplitude in eq.(4) as well, at tree level and without pole terms. An analogous situation appears in the calculation of the decay $K^+ \to \pi^+ e^+ e^-$. In CHPT, the predicted decay rate[9a] is a function of the combination of $0(p^4)$ coupling constants

$$\omega_+ = \frac{16\pi^2}{3} (12L_9 - w_1 - 2w_2) + \frac{1}{3} \log \frac{M_\rho^2}{M_K m_\pi} . \tag{8}$$

This dependence is plotted in Figure 2 below :

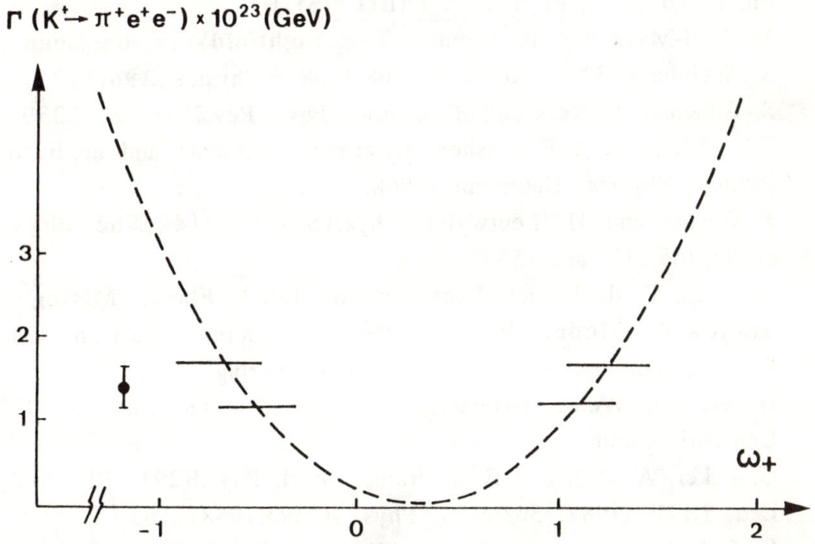


Fig.2

Here, the analog of the VMD contribution discussed above[15],[14] is in the constant $L_9$, which is phenomenologically known $[L_9(M_\rho) = (6.9 \pm 0.7) * 10^{-3}$ from ref. 6b]. The direct weak couplings $w_1$ and $w_2$ are the analog of the $\omega$ in eq.(7). It appears from fig.2 that setting $w_1 = w_2 = 0$ would lead to a huge branching ratio for $K^+ \to \pi^+ e^+ e^-$ totally excluded by experiment (the point with error bars in Fig.2). It is likely that similar cancellations among the various $O(p^6)$ terms contributing to $K_L \to \pi^0 \gamma\gamma$ occur, bringing the whole $O(p^6)$ amplitudes down to the order of magnitude expected from naïve chiral power counting[9c].

Theoretical progress along the lines discussed in this talk can still be made, but it will eventually require more experimental input than the one which is available at present. In that respect, some information on the Dalitz plot of $K_L \to \pi^0\gamma\gamma$ and for $K^+ \to \pi^+\gamma\gamma$ would be most welcome.

**Acknowledgments**

The Blois Conference on CP Violation was a memorable event. I wish to thank J. Tran Thanh Van and the organizing committee for this opportunity to do physics in such an excellent environment.

**References**

[1] Ph. Bloch et al., Phys. Lett. **56B(1975)**201

[2] M. Gell-Mann and Y. Neeman, **The EightfoldWay**, Benjamin (1954)

[3] S. Weinberg, Phys. Rev.**166**(1968)1568 ; Physica A**96**(1979)327

[4] S. Coleman, J. Wess and B. Zumino, Phys. Rev.**177**(1969)2239

[5] S.L. Adler and R.F. Dashen, **Current Algebras** and applications to Particle Physics, Benjamin (1968)

[6] J. Gasser and H. Leutwyler, Phys.**158**(1984)142;Nucl.Phys.**B250** (1985)465,517 and 539

[7] See e.g., E. de Rafael, **Lectures on Quark Flavor Mixing in the Standard Model**, Proc. of the 1984 Latin American School of Physics, Santiago, Chile 1984, Springer Verlag

[8] H. Georgi, **Weak Interactions and Modern Particle Theory.** Benjamin/Cummings 1984

[9] G. Ecker, A. Pich and E. de Rafael, Nucl. Phys.**B291**(1987)692;Phys. Lett. **189B** (1987) 363;Nucl. Phys. **B 303**(1988) 665

[10] G. D'Ambrosio and D. Espriu, Phys.Lett. **B175**(1986)237; J.L. Goity, Z. Phys. **C34**(1987)341

[11] L. Cappiello and G. D'Ambrosio, INFN-Napoli preprint (June 1987)

[12] J.F.Donoghue,B.R.Holstein and G.Valencia,Phys.Rev. **D35** (1987)2769

[13] F. Gilman CP Violation in Rare K Decays, talk in these Proceedings

[14] J.M. Flynn and L. Randall, UCB-PTH-88-21 ; LBL-26008 (1988)

[15] L.M. Sehgal, Phys. Rev. **D38** (1988) 808

[16] G. Ecker, A. Pich and E. de Rafael, unpublished ; A. Pich, Proc. of 1988 Amer. Soc. Storrs (1988)

[17] L.M. Sehgal, PITHA 89/16 (1989)

[18] P. Ko and J.L. Rosner, IASSNS-HEP 89/16 ; EFI 89-13 (1989)

[19] G. Ecker, J. Gasser, A. Pich and E. de Rafael, CERN-TH.5185/88 (to appear in Nucl. Phys. B)

[20] G. Ecker, J. Gasser, H. Leutwyler, A. Pich and E. de Rafael, Phys. Lett. **223B** (1989) 425

[21] J.J. Sakurai, **Currents and Mesons**, University of Chicago Press, 1969

[22] R.S. Chivukula and J.M. Flynn, Phys. Lett. **168B** (1986)127

[23] G. Ecker, A. Pich and E. de Rafael, to be published.

*Alice Bermarchick-Tuts singing at the "Château de Chambord".*

# Rare decays : BNL 878 Results

Smith A. J. Stewart
Princeton University
Dept of Physics
PRINCETON NJ 08540
USA

No written contribution received.

# LEPTON FLAVOR VIOLATION IN RARE KAON DECAYS

Jack L. Ritchie
*Department of Physics*
*University of Texas, Austin, TX 78712*

**Abstract**

The motivation for searches for separate lepton number violating processes in the kaon system are discussed and the status of recent experiments is described. Major emphasis is placed on BNL Experiment 791.

The Standard Model, consisting of the Weinberg-Salam theory of the electroweak interaction and the theory of the strong interaction known as Quantum Chromodynamics, accurately predicts a wide range of phenomena and is not presently challenged by any experimental result. Despite this success, the theory is generally considered to be incomplete, perhaps a kind of low energy limit of some grander theory. The reasons for this view can be traced to several unsatisfying features of the Standard Model:

- There are 19 free parameters in the theory.
- There is no explanation of why the quarks and leptons appear in three generations, or why there should not be more.
- There is no explanation of how the quarks and leptons are related.
- There is a technical difficulty known as the "gauge heirarchy problem," which refers to the need for unnatural fine-tuning of parameters.
- There is no unification of the strong and electroweak interactions (and gravity is nowhere to be seen).

The effort to find the grander theory has been the focus of tremendous theoretical effort, resulting in several candidates: supersymmetry, technicolor, right-left symmetry, horizontal symmetry, and a host of others. Typically, each of these addresses one or more of the problems listed above. But in no case is there a shread of experimental support. Therefore, one of the highest priorities for experimentalists is to find evidence for physics beyond the Standard Model. One of the most promising places to look for such evidence is in the very rare decays of familiar particles. The subject of this report is the search for such evidence in the decays of the neutral long-lived kaon, $K_L$. Other similar searches, for example in rare muon decays, are also important.

Since the decay $K_L \rightarrow \mu e$ violates separate lepton number conservation (i.e., both muon number and electron number are not conserved), its observation would be clear evidence for physics beyond the Standard Model. In most extensions of the Standard Model, the process would occur at some level. But instead of discussing individual models, it is more satisfying to state a general motivation—that a search for a small branching ratio probes a high mass scale. This can easily be seen by comparing the ratio of decay rates of the processes shown in Figure 1. The diagram on the left shows $K_L \rightarrow \mu e$, mediated by a heavy particle $H$ with mass $M_H$. The diagram on the right show the familiar decay $K^+ \rightarrow \mu^+\nu$. To a good approximation, phase space and helicity factors cancel in the ratio, so that

$$\frac{\Gamma(K_L \rightarrow \mu e)}{\Gamma(K^+ \rightarrow \mu^+\nu)} \simeq \left|\frac{ff'/M_H^2}{g^2 sin\theta_c/M_W^2}\right|^{1/2},$$

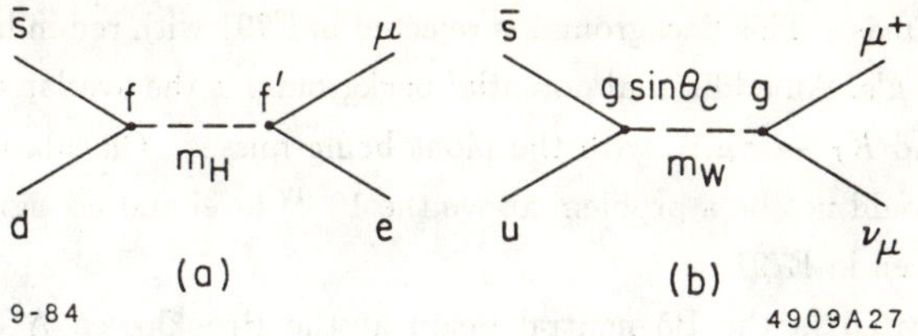


Figure 1: Diagrams for the decays $K_L \to \mu e$ (left) and $K^+ \to \mu^+ \nu$ (right).

where $g$ is the electroweak coupling, $\theta_c$ the Cabibbo angle, and $M_W$ the mass of the $W$ boson. Plugging in known quanitities and rearranging the equations leads to

$$M_H \simeq 20\ TeV \left( \frac{1 \times 10^{-8}}{Br(K_L \to \mu e)} \right)^{1/4} \left| \frac{ff'}{g^2} \right|^{1/2} .$$

Assuming couplings similar to the electroweak coupling, a branching ratio of $10^{-10}$ corresponds to a mass around 60 TeV. No presently forseeable accelerator can reach such energies.

Several groups are actively studying a variety of rare kaon decays (at BNL, KEK, Fermilab, and CERN). A vigorous program at BNL includes four groups focusing on four different rare modes: E777[1] on $K^+ \to \pi^+ \mu^+ e^-$, E791[2] on $K_L \to \mu e$ (and the topologically similar $K_L \to \mu\mu$ and $K_L \to ee$), E787[3] on $K^+ \to \pi^+ \nu \bar{\nu}$ (see Stew Smith's paper, this proceedings), and E845[4] on $K_L \to \pi^o e^+ e^-$ (see Mike Schmidt's paper, this proceedings). This report will describe E791 and report on results from data taken in the Spring of 1988. A run only recently completed should more than double our sensitivity, but it is too early to show the new data.

Before describing the E791 detector, it is usef l to discuss the backgrounds for $K_L \to \mu e$. The most important potential backgrour comes from the decay $K_L \to \pi e \nu$, followed by $\pi \to \mu \nu$ in flight (or alternatively, simple misidentification of the $\pi$ as a $\mu$). If the energy of the $\nu$ in the $K_{e3}$ decay is close to zero, the event can be confused with $K_L \to \mu e$. Rejection of this background depends on the invariant mass distribution of the background process having an end-point 8.4 MeV below the mass of the $K_L$. Therefore, precise event reconstruction is essential. To achieve this, multiple Coulomb scattering must be minimized; in E791, no unnecessary material is placed upstream of the most downstream drift chamber. Also, two consecutive dipole magnets allow two independent momentum measurements, so that $\pi \to \mu \nu$ decay in flight in the spectrometer can be rejected. The next most important potential background is the decay $K_L \to \pi e \nu$ in which the $\pi$ is misidentified as an $e$ and the $e$ is misidentified as a $\mu$. If this occurs, then in the case where the misidentified $\pi$ has large momentum compared to the electron, the event can reconstruct with invariant mass equal to (or above) the $K_L$ mass. This background is rejected in E791 with redundant identification of both $e$'s and $\mu$'s. An additional potential background is the overlap of two $K_L$ decays ($K_L \to \pi e \nu$ and $K_L \to \pi \mu \nu$), with the pions being missed. Calculations indicate this background should not be a problem above the $10^{-12}$ level and no sign of this problem has yet been seen in E791.

E791 is located in the B5 neutral beam at the Brookhaven AGS. Protons with 24 GeV in energy are incident on a one interaction length copper target. A neutral beam is defined by a series of collimators centered at a 2.75 degree angle from the incident proton beam direction. This reduces the neutron contamination in the beam, so that the $n$ to $K_L$ ratio is roughly 3:1. Two dipole sweeping magnets remove charged particles from the beam. The "decay volume" is a region extending from roughly 10 meters from the target to the upstream most drift chamber at 18 meters from the target. The target region, the collimation channel and the decay volume are all under vacuum.

The spectrometer is shown in Figure 2. Tracking is performed by 5 drift chamber modules (each module makes 2 $x$ and 2 $y$ measurements). The regions between drift chambers are filled with helium to reduce multiple scattering and particle interactions. Each of the two dipole magnets provides $\Delta p_T \simeq 300$ MeV, but with opposite sign. Downstream of the final drift chamber, a finely segmented scintillator hodoscope, a gas

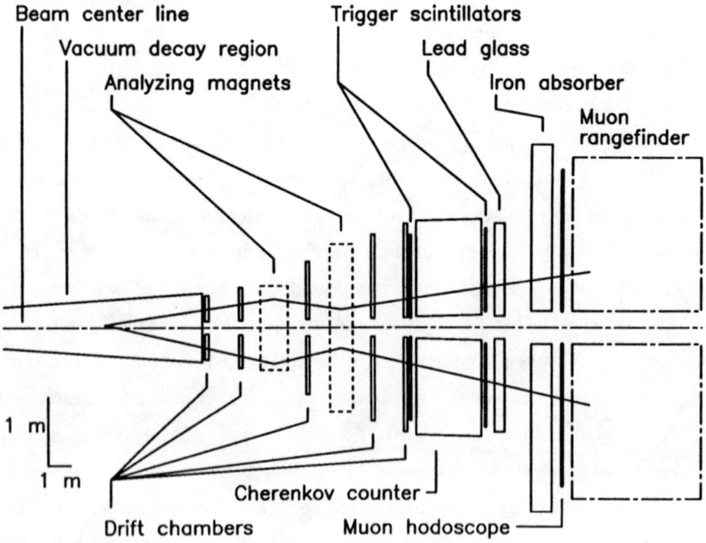


Figure 2: Plan view of the E791 spectrometer.

threshold Cerenkov counter, another hodoscope and a large lead glass array follow in sequence. The scintillation counter hodoscopes provide the signals used in the lowest level (fast logic) trigger. The Cerenkov counter also provides a fast signal, corresponding to the presence of an electron, which is used in the low level trigger. The lead glass yields an electron energy measurement which is used for offline $\pi : e$ discrimination. A meter of steel follows the lead glass to stop all particles except muons. Behind the steel, a segmented scintillation hodoscope provides a fast muon signal for the low level trigger. Finally, muons are stopped in a segmented absorber stack with large proportional wire chambers spaced throughout the stack. This "rangefinder"[5] provides a muon range measurement which corresponds to a 10% measurement of momentum and is important in the offline muon identification.

High rate capability is achieved by fully-custom high speed front-end electronics[6,7], a massively parallel readout architecture[8], and a multi-level trigger system. Trigger decisions are made in two stages. A low level fast logic trigger is based on the scintillation counter hodoscopes, the Cerenkov counter, the muon hodoscopes, and meantimer signals from the upstream drift chambers. A high level trigger is based on rudimentary event reconstruction using drift chamber hits. This is carried out in a bank of eight SLAC 3081/E processors.[9]

The decay $K_L \rightarrow \pi^+\pi^-$ has the same topology as $K_L \rightarrow \mu e$, is copious by E791 standards, and its branching ratio is well known. It is useful as an indicator of detector performance and a heavily prescaled sample is used as the branching ratio normalization. A useful quantity, denoted by $\Theta_K$, is the angle the reconstructed $K_L$ makes with respect to the line from the target to the decay vertex. This quantity will be zero (within measurement resolution) for decays with no missing particles, but will usually be much larger in cases where there is missing transverse momentum due to missed neutrinos. Figure 3 (left box) shows $\Theta_K^2$ versus $\pi\pi$ invariant mass for a sample of $K_L \rightarrow \pi^+\pi^-$ events. A gaussian fit to the mass distribution after a cut requiring $\Theta_K < 1$ mrad, shown in Figure 3 (right box), yields a mass resolution of 1.4 MeV.

Similar distributions can be made for candidates for $K_L \rightarrow \mu\mu$, $K_L \rightarrow \mu e$, and $K_L \rightarrow ee$. Particle identification information from the Cerenkov counter, lead glass array and muon hodoscopes and muon rangefinder are used in selecting these candidates. Figure 4 shows 87 $K_L \rightarrow \mu\mu$ candidates. From this sample, a new measurement of the $K_L \rightarrow \mu\mu$ branching ratio will be made.[10]

Figure 5 shows the corresponding plots for $K_L \rightarrow \mu e$. No events are observed with invariant mass near the $K_L$. From these data, a preliminary upper limit of $3\times10^{-10}$ (90% confidence level) has been set. A final result will be published[11] shortly. No candidates

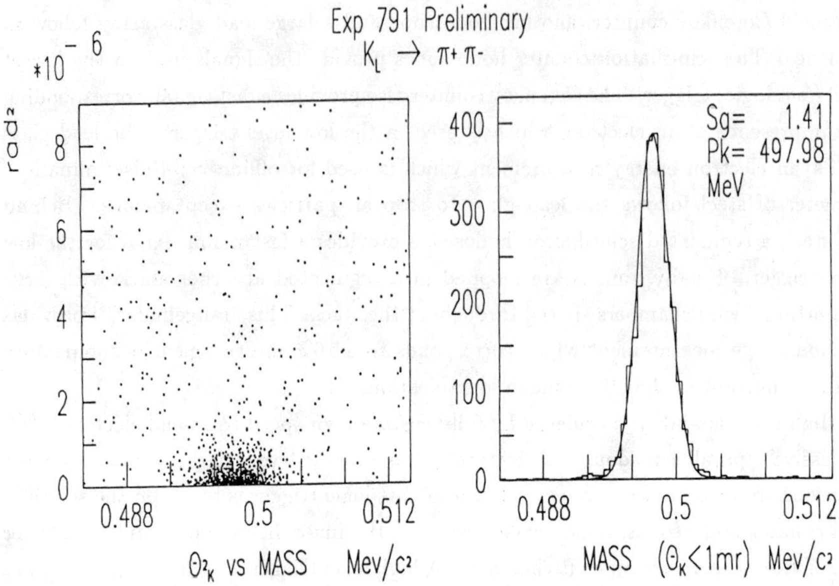


Figure 3: $K_L \to \pi^+\pi^-$ events.

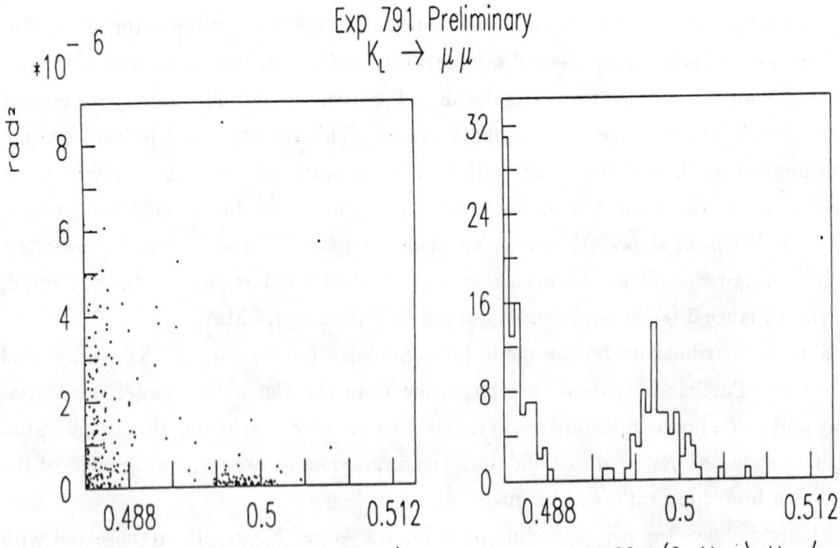


Figure 4: $K_L \to \mu^+\mu^-$ events.

for the decay $K_L \to ee$ are observed. This channel is particularly background free; no events even appear on a plot covering the same mass and $\Theta_K^2$ range as Figure 5.

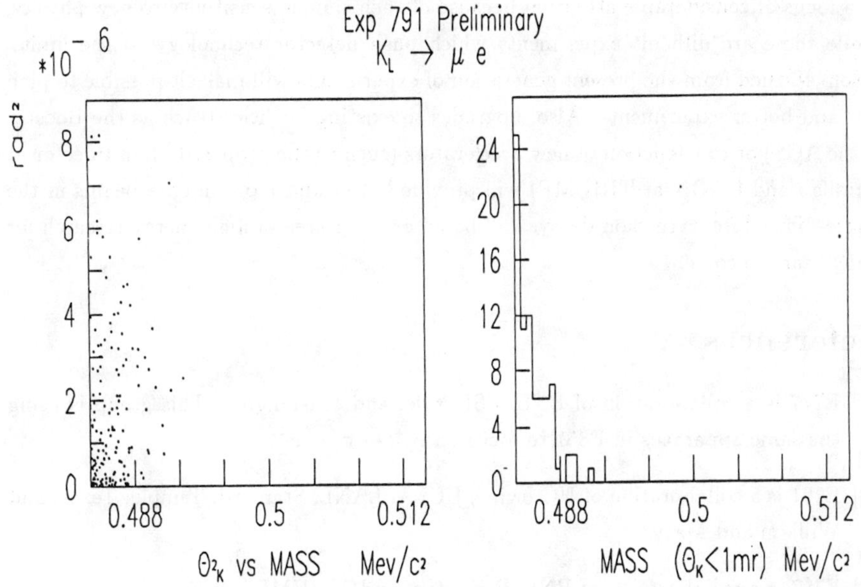


Figure 5: $K_L \to \mu e$ candidates. No events appear at the correct mass.

These results represent the best so far on these decay modes. The KEK 137 experiment has reported 54 $K_L \to \mu\mu$ events and set a limit on $K_L \to \mu e$ of $4.3 \times 10^{-10}$.[12] More recent data from E791, for which the analysis is in an early stage, should represent an improvement of a factor of two or better over the data described here. Also, E791 will take data again beginning late this year.

A complimentary process to $K_L \to \mu e$ is $K^+ \to \pi^+ \mu e$. A comparison of this process to $K_L \to \pi^\circ \mu\nu$, following along the lines of the earlier argument for $K_L \to \mu e$, can be used to show that

$$M_H \simeq 8\ TeV \left( \frac{1 \times 10^{-8}}{Br(K^+ \to \pi^+ \mu e)} \right)^{1/4} \left| \frac{ff'}{g^2} \right|^{1/2}.$$

Therefore, $K^+ \to \pi^+ \mu e$ is somewhat less sensitive to high mass scales than $K_L \to \mu e$ for a $V - A$ interaction. However, this process probes the scaler or vector part of a new interaction, while $K_L \to \mu e$ is sensitive to the pseudoscaler or axial vector part. It is important therefore to do both experiments. The recently completed E777 at BNL has set a new upper limit on $K^+ \to \pi^+ \mu^+ e^-$ of $3 \times 10^{-10}$ (90% confidence level). The best limit on the other sign process, $K^+ \to \pi^+ \mu^- e^+$, is $6.9 \times 10^{-9}$ from an experiment performed several years ago at the CERN PS.[13]

Over the last few years, significant progress has been made in improving the sensitivity of experiments to rare or forbidden kaon decays. These experiments will continue to be a focus of considerable attention because of their unique sensitivity to new physics. While these are difficult experiments which push detector technology to the limits, lessons learned from the present generation of experiments will make it possible to plan new and better experiments. Also, upgrades to existing facilities (such as the Booster at the AGS) or construction of new accelerators (such as the proposed Main Injector at Fermilab and KAON at TRIUMF) will provide better and more intense beams in the future. Therefore, rare kaon decays will be an exciting area of high energy research for many years to come.

## References

[1] E777 is a collaboration of BNL, PSI, Yale, and Washington. This group is using the same apparatus in E851 to focus on $K^+ \to \pi^+ e^+ e^-$.

[2] E791 is a collaboration of UC Irvine, UCLA, LANL, Stanford, Temple, Texas, and William and Mary.

[3] E787 is a collaboration of BNL, Princeton, and TRIUMF.

[4] E845 is a collaboration of BNL, Yale, and Vassar.

[5] J. Frank *et al.*, "A Tracking Rangefinder for Muons from Kaon Decay," IEEE Trans. in Nucl. Science **36**, 79(1989).

[6] R. D. Cousins, C. Friedman, and P. L. Melese, "32-channel Digital 6-bit TDC with 2.5 ns Least Count," IEEE Trans. in Nucl. Science **36**, 646(1989).

[7] K. A. Biery, D. A. Ouimette, and J. L. Ritchie, "A Fast Integrating Eight-bit Bilinear ADC," IEEE Trans. in Nucl. Science **36**, 650(1989).

[8] R. D. Cousins et al., "Fast Parallel Pipelined Readout Architecture for a Completely Flash Digitizing System with Multilevel Trigger," Nucl. Instr. and Meth. **A277**, 517(1989).

[9] P. F. Kunz *et al.*, "The 3081/E Processor," SLAC-PUB-3332, April, 1984.

[10] C. Mathiazhagan *et al.*, "Measurement of the Branching Ratio for the Decay $K_L \to \mu\mu$," submitted to Phys. Rev. Lett.

[11] C. Mathiazhagan *et al.*, "New Experimental Limits on $K_L \to \mu e$ and $K_L \to ee$ Branching Ratios," submitted to Phys. Rev. Lett.

[12] T. Inagaki *et al.*, "Search for $K_L \to \mu e$ and $K_L \to ee$ Decays," KEK Preprint 89-25, May, 1989.

[13] A. M. Diamant-Berger *et al.*, "Study of Some Rare Decays of the $K^+$ Meson," Phys. Lett. **62B**, 285(1976).

*Stewart Smith, André Martin, Penny and Alain Martin.*

# A DEDICATED SEARCH FOR $K_L \to \pi^0 e^+ e^-$

**BNL Experiment 845**

M.P. Schmidt, R.K. Adair, H.B. Greenlee, H. Kasha, E. B. Mannelli, K.E. Ohl, and M.R. Vagins
*Department of Physics*
*Yale University, New Haven, CT 06511, U.S.A.*

E. Jastrzembski, R.C. Larsen, L.B. Leipuner, and W.M. Morse
*Department of Physics*
*Brookhaven National Laboratory, Upton, NY 11973, U.S.A.*

C.B. Schwarz
*Department of Physics and Astronomy*
*Vassar College, Poughkeepsie, NY 12601, U.S.A.*

ABSTRACT

We report on the progress of Brookhaven AGS experiment 845, a dedicated search for the CP invariance violating decay $K_L \to \pi^0 e^+ e^-$. The detector is described and prospects for results from data recently collected are reviewed.

## Introduction

Continuing our program of searching for forbidden or rare $K^0$ decays, which established 90% confidence level limits of $1.9 \times 10^{-9}$ for $K_L \to \mu^\pm e^\mp$,[1] $1.2 \times 10^{-9}$ for $K_L \to e^+e^-$,[2] and $3.2 \times 10^{-7}$ for $K_L \to \pi^0 e^+e^-$,[2] we have designed and have run an experiment at the Brookhaven AGS dedicated to the search for $K_L \to \pi^0 e^+ e^-$.

The decay $K \to \pi e^+e^-$ is a strangeness changing neutral current and is highly suppressed by the GIM mechanism. The decay $K^+ \to \pi^+ e^+ e^-$ has been observed with a branching fraction of $(2.7 \pm 0.5) \times 10^{-7}$ and is currently under study at Brookhaven.[3] The decay $K_S \to \pi^0 e^+ e^-$ is expected to occur at about the same rate, with the theoretical uncertainty being a factor of 2-3 either way. The decay $K_L \to \pi^0 e^+e^-$ however violates CP invariance, assuming single $\gamma$ or $Z^0$ exchange. In the context of a superweak theory the rate for $K_L \to \pi^0 e^+ e^-$ would be $\epsilon^2$ times the $K_S \to \pi^0 e^+ e^-$ rate, as a result of *indirect* CP invariance violation in $K_L - K_S$ mixing, giving $\mathrm{Br}(K_L \to \pi^0 e^+ e^-) \sim 6 \times 10^{-12}$.

A *direct* CP invariance violating amplitude may also be important and estimates made in the context of the Kobayashi-Maskawa 6 quark (KM) model lead to a predicted range: $2 \times 10^{-12} < Br(K_L \to \pi^0 e^+ e^-) < 3 \times 10^{-11}$. Thus the direct CP violating contribution is expected to be comparable to the indirect CP violating contribution[4], in sharp contrast to the case for two pion decays of the $K_L$. In that case $\epsilon'/\epsilon$, which measures the ratio of direct to indirect contributions, appears to be at most a few parts in a thousand[5] if it is nonzero at all.

A very sensitive measurement of $K_L \to \pi^0 e^+ e^-$ that isolated the (direct) $K_2$ and (indirect) $K_1$ decay amplitudes would be an important test of the KM model of CP nonconservation. This would also require resolution of a controversy[4] over the relative importance of an additional CP conserving contribution to $K_L \to \pi^0 e^+ e^-$ through the $\pi^0 \gamma\gamma$ intermediate final state.

The present limit[6] of $4 \times 10^{-8}$ for $K_L \to \pi^0 e^+ e^-$ is many orders of magnitude above any Standard Model prediction. This provides an excellent window to search for light scalar particles that couple to $e^+e^-$ or to detect the effects of nonstandard models of CP invariance violation.[7]

## The E845 Detector

In order to perform a dedicated search for $K_L \to \pi^0 e^+ e^-$, the Yale-BNL E780 detector[8], which was optimized for the $K_L \to \mu^\pm e^\mp$ and $K_L \to e^+e^-$ search, was reconfigured for E845 as shown in Figure 1. The E780 detector was in effect collapsed along the beam line and the analysis magnet aperture was opened to 91 cm (with $\Delta p_t = 114$ MeV/c). This design resulted in an increase of the acceptance for $K_L \to \pi^0 e^+ e^-$ decays by a factor of about fifty (to $\sim$1% for $p_K > 4$ GeV/c). In addition the decay zone length was increased by a factor of about two (to $\sim$ 6 m).

The larger acceptance of E845 is important not only because it allows for an increase in sensitivity, but also because it increases the immunity to possible backgrounds resulting from the overlap of two decays — for example $K_L \to \pi e \nu$, with the pion misidentified as an electron, in coincidence with *only two* gamma rays from another $K$ decay such as $K_L \to 3\pi^0$. The larger geometric acceptance of the detector decreases the likelihood of missing all of the extra gamma rays. In fact the ratio of signal to background is proportional to the square of the geometric acceptance of the detector.[9] The detector has good acceptance for $e^+e^-$ pair masses above the $\pi^0$ mass as shown in Figure 2.

In changing over from E780 to E845, the neutral beam line production angle was increased from 0° to 2° so as to increase the kaon to neutron ratio. The neutral beam is derived from the interactions of the 24 GeV/c primary proton beam with an 18 cm long copper target. The divergence of the beam, 15 mr vertically by 2.3 mr horizontally, is defined by a series of brass collimators embedded in heavily shielded sweeping magnets. About 10 m of shielding and $\sim 6$ Tesla-meters of sweeping field separate the target from the beginning of the decay region. The final 1.5 m of shielding consists of boron loaded concrete used to suppress the flux of low energy neutrons produced in the target, collimators and primary proton beam dump. A 2.5 cm lead plug is used to remove gamma rays from the target. The neutral beam is transported in vacuum ($\sim 70\mu$m Hg) from the defining collimators to the end of the decay region which is terminated by a window consisting of 4.3 mm of Kevlar and 1.3 mm of mylar.

In order to maximize the acceptance, the intense neutral beam ($\sim 3\times10^8$ neutrons/pulse) passes through the active volumes of the chambers in the spectrometer. There are four sets of chambers positioned upstream and downstream of the analysis magnet. The chambers have 3.2 mm drift cells and 6.4 mm anode to cathode plane separation, and each set of chambers provides four measurements ($X, X', Y$ and $Y'$) over a 1 m$^2$ region. The chambers are operated with a gas mixture of 79% argon, 15% isobutane and 6% methylal. Experience with these chambers in E780 has shown that space points on the trajectories can be determined with good resolution ($\sim 250\mu$m) and high efficiency ($\sim 99\%$). Each chamber set represents $< 10^{-3}$ of an interaction length in material.

The electrons of interest (1 GeV/c $< p_e <$ 8 GeV/c) are identified by an atmospheric pressure hydrogen Čerenkov counter as well as by their energy deposition in a lead glass calorimeter. The Čerenkov counter has a nominal 2 m radiator volume that passes through the magnet gap. The counter has four cells, hereafter referred to as quadrants. Each quadrant is viewed by a 12.7 cm diameter EMI-9823B quartz window photomultiplier tube by reflection from a front surface aluminized spherical mirror. Cones of simple construction covered with aluminized mylar are also used to increase the light collection at the phototube. The efficiency of each cell is about 95% as determined from $K_{e3}$ events in the data sample.

Gamma ray positions and energies are measured in the lead glass array, and $\pi^0$s are reconstructed assuming the decay vertex established by the electron trajectories. The lead glass array consists of 244 blocks of Schott F2 glass, each of dimension 6.4 cm $\times$ 6.4 cm $\times$ 46 cm, that are arranged in a square (1 m$^2$) array with a 12.7 cm $\times$ 38 cm hole in the center to pass the neutral beam. In addition 64 larger blocks define a fiducial border around the periphery of the central array. Based on experience from E780 a spatial resolution of $\sim$ 6 mm and an energy resolution of $\sim$ 9% is expected for 1 GeV/c gamma rays.

Accepted events are required to have two electron tracks *and* a minimum of 4 GeV of energy visible in the lead glass array. An electron is identified at the trigger level by the activation of one Čerenkov quadrant, an appropriate $E$ scintillation counter and >1 GeV of energy in that quadrant of the lead glass, as determined by a fast analog summing network. In addition one of the electrons is required to activate one of the upstream scintillation counters ($F$) in the appropriate quadrant. Pions and muons are vetoed at the the trigger level by a bank of scintillation counters located after a 15 cm thick lead wall located downstream of the lead glass array. Events with wide angle gamma rays are largely vetoed by scintillation counters (covered with 3 radiation lengths of material) located around the opening in the evacuated decay tank window and outside the fiducial face of the lead glass array. The trigger logic is implemented in ECL programmed array devices (PAL).

The normalization for the experiment (counting the number of $K$ decays) will be accomplished using the decay $K_L \to 3\pi^0$ with one $\pi^0$ suffering a Dalitz decay. In addition unbiased triggers, that is with no particle identification or energy requirements, are also collected concurrently, prescaled by a factor of 10,000. From this sample $K_L \to \pi^+\pi^-$ and $K_L \to \pi^+\pi^-\pi^0$ decays can also be observed providing a cross check for the normalization. The unbiased triggers also provide data with which to establish and monitor the performance of specific elements of the detector (*e.g.* the Čerenkov counter efficiency).

The ability of the detector to distinguish signal from background depends on the resolution of the charged particle spectrometer and the electromagnetic calorimeter, the quality of the electron identification, and the time resolution of the detector. Expectations for the detector resolution are: $\sigma(m_{\pi^0}) \sim 10$ MeV/c$^2$, $\sigma(m_{\pi^0 ee}) \sim 12$ MeV/c$^2$, and $\sigma(\theta_K) \sim 1.3$mr, where $\theta_K$ is the angle between the vectors that define the reconstructed kaon momentum and the distance from the target to the reconstructed vertex. A $\pi - e$ rejection ratio of a few times $10^4$ is expected through combined use of the Čerenkov counter, the response of the lead glass array, and the lead wall veto. For good background rejection, time and pulse height information is recorded for trigger counters and calorimeter elements.

**Data Acquisition**

The data acquisition system for this experiment is one that has been developed[10,8] and used in a series of experiments at the Brookhaven AGS (E735, E749, E780) by the Yale-BNL collaboration. It employs a water cooled variant of FastBus, one that supports a subset of the standard protocol. The present system has six interconnected crates of custom built electronics with about 3600 channels in all.

The acquisition scheme operates as follows: given a trigger a programmable hardwired processor (Fast List Instruction Processor) retrieves data over FastBus cable segments from front-end devices (TDCs for the drift chambers, ADCs for the lead glass, etc.) residing on separate FastBus crate segments. This data, typically consisting of about 100 (32 bit) words, is transferred at about 100 nsec per word into a 2 Mbyte barrel (FIFO) memory. Data in this dual-ported memory is asynchronously accessed by an LSI-11/73 processor that formats and checks the data for the host PDP-11/44.

Two devices on FastBus rapidly abort the acquisition of events that do not pass certain broad, selectable criteria. One of these devices, called a TDC Spy, rejects events having too many or too few hits in the drift chambers. The other device, an ADC Spy, rejects events with less than 4 GeV of energy deposited in the lead glass array.

A special (homemade) interface between the Q-bus resident LSI-11/73 and the Unibus resident PDP-11/44 allows events to be transferred between the two computers by a path separate from the heavily used FastBus. This interface has two 'windows' each acting as a buffer for up to four formatted events. Each computer can access one of the windows at a time, with the windows switched on request of the host.

On-line selection of events and tape writing is controlled by the host PDP-11/44. Wire address (not time) information is used from the $X$ and $Y$ chambers in a search for two track candidates that point back to the decay region. The on-line program also monitors the performance of the detector and provides the experimenters with numerous diagnostic aides.

Additional electronics was built for E845 to improve the time resolution of the detector. A 40 channel TDC with least count of 120 psec was built the trigger counters ($F, E$ and $\check{C}$). Also the pulse height of the $F$ and $\check{C}$ counters was recorded in FastBus ADCs to aid in the (off-line) rejection of $K_L \to 2\pi^0$ and $3\pi^0$ decays with two $\pi^0$s suffering highly asymmetric Dalitz decays. In such events the charged spectrometer might only detect two complete tracks, but the Dalitz pairs might be revealed by pulse heights found in the upstream detectors.

The integration time for the analog to digital conversion of the lead glass pulse heights is $\sim 100$ nsec, and this is long compared to the time resolution in the rest of the detector. For E845 the analog signals from the lead glass were discriminated (using modified E780/845 drift chamber amplifier-discriminator cards) at a low threshold (equivalent to $\sim 300$ MeV). These discriminated signals were each fed into two channels of (E780/845 drift chamber) FastBus TDCs that have a 4 nsec least count, with a relative delay of 2 nsec between the inputs to the two channels. For the 600 channel system an effective least count of 2.3 nsec was obtained.

Figure 3 presents the invariant mass distribution for pairs of reconstructed gamma rays from a preliminary analysis of E845 data for $K_L \to \pi^+\pi^-\pi^0$ decays. Events have been selected from unbiased trigger data with two oppositely charged tracks that satisfy the kinematic constraints ($p_t$ of the $\pi^+\pi^-$ pair vs. the invariant mass of the $\pi^+\pi^-$ pair) consistent with $K_L \to \pi^+\pi^-\pi^0$ decay. In one case no timing constraint is placed on the gamma rays; in the other case the gamma rays are required to be 'in time' to $\pm 4$ nsec with the rest of the event. The event time is defined by the timing of the scintillation trigger counters and the gamma ray time is determined by the lead glass block with the largest pulse height in the cluster. The importance of the timing criteria is obvious. The timing resolution has been found to be $\sim 1.4$ nsec or better throughout the detector.

**Present Status**

The current AGS running period, which ends during this conference, has been a very successful one for E845. The experiment was approved in February 1988 and installed in the interim between the end of E780 (June 1988) and the beginning of the current run (January 1989). The detector was commissioned in January and after climbing the usual learning curve a stable running condition was established. About $10^{12}$ protons per pulse were taken on target resulting in about 500 fast triggers per pulse and about 6 MHz singles rates at the face of the lead glass array. About 80 events were written to tape per pulse after passing on-line track editing. In all some 700 (6250 bpi) tapes were collected — roughly 100 million events.

Based on previous experience with the E780 $K_L \to \pi^0 e^+e^-$ search, an initial version of the first pass data reduction program has been written and is currently under evaluation and refinement. The intent is to perform the first pass reduction on the Brookhaven IBM-3090 and it is estimated that about 1000 CPU hours will be needed.

Very preliminary estimates indicate a sensitivity to a branching fraction at the level of several times $10^{-10}$ for $K_L \to \pi^0 e^+e^-$. This will represent more than an order of magnitude improvement over the current limit. Similar sensitivities are expected for other rare decay modes of the $K_L$ to electromagnetic final states such as $K_L \to e^+e^-$ and the Dalitz decays $K_L \to e^+e^-\gamma$ and $K_L \to e^+e^-e^+e^-$.

The decay $K_L \to e^+e^-$ is a flavor changing neutral current which is expected to occur in the Standard Model at $\sim 10^{-4}$ the rate for $K_L \to \mu^+\mu^-$ by virtue of the electron to muon mass ratio. The Dalitz decay modes are of interest because of the interplay required between the strong, weak and electromagnetic interactions. Several thousand $K_L \to e^+e^-\gamma$ events will be observed, to be compared with the world sample of a handful of events; the Kroll-Wada prediction for the $K_L \to e^+e^-\gamma$ branching fraction is $9 \times 10^{-6}$. It will be possible to make a detailed comparison of the $e^+e^-$ pair distribution with the Kroll-Wada predictions. These detailed studies can reveal information on the contribution of the so called *penguin* diagrams in $K$ decays.[11] The double Dalitz decay $K_L \to e^+e^-e^+e^-$ has yet to be observed; the Kroll-Wada prediction for the branching fraction is $3.6 \times 10^{-8}$.

**Acknowledgements**

We wish to thank the management and staff at Brookhaven for the strong support that made the changeover from E780 to E845 possible. We wish to acknowledge the contributions to E845 by M. Mannelli and S.F. Schaffner resulting from their efforts on E780. This research is supported by the U.S. Department of Energy under contracts No. DE-AC02-76ER03075 and DE-AC02-76CH00016. One of us (M.P.S.) is supported in part by the Alfred P. Sloan Foundation.

## References

[1] S.F. Schaffner *et al.*, Phys. Rev. **D39** (1989) 990.

[2] E. Jastrzembski *et al.*, Phys. Rev. Lett. **61** (1988) 2300.

[3] Brookhaven AGS experiments 777 and 851, M. Zeller, spokesman.

[4] C.O. Dib, I. Dunietz, and F.J. Gilman, Phys. Rev. **D39** (1989) 2639.

[5] K. Peach and B. Winstein, these proceedings.

[6] G.D. Barr *et al.*, Phys. Letts. **B214** (1988) 303; L.K. Gibbons *et al.*, Phys. Rev. Letts. **61** (1988) 2661.

[7] J. Flynn and L. Randall, LBL-26310, 1988.

[8] M. Mannelli (1988), E.B.M. Mannelli (1989), and S.F. Schaffner (1989), Yale University doctoral dissertations, unpublished.

[9] Brookhaven AGS proposal 845.

[10] *Data Processing in AGS Experiment 780*, W.M. Morse, *et al.* Proceedings of the Workshop on Triggering, Data Acquisition and Computing for High Energy/High Luminosity Hadron-Hadron Colliders, Fermilab, 1985. Editors B. Cox, R. Fenner, and P. Hale.

[11] L. Bergström, E. Massó and P. Singer, Phys. Letts. **131B** (1983) 229.

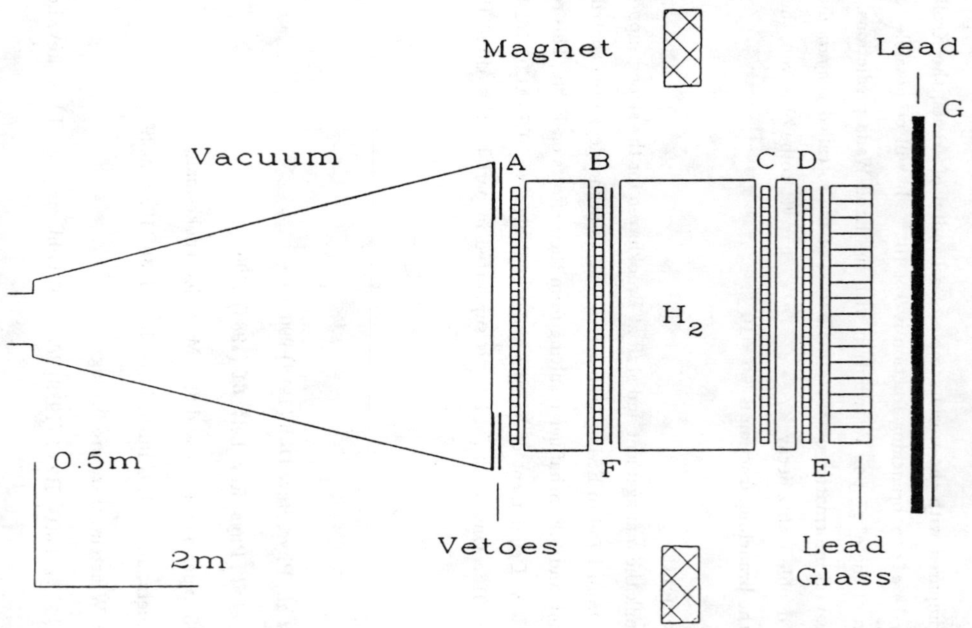


**FIGURE 1. A schematic plan view of the Brookhaven E845 detector. The neutral beam enters from the left.**

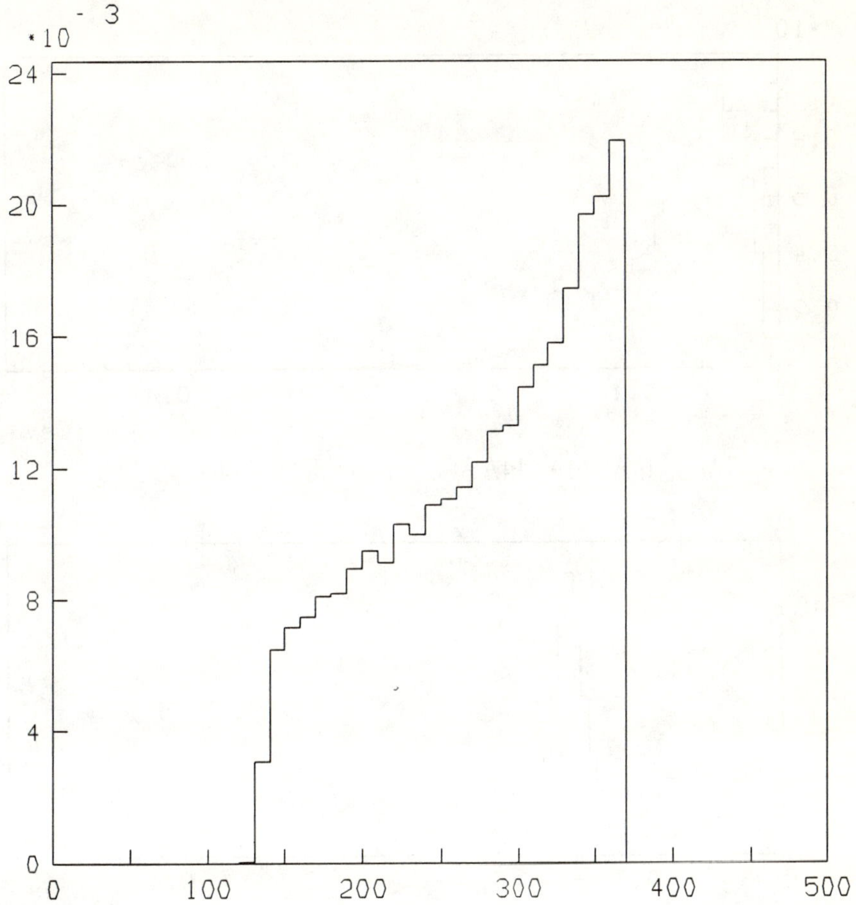


**FIGURE 2. The acceptance of the detector to $K_L \rightarrow \pi^0 e^+ e^-$ for $p_K > 4$ GeV/c as a function of the $e^+e^-$ pair mass (in MeV/c$^2$).**

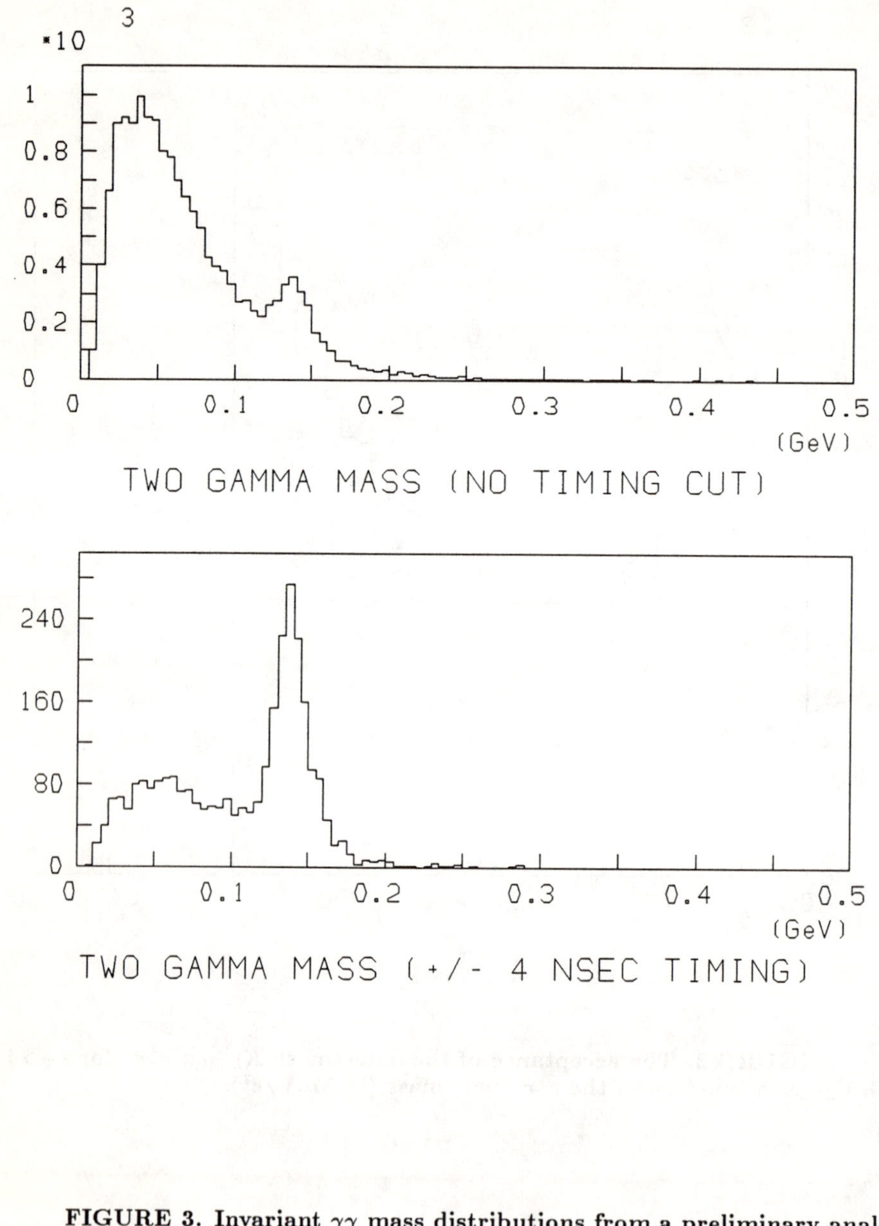


**FIGURE 3. Invariant $\gamma\gamma$ mass distributions from a preliminary analysis of BNL-845 data for $K_L \rightarrow \pi^+\pi^-\pi^0$ decays; with and without timing constraints on the gamma rays.**

# STUDY OF SOME K°RARE DECAY PROCESSES IN NA31

**Etienne Augé**

**Laboratoire de l'Accélérateur Linéaire - Orsay**

## Abstract

The large amount of data accumulated by the NA31 collaboration in order to measure Direct CP Violation in the $K^\circ \to 2\pi$ decays can also be used to study some rare $K_L$ and $K_S$ decays. A search for the decay $K_L \to \pi^\circ e^+e^-$ is presented, together with a search for the sequence $K_L \to \pi^\circ H^\circ$ followed by $H^\circ \to e^+e^-$, where $H^\circ$ would be a scalar boson. The prospects for studying the decays $K_L \to \pi^\circ\gamma\gamma$ and $K_L \to \gamma e^+e^-$ are presented. A new limit on the decay $K_S \to e^+e^-$ is also given.

### 1) Motivation

The interest of a search for some yet unobserved $K^0$ decays has been discussed extensively [1]. These rare decays can be grouped in classes corresponding to different physics motivations.

• The first class corresponds to decays which are highly suppressed in the Standard Model. Because of some new physics, this suppression might not hold, and the corresponding final state might be observed at a very different rate than the one predicted within the Standard Model. Moreover, putting upper limits on these exotic decay modes provides constraints on possible extensions of the Standard Model. For example, an upper limit on the decay $K_S^0 \rightarrow e^+e^-$ around $10^{-10}$ would not allow Flavor Changing Neutral Currents to play a significant role up to energy above a TeV. In the same spirit, direct CP violation in $K^0 \rightarrow 2\pi$ is a good place to learn new physics, because of the smallness of the Standard Model prediction.

• A second class of interesting decay modes allows a better understanding of the Standard Model (if no new physics is discovered). Some more information about quark mixings, in connection with CP violation would be especially interesting. This could be obtained from the decay $K_L^0 \rightarrow \pi^0 e^+e^-$. Although there is some controversy on the predictions [2], the CP conserving amplitude, which is second order in electromagnetic interactions (fig.1), is thought to be small compared to the CP violating ones. Moreover, the direct CP violation (through an electromagnetic penguin graph) should be comparable to the CP violation from states mixing. The situation is therefore much cleaner than in the $2\pi$ case, where the CP violation from states mixing dominates. Unfortunately, the branching ratio is only $10^{-11}$, still out of reach for the present experiments.

• Some experimental information allowing a better understanding of strong interactions in the non-perturbative domain of QCD is also missing. This reflects in particular into the large theoretical uncertainty on the quantities $\epsilon$ and $\epsilon'$ describing CP violation. Some particular decay processes such as $K_S^0 \rightarrow \gamma\gamma$ and $K_L^0 \rightarrow \pi^0\gamma\gamma$ are especially suited to test the various models [3]. Measuring the latter mode would also clearly settle the controversy about the CP conserving amplitude in $K_L^0 \rightarrow \pi^0 e^+e^-$ (fig. 1). The $K_L^0 \rightarrow \gamma e^+e^-$ (Dalitz decay) also belongs to this class, since the shape of the $e^+e^-$ mass spectrum is related to the amplitude of the penguin graphs in $K^0 \rightarrow 2\pi$ [4].

• Finally, the Standard Model Higgs boson might be discovered in the decay $K_L^0 \rightarrow \pi^0 H^0$, if it was light enough. The decay $H^0 \rightarrow e^+e^-$ would be dominant for Higgs masses between $2m_e$ and $2m_\mu$.

### 2) Data

In order to measure the ratio $\epsilon'/\epsilon$ with the desired precision of 0.1%,the NA31 collaboration [5] needed a large number ($10^9$) of $K_L^0$ to decay in the experimental fiducial region,with very clean beam conditions. The number of $K_S^0$ needed being much smaller ($10^7$), our sensitivity is better for $K_L^0$ than for $K_S^0$. The on-line event selection accepted final states with 2 to 4 electromagnetic showers, no missing transverse momentum, and no muon. Athough this was intended essentially to select $\pi^0\pi^0$ final states, all the decay modes mentionned above are also selected.

Data taken in 1986 have been extensively studied.Data taken in 1988 are being analysed, and new data will be collected during the second half of 1989.

### 3) Search for $K_L^0 \rightarrow \pi^0 e^+e^-$

This study has been performed with 1986 data only, and the result is published [6]. The final state contains 4 electromagnetic showers, like $\pi^0\pi^0$. The impact points and energies are measured with a Liquid Argon Calorimeter (LAC - fig.2). The two electron tracks are measured with two Drift Chambers.

We have imposed strict fiducial cuts to ensure that each shower was correctly measured. For example, we have rejected events with two showers aligned in one direction. We have rejected events when the energy barycenter in LAC was too far from the beam spot (i.e. when some transverse momentum was missing),and events with a fifth shower. Finally, we have used the information on lateral and longitudinal shower profiles to reject $\pi^+\pi^-\pi^0$ final states.

We estimate the longitudinal position of the decay vertex in two ways. We use the intersection point of the two electron tracks ("charged vertex") to compute the 4-particle invariant mass. We use the point where the 4-particle invariant mass is the nominal kaon mass ("neutral vertex") to compute the $\gamma\gamma$ invariant mass. In this way, the correlation between the 4-particle mass and the $\gamma\gamma$ mass is negligible for good events. Conversely, some correlation is observed for $\pi^+\pi^-\pi^0$ background events (fig.3). The ellipse drawn on figure 3 corresponds to three standard deviations in each of the two masses. In real data, a single event is found in that ellipse.

We anticipate two main background sources. The first one comes from decays into $\pi^0\pi^0$ where each photon from a $\pi^0$ converts into an $e^+e^-$ pair. The most probable place for this conversion corresponds to the window closing the decay region, which represents 0.004 $X_0$ (fig.4). Since we have no magnetic field, each pair is seen as a single electron track. These double conversion events should have a "$e^+e^-$" invariant mass equal to the nominal $\pi^0$ mass. This is indeed the case for the single observed candidate. Since the amount of data used corresponds to 70,000 measured $K_L^0 \to \pi^0\pi^0$, we expect 1.5 such events. The second background source comes from $K_L^0 \to \pi^0\pi^0$, where one of the $\pi^0$ subsequently decays into $\gamma e^+e^-$ (Dalitz decay), and the photon is missed. This background is negligible, as will be clear later.

The simulation used to compute the detector acceptance has been extensively tested on $\pi^0\pi^0$ and $\pi^+\pi^-$ final states.The only additionnal test needed for the present study concerns event acceptance as the angle between the two electron tracks becomes small. We have checked that the simulation correctly reproduces the $e^+e^-$ mass distribution for $K_S^0 \to \pi^0\pi^0$ followed by the decay into $\gamma e^+e^-$ for one of the $\pi^0$s (fig. 5). The acceptance varies slowly with the mass, and remains above 10% down to a mass of 10 $MeV/c^2$. These events, called $(\pi^0\pi^0)_D$ in the following, were selected by a dedicated trigger. The sample used here correspond to 1986 and 1988 data together, and to slightly looser fiducial cuts.

The average acceptance for the expected $K_L^0 \to \pi^0 e^+e^-$ signal is 13.1%. We use the observed the 70,000 observed $K_L^0 \to \pi^0\pi^0$ decays to normalise our result. The average acceptance within the chosen cuts is 14.3%. We can therefore conclude

$$Br(K_L^0 \to \pi^0 e^+e^-) < 4 \cdot 10^{-8} \quad (90\% \ Confidence \ Level)$$

### 4) Search for the decay $K_L^0 \to \pi^0 H^0$ followed by $H^0 \to e^+e^-$

This study, which is similar to the previous one, has been performed on 1986 and 1988 data samples ($\approx$ 150,000 $K_L^0 \to \pi^0\pi^0$ events in total). We have chosen looser fiducial cuts to improve the acceptance at low $e^+e^-$ mass (this was not critical for the previous study), and we have allowed for a measurable $H^0$ distance of flight. This distance is indeed expected to be $c\tau \approx 1\ cm \cdot (100MeV/m_H)$ in the minimal Standard Model.

As a measure of the longitudinal position of the K decay vertex,we choose the point such that the $\gamma\gamma$ invariant mass is the nominal $\pi^0$ mass (fig.6). The direction of the $K_L^0$ is approximated by the line joining the target to the barycenter of the 4 showers in the calorimeter. The direction of the $e^+e^-$ pair is reconstructed from the two electron tracks and energies. We define a "miss distance" to be the distance between the $e^+e^-$ direction and the $K_L^0$ direction in the plane transverse to the beam axis at the K vertex. Both projections of this distance (x and y) should be compatible with 0 within 2.5 standard deviations ($\sigma$). This quantity is computed for each event. The cut has been checked using $K_S^o$ and $K_L^o \to (\pi^o\pi^o)_D$ events (see fig. 7). We then cut at two standard deviations around the nominal $K^o$ mass (fig. 8), and we are left with three events (including the $\pi^o e^+e^-$ candidate mentionned in the previous section).

The expected background again comes from "double conversion events" and from $\pi^o\pi^o$ events where one $\pi^o$ decays in to $\gamma e^+e^-$ with the $\gamma$ unobserved. We expect 2.6 events from the first source and 0.7 fom the second. Figure 9.a shows the $e^+e^-$ pair invariant mass as a function of the $H^o$ flight path for the three selected events. One can see that two of them indeed have the nominal $\pi^o$ mass,whereas the third one has a much smaller mass, as expected for the Dalitz decay of a $\pi^o$. All three of them have a flight path compatible with 0.

The background estimate has been checked with $K_S^o$ data, where the $\pi^o\pi^o$ final states are plenty. We then expect 20 double conversion events and 5 events with a $\pi^o$ Dalitz decay. We observe 31 events (fig.9.b).

The acceptance has been computed from the simulation for various masses and lifetimes for the Higgs boson. We have used the observed $(\pi^o\pi^o)_D$ events to normalise. For each value of the Higgs boson mass and life time, we compute a 90% confidence upp.- limit for the product of the branching ratios :

$$Br(K_L^o \to \pi^o H^o) \cdot Br(H^o \to e^+e^-)$$

Figure 10 shows contours of equal upper limit in the mass-life time plane. The dashed line on that figure represents the relation between the mass and the life time expected for the Higgs boson of the minimal standard model. Taking this relation into account, and given the fact that in the mass region considered the $H^o \to e^+e^-$ decay mode should be dominant, we conclude that the 90% confidence level upper limit for the branching ratio $Br(K_L^o \to \pi^o H^o)$ ranges from $10^{-7}$ for $m_H = 15MeV/c^2$ to $10^{-8}$ for $m_H$ above 100 $MeV/c^2$. This may be compared with some theoretical predictions [7],[8],[9] in the range $10^{-6} - 10^{-3}$ (fig. 11). We conclude that,unless the top quark mass precisely coincides with a specific value (which is most un-natural), the Higgs boson of the minimal standard model is excluded in the range from 10 to 200 $MeV/c^2$.

### 5) Measurement of the decay $K_S^o \to \gamma\gamma$

This decay has been seen for the first time by NA31 [10]. The measurement gives :

$$\frac{\Gamma(K_S^o \to \gamma\gamma)}{\Gamma(K_L^o \to \gamma\gamma)} = 2.3 \pm 1.0 \pm 0.4$$

which translates into : $Br(K_S^o \to \gamma\gamma) = (2.4 \pm 1.2)10^{-6}$. This results agrees nicely with the predicted value of $2 \cdot 10^{-6}$ [11].

### 6) Measurement of the decay $K_L^o \to \pi^o\gamma\gamma$

The predicted branching ratio for this decay mode is $6.8 \cdot 10^{-7}$ [3], and only an upper limit is available from experiment [12]. To measure this decay mode, one indeed has to fight against the tremendous background from $K_L^o \to 3\pi^o$ decays. Given the excellent granularity and excellent resolution of the NA31 calorimeter [13], we anticipate being able to isolate around 20 such events on a background of 20 events. If necessary, the peculiar invariant mass spectrum predicted for the $\gamma\gamma$ pair might help reducing the background. Work on the subject is in progress.

### 7) Search for the decay $K_S^o \to e^+e^-$

This search has been performed with 1986 data.No event has been found within the fiducial cuts, despite the good acceptance (50%). We obtain the preliminary result :$Br(K_S^o \to e^+e^-) < 5 \cdot 10^{-7}$, which is very far from the Standard Model predictions ($10^{-14}$).

**8) Measurement of the decay $K^o_L \to \gamma e^+ e^-$**

The branching ratio for this decay mode is predicted to be $10^{-5}$. We observe approximatively 800 such decays (1988 data) on a very low background. A significant measurement of the $e^+e^-$ invariant mass spectrum at high mass should therefore be possible. Some work is in progress.

**Conclusion**

Interesting informations about hadronic matrix elements are expected from the measurement of the decay modes $K^o_L \to \pi^o\gamma\gamma$ and $K^o_L \to \gamma e^+e^-$, together with the already measured $K^o_S \to \gamma\gamma$ mode. Upper limits have been obtained for the decay modes $K^o_L \to \pi^o e^+e^-$ and $K^o_S \to e^+e^-$. No signal was indeed expected within the Standard Model, since the predictions lie much below the obtained limits. In the end, we find that a Higgs boson with mass between 10 and 200 $MeV/c^2$ is very unlikely within the minimal Standard Model.

I would like to aknowledge major contributions from H.N. Nelson, G.D. Barr, H. Rohrer, L. Gatignon and D. Cundy. I thank A. Schaffer for a critical reading of the text.

**References**

1) A. Pich - CERN preprint TH5194/88
W.J. Marciano - BNL preprint 42381
F. Gilman - these proceeding
E. De Rafael - these proceedings
2) Ecker, Pich, De Rafael - Nucl. Phys. B291 (1987) 692
Seghal - Phys. Rev. D38 (1988) 808
Donoghue, Holstein, Valencia - Phys. Rev. 35 (1987)2769
Gilman, Wise - Phys. Rev. 21 (1980) 3150
3) Pich, De Rafael - Phys. Lett. 189B (1987) 363
Seghal - Aachen preprint PITHA 89/16 (1989)
4) Bergstrom, Masso, Singer - Phys. Lett. 131B (1983) 229
5) Burkhardt et al - Phys. Lett. 206B (1988) 169
6) Barr et al - Phys. Lett. 214B (1988) 303
7) Willey - Phys. Lett. 173B (1986) 480
8) Chivukula, Manohar - Phys. Lett. 207B (1988) 123
9) Pham, Sutherland - Phys. Lett. 151B (1985) 444
10) Burkhardt et al - Phys. Lett. 199B (1987) 139
11) L.L. Chau, H.Y. Cheng - Phys. Rev. Lett. 54 (1985) 1768
G. D'Ambrosio, D. Espriu - Phys. Lett. 175B (1986) 237
J.L. Goity - Zeit. für Phys. C34 (1987) 341
12) E 731 collaboration, presented by R. Patterson - these proceedings
13) Burkhardt et al - Nucl. Inst. and Meth. 268A (1988)

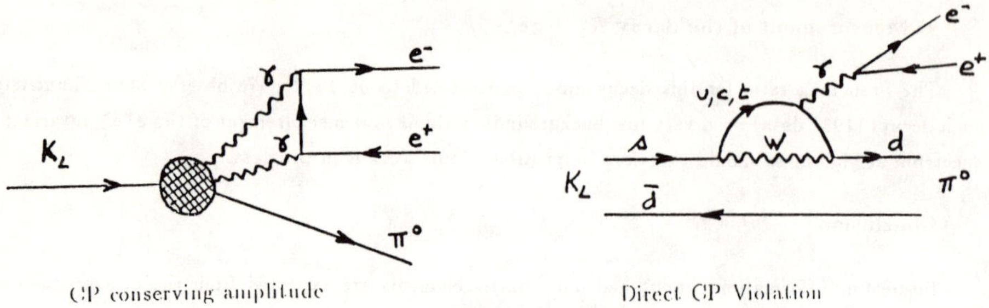


Figure 1 : Contributions to the decay $K_L^0 \to \pi^0 e^+ e^-$

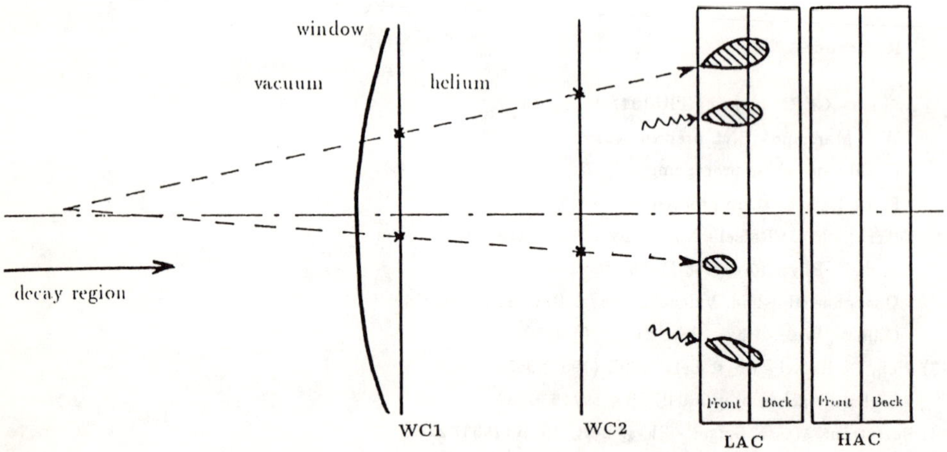


Figure 2 : Aspect of the $\pi^0 e^+ e^-$ final state in the apparatus

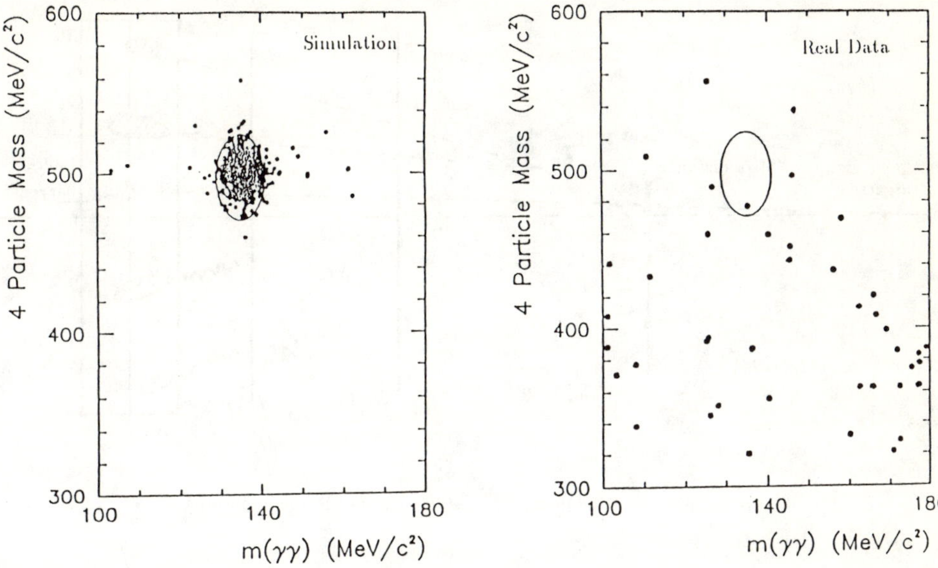


Figure 3 : Selected $K_L^0 \to \pi^0 e^+ e^-$ candidates

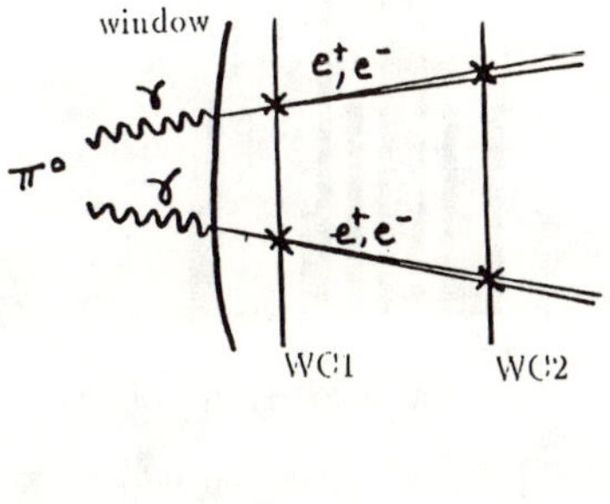


Figure 4 : Background from Double Conversion

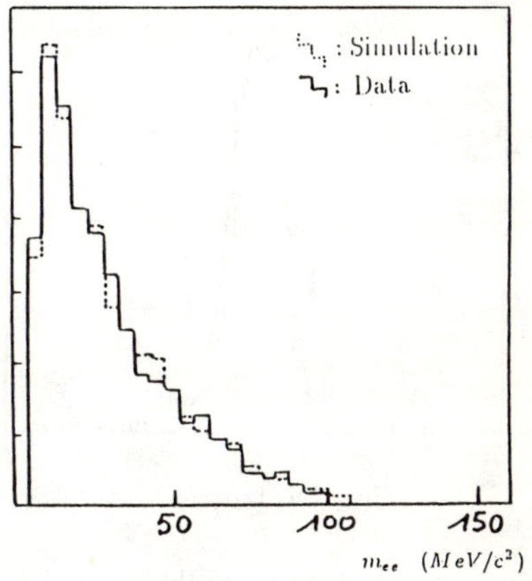


Figure 5 : (ee) mass spectrum for $K_S^0 \to \pi^0 \pi^0$ $\to \gamma\gamma$, $\to \gamma e^+ e^-$

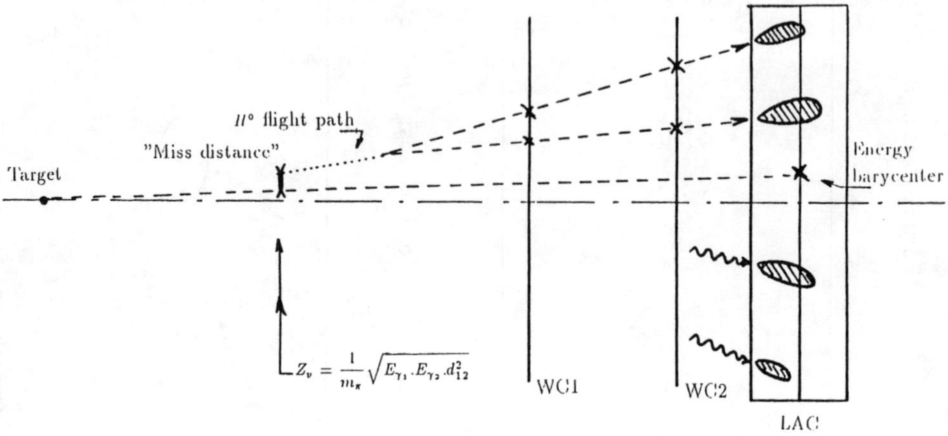


Figure 6 : Aspect of the decay $K_L^o \rightarrow \pi^o H^o \rightarrow \pi^o e^+ e^-$ in the apparatus

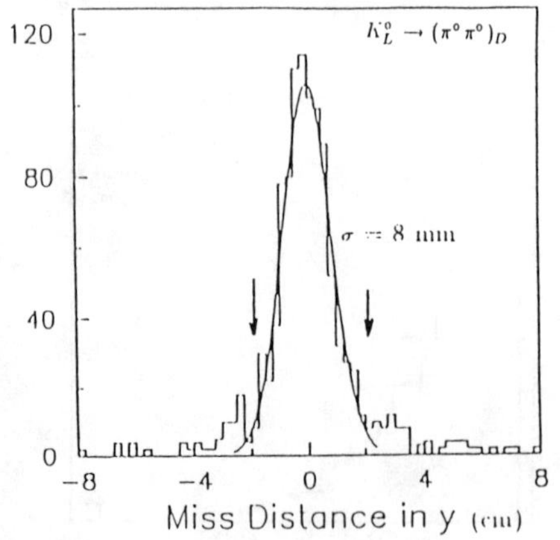


Figure 7 : Check of the cut on the Miss distance

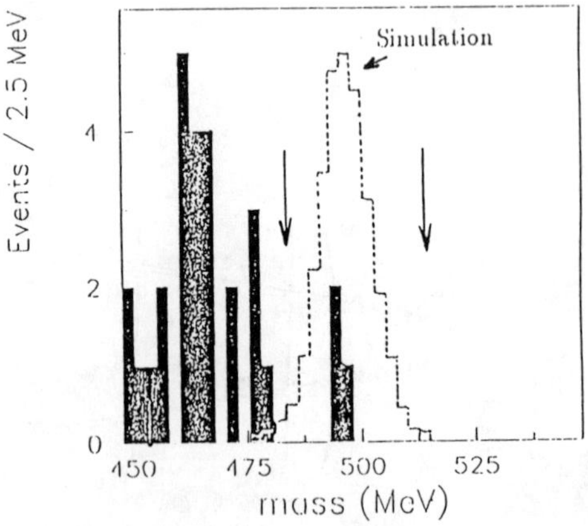


Figure 8 : Total invariant mass spectrum

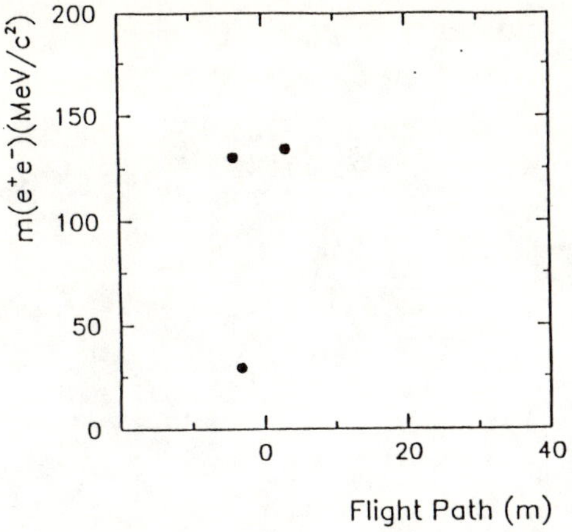


Figure 9.a : $K_L^0$ candidates

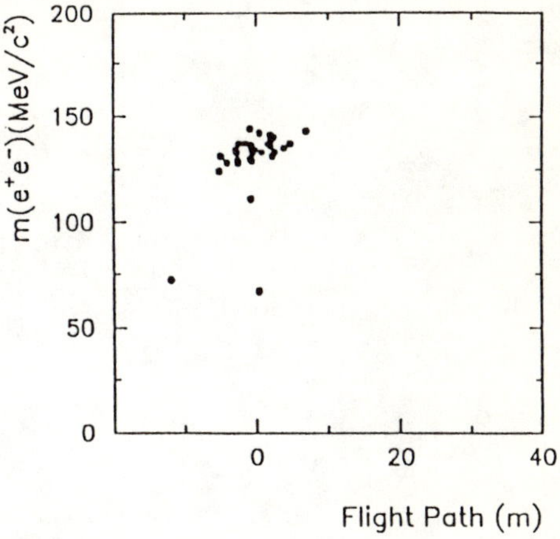


Figure 9.b : $K_S^0$ candidates

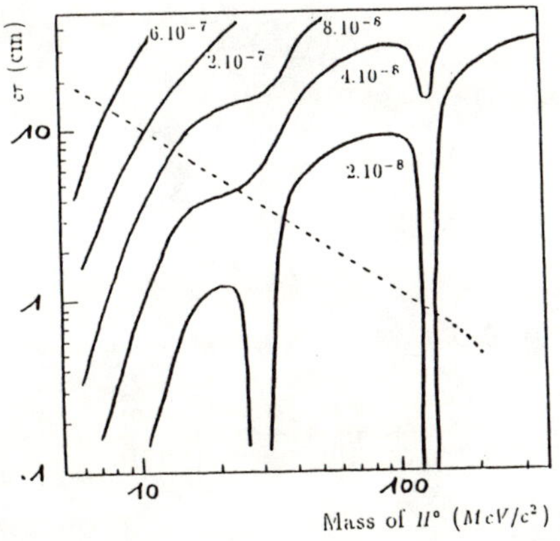


Figure 10 : Upper limit for $Br(K_L^0 \to \pi^0 H^0) \cdot Br(H^0 \to e^+e^-)$ (90% confidence level).

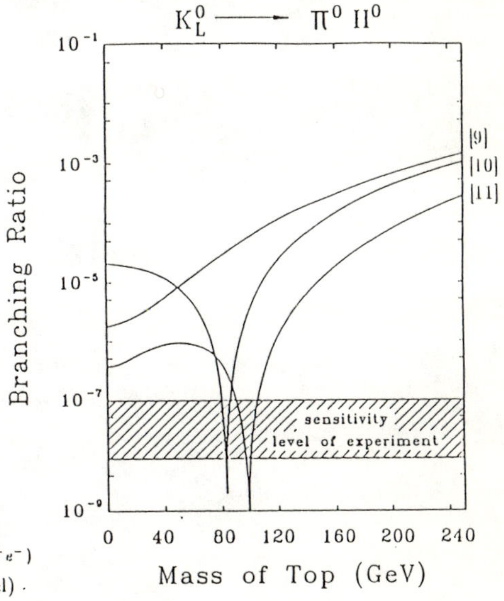


Figure 11 : Theoretical predictions for $Br(K_L^0 \to \pi^0 H^0)$

# SEARCHES FOR $K_{L,S} \to \pi^0 e^+ e^-$ AND $K_L \to \pi^0\gamma\gamma$ BY FERMILAB E-731

J. R. Patterson
The Enrico Fermi Institute and the Department of Physics,
The University of Chicago

## ABSTRACT

Searches for the rare decay modes $K_{L,S} \to \pi^0 e^+ e^-$ and $K_L \to \pi^0\gamma\gamma$ were performed using 20% of the data from Fermilab experiment E-731. The decay to $\pi^0 e^+ e^-$ is a possible avenue of observation of direct CP violation, and the decay to $\pi^0\gamma\gamma$ is of interest for its contribution to the $\pi^0 e^+ e^-$ decay and as a test of chiral perturbation theory. The results are $BR(K_L \to \pi^0 e^+ e^-) < 4.2 \times 10^{-8}$, $BR(K_S \to \pi^0 e^+ e^-) < 4.5 \times 10^{-5}$, and $BR(K_L \to \pi^0\gamma\gamma) < 2.7 \times 10^{-6}$, all with 90% confidence.

**Introduction**

In this talk I will present the results of searches for the rare kaon decays $K_{L,S} \to \pi^0 e^+ e^-$ and $K_L \to \pi^0 \gamma\gamma$ by the E-731 experiment at Fermilab. E-731 is a collaboration between the University of Chicago, Elmhurst College, Fermilab, Princeton University and the CEN Saclay Laboratory, and was designed to measure the CP violation parameter Re($\varepsilon'/\varepsilon$). During the experimental run, which lasted from August, 1987 to February, 1988, about one billion $K_L$ decayed in the apparatus. So far 20% of the data collected have been analyzed. Aside from the two pion decays collected for the Re($\varepsilon'/\varepsilon$) measurement, the trigger accepted possible decays to a number of rare decay modes listed in Table I, and when analysis of the full data set is complete, we expect to present our findings on all of these modes. Because the $\pi^0 e^+ e^-$ and $\pi^0 \gamma\gamma$ mode are of special interest, searches for these decays have been completed using the partial data set.

Table I

| Mode | PDG Branching Fraction[1] |
|---|---|
| $K_L \to \pi^0 e^+ e^-$ | $< 2.3 \times 10^{-6}$ |
| $K_S \to \pi^0 e^+ e^-$ | no existing limit |
| $K_L \to \pi^0 \gamma\gamma$ | $< 2.4 \times 10^{-4}$ |
| $K_L \to \pi^+ \pi^- \gamma$ | $(4.4 \pm 0.3) \times 10^{-5}$ |
| $K_L \to \pi e \nu \gamma$ | $(1.3 \pm 0.8) \times 10^{-2}$ |
| $K_L \to \pi^0 \pi e \nu$ | $(6.2 \pm 2.0) \times 10^{-5}$ |
| $K_L \to e^+ e^- \gamma$ | $(1.7 \pm 0.9) \times 10^{-5}$ |
| $K_L \to e^+ e^- e^+ e^-$ | $< 2.6 \times 10^{-6}$ |
| $\pi^0 \to e^+ e^-$ | $(1.8 {}^{+0.7}_{-0.6}) \times 10^{-7}$ |
| $K_L \to \pi^0 \pi^0 \gamma\gamma$ | no existing limit |

The importance of the $K_L \to \pi^0 e^+ e^-$ decay has already been described in the theoretical talks of Fred Gilman and Edouardo de Rafael. It can occur either via CP conserving or CP violating channels. In contrast to the $\pi\pi$ modes, however, direct CP violation is expected to contribute about the same amount to the decay amplitude as CP violation due to mixing. Thus it provides an alternative window for the observation of direct CP violation. Standard Model calculations generally predict that the branching ratio for the decay will lie in the $10^{-11}$ range[2]. Predictions for $K_S \to \pi^0 e^+ e^-$ vary from $5 \times 10^{-9}$ to $5 \times 10^{-10}$.

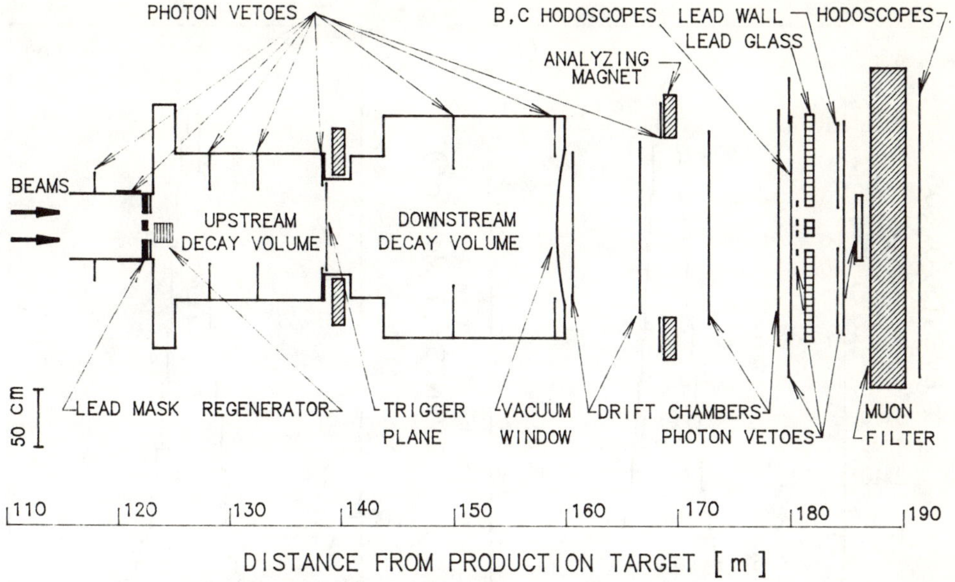


**Figure 1.** Detector schematic, elevation view.

**Detector**

A schematic representation of the apparatus used for the experiment is shown in Figure 1. It is composed of an evacuated decay volume followed by a four drift chamber spectrometer and an 804 block lead glass electromagnetic calorimeter. The momentum resolution of the spectrometer is $(\Delta p/p)^2 = (0.6\%)^2 + (0.016\%\ p[\mathrm{GeV/c}])^2$, and the energy resolution of the calorimeter is about $1.5\% + 5\%/\sqrt{E[\mathrm{GeV}]}$ for electrons, and about 1% worse, added linearly, for photons. In addition, there are 11 planes of counters distributed along the apparatus, each collaring the fiducial volume, which are used to detect escaping particles.

**Search for $K_{L,S} \to \pi^0 e^+ e^-$**

For the purpose of measuring $\mathrm{Re}(\varepsilon'/\varepsilon)$ the experiment was triggered on either two tracks originating in the upstream portion of the decay volume (the "two track" trigger), or on four electromagnetic showers in the lead glass with total energy greater than about 30 GeV, and no charged particles in the upstream segment of the decay volume (the "four cluster" trigger). $\pi^0 e^+ e^-$ decays in the upstream portion of the decay volume satisfied the first of these, while those in the downstream portion satisfied the second. Events with two charged tracks and four clusters in the calorimeter were reconstructed as follows: the tracks were matched with two of the clusters in the lead glass, and were identified as electrons

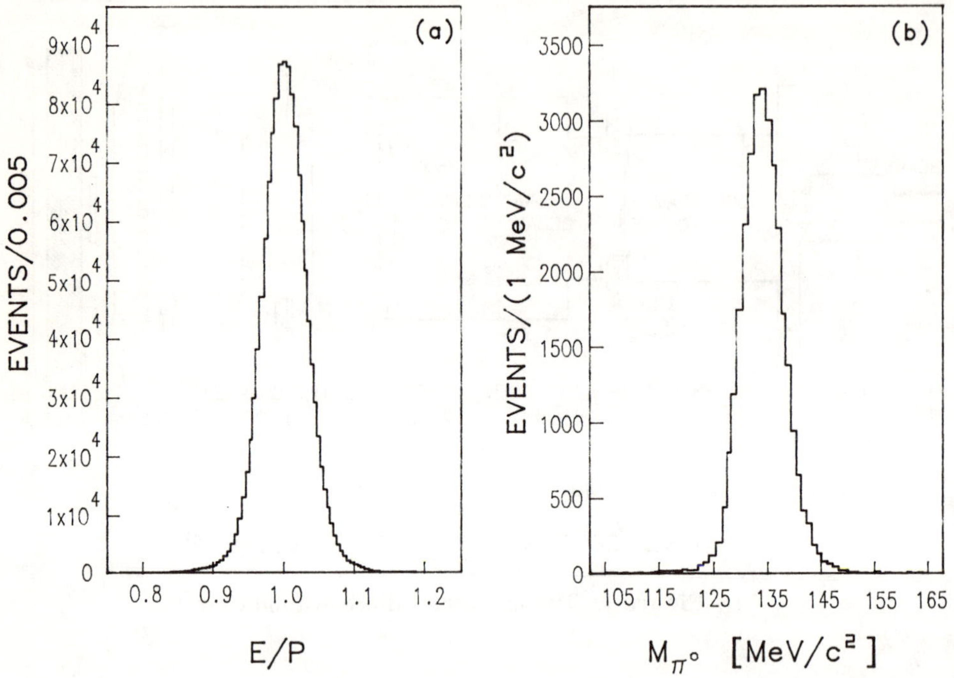


**Figure 2.** (a) E/p in the lead glass for electron calibration data. The resolution is about 4%; (b) Two photon invariant mass for $\pi^+\pi^-\pi^0$ events.

if the ratio of energy measured in the calorimeter to track momentum (E/p) was close to one. The E/p distribution of a sample of independently identified electrons is shown in Figure 2a. For each track it was required that $0.85 < E/p < 1.15$, a cut which retains 95% of true electrons, but only about 2% of hadrons. The decay vertex was then identified by finding the position of closest approach of the two tracks. This was used along with the energies and positions of the two remaining clusters to calculate the two photon invariant mass, $m_{\gamma\gamma}$, which was required to be within $\pm 5$ MeV of the $\pi^0$ invariant mass. For comparison, $m_{\gamma\gamma}$ of $\pi^+\pi^-\pi^0$ events similarly reconstructed is shown in Figure 2b.

For the remaining $\pi^0e^+e^-$ candidates Figure 3b shows the squared transverse momentum of the decay products with respect to that of the original kaon ($P_t^2$) plotted against invariant mass ($m_{\pi ee}$). For comparison, the corresponding plot is shown for $\pi^+\pi^-\pi^0$ decays in Figure 3a. No events are observed in the signal region defined by $0.489\ \mathrm{MeV/c^2} < m_{\pi ee} < 0.507\ \mathrm{MeV/c^2}$ and $P_t^2 < 200\ (\mathrm{MeV/c})^2$ which is expected to contain 95% of good $\pi^0e^+e^-$ decays. The number and distribution of the three events observed in the much broader window $0.470\ \mathrm{MeV/c^2} < m_{\pi ee} < 0.525\ \mathrm{MeV/c^2}$ and $P_t^2 < 1000\ (\mathrm{MeV/c})^2$ are consistent with radiative $K_L \to \pi e\nu$ decays with an accidental photon superimposed.

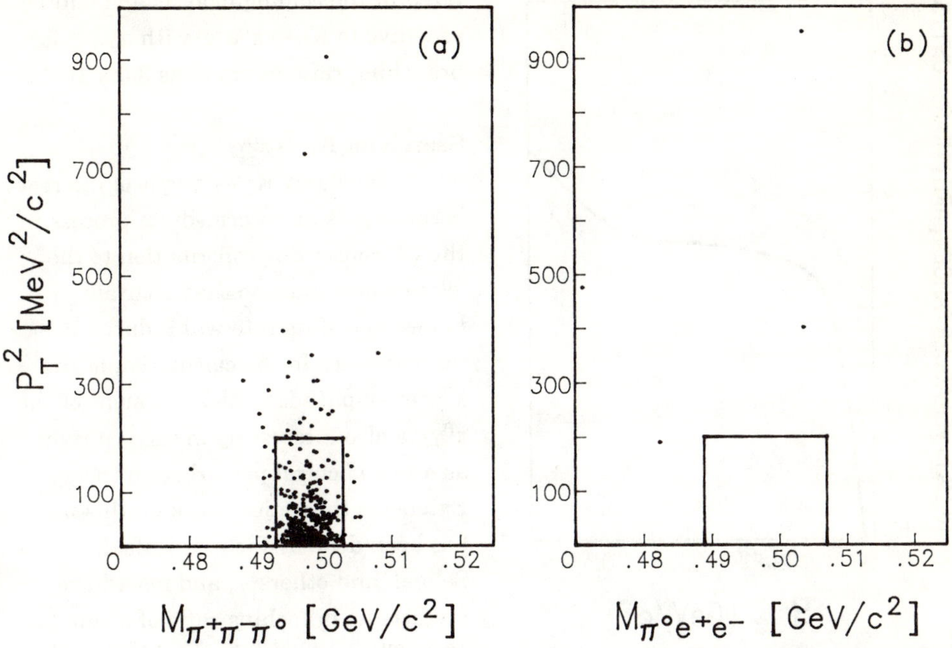


**Figure 3.** The squared transverse momentum versus invariant mass for (a) $K_L \to \pi^+\pi^-\pi^0$ events, and (b) candidate $K_L \to \pi^0 e^+ e^-$ events. In (b), 24 events are above the top of the plot. 95% of signal events are expected to lie within the inscribed boxes. There are no background events for the $K_S$ decay.

The acceptance for $\pi^0 e^+ e^-$ decays was calculated using a detailed monte carlo developed for use in the Re($\varepsilon'/\varepsilon$) analysis, and its dependence on the invariant mass of the electron-positron pair is shown in Figure 4. The absolute tracking efficiency was checked by verifying that the measured ratio of $\pi^0\pi^0$ and $\pi^+\pi^-$ branching fractions was correct within the desired accuracy. For events satisfying the "four cluster" trigger the overall acceptance is 9.5%. When normalized to the 59,000 observed $K_L \to \pi^0\pi^0$ decays for which the acceptance is about 10%, this subset of the data sample leads to an upper limit on the branching ratio of $4.7 \times 10^{-8}$. The data satisfying the "two track" trigger were normalized to 71,000 observed $K_L \to \pi^+\pi^-$ and 224,000 $K_S \to \pi^+\pi^-$ decays, yielding upper limits on the $K_L$ and $K_S$ branching ratios of $4.1 \times 10^{-7}$ and $4.5 \times 10^{-5}$ respectively. Combining the limits from the two triggers gives the results BR($K_L \to \pi^0 e^+ e^-$) < $4.2 \times 10^{-8}$ and BR($K_S \to \pi^0 e^+ e^-$) < $4.5 \times 10^{-5}$, with 90% confidence.

This result[3] is about an order of magnitude more sensitive than the upper limit reported by Brookhaven E845[4], and is comparable in sensitivity to that of CERN NA31[5]. The remaining 80% of the data from this experiment will be ana-

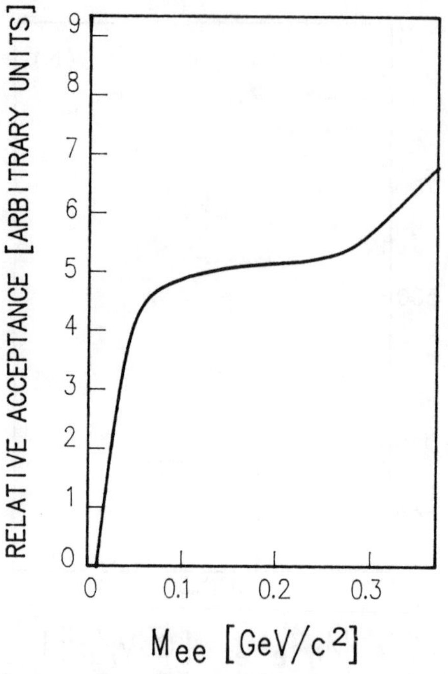


**Figure 4**. The relative acceptance of the $K_L \rightarrow \pi^0 e^+ e^-$ decay versus the $e^+e^-$ invariant mass for the neutral trigger.

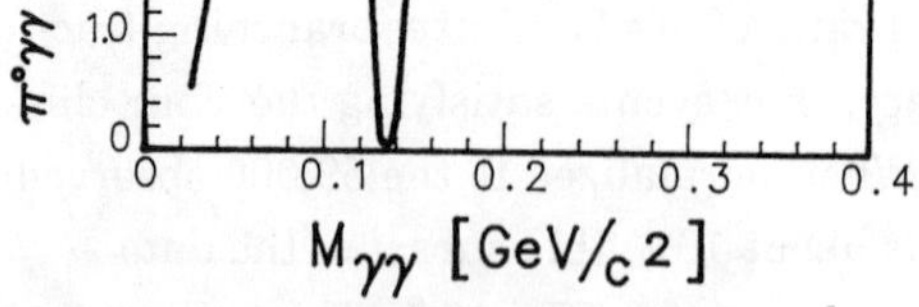


**Figure 5**. The acceptance of the $K_L \rightarrow \pi^0\gamma\gamma$ decay versus the invariant mass of the two photons.

lyzed in the coming months and will be sensitive to $K_L \rightarrow \pi^0 e^+ e^-$ with a branching ratio as small as $3.5 \times 10^{-9}$.

**Search for $K_L \rightarrow \pi^0\gamma\gamma$**

The decay $K_L \rightarrow \pi^0\gamma\gamma$ is of interest because it is an intermediate process in the CP conserving contribution to the $\pi^0 e^+ e^-$ decay rate, so that accurate knowledge of its rate will reduce uncertainties in the calculation of the $\pi^0 e^+ e^-$ amplitudes. Measurement of the $\pi^0\gamma\gamma$ is also interesting in its own right as a test of chiral perturbation theory. Extensive theoretical work on this mode has been done by Pich, de Rafael, Sehgal, and others[6,7], and most have predicted a branching ratio of about $7 \times 10^{-7}$. The best experimental limit is due to Banner, Cronin, Liu and Pilcher[8] and is $2.4 \times 10^{-4}$.

The "four cluster" trigger was used in the search for $K_L \rightarrow \pi^0\gamma\gamma$. Candidate events were reconstructed as follows: first the position of the decay vertex was calculated from the four final state photon energies and positions using the kaon mass as a constraint. Next the two photons originating from the pion were identified by choosing that pair with invariant mass closest to that of the pion. Finally, it was required that the remaining two photons be inconsistent with pion decay from the same vertex.

There is a potentially large background to the $\pi^0\gamma\gamma$ decay mode due to misreconstructed $K_L \rightarrow \pi^0\pi^0\pi^0$ decays. In particular, wide angle photons which miss the lead glass and photon showers which fuse in the lead glass can reduce the number of clusters observed from six to four. Background from such events has characteristics which can be used to reduce its size. First, photons escaping from the fiducial volume of the detector are generally detected by one of the 11 photon veto counter planes. Second, the center of energy at the lead glass of events with one or more missing photons is displaced, and finally, because some of the final state energy is missing, the reconstructed decay vertex is artificially shifted downstream. The decay vertex of background events due to overlapping photon showers also tends to be in the downstream portion of the decay region. In addition, fused clusters differ in shape from clusters composed of a single photon shower.

The final cuts were as follows: the invariant mass of the pair of photons from pion decay was required to be within 5 MeV of the true pion mass, while that of the other photon pair was required to be at least 12 MeV away from the pion mass. Kaons with energies between 40 and 150 GeV were accepted, and their center of energy at the lead glass was required to lie within the $K_L$ beam. To reduce $\pi^0\pi^0\pi^0$ background only decays occurring within 124 meters of the target were accepted (the upstream cut was at 110 meters, the point at which the detector acceptance becomes significant), and events with significant energy deposit in the photon veto counters, or with a cluster shape inconsistent with that expected for a single photon were rejected. Finally, events with charged particles were cut, as were events which were consistent with the $K_L \rightarrow \pi^0\pi^0$ hypothesis. After these cuts 130 candidate $\pi^0\gamma\gamma$ events remain.

The acceptance as a function of $m_{\gamma\gamma}$ is shown in Figure 5. It is about 4% everywhere above about 80 MeV/$c^2$, except at the pion mass where the cut has been applied. The observed $m_{\gamma\gamma}$ distribution is shown in Figure 6a, along with that predicted theoretically by Ecker, Pich and de Rafael[6]. The peak at about 0.27 GeV/$c^2$ is due to $\pi^0\pi^0\pi^0$ decays in which both photons from two of the pions fuse together: each photon cluster from one $\pi^0$ lies on top of a photon cluster from a second $\pi^0$. In order to eliminate most of this "double fusion" background while preserving 60% of the predicted signal, the limit is calculated considering only events with $m_{\gamma\gamma}$ greater than 0.3 GeV/$c^2$. With this cut there are 24 candidate events. If all of these events are assumed to be $\pi^0\gamma\gamma$ decays, then the corresponding upper limit on the branching ratio is 6 x $10^{-6}$. According to predictions, three or four of these events are due to real $\pi^0\gamma\gamma$ decays.

The limit can be much improved, however, if the origin of these candidates is understood, and for this we rely on a monte carlo simulation of $K_L \rightarrow \pi^0\pi^0\pi^0$ decays in our detector. The same $\pi^0\pi^0\pi^0$ monte carlo is used to simulate the background to the $K_L \rightarrow \pi^0\pi^0$ signal in the Re($\varepsilon'/\varepsilon$) analysis, and a sample distribution from that application is shown in Figure 7. The monte carlo duplicates the shape of the background distribution well, and the normalization, which is absolute, is correct to better than 10% of itself. Figure 6b shows the $K_L \rightarrow \pi^0\pi^0\pi^0$ monte carlo $m_{\gamma\gamma}$ distribution superimposed on the distribution for candidate $\pi^0\gamma\gamma$ events. The small background due to misreconstructed $\pi^0\pi^0$ decays as simulated by a $\pi^0\pi^0$

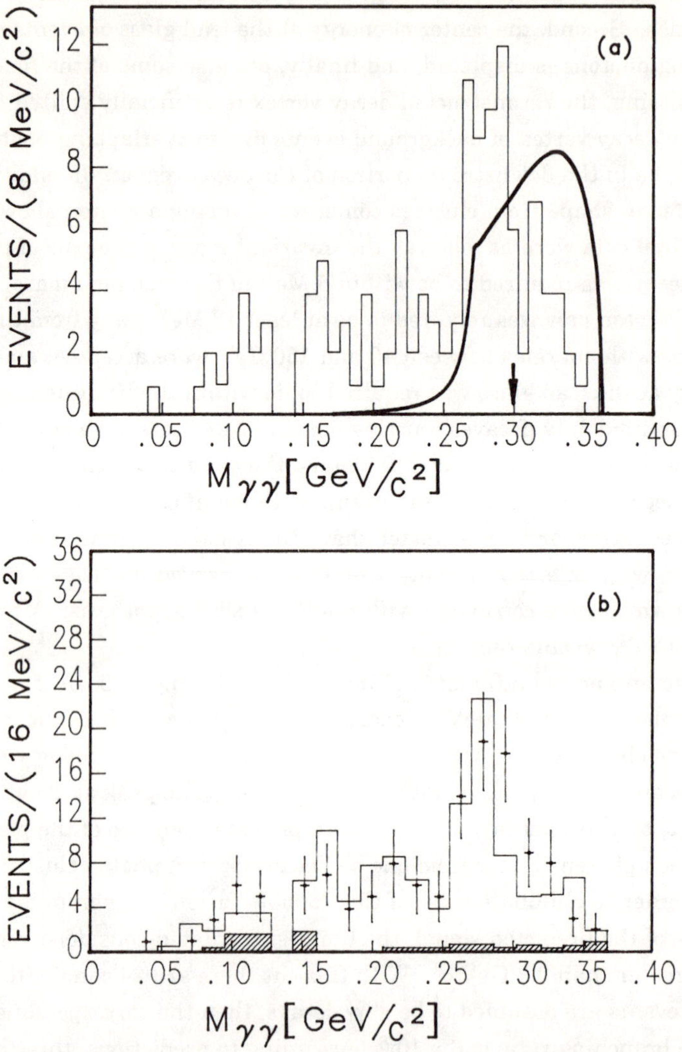


**Figure 6**. The invariant mass of the two photons. The histograms are the candidate events. (a) the solid curve shows the distribution predicted by chiral perturbation theory for a branching ratio of $10^{-5}$, and the arrow shows the position of the cut. (b) The crosses represent the sum of events from the $K_L \to \pi^0\pi^0\pi^0$ and $K_L \to \pi^0\pi^0$ monte carlos, absolutely normalized. The shaded histogram shows the contribution from $K_L \to \pi^0\pi^0$.

monte carlo is also shown. Figure 8 shows the decay vertex distribution of the data and the $K_L \to \pi^0\pi^0\pi^0$ monte carlo.

For $m_{\gamma\gamma} > 0.3$ GeV/c$^2$ there are $20.8 \pm 3.2 \pm 2.5$ background events due to misreconstructed $K_L \to \pi^0\pi^0$ and $K_L \to \pi^0\pi^0\pi^0$ decays, where the first error is statistical and the second is systematic, while $24 \pm 4.9$ events are observed. Based on a maximum liklihood fit, we conclude that there are fewer than 13.9 $\pi^0\gamma\gamma$ signal events in

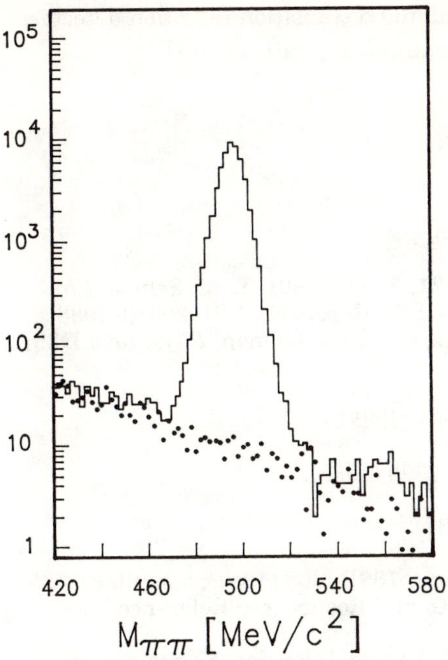


**Figure 7**. The $4\gamma$ invariant mass distribution for $K_L \rightarrow \pi^0\pi^0$ decays (histogram), and for the $K_L \rightarrow \pi^0\pi^0\pi^0$ monte carlo (dots). The monte carlo is absolutely normalized to the data, and has about twice the statistics.

that mass region, with 90% confidence. Normalizing to 11,000 $K_L \rightarrow \pi^0\pi^0$ decays in the same region and analyzed with similar cuts, for which the acceptance is 5.8%, leads to an upper limit on $BR(K_L \rightarrow \pi^0\gamma\gamma)$ of 2.7 x $10^{-6}$ with 90% confidence. If the limit is calculated accepting all $m_{\gamma\gamma}$ away from the double fusion peak which spans the region from 0.264 to 0.294 GeV/c$^2$, then the resulting limit is 4.4 x $10^{-6}$ with 90% confidence[9]. The first of these limits represents an improvement on the previous experimental limit by a factor of about 90.

**Future Prospects**

Upon completion of E-731 data analysis we can expect even more stringent limits on $\pi^0\gamma\gamma$ and $\pi^0e^+e^-$. Further improvement on rare decay limits can be expected from FNAL E-799. By increasing the kaon flux, tuning the

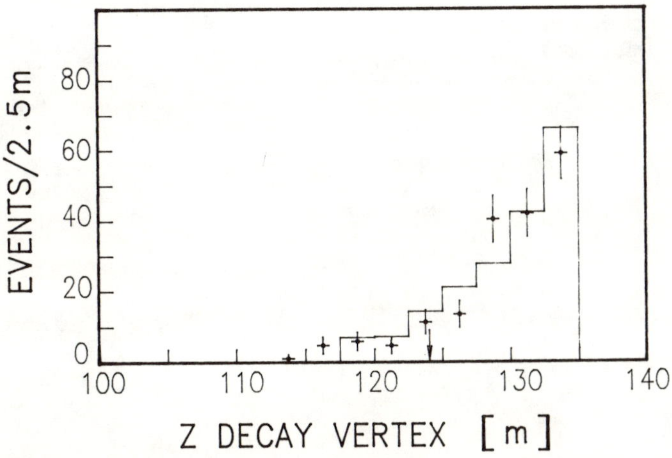


**Figure 8**. The position of the decay vertex along the beam axis for candidate $K_L \rightarrow \pi^0\gamma\gamma$ decays (histogram) and for the $K_L \rightarrow \pi^0\pi^0\pi^0$ monte carlo (crosses) for events with $m_{\gamma\gamma} > 0.3$ GeV/c$^2$. An arrow indicates the position of the cut.

E-731 detector to increase acceptance, and adding a transition radiation detector, E-799 will be sensitive to $\pi^0e^+e^-$ decays with a branching ratio of $10^{-11}$.

---

REFERENCES


1 Particle Data Group, *Phys. Lett.* **204B** (1988).

2 F. J. Gilman and M. B. Wise, *Phys. Rev.* **D21**, 3150 (1980); L. M. Sehgal, *Phys. Rev.* **D38**, 808 (1988); J. Flynn and L. Randall, LBL Report No. LBL-26310, 1988 (unpublished); Claudio 0. Dib, Isard Dunietz, and Fred Gilman, *Phys. Rev.* **D39**, 2639 (1989).

3 L. K. Gibbons, et al., *Phys. Rev. Lett.* **61**, 2661 (1988).

4 E. Jastrzembski, et al., *Phys. Rev. Lett.* **61**, 2300 (1988).

5 G. D. Barr, et al., *Phys. Lett.* **214B**, 303 (1988).

6 G. Ecker, A. Pich and E. de Rafael, *Phys. Lett.* **189B**, 363 (1987); a similar invariant mass distribution is predicted by Ko and Rosner (see Reference 7.)

7 L. M. Sehgal, *Phys. Rev.* **D6**, 367 (1972); P. Ko and J. Rosner, Enrico Fermi Institute Report No. EFI 89-13, 1989 (to be published in *Phys. Rev.* **D40**, 1989); T. Morozumi and H. Iwasaki, KEK Report No. KEK-TH-206, 1988 (unpublished).

8 M. Banner, et al., *Phys. Rev.* **188**, 2033 (1969).

9 V. Papadimitriou, et al., *Phys Rev. Lett.* **63**, 28 (1989); the analysis will be described in detail in the Ph.D. thesis of V. Papadimitriou, The University of Chicago, 1990.

# K Factory

BOCK Greg
FERMILAB
Dept of Physics
BATAVIA IL 60510
USA

No written contribution received.

# CP VIOLATION IN THE B SYSTEM

# CP VIOLATION IN THE B SYSTEM

A. I. SANDA

*Rockfeller University New York, N.Y. 10021, USA*

## ABSTRACT

We discuss a strategy for studying CP violation in the B system using the framework of the unitarity triangle. A special attention is placed on obtaining the parameters of the triangle with minimal of strong interaction ambiguities.

# 1. INTRODUCTION

I can not resist the oppertunity to add my personal historical remark: I was a student at University of Illinois during the period of 1962-1965. I took a mechanics cource in the spring of 1963 from Al Abasian. It was a fun cource. It was fun in the sense that it was held at 8 am, and the professor came half asleep. Soon after the begining of the cource, I was convinced that the professor was either totally absent minded or he must have been awake all night, before each mechanics class. I was not aware of all the excitement that was going on in the field of particle physics at that time. Listening to the previous talks in this conference, I am now conviced that it was the latter.

The standard model of electroweak interaction together with the three families of quarks and leptons, and the Kobayashi-Maskawa (KM) matrix explains nearly all experimental observations. It does not, however, explain CP violation. The KM description of CP violation is merely a convenient parametrization. In fact, the origin of CP violation is likely to be outside of the standard model. If one takes the scenario of a grand-unified theory together that of the big-bang cosmology, baryon number asymmetry observed today was be generated when the universe cooled down out of equilibrium. Both baryon number changing interaction as well as CP violation is necessary to realize this possibility. It is not unreasonable to guess that the CP violation observed in K meson decay is a relic of that responsible for baryon number asymmetry generated at the early stage of the universe.

Thus the study of CP violation is a best bet for a guide to physics beyond the standard model - the study of B meson decays is the best bet for cracking the mystery of CP violation.

It is interesting to reflect upon the status of CP violation in particle physics now, and compare it to that soon after its discovery. Fig. 1 shows a cartoon which was shown by Cabibbo[1]

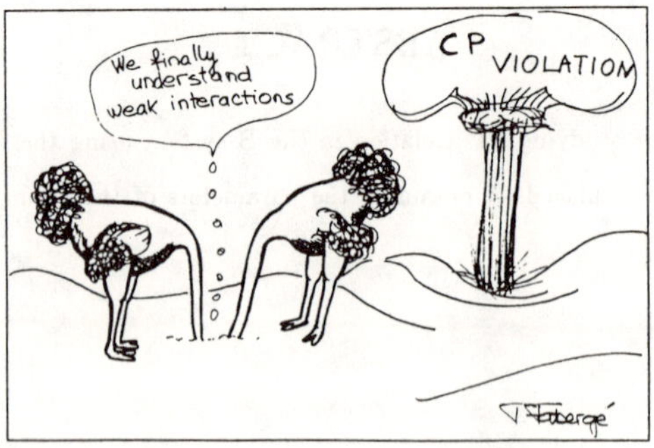


Fig. 1. The way it is today and the way it was back in 1966.

back in 1966. At that time, physicists thought that weak interaction was in good shape since they understood how to calculate radiative correction to $\mu$ decay. On the contrary, CP violation was comparable to an atomic bomb blowing up in the back ground. There was no understanding. We made a tremendous progress in understanding weak interactions since then. Similarly, we now understand CP violation a lot better. However, while weak interaction has gained a status comparable to QED, there is no fundermental understanding of CP violation. Relatively speaking, it is still an atomic bomb. This is a field in which there is a best prospect for progress.

In this talk we outline an approach to study CP violation in the B system through the unitarity triangle. We then go on to discuss, very briefly, the rare decays.

Here we stress just one point: The asymmetry

$$\frac{\Gamma(B \to f) - \Gamma(\bar{B} \to f)}{\Gamma(B \to f) + \Gamma(\bar{B} \to f)} \tag{1.1}$$

for CP eigenstate f is predicted precisely in terms of the KM parameters alone. Thus it is not too outragious to state that: we can push the standard model to its limits, just as $g-2$ measurements keep on pushing QED to its limits. Even a small deviation between theory and experiment is a signiture for physics beyond the standard model. Also, if we were to search for the true origin of CP violation, it is necessary to search for a fundermental theory which contains the standard model in its low energy limit. It is, therefore, necessary to gain more confidence as to the correctness of the standard model.

## 2. OVERVIEW

There have been considerable progress in experimental studies of CP violation during the past year. First, non-vanishing[2] $\epsilon'/\epsilon$ has been reported:

$$\epsilon'/\epsilon = (3.3 \pm 1.1) \times 10^{-3}. \tag{2.1}$$

Second, a new upper limit for the neutron electric dipole moment (edm) has been anounced[3]

$$edm = (-6 \pm 4 \pm 2) \times 10^{-26}. \tag{2.2}$$

The nonvanishing $\epsilon'/\epsilon$ will rule out the super weak model of CP violation.[4] If CP violating interaction is due to an exchange of Higg's boson, it tends to generate a large edm. Since this type of models have several free parameters, it is difficult to mathematically exclude them. The simplest version of the model which gives a precise lower bound for the edm[5]

$$edm > 10^{-25} \tag{2.3}$$

is excluded by the measurment. The Left-Right symmetric models also have several free parameters and not much can be said about their fate at this time.

**The KM model is not without ambiguities. The parameters associated with the strong interaction matrix elements of weak operators as well as the top quark mass $m_t$ give uncertainties and thus it is not possible to make precise predictions. The nonvanishing $\epsilon'/\epsilon$ as well as the new upper limit for the edm, however, are very encouraging for the KM model. The present status of the various models are summarized in table I together with their experimental prospects.**

How can we put these models to further experimental test? It is likely that experiments can be sensitive to the branching ratio for the CP violating decay at the level of[6]

$$Br(K_L \to \pi^\circ + e^+ + e^-) \sim 10^{-11} \tag{2.4}$$

in the next few years. While this channel may offer a best opertunity to confirm nonvanishing $\epsilon'$, the direct CP violation[7], it is very difficult to devise a quantitative check of the standard model.

Note that certain CP asymmetries in B meson decays are as large as $(10 - 60)$ percent.

| | edm | $\epsilon'/\epsilon$ | Br$(K_L \to \pi^\circ + e^+ + e^-)$ | B asymmetry |
|---|---|---|---|---|
| EXP | $(-6\pm4\pm2) \times 10^{-26}$ | $(3.3 \pm 1.1) \times 10^{-3}$ | *expectedsensitivity* $10^{-11}$ | ? |
| Spont. CP | $> 10^{-25}$ | $\leq .05$ | ? | ? |
| L-R | $O(10^{-26})$ | many parameters | ? | small |
| KM | $< 10^{-31}$ | $(8 - 1) \times 10^{-3}$ | $O(10^{-11})$ | $.12 - .60$ |
| Sup. Weak | 0 | 0 | 0 | 0 |

Table 1. A brief summary of the present experimental and theoretical status of CP violation.

## 3. UNITARITY TRIANGLE

CP observables can be discussed conveniently in terms of a unitarity triangle[8]. Consider a unitarity constraint on the KM matrix elements:

$$V_{td}^* V_{tb} + V_{cd}^* V_{cb} + V_{ud}^* V_{ub} = 0. \tag{3.1}$$

This relation can be graphically represented as:

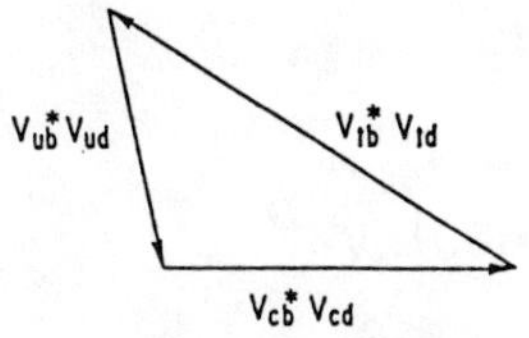


Fig. 2 The unitarity triangle representing the unitarity relation. Measurement of the triangle forms the basis for our study of CP violation.

The area of the triangle is equal to the imaginary part of CP rephrasing invariants[9]:

$$V_{ij}V_{kl}V_{kj}^{*}V_{il}^{*} \tag{3.2}$$

for any i, j, k, l $= 1 \sim 3$. These are invariant under the transformation which changes the phases of the quark fields:

$$q_n \rightarrow e^{i\phi_n} q_n \tag{3.3}$$

where n= i, j, k, l. It can be shown that this quantity has a unique imaginary part independent of the indices:

$$Im(V_{ij}V_{kl}V_{kj}^{*}V_{il}^{*}) = A^2\lambda^6\eta \tag{3.4}$$

where the right hand side is defined by the Wolfenstein representation of the KM matrix

$$KM = \begin{pmatrix} 1 & \lambda & A\lambda^3(\rho - i\eta) \\ -\lambda & 1 & A\lambda^2 \\ A\lambda^3(1-\rho+i\eta) & -A\lambda^2 & 1 \end{pmatrix} \tag{3.5}$$

It is convenient to rescale the triangle by $A\lambda^3$ and obtain:

For various constraints on the $\rho - \eta$ plane, see below.

Within the context of the standard model, study of CP violation is reduced to determining the triangle. At present, we know the following about it[8]:

1. $V_{ub}/V_{cb}$

From the constraint[10]

$$V_{ub}/V_{cb} < .16 \tag{3.6}$$

we obtain:

$$\sqrt{\rho^2 + \eta^2} < .6 \tag{3.7}$$

which is the area bounded by a semi-circle in Fig. 3.

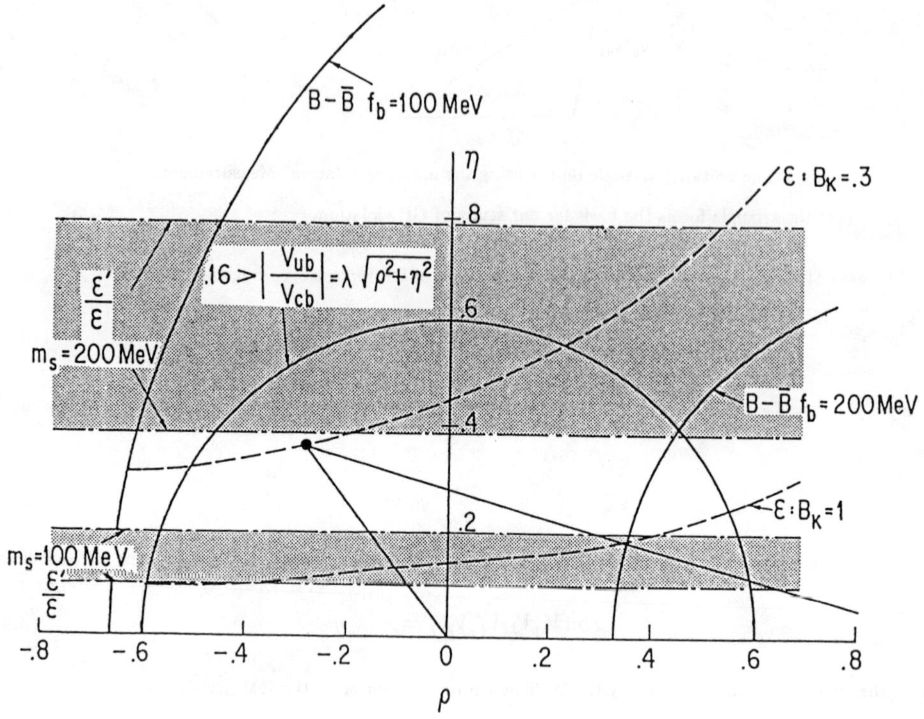


Fig. 3 A rescaled unitarity triangle on the $\rho - \eta$ plane.

2. $\epsilon$ from K decay

The expression for $\epsilon$ is given by[11]

$$|\epsilon| \approx 1.5 B_K \eta(\frac{5}{4} - \rho)(\frac{m_t}{m_W})^{4/3} \times 10^{-2} \approx 2.2 \times 10^{-3} \tag{3.8}$$

where $B_K$ represents the ambiguity in evaluating the matrix element

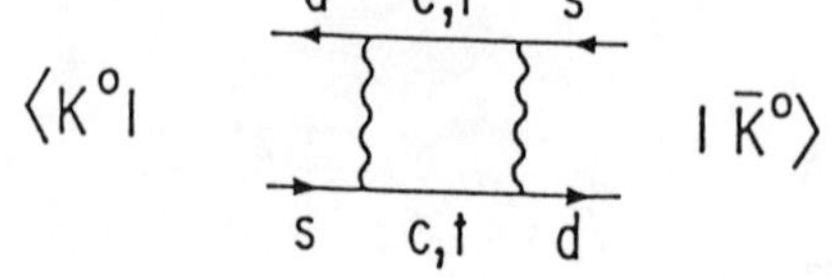


Fig. 4 the matrix element relevant for obtaining $\epsilon$.

For simplicity, we show the constraint given by Eq(3.8) for $m_t = m_W$ in Fig. 4. We have taken $B_K = .3$ and 1.

3. $\epsilon'/\epsilon$

$\epsilon'$ is obtained by evaluating the matrix element of the penguin operator as shown in Fig. 5.

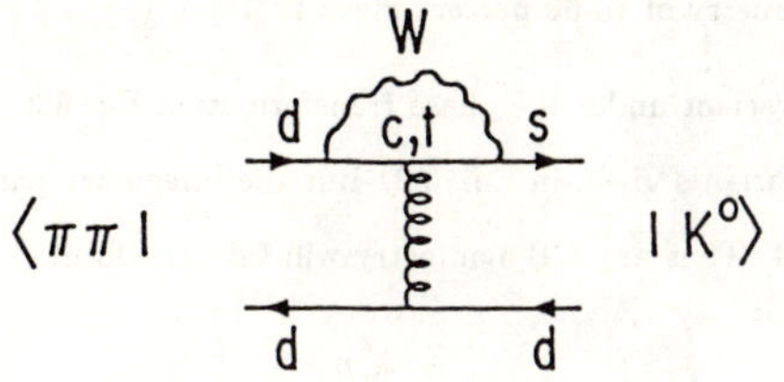


Fig. 5 The matix element which gives direct CP violation.

Evaluating the matrix element using $1/N$ approximation to the chiral purterbation theory, the following expression has been obtained [11]

$$| \epsilon'/\epsilon | \approx 1.1\eta(150MeV/m_s)^2 \times 10^{-2}. \tag{3.9}$$

For a given strange quark mass $m_s$, the experimental value for $\epsilon'/\epsilon$ will give a horizontal band on the $\rho - \eta$ plane. Bands corresponding to the NA31 result and $m_s = 100$ and 200 MeV are shown on Fig. 3.

4. $B - \bar{B}$ mixing.

The matrix element relevant to the evaluation of $M_{12}$ for the B system is given in Fig. 6.

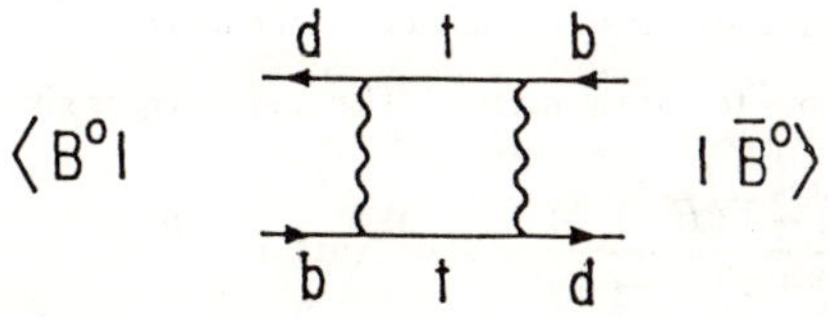


Fig. 6 The box diagram giving $M_{12}$, the $B - \bar{B}$ transition matrix element.

The expression for $x_B = \Delta M/\Gamma$ is approximated by [8]

$$x_B \approx 1.8R_B(\frac{f_B}{150MeV})(\frac{m_t}{m_W^2})^{.84}((1-\rho)^2 + \eta^2) \tag{3.10}$$

The constraint obtained from this equation is a circle centered around $(1, 0)$. In Fig. 3 we present constraints corresponding to $m_t = m_W$ and $f_B = 100 - 200MeV$.

In a qualitative sense, these are valuable restrictions on the $\rho - \eta$ plane. It is far from restricting the unitarity triangle, however. Below we shall see that the study of B meson decays will provide much better constraint on the $\rho - \eta$ plane.

## 4. WHY ARE ASYMMETRIES SO LARGE IN THE B SYSTEM?

Before we discuss the asymmetry in the B meson system, we present here a physical argument which allows one to understand the gigantic asymmetry of 10-60 percent given in Table 1.

Any CP asymmetry must be invariant under the phase transformation Eq. 3.3. Thus it is proportional to the imaginary part of one of the invariants given in Eq. 3.2. But the imaginary part of all the invariants are equal - the right hand side of Eq. 3.4. Thus any CP aymmetry will take the form

$$Asym = \frac{A^2\lambda^6\eta}{(denominator)} \tag{4.1}$$

Since the numerator is more or less fixed, the magnitude of an asymmetry depends on the size of the denominator.

For K meson CP asymmetry, the denominator is the K decay width, which is $0(\lambda^2)$. Thus the asymmetry is no larger than $O(A^2\lambda^4\eta)$. On the contrary, for B meson decays, the widths of the major decay modes are $\approx O(\lambda^4)$. Also for B decay asymmetry which requires $B - \bar{B}$ mixing, the parameter $x_B \approx .7$ inspite of the fact that it is $O(\lambda^2)$. Thus it can be as large as:

$$Asym = \frac{A^2 x_B \lambda^4 \eta}{\lambda^4} \approx 0(\eta) \tag{4.2}$$

As we see in Fig. 3, $\eta$ can be quite sizable.

## 5. B DECAY ASYMMETRY

The CP asymmetry for various channels are discussed in detail in the literature.[12] In this talk, we restrict our selves to the decays of neutral B mesons to CP eigenstate f. The asymmetry is given by

$$\frac{\Gamma(B \to f) - \Gamma(\bar{B} \to f)}{\Gamma(B \to f) + \Gamma(\bar{B} \to f)} = sin(\Delta mt) Im(\frac{p}{q}\rho) \tag{5.1}$$

where

$$\frac{q}{p} = [\frac{M_{12}^* - \frac{i}{2}\Gamma_{12}^*}{M_{12} - \frac{i}{2}\Gamma_{12}}]^{\frac{1}{2}} \tag{5.2}$$

and

$$\rho = \frac{A(B \to f)}{A(\bar{B} \to f)} \tag{5.3}$$

Then we show below that, for certain CP eigenstate f, we can write

$$Im(\frac{p}{q}\rho) = sin(2\alpha) \tag{5.4}$$

where $\alpha$ is one of the angles of the triangle shown in Fig. 2.

This result shows that it is independent of hadronic uncertainties. To what extent is this result exact? For this discussion it is convenient to prove a theorem.

Theorem: Suppose there is only one operator $\Theta$, made out of quark fild operators, which has non-vanishing matrix element $\langle f \mid \Theta \mid B \rangle$ responsible for $B \to f$ decay. Also, suppose it is a good approximation to write[13] $\frac{q}{p} = (M_{12}^*/M_{12})^{\frac{1}{2}}$. Then:

$$Im(\frac{p}{q}\rho) = sin(2\alpha) \tag{5.5}$$

where $\alpha$ is determined purely from the KM matrix.

Proof: First note that $\frac{p}{q} = arg(2V_{td}V_{tb}^*)$. To obtain $\rho$, write the amplitude

$$A(B \to f) \sim \langle f \mid \Theta \mid B \rangle$$

$$= \sum_{quarkstates} \frac{\langle f \mid q'\bar{q}' \rangle \langle q'\bar{q}' \mid \Theta \mid q\bar{q} \rangle \langle q\bar{q} \mid B \rangle}{(E_f - E_{q'\bar{q}'})(E_{q\bar{q}} - E_B)} \tag{5.6}$$

where we have incerted complete set of quark states with appropriate energy denominators. Simmilarly,

$$A(\bar{B} \to f) \sim \langle f \mid \Theta \mid \bar{B} \rangle \tag{5.7}$$

$$= \sum_{quarkstates} \frac{\langle f \mid \bar{q}'q' \rangle \langle \bar{q}'q' \mid \Theta \mid \bar{q}q \rangle \langle \bar{q}q \mid \bar{B} \rangle}{(E_f - E_{\bar{q}'q'})(E_{\bar{q}q} - E_{\bar{B}})} \tag{5.8}$$

If there is only one operator $\Theta$ giving nonvanishing contribution to $\langle q'\bar{q}' \mid \Theta \mid q\bar{q} \rangle$, it is qiven by the quark model amplitude which depends on the KM matrix elements. Then the expression simplifies:

$$\frac{A(B \to f)}{A(\bar{B} \to f)} = \frac{\langle q'\bar{q}' \mid \Theta \mid d\bar{b} \rangle}{\langle \bar{q}'q' \mid \Theta \mid b\bar{d} \rangle .} \tag{5.9}$$

For $B \to \pi\pi$ decay we see that $\rho = arg(2V_{ub}V_{ud}^*)$ and we obtain

$$Im(\frac{p}{q}\rho) = sin[2arg(V_{td}V_{tb}^*V_{ub}V_{ud}^*)]. \tag{5.10}$$

If we take $V_{tb} = V_{ud} = 1$, we have: $Im(\frac{p}{q}\rho) = sin(2arg(V_{td}V_{ub}))$ so that the asymmetry is related to one of the angle of the triangle shown in Fig. 5. Similarly, one can show that the asymmetry for $B \to \psi K_S$ decay is related to the argument of $V_{cb}V_{td}$.

Note that the final state interaction as well as the initial wave function dependence cancel in the ratio $\rho$. If there are more than one operator contributing to the decay, their relative magnitude becomes important. This will depend on the detail of the strong interaction.

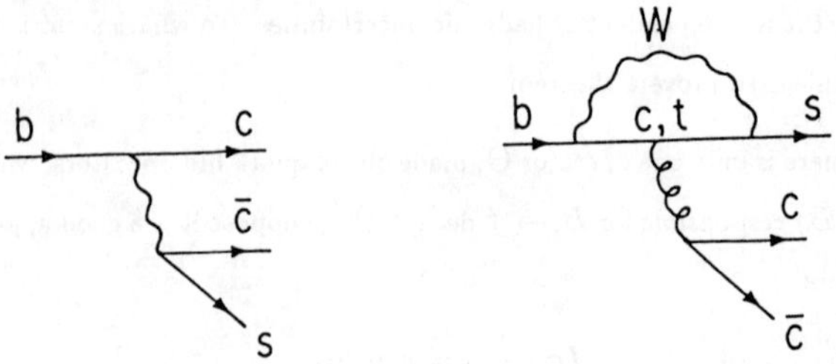


Fig. 7 Two competing amplitudes which may introduce some ambiguity in $\rho$ for $B \to \psi K_S$.

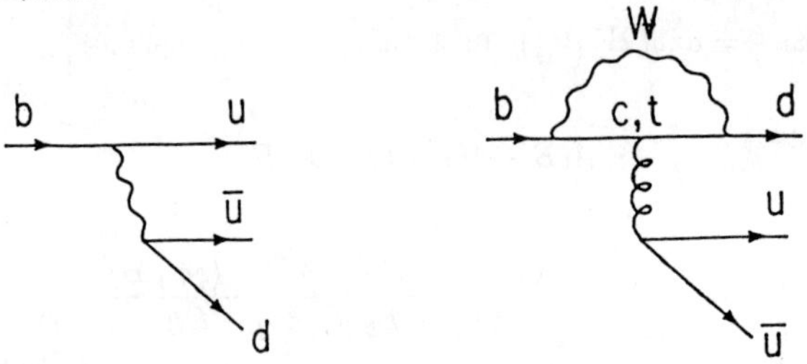


Fig. 8 Two competing amplitudes which may introduce some ambiguity in $\rho$ for $B \to \pi\pi$.

Let us give a specific example. $B \to \psi K_S$ decay is caused by

The penguin operator can not compete with the tree graph, although the KM factors for both diagrams are expected to be similar. The emission of $c\bar{c}$ state from a gluon interaction is exprected to suppress the penguin graph by at least a factor of 100. We thus expect the four fermi operator to completely dominate, and the condition of the theorem is satisfied. Such argument may not work as well for $B \to \pi\pi$ decay.

The KM factor for both diagrams are again expected to be comparable. But, it is relatively easy to emit a light $q\bar{q}$ pair through a gluon interaction. We expect the amplitude to be down by about $\frac{\alpha_S}{\pi}$.

Gronau has concluded that the theoretical uncertainty in predicting the asymmetry is less than a percent for the $B \to \psi K_S$ asymmetry, and at most 20 percent for the $B \to \pi\pi$ asymmetry.[14]

# 6. THE EFFECT OF TAGGING

So far we have been pretending that we have access to separated $B^\circ$, and $\bar{B}^\circ$ beams. In reality we have a pair of B's from

$$\Upsilon(4S) \to B\bar{B} \tag{6.1}$$

or

$$\Upsilon(4S+) \to B\bar{B}^* \to B\bar{B}\gamma \tag{6.2}$$

where the + sign indicates the beam energy slightly above $\Upsilon(4S)$. Note that $B\bar{B}$ pair from Eq.(6.1), and Eq.

(6.2) are P wave and S wave, respectively. Inorder to detect CP violating B decay, define:

$$\Gamma^{\pm}_{\binom{\bar{l}}{l}} = \Gamma([B\bar{B}]^{\pm} \rightarrow f_t + \binom{\bar{l}}{l}_{t'} + \binom{\nu}{\bar{\nu}} + ...), \tag{6.3}$$

where $[B\bar{B}]^{\pm}$ is a time dependent oscillating $B - \bar{B}$ state with the relative angular momentum $\binom{even}{odd}$. The times t and t' denote the time at which the B or $\bar{B}$ meson decayed to f and $\binom{\bar{l}}{l} + \binom{\nu}{\bar{\nu}} + ...)$, respectively.[15] The asymmetry

$$Asym_{\pm} = \frac{\Gamma^{\pm}_{\bar{l}} - \Gamma^{\pm}_{l}}{\Gamma^{\pm}_{\bar{l}} + \Gamma^{\pm}_{l}}$$

$$= sin(\Delta m(t \pm t'))sin(\frac{p}{q}\rho) \tag{6.4}$$

Note that the asymmetry is odd under the decay time of two B's. Unfortunately, presently available technology is not capable of recording the decay position of two B's if $\Upsilon(4S)$ is at rest. Thus $Asym_{-}$ is not accessable in the this case, since one effectively integrates over time. While $Asym_{+}$ does not vanish under this integration, one sacrifices the event rate by going off the $\Upsilon(4S)$ resonance.

The way to solve this dilema is to produce $\Upsilon(4S)$ with momentum, so that B's leave measuable gaps. To accomplish this, one has to either build an asymmetric $e^+e^-$ collider with large enough luminosity or learn to study B's in a hadronic colliders.

# 7. SIZE OF EXPECTED ASYMMETRY

In order to estimate the size of the expected asymmetry, we plot $sin(2\alpha)$ where $\alpha$ is the anlgle which is defined by (0,0), (1,0) and $(\rho, \eta)$, for $\rho = -.6 \sim +.6$. We think we are being conservative in choosing one of the vertex to lie on $\eta > .1$. The resulting asymmetry for f= $\psi K_S$and $\pi\pi$ are shown in Figs. 8 and 9, respectively.

We first note that if the asymmetry turns out to be at the higher end of our estimated region, one needs only few $10^{32}cm^{-2}sec^{-1}$ luminosity to measure the CP asymmetry. One has to be some what cautious, however, considering the fact that our luminosity estimate is only applicable for a perfect detector. More realistic estimate for a specific detector and accerlerator configuration is necessary. A bench mark luminosity to shoot at may be $10^{34}cm^{-2}sec^{-1}$.

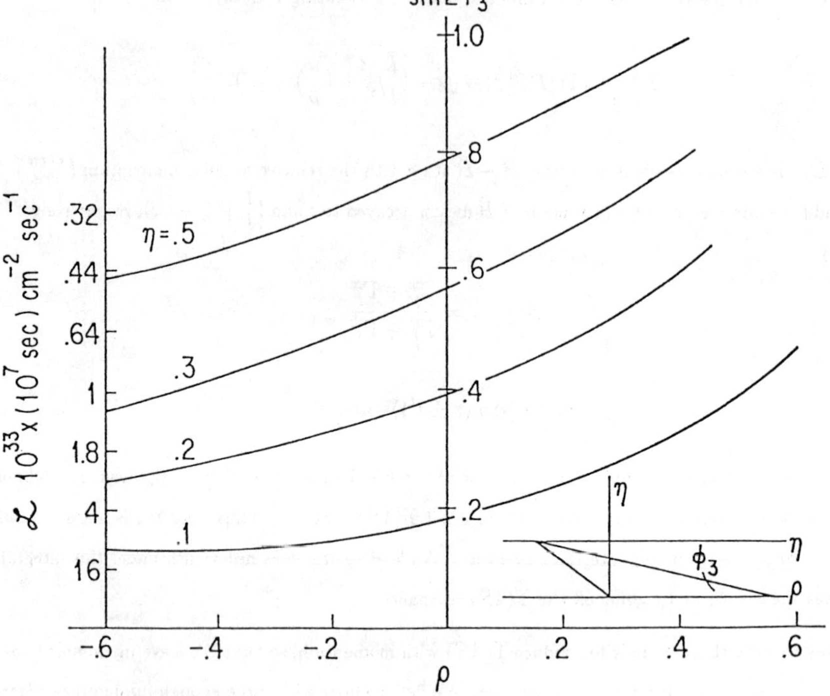


Fig. 9 Expected asymmetry and necessary luminosity for $\psi K_S$ decay mode. The luminosity given here corresponds to asymmetric collider configuration discussed in Ref. 16. The necessary luminosity for other configurations can be estimated by scaling the numbers given here by an appropriate ratios given in the same reference.[16]

## 8. GETTING AT THE TRIANGLE

In principle, the unitarity triangle represents the relationship between six physical observables. At present, only the size of the base, $\mid V_{cb} \mid$ is known.

As we discussed above, two of the angles are related to the asymmetry for $f = \psi K_S$, and $\pi\pi$. The third angle is related to the asymmetries for $B_s$ decay into CP eigenstates.

If both $B_d - \bar{B}_d$ and $B_s - \bar{B}_s$ oscillation parameters $x_d$ and $x_s$ are measured, the ratio:

$$\frac{x_s}{x_d} \approx (\frac{\mid V_{ts} \mid}{\mid V_{td} \mid})^2 \tag{8.1}$$

gives one of the sides as $V_{ts} = -A\lambda^2$. Here the $\approx$ sign implies the assumed equality $\Gamma_d = \Gamma_s$, and cancellation of the hadronic factors in evaluating the ratio of the matrix elements. Both of these approximations are expected to be very good.

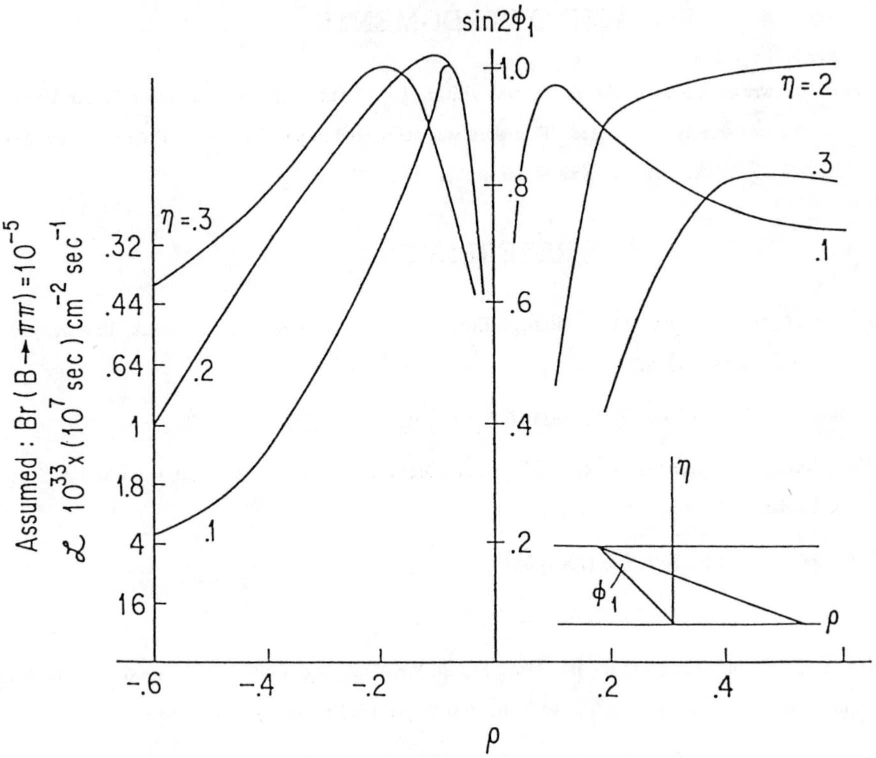


Fig. 10 Expected asymmetry and necessary luminosity for $\pi\pi$ decay mode.

Probably $V_{ub}$ is the most difficult one to obtain free of hadronic ambiguities. One can consider measuring[17]

$$\frac{\Gamma(B \to \pi e^+ \nu)}{\Gamma(B \to \bar{D} e^+ \nu)}. \tag{8.2}$$

Whether this indeed determines $V_{ub}$ to the accuracy we need remaines to be studied.

## 9. CONCLUSIONS

We have discussed a strategy for studying CP violation in the B system in terms of the unitarity triangle. Any inconsistency among the 6 observables represented by the triangle, 3 sides and 3 angles, implies a deviation from the standard model. The CP asymmetries in $\psi K_S$ and $\pi\pi$ modes are directly related to the two angles of the triangle. A measurement of $x_s/x_d$ together with $|\ V_{cb}\ |$ give two sides of the triangle. These four quantities give already one consistency check on the triangle. We expect $|\ V_{ub}\ |$ will give an aditional constraint. Extraction of $|\ V_{cb}\ |$, however, requires understanding of the strong interaction dynamics. Whether one can measure the third angle which is associated with the decay asymmetry of $B_s$ decay remains to be studied. Measurement of asymmetries as well as $x_s$ requires detecting the decay vertex of the B mesons.

# ACKNOWLEDGMENTS

This work was written up while the author was visiting KEK. The hospitality extended to him by the KEK staff members was greatly appreciated. This work was supported in part by the DOE Contract number DE-AC-02-87ER-40325 TASKB, and the NSF Grant number INT-8613131.

# REFERENCES


1. Proceedings of the XIIIth Int. Conf. in High Energy Physics. Edited by C. E. Mauk, University of California Press, Berkeley, (1967).
2. H. Burkhardt at. al, Phys. Lett. **206**, (1988) 169.
3. B. Heckel, Second Int. Symp. on the 4th Family, Santa Monica, Feb. 89, Edited by D. Cline and A. Soni. New York Academiy of Science (New York).
4. L. Wolfenstein, Phys. Rev. Lett. **13**, (1964) 562.
5. I.I. Bigi, A. I. Sanda, Phys. Rev. Lett. **58**, (1987) 1604.
6. See Konisberg, The proceedings of the Int. Workshop on Weak Int. and Neutrino's. Ginosar Israel Edited by P. Singer and G. Eilam, (to be published in Nuclear Physics B Proceedings Suppliments Section).
7. C. O. Dib, I. Dunietz, and F. J. Gilman, Phys. Rev. D **39**, 2639 (1989), and references therin.
8. C. Hamzaoui, J. Rosner, and A. I. Sanda Fermilab workshop edited by N. Lockyerd, and J. Slaughter (1988).
9. C. Jarlskog, Phys. Rev. Lett. **55**, 1039 (1985);Z. Phys. **C29**, 491 (1985).
10. See for example, S. Stone, Proceedings of the XIV Int. Symp. in Multiparticle Dynamics, Arles, France, 1988; D. Kreinick, Proceedings of the XXIV Int. Conf. on High Energy Physics, Editors: R. Kotthaus and J. Kuhn, Springer-Verlag Berlin 1989.
11. See, for example, A. J. Buras, and J-M. Gerard, Phys. Lett. **203B**, 272 (1988)
12. K. J. Foley et. al. p. 701 Edited by R. Donaldson and M. G. D. Gilchriese, Experiments, Detectors, and Exp. Areas for the Super Collider, World Scientific Publishing Co., Singpore (1988).
13. It is a good approximation of a heavy quark system.
14. See M. Gronau, The proceedings of the Int. Workshop on Weak Int. and Neutrino's. Ginosar Israel Edited by P. Singer and G. Eilam, (to be published in Nuclear Physics B Proceedings Suppliments Section).

15. Note that this is different from usual definition of t and t', for example, given in Ref. 12. For derivattion of the time dependence for Eq. 6.4, see D. Hitlin, T. Nakada, and A. I. Sanda, Proceedings of Snowmass workshop on Physics for 1990's.

16. G. Feldman's summary of $e^+e^-$ collider B factory, Proceedings of Snowmass workshop on Physics for 1990's.

17. Some references on this topic are: B. Grinstein, M. B. Wise, and N. Isgur CALT-68-1311; C. A. Dominguez Phys. Lett. **197B**, 423 (1987); M. Wirbel, B. Stech, and M. Bauer, Z. Phys. **C29**, 637 (1985).

# BETA - The $B$-Meson Factory Project at PSI

Klaus R. Schubert
Institut für Experimentelle Kernphysik
Universität Karlsruhe

Abstract

The Paul Scherrer Institute (PSI) at Villigen, Switzerland has proposed to construct a $B$-Meson Factory, i.e. an $e^+e^-$ collider for centre-of-mass energies around 10 $GeV$ with very high luminosity. The proposal is based on a double storage ring with 640 $m$ circumference, now called BETA, and aims at a luminosity of some $10^{33}/cm^2/s$. The machine design, a detector study, and the physics goals are described. Two of these goals, the precise measurement of the CKM matrix element $|V_{ub}|$ and the search for CP violation in the decays $B^0$, $\overline{B}^0 \rightarrow J/\Psi K_s^0$, are presented with detailed Monte Carlo studies.

Most of our present knowledge on $B$ mesons, apart from lifetime measurements at PEP and PETRA, originates from the $e^+e^-$ storage rings DORIS and CESR operating on the $\Upsilon(4S)$ resonance: existence and mass of $B^0$ and $B^+$, rates of weak decays into nonleptonic and semileptonic final states with charm, determination of the CKM matrix element $|V_{cb}|$ from these rates, limits on decay rates for final states without charm and corresponding limits on $|V_{ub}|$, observation of $B^0\overline{B}^0$ mixing yielding a value for $|V_{td}| \cdot m_t$, limits on $b \to s$ penguin rates ... Table 1 gives the present average and peak luminosities of these two storage rings at $\sqrt{s} = m(\Upsilon 4S)$ as well as the $B\overline{B}$ production rates.

| | $\mathcal{L}_{peak}$ | $\mathcal{L}_{av}$ | $\dot{n}(\Upsilon 4S)$ |
|---|---|---|---|
| DORIS | $3 \cdot 10^{31}/cm^2/s$ | $100/pb/y$ | $10^5/y$ |
| CESR | $10^{32}/cm^2/s$ | $500/pb/y$ | $5 \cdot 10^5/y$ |

Table 1: Present luminosities and $\Upsilon(4S) = B\overline{B}$ production rates.

Both storage rings are old, DORIS was completed in 1974 and modified in 1977 and 1982, CESR completed in 1979 and continuously upgraded between 1985 and today, and both were not conceived for maximum luminosity at the threshold energy for $B$ meson production. A new and optimized ring could reach $\mathcal{L}_{peak} = 10^{33} - 10^{34}/cm^2/s$, corresponding to $10^7 - 10^8 B\overline{B}$ pairs per year, as first demonstrated in a study and motivation paper in 1986 [1]. Discussion of this paper and the worldwide increase of interest in $B$ meson physics has led the Paul Scherrer Insitute (PSI) at Villigen, Aargau, Switzerland to propose [2] in July 1988 the construction of such a $B$-Meson Factory.

There are plenty of topics which motivate the proposal, some of these are:

- Missing states in $b\bar{b}$ spectroscopy,
- Missing states in $c\overline{q}$, $cqq$, $b\overline{q}$, and $bqq$ spectroscopy,
- Lifetimes and decay fractions of the four quasi-stable charmed baryons,
- The missing one-prong problem in $\tau$ decays,
- Decay parameters of the $\tau$ and search for parity violation therein,
- More sensitive measurements of the $\tau$-neutrino mass,
- Search for rare $\tau$ decays with decay fractions $\mathcal{B} < 10^{-4}$,
- Measurement of charmed meson decay fractions $\mathcal{B} < 10^{-2}$,
- Decay fractions of $B$ mesons with 5 – 10 % precision,
- Separate lifetimes for $B^+$ and $B^0$,
- More precision on $|V_{cb}|$,
- $|V_{ub}|$ from $B \to \pi l\nu$, $\rho l\nu$ ...
- $f_B$ from $B \to \tau\nu$,
- More precision on the $B^0\overline{B}^0$ mixing parameter $x_0$,
- Existence, lifetime (two lifetimes ?) and decay fractions of the $B_s$ meson,
- Measurement of the $B_s\overline{B}_s$ mixing parameter $x_s$,
- CP violation in $B$ meson decays,
- Search for CP violation surprises in rare $D$ and $\tau$ decays ...

This list, by far not complete, contains two equally important topics for the central question of this Blois Conference: *What have we learnt from the 1964 discovery of Christenson, Cronin, Fitch, and Turlay?* A determination of $|V_{ub}|$, with 10 % precision or better, in charmless $B$ meson decays is as important as the direct search for CP violation in $B$ meson decays. I will come back to these two items at the end of the paper and just recall here that the standard electroweak theory with three families of fermions and with minimal Higgs assumptions on the mass generating mechanism is able to describe the observed CP violations in the $K^0\overline{K}^0$ system if $V_{ub}$ is of the order of $0.004 \cdot e^{-3i\pi/4}$.

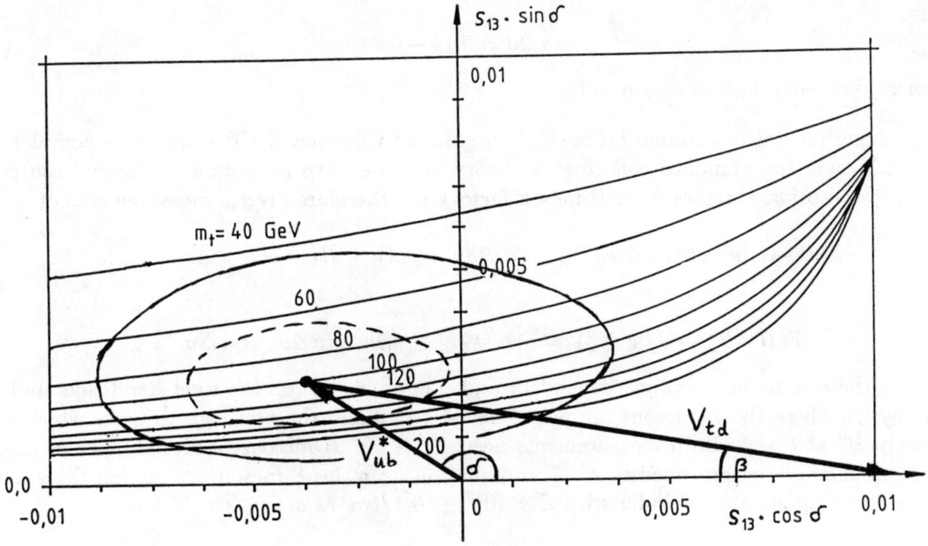


Fig.1: The "CP triangle" of the CKM matrix with approximate one- (dashed ellipse) and two- (solid ellipse) standard deviation contours [7]. The solid lines relate $V_{ub}^*$ to the observed value of $\epsilon(K^0)$ for different $m_t$ values [8].

The CKM matrix [3] has five well-measured matrix elements, and seven are well known if unitarity is imposed [4]. In the Wolfenstein parametrization [5], these seven are essentially real and the matrix has the form

$$V_{CKM} = \begin{pmatrix} V_{ud} & V_{us} & V_{ub} \\ V_{cd} & V_{cs} & V_{cb} \\ V_{td} & V_{ts} & V_{tb} \end{pmatrix} \approx \begin{pmatrix} 1 & \lambda & ? \\ -\lambda & 1 & \lambda^2 \\ ? & -\lambda^2 & 1 \end{pmatrix}$$

with $\lambda = 0.22$ and $V_{ub} = |V_{ub}| \cdot e^{-i\delta}$. The two unknown elements are related by unitarity,

$$V_{ub}^* + V_{td} \approx \lambda^3 \approx 0.01 \ . \tag{1}$$

For many discussions on CP violation it is very convenient to draw this relation as a triangle in the complex plane [6].

Fig. 1 shows the "CP triangle" with approximate one- and two- standard deviation contours for $V_{ub}^*$ as obtained in a fit [7] using experimental information on the rate limit $\Gamma(b \to ul\nu)$, the $B^0\overline{B}^0$ mixing parameter $x_0$, the CP violation parameters $\epsilon(K^0)$ and $\epsilon'(K^0)$, and the

$m_t$ limit as obtained from radiative corrections to $m(Z^0)/m(W^{\pm})$. The solid lines in the figure give $V_{ub}^*$ values as required to explain $\epsilon(K^0)$ for top quark mass values $m_t$ between 40 and 200 $GeV$ and for $B_K = 2/3$ in the standard electroweak theory [8]. The fit uses $B_K = 2/3 \pm 1/6$ and results in $|V_{ub}| = (0.4 \pm 0.2) \cdot \lambda^3$, $\delta = (140^{+20}_{-50})^o$, $m_t = (110 \pm 35)\ GeV$. With approximately 95% confidence, we know

$$|V_{ub}| = 0.001 - 0.009, \tag{2}$$

and

$$sin\ 2\beta = 0.14 - 0.84, \tag{3}$$

where $\beta = -arg\ V_{td}$, as shown in fig. 1.

The experimental determination of $|V_{ub}|$ and $sin\ 2\beta$ will show if CP violation is completely embedded in the standard electroweak theory or if we have to assume additional sources. Future top-priority topics for a $B$-meson factory are therefore precise measurements of

$$\Gamma(B \to \rho l \nu,\ \pi l \nu \ldots) \ = \ |V_{ub}|^2 \ \cdot \ Phasespace \ \cdot \ Hadronic\ Engineering, \tag{4}$$

and

$$\Gamma(B^0, \overline{B}^0 \to J/\Psi\ K_s^0) \ = \ const \cdot e^{-\Gamma t} \cdot [1 \mp sin\ 2\beta \cdot sin(\Delta m \cdot t)], \tag{5}$$

where the rate in eq. 5 is time-dependent and the minus sign on the right-hand side applies to decays where the $B$ mesons are purely $B^0$ at $t = 0$ and the plus sign to those which are purely $\overline{B}^0$ at $t = 0$. Both measurements need $10^7$ – $10^8$ $B$ mesons produced. The cleanest measurements are performed in $e^+e^-$ annihilation near the $B$-meson production threshold, thus requiring an $e^+e^-$ collider with $\mathcal{L} = 10^{33}$ – $10^{34}/cm^2/s$ near $\sqrt{s} = 10.6\ GeV$.

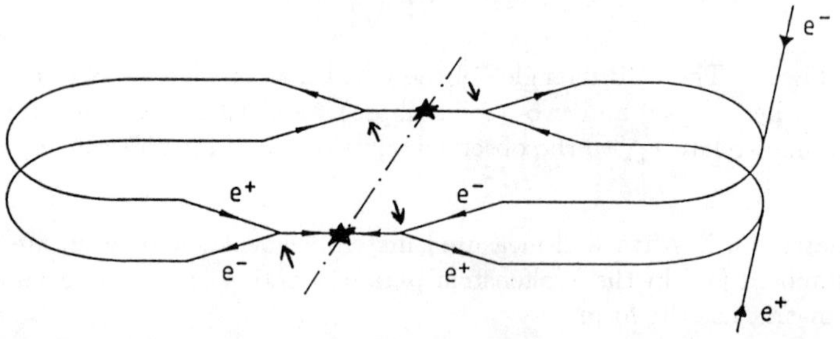


Fig. 2: Double storage ring for separate storage of $e^+$ and $e^-$ bunches.
The stars indicate the interaction points, the arrows the separation regions.

A fast and safe design for such a collider is a storage ring, the luminosity of which is given by

$$\mathcal{L}_{SR} \ = \ \frac{N_b N_1 N_2 f}{4\pi\sigma_x^*\sigma_y^*}, \tag{6}$$

where $N_b$ is the number of stored bunches, $N_1$ and $N_2$ are the numbers of $e^+$ and $e^-$ per bunch, $f$ is the revolution frequency of one bunch, and $\sigma_x^*$ and $\sigma_y^*$ are horizontal and vertical beam widths in the interaction point. The density of particles in one bunch is limited by beam-beam interaction. This can either be quantified by the tune shift $\Delta Q$ or by the oberservation [9] that the disruption angle $\vartheta_{dis}$ — the deflection angle of a particle in one bunch caused by the electromagnetic field of the other bunch — has to be smaller than the natural angular spread $\vartheta^* = \epsilon/\sigma^*$ in the interaction point. With $\sigma_x^* = \sigma_y^* = \sigma^*$ and $\epsilon_x = \epsilon_y = \epsilon$, this condition leads to the following limitation on $N_1 = N_2 = N$:

$$\vartheta_{dis} = \frac{r_e N}{\gamma \sigma^*} \leq \frac{\epsilon}{\sigma^*}, \tag{7}$$

where $r_e$ is the classical electron radius and $\gamma = E_e/m_e$. Inserted into eq. 6, this gives

$$\mathcal{L}_{SR} \leq \frac{\gamma^2}{4\pi \, r_e^2} \cdot (N_b f) \cdot \frac{\epsilon}{\beta^*} , \tag{8}$$

where $\beta^* = \sigma^{*2}/\epsilon$. Maximum luminosity requires large emittance $\epsilon$, small $\beta$-function value $\beta^*$ in the interaction point, and a rapid sequence of bunches $(N_b f)$. Storage of many bunches which interact in only one or two points around the circumference leads to the concept of a double storage ring as sketched in fig. 2.

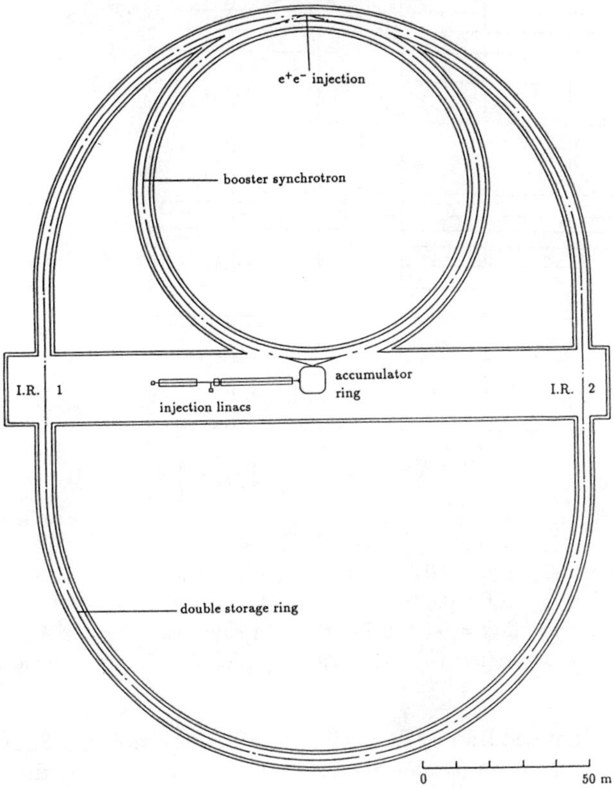


Fig. 3:

Layout of the proposed $B$-Meson Factory BETA at PSI. For details see ref. [2].

The parameter choice of $\beta^*$, $\epsilon$, and $(N_b f)$ is limited by the following considerations: Accelerating with a RF system of 500 $MHz$ leads to bunch lengths of $\sigma_z \approx 2\ cm$. The minimum $\beta^*$ values are of that order, so $\beta_x^*$, $\beta_y^* \geq 2\ cm$. The vacuum tube radius should not be larger than about 5 $cm$. Storing particles up to 10 $\sigma$ and $\beta_{max} < 30\ m$ leads to $\epsilon_x$, $\epsilon_y \leq 10^{-6} m$. Separation of $e^+$ and $e^-$ before and after the interaction points (see fig. 2) requires a bunch distance $d = c/(N_b f) \geq 10\ m$. Standard magnet optics leads to $\epsilon_y = k \cdot \epsilon_x$ with $k \leq 0.03$. At $\sqrt{s} = 10\ GeV$, where $\gamma = 10^4$, we obtain

$$\mathcal{L}_{max} \approx \frac{c\,\gamma^2\,k\,\epsilon_x}{4\pi\,r_e^2\,d\,\beta_y^*} = \frac{\epsilon_x}{10^{-6}\ m} \cdot \frac{k}{0.03} \cdot \frac{10\ m}{d} \cdot \frac{2\ cm}{\beta_y^*} \cdot \frac{5 \cdot 10^{33}}{cm^2 s}\ . \tag{9}$$

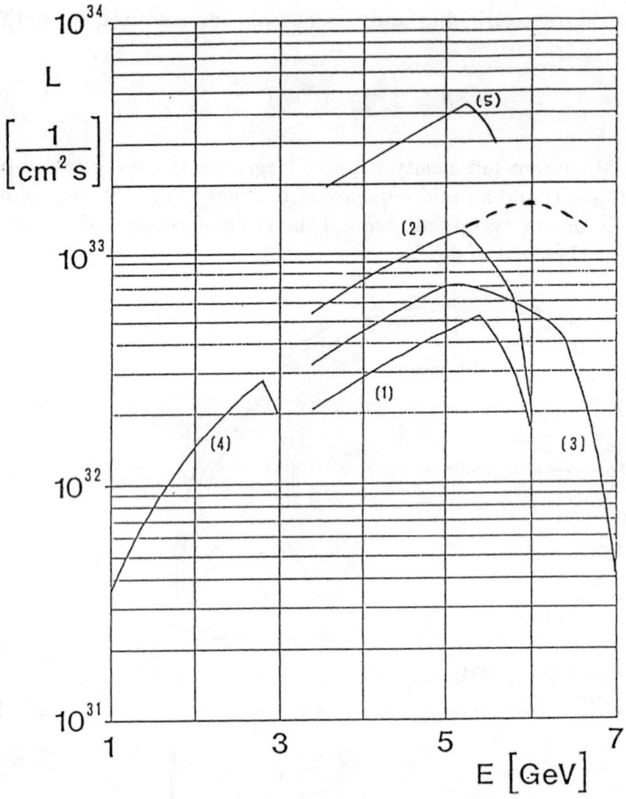


Fig. 4: Luminosity curves of BETA.
(1): $\beta_y^* = 3\ cm$, $\Delta Q = 0.04$, 10 bunches, 10 cavities;
(2): $\beta_y^* = 2\ cm$, $\Delta Q = 0.03$, 10 bunches, 10 cavities;
(3): $\beta_y^* = 2\ cm$, $\Delta Q = 0.04$, 6 bunches, 14 cavities;
(4): $\beta_y^* = 3\ cm$, $\Delta Q = 0.03$, 2 cavities; (5): $\beta_y^* = 1.5\ cm$, $\Delta Q = 0.05$, 18 bunches.

The 1988 layout of BETA, the proposed BEautyTAufactory at PSI, is shown in fig. 3. Design details can be found in ref. [2]. The main parameters are: 2 rings on top of each other with 60 $cm$ distance; 648 $m$ circumference; 56 horizontal and 12 vertical bending magnets per ring; horizontal bending radius 41 $m$; energy loss per turn 1.4 $MeV$ at 5 $GeV$; energy spread of $e^+e^-$ collisions in the center-of-mass system $\pm$ 4.4 $MeV$ at $\sqrt{s} = 10\ GeV$; horizontal beam emittance $5.5 \cdot 10^{-7}\ m$ at 5 $GeV$; 48 $mA$ current per bunch; synchroton radiation loss 100 $kW$ per bunch at 5 $GeV$; installed RF power $\approx 1.2\ MW$ per ring if operating with 10 bunches of 48 $mA$ each at 5 $GeV$.

An important lesson from the operation of DORIS-I is the fact that $e^+e^-$ collisions with a non-zero crossing angle lead to serious transverse beam instabilities. BETA is therefore planned to run with zero collision angle requiring electrostatic or RF-magnetic beam separators before and after the collision points. Both schemes are under investigation at PSI. Fig. 4 shows the expected luminosities under five different conditions. Curve 1 should be reached soon after the start of machine operation, curve 2 after some machine development, and curve 5 after experience and upgrade. Curves 1 to 3 are achieved with $\approx 1.2\ MW$ installed RF power per ring, 5 and the high-energy extension of 2 need more. Curve 4 is for optimized conditions in order to operate the machine on the $J/\Psi$ or the $\Psi(2S)$ resonance.

The investment for the storage ring project, including all injection machines and buildings, but excluding manpower, is estimated to be 130 $MSFr$ at 1988 prices. For details, see ref. [2]. The estimated manpower is shown in fig. 5. Design and construction requires 420 Menyears; operation requires a permanent staff of about 50. The proposed timetable is 2 more years of $R\&D$ and design, 3 – 4 years of construction, and about 1 year of running-in. A decision of the Swiss Schulrat, the board of Swiss federal schools and their annex institutes, is expected for September 89, a decision of the Swiss Bundesrat (government) in January 90, and of the Swiss Nationalrat (parliament) towards the end of 90. Fig. 5 is based on this schedule.

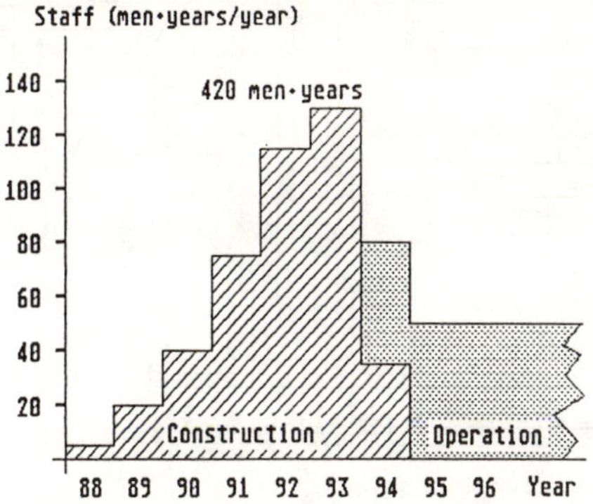


Fig. 5:

Manpower profile for construction and operation of BETA.

BETA will have two interaction areas for one experiment each. The Crystal Barrel Collaboration [10], now working with $\bar{p}p$ annihilation at LEAR (CERN), has expressed interest to move their detector into one of the areas for $J/\Psi$ and $\psi(2S)$ decay studies as well as for studies of two-photon reactions. The other area should be equipped with a new and optimally designed universal detector for investigating all the topics which I listed at the beginning of the paper including the precise measurement of $|V_{ub}|$ and the search for CP violation in various $B$ meson decay channels. The proposal for BETA [2] contains a detailed study for such a detector. It has been worked out in 1987/88 by 45 physicists from 20 institutions in Switzerland, Germany, France, and Poland. Fig. 6 shows a sketch of the detector. The necessary investment, again excluding manpower, is estimated to be 50 $MSFr$.

The main features of the studied Universal Detector [2] are as follows: a Beryllium beam tube of 1.5 $cm$ radius and of 180 $\mu m$ thickness; double-side-readout microstrip Silicon detectors and a precision drift chamber up to $r = 20\ cm$; a main drift chamber of the Jet Chamber type up to $r = 64\ cm$ with a length of 160 $cm$; a Ring Image Cherenkov Counter from $r = 65\ cm$ to $r = 85\ cm$ and in the endcaps aiming at $\pi/K$ separation up to 2.5 $GeV/c$ with 4 $\sigma$ and $p/K$ separation up to 4.5 $GeV/c$; a CsI electromagnetic calorimeter of 18 radiation lengths depth with special readout efforts for good resolution at low photon energies $E_\gamma = 20 - 100\ MeV$; a superconducting coil with a solenoidal field of 1.5 $T$; finally 4 layers of muon chambers for good muon identification with very low $\pi$ and $K$ punchthrough. The detector aims at resolutions for momentum of charged particles and for direction and energy of photons at least as good as in the CLEO-II detector [11], for direction of charged particles better than in CLEO-II, and at a charged particle identification which is much better than there.

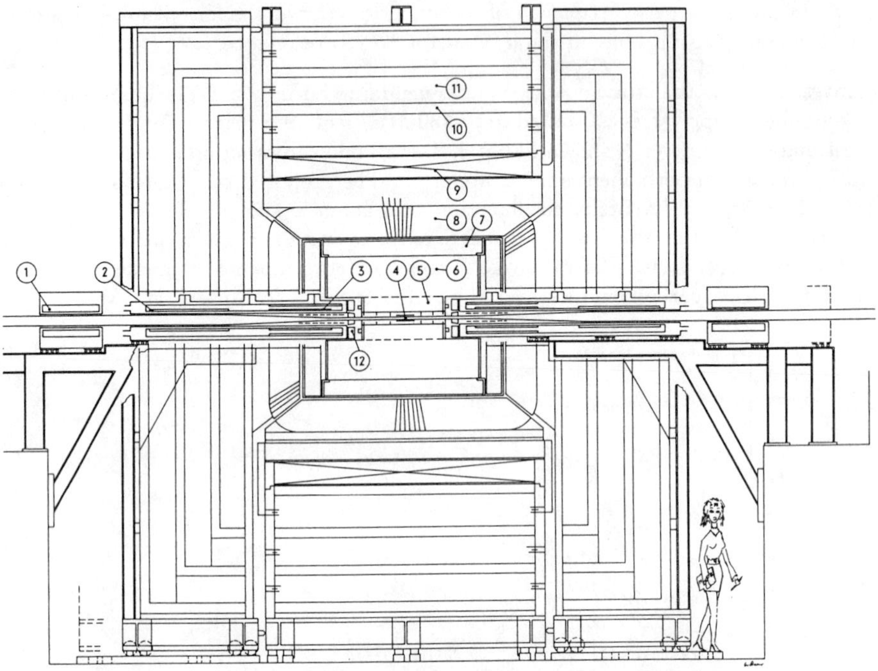

Fig. 6: Sketch of a detector study [2] for BETA. This Universal Detector consists of a Silicon Vertex Detector (4), a Precision Tracking Chamber (5), a Main Tracking Chamber (6), a Ring Image Cherenkov Counter (7), a CsI Electromagnetic Calorimeter (8), a Superconducting Coil (9), Muon Chambers (10) between Iron Absorbers (11), and High Density Forward Calorimeters (12). Also shown are the Mini-$\beta$ Quadrupoles (1,2,3).

Performance figures of merit for the detector have been obtained by extensive Monte Carlo simulation calculations; they can be found in ref. [2]. Let me present here only the more recent results for two pilote reactions on $V_{ub}$ and on $sin\ 2\beta$.

A full Monte Carlo study of the search for the $V_{ub}$-induced decay $B^+ \rightarrow \rho^0 l^+ \nu$ has just been finished [12, 13]. $B$ meson pairs without additional particles are produced on the $\Upsilon(4S)$ resonance,

$$e^+e^-\ (\sqrt{s} = 10580\ MeV)\ \rightarrow\ \Upsilon(4S)\ \rightarrow\ B^\pm\ B^\mp\ \rightarrow\ (\rho^0 l^\pm \nu)\ (hadrons)\ ,$$

and $\rho l\nu$ decays of the one $B$ are "tagged" by completely reconstructing the other $B$. The use of all known exclusive decay channels leads to a tagging efficiency of only about 1 %. If the $\rho l\nu$ candidates are tagged, however, by summing energies and momenta of all charged and neutral particles which are observed in the event in addition to $\rho$ and $l^{\pm}$, one can obtain $\eta_{tag} \approx 10\%$. True $\rho l\nu$ decays are identified by their missing mass which has to fulfill $MM^2 = m_\nu^2 = 0$. In contrast to the missing mass method as introduced by ARGUS [14] for the reconstruction of $B^0 \to D^{*-}l^+\nu$ assuming $\vec{p}(B) = \vec{0}$, tagging allows to use $\vec{p}(B) = -\vec{p}(B_{tag})$. This reduces the missing mass resolution from $\sigma(MM^2) = \pm 0.8\ GeV^2$ to $\pm 0.1\ GeV^2$ in $\rho l\nu$ decays.

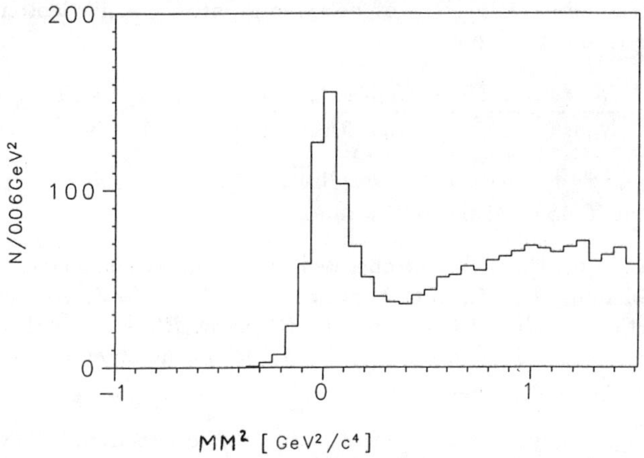


Fig. 7: Signal and background of tagged $B^+ \to \rho^0 l^+\nu$ decays ($l = e$ and $\mu$) if $\mathcal{B}(B^+ \to \rho l\nu) = 2.0 \cdot 10^{-4}$, as obtained in a Monte Carlo study [12, 13] for the Universal Detector at BETA.

The background of miss-tagged $\rho l\nu$ candidates in this new method corresponds to a decay fraction level of $\mathcal{B}(\rho l\nu) \approx 1 \cdot 10^{-5}$. Fig. 7 shows Monte Carlo signal and background of $B^+ \to \rho^0 l^+\nu$ from $10^7$ $B^+B^-$ pairs, including background from $B^0\overline{B}^0$ and continuum, assuming $\mathcal{B}(B^+ \to \rho^0 l^+\nu) = 2 \cdot 10^{-4}$. The present upper limit from the ARGUS experiment is $1.0 \cdot 10^{-3}$ [13]. The situation in fig. 7 allows to determine $\mathcal{B}(\rho^0 l^+\nu) = (2 \pm 0.2) \cdot 10^{-4}$ from electrons and muons together. The extraction of $|V_{ub}|$ from such a result would presently be limited by a large discrepancy in the theoretical predictions relating $\Gamma(\rho l\nu)$ and $|V_{ub}|$ [15, 16, 17]. The observation of events and their $q^2$ dependence, however, would then allow to adjust the models.

The second studied pilot reaction is CP violation in the decays $B^0$, $\overline{B}^0 \to J/\Psi\ K_s^0$. This decay channel has some clear advantages: (1) The order of magnitude of the decay fraction is known; the four observed events [18, 19] correspond to $[\mathcal{B}(B^0 \to J/\Psi\ K_s^0) + \mathcal{B}(\overline{B}^0 \to J/\Psi\ K_s^0)]/2 \approx 4 \cdot 10^{-4}$. (2) The standard electroweak theory relates the observable particle-antiparticle asymmetry $A$ in this decay to the angle $\beta$ in the CP triangle of fig.1 independent of hadronic matrix element uncertainties. (3) No other decay channel discussed so far has a better figure of merit $\mathcal{M} = (1 - A^2)/(A^2 \cdot \mathcal{B})$ which determines the number of necessary $B\overline{B}$ pairs for observing CP violation in $B$ decays, i. e. all other channels require more $B\overline{B}$ pairs in the frame of the standard theory.

The observed asymmetry [20] from a coherent two-particle state $B\overline{B}$ is

$$A_{obs} = \frac{N(\Psi K_s;\ \overline{T})\ -\ N(\Psi K_s;\ T)}{N(\Psi K_s;\ \overline{T})\ +\ N(\Psi K_s;\ T)} = A_0 \cdot sin[\Delta m(t_1 \pm t_2)]\ , \tag{10}$$

as derived from eq. 5 with $A_0 = sin2\beta$, $t_1$ the time between $B^0\overline{B}^0$ production and the $\Psi K_s$ decay, and $t_2$ the time between $B^0\overline{B}^0$ production and the tagging decay. The plus sign between $t_1$ and $t_2$ is for a two particle state with $C = P = +1$. The minus sign is for $C = P = -1$ as for a $B^0\overline{B}^0$ pair from the $\Upsilon(4S)$. The decay $\overline{T}$ is a $\overline{B}^0$ decay at $t = t_2$, e. g. into a negative lepton and anything, T a $B^0$ decay, e. g. into a positive lepton. If integrated over all decay times $t_1$ and $t_2$ we obtain

$$A_{obs,int} = \frac{N_{int}(\Psi K_s;\ \overline{T})\ -\ N_{int}(\Psi K_s;\ T)}{N_{int}(\Psi K_s;\ \overline{T})\ +\ N_{int}(\Psi K_s;\ T)} = \begin{cases} 0 & \text{for } C = -1, \\ r \cdot A_0 & \text{for } C = +1, \end{cases} \tag{11}$$

where $r = 2x_0/(1 + x_o^2)^2$; $r = 0.62$ for $x_0 = x(B^0) = 0.72$. This result leads to two possible strategies, the moving $\Upsilon(4S)$ and the $BB^*$ strategy.

In the $BB^*$ method [20, 21], BETA is operated in an energy-symmetric way at $\sqrt{s} =$ 10.63 $GeV$, i. e. colliding $e^+$ of 5.315 $GeV$ against $e^-$ of 5.315 $GeV$. At this energy there is abundant production of $BB^*$ events. After the $B^*$ decay, $B^* \to B\gamma$ with $\mathcal{B} = 100\%$, the remaining $B\overline{B}$ system is in a state with $P = C = +1$, yielding an observable rate asymmetry even without any vertex information.

For measuring the asymmetry it is essential to separate $BB^*$ events from $B\overline{B}$ events produced at the same energy. A Monte Carlo study [22] leads to the result that this is perfectly achieved using the different kinematics of the two event tpyes even without detection of the 50 $MeV$ gamma ray. By reconstructing the invariant $\Psi K_s$ mass from the observed decay products $\Psi \to l^+l^-$ and $K_s \to \pi^+\pi^-$ with the beam-energy constraint $E_B = \sqrt{s}/2$, one finds the right mass $m_B$ for $\Psi K_s$ decays from $B\overline{B}$ events and a shifted mass for decays from $BB^*$ events as shown in fig. 8.

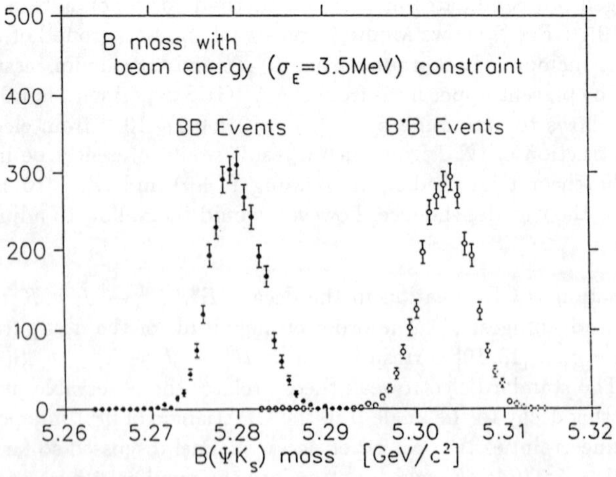


Fig. 8: Beam-energy-constrained mass of $B \to \Psi K_s$ decays from $B\overline{B}$ and $BB^*$ events at $\sqrt{s} = 10.63$ $GeV$ in energy-symmetric $e^+e^-$ collisions [22].

The separation in fig. 8 is obtained with $\sigma(\sqrt{s}) = \pm 4.9$ $MeV$ as expected in BETA at $\sqrt{s} =$ 10.6 $GeV$. Selecting $BB^*$ events with flavour tag allows then to observe an asymmetry

$A = A_{obs,int}(C = +1)$ with a statistical significance of $S$ standard deviations if one starts with the following number of $BB^*$ events:

$$N(BB^*) = \frac{S^2(1-A^2)}{2A^2} \cdot \frac{\sigma(BB^*)}{\sigma(B^0B^{0*})} \cdot \frac{1}{\mathcal{B}_1 \cdot \mathcal{B}_2 \cdot \mathcal{B}_3} \cdot \frac{1}{\eta_{rec} \cdot \eta_{tag}}, \tag{12}$$

where $A = r \cdot A_0$ as in eq. 11, $\mathcal{B}_1 = \mathcal{B}(B^0 \to J/\Psi K_s)$, $\mathcal{B}_2 = \mathcal{B}(J/\Psi \to l^+l^-)$, $\mathcal{B}_3 = \mathcal{B}(K_s \to \pi^+\pi^-)$, $\eta_{rec}$ the reconstruction efficiency of $J/\Psi K_s$ decays, and $\eta_{tag}$ the reconstruction efficiency for the tag channels $T$ and $\overline{T}$.

We find $N(BB^*) = 1 \cdot 10^7$ with $S = 3$, $A = 0.62 \cdot sin2\beta$, $sin2\beta = 0.40$ as in fig. 1, $\mathcal{B}_1 = 5 \cdot 10^{-4}$, $\eta_{rec} = 0.65$, and $\eta_{tag} = 0.44$.

The moving $\Upsilon(4S)$ method requires energy-asymmetric operation of BETA,

$$e^+(E_1)\, e^-(E_2) \to \Upsilon 4S(\beta\gamma \neq 0) \to B^0\overline{B}^0 .$$

In this case, the rate asymmetry becomes observable if the distance $l_2 - l_1 = c\beta\gamma(t_2 - t_1)$ between the $B \to \Psi K_s$ decay vertex and the decay vertex of the tagging $B$ is resolved. Fig. 9 shows the expected significance $S = A/\sigma_A$ on the rate asymmetry $A$ for 250 reconstructed events $\Upsilon(4S) \to B^0\overline{B}^0 \to (J/\Psi K^0_s)(l^\pm \; or \; K^\pm)$ as a function of the energy of the more energetic beam assuming, as in the $BB^*$ method, $sin2\beta = 0.40$. The three curves in fig. 9 are for three different assumptions for the experimental resolution on $l_2 - l_1$. Also shown is the geometrical acceptance of six particles in the Universal Detector of BETA. An asymmetry of 7 *GeV vs.* 4 *GeV* is attractive for this method of CP-violation search.

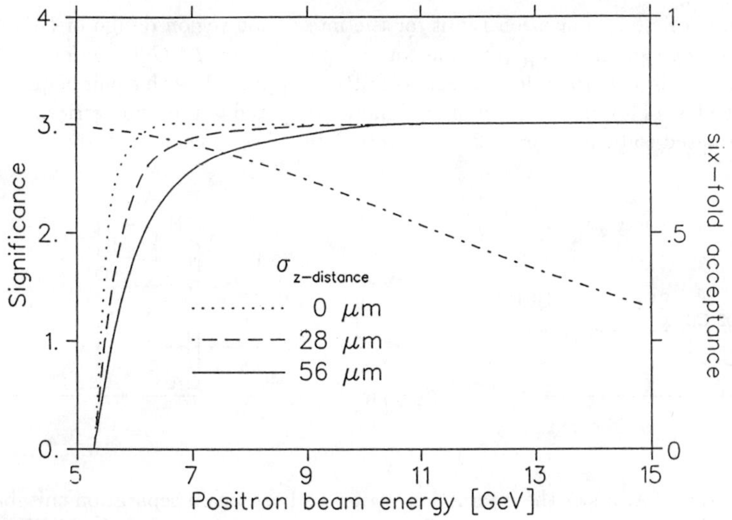


Fig. 9: Significance $S = A/\sigma_A$ of the CP asymmetry $A$ in $B^0 \to J/\Psi K^0_s$ decays in energy-asymmetric $e^+e^-$ collisions as function of the energy of the higher momentum beam. Also shown is the geometrical acceptance for six particles.

A detailed Monte Carlo study [22] with this asymmetry and with the properties of the Universal Detector as described before and in ref. [2] leads to a $\Psi K_s$ reconstruction efficiency

$\eta_{rec} = 0.60$ and a vertex resolution of $\sigma(l_1) = 14\ \mu m$. The flavour of the tagging $B$-meson is found from a high momentum lepton or a charged Kaon from the subsequent $D$-decay. The vertex of the tagging $B$ is determined by using all tracks from this $B$. This method, owing to the modest boost, has a bias of only 6 $\mu m$ and has a resolution of about $\pm 30\ \mu m$. The tagging efficiency is found to be $\eta_{tag} = 0.30$. The Gaussian resolution on $l_2 - l_1$ is 35 $\mu m$. Using this resolution, the observable CP asymmetry $A_{obs,asym}$ defined as

$$A_{obs,asym} = \frac{(N_1 + N_4) \ - \ (N_2 + N_3)}{N_1 + N_2 + N_3 + N_4}$$

with $N_1 = N(\Psi K_s$ and $B^0$ decaying after $\Psi K_s)$, $N_2 = N(\Psi K_s$ and $\overline{B}^0$ after $\Psi K_s)$, $N_3 = N(\Psi K_s$ and $B^0$ before $\Psi K_s)$, and $N_4 = N(\Psi K_s$ and $\overline{B}^0$ before $\Psi K_s)$, is found to be

$$A_{obs,asym} \ = \ A \ = \ 0.90 \cdot \frac{x_0}{1 + x_0^2} \cdot sin2\beta \ = \ 0.42 \cdot sin2\beta \ .$$

The required number of produced $\Upsilon(4S)$ mesons for observing $A$ with a significance $S = A/\sigma_A$ is given by the analoguous expression to eq. 12,

$$N(\Upsilon 4S) \ = \ \frac{S^2(1 - A^2)}{2A^2} \cdot \frac{1}{\mathcal{B}(\Upsilon 4S \to B^0\overline{B}^0)} \cdot \frac{1}{\mathcal{B}_1 \cdot \mathcal{B}_2 \cdot \mathcal{B}_3} \cdot \frac{1}{\eta_{rec} \cdot \eta_{tag}} \ . \qquad (13)$$

Assuming $\mathcal{B}(\Upsilon 4S \to B^0\overline{B}^0) = 0.50$ and $\mathcal{B}_1$ and $sin2\beta$ as in the $BB^*$ method in eq. 12, we obtain $N(\Upsilon 4S) = 4 \cdot 10^7$ for a significance of 3 standard deviations.

Fig. 10 shows a first optimization result for the interaction region design of BETA suitable as well for energy-symmetric operation as for an operation with 4 $GeV$ against 7 $GeV$. This design contains split electrostatic separators (SEP1) and (SEP 2) with a pair of quadrupoles in between. The loss of luminosity in the 4 $vs$. 7 mode compared to the symmetric 5.3 $vs$ 5.3 $GeV$ mode is estimated to be a factor of 2 to 3. More detailed optimization studies are underway.

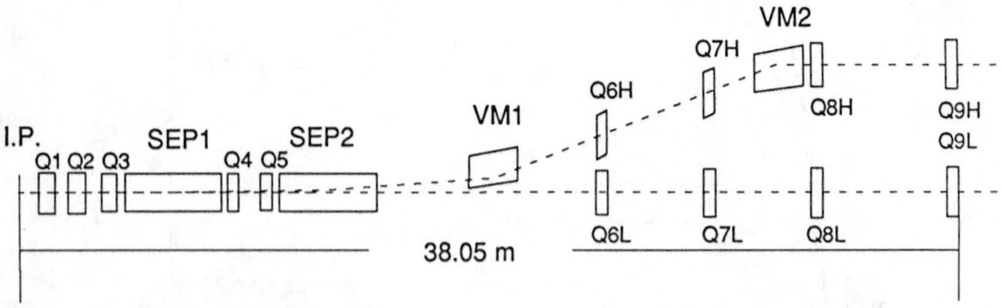


Fig. 10: Layout of the interaction region and the beam separation suitable as well for energy-symmetric as for energy-asymmetric operation of BETA.

When comparing the two strategies for measuring the CP violation in the $\Psi K_s$ decay mode, one has to take into account the differences in luminosity $\mathcal{L}$ and in production cross section $\sigma$. Both factors are not too well known at present. Assuming $\mathcal{L}_{sym}/\mathcal{L}_{asym} = 2.5$ and $\sigma(BB^*\ at\ 10.63\ GeV)/\sigma(\Upsilon 4S) = 0.25$ [23] leads to the conclusion that about two times less running time is necessary in an energy-symmetric operation at $\sqrt{s} = 10.63\ GeV$ than in the enery-asymmetric operation on the $\Upsilon(4S)$ resonance to observe an effect with the same significance. Both ways look equally promising, and there is no need to define the best search strategy now. The flexibility of a double ring machine like BETA allows to decide this question after the determination of the unknown factors in cross section and luminosity.

To conclude, a $B$-Meson Factory with $\mathcal{L} = 10^{33} - 10^{34}/cm^2/s$ and with the option of energy-asymmetric operation is technically feasable. The planning of BETA at PSI is well advanced and a final decision for construction is expected in 1990. The rich physics program includes $|V_{ub}|$ and $A(B^0, \overline{B}^0 \rightarrow J/\Psi K_s^0)$, both equally important for understanding the origin of CP violation.

I would like to acknowledge many useful discussions with R. Eichler, T. Nakada, T. Ruf, and K. Wille, and the typesetting help of W. Weißmann.

References:

[1] R.Eichler, T.Nakada, K.R.Schubert, S.Weseler, and K.Wille, SIN-PR-86-13 (1986)
[2] "Proposal for an Electron Positron Collider for Heavy Flavour Particle Physics and Synchrotron Radiation", PSI-PR-88-09 (July 1988)
[3] N.Cabibbo, Phys.Rev.Lett. 10(1963)531
M.Kobayashi and T.Maskawa, Progr.Theor.Phys. 49(1973)652
[4] For recent reviews see e.g. H.Harari, these Proceedings,
K.Kleinknecht, Proc.XXIV.Int.Conf.High Energy Phys., Munich 1988, p.98, and K.R.Schubert [7]
[5] L.Wolfenstein, Phys.Rev.Lett. 51(1984)1945
[6] J.D.Bjorken, presented at the Workshop on Experiments, Detectors, and Experimental Areas for the Supercollider, July 1987
C.Jarlskog and R.Stora, Phys.Lett.B 208(1988)268, and references therein
[7] K.R.Schubert, Karlsruhe preprint IEKP-KA/88-4, to be published in Proc.Conf. Phenomenology in High Energy Physics, ICTP Trieste 1988
[8] A.Buras and J.M.Gerard, Phys.Lett.B 203(1988)272, and references therein
[9] K.Brown, SLAC-PUB-4366 (1987)
[10] E.Aker et al., CERN Proposal PS197 (1985)
[11] CLEO-II Proposal, Cornell preprint CLNS 85/634 (1985)
[12] T.Ruf, Karlsruhe preprint IEKP-KA/89-2 (1989)
[13] T.Ruf, Dr.rer.nat. Thesis, Karlsruhe University, July 1989
[14] H.Albrecht et al. (ARGUS), Phys.Lett. B 197(1987)452
[15] M.Wirbel, B.Stech, and M.Bauer, Z.Physik C 29(1985)637
[16] J.G.Körner and G.A.Schuler, Z.Physik C 38(1988)511
[17] N.Isgur, D.Scora, B.Grinstein, and M.B.Wise, Phys.Rev. D 39(1989)799
[18] H.Albrecht et al. (ARGUS), Phys.Lett. B 199(1987)451
[19] I.J.Kim (CLEO), presented at the XXIV.Int.Conf.High Energy Phys., Munich 1988
[20] I.Bigi and A.Sanda, Nucl.Phys. B 193(1981)85
[21] S.Stone, Mod.Phys.Lett. A 3(1988)541
[22] T.Nakada, PSI unpublished reports PSI/TN-1/88, PSI/TN-2/88, and PSI/TN-3/88 (1988),
R.Eichler, K.Gabathuler, E.Lorenz, T.Nakada, T.Ruf, K.R.Schubert, and K.Wille, PSI-PR-88-22 (1988)
[23] S.Ono, A.Sanda, and N.Törnqvist, Phys.Rev. D 24(1986)186

# B Meson Factory

Hitlin David
California Institute of Technology
Lauritsen Lab. of Physics
PASADENA CA 91125
USA

No written contribution received.

*Arthur Schaffer and Guy Wormser.*

# BEAUTY PHYSICS AT $e^+ e^-$ COLLIDERS

G. WORMSER
Laboratoire de l'Accélérateur Linéaire - ORSAY -

**ABSTRACT** : Beauty physics to be performed in the next decade at the resonances Υ(4S) and $Z^0$ are compared. Large similarities are found in the physics program and the reconstruction techniques of $Z^0$ and asymmetric Υ(4S) colliders. The physics potential of the latter is found to be superior at equal luminosity to a symmetric machine, provided a large enough boost (≥.5). $Z^0$ machines will probably be the main source of the rich $B_s^0$ physics during that period.

## INTRODUCTION

There are two obvious places to study B physics at $e^+ e^-$ colliders : the Y(4S) resonance, where a pair of neutral or charged B mesons are produced at rest in the Y(4S) rest frame, and the $Z^0$ resonance, where are produced pairs of back-to-back jets containing B particles. Both settings offer a large beauty cross-section, a large signal to noise ratio (around 25 %). Other important advantages or limitations are different for the 2 situations and are summarized in section 1.

These differences between the two approaches will then be discussed in the framework of 4 important topics for B physics, namely :

(i) lifetime measurements
(ii) $b \to u$ transitions
(iii) $B_s^0$ physics
(iv) CP violation.

These topics are clearly ordered with increasing difficulty but also with increasing interest.

Another important difference to keep in mind while comparing B physics at the Y(4S) and at the $Z^0$ is that a large experience has been gained during the past ten years on Y(4S) physics, resulting in many clever reconstruction techniques, while $Z^0$ physics has not even started. We will therefore concentrate on the intrinsic capabilities of the 2 methods and we will not try to rely too heavily on a "realistic" simulation of B physics at LEP.

## I – GENERAL COMPARISON OF Y (4 S) AND $Z^0$ PHYSICS

The main features are summarized in table 1.

From lines (a) or (b), one can see that the $B\bar{B}$ samples available from both sources should be very comparable in the absence of unexpected breakthrough in machine luminosity for the next decade. This will only be true for $B_d^0$ and $B^\pm$ particles and $Z^0$ machines will be the dominant source of $B_s^0$ and $\Lambda_b$ particles.

The existence of kinematic constraints on the Y(4S) leads to very good mass resolutions, clearly desirable for exclusive studies, while on the $Z^0$, exclusive studies will have to be limited to background free channels, such as semileptonic decays or channels involving a $J/\psi$.

The main drawback of a symmetric Y(4S) machine is the negligible mean flight path of the B particles, which makes impossible any time dependent study. This can be overcome by the advent of asymmetric Y(4S) machines[1]. In the other hand, the large flight path of B particles at $Z^0$ machines will serve, not only to measure B properties but also get almost pure samples of B events on an event-by-event basis.

## II - LIFETIME MEASUREMENTS

While the average lifetime of beauty particles is now known with a 10 % accuracy from measurements in $e^+ e^-$ continum around 30 GeV[2], very little is known of individual lifetime of the various beauty particles : $B^0_d$ , $B^+$, $B^0_s$ , $\Lambda_b$ . In the framework of the spectator model, all these lifetimes should be the same. It is therefore interesting to try to measure possible departures from such a situation.

On a symmetric Y(4S) machine, indirect information can be gathered by measuring the ratio of $B^+$ and $B^0$ semileptonic ratios. With the reasonable assumption that semileptonic widths are equal, this ratio is equal to the lifetime ratio $\tau_{B^+} / \tau_{B^0_d}$ . Argus and Cleo preliminary results show that $B^0$ and $B^+$ lifetime are equal within 20 %[3]. A more precise and direct measurement will certainly be the first output of an Y(4S) asymmetric machine where well known techniques can be used to reconstruct exclusively $B^0$ and $B^+$ mesons. In addition, thanks to the detailed vertex information which will lead to a better background rejection, more inclusive methods can be used. As an exemple, $D^{*+}$ particles will be reconstructed using only the slow pion in the decay $D^{*+} \rightarrow \pi^+ D^0$. This is illustrated in figure 1 for the semi leptonic decay $B^0_d \rightarrow D^{*+} e^- \nu_e$ where a clear missing mass peak is observed when using only the lepton and the slow pion. These high efficiency reconstruction techniques will enable also absolute branching ratio measurements because of the large sample of tagged $B^0$ decays.

On the $Z^0$, several analyses will be performed, ranging from completely exclusive channels to very inclusive ones. $D\pi$ or $D^*\pi$ peaks will yield only around 20 events per $10^6$ $b\bar{b}$ events, a factor 2 worse than $\psi K$ or $\psi K\pi$ channels. More promising is the method based on inclusive $\psi$ reconstruction. The idea is to count the charge at the secondary vertex, the B vertex itself in that case, to distinguish $B^0_d$ from $B^+$. It seems to be possible to reach a 80 % purity for charged and neutral samples[4]. This method does require a good microvertex detector but no hadron identification. Conversely, exclusive semileptonic decays such as $D^{*+} e^- (\nu e)$ require complete charm reconstruction but are less demanding on vertex reconstruction. Samples of ~ 1000 events per $10^6$ $b\bar{b}$ could be selected for $B^0$ or $B^+$ modes[5]. For $B_s$ and $\Lambda_b$, the first glance could be provided by a few events obtained in background free modes like $\psi\,\phi$ or $\psi\Lambda$, or by more inclusive correlations between leptons and $D_s$ or baryons.

Let us precise what we mean by a good or a very microvertex detector. As an example, we consider the Si vertex detector of the Cern DELPHI experiment. In its present version (fig. 2a), it comprises 2 layers of 25 μm pitch strips. Since this detector azimuthal leads to a longitudinal precision equal to 10 % of the mean flight distance of a B meson we consider it a good microvertex detector. A very good microvertex is correspond to the upgraded version of this detector, foreseen in 1991, with now R–φ and R–z readout, three layers and smaller radius for the first plane (fig. 2b). Only this version will allow charge counting at a secondary vertex with a reasonable precision[5].

## III – B → U TRANSITIONS

The transition amplitude $V_{bu}$ is a crucial element in the Kobayashi-Maskawa matrix, since no CP violation can occur in the Standard Model if it is equal to 0. Therefore, we can proceed in two steps : first gather evidence that $V_{bu}$ is not 0, then measure it precisely. The experimental techniques to achieve these 2 goals can vary widely and are not the same from machine to machine.

The various methods are listed in Table 2.

The simplest method to prove that $V_{bu}$ is not zero is certainly the exclusive reconstruction of B mesons decaying is charmless final state such as $\pi^+ \pi^-$ or $p\bar{p}$. Unfortunately, branching ratios into two body modes are very small and present upper limits do not yet constraint $V_{bu}$ significantly.

The most promising method is still today the search for leptons past the charm end point on inclusive semileptonic decays. Cleo[6] already reported evidence for such an effect. This method can only be applied at Y(4S) machines. When the luminosity will be sufficient, exclusive b → u semileptonic transitions such as $B^0 \to e^- \pi^+ \nu_e$ or $B^- \to e^- \rho^+ \nu_e$ will allow a good measurement of $V_{bu}$.

On the $Z^0$ peak, inclusive b → u semileptonic decays will be searched using a powerful microvertex detector to tag B decays with only 1 secondary vertex requiring that all tracks coming from the semileptonic decay come from the same point. This method has the advantage of being sensitive to all b → u semileptonic decay but requires at least a factor 20 rejection of charmed decay which can only be possible with a very good microvertex detector.

Another possibility is the observation on the leptonic decay $B^+ \to \tau^+ \nu_e$ whose branching ratio is $-10^{-5}$. This can only be performed on $Z^0$ machines where decay products of the 2 B mesons are well separated.

On a asymmetric machine, the situation is ideal in the sense that both Y(4S) and $Z^0$ approaches can be performed, but the required vertex precision is such that a large boost is required.

## IV – $B_s^0$ PHYSICS

The study of the properties of the $B_s^0$ meson is a new and exciting field, almost totally untouched up to now. While $B_s^0$ mesons have never been reconstructed, some of its properties are precisely predicted by the standard model and therefore need to be checked. The most interesting one is of course, $x_s$, the oscillation frequency between $B_s$ and $\bar{B}_s$, which should lie around 10[7]. Such rapid oscillations can only be measured by observing the time evolution of the $B_s - \bar{B}_s$ system, thus we will only consider 2 cases : an Y(4S) asymmetric machine and a $Z^0$ machine.

The measurement will proceed as follows :

(a) reconstruct at least partially a $B_s$ (presumably through its semileptonic decay $B_s \rightarrow D_s\ \ell\ \nu_e$)

(b) tag the other B to find out if it is a B or a $\overline{B}$

(c) measure the proper lifetime for the $B_s$ meson

(d) deduce $x_s$ from proper lifetime distributions with or without observed mixing.

For point (a) and (b), experimental techniques should be very similar on both machines. $D_s$ exclusive reconstruction efficiency will always be low, due to the small branching ratios of easily accessible modes ($\phi\pi$, $K^*K$). An excellent calorimeter on a Y(4S) machine would be very rewarding to increase this efficiency due to the large $D_s$ branching ratios in modes involving neutrals ($\eta\pi$, $\eta'\pi$)[8]. Otherwise, one will have to rely on particle identification and tag as $D_s$ charm decays with 0 or 2 strange particles. In any case, it is not necessary to reach a pure sample of $B_s$ since the oscillation frequency should be very different for $B_d$ and $B_s$.

Tagging efficiency will rely as usual on leptons, exclusively reconstructed charmed mesons or $K^{\pm}$ tagging, which can be performed on both machines with a comparable efficiency, around 40 %.

Two quantities need to be measured to extract the proper lifetime : the $B_s$ energy and the flight distance. On a $Z^0$ machine, the flight distance will be given by the distance between a well measured primary vertex and the $B_s$ secondary vertex which will be taken as the intersection of the lepton and the totally or partially reconstructed $D_s$. This can be achieved with a precision of $f = \frac{\nu}{\sigma_\nu} = 300\ \mu m$ to be compared with the mean flight distance $\nu \sim 2.2$ mm. [This figure of merit $f = \sim 10$ allows to be sensitive to $x_s$ up to 15]. A comparable result can only be achieved on a asymmetric machine when the boost is sufficiently high grows almost linearly with the bost since $\sigma$ is almost boost independent in the interesting range. Therefore, a minimum boost is required to be sensitive to the high $x_s$ region ($x_s \geq 10$) which is the favoured one[9].

It is important to realize that a combined measurement of $V_{bu}$ and $x_s$ gives access to the CP violating phase in the standard model[5].

On a $Z^0$ machine, the $B_s^0$ energy will not be directly measured since a missing neutrino will always be present in the semileptonic decay method used here but it has been shown[5] than replacing the exact $B_s$ energy by its mean on value does not introduce a significant loss of precision up to 2 or 3 lifetimes. Furthermore, this algorithm can be refined by taking into account the amount of energy observed at the primary vertex which is strongly anticorrelated with the $B_s^0$ energy.

In summary, the interesting $x_s$ range –between 10 and 20– can be covered with a $Z^0$ machine provided its luminosity reaches 10 to 50 $Mb\overline{b}$ events.

$B_s$ physics on a 10 GeV machine has to be performed on the Y(5S) resonance, whose cross section is ~ 6 times smaller than Y(4S). It is not known yet how many $B_s$ are produced at the Y(5S) nor in which mode ($B_s$ - $B_s$ , $B_s^*$ - $B_s$ or $B_s^*$ - $B_s^*$). Preliminary evidence by CUSB[10] indicate however that this fraction could be high (~ 30 %). The necessary running time on the Y(5S) resonance would then correspond to the one needed to accumulate ~ 50 M $b\bar{b}$ on the Y(4S).

## V - CP VIOLATION

We will only discuss here CP violation effects in the decay $B_d^0$ ( $\bar{B}_d^0$ ) → f, where f is a CP eigenstate mode such as $\psi K^0$ , $\pi^+ \pi^-$ , $D^+ D^-$ ,... Those decays have been recognized as the most promising ones[11].

The measurement will proceed as follows :

(a) measure a CP eigenstate

(b) tag the other side as B or $\bar{B}$

(c) compare the time evolution of the system f–B and f–$\bar{B}$.

Asymmetries up to 10-50 % could be observed. Therefore, 1000 such decays are needed, which in turn correspond to roughly $10^8$ $b\bar{b}$ events.

No large differences are expected between the different machine types for points (a) since CP eigenstates modes appear to be background free. The same is also true for point (b) as mentioned in the preceding section.

Concerning point (c), the situation is very different for symmetric and asymmetric low energy machines. Since no time evolution is visible for a symmetric machine at the Y(4S), one has to rely on integrated measurements. It is now well known that no CP time integrated asymmetry can be observed on the Y(4S), due to the initial quantum state L = – 1 of the 2 B mesons. The best way out consists in changing the c.m. energy from the Y(4S) to the $B^*B$ threshold. In that case, the initial states is a L = + 1 state and time integrated measurements are possible. However, there is a loss of ~ 4 due to the ratio of cross-sections between the Y(4S) ad the $B^*B$ threshold.

The precision required on the time measurement asymmetric Y(4S) colliders is not very stringent, since, in first approximation, it is sufficient to know which B decayed first. A figure of merit, $f = v/\sigma_v$ , of 3 or 4 is therefore adequate, althrough a loss of sensitivity of a factor 2 has to be expected compared to a higher f, where a detailed fit to the time evolution can be performed[12]. This will easily be done on a $Z^0$ machine, where the only problem consists in accumulating between 100 and 500 M $Z^0$ to reach the desired sensitivity.

## CONCLUSION

We have seen the similarities between the physics program and the reconstruction techniques of a $Z^0$ machine and a asymmetric Y(4S) collider. Their physics possibilities are superior to an Y(4S) symmetric machine provided :

(i) their luminosity is comparable. The boost optimization of an asymmetric machine depends essentially of the variation of the luminosity with the boost. If the luminosity can be made boost-invariant, then it is clear that the optimum boost corresponds to $\beta\Gamma$ between .7 and 1, allowing the figure of merit $f = v/\sigma_v$ to lie between 5 and 10. Otherwise, a lower boost will have to be chosen with a minimum value around .3, which will only allow CP violation study to be performed and with an efficiency reduced by a factor ~ 2[13].

For a symmetric Y(4S) machine, every effort has to be made to ensure a luminosity around $10^{34}$ $cm^{-2}$ $s^{-1}$. Only in that range, CP violation effects could be detected in the $BB^*$ system.

(ii) the detector is optimized for an asymmetric configuration. This mean a state-of-the-art microvertex detector, increased coverage in the forward region and good particle identification. Emphasis will be less important on electromagnetic calorimeter since neutrals cannot be tracked back to a secondary vertex.

For $Z^0$ machines, the luminosity threshold is around $10^{32}$ $cm^{-2}s^{-1}$. It is now hoped that in a few years. LEP will reach the luminosity and this enable a rich harvest in B physics especially in $B_s$ physics. The present LEP detectors will have to be upgraded, that is equipped with better microvertex detectors and smaller beam pipes. They all seem to be able to cope with the increased event rate.

In summary, B physics in the next decade could well come from three different machines : a symmetric Y(4S) machine with a luminosit between $10^{33}$ ad $10^{34}$ $cm^{-2}s^{-1}$, the first asymmetric Y(4S) machine with a luminosity around $10^{33}$ $cm^{-2}s^{-1}$ and LEP with a luminosity of $10^{32}$ $cm^{-2}s^{-1}$. In this scenario, most aspects of the ambitious B physics program discussed here will be very nicely covered.

## ACKNOWLEGMENTS

I would like to acknowledge here many useful discussions held with my Delphi colleagues P. Roudeau and D. Treille and the work done on asymmetric B factories with R. Aleksan and J. Chauveau.

## FIGURES

1 • Inclusive $B^0 \rightarrow D^{*+} e - \nu_e$ ; $D^{*+} \rightarrow \pi_s^+ D^0$ reconstruction.

(a) " $D^*$ " e missing mass at a symmetric machine or a low boost asymmetric machine. "$D^*$ " is the kinematically reconstructed $D^*$ from the $\pi_s$ alone.

(b) " $D^*$ " e missing mass at an asymmetric machine ($\beta\Gamma \sim 1$), requiring detached $\pi - e$ vertex by 400 $\mu$m.

2 • Microvertex upgrade foreseen by the Delphi experiment.

3 • $B_s$ oscillations as they can measured at LEP (from [5]).

## REFERENCES

[1] P. Oddone, Proceedings of the UCLA Workshop : Linear Collider $B\bar{B}$ Factory Conceptual Design, ed. by D. Stork, p.243 (1987) and R. Siemann et al., SLAC-PUB-4839 (1989).

[2] R. A. Ong, Proceedings of the 1987 Heavy Flavour Symposium, Stanford, (1987), E. Bloom ed.

[3] K. Berkelman, Proceedings of the 1989 Heavy Flavour Symposium, Cornell University (1989).

[4] G. Wormser, Third Mark II Workshop on SLC physics, SLAC-315 (1987).

[5] P. Roudeau, LAL 89/21.

[6] D.Kreinick and M. Danilov, Proceedings of the 1989 Lepton-Photon Conference, Stanford University, (1989).

[7] H. Harari, these proceedings.

[8] G. Wormser et al., Phys. Rev. Letters 61 (1988) 1057.
G. Wormser LAL 89/27.

[9] D. Hitlin, these proceedings.

[10] J. Lee-Franzini, Proceedings of the XXIV Rencontres de Moriond, 5-12 March 1989.

[11] I. I. Bigi and A. Sanda, Nucl. Phys. B 193, 85 (1981), Nucl. Phys. B281, 41 (1987).

[12] Report of the B-Factory group, Physics and techniques, Snowmass 88, SLAC-PUB-4838.

[13] T. Nakada, these proceedings and PSI-PR-89-13 (1989).

**TABLE 1**

| | | Y(4S) | | $Z^0$ |
|---|---|---|---|---|
| (a) | $\sigma (b\bar{b})$ | 1 nb | | 6.5 nb |
| (b) | Foreseen luminosity in 1990/1995 ($cm^{-2}\ s^{-1}$) | $10^{32}/10^{34}$ | | $10^{31}/10^{32}$ |
| (c) | $\frac{\sigma (b\bar{b})}{(\sigma (tot))}$ | .25 | | .22 |
| (d) | b particles produced | 50 % $B_d^0, \bar{B}_d^0$, 50 % $B^+B^-$ <br> ~ 30 % $B_s \bar{B}_s$ on Y(5S) | | 40 % $B_d^0$, 40 % $B^+$, 15 % $B_s$, 10 % $\Lambda_b$ |
| (e) | Multiplicity at primary vertex | 0 | | ~ 5 charged particles |
| | | Symmetric | Asymmetric with a $\beta\Gamma$ boost | |
| (f) | B momentum (GeV) | 5 | $5 (1 + \beta\Gamma)$ | ~ 35 |
| (g) | Particle Id Range (GeV) | 0 – 2 | 0 – 3 | 2–20 |
| (h) | Mass resolution | 2–3 MeV | 3 MeV | 50 MeV |
| (i) | Mean flight path | ~ 20 μm | $\beta\Gamma$ x 300 μm | 2.2 mm |
| (j) | Longitudinal vertex resolution | 50 μm | 50 μm | 2 50 μm |
| (k) | Figure of merit $f = v/\sigma_v$ | .4 | 6 x $\beta\Gamma$ | ~ 10 |

## TABLE 2

| Machine / Methods | Sym. Y(4S) | A sym. Y(4S) | $Z^0$ | Fixed Target | Comments: Theo. | Comments: Exp. |
|---|---|---|---|---|---|---|
| Exclusive peaks ($\pi^+\pi^-$, $p\bar{p}$ ,...) | +++ | +++ | ++ | ++++ | Only NON–ZERO strong model dependance | Best use of fixed target capabilities |
| Inclusive semileptonic end point | +++ | +++ | — | — | Limited model dependance | Probably first reliable measurement |
| Exclusive semileptonic decay ($\pi e \nu_e$,..) | ++ | + | + | — | No model dependance | Unfortunately very small branching ratios |
| Inclusive semileptonic | + | ++++ | +++ | — | Limited model dependance | Best use of asym (Y(4S)) colliders |
| Leptonic decays | — | ++ | ++++ | — | No model dependance | Low branching ratio but important. Can only be done at $Z^0$ machines. |

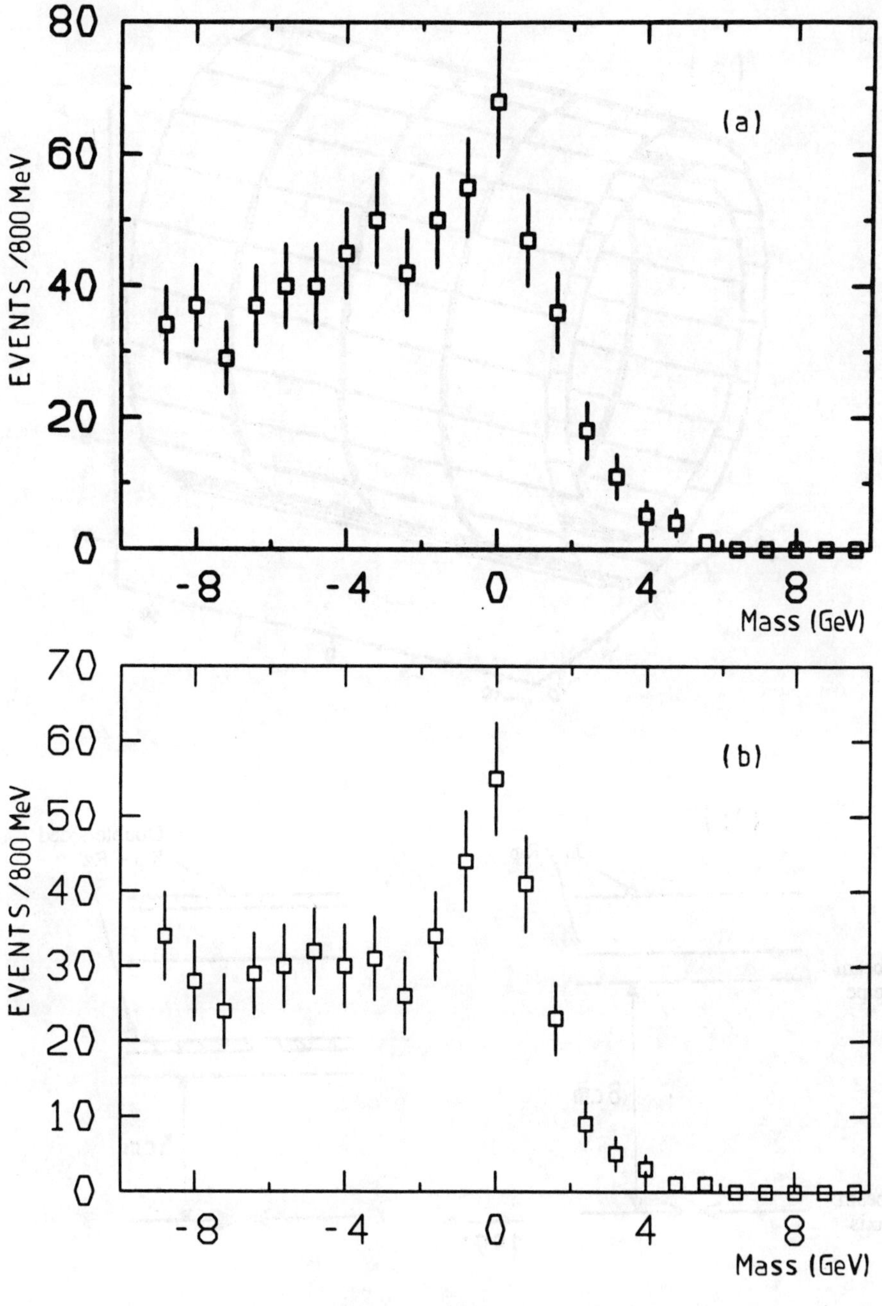
80
60
40
20
0
EVENTS /800 MeV
(a)
-8
-4
0
4
8
Mass (GeV)
70
60
50
40
30
20
10
0
EVENTS /800 MeV
(b)
-8
-4
0
4
8
Mass (GeV)

Figure 1

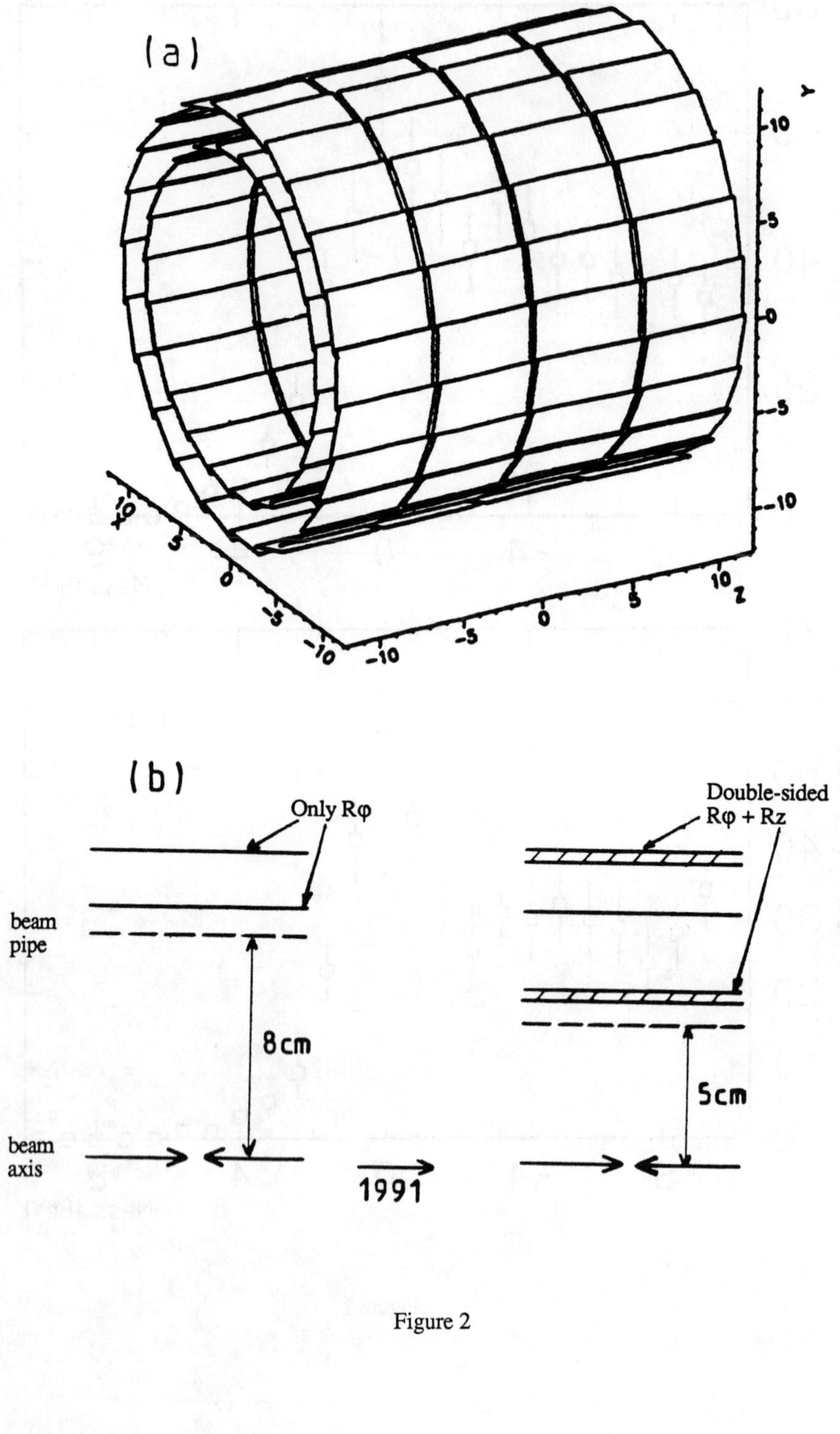
(a)
Y
10
5
0
-5
-10
X
10
5
0
-5
-10
-10
-5
0
5
10
Z
(b)
Only Rφ
Double-sided
Rφ + Rz
beam
pipe
8cm
5cm
beam
axis
1991

Figure 2

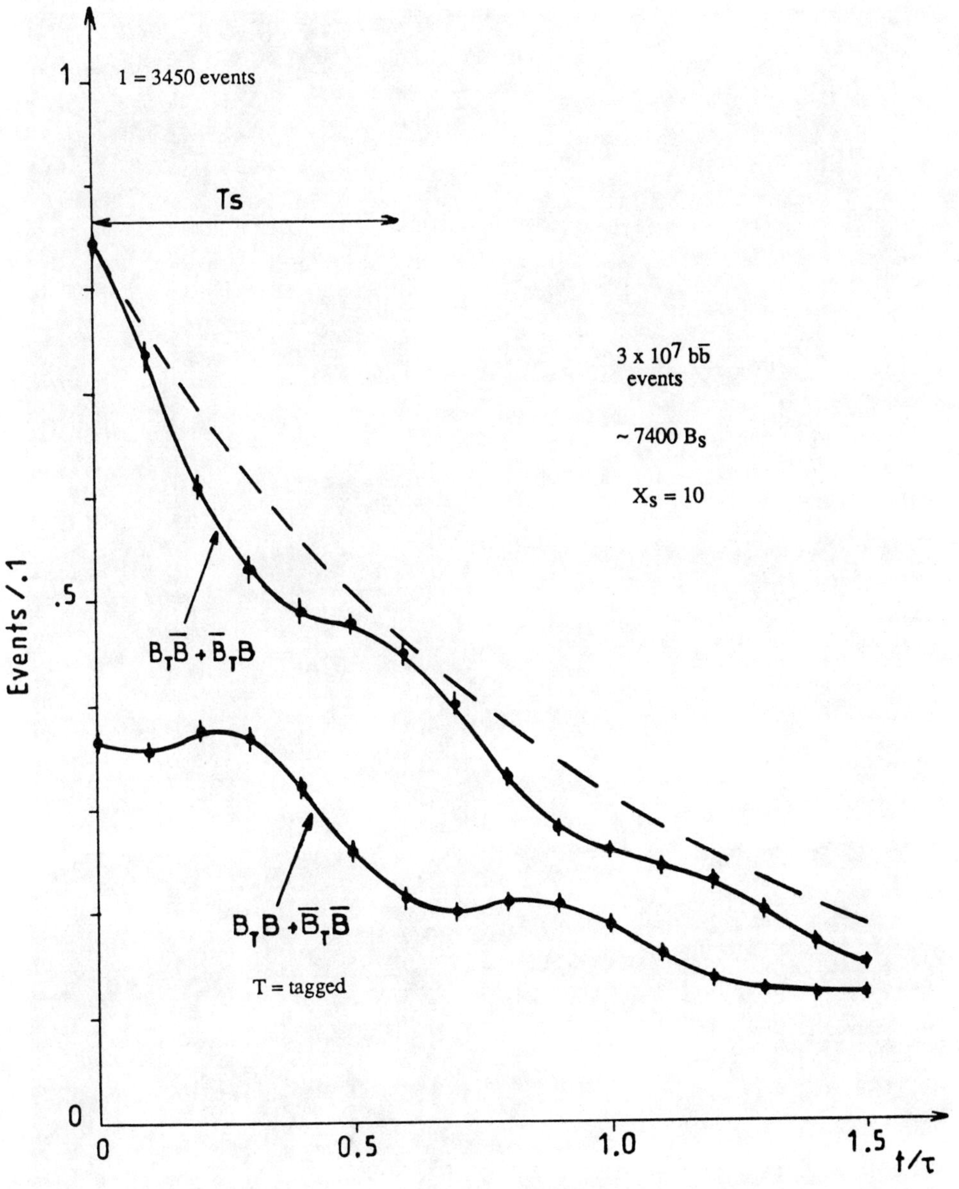
1
.5
0
1 = 3450 events
Ts
3 x 10^7 bb̄
events
~ 7400 Bs
Xs = 10
Events /.1
B_T B̄ + B̄_T B
B_T B + B̄_T B̄
T = tagged
0
0.5
1.0
1.5
t/τ

Figure 3

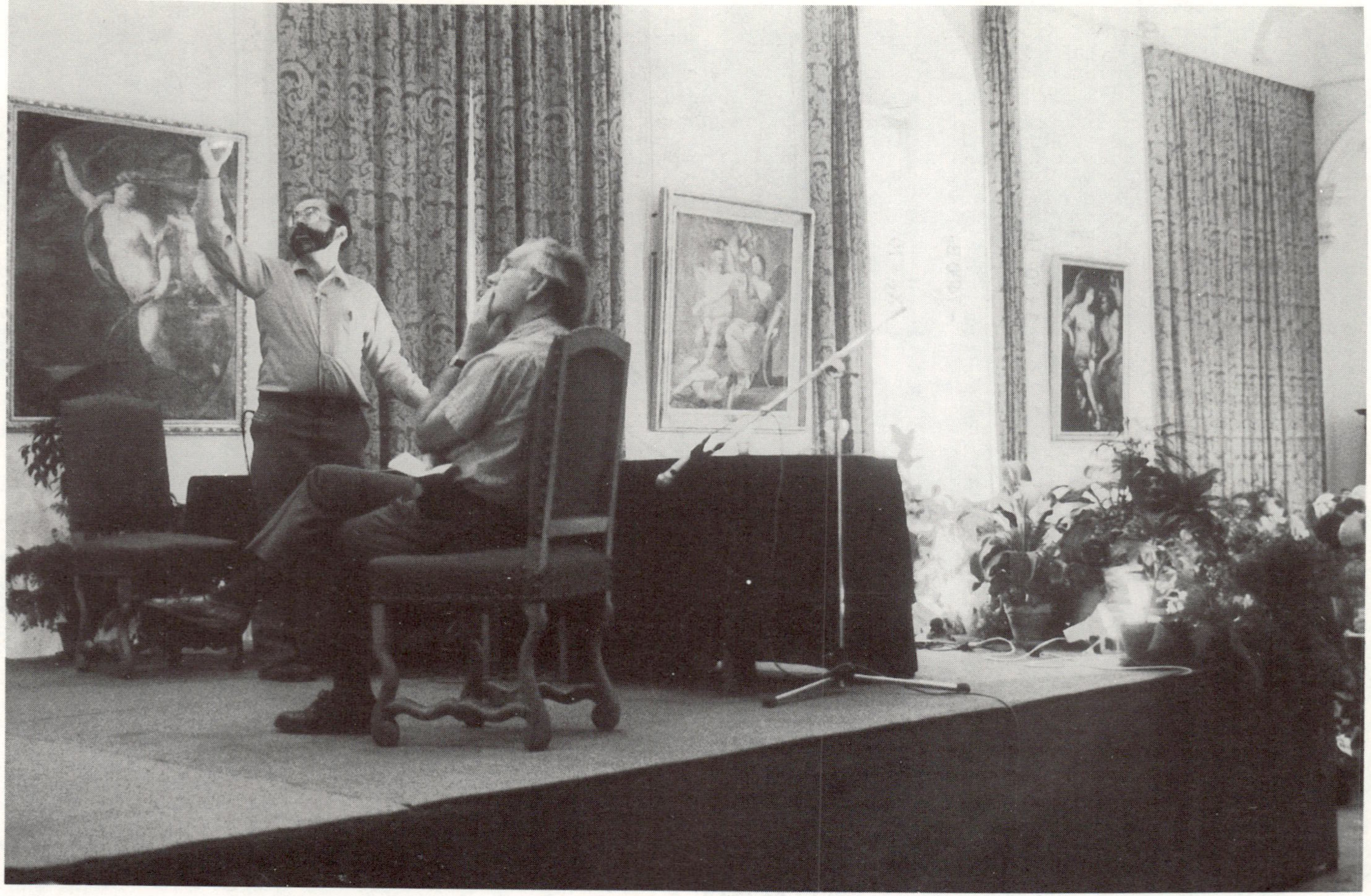

*Larry McLerran and Val Fitch.*

# CP VIOLATION AND COSMOLOGY

*Jean Audouze and Hubert Curien, Minister of Research and Technology.*

# Cosmology and Particles

Audouze Jean
Institut d'Astrophysique de Paris
98 bis Bd Arago
75014 PARIS
France

No written contribution received.

*Edward Kolb and Cynthia Sikivie.*

# CP VIOLATION IN THE EARLY UNIVERSE

Edward W. Kolb
NASA/Fermilab Astrophysics Group
Fermi National Accelerator Laboratory
Batavia, Illinois 60510 USA

## ABSTRACT

**Application of fundamental physical law is a crucial step in the attempt to understand the early (as well as the present) Universe. The discovery of CP violation has provided a key element in the framework of theories that explain why the present Universe seems to consist of matter, rather than equal amounts of matter and antimatter. In this talk I will discuss the role of CP violation in the generation of a cosmological baryon asymmetry.**

**Introduction**

Although CP violation was only discovered in 1964, the evidence that CP is not an exact symmetry has been present for the last 13.6 billion years or so in the form of a cosmological baryon asymmetry [1]. Antimatter is rare on earth, as evidenced by the heroic efforts of accelerator physicists to accumulate antimatter at Fermilab and CERN. Antimatter is also rare throughout the Universe. We all know there are no antimatter bodies in the solar system. Since we all recall that Neil Armstrong survived his "one small step for mankind," the moon must be matter; if not, it would have been one small step indeed. Solar cosmic rays are evidence that the sun is also composed of matter. Planetary probes have visited eight of the planets, dramatically demonstrating that the entire solar system is made of matter.

Beyond the solar system, cosmic rays provide samples of material from throughout the entire galaxy, and probably from distant galaxies as well. Although antiprotons are seen in cosmic rays (at about the $10^{-4}$ level compared to protons [2]), their numbers are consistent with the assumption that the antiprotons are secondaries. The flux of anti-nuclei (in particular anti-$^4$He) is less than $10^{-5}$ that of nuclei, and there is no clear evidence for the detection of an antinucleus. Cosmic rays prove that the asymmetry of matter and antimatter extends (at least) throughout our galaxy. The evidence for the matter–antimatter asymmetry beyond out galaxy is more indirect, but nevertheless convincing. Clusters of galaxies contain gas. If a cluster consisted of both matter galaxies and antimatter galaxies, one would expect to detect x- and $\gamma$-ray emission from matter–antimatter annihilations between the galaxies and the gas. The absence of evidence for such annihilations suggests that clusters of galaxies are either all baryons or all antibaryons. There is little or no information on scales larger than clusters of galaxies. A review of the evidence against the existence of antimatter in the Universe is given in [3].

There is another reason to suspect that the Universe is all matter (or all antimatter). In the very early Universe matter and antimatter should have been in thermal contact, and reactions like matter + antimatter $\leftrightarrow$ radiation should occur rapidly. The nucleon–antinucleon annihilation cross section is so large that in a Universe with equal amounts of matter and antimatter, annihilation is so efficient that by the time the annihilations "freeze out" at temperatures of about 38 MeV [4], the amount of matter and antimatter surviving would be too small to account for the observed mass density of the Universe. The only way to have enough matter survive the annihilation epoch is to have an excess of matter over antimatter, so annihilations effectively stop when the antimatter is "used up" before all the matter has annihilated.

The most reasonable conclusion is that at high temperatures, $T \gtrsim 38$ MeV, corresponding to early times, the Universe possessed an asymmetry between the number of baryons and the number of antibaryons. This asymmetry can be quantified by something known as the "baryon number of the Universe," the ratio of the number density of baryons divided by the entropy density, $B \equiv n_B/s$, where $n_B$ is the Baryon number number density, the difference between the number density of baryons and the number density of antibaryons ($n_B \equiv n_b - n_{\bar{b}}$) and $s$ is the present entropy density of the Universe. If there are three species of massless neutrinos, today $s \simeq 3000\ \mathrm{cm}^{-3}$. If today $n_{\bar{b}} \ll n_b$, then $n_B \simeq n_b$. Estimates for $n_b$ are uncertain because the baryonic mass density of the Universe is uncertain. $n_B$ can be expressed in terms of $\Omega_B h^2$, where $\Omega_B$ is the fraction of the critical density in baryons and $h$ is Hubble's constant in units of 100 km $\mathrm{s}^{-1}$ $\mathrm{Mpc}^{-1}$. The best limit on $\Omega_B h^2$ comes from considerations of primordial nucleosynthesis [5], and constrains $\Omega_B h^2$ to be in the range $0.011 \lesssim \Omega_B h^2 \lesssim 0.037$. Therefore the baryon number of the Universe must be in the range

$$4 \times 10^{-11} \lesssim B \equiv n_B/s = 3.81 \times 10^{-9} \Omega_B h^2 \lesssim 1.4 \times 10^{-10}. \tag{1}$$

In the absence of B-nonconserving interactions and if the Universe maintained a state of near equilibrium resulting in constant total entropy, $B$ will remain constant in the course of the cosmological evolution. Although the baryon asymmetry is "maximal" today (no antibaryons), at temperatures $T \gtrsim m_p \simeq 1$ GeV, quark-antiquark pairs were present in large numbers in the thermal background, and the baryon asymmetry we see today corresponded to a very small quark–antiquark asymmetry at early times,

$$\frac{n_q - n_{\bar{q}}}{n_q} \lesssim 10^{-7}. \tag{2}$$

This means that for every 10 million antiquarks, there were 10 million plus one quarks. A very curious situation that demands a physical explanation.

**Drift and Decay**

The basic ingredients for the generation of a cosmological baryon asymmetry from an initial state of perfect symmetry were outlined by Sakharov [6]. The three major ingredients are 1) Baryon Number violation, 2) C and CP violation, and 3) Non-equilibrium conditions. The first two ingredients are obvious. If baryon number is an exact symmetry of nature, the present baron asymmetry must simply reflect initial conditions. Since $B$ is odd under C and CP, any change in $B$ requires C and CP non-invariance. The third condition is somewhat more subtle. The easiest way to understand it is to realize that the baryon number is also odd under T, and non-equilibrium conditions are necessary to break T invariance by providing an arrow of time. Another way to understand the necessity of the third condition is to note that in equilibrium the entropy is maximal, and if baryon number can be violated, the maximal entropy state will have $B = 0$.

| particle | | final state | branching ratio | B |
|---|---|---|---|---|
| $X$ | $\rightarrow$ | $qq$ | $r$ | 2/3 |
| $X$ | $\rightarrow$ | $\bar{q}l$ | $1-r$ | $-1/3$ |
| $\bar{X}$ | $\rightarrow$ | $\bar{q}\bar{q}$ | $\bar{r}$ | $-2/3$ |
| $\bar{X}$ | $\rightarrow$ | $q\bar{l}$ | $1-\bar{r}$ | 1/3 |

**Table 1: Final states and branching ratios for $X$ and $\bar{X}$.**

The three conditions may in fact be realized in the eraly Universe. The second condition is easiest to imagine. C is violated (maximally) by the weak interactions, and we are all here because CP is not an exact symmetry of nature. Although the CP violation needed in the early Universe may be (and probably is) unrelated to CP violation in ther Kaon system (see [7]), the philosophy that once the toothpaste is out of the tube it is hard to get it back in, implies that there is no reason to expect that CP will be conserved in the early Universe. The third ingredient is easily provided by the expansion of the Universe. There is absolutely no evidence for baryon number non-conservation, so the first ingredient might be the hardest to obtain. Nevertheless, baryon number is violated by non-perturbative effects in the standard electroweak model, and there is every expectation that it is violated in the perturbative sector as well. Almost all grand unified theories (GUTs) incorporate baryon number non-conservation.

To see how the three ingredients operate, conside the following simple model. Assume that there is some massive particle denoted as $X$, which decays to two possible final states, one with two quarks, the other with an antiquark and a lepton. The decays of the $X$ boson and its antiparticle are shown in Table 1. Note that CPT implies that the total decay rates of the $X$ and $\bar{X}$ are equal. C and CP will obtain if the branching ratio of the $X$ to the $qq$ final state is not the same as the branching ratio of the $\bar{X}$ to the $\bar{q}\bar{q}$ final state (i.e., $r \neq \bar{r}$).

Now for the simple procedure. Imagine a box containing equal numbers of $X$ and $\bar{X}$, i.e., symmetric initial conditions. The baryon number produced by decay of an $X$ is $B_X = r(2/3) + (1-r)(-1/3)$, and the baryon number produced by the decay of a $\bar{X}$ is $B_{\bar{X}} = \bar{r}(-2/3) + (1-\bar{r})(1/3)$. The net baryon number produced by the decay of an $X$, $\bar{X}$ pair is

$$\epsilon \equiv B_X + B_{\bar{X}} = r - \bar{r}. \tag{3}$$

The baryon number produced vanishes, of course, if C or CP is conserved ($r = \bar{r}$) or if baryon number is conserved.

This very simple picture has all the ingredients of the drift and decay scenario. In the expanding Universe the initial abundance of $X$, $\bar{X}$ bosons is thermal: $n_X = n_{\bar{X}} \sim n_\gamma$. Provided that the $X$ is massive enough, the interactions that decrease the number of $X$, $\bar{X}$'s are ineffective for $T < m_X$, and $X$, $\bar{X}$ bosons become overabundant as $T$ falls below the $X$-boson mass. Eventually, when $T \ll m_X$, $X$, $\bar{X}$ bosons freely decay without "back" reactions, since inverse decays and other B-violating reactions are suppressed by the low temperature. The resulting baryon number-to-entropy ratio is

$$n_B/s \sim \epsilon n_X/g_* n_\gamma \sim \epsilon g_*. \tag{4}$$

## GUTS

Most of our present scenarios for the generation of the baryon asymmetry were developed in the late 70's and early 80's, and were motivated by the interest in GUTs. Baryon number is violated in almost all GUTs, mediated by some massive ($M \gtrsim 10^{14}$ GeV) vector or scalar boson. Thus, in the early realizations of GUTs, baryon number violation at low energy was characterized by an effective four-fermion "Fermi" interaction, with a coupling constant $G_X \sim m_X^{-2} \sim 10^{-30}$ GeV$^{-2}$, where $m_X$ is the mass of the baryon-number violating boson (or equivalently the mass scale for unification). However, if the Universe was ever at temperatures greater than $m_X$, baryon-number violationg reactions should be just as "strong" as other reactions, in the same way that at energies above $m_Z$, weak interactions have about the same strength as electromagnetic interactions. Thus if GUTs are correct in any way, shape, or form, baryon number should have been violated in the early Universe.

There are two points that must be noted. First, current limits to the lifetime of the proton are in excess of $10^{31}$ to $10^{32}$ yr for many of the branching modes, while the simplest GUTs predict something more like $10^{29}$ yr. There is no direct laboratory evidence for baryon number violation. However, neither is there an understanding for why baryon number should be an exact symmetry of the perturbative sector. There are also GUTs that have an unstable proton, but with lifetime beyond the reach of current experiments. For the remainder of this section I will assume that there is perturbative baryon number violation, and that it is responsible for the baryon asymmetry. For the interesting and exciting possibility that non-perturbative effects are responsible, see [8].

The second point is that in the standard picture, the baryon asymmetry must develop very early in the evolution of the Universe because the mass of the $X$ bosons must be very large. Recall that for a radiation-dominated Universe

$$t = 0.3 g_*^{-1/2} m_{Pl}/T^2, \tag{5}$$

where $g_*$ is the "effective" (sum over bosons and 7/8 times the sum over fermions) number of spin degrees of freedom in relativistic particles. For "simple" GUTs like $SU(5)$ or $SO(10)$, $g_*$ is of order 200. Therefore, using $g_* = 100$ and $T = m_X = 10^{15}$ GeV, the time of baryogenesis must be approximately $t = 3.6 \times 10^{-13}$ GeV$^{-1}$ $= 2 \times 10^{-37}$ sec.

Before discussing the dynamics of baryogenesis, I would like to describe in more detail how CP violation enters. To see explicitly how C, CP violation enters, consider a system with two superheavy bosons, $X$ and $Y$, with baryon number violating decays. The generalization of $\epsilon$ defined in (3) is

$$\begin{aligned} \epsilon_X &= \sum_f B_f \frac{\Gamma(X \to f) - \Gamma(\bar{X} \to \bar{f})}{\Gamma_X} \\ \epsilon_Y &= \sum_f B_f \frac{\Gamma(Y \to f) - \Gamma(\bar{Y} \to \bar{f})}{\Gamma_Y}, \end{aligned} \tag{6}$$

where the sum runs over all final states $f$, each with baryon number $B_f$, and $\Gamma_X$ and $\Gamma_Y$ is the total $X$ and $Y$ decay widths. For simplicity, assume there are but two final states for $X$ and $Y$ decay, and that the interaction Lagrangian is given by

$$\mathcal{L} = g_1 X i_2 i_1^\dagger + g_2 X i_4 i_3^\dagger + g_3 Y i_1 i_3^\dagger + g_4 Y i_2 i_4^\dagger + \text{h.c.} \tag{7}$$

where $i_1$, $i_2$, $i_3$, $i_4$ are fermions and $g_i$ are coupling strengths (which must be complex). The model leads to the decay processes: $X \to \bar{i}_1 + i_2$, $\bar{i}_3 + i_4$, and $Y \to i_1 + \bar{i}_3$, $i_2 + \bar{i}_4$.

The lowest order processes cannot contribute to $\epsilon$, as $\Gamma(X \to \bar{i}_1 i_2) = |g_1|^2 I_X = \Gamma(\bar{X} \to i_1 \bar{i}_2) = |g_1^*|^2 I_{\bar{X}}$ where the kinematic factors $I_X = I_{\bar{X}}$ arise from the phase space integrals. The first non-zero contribution to $\epsilon$ comes from the interference of the lowest order graphs with the one-loop corrections. These interference terms are given by

$$\begin{aligned} \Gamma(X \to \bar{i}_1 i_2) &= g_1 g_2^* g_3 g_4^* I_{XY} + (g_1 g_2^* g_3 g_4^* I_{XY})^* \\ \Gamma(\bar{X} \to i_1 \bar{i}_2) &= g_1^* g_2 g_3^* g_4 I_{XY} + (g_1^* g_2 g_3^* g_4 I_{XY})^*, \end{aligned} \tag{8}$$

where the phase-space factors $I_{IJ}$ ($I$, $J = X, Y$) include the kinematic factors arising from integrating over the internal momentum loop due to $J$ exchange in $I$ decay. If the intermediate particles ($i_1$, $i_2$, $i_3$, $i_4$) in the loop are kinematically allowed to propagate on shell, as will be the case if $X$, $Y$ are superheavy bosons and $i_{1-4}$ are light quarks and leptons, the quantity $I_{IJ}$ will be complex. As I will show shortly, the complexity of $I_{IJ}$ is crucial to baryogenesis.

The difference between $X \to \bar{i}_1 i_2$ and $\bar{X} \to i_1 \bar{i}_2$ is given by

$$\begin{aligned} \Gamma(X \to \bar{i}_1 i_2) - \Gamma(\bar{X} \to i_1 \bar{i}_2) &= 2i I_{XY} \mathrm{Im}(g_1 g_2^* g_3 g_4^*) \\ &\quad + 2i I_{XY}^* \mathrm{Im}(g_1^* g_2 g_3^* g_4) \\ &= 4 \mathrm{Im} I_{XY} \mathrm{Im}(g_1^* g_2 g_3^* g_4). \end{aligned} \tag{9}$$

After a similar calculation for the other decay mode,

$$\epsilon_X = \frac{4}{\Gamma_X} \mathrm{Im} I_{XY} \mathrm{Im}(g_1^* g_2 g_3^* g_4) \left[ (B_{i_4} - B_{i_3}) - (B_{i_2} - B_{i_1}) \right]. \tag{10}$$

Repeating the same calculations for the $Y$ decay modes, the result is $\epsilon_Y = -\epsilon_X$.

From this exercise we see that 3 things are required in order to have $\epsilon \neq 0$ (and therefore successful baryogenesis). First, there must be (at least) *two* baryon-number violating bosons, each with mass greater than the sum of the (fermion) masses in the internal loops; otherwise, $\mathrm{Im}I_{XY} = 0$. Second, C, CP violation arises from the interference of loop diagrams with the tree graph, and manifests itself in complex coupling constants. Thus, in general we expect $\epsilon$ to be of the order of $\alpha^N$, where $\alpha$ characterizes the coupling constant of the loop particle(s) and $N$ is the number of loops in the lowest-order diagram that interferes with the tree graph to give rise to $\epsilon \neq 0$. Note that for this reason, $\epsilon \sim \alpha^N$ is expected to be small. Third, the $X$ and $Y$ particles in the above example must not be degenerate in mass, or the baryon number produced by $X$ decays will precisely cancel that produced by $Y$ decays.

Much more detail about the role of CP violation in baryogenesis will be given in [7]. Henceforth I will assume that CP violation is present (in the form of some complex coupling constant) and I will simply parameterize it by $\epsilon$. Of course, we expect epsilon to be small, $\epsilon \ll 1$.

The detailed study of the origin of the baryon asymmetry requires some treatment of the interactions of all the particles present in the Universe. The techniques for the reaction networks in the curious case where CP is violated were developed by Kolb and Wolfram, and by Fry, Olive, and Turner [9]. In these toy models, which typically have one massive particle that decays into baryons or antibaryons, it was found that the resulting baryon number depended upon a parameter $K$, defined as

$$K \equiv \left(\frac{\Gamma_D}{H}\right)_{T=m_X} = \left(\frac{\alpha m_X}{(\pi^2 g_*/30)^{1/2} T^2/m_{Pl}}\right)_{T=m_X} \sim \frac{\alpha m_{Pl}}{m_X}, \tag{11}$$

where $\Gamma_D$ is the $X$ decay width, $H$ is the expansion rate of the Universe, $H = g_*^{1/2} T^2/m_{Pl}$, and $\alpha$ is the GUT fine structure constant.

If $K$ is small (much less than one), the $X$ boson will decay when the Universe is at a temperature much less than $m_X$, and inverse reactions will not destroy the baryon number produced by $X$ decays. If $K$ is large (much greater than one) the $X$ boson will decay when the Universe is still hot enough that inverse reactions involving real and intermediate $X$ bosons will eradicate the baryon number produced by the $X$ boson decays as equilibrium is approached (recall that non-equilibrium conditions are required for baryogenesis). $K$ is said to determine the effectiveness of decays in producing the baryon excess. Note that the requirement that $K \ll 1$ sets a requirement on the $X$ boson mass. If the $X$ mass is too low, than $K$ will be large, and baryogenesis ineffective. This is the problem faced by all low-temperature scenarios.

Once the techniques for developing the Boltzmann equations to study baryogenesis were developed, they were applied to "realistic" GUTs like $SU(5)$ and $SO(10)$ [10], and even to supersymmetric GUTs [11]. What we have learned after the dust settled is that the standard "out-of-equilibrium" decay scenario provides an attractive framework in understanding the origin of the baryon asymmetry. Although I did not nemtion $B-L$ violation, in most models a lepton number also arises, and in models where $B-L$ is not a global symmetry, a non-zero value of $B-L$ often results. This last point may be of relevance for sphaleron considerations.

Although it does provide an attractive framework, it is not without problems. One of the fundamental problems of relevance for this conference is that the CP violation required for baryogenesis does not seem to be related to either the Kaon system or neutron electric dipole moment. Another problem is that even in the most favorable of circumstances the produced baryon number is small, leaving little room for entropy production that might dilute the baryon number. Thus inflation (which produces entropy) must occur *before* baryogenesis, which presents a difficulty for models of inflation. Another inflation-related problem is that for inflation to succeed, the inflaton must be very weakly coupled, which results in low reheat temperatures. If baryogenesis occurs after inflation, a low starting temperature might result.

**Alternatives**

Here I will mention 5 alternatives to the drift and decay scenarion. Almost all of them incorporate one or more features of the standard picture discussed above.

- *Spahaerons:* As the talk following mine is devoted to this topic, I will say nothing about them, and simply refer to [8].

- *Primordial Black Holes:* The fact that "black holes have no hair" implies that the baryon number of a black hole is undetermined. If black holes are hot enough, they can radiate $X$ bosons, which can decay and produce the asymmetry [12].

- *Low-Temperature Scenarios:* The discussion above assumed that the massive perticle responsible for baryogenesis was a particle with $SU(3) \otimes SU(2) \otimes U(1)$ interactions. The key to any low-temperature scenario is to keep the massive particle out of equilibrium at low temperatures. This is only possible if the particle is an $SU(3) \otimes SU(2) \otimes U(1)$ singlet. Such models where the particle is a massive Majorana fermion was proposed by [13].

- *Spontaneous Baryogenesis:* A proposal to account for the baryon asymmetry through a temporary dynamical violation of CPT is one of the most radical scenarios [14]. The basic idea is to assume that there is some scalar field $\phi$ called the ilion that appears in some effective Lagrangian coupled to the baryon number current: $\mathcal{L}_{\rm eff} = f_\phi^{-1}\partial_\mu\phi j^\mu_B$, (cf. the axion!). This term breaks CPT by shifting the energy of a baryon relative to an antibaryon, and in equilibrium $n_B \neq 0$. Presumably there is some potential for $\phi$ and eventually $\dot{\phi}$ is damped, $\partial\phi \to 0$ and the offending term disappears. But as it disappears, it leaves behind the baryon number of the Universe.

• *Squark/Slepton Condensates:* Afflect and Dine have pointed out that in many supersymmetric models, there are "flat" directions in the scalar potentials corresponding to vacuum expectation values (VEVs) for squarks and sleptons [15]. As the squark and slepton fields relax to their ground state (assuming that at high temperatures they are displaced from their ground states) the coherent oscillations of the squark/slepton fields will produce a baryon number.

**Acknowledgements**

This work was supported by the DOE and NASA (grant number NAGW–1340).

**References**

1. For a review of baryogenesis, see E. W. Kolb and M. S. Turner, *Ann. Rev. Nuc. Part. Sci.* **33**, 645 (1983); and in *The Eraly Universe* (Addison–Wesley, Redwood City, Calif., 1989), chapter 6.

2. E. A. Bogomolov, et al., *Proc. of the 16th Intl. Cosmic Ray Conference* (Kyoto, 1979), Vol. 2, p. 330; R. L. Golden, et al., *Phys. Rev. Lett.* **43**, 1196 (1979); M. V. K. Apparao, *Can. J. Phys.* **46**, S654 (1968); A. Buffington, et al., *Ap. J.* **248**, 1179 (1981); S. P. Ahlen, et al., *Phys. Rev. Lett.* **61**, 145 (1988); R. Streitmatter, et al., (LEAP collaboration), in preparation (1989).

3. G. Steigman, *Ann. Rev. Astron. Astrophys.* **14**, 339 (1976).

4. E. W. Kolb and M. S. Turner, *The Eraly Universe* (Addison–Wesley, Redwood City, Calif., 1989)

5. See J. Audouze in these proceedings.

6. A. D. Sakharov, *JETP Letters* **5**, 24 (1967).

7. See R. Peccei, these proceedings.

8. See L. McLerran, these proceedings.

9. E. W. Kolb and S. Wolfram, *Nucl. Phys.* **B172**, 224 (1980); J. N. Fry, K. A. Olive, and M. S. Turner, *Phys. Rev. D* **22**, 2953 (1980).

10. Boltzmann equations for $SU(5)$ have been derived by J. N. Fry, K. A. Olive, and M. S. Turner, *Phys. Rev. D* **22**, 2953 (1980); **22**, 2977 (1980), and for both $SU(5)$ and $SO(10)$ by J. A. Harvey, E. W. Kolb, D. B. Reiss, and S. Wolfram, *Nucl. Phys.* **B201**, 16 (1982).

11. Baryogenesis in supersymmetric unified models has been studied by E. W. Kolb and S. Raby, *Phys. Rev. D* **27**, 2990 (1983); J. N. Fry and M. S. Turner, *Phys. Lett.* **125B**, 379 (1983); H. E. Haber, *Phys. Rev. D* **26**, 1317 (1982).

12. Alternative baryogenesis scenarios involving primordial black holes have been considered by M. S. Turner, *Phys. Lett.* **B89**, 155 (1979); D. Lindley, *Mon. Mot. R. astron. Soc.* **196**, 317 (1981); A. D. Dolgov, *Zh. Eksp. Teor. Fiz.* **79**, 337 (1980); A. F. Grillo, *Phys. Lett.* **94B**, 364 (1980); and L. L. Krauss, *Phys. Rev. Lett.* **49**, 1459 (1982).

13. Models in which the baryon asymmetry evolves at very low temperature ($T \ll 10^{14}$ GeV) have been discussed by S. Dimopoulos and L. Hall, *Phys. Lett.* **196B**, 135 (1987); M. Claudson, L. Hall, and I. Hinchliffe, *Nucl. Phys.* **B241**, 309 (1984).

14. A. Cohen and D. Kaplan, *Phys. Lett.* **199B**, 251 (1987).

15. I. Affleck and M. Dine, *Nucl. Phys.* **B249**, 361 (1985).

# CP Violation and Baryogenesis

**R. D. Peccei†**
Department of Physics, UCLA, Los Angeles, CA. 90024

## ABSTRACT

After reviewing the conditions for obtaining baryogenesis in the early universe, I discuss why, in general, one does not expect the CP phase of baryogenesis to be connected to that giving rise to CP violation in Kaon decays. Two special cases where this connection exists are, however, also examined.

† On leave from DESY, Hamburg, Fed. Rep. Germany

## I. Baryogenesis and CP Violation

A few years after the discovery of CP violation [1], which we are celebrating in this Symposium, A. Sakharov [2] in a seminal paper outlined the necessary ingredients for dynamical baryogenesis. These are:

i) The theory must possess baryon number violating interactions.

ii) These B-violating processes must go out of equilibrium sometime during the history of the universe.

iii) C and CP must, additionally, be violated.

The first point above is obvious, while the second will be discussed in detail by Kolb [3] in this Symposium. * I will, therefore, focus on the last point above, and most particularly, on the CP violating aspects of it. For most of my discussion I will assume that the universe baryon asymmetry originated from the baryon violating interactions present in a grand unified theory (GUT). However, I shall try to keep my treatment rather general since, as is well known, there is no positive evidence for any given GUT. Indeed the absence of proton decay can be used to rule out the simplest versions of $SU(5)$.

To illustrate the issues involved let me consider an excitation $\chi$ which has at least two decay channels, with different final baryon numbers, and which is responsible for the asymmetry. An example for $\chi$ is provide by the Higgs **5** of $SU(5)$ which can decay into $q^c q^c$ with $B = -\frac{2}{3}$ and $q\ell$ with $B = \frac{1}{3}$. However, the discussion that follows is quite general. For baryogenesis to begin, the number density of $\chi$ must go out of equilibrium. This will occur when the rate for $\chi$ interactions [decays, inverse decays and scattering processes] becomes much slower than the universe's expansion rate [3]: $\Gamma_\chi << H$. One can then calculate the difference in the density of baryons and antibaryons relative to photons directly in terms of the baryon asymmetry produced in $\chi$ decays. Since the density ratio is an invariant in an expanding universe** and at present our universe is baryon dominated, one has:

$$\frac{n_B - n_{\bar{B}}}{n_\gamma} \simeq \frac{n_B}{n_\gamma} = A\, \Delta B_\chi \tag{1}$$

* Roughly speaking, it is clear that unless one blocks the inverse processes that destroy any baryon asymmetry one creates, no asymmetry can be established. Dynamically, this blocking can only occur if the relevant interactions go out of equilibrium.

** This ignores reheating effects which change $n_\gamma$, but these effects can be accounted for readily and will be omitted here.

Here A is a kinematical/dynamical factor which is model dependent and is a function of the precise way in which the $\chi$ population goes out of equilibrium. Typically, however, $A \sim 10^{-1} - 10^{-2}$ [3].

The baryon asymmetry produced in $\chi$ decays is, by definition,

$$\Delta B_{\chi} = \sum_f B_f \frac{[\Gamma(\chi \to f) \; - \; \Gamma(\bar{\chi} \to \bar{f})]}{\Gamma_{tot}\,(\chi)} \tag{2}$$

Here $B_f$ is the baryon number of the final state $f$ and, as usual, $\bar{\chi}$ and $\bar{f}$ are the antiparticles of $\chi$ and $f$. The baryon number weighting above is important since the CPT theorem, plus unitarity, implies that

$$\Gamma_{tot}(\chi) = \sum_f \Gamma(\chi \to f) = \sum_{\bar{f}} \Gamma(\bar{\chi} \to \bar{f}) = \Gamma_{tot}\,(\bar{\chi}) \tag{3}$$

If CP (or C) is conserved, however, one has not only equality for the total decay rates but also for the partial rates.

$$\Gamma(\chi \to f) = \Gamma(\bar{\chi} \to \bar{f}) \quad [CP \;\; \text{conservation}] \tag{4}$$

Hence, in this case, the baryon number weighting in Eq. (2) is irrelevant and $\Delta B_{\chi} = 0$. However, if CP (and C) is violated then the partial rates can differ and $\Delta B_{\chi}$ need not vanish.

Let me, for simplicity, consider the case when $\chi$ has only two decay channels, with final baryon numbers $B_1$ and $B_2$ and denote by $r_i$ the partial decay rate branching ratio:

$$r_i = \frac{\Gamma(\chi \to f_i)}{\Gamma_{tot}(\chi)} \quad . \tag{5}$$

Then, even though $r_i \neq \bar{r}_i$, the sum of the partial decay rates for $\chi$ and $\bar{\chi}$ is equal: $r_1 + r_2 = \bar{r}_1 + \bar{r}_2 = 1$. Thus $\Delta B_{\chi}$ can be written simply as

$$\Delta B_{\chi} = (B_1 - B_2)(r_1 - \bar{r}_1) \tag{6}$$

In practice there are two additional conditions, besides C and CP violation, which are necessary for a calculation of $r_1 - \bar{r}_1$ not to give a vanishing result [4]. One must compute the branching ratios beyond lowest order and a physical discontinuity must ensue in this process. Both of these conditions are easily understandable. Obviously, a lowest order calculation will not do. Rates involve absolute values of decay amplitudes and no difference between $r_1$ and $\bar{r}_1$ can exist if these amplitudes are the complex conjugate of each other. This is also the reason why one needs a discontinuity to appear. Without such a discontinuity again the amplitude for $r_1$ and $\bar{r}_1$ are the complex conjugate of each other and $r_1 = \bar{r}_1$.

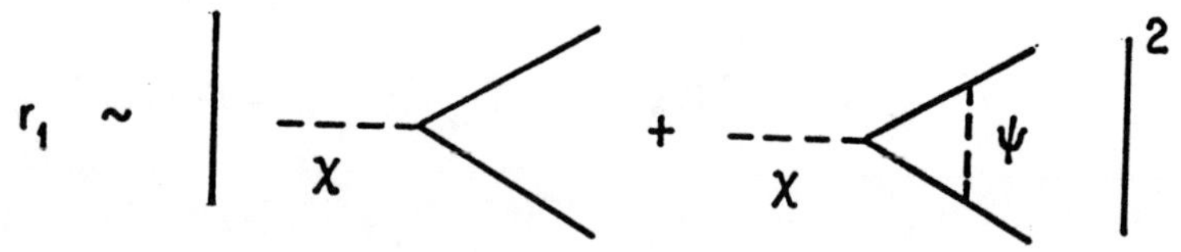


Fig. 1: Contributions to $r_1$

I illustrate the above considerations pictorially in Fig. 1, where the branching fraction $r_1$ is computed including a next to lowest order correction, involving the exchange of a state $\psi$. Let me write, schematically,

$$r_1 = |\gamma_0 + \gamma_1 I(M_\chi^2 - i\epsilon)|^2 \tag{7}$$

where $\gamma_0$ and $\gamma_1$ are (products of) complex coupling constants and I is the contribution of the loop integral. Then the branching fraction for the decay of $\bar{\chi}$ into the corresponding antiparticles is just

$$\bar{r}_1 = |\gamma_0^* + \gamma_1^* I(M_\chi^2 - i\epsilon)|^2 = |\gamma_0 + \gamma_1 I(M_\chi^2 + i\epsilon)|^2 \tag{8}$$

Whence

$$r_1 - \bar{r}_1 = 4\ Im(\gamma_0\gamma_1^*) Im\ I(M_\chi^2 - i\epsilon) \tag{9}$$

Eq. (9) illustrates clearly the necessity of going beyond leading order in computing the rates and the requirement of having a non trivial discontinuity.

There are a number of features concerning baryogenesis, which one can glean from Eq. (9):

i) It is clear from this equation that the actual value for $\Delta B_\chi$, and hence the value of $\frac{n_B}{n_\gamma}$, depends not only on the CP phases of the underlying theory, but it also depends on the underlying dynamics through the discontinuity of I. In particular, since one must consider all processes that lead to an asymmetry, it is not hard to show that there are further restrictions. For instance, if the two states $\chi$ and $\psi$ in Fig. 1 have decays and mutual interactions which can contribute to $\Delta B$, one can show that unless both $\chi$ and $\psi$ have $B$ violating interactions, then $\Delta B = 0$ [5]. Similarly, there is no asymmetry if $M_\chi = M_\psi$, because in this case $\Delta B_\chi = -\Delta B_\psi$ [5]. One concludes, unfortunately, that to predict the value or even the sign of $\Delta B$ requires making dynamical inputs in a largely unknown theory - inputs that we often do not have.

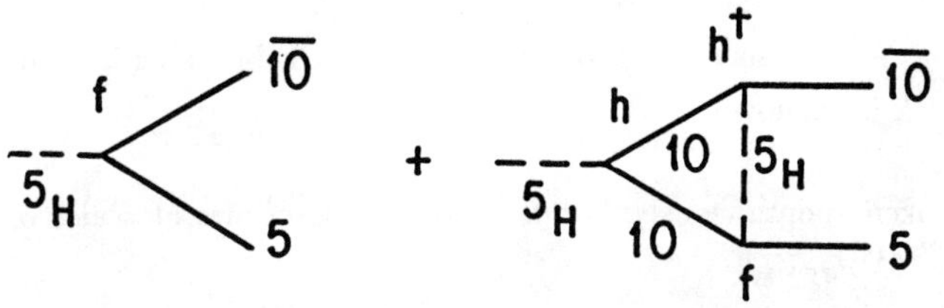


Fig. 2: Decay of a **5** Higgs to quarks and leptons in the $\bar{\mathbf{10}}$ and **5** representations of $SU(5)$.

ii) It is possible that, for simple enough theories, $Im(\gamma_0\gamma_1^*) = 0$, so that there is no net baryon asymmetry. This is true, for instance in GUTs based on $S0(10)$, when the breakdown preserves a left-right symmetric subgroup. In these cases [6] one can still define what amounts to a conserved C operator in the broken theory and no relative phase appear between $\gamma_0$ and $\gamma_1$. A perhaps more familiar example is the case when $\chi$ is a **5** Higgs field in $SU(5)$. In this case there are two couplings of this field to quarks and leptons and $\psi = \chi$. However, as can be seen from Fig. 2,

$$Im(\gamma_0\gamma_1^*) = Im(ff^\dagger hh^\dagger) = 0 \tag{10}$$

and there is no net baryon asymmetry [7].

**II. Is the Baryogenesis CP Phase the Same as that in Kaon Decays?**

The above discussion makes it clear that is is not really possible to give a quantitative computation of $\frac{n_B}{n_\gamma}$, even in GUTs where all the conditions for establishing the asymmetry obtain [8]. Qualitatively, however, at least for a Higgs generated asymmetry, numbers of $0(10^{-6})$ for $Im\gamma_0\gamma_1^*$ and of $0(10^{-3})$ for Im ï are not unreasonable, so that values of $\frac{n_B}{n_\gamma} \sim 10^{-10}$ (as is experimentally observed) are perfectly sensible. In any case, it is interesting to ask whether the CP phase - which I will call $\delta_B$ - connected with baryogenesis is, or is not, connected with the CP phase $\delta_K$, responsible for CP violating phenomena in the $K$ - system.* As I will discuss below, basing myself on some work done long ago with Masiero and Mohapatra [9], the answer, unfortunately, appears to be **no**.

To answer the question whether $\delta_B$ and $\delta_K$ are one and the same, it is useful first to distinguish various different options for the source of CP violation in the gauge theories describing the fundamental interactions. These are:

i) CP is broken spontaneously and the scale of the CP breaking is around that of the electroweak breakdown.

ii) CP is broken spontaneously, but this breakdown occurs at scales of the order of $M_{GUT} \sim 10^{15}$ $GeV$.

iii) CP is intrinsically broken by the Yukawa couplings $\Gamma_{Hf\bar{f}}$ of the Higgs fields(s) to the fermions.

iv) CP is intrinsically broken by both Yukawa couplings $\Gamma_{Hf\bar{f}}$ and Higgs self couplings $\lambda_H$.

The first option above is the easiest to discuss, since one can readily dismiss it if the baryon violating phenomena occur at a scale of $0(M_{GUT})$. In this case, the $B$ asymmetry is supposed to be generated at temperatures T which are also of $0(M_{GUT})$. However,

* I am talking here somewhat loosely about a phase $\delta_B$ and a phase $\delta_K$. In both cases there may be more than one CP violating phases contributing.

at such high temperatures the Higgs vacuum expectation values which violated CP spontaneously surely must vanish. Hence, no baryon asymmetry is generated at all!

It is possible, however, to imagine that the universe baryone asymmetry was generated at the electroweak scale, due to weak instanton effects [10]. Although this is an interesting possibility, which I will discuss later on, it is not bereft of problems. It seems, therefore, a little perverse to mix it in with low energy spontaneous CP violation, which has its own difficulties. In fact, independently of whether or not one can generate a baryon asymmetry in this scenario, low energy spontaneous CP violation is probably not tenable by itself because of the domain wall problem. As Kobzarev, Okun and Zeldovich [11] pointed out, in the case of spontaneous CP violation one generates domains with different CP values in the universe, as one goes through the phase transition temperature. The walls separating these domains have a large surface density and dissipate very slowly [$\rho_{wall} \sim T$] as the universe expands. Thus eventually the energy density in the domain walls would end up exceeding the universe critical density. This cosmological problem makes considering spontaneous CP violation difficult, unless one can imagine inflating the domains away. However, this requires that the CP violating phase transition occur at high, rather than low scales.

If CP violation is spontaneous, but the breakdown occurs at scales of $0(M_{GUT})$, one can presume that an inflationary period got rid of the domain wall problem [we live in one domain!]. Even in this case, however, one does not expect that $\delta_B$ and $\delta_K$ are the same. The reason for this is rather clear. CP violation in these models is intimately tied to the Higgs potential and this potential varies with temperature. Thus one has different minimization equations to solve as a function of temperature for the CP phase:

$$\frac{\partial V(\delta;T)}{\partial\delta} = 0 \tag{11}$$

and $\delta = \delta(T)$. Under these circumstances there is no reason why $\delta_B \simeq \delta(M_{GUT})$ should be equal to $\delta_K \simeq \delta(M_W)$.

Let me illustrate the above with a simple model based on $SU(2)_R \times SU(2)_L \times U(1)_{B-L}$ [12], in which one has three kinds of Higgs fields [with associated vacuum expectation values]:

$$\Delta_R \sim (1,0) \qquad <\Delta_R> = V_R e^{i\delta_R} \tag{12a}$$

$$\Delta_L \sim (0,1) \qquad <\Delta_L> = V_L e^{i\delta_L} \tag{12b}$$

$$\Phi \sim (\frac{1}{2},\frac{1}{2},) \qquad <\Phi> = \kappa e^{i\delta} \tag{12c}$$

The Higgs potential for the model has a minimum which admits a hierarchy [13]

$$V_R >> \kappa >> V_L \quad ; \quad \kappa^2 \sim V_R V_L \tag{13}$$

and one finds * that the associated phases, which signal spontaneous CP violation, obey an inverse hierarchy

$$\delta_L \sim 0(1) \quad ; \quad \delta \sim \frac{V_L}{V_R}\delta_L \quad ; \quad \delta_R \sim \left(\frac{V_L}{V_R}\right)^2 \delta_L \tag{14}$$

Thus in this model, at $T = 0$ one finds $\delta_R \simeq 0$. However, as one increases the temperature beyond $V_L$ and $\kappa$ only the $SU(2)_R$ group is broken. The relevant minimization equation for $\delta_R$, for $T >> \kappa, V_L$, is simply of the form

$$A + B\cos\delta_R = 0 \tag{15}$$

with $A$ and $B$ being coefficents of order unity. Hence, at high temperatures $\delta_R(T >> \kappa, V_L) \sim 0(1)$, which is quite different from $\delta_R(T = 0) \simeq 0$.

If CP is explicitly broken in the theory by the appearance of phases in Yukawa and Higgs self couplings the situation is not much better, as far as relating $\delta_B$ and $\delta_K$ goes. Consider first the case in which one worries only about the phases entering in the Yukawa couplings. It is well known in this case that, in the low energy theory, one can trade the complex matrices $\Gamma^u$ and $\Gamma^d$, coupling the doublet Higgs field $\phi$ to charge $\frac{2}{3}$ and charge $-\frac{1}{3}$ quarks respectively, into a complex mixing matrix V describing the coupling of these states to the W bosons. For $n$ generations this matrix $V$ - the Cabibbo Kobayashi Maskawa matrix [14] - has $\frac{1}{2}(n-2)(n-1)$ phases. Thus for 3 generations there is a single phase $\delta$ which is responsible for the low energy CP violation seen in the Kaon system: $\delta_K = \delta$. However, in GUTs, even when one has only 3 generations, there are always **more** phases than those in the low energy theory, so $\delta_B \neq \delta_K$.

Let me illustrate the above by considering an example based on the simplest form of an *SU*(5) *GUT*.** I shall denote the Higgs field by $\mathbf{5_H}$ and the two fermion representation of the quarks and leptons by their dimension: $\mathbf{\bar{10}}$, $\mathbf{5}$. Then, as shown in Fig. 3, there

* One needs technically to have more than one $\Delta_L$ and $\Delta_R$ fields for non trivial phases to exist [12].

** For these purposes I ignore the fact that this theory is not viable, since it predicts too rapid proton decay.

are two relevant Yukawa couplings: $f_{ij}$ and $h_{ij}$. For $n$ generations the matrix $f_{ij}$ will contain $n^2$ phases, while $h_{ij}$, since $h = h^T$, contains $\frac{1}{2}n(n+1)$ phases. As in the case of the low energy theory, not all these phases are physical. Indeed one can [9]:

i) Remove $n^2$ phases by diagonalizing $f$ via a bi-unitary transformation

$$f = U^\dagger f_{diag}\, V \tag{16}$$

This redefines new fermion fields

$$\mathbf{5} \rightarrow V\mathbf{5} \;\; ; \;\; \mathbf{10} \rightarrow U\mathbf{10} \tag{17}$$

and changes $h$ to $h' = U^T\, hU = h'^T$

ii) One can further remove $n$ phases from $h'$ by doing an extra transformation on the fermions:

$$\mathbf{5} \rightarrow W\mathbf{5} \;\; ; \;\; \mathbf{10} \rightarrow W\mathbf{10} \tag{18}$$

with $W_{ij} = \delta_{ij}e^{i\delta_i}$ since this transformation does not affect $f_{diag}$.

After these transformations, in the $SU(5)$ $GUT$ one has left $\frac{1}{2}n(n-1)$ physical phases. Thus for 3 generations one has 3 GUT phases $\delta_{B_1}, \delta_{B_2}$ and $\delta_{B_3}$, compared to one low energy phase $\delta_K = \delta$.

Fig. 3: Definition of the GUT Yukawa couplings in $SU(5)$

I have mentioned earlier that $SU(5)$ with just one $\mathbf{5_H}$ of Higgs is not a realistic model for the universe baryon asymmetry because $Im\gamma_0\gamma_1^*$ vanishes, as is illustrated schematically in Fig. 4a. Nevertheless, one can obtain a non vanishing $\Delta B_\chi$ by considering a higher order correction to the graph of Fig. 4a. While the contribution of Fig. 4a is proportional to $ImTr(ff^\dagger hh^\dagger)$. and hence vanishes, that of Fig. 4b is proportional to $ImTr(h^\dagger ff^\dagger hf^\dagger fh^\dagger h)$ and gives a finite, albeit probably very small result. Assuming that $h$ and $f$ are typical Yukawa couplings, Ellis et al [7] estimated that for this case $\frac{n_B}{n_\gamma} \sim 10^{-18}$. However, given a large top quark mass one may actually obtain a considerably larger asymmetry from Fig. 4b. Indeed, long ago, Segre' and Turner [15] argued that with 4 generations this graph may suffice to give $\frac{n_B}{n_\gamma} \sim 10^{-10}$. Clearly, it is difficult to argue very cogently for this possibility, particularly since we only have negative indications on $SU(5)$ due to the long proton lifetime! Nevertheless, even if Fig. 4b were to give the right order of magnitude for $\frac{n_B}{n_\gamma}$, it is clear that $\frac{n_B}{n_\gamma}$ is a function of the **three** $SU(5)$ phases $\delta_{B_i}$, which are not directly connected to $\delta_K = \delta$. Furthermore, more "realistic" GUTs end up by having more parameters (and phases) than the minimal $SU(5)$ model. Thus also here there is no direct connection between $\delta_B$ and $\delta_K$.

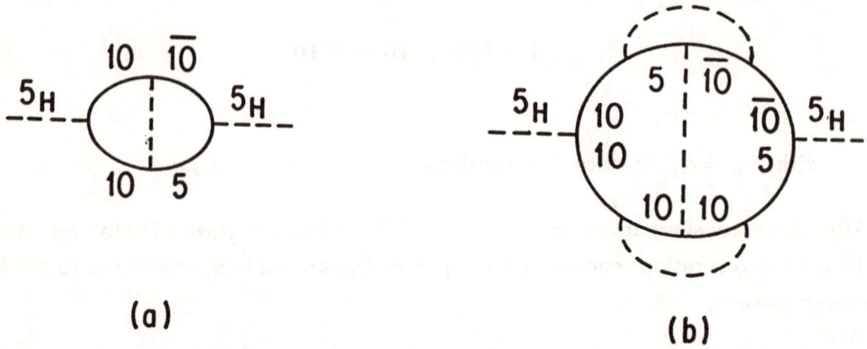


Fig. 4: Graphs whose discontinuity does not (a), and does (b), contribute to $\frac{n_B}{n_\gamma}$ for the minimal $SU(5)$ model with just one $\mathbf{5_H}$.

As a last example let me consider the case in which, in addition to having complex Yukawa couplings, one has also complex Higgs self couplings.* Obviously in these circumstances one also does not expect that $\delta_B$ and $\delta_K$ be related. By the same kind of arguments as before, one can identify many more GUT phases $\delta_{B_i}$ than the unique low energy phase $\delta$ of the 3 generation Cabibbo Kobayashi Maskawa model. So this case is no different than the one with purely complex Yukawa couplings. I mention it here

* For a sufficiently complicated Higgs sector, renormalizability always will require that certain Higgs couplings be complex, if CP is not conserved

separately only because of a mathematically false - but probably physically true - claim made long ago by Ellis Gaillard Nanopolous and Rudaz (EGNR) [16] connected with this type of models. Some of the arguments involved in the discussion related to this claim, we shall see, are quite helpful to clarify the issue of the relation of $\frac{n_B}{n_\gamma}$ to other CP violating phenomena. Hence they are useful to address here.

In the early 1980's, EGNR [16] claimed to have established a "quasi" theorem relating $\frac{n_B}{n_\gamma}$ to the electric dipole moment of the neutron, $d_n$. One of the main reasons of the paper which I wrote with Masiero and Mohapatra [9] was to show that this "quasi" theorem is in fact, false. Although this contention remains true, with some hindsight I have to admit that there is considerable more "physical" truth today about this false theorem than we were willing to grant in 1981! Let me first briefly summarize the gist of the EGNR "theorem". What was noticed in [16] was that the electric dipole moment of the neutron follows directly from the complex part of the $u$ and $d$ quark self energy $\sum(p)$, due to Higgs exchange. Graphically, these contributions are quite similar to those that appear in the discontinuity of Fig. 4, apart from the presence of an external Higgs field. By these means EGNR obtained an inequality [which I have updated by taking $m_t \geq 60\ GeV$] between $d_n$ and $\frac{n_B}{n_\gamma}$:

$$d_n \gtrsim 3 \times 10^{-17} \frac{n_B}{n_\gamma} (ecm) \tag{19}$$

This result is very interesting since for $\frac{n_B}{n_\gamma} \sim 10^{-10}$ it gives a dipole moment for the neutron much above that expected in the Cabibbo Kobayashi Maskawa model [17] and near to a value which is possibly observable experimentally. Nevertheless, since there are, in general, different phases in the GUT theory and in the low energy theory the result (19) can at best be only approximate.

In [9] we constructed a simple example which illustrates both why one can expect something like the EGNR "semi" theorem and why, however, this result cannot be strictly true. I will close this section with this example. However, I will return in the final section once again to argue in favor of the EGNR conclusion!

Consider [9] again $SU(5)$, for simplicity with one family, but with two Higgs fields: $\mathbf{5_H}$ and $\mathbf{5'_H}$ Since there are two Higgs fields, there is now an asymmetry from the graphs of Fig. 4a, where what is exchanged is the other Higgs field. If one lets $f, f'$ and $h, h'$ be the couplings of the Higgs fields with the fermions then

$$\frac{n_B}{n_\gamma} \sim Im\, f f^{*\prime} h' h^* \tag{20}$$

The argument of EGNR [16] uses the fact that $h$ and $h'$ contribute to $m_u$, while $f$ and $f'$ contribute to $m_d$, so that analogous phases to those in (20) appear in $d_n$. However, this is not quite correct since an **additional** phase enters in the low energy theory, through the Higgs potential [9]. Since CP is violated explictly, the coupling coefficients of the low energy Higgs potential $V(\phi, \phi')$ are necessarily complex. Thus the general solution at the minimum of the potential will involve a relative phase:* $<\phi>= V$ ; $<\phi'>= V'e^{i\delta_H}$ and

$$\begin{aligned} m_u &= hV + h'V'e^{i\delta_H} \\ m_d &= fV + f'V'e^{i\delta_H} \end{aligned} \tag{21}$$

As a result $d_n$ will depend not only on the "hard" phases contained in $f, f', h$ and $h'$, but also on the dynamical Higgs phase $\delta_H$. For $T \sim M_{GUT}$, of course, this phase - along with the vacuum expectation values $V$ and $V'$ - disappears. So again there is no strict low energy - high energy connection.

## III. Models Where the Baryogenesis CP Phase and the Low Energy CP Phase are Related

Having given numerous illustrations in the last section of why, in general, $\delta_B \neq \delta_K$ it is interesting to contemplate if there are any circumstances in which these phases could be one and the same. I know of only two examples where this may indeed by the case. One example is provided by a class of models which have spontaneous CP breaking at the GUT scale, but where the $\delta_B$ dynamical phases are not suppressed at low temperatures. In these models one is able to avoid the decoupling of this high energy phase from the low energy theory, so that indeed $\delta_K = \delta_B$. The second example is provided by the exciting possibility that $\frac{n_B}{n_\gamma}$ is generated at the electroweak scale, rather than at a GUT scale [10]. In this symposium L. McLerran [18] has discussed the still controversial dynamics associated with this option, so I shall limit myself here only to considerations germane to the CP phase issue.

Numerous authors [19] have indicated ways in which one can avoid the decoupling of phases spontaneously generated at high energy from the low energy theory. This is somewhat tricky since the decoupling originates in the hierarchy of vacuum expectation values in the Higgs potential [c.f. the $SU(2)_L \times SU(2)_R \times U(1)_{B-L}$ example of the previous section]. The way to get around this natural decoupling is to isolate the GUT scale CP violating sector from the low energy sector. However, one must then provide some bridge to transmit the $\delta_B$ phase to the low energy theory. This is illustrated schematically in Fig. 5. Typical bridges are provided by radiative effects or alternatively

* Strictly speaking one needs also some singlet field for a phase to appear

one can pursue the connection between the two sectors via some super heavy fermions. To provide isolation the phase $\delta_B$ is generally associated with the vacuum expectation value of some $SU(2) \times U(1)$ singlet field and CP is spontaneously broken at the GUT scale.

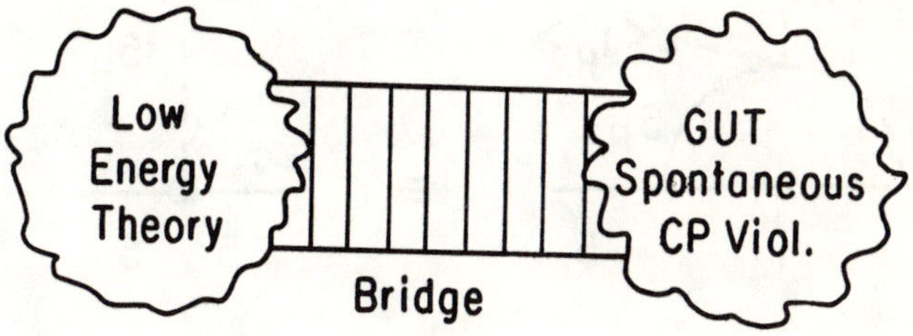


Fig. 5: Schematic of models which avoid decoupling of the spontaneously generated CP phase $\delta_B$.

I will illustrate how all of this works with a "simple" example due to Barr [20]. The main raison d'etre of this example is to provide a soft CP solution to the strong CP problem, as Barr himself has discussed at this meeting [21]. Nevertheless, this example also provides a flavor of the intricacies one requires if one wants to connect $\delta_B$ and $\delta_K$. Barr [20] adds to the minimal $SU(5)$ menagerie of fermions and Higgs fields [$\mathbf{5}, \bar{\mathbf{10}}$ and $\mathbf{5_H}$] two new Higgs fields [$\mathbf{15_H}$ and $\mathbf{1_H}$] plus a superheavy singlet fermion: **1**. The vacuum expectation of the $\mathbf{1_H}$ gives the spontaneous CP violating phase

$$< \mathbf{1_H} > \sim M_{GUT} e^{i\delta_B} \tag{22}$$

while the $\mathbf{15_H}$ provides both a net baryon asymmetry $\Delta B$ (through the baryon violating decay of a leptoquark $\rho$ in this representation) and a means of generating $\epsilon$ (via the decay of a diquark $\eta$ also present in the $\mathbf{15_H}$).

The singlet fermion and a quartic Higgs coupling $\lambda$ between $\mathbf{1_H 15_H 5_H 5_H}$ provide the necessary bridges between the high and low energy theory. As illustrated in Fig. 6 the spontaneous CP phase $\delta_B$ - which is the only phase in this theory - gives an effective (higher order) complex coupling between the $\mathbf{15_H}$ and the fermions in the **5**. This complex coupling is responsible both for $\Delta B$ and for $\epsilon$, as shown in Fig. 7. So here is an example where really $\delta_K$ and $\delta_B$ are one and the same. However, the resulting model is superweak, so that $\epsilon'$ must vanish. Furthermore, although the model is straightforward its simplicity is not manifest, at least to me!

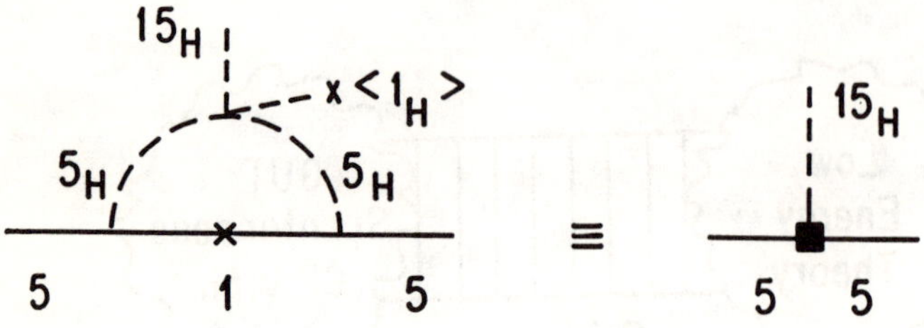


Fig. 6: Effective $\mathbf{15_H 55}$ coupling generated through a radiative bridge.

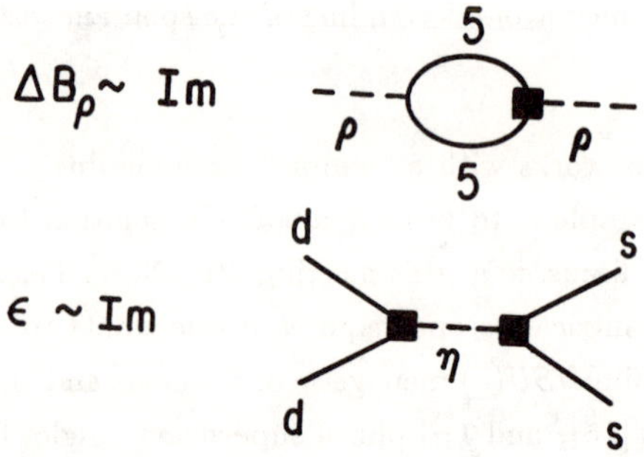


Fig. 7: $\Delta B_\rho$ and $\epsilon$ contributions arising from the effective $\mathbf{15_H 55}$ coupling.

The second example - where $\delta_B = \delta_K$ occurs - arises if the universe baryon asymmetry is generated directly by the electroweak interactions. Because in the usual electroweak theory the fermions are in chiral representations, the total fermion number of the theory - which corresponds to $B+L$ - is not conserved due to the existence of a chiral anomaly:

$$\partial_\mu [B+L]^\mu = \frac{N_g g_2}{32\pi^2} W_a^{\mu\nu} \tilde{W}_{a\mu\nu} \tag{23}$$

At $T = 0$ the effects of this anomaly are exponentially suppressed [22]. However, Kuzmin, Rubakov and Schaposhnikov [10] have argued that at temperatures near the electroweak phase transition this exponential suppression no longer exists. Hence, if these anomaly driven $B$ violating interactions could get out of equilibrium at $T \sim M_W$, one might be able to generate the observed $\frac{n_B}{n_\gamma}$ directly using the standard model, without having to appeal to GUTs!

There is presently some controversy regarding whether one can or one cannot really generate an asymmetry via the anomaly in (23). I do not want to enter into these matters here, as McLerran [18] has covered this topic in detail. Again the crucial issue is really whether one can ever go out of equilibrium for sufficiently long to generate a reasonable asymmetry. I do want, however, to make a comment on the CP phase. Obviously, $\delta_B = \delta_K$ a fortiori, since one is dealing now only with the low energy theory. However, in the standard 3 generation Cabibbo Kobayashi Maskawa model it appears - at least naively - that the effect is too small. The argument, which is due to Schaposhnikov [23], is as follows. Since the effect is anomaly connected one should look for a CP odd piece to the $W$ propagator, for an order of magnitude estimate. Because of the GIM mechanism, however, this contribution is non vanishing only at the 14th order [23]. [A typical graph is shown in Fig. 8]. If this is so, the magnitude of the CP violating effect is obviously too small to produce a reasonable value for $\frac{n_B}{n_\gamma}$. Indeed the CP violating effect from Fig. 8 is of order

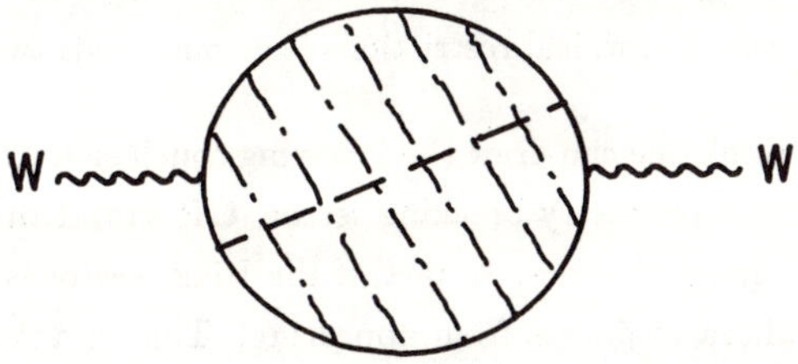


Fig. 8: Graph contributing to CP violation in the $W$ propagator.

$$\Delta \sim \sin^2\Theta_1 \sin\Theta_2 \sin\Theta_3 \sin\delta \left[(m_t^2 - m_c^2)(m_t^2 - m_u^2)\right.$$
$$\left.(m_c^2 - m_u^2)(m_b^2 - m_d^2)(m_b^2 - m_s^2)(m_s^2 - m_d^2\right] \frac{(m_t^2 - m_b^2)}{M_W^{14}} \simeq 10^{-18} \qquad (24)$$

However, as Ringwald [24] has emphasized to me, this may just be too pessimistic an estimate. The actual processes which give rise to $(B+L)$ - violation must include a zero mode for **all** the fermions in the theory. Thus, it is not immediately obvious that the asymmetry itself should be GIM suppressed, since all the quarks must participate. Clearly a clarification of this matter is important, since if the estimate (24) obtains low energy baryogenesis would only be of academic interest.

## IV. Conclusion: A Plethora of Phases

I would like to close with a few remarks, which summarize and distill some of the elements in the above discussion. I have tried, by means of various examples, to make two principal points:

i) Except for very constrained scenarios, there is no direct relation between the CP violating phenomena responsible for $\frac{n_B}{n_\gamma}$ and the violation of CP observed 25 years ago for the first time by Christensen et al [1].

ii) Nevertheless, one should not feel that these phenomena are unrelated. They occur in general because the underlying gause theory **permits** the existence of CP violating phases. These phases are not directly related, but otherwise they are not constrained. As Harari [25] has emphasized in this meeting, $\epsilon$ and $\epsilon'$ in the standard model are not small because the CP phase $\delta = \delta_K$ is small, but rather because the intrageneration mixing angles are small. Similarly, using typical Yukawa couplings but arbitrary phases $\delta_{B_i}$, one can get numbers for $\frac{n_B}{n_\gamma}$ of $0(10^{-10})$ [3], provided there are no unnatural kinematical restrictions [e.g. one needs two $5_H$ in SU(5)].

From these two points I think one can draw the following conclusion [26]. Provided that there is an associated Higgs symmetry breaking sector, **CP violation is a perfectly natural occurrence** in gauge theories. In fact, if the Higgs sector is sufficiently rich, one cannot avoid, a plethora of phases from appearing. This means that, in general, different CP violating phenomena will not be strictly connected. Nevertheless, the order of magnitude of these phenomena should be roughly understandable since their pattern will be mostly determined by the masses and mixing angles in the theory, rather than by the magnitude of the phases themselves. From this point of view, the false EGNR "theorem" is probably correct physically! Thus from Eq (19), provided the dipole moment is not kinematically restricted [as is the case in the 3 generation Cabibbo Kobayashi Maskawa model], it is very natural to expect that $d_n \sim 10^{-26}$ ecm. However, the various CP phases $\delta_{d_n}, \delta_K$ and $\delta_B$ are likely to be distinct.

## Acknowledgements

I am grateful to Thran Than Van for the very nice hospitality at Blois.

**References**

[1] J. H. Christensen, J. W. Cronin, V. L. Fitch and R. Turley, Phys. Rev. Lett. 13 (1964)138

[2] A. Seklarov, Pisma Zh. Eksp Teor. Fiz. 5 (1967) 32 [JETP Lett 5 (1967) 24]

[2] E. W. Kolb, these Proceedings

[3] See for example, D. V. Nanopolous and S. Weinberg, Phys. Rev. D20 (1979) 2484

[4] J. A. Harvey, E. W. Kolb, D. B. Reiss and S. Wolfran, Nucl. Phys. B201 (1982) 16

[5] Y. Yanagida and M. Yoshimura, Nucl. Phys. B168 (1980) 534

[6] J. Ellis, M. K. Gaillard and D. V. Nanopolous, Phys. Lett. 80B (1979) 360; 82B (1979) 464 (E); S. Barr, G. Segre' and A. Weldon, Phys. Rev. D20 (1979) 2494

[7] M. Yoshimura, Phys. Rev. Lett. 41 (1978) 281; 42 (1979) 746 (E); Phys. Lett. 88B (1979) 294; A. Yu. Ignatiev, N. V. Krasnikov, V. A. Kuzmin and A. N. Tavkhelidze, Phys. Lett 76B (1978) 436; S. Dimopolous and L. Susskind, Phys. Rev. D18 (1978) 4500; Phys. Lett 81B (1979) 415; D. Toussaint, S. B. Treiman, F. Wilczek and A. Zee, Phys. Rev. D19 (1979) 1036; J. Ellis, M. K. Gaillard and D. V. Nanopolous, Phys. Lett. 80B (1979) 360; 82B (1979) 464 (E); S. Weinberg, Phys. Rev. Lett. 42 (1979) 850

[8] A. Masiero, R. N. Mohapatra and R. D. Peccei, Phys. Lett. 108B (1982) 111

[9] V. A. Kuzmin, V. A. Rubakov and M. E. Schaposhnikov, Phys. Lett. 155B (1985) 36

[10] I. Kobzarev, L. Okun and Ya. B. Zeldovich, Phys. Lett. 50B (1974) 340

[11] A. Masiero, R. N. Mohapatra and R. D. Peccei, Nucl. Phys. B192 (1981) 66

[12] R. N. Mohapatra and G. Senjanovic, Phys. Rev. D20 (1979) 3390

[13] M. Kobayashi and T. Maskawa, Prog. Theor. Phys. 49 (1973) 652

[14] G. Segre' and M. S. Turner, Phys. Lett. 99B (1981) 339

[15] J. Ellis, M. K. Gaillard, D. V. Nanopolous and S. Rudaz, Phys. Lett. 99B (1981) 101

[16] E. P. Shabalin, Yad. Fiz. 28 (1978) 151; J. Ellis and M. K. Gaillard, Nucl. Phys. B150 (1979) 141; B. F. Morel, Nucl. Phys. B157 (1979) 23

[17] M. McLerran, these Proceedings

[18] D. Chang, R. N. Mohapatra and G. Senjanovic, Phys. Rev. Lett. 53 (1984) 1419; G. Branco and A. I. Sanda, Phys. Lett. 135B (1984) 383; S. M. Barr, Phys. Rev. D34 (1986) 1567; A. Nelson, Phys. Lett. 136B (1983) 387; 143B (1984) 165

[19] S. M. Barr, Phys. Rev. D34 (1986) 1567

[20] S. M. Barr, these Proceedings

[21] G. 't Hooft, Phys. Rev. Lett. 37 (1976) 8

[22] M. E. Schaposhnikov, Nucl. Phys. B287 (1987) 757; B299 (1988) 797

[23] A. Ringwald, private communication

[24] H. Harari, these Proceedings

[25] This point has been strongly emphasized to me by S. Weinberg, private communication

# Can the Baryon Number of the Universe Arise at the Electroweak Phase Transition?

**Larry McLerran**

Theoretical Physics Institute
School of Physics and Astronomy
University of Minnesota
*116 Church St.*
*Minneapolis, Mn. 55455*

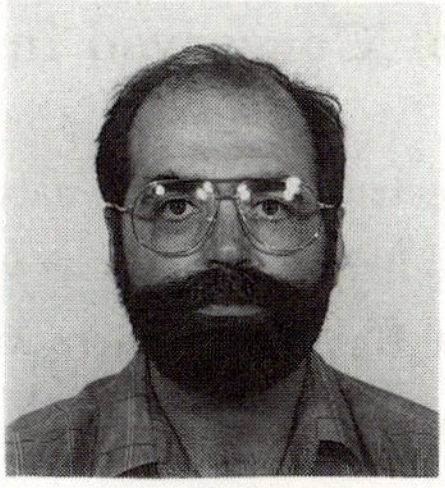

**Abstract**

Baryon number non-conservation in electroweak theory is discussed. The rate of non-conservation is argued to be larger than the expansion rate of the universe in cosmology, and is not in contradiction with instanton estimates. Various schemes for generating the baryon asymmetry of the universe at the electroweak transition are presented.

# 1 Introduction

In this talk, I will try to convince you of the following points:

1. Baryon number is not conserved in electroweak theory.
2. The rate of non-conservation is large enough so that there are physical consequences.
3. There is no contradiction between estimates of baryon number non-conservation which are large at high temperature and the unobservably small rate at zero temperature.
4. It is difficult, but perhaps possible, to generate the baryon asymmetry of the universe at the electroweak phase transition, that is, at a temperature of order of 100's of GeV's.

# 2 Baryon Number is Not Conserved in Electroweak Theory

The equations of motion for electroweak theory naively conserve baryon number. This should be true since we would like the rate to be very small, and since the scale of electroweak physics is a mere 100 GeV, and if there was explicit violation of baryon number, it should be observably large.

When the equations of motion of electroweak theory are quantized, baryon number is no longer conserved. This is a consequence of the Adler-Bardeen anomaly.[1]-[3]. Baryon number is violated as

$$\partial_\mu J^\mu_B = \frac{g^2}{32\pi^2} tr F^{\mu\nu} F^d_{\mu\nu} = \partial_\mu K^\mu_{CS} \tag{1}$$

In this equation, $F^{\mu\nu}$ is the field strength of the electroweak W and Z fields,

$$F^d_{\mu\nu} = \frac{1}{2}\epsilon_{\mu\nu\lambda\sigma} F^{\lambda\sigma} \tag{2}$$

The product $FF^d$ can be written as the divergence of the Chern-Symons current where

$$K^\mu_{CS} = \frac{1}{16\pi^2}\epsilon_{\mu\nu\lambda\sigma}\, Tr(F^{\nu\lambda}A^\sigma - \frac{2}{3}A^\nu A^\lambda A^\sigma) \tag{3}$$

The fields $A$ are the W and Z fields in a matrix basis

$$A = \tau^a \cdot A^a \tag{4}$$

Therefore the Adler-Bardeen anomaly requires that the change of baryon number be equal to the change in Chern-Symons number,

$$\Delta N_B = \Delta Q_{CS} = \int_{t_1}^{t_2} dt \int d^3x \; \partial_\mu K^\mu_{CS} \tag{5}$$

The change in Chern-Symons charge is a gauge invariant quantity (but the the Chern-Symons charge density is not).

It is not at all obvious that the anomaly generates baryon number violation at a sufficiently small rate that it would have escaped detection. We will later see that this is the case, where we shall argue that in weak coupling, the rate of change of topological charge at zero temperature is of order $e^{-4\pi/\alpha_W} \sim 10^{-170}$ and can be ignored.

The existence of a topological charge $Q_{CS}$ is a consequence of the underlying vacuum structure of electroweak theory. As a function of $\Delta Q_{CS}$, the minimal energy or effective potential energy is multiply periodic. This is shown in Fig. 1.

With such a multiply periodic structure, it is straightforward to gain some physical insight into the nature of the baryon non-conservation anomaly.[4] Consider Figures 2a - 2c. On the left hand side of each figure, there is a plot of the effective potential. On the right hand side, there is a plot of the fermi levels. In Fig. 2a, we imagine being at the minimum of the effective potential labeled by $x$ which corresponds to $\Delta Q_{CS} = 0$. In this case, the fermi levels are assumed to be unoccupied for energies greater than zero. For all energies less than zero, the fermi levels are occupied. Particles are interpreted to be occupied positive energy states and anti-particles unoccupied negative energy states.

The fermi level picture is a consequence of quantizing the Dirac equation. What we will show here is that the picture of multiply periodic vacuum structure, together with fermi levels for the Dirac equation lead to the anomaly.

Before proceeding further, we should mention that the minima at integer values of $\Delta Q_{CS}$ are all gauge transforms of one another. Therefore the spectra of the Dirac equation corresponding to these different minima must be identical. As we shall soon see, if we continuously deform the fields from one minimum to another, the occupation numbers of positive energy states is no invariant. Transitions from one minimum to another may therefore produce particles.

In Fig. 2b, we have deformed the W and Z fields from their minima to a maxima separating the two minima. In the process of this variation, the fermi levels move. As shown by Manton[5] (for not too strong a scalar coupling $\lambda$ [6]) the configuration which sits at the top of the barrier corresponds to a level crossing of a fermion at zero energy.

The configuration of W, Z and Higgs fields at the top of the barrier between the two minima is called the sphaleron. It was first constructed for electroweak theory by Klinkhammer and Manton.[7] The sphaleron is a solution of the static classical equations of motion. Since it corresponds to a top of an energy barrier, it is classically unstable under time dependent perturbations. By making small fluctuations around the sphaleron, it is also possible to construct the shape of the potential in the neighborhood of the top of the barrier. The sphaleron has Chern-Symons charge of 1/2.

In Fig 2c, the W, Z and Higgs fields have been deformed to the minimum to the right of the original minimum. The fields are now gauge transforms of the fields near the original minimum, and therefore the spectrum of fermi levels is the same. The only difference is that in the deformation, a filled energy level which originally had negative energy has gained positive energy. A particle has been produced.

This adiabatic level crossing is a general feature of electroweak theory. The relation between it and the original computation of the anomaly is not obvious. The original computation required a careful use of ultraviolet regulators. In this level crossing picture, it is necessary to know that such a regulation scheme exists, since we have ignored what is happening at the bottom of the fermi sea.

The non-conservation of baryon number is related to transitions between degenerate minima of the effective potential with differing topological charge. At zero temperature, the only way such transitions can take place is by quantum mechanical tunneling, as shown in Fig. 3. This tunneling process was first computed by t' Hooft[8]. In the computation of the rate, a classical solution of the equations of motion is used which is called an instanton,[9] which corresponds to the classical solutions used in the WKB method of ordinary quantum mechanics. The process computed by t' Hooft is therefore often referred to as instanton mediated.

The result of t' Hooft's computation is that the rate is

$$Rate \sim e^{-4\pi/\alpha_w} \sim 10^{-170} \tag{6}$$

In this formula, there are no dimensional factors. Remember that even if we measure the space time volume of the universe in terms of the W boson compton wavelength, it is only

$$Vt_{universe} \sim 10^{168} \tag{7}$$

Since there are in fact very few protons per Compton volume of the W boson, the rate given by t' Hooft is so small that never in the history of universe was it even marginally likely that an instanton generated a baryon number change. Nevertheless, the computation does prove that the rate of baryon number change in electroweak theory is non-zero.

The process generated by instantons is visualized in Fig. 4. This amplitude violates both baryon number and lepton number by 3 units. The sum of baryon number plus lepton number is not conserved, but the difference $B+L$ is conserved. This is generally true in electroweak theory. There is a corresponding anomaly for the lepton current which is identical with that of the baryon current. Note also, that each generation of lepton and baryon is produced in the process. The electromagnetic and color charges are also conserved.

Even if such a process as shown in Fig. 4 existed at a large rate in electroweak theory, it would be very difficult to detect experimentally. Most proton decay experiments are designed to measure only single proton decays. Requiring a three body decay in a nucleus would surely reduce the rate due to the small probability that three nucleons overlap. In addition, requiring that only light mass quarks are produced would involve strong suppression due to weak mixing angles. Finally, there would in general be much missing energy in the event due to the large probability of emitting a neutrino.

## 3 The Rate of Baryon Number Non-conservation at Finite Temperature

At finite temperature, an entirely new method of generating baryon number change arises. There is some probability that there is a thermal fluctuation with an energy higher than the height of the barrier between two degenerate minima. In this case, the fluctuation can classically traverse the maximum and enter the new minimum, as shown in Fig. 5. The probability that the thermal fluctuation has an energy higher than the top of the barrier is[10]

$$Rate \sim e^{-E_{sph}/T} \tag{8}$$

As discussed above, if we know the sphaleron solution of the classical equations of motion with topological charge $Q_{sph} = 1/2\ \Delta Q_{cs}$, then we know the height of the barrier $E_{sph}$. The energy of the sphaleron comes from a classical solution, and classical solutions correspond to the limit of very many quanta $\langle N \rangle \sim 1/\alpha_w$, so that the energy is of order

$$E_{sph} \sim M_W/\alpha_W \sim 1 - 10\ TeV \tag{9}$$

The typical energy per quanta is of order $M_W$, and therefore the size of the sphaleron is of order

$$R_{sph} \sim 1/M_W \tag{10}$$

We can visualize the process of making a classical transition over the top of the barrier by the diagram of Fig. 6. The diagram is identical to that of the instanton process, and the conservation laws associated with the event are identical. The sphaleron process may be thought of as analogous to an instanton process, except that we imagine the sphaleron process taking place in real time at finite temperature.

The above estimate of the probability that there is a thermal fluctuation with an energy larger than the height of the barrier clearly gives an overestimate of the rate. We must also know the probability that the thermal fluctuation is in some sense near the top of the barrier. For example, if we are lost in a forest in the high Himalayas, we may be much higher than a mountain pass in the Rockies, but we may never find our way to it. Put another way, the function space in which we imagine our thermal fluctuation is multi-dimensional and complicated. It is not obvious what weight a thermal fluctuation has for inducing a baryon number changing transition.

There is a reliable method for computing thermal transition processes if the coupling is weak.[11]-[12]. We must first ask if there is a range of temperatures where weak coupling methods are reliable. First we must recall that at high temperature, the mass of the W boson becomes temperature dependent $M_W \rightarrow M_W(T)$.[13] At some temperature, the masses of the W and Z bosons approach zero, and the broken symmetry of electroweak theory is restored.

In general, it is possible to show that perturbation theory breaks down in electroweak theory when $T \geq M_W(T)/\alpha_W$.[14]-[15] This is because infrared integrations over thermal propagators become so singular that the factors of alpha generated at each order of the loop expansion are canceled. In addition, the use of the Langer-Affleck formalism for computing thermal transitions over the barrier is only valid in the high temperature limit, $T >> M_W(T)$. This latter case is not much of a restriction since only in this limit are thermal transitions larger than quantum tunneling, that is $e^{-M_W/T} \geq e^{-4\pi/\alpha_W}$. We therefore conclude that there is a region of temperatures where the rate of baryon number changing transitions may be reliably computed in electroweak theory,

$$M_W(T) << T << M_W(T)/\alpha_W \tag{11}$$

In the above range of temperatures, the rate of baryon number change has been computed in electroweak theory for $\lambda/g^2 \sim 1$ where $\lambda$ is the four Higgs coupling strength.[16]-[17] The rate is large compared to the expansion rate of the universe in cosmology,

$$\Gamma_{\Delta B}/\Gamma_{universe} \geq 10^9 \; for \; T \sim 100 \; GeV \tag{12}$$

The rate may be estimated also for $\lambda/g^2$ in other ranges and for this ratio in a reasonable range around 1, the rate is large.[18]

For temperature $T \geq M_W(T)/\alpha_W$, it is not possible to do reliable weak coupling estimates. It is however possible to make estimates based on scaling arguments or approximate classical solutions.[16]-[19] The rate here is also many orders of magnitude larger than the expansion rate of the universe.

## 4 The Problem with Instantons

The sphaleron estimates of the rate of baryon number change have been criticized by several authors.[20]-[21] The paper by Ellis et. al. points out several puzzling apparent paradoxes between instanton computations and those of sphalerons, and we will address these apparent puzzles here. The resolution of these apparent paradoxes sheds some interesting light on the mechanism of sphaleron transitions.[16]. We will not discuss the second set of criticisms due to Cohen et. al., as this would go beyond the scope of this talk, except to say that this paper has, in my opinion, at least two essential mistakes which invalidate the conclusions.

The first set of apparent problems with the sphaleron computation is that the sphaleron is a large classical object. It has $\langle N \rangle \sim 1/\alpha_W$ quanta in it and is large compared to its Compton wavelength. It should be hard to make such a fat, fluffy object.

In fact, when the method of Langer-Affleck is applied to the computation of the rate of sphaleron processes, one estimates the probability of this production. There is a suppression, roughly proportional to $\alpha_W^4$, which results. There are some large numerical factors however which enhance the rate. When all the factors are combined together, the overall suppression is about $10^{-5}$. This suppression is not enough to make the rate irrelevant cosmologically since the rate of expansion of the universe is slow compared to the natural time scale set by thermal processes $T/M_{planck} \sim 10^{-16}$.

It is easy to get confused by this issue nevertheless. If we think in terms of exclusive probabilities, the probability to have only a sphaleron in a box the size of a sphaleron compared to that to have ordinary thermal fluctuations is very small $P \sim e^{-T^3/M_W^3}$ The question we are asking is however an exclusive one. We ask what is the probability that there is a sphaleron transition within the volume of the sphaleron plus all thermal excitations within the same volume. The sphaleron is weakly interacting, and the probability that a highly excited Z or W boson interacts with it during its lifetime is small, that is, the sphaleron is transparent to the high energy modes and one can compute using inclusive probabilities. This probability is not a priori small.

A more difficult instanton argument to evade is provided by the following "theorem". Consider the baryon number changing expectation value $\langle qqql \rangle$ where $q$ is a quark field and $l$ is a lepton field. We will consider for purposes of this argument a theory with only one generation of quarks and leptons. We will consider this expectation value at finite temperature. This expectation value gives the amplitude for a baryon number changing process.

We can estimate this expectation value using the path integral,

$$\langle qqql \rangle = \frac{\int qqqle^{-S}}{\int e^{-S}} \tag{13}$$

This amplitude may be computed in Euclidean space, and later analytically continued to Minkowski space. The numerator of the above ratio of path integrals is nonzero only if

$$\int d^4x \; FF^d \neq 0 \tag{14}$$

Now we also have

$$F^2 \geq FF^d \tag{15}$$

and the Higgs part of the action is positive definite. Therefore

$$S \geq S_{inst} \tag{16}$$

and

$$\langle qqql \rangle \leq e^{-2\pi/\alpha_W} \tag{17}$$

The conclusion seems to be that baryon number processes are exponentially suppressed in contradiction with the sphaleron arguments.

How can this argument be wrong? One possibility is that instanton computations may break down in a weak coupling expansion for $E \geq M_W/\alpha_W$, and that the estimate above is simply incorrect.

Even if the above argument is correct, it does not prove that all baryon number changing processes are suppressed.[22] Recall that the sphaleron is composed of many quanta. We would expect at high temperature that the dominant matrix elements involve of order $1/\alpha_W$ quanta. If these matrix elements are Poisson distributed in the number of participating quanta

$$P_N = e^{-\langle N\rangle} \frac{\langle N\rangle^N}{N!} \tag{18}$$

so that while the total probability is of order one, the probability of a few prong event is exponentially suppressed

$$P_1 \sim e^{-\langle N\rangle} \tag{19}$$

If the above situation is true, then it must also be true that if instanton methods are used to compute these multi-particle processes, the weak coupling expansion around the instanton will break down, since instanton processes always have a prefactor of $e^{-2\pi/\alpha_W}$. This has recently been shown to be true by Ringwald.[23].

It is possible to use Ringwald's method to compute the total cross section for quark-quark scattering to inclusively produce baryon number change. This cross section may be only computed in the limit that it is exponentially suppressed, so that perturbation theory breaks down when the result becomes interesting. Due to the contribution of multi-particle final states, the cross section is growing incredibly rapidly. For $E << M_W/\alpha_W$, Ringwald's computations gives

$$\sigma \sim e^{-4\pi/\alpha_W} e^{\alpha_W (E/M_W)^2} \tag{20}$$

Ringwald clearly shows that at high energy, which in his computation corresponds to the production of large numbers of Higgs, W's and Z's, the instanton approximation breaks down. This also allows the remote possibility that the breakdown in perturbation theory may also allow for the production of baryon number in pp collisions at SSC energies. This is not necessary however for large thermal rates since at finite temperature there may be large number of quanta in the initial state, unlike the case in particle collisions.

We can also see in an explicit quantum mechanics example that a large suppression associated with instantons does not imply a similar suppression for sphaleron processes.[22] Consider the pendulum shown in Fig. 7. In a gravitational field, the potential for the pendulum is periodic as is shown in Fig. 8. The instantons of this theory are classical solutions which in Euclidean space interpolate between $\theta = 0$ and $\theta = 2n\pi$ in a time 1/T. The sphaleron is the classical solution which sits on top of the barrier, $\theta = 2(n + 1/2)\pi$.

We can easily compute both the sphaleron and instanton contributions at high temperature, where the potential can be ignored. The sphaleron contribution is $e^{-E_{sp}/T} \sim 1$. The instanton classical solution is

$$\theta = 2\pi t T \tag{21}$$

The Euclidean action is

$$S \sim \int_0^\beta dt \frac{1}{2} \dot{\theta}^2 \sim 2\pi^2 T \tag{22}$$

Therefore at high temperatures, the instanton contribution is greatly suppressed, $e^{-2\pi^2 T}$

In the case of the simple pendulum at high temperature, we of course know the correct answer. The highly excited pendulum will roll over many times. The sphaleron therefore predicts qualitatively correct results where the instanton naively gives incorrect intuition.

## 5 Sphalerons and the Baryon Asymmetry of the Universe

We have seen that sphalerons generate rapid baryon number changing processes in cosmology at temperatures larger than $T \geq 100\ GeV$. These processes at the least greatly reduce any baryon asymmetry produced at a higher temperature scale in a GUT.

Sphalerons however change only $B + L$. They in fact reduce the asymmetry to such a large extent that any GUT which only has $B + L$ violation can almost certainly not produce enough baryon asymmetry.[24]–[25]. It is always possible to produce a baryon asymmetry through $B - L$ violations, and there are viable GUT models which do this. It is also possible in some models to produce the baryon asymmetry by GUT mechanisms at temperatures below the electroweak transition.[26]

An interesting question is whether the sphalerons can produce an asymmetry of the correct magnitude at the electroweak phase transition. This is difficult to do, since the Sakharov conditions must be satisfied[27]

1. C non-conservation.

2. Out of equilibrium.

3. CP non-conservation.

The first Sakharov condition is easy to satisfy in electroweak theory, since C is violated in the sphaleron process itself.

The second condition that the system be out of equilibrium requires that the electroweak phase transition be first order. This can be arranged in the standard model if the theory has a very light mass Higgs, that is, one is in the Coleman-Weinberg limit.[28] Another possibility is that there are multiple Higgs bosons, that is one goes out of the framework of the standard model. In such a multiple Higgs model, first order phase transitions are easy to arrange.

The essential problem for making a viable scenario for producing the observed baryon asymmetry is having an acceptable strength for CP violation. In the standard model, CP violation may have an intrinsically large strength. Processes which violate CP must however have a large number of quark mass insertions. As reviewed in the talk by Peccei, this leads to an unacceptably small asymmetry $N_B/S \leq 10^{-21}$.[29]

There are a variety of ways of evading this conclusion. Perhaps the simplest is to either have more generations of massive fermions or to have multiple massive Higgs doublets with CP violation in this sector of the theory.

There is also a way to generate an acceptable asymmetry in the standard model, but it requires some as yet not understood dynamics. Shaposhnikov has argued that it may be possible that at some temperature, CP may be spontaneously broken through a condensation of Chern-Symons charge.[30] At the electroweak transition, the Chern-Symons charge is converted to baryon number.

The scenario of Shaposhnikov is in principle possible, but there is not yet a dynamical model where this effect has been convincingly demonstrated. It also seems that the dynamical violation of Chern-Symons number can only happen in a finite range of temperature. At high temperature, the infrared dynamics of electroweak theory becomes that of a 3 dimensional pure Yang-Mills theory. This theory is confining, and it is not expected that there is a spontaneous violation of CP.

Another possibility is that at high temperatures $T \geq 10\ GeV$ there may be U(1) CP violation in the strong interaction sector of the theory. Recall that in axion cosmology, the strong $\Theta$ angle is non-zero for temperatures larger than 10 $GeV$.[31]. At lower temperatures, the instantons of QCD generate a potential for the $\Theta$ angle. It begins to oscillate around zero, so that CP violation on the average disappears, and eventually the coherent oscillations result in axionic dark matter.

If the $\Theta$ angle is of order one at the $T \geq 100\ GeV$, it is possible to make a model where the baryon asymmetry is generated.[32] This requires however extra multiplets of massive Higgs bosons. These Higgs bosons must have couplings of order $g_W$ to the quarks, and no expectation value at zero temperature.

In summary, it may be possible to generate the observed baryon asymmetry at the electroweak phase transition. This however either involves as yet not understood dynamics, or a generalization of the standard model to multiple Higgs doublets or including extra heavy generations of fermions. It seems it is not possible to generate such an asymmetry entirely within the context of the standard model without invoking as yet not understood dynamics.

## 6 Acknowledgments

I would like to thank Tran for the wonderful organization of this meeting.

## References

[1] S. Adler, *Phys. Rev.* **177**, 2426 (1969).

[2] J. S. Bell and R. Jackiw, *Nuovo Cimento* **51**, 47 (1969).

[3] W. Bardeen, *Phys. Rev.* **184**, 1841 (1969).

[4] J. Ambjorn, J. Greensite, and C. Peterson *Nuc. Phys.* **B221**, 381 (1983).

[5] N. Manton, *Phys. Rev.* **D28**, 2019 (1983).

[6] L. Yaffe, University of Washington Preprint (1989).

[7] F. Klinkhammer and N. Manton, *Phys. Rev.* **D30**, 2212 (1984).

[8] G. 't Hooft, *Phys. Rev. Lett.* **37**, 8 (1976).

[9] A. Belavin et. al., *Phys. Lett.* **59B**, 85 (1975).

[10] V. Kuzmin, V. Rubakov and M. Shaposhnikov, *Phys. Lett.* **155B**, 36 (1985).

[11] J. Langer, *Ann. Phys.* **41**, 108 (1967); **54**, 258 (1969).

[12] I. Affleck, *Phys. Rev. Lett.* **46**, 388 (1981).

[13] D. A. Krizhnits and A. Linde, *Ann. Phys.* **101**, 195 (1976).

[14] A. D. Linde, *Phys. Lett.* **96B**, 289 (1980).

[15] D. Gross, R. Pisarski and L. Yaffe, *Rev. Mod. Phys.* **53**, 43 (1981).

[16] P. Arnold and L. McLerran, *Phys. Rev.* **D36**, 581 (1987).

[17] A. Ringwald, *Phys. Lett.* **201B**, 510 (1988).

[18] L. Carson and L. McLerran, University of Minnesota Preprint (1989).

[19] S. Khlebnikov and M. Shaposhnikov, *Nuc. Phys.* bf B308, 885 (1988)

[20] J. Ellis, R. Flores, S. Rudaz, and D. Seckel, *Phys. Lett.* **194B**, 241 (1987).

[21] A. Cohen, S. Dugan and A. Manohar, *Phys. Lett.* **222B**, 91 (1989).

[22] P. Arnold and L. McLerran, *Phys. Rev.* **D37**, 1020 (1988).

[23] A. Ringwald, DESY Preprint (1989).

[24] V. Kuzmin, V. Rubakov and M. Shaposhnikov, *Phys. Lett.* **191B**, 171 (1987).

[25] E. W. Kolb and M. Turner, *Mod. Phys. Lett* **A2**, 285 (1987).

[26] For a review see E. W. Kolb in the proceedings of this meeting.

[27] A. D. Sakharov, *Zhetf Pisma* **5**, 32 (1967).

[28] S. Coleman and E. Weinberg, *Phys. Rev.* **D7**, 888 (1973).

[29] For a review see R. Peccei in the procedings of this meeting.

[30] M. Shaposhnikov, *JETP Lett.* **44**, 4651 (1986); *Nuc. Phys.* **B287**, 757 (1987); *Nuc. Phys.* **B299**, 797 (1988).

[31] For a review see J. Kim, *Phys. Rep.* **150**, 1 (1987)

[32] L McLerran, *Phys. Rev. Lett.* **62**, 1075 (1989).

**Figures:**

Fig. 1 The minimal energy, or effective potential, as a function of $\Delta Q_{CS}$. The effective potential is multiply periodic.

Fig. 2a The effective potential and the fermi levels for W, Z and Higgs fields near the minimum with zero topological charge. 2b The W, Z and Higgs fields have been deformed to near the top of the energy barrier. The fermi levels move, and an occupied state crosses zero energy. 2c The W, Z, and Higgs fields are now near the next minima and the topological charge has changed by one unit. An occupied level has crossed zero energy and is now at the position of the first unoccupied level near the previous minimum.

Fig. 3 Tunneling through the effective potential from one minimum to another.

Fig. 4 The amplitude for baryon number change generated by instantons.

Fig. 5 Transitions across the maximum between two degenerate minima of the effective potential for a thermal fluctuation.

Fig. 6 A sphaleron induced baryon number violating process.

Fig. 7 A simple pendulum.

Fig. 8 The multiply periodic potential of the simple pendulum in a gravitational field.

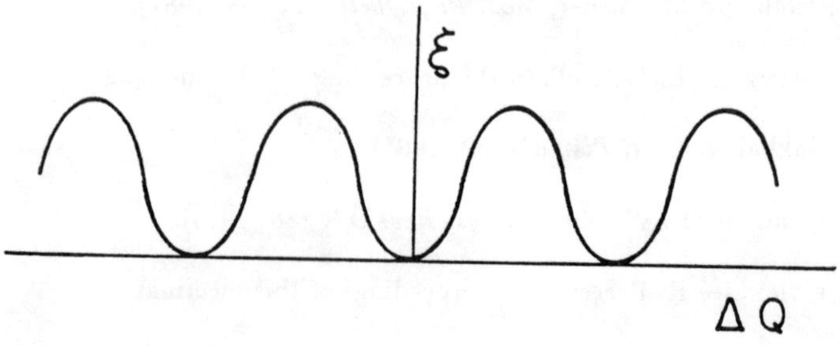


Figure 1

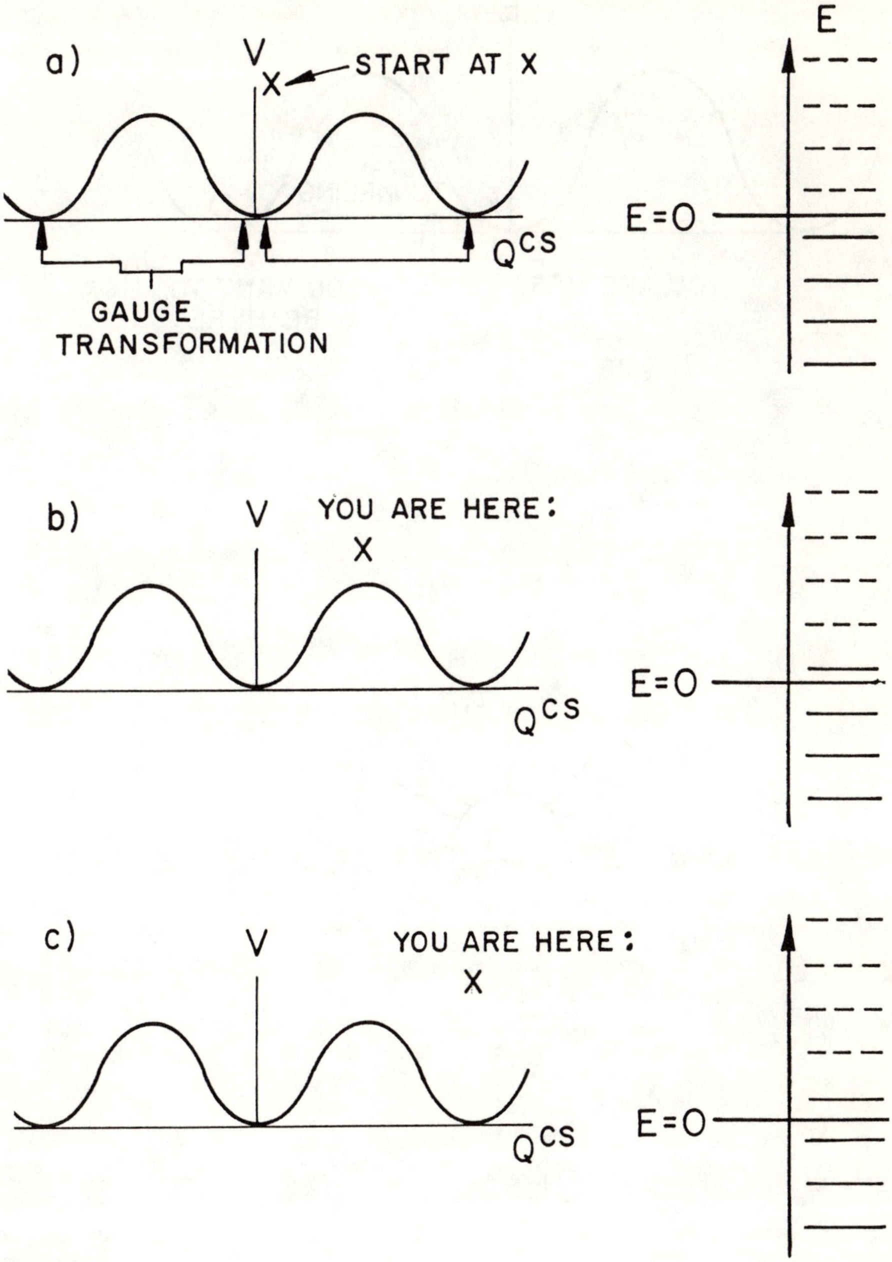
a)
V
X
START AT X
Q^CS
GAUGE
TRANSFORMATION
E
E=O
b)
V
YOU ARE HERE:
X
Q^CS
E=O
c)
V
YOU ARE HERE:
X
Q^CS
E=O

Figure 2

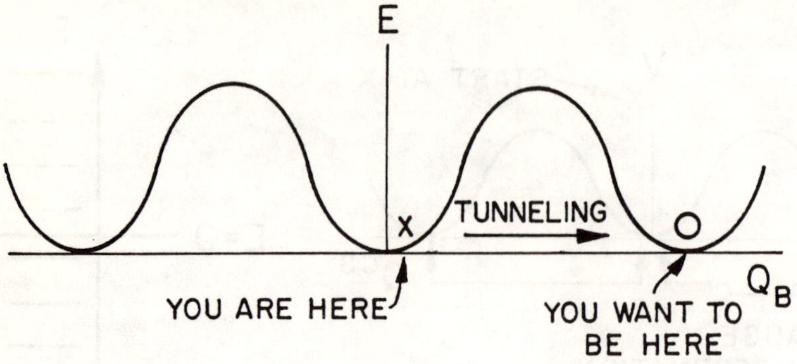


Figure 3

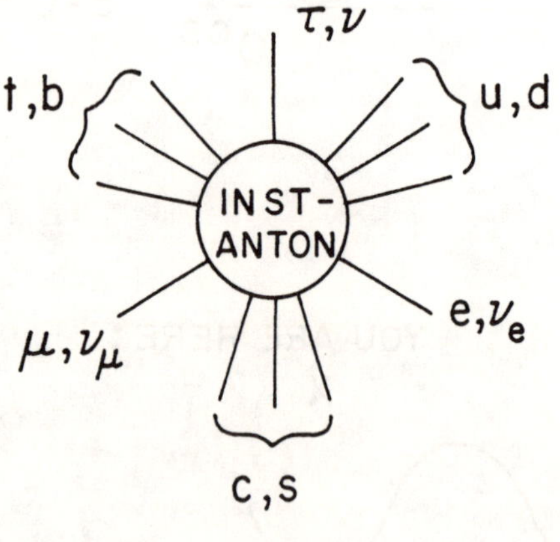


Figure 4

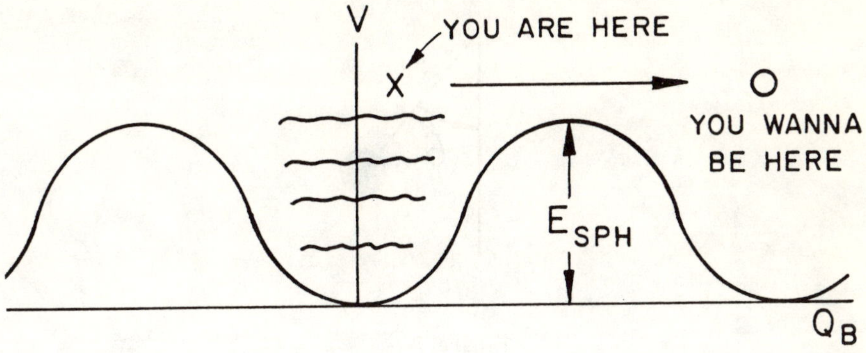


Figure 5

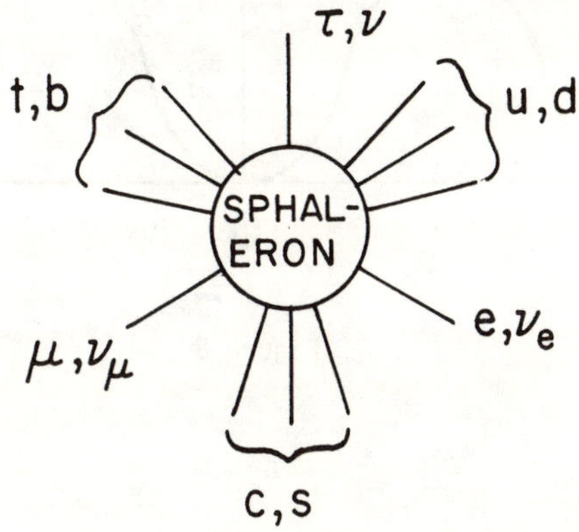


Figure 6

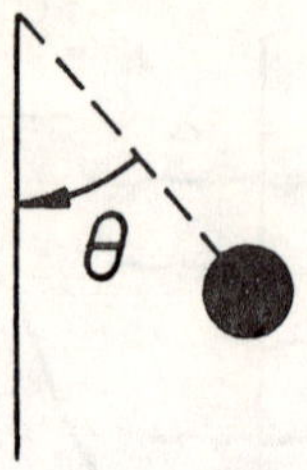


Figure 7

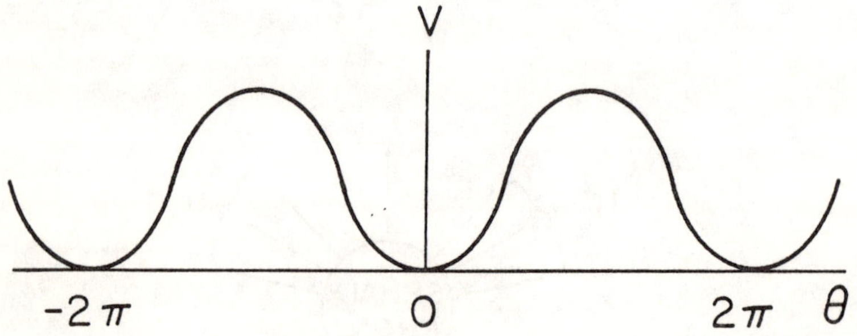


Figure 8

# NON-EUCLIDEAN QUANTUM COSMOLOGY

**Andrei Linde**

CERN, CH-1211 Geneva 23, Switzerland

## ABSTRACT

We investigate an interplay between elementary particle physics, quantum cosmology and inflation. We argue in particular that the absence of intersection of parallel lines in the observable part of the universe can be considered as a first experimental evidence of the baryon non-conservation and as a second experimental evidence of CP violation. Stochastic approach to quantum cosmology is developed. The results obtained within this approach are compared with the results obtained in the context of Euclidean quantum cosmology. It is argued that if the present vacuum energy density $\rho_v$ exceeds some extremely small critical value $\rho_c$ ($\rho_c \sim 10^{-10^7} g.cm^{-3}$ for chaotic inflation in the theory $\frac{m^2}{2}\phi^2$), then the lifetime of mankind in the inflationary universe should be finite, even though the universe as a whole will exist without end. A possible way to justify the anthropic principle in the context of the baby universe theory and to apply it to the evaluation of masses of elementary particles, of their coupling constants and of the vacuum energy density is also discussed.

# 1 Introduction

With the invention of unified theories of strong, weak, electromagnetic and gravitational interactions, elementary particle physics has entered a very interesting and unusual stage of its development. It appears that in the context of these theories we can predict much more than we can actually verify by the standard experimental methods of high energy physics.

The energy scale at which the unified nature of all four fundamental interactions is expected to become manifest is not very different from the Planck mass $M_P$, about $10^{-5}$ grams, where quantum gravity effects become important. (The Planck mass is that mass for which the Compton wavelength $l_P$, about $10^{-33}$ cm, equals the Schwarzschild radius.) Its rest energy $M_P c^2$, about $10^{19}$ GeV, corresponds to the kinetic energy of a small airplane. By contrast, the 80-km-circumference Superconducting Super Collider the Americans hope to build in the near future will accelerate particles up to $10^4$ GeV. The largest accelerator ring one could build on Earth, with a circumference of 40.000 km, could not accelerate particles beyond about $10^8$ GeV, a center-of-mass energy occasionally to be seen in cosmic-ray collisions. But this still leaves us 12 orders of magnitude short of the energy necessary for a direct test of the unified theories. Of course, there are some indirect tests, such as the searches for proton decay and for supersymmetric partners of ordinary particles. But trying to get a correct theory of all fundamental interactions with only such low-energy experiments is like looking for the correct unified electroweak theory by using nothing but radiotelescopes. (Note that $M_P/E_W \sim E_W/E_\gamma$, where $E_W \sim 10^2$ GeV is the unification energy scale in the theory of weak and electromagnetic interactions, $E_\gamma \sim 10^{-5}$eV is the typical energy of photons in a radiowave).

The only accelerator that could ever produce particles energetic enough for a direct testing of the unified theories of all fundamental interactions is our universe itself. The Big Bang scenario as it stood ten years ago, which I will call the hot-universe theory, asserts that the universe was born at some moment $t = 0$ about 15 billion years ago, in a state of infinitely large density $\rho$. Of course, one cannot really speak of classical space-time for the earliest moments, when $kT/c^2$ was greater than the Planck mass, and $\rho$ exceeded $M_P/l_P^3$, making quantum fluctuations of the metric predominant. (We will use a convenient unit system that sets $k, c$ and $\hbar$ all equal to 1, so that $l_P = 1/M_P$ and the Planck density is $M_P^4$, roughly $10^{94} g/cm^3$.) It is just at such times, when the average particle energy exceeded $M_P$, that the unity of all four fundamental interactions would have been manifest.

With the rapid expansion of the universe, the average energy of particles, given by the temperature, decreases rapidly, and the universe becomes cold. The temperature falls as the reciprocal of $a(t)$, the scale factor, or "radius", of the universe. This means that particle interactions at extremely large energies can have occurred only at the very early stages of the evolution of the universe. One might think it very difficult to extract useful and reliable information from the unique experiment carried out about $10^{10}$ years ago. Thus, it came as a great surprise to those who study elementary particles that the investigation of physical processes at the very early stages of the universe can rule out most of the existing unified theories.

For example, all the grand unified theories predict the existence of superheavy stable particles carrying magnetic charge: magnetic monopoles. These objects have a typical mass $10^{16}$ times that of the proton. According to the standard hot-universe theory, monopoles should appear at the very early stages of the universe, and they should now be as abundant as protons. In that case the mean density of matter in the universe would be about 15 orders of magnitude higher than its present value of about $10^{-29} g/cm^3$ [1].

It was shown that all theories with spontaneous breaking of a discrete symmetry (including the simplest $SU(5)$ model, the models of spontaneous $CP$ violation, most of the theories with axions, etc.) lead to the existence of superheavy domain walls, which would be in a drastic contradiction with observational cosmology [2]. It was shown that in most of the theories based on $N = 1$ supergravity the primordial abundance of gravitinos contradicts cosmological data by about 10 orders of magnitude [3], whereas the energy density stored in the so-called Polonyi fields contradicts cosmological data by 15 orders of magnitude for the simplest models [4], and by 6 orders of magnitude for the no-scale models [5]. Several models based on superstring theory lead to a cosmologically unacceptable type of axions [6]. Most of the Kaluza-Klein theories based on $N = 1d = 11$ supergravity predict the present vacuum energy to be $O(-M_P^4)$, which would contradict the observational data by 125 orders of magnitude. The situation with superstring theories seemed, at first glance, to be somewhat better. To be fair, however, one must say that no sufficiently good cosmological model based on superstrings has been suggested so far, though some interesting ideas have been already proposed, see e.g. [7].

Despite many efforts, some of the problems listed above still remain unsolved, which probably means that the corresponding theories are actually unrealistic. Fortunately, however, the problems of primordial monopoles, gravitinos, domain walls and some other important cosmological problems such as the flatness problem, the homogeneity and isotropy problem, etc. can be solved simultaneously in the context of a relatively simple scenario of the universe evolution - the inflationary universe scenario [8]-[12]. This scenario makes it possible also to remove some constraints on the parameters of the elementary particle theory, for example, the constraint on the axion mass $m_A \geq 10^{-5} eV$ ($f_A \leq 10^{12} GeV$) [13]. The invention of this scenario has modified considerably the standard cosmological paradigm. One of the most important modifications was performed within the last three years. Therefore, we will start with the discussion of the "standard" inflationary universe scenario. Then we will discuss some recent developments in the inflationary cosmology [14,15] and some new trends in the theory of baryogenesis and in the theory of galaxy formation. We will finish with the discussion of exciting possibilities related to baby-universe theory, Anthropic Principle and the cosmological constant problem [16]-[26].

# 2 The "Standard" Inflationary Scenario

## 2.1 The stage of inflation

Historically, there were several different versions of the inflationary universe scenario. At present, it seems that the simplest and, simultaneously, the most general one is the chaotic inflation scenario [12]. To describe it, let us consider the simplest model based on the theory of a massive non-interacting scalar field $\phi$ with the Lagrangian

$$L = \frac{M_P^2}{16\pi} R + \frac{1}{2}\partial_\mu \phi \partial^\mu \phi - \frac{m^2}{2}\phi^2 \tag{1}$$

Here $M_P^{-2} = G$ is the gravitational constant, $M_P \sim 10^{19} GeV$ is the Planck mass, $R$ is the curvature scalar, $m$ is the mass of the scalar field $\phi, m \ll M_P$. If the classical field $\phi$ is sufficiently homogeneous in some domain of the universe (see below), then its behaviour inside this domain is governed by the equations

$$\ddot{\phi} + 3H\dot{\phi} = -dV/d\phi, \tag{2}$$

$$H^2 + \frac{k}{a^2} = \frac{8\pi}{3M_P^2}(\frac{1}{2}\dot{\phi}^2 + V(\phi)). \tag{3}$$

Here $V(\phi)$ is the effective potential of the field $\phi$ (in our case $V(\phi) = \frac{1}{2}m^2\phi^2$), $H = \dot{a}/a, a(t)$ is the scale factor of the locally Friedmannian universe (inside the domain under consideration), $k = +1, -1$, or 0 for a closed, open or flat universe, respectively. If the field $\phi$ initially is sufficiently large ($\phi \geq M_P$), then the functions $\phi(t)$ and $a(t)$ rapidly approach the asymptotic regime

$$\phi(t) = \phi_0 - \frac{mM_P}{2(3\pi)^{1/2}}t, \tag{4}$$

$$a(t) = a_0 exp(\frac{2\pi}{M_P^2}(\phi_0^2 - \phi^2(t))). \tag{5}$$

According to (4) and (5), during a time $\tau \sim \phi/mM_P$ the value of the field $\phi$ remains almost unchanged and the universe expands quasi-exponentially:

$$a(t + \Delta t) \sim a(t) exp(H\Delta t) \tag{6}$$

for $\Delta t \leq \tau = \phi/mM_P$. Here

$$H = \frac{2\pi^{1/2}}{\sqrt{3}}\frac{m\phi}{M_P} \tag{7}$$

Note that $H \gg \tau^{-1}$ for $\phi \gg M_P$.

The regime of quasi-exponential expansion (inflation) occurs for $\phi \geq \frac{1}{5}M_P$. For $\phi \leq \frac{1}{5}M_P$ the field $\phi$ oscillates rapidly, and if this field interacts with other matter fields (which are not written explicitly in eq. (1)), its potential energy $V(\phi) \sim \frac{m^2\phi^2}{2} \sim m^2M_P^2$ is transformed into heat. The reheating temperature $T_R$ may be of the order $(mM_P)^{1/2}$ or somewhat smaller, depending on the strength of the interaction of the field $\phi$ with other fields. It is important that $T_R$ does not depend on the initial value $\phi_0$ of the field $\phi$. The only parameter which depends on $\phi_0$ is the scale factor $a(t)$, which grows $\exp((2\pi/M_P^2)\phi_0^2)$ times during inflation.

If, as is usually assumed, a classical description of the universe becomes possible only when the energy-momentum tensor of matter becomes smaller than $M_P^4$, then at this moment $\partial_\mu\phi\partial^\mu\phi \leq M_P^4$ and $V(\phi) \leq M_P^4$.

Therefore, the only constraint on the initial amplitude of the field $\phi$ is given by $\frac{1}{2}m^2/\phi^2 \leq M_P^4$. This gives a typical initial value of the field $\phi$:

$$\phi_0 \sim \frac{M_P^2}{m} \tag{8}$$

Let us consider for definiteness a closed universe of a typical initial size $O(M_P^{-1})$. It can be shown that if initially $\partial_\mu\phi\partial^\mu\phi$ becomes much smaller than $V(\phi)$, the evolution of the universe becomes describable by (2)-(7), and after inflation the total size of the universe becomes larger than

$$l \sim M_P^{-1} exp(\frac{2\pi}{M_P^2}\phi_0^2) \sim M_P^{-1} exp(\frac{2\pi M_P^2}{m^2}) \tag{9}$$

For $m \sim 10^{-6} M_P$ (which is necessary to produce density perturbations $\frac{\delta \rho}{\rho} \sim 10^{-5}$, see below)

$$l \sim M_P^{-1} \exp(2\pi 10^{12}) \geq 10^{10^{12}} cm, \tag{10}$$

which is much greater than the size of the observable part of the universe $\sim 10^{28} cm$.

After such a large inflation the term $k/a^2$ in (3) becomes negligibly small compared with $H^2$, which means that the universe becomes flat and its geometry locally Euclidean. This implies that the total density of the universe $\rho$ becomes almost exactly equal to the critical density $\rho_c = \frac{3M_P^2}{8\pi} H^2$, i.e.

$$\Omega = \frac{\rho}{\rho_c} \sim 1. \tag{11}$$

For similar reasons the universe becomes locally homogeneous and isotropic. The density of all 'undesirable' objects (monopoles, domain walls, gravitinos) created before or during inflation becomes exponentially small, and they never appear again if the reheating temperature, $T_R$, is not too large.

We should like to emphasize that for a realisation of this scenario it is sufficient that initially $\partial_\mu \phi \partial^\mu \phi \leq V(\phi) \sim M_P^4$ in a domain of a smallest possible size $l \sim M_P^{-1}$. Since $\partial_\mu \phi \partial^\mu \phi \leq M_P^4, V(\phi) \leq M_P^4$ in any classical spacetime, the above-mentioned initial conditions are quite natural. For a more detailed discussion of initial conditions which are necessary for inflation see [28].

## 2.2 Baryosynthesis

The discovery of a possible mechanism of generation of the baryon asymmetry of the universe [30] was one of the most important events in modern cosmology. The existence of such a mechanism is very important also for the realization of the inflationary universe scenario. Indeed, in the pre-inflationary cosmology one could just assume that the universe had the excess of particles over antiparticles *ab initio*. However, after inflation, the pre-existing baryon density becomes exponentially small, and it should be created anew.

As was emphasized by Zeldovich, cosmological observations, which show that the universe is flat, homogeneous and isotropic, and which imply therefore that it should have been inflationary, serve at present as the sole experimental evidence of the nonconservation of baryon number. Let us slightly generalize this statement and explain its meaning.

One of the reasons why pre-inflationary cosmology was not satisfactory can be explained as follows. When we are going to school, our teachers tell us that parallel lines never intersect. In the university professors tell us that parallel lines can intersect (or diverge from each other at infinity). However, nobody have ever told us why nobody have ever seen parallel lines intersecting. Meanwhile in the context of general relativity the fact that space in our part of the universe is almost exactly Euclidean should be considered as a surprising experimental result which needs explaination. Indeed, the only natural parameter of dimensionality of length which exists in quantum cosmology (which therefore could give us a typical radius of curvature of space) is the Planck length, $l_p \sim 10^{-33}$ cm, but it is 60 orders of magnitude smaller than the size of the observable part of the universe, and in any case it is much smaller than a typical size of a table, where properly driven parallel lines never intersect !

The only physical reason known to us at present which explains why parallel lines do not intersect is given by inflation. In all realistic models of inflation the size of the nearly Euclidean domains of the universe is many orders of magnitude bigger than the size of the observable part of the universe; see eq. (10) which gives a typical distance between the points where two parallel lines could intersect (e.g. the distance between "the North and the South poles" in a closed universe of a typical initial size $M_p^{-1}$.

We are not going to discuss now many other observational data (homogeneity and isotropy of the universe, absence of primordial monopoles etc.) which at present can be explained only within the inflationary cosmology, see e.g. [29]. What we are going to tell now is that all these experimental data strongly indicate that the very early universe was inflationary. In the standard hot universe theory the universe could have all its baryons from the very beginning. In the inflationary cosmology it is impossible, since the density of baryons existing in the universe *ab initio* after inflation becomes vanishingly small.

Therefore without the baryon nonconserving processes the universe after inflation would contain equal density of baryons and antibaryons, and after their annihilation the universe would contain almost no baryonic matter at all. The only possibility to produce excess of baryons over antibaryons after inflation is related to nonequilibrium processes with C and CP violation and with the baryon number nonconservation [30]. Thus, without inflation one has no theoretical explaination why parallel lines do not intersect, and without baryon number nonconservation and CP violation one cannot have a consistent inflationary cosmology. This means that the absence of intersection of parallel lines in the observable part of the universe can be considered as a first experimental evidence of the baryon nonconservation and as a second experimental evidence of CP violation...

The first mechanism of cosmological baryon production [30] was based on the theory of superheavy particle decay in an expanding universe. Implementation of this mechanism in the context of the inflationary cosmology proved to be not quite trivial. The universe soon after inflation is very far from thermodynamic equilibrium, which is good for baryosynthesis [31]. However, if the reheating temperature $T_R$ is much smaller than $10^{14} - 10^{15} GeV$, then the standard mechanism of baryosynthesis does not work, or at least becomes very inefficient. Therefore, it would be good to have an efficient mechanism of low-temperature baryon production. The search for such a mechanism was also stimulated by a very important paper by Kuzmin, Rubakov and Shaposhnikov [32], in which it was shown that non-perturbative effects near the critical temperature of the phase transition in the Glashow-Weinberg-Salam model (i.e. at $T \sim 200 GeV$) can wash out any excess of baryons produced earlier by B-L conserving processes.

A detailed discussion of this effect, of the possible ways to circumvent the above-mentioned difficulty and to generate baryon asymmetry of the universe during the epoch of the electroweak phase transition is contained in the talk by Shaposhnikov at this conference.

Within the last three years many other mechanisms of low-temperature baryon production were proposed. One of the most interesting mechanisms was suggested by Affleck and Dine [33]. It is based on the observation that in supersymmetric theories baryon and lepton charges can be carried not only by quarks and leptons, but by classical squark-slepton (scalar) fields as well. This mechanism can be naturally implemented [34] in the context of the chaotic inflation scenario, since in this scenario large classical scalar fields either can exist from the very beginning of the universe evolution, or they can be generated by quantum effects to be discussed below. An important feature of this scenario is that the final stage of the baryon production process occurs at a temperature $T \sim 100$ GeV [34], and, therefore, the baryons produced by this mechanism do not disappear due to the nonperturbative effects discussed above. Realization of this mechanism in some supergravity- and superstring-inspired models was considered in a series of papers [35]. Other interesting models of the low-temperature baryosynthesis were considered in [36].

## 2.3 Scalar field fluctuations, perturbations of density and galaxy formation

According to quantum field theory, empty space is not entirely empty. It is filled with quantum fluctuations of all types of physical fields. These fluctuations can be regarded as waves of physical fields with all possible wave-lengths, moving in all possible directions. If the values of these fields, averaged over some macroscopically large time, vanish, then the space filled with these fields seems to us empty and can be called the vacuum.

In the exponentially expanding universe the vacuum structure is much more complicated. The wave-lengths of all vacuum fluctuations of the scalar field $\phi$ grow exponentially in the expanding universe. When the wavelength of any particular fluctuation becomes greater than $H^{-1}$, this fluctuation stops propagating, and its amplitude freezes at some nonzero value $\delta\phi(x)$ because of the large friction term $3H\dot{\phi}$ in the equation of motion of the field $\phi$. The amplitude of this fluctuation then remains almost unchanged for a very long time, whereas its wavelength grows exponentially. Therefore, the appearance of such a frozen fluctuation is equivalent to the appearance of a classical field $\delta\phi(x)$ that does not vanish after averaging over macroscopic intervals of space and time.

Because the vacuum contains fluctuations of all wavelengths, inflation leads to the creation of more and more new perturbations of the classical field with wavelengths greater than $H^{-1}$. The average amplitude of such perturbations generated during a time interval $H^{-1}$ (in which the universe expands by a factor of e) is given by

$$|\delta\phi(x)| \approx \frac{H}{2\pi}. \tag{12}$$

Perturbations of the field lead to adiabatic perturbations of density $\delta\rho \sim V'(\phi)\delta\phi$, which after inflation grow, and acquire the amplitude [37,29]

$$\frac{\delta\rho}{\rho} = \frac{48}{5}\sqrt{\frac{2\pi}{3}}\frac{V_\phi^{3/2}}{M_P^3 V'(\phi)}, \tag{13}$$

where $\phi$ is the value of the classical field $\phi(t)$ (4), at which the fluctuation we consider has the wavelength $l \sim k^{-1} \sim H^{-1}(\phi)$ and becomes frozen in amplitude. In the theory of the massive scalar field with $V(\phi) = \frac{m^2}{2}\phi^2$

$$\frac{\delta\rho}{\rho} = \frac{24}{5}\sqrt{\frac{\pi}{3}}\frac{m}{M_P}(\frac{\phi}{M_P})^2. \tag{14}$$

Taking into account of (4), (5) and also of the expansion of the universe by about $10^{30}$ times after the end of inflation, one can obtain the following result for the density perturbations with the wavelength $l(cm)$ at the moment when these perturbations begin growing and the process of the galaxy formation starts:

$$\frac{\delta\rho}{\rho} \sim 0.8\frac{m}{M_P}\ln l(cm). \tag{15}$$

At a galaxy scale ($l \sim 10^{21} - 10^{22} cm$)

$$\frac{\delta\rho}{\rho} \sim 40\frac{m}{M_P}, \tag{16}$$

which gives a desirable amplitude $\frac{\delta\rho}{\rho} \sim 10^{-4} - 10^{-5}$ for $m \sim 2.10^{-7} - 2.10^{-6} M_p$. In what follows we will assume that $m \sim 10^{-6} M_P$.

# 3 Self-Reproducing Universe and Life After Inflation

In our previous investigation we only considered local properties of the inflationary universe, which is quite sufficient for a description of the observable part of the universe of the present size $l \sim 10^{28}$ cm. Indeed, in accordance to (15) our universe remains relatively homogeneous on a scale

$$l \leq l^* \sim exp(2\pi \frac{M_P}{m})cm \sim 10^{3.10^6} cm. \tag{17}$$

The corresponding density perturbations were formed at the time, when the scalar field $\phi(t)$ was bigger than $\phi^*$, where

$$\phi^* \sim M_P \sqrt{\frac{M_P}{m}} \sim 30 M_P, \tag{18}$$

see (14). (Note that $V(\phi^*) \sim \frac{m^2}{2}(\phi^*)^2 \sim m M_P^3 << M_P^4$). On a scale $l > l^*$ the universe becomes extremely inhomogenous due to quantum fluctuations produced during inflation. We are coming to a paradoxical conclusion, that the global properties of the inflationary universe are determined not by classical but by quantum effects.

Let us try to understand the origin of such a behaviour of the inflationary universe.

A very unusual feature of the inflationary universe is that processes separated by distances $l$ greater than $H^{-1}$ proceed independently of one another. This is so because during exponential expansion any two objects separated by more than $H^{-1}$ are moving away from each other with a velocity $\nu$ exceeding the speed of light. (This does not contradict special relativity because $\nu$ is not the speed of any signal; it is just the rate at which the general expansion of the universe separates two distant points.) As a result, any observer in the inflationary universe can see only those processes occurring nearer than $H^{-1}$.

An important consequence of this general result is that the process of inflation in any spatial domain of radius $H^{-1}$ occurs independently of any events outside it. Any two inflationary domains displaced by more than $H^{-1}$ cannot collide or eat one another, or do each other any damage. Their expansion is due not to the annexation of the territory of their neighbors, but rather to the peaceful (and very rapid) growth in their own volume, as allowed by general relativity. In this sense any inflationary domain of initial size exceeding $2H^{-1}$ can be considered as a separate mini-universe, expanding independently of what occurs outside it.

To investigate the behavior of such a mini-universe, with an account taken of quantum fluctuations, let us consider an inflationary domain of initial size roughly $H^{-1}$ containing a sufficiently homogeneous field whose initial value $\phi$ greatly exceeds $M_P$. Eq.(4) tells us that during a typical time interval $\Delta t = H^{-1}$ the field inside this domain will be reduced by

$$\Delta\phi = \frac{M_P^2}{4\pi\phi}. \tag{19}$$

By comparison of (12) and (19) one can easily see that if $\phi$ is much less than $\phi^* \sim M_P\sqrt{\frac{M_P}{m}}$, the decrease of the field $\phi$ due to its classical motion is much larger than the amplitude of the quantum fluctuations $\delta\phi$ generated during the same time. But for large $\phi$ (up to the classical limit of $10^4 M_P$), $\delta\phi(x)$ will exceed $\Delta\phi$, i.e. the Brownian motion of the field $\phi$ becomes more rapid than its classical motion. Because the typical wavelength of the fluctuation field $\delta\phi(x)$ generated during this time is $H^{-1}$, the whole domain volume after $\Delta t$ will effectively have become divided into $e^3$ separate domains (mini-universes) of diameter $H^{-1}$. In almost half of these domains the field $\phi$ grows by $|\delta\phi(x)| - \Delta\phi$, which is not very different from $|\delta\phi(x)|$ or $H/2\pi$, rather than decreases. During the next time interval $\Delta t = H^{-1}$ the field grows again in half of these mini-universes. It can be shown that the total physical volume occupied by a permanently growing field $\phi$ increases with time like $exp(3 - \ln 2)Ht$, and the total volume occupied by a field that does not decrease grows almost as fast as $\frac{1}{2}e^{3Ht}$.

Because the value of the Hubble constant $H(\phi)$ is proportional to $\phi$, the main part of the physical volume of the universe is the result of the expansion of domains with nearly the maximal possible field value, $M_P^2/m$, for which $V(\phi)$ is close to $M_P^4$. There are also exponentially many domains with smaller values of $\phi$. Those domains in which $\phi$ eventually becomes smaller than about $30M_P$ give rise to the mini-universes of *our* type. In such domains, $\phi$ eventually rolls down to the minimum of $V(\phi)$, and these mini-universes are subsequently describable by the usual Big Bang theory. However a considerable part of the physical volume of the entire universe remains forever in the inflationary phase [14,15].

Thus in our scenario the universe, in which there was initially at least one domain of a size on the order of $H^{-1}$ filled with a sufficiently large and homogeneous field $\phi$, unceasingly reproduces itself and becomes immortal. One mini-universe produces many others, and this process goes on without end, even if some of the mini-universes eventually collapse.

But this means that it is vanishingly improbable that our mini-universe would have been the first in the sequence of all mini-universes. Moreover, it no longer seems necessary to assume that there actually was some first mini-universe appearing from nothing or from an initial singularity at some moment $t = 0$ before which there was no space-time at all.

From general topological theorems about singularities in cosmology it does not actually follow that our universe was created *as a whole* at some moment before which the universe did not exist. The usual assumption that the whole universe appears from the unique Big Bang singularity at $t = 0$ is based on the implicit assumption that the universe *as a whole* is sufficiently homogeneous. Indeed, the observable part of our universe is very homogeneous. Observed density fluctuations are less than a part in a thousand and there has been no reason to expect that the universe is inhomogeneous on a larger scale beyond the $10^{10}$-light-year horizon. In a homogeneous universe one can use the density $\rho(t)$ as a measure of time. In that case it can be shown that the universe appears *as a whole* from a singularity at $t = 0$. The initial density $\rho(0)$ is infinite, and it becomes possible to describe the whole universe in terms of classical space-time after the Planck time $M_P^{-1}$ (about $10^{-43}$ seconds), when the energy density everywhere simultaneously becomes smaller than the Planck density $M_P^4$.

With the invention of the inflationary scenario the situation changes drastically. At present only inflation can explain why the observable part of the universe is so homogeneous, but from inflation it also follows that on a much larger scale the universe is extremely *in*homogeneous. In some parts of the universe the energy density $\rho$ is now of the order of $M_P^4$, 125 orders of magnitude higher than the $10^{-29}$ or $10^{-30} g.cm^{-3}$ we can see nearby. In such a scenario there is no reason to assume that the universe was initially homogeneous and that all its causally disconnected parts started their expansions simultaneously.

If the universe is infinitely large (like the Friedmann open or flat universe), then it cannot have had a single beginning; a simultaneous creation of infinitely many causally disconnected regions is totally improbable. Therefore the universe cannot be infinite *ab initio*, or it must exist eternally as a huge self-reproducing entity. Some of its parts appear at different times from singularities, or may die in a singular state. New parts are constantly being created from the space-time foam when $V(\phi)$ exceeds the Planck density, or they may revert to the foamlike state again as a result of large fluctuations in $\phi$. But the evolution of the universe as a whole has no end, and it may have had no beginning.

## 4 Inflation and the Wave Function of the Universe

One of the most ambitious approaches to cosmology is based on the investigation of the Wheeler-DeWitt equation for the wave function $\Psi$ of the universe [39]. However, this equation has many different solutions, and *a priori* it is not quite clear which of these solutions describes our universe.

A very interesting idea was suggested by Hartle and Hawking [40]. According to their work, the wave function of the ground state of the universe with a scale factor $a$ filled with a scalar field $\phi$ in the semi-classical approximation is given by

$$\psi_0(a,\phi) \sim exp(-S_E(a,\phi)). \tag{20}$$

Here $S_E(a,\phi)$ is the Euclidean action corresponding to the Euclidean solutions of the Lagrange equation for $a(\tau)$ and $\phi(\tau)$ with the boundary conditions $a(0)=a, \phi(0)=\phi$. The reason for choosing this particular solution of the Wheeler-DeWitt equation was explained as follows. Let us consider the Green's function of a particle which moves from the point $(0,t')$ to the point $\mathbf{x},t$:

$$\begin{aligned} < \mathbf{x},0|0,t' > &= \sum_n \psi_n(\mathbf{x})\psi_n(0)exp(iE_n(t-t')) \\ &= \int d\mathbf{x}(t)exp(iS(\mathbf{x}(t))), \end{aligned} \tag{21}$$

where $\Psi_n$ is a complete set of energy eigenstates corresponding to the energies $E_n \geq 0$.

To obtain an expression for the ground-state wave function $\Psi_0(\mathbf{x})$, one should make a rotation $t \to -i\tau$ and take the limit as $\tau \to -\infty$. In the summation (21) only the term $n=0$ with $E_0=0$ survives, and the integral transforms into $\int dx(\tau)\exp(-S_E(\tau)))$. Hartle and Hawking have argued that the generalisation of this result to the case of interest in the semiclassical approximation would yield (20).

The gravitational action corresponding to the Euclidean section $S_4$ of de Sitter space $dS_4$ with $a(\tau) = H^{-1}(\phi)\cos H\tau$ is negative,

$$S_E(a,\phi) = -\frac{1}{2}\int d\eta[(\frac{da}{d\eta})^2 - a^2 + \frac{\Lambda}{3}a^4]\frac{3\pi M_P^2}{2} = -\frac{3M_P^4}{16V(\phi)} \tag{22}$$

Here $\eta$ is the conformal time, $\eta = \int dt/a(t), \Lambda = 8\pi V/M_p^2$. Therefore, according to [40],

$$\psi_0(a,\phi) \sim \exp(-S_E(a,\phi)) \sim \exp(\frac{3M_P^4}{16V(\phi)}). \tag{23}$$

This means that the probability $P$ of finding the universe in the state with $\phi = const, a = H^{-1}(\phi) = (3M_P^2/8\pi V(\phi))^{1/2}$ is given by

$$P(\phi) \sim |\psi_0|^2 \sim \exp(\frac{3M_P^4}{8V(\phi)}). \tag{24}$$

This expression has a very sharp maximum as $V(\phi) \to 0$. Therefore the probability of finding the universe in a state with a large field $\phi$ and having a long stage of inflation becomes strongly diminished.

There exists an alternative choice of the wave function of the universe. It can be argued that the analogy between the standard theory (21) and the gravitational theory (22) is incomplete. Indeed, there is an overall minus sign in the expression for $S_E(a,\phi)$ (22) which indicates that the gravitational energy associated with the scale factor $a$ is negative. (This is related to the well-known fact that the total energy of a closed universe is zero, being a sum of the positive energy of matter and the negative energy of the scale factor $a$.) In such a case, to obtain $\psi_0$ from (22) one should rotate $t$ not to $-i\tau$, but to $+i\tau$ which leads to [41]

$$P \sim |\psi_0(a,\phi)|^2 \sim \exp(-2|S_E(a,\phi)|) \sim \exp(-\frac{3M_P^4}{8V(\phi)}) \tag{25}$$

Actually, this result is valid only if the evolution of the field $\phi$ is very slow, so that this field acts only as a cosmological constant $\Lambda(\phi) = 8\pi V(\phi)/M_P^2$ in (22). Fortunately, this is indeed the case during inflation. Later the same result was obtained by another method, devised by Zeldovich and Starobinsky [42], Rubakov [43], and Vilenkin [44]. This result can be interpreted as the probability of quantum tunnelling of the universe from $a = 0$ (from 'nothing') to $a = H^{-1}(\phi)$. In complete agreement with our previous argument (see eq. (8)), eq.(25) shows that a typical initial value of the field $\phi$ is given by $V(\phi) \sim M_P^4$ (if one does not speculate about the possibility that $V(\phi) >> M_P^4$), which leads to a very long state of inflation.

It must be said that there is no rigorous derivation of either eq. (23) or (25), and the physical meaning of creation of everything from 'nothing' is far from clear. Therefore a deeper understanding of the physical processes in the inflationaly universe is necessary in order to investigate the wave function of the universe $\Psi_0(a,\phi)$ and to suggest a correct interpretation of this wave function. With this purpose we shall try to investigate the global structure of the inflationary universe, and go beyond the minisuperspace approach used in the derivation of (23) and (25). This can be done with the help of a stochastic approach to inflation [45, 15], which is a more formal way to investigate the Brownian motion of the scalar field $\phi$ studied in the previous section.

The evolution of the fluctuating field $\phi$ in any given domain can be described with the help of its distribution function $P(\phi)$, or in terms of its average value $\bar{\phi}$ in this domain and its dispersion $\Delta = (\langle\delta\phi^2\rangle)^{1/2}$. However, one will obtain different results depending on the method of averaging: one can consider the distribution $P_c(\phi)$ over the non-growing coordinate volume of the domain (i.e. over its physical volume at some initial moment of inflation), or the distribution $P_P(\phi)$ over its *physical* (proper) volume, which grows exponentially at a different rate in different parts of the domain. It can be shown that the dispersion of the field $\phi$ in the coordinate volume $\Delta_c$ is always much smaller than $\bar{\phi}_c$ for $V(\bar{\phi}_c) << M_P^4$. Therefore the evolution of the averaged field $\bar{\phi}_c$, can be described approximately by (2)-(5). However, if one wishes to know the resulting spacetime structure and the distribution of the field $\phi$ after (or during) inflation, it is more appropriate to take an average $\bar{\phi}_P$ over the physical volume, and in some cases the behaviour of $\bar{\phi}_P$ and $\Delta_P$ differs considerably from the behaviour of $\bar{\phi}_c$ and $\Delta_c$.

The Brownian motion of the field $\phi$ can be described by the diffusion equation

$$\frac{\partial P_c}{\partial t} = \frac{\partial}{\partial\phi}[\frac{\partial(\mathcal{D}P_c)}{\partial\phi} + \frac{P_c}{3H}\frac{\partial V}{\partial\phi}], \tag{26}$$

where the coefficient of diffusion $\mathcal{D} = H^3/8\pi^2$. This equation for the case $H(\phi) = $ const. was first derived by Starobinsky [45]; for a more detailed derivation see [46]. For the special case $\frac{\partial V}{\partial \phi} = 0$ this equation was obtained by Vilenkin [47].

The stationary solution ($\partial P_c/\partial t = 0$) would be [45]

$$P_c \sim exp(3M_P^4/8V(\phi)) \tag{27}$$

which is equal to the square of the Hartle-Hawking wave function of the universe (35). At the first glance, this result is a direct confirmation of the Hartle-Hawking prescription for the wave function of the universe. (In fact, this is the *only* more or less rigorous confirmation of the Hartle-Hawking prescription which is known to us at present).

However, in all realistic cosmological theories, in which $V(\phi) = 0$ in its minimum the distribution (27) is not normalizable. The source of this difficulty can be easily understood: any stationary distribution may exist only due to a compensation of a classical flow of the field $\psi$ downwards to the minimum of $V(\phi)$ by the diffusion motion upwards. However, diffusion of the field $\phi$ discussed above exists only during inflation, i.e. only for $\phi \geq M_P, V(\phi) \geq V(M_P) \sim m^2M_P^2 \sim 10^{-8}M_P^4$. Therefore eq. (27) would correctly describe the stationary distribution $P_c(\phi)$ in the inflationary universe only if $V(\phi) \geq 10^{-8}M_P^4$ in the absolute minimum of $V(\phi)$, which, of course, is unrealistic [15]. Therefore, at present, we do not know how one can use the Hartle-Hawking wave function (20), (23) for the description of the inflationary universe.

Now let us study a general non-stationary solution of (26) describing the time-evolution of the initial distribution $P_c(\phi, t = 0) = \delta(\phi - \phi_0)$. This solution for the theory $\frac{m^2}{2}\phi^2$ is [15].

$$P_c(\phi, t) = exp[-\frac{3(\phi - \phi(t))^2 M_P^4}{2m^2(\phi_0^4 - \phi^4(t))}], \tag{28}$$

which shows that at the first stage of the process during the time $\tau \leq \frac{\phi_0 2(3\pi)^{1/2}}{mM_P}$ (see (4)), the maximum of the distribution $P_c(\phi, t)$ almost does not move, whereas the dispersion linearly grows. Then, at $t >> \tau$, the maximum of $P_c(\phi, t)$ moves to $\phi = 0$, just as the classical field $\phi(t)$ (4). However, during the first period of time the volume of domains filled by the field $\phi$ increases approximately by $exp(3H(\phi)\tau) \sim exp(c_1\frac{\phi_0\phi}{M_P^2})$ where $c_1 = 0(1)$. After this time the distribution $P_p(\phi, \tau)$ looks (approximately) as follows

$$P_p(\phi, \tau) \sim P_c(\phi, \tau)exp(3H(\phi)\tau) \sim exp[-\frac{\phi^2 M_P^4}{c_2 m^2 \phi_0^4} + c_1\frac{\phi_0\phi}{M_P^2}], \tag{29}$$

where $c_2 = 0(1)$. One can easily verify that, if $\phi_0 > \phi^* \sim M_P\sqrt{\frac{M_P}{m}}$, then the maximum of $P_p(\phi, t)$ during the time $\tau$ becomes shifted to some field $\phi$ which is bigger than $\phi_0$. This just corresponds to the process of eternal self-reproduction of the inflationary universe studied in the previous section. For completeness we shall mention here another solution of eq. (26). If the initial value of the field $\phi$ is very large, $\phi_0 \geq M_P^2/m$, i.e. if one starts with the spacetime foam with $V(\phi_0) \geq M_P^4$, then the evolution of the field $\phi$ in the first stage (rapid diffusion) becomes more complicated (the naively estimated dispersion $\Delta_c^2 \sim H^3\Delta t$ soon becomes greater than $\phi_0^2$). In this case the distribution of the field $\phi$ is not Gaussian. The solution of eq. (26) at the stage of diffusion from $\phi_0$ to some field $\phi$ with $V(\phi) << M_P^4$ is given by

$$P_c(\phi) \sim exp(-\frac{3\sqrt{3}\pi M_P^3}{m^3\phi t}). \tag{30}$$

This solution describes quantum creation of domains of a size $l \geq H^{-1}(\phi)$, which occurs due to the diffusion of the field $\phi$ from $\phi_0 \geq M_P^2/m$ to $\phi << \phi_0$. Direct diffusion with formation of a domain filled with the field $\phi$ is possible only during the time $= c(2\sqrt{3}\pi\phi/mM_P), c = 0(1)$. At larger times a more rapid process is a diffusion to some field $\tilde{\phi} > \phi$ and a subsequent classical rolling down from $\tilde{\phi}$ to $\phi$. Therefore one may interpret a distribution $P_c(\phi)$ formed after a time $t = c(2\sqrt{3}\pi\phi/mM_P)$ as a probability of a quantum creation of a mini-universe filled with a field $\phi$ [15],

$$P_c(\phi) \sim exp(-c\frac{3M_P^4}{2m^2\phi^2}) \sim exp(-2c\frac{3M_P^4}{8V(\phi)}), \tag{31}$$

which is in agreement with the previous estimate for the probability of quantum creation of the universe, eq. (26) [41-44].

These results imply that life in the inflationary universe will never disappear. Unfortunately, this conclusion does not automatically mean that one can be very optimistic about the future of mankind. Even though new domains of our type are permanently produced during inflation, later on each particular domain either will become practically empty and unsuitable for life or it will evolve into a huge black hole and collapse [27]. The only possible strategy of survival which we see at the moment is to travel from old domains to the new ones, which are displaced near the self-reproducing inflationary domains and always have large enough density of baryons. In the worst case, if we will be unable to travel to such distant places ourselves, we can try to send some information about us, our life and our knowledge, and maybe even stimulate development of such kinds of life there, which would be able to receive and use this information. In such a case one would have a comforting thought that even though life in our part of the universe will disappear, we will have some inheritors, and in this sense our existence is not entirely meaningless. (At least it would be not worse than what we have here now.)

In order to check whether such a possibility does actually exist, it would be necessary to perform a detailed investigation of the global structure of the universe, and especially of its causal structure, which becomes very nontrivial in the context of inflationary cosmology [15]. One should answer the question whether it is actually possible to travel or to send a signal from our part of the universe to a vicinity of a nearby self-reproducing inflationary domain, and to do it in such a way that after a long time which is necessary for the signal to arrive the density of matter near this domain will be sufficiently large for the existence of life there. At present we do not know any no-go theorems which would imply that it is generally impossible and that one cannot elaborate any successful strategy of survival of life in our part of the universe. One of the main aims of this paper is to show that the answer to this question may depend crucially on the value of the vacuum energy density, i.e. on the absence or presence of a tiny cosmological constant in the Einstein equations.

Namely, according to [14, 15], the self-reproducing domains appear in the inflationary universe independently of the existence or nonexistence of a small energy density $\rho_v$ of the vacuum state in which we live now (if it is not negative and as big as - $10^{-6}M^4{}_p$). However, the causal structure of the universe, the possibility of communication between its different domains and the possibility of the existence of life there depend crucially on the value of $\rho_v$. For example, if $\rho_v$ is negative and its absolute value is much bigger than the present energy density of matter $\rho_o \sim 10^{-29}g.cm^{-3}$, each domain of the universe after inflation and a sufficiently long subsequent period of expansion behaves as an anti de Sitter space and collapses at a time $t \sim M_p\rho_v^{-1/2} \ll M_p\rho_o^{-1/2} \sim 10^{10}$ years, i.e. much earlier than the time necessary for the development of life of our type. Note, that this process occurs locally, at different times in different places of a self-reproducing universe. Therefore one may escape the local collapse by travelling towards the self-reproducing domains. However, since the typical distance between our place and a nearby self-reproducing domain is extremely large, such a journey is possible only if the lifetime of the anti de Sitter domain $t \sim M_p \rho_v^{-1/2}$ is also extremely large and, consequently, the absolute value of $\rho_v$ is extremely small (see below).

If the vacuum energy density is positive and bigger than $10^{-27} g.cm^{-3}$, formation of galaxies of our type becomes impossible [48]. Smaller values of $\rho_v$ do not affect galaxy formation, but do affect the causal structure of the universe. For example, the univesre with $\rho_v > \rho_0$ at the time $t > 10^{10}$ years behaves as de Sitter space with the event horizon of the size $H^{-1} \sim 10^{28} cm$ (i.e. smaller than the particle horizon $\sim t$ now). In such a universe the distance between the galaxies grows as $e^{Ht}$, and after a time of the order $H^{-1} < 10^{10}$ years there would remain only one (our own) galaxy inside our event horizon. (The galaxy itself is not stretched by the universe expansion since it is a gravitationally bounded system with the energy density much bigger than $\rho_o$.) It is well known that in the exponentially expanding universe one can travel or send any information only to those regions which were inside the event horizon at the moment when the information was sent. This means that soon after the beginning of the exponential expansion of the universe with $\rho_v > \rho_0$ there will be no possibility to travel or send information to other galaxies, and it is certainly impossible to send a signal to any self-reproducing inflationary domain, since a typical distance from our place to any such domain is much bigger than $H^{-1}$.

This result can be considerably strengthened in all realistic models of inflation. To do it one should first make an estimate of a typical distance $l^*$ from our place to a nearby inflationary self-reproducing domain. This can be easily done, since there should be many such domains in each region of initial size $H^{-1}(\phi)$ containing field $\phi > \phi^*$ (18), and there should be no such domains inside the regions of initial size $H^{-1}$ containing field $\phi < \phi^*$. For the theory $m^2\phi^2/2$ this gives, according to eq.(5),

$$l^* \sim H^{-1}(\phi^*)\, exp(\frac{2\pi\phi^{*2}}{M^2{}_p})\, (\frac{T_R}{T_o}). \tag{32}$$

Here $H^{-1}(\phi^*) = M_p(\frac{3}{8\pi V(\phi^*)})^{1/2}$, $T_R$ is the reheating temperature of the universe after inflation, $T_o$ is the present temperature of the blackbody radiation in the universe. The last factor appears due to the additional stretching of the universe when the temperature falls down from $T_R$ to $T_o$. Typically this factor is of the order of $10^{30}$ and $H^{-1}(\phi^*)$ is of the order of $10^{-30}$ cm. Therefore from (18), (32) it follows that

$$l^* \sim exp(2\pi M_p/m)\ cm \sim exp((2\pi)10^6)\ cm. \tag{33}$$

for $m \sim 10^{-6} M_p$. The universe becomes exponentially expanding and acquires an event horizon $H^{-1}(\rho_v) = M_p(\frac{3}{8\pi\rho_v})^{1/2}$ at the late stages of its evolution, when the energy density of matter, which decreases as $t^{-2}$, becomes smaller than the vacuum energy density $\rho_v$. It occurs at the time which is bigger than the present age of the universe $t_o \sim 10^{10}$ years by the factor $(\frac{\rho_o}{\rho_v})^{1/2}$. At that time the distance $l^*$ from us to the nearby inflationary self-reproducing domain grows by a factor $(\frac{\rho_o}{\rho_v})^{1/3}$, since the scale factor of a cold matter dominated universe grows as $t^{2/3}$. The causal connection between us and this region will be possible only if this distance will remain smaller than $H^{-1}(\rho_v)$,

$$M_p\left(\frac{3}{8\pi\rho_v}\right)^{1/2} > M_p\left(\frac{3}{8\pi V(\phi^*)}\right)^{1/2} exp\left(\frac{2\pi\phi^{*2}}{M^2{}_p}\right)\left(\frac{T_R}{T_o}\right)\left(\frac{\rho_o}{\rho_v}\right)^{1/3}, \tag{34}$$

which implies that the causal connection is possible only if $\rho_v$ is smaller than some critical value $\rho_c$, where

$$\rho_c \sim \frac{V^3(\phi^*)}{\rho^2{}_o}\left(\frac{T_o}{T_R}\right)^6 exp\left(-\frac{12\pi M_p}{m}\right) \sim 10^{-10^7} g.cm^{-3}. \tag{35}$$

Here $V(\phi^*) \sim mM^3{}_p \sim 10^{-6}M^4{}_p$, $\rho_o \sim 10^{-29}g.cm^{-3} \sim 10^{-125}M^4{}_p$, $T_R \sim 10^{10}GeV$ . One can easily see that the constraint (35) on the value of the vacuum energy density $\rho_v$ is very stringent and practically does not depend on the particular values of $V(\phi^*), \rho_o$ and $T_R$ .

Returning to the case of a negative vacuum energy, it can be shown by methods similar to those used above that in this case as well a causal contact between our part of the universe and a nearby self-reproducing inflationary domain is impossible if $|\rho_v| > \rho_c$ , since in this case the time which is necessary for the contact is bigger than the time before the local collapse of our part of the universe.

These results are rather unexpected. It was clear *a priori* that the existence of a nonvanishing vacuum energy may slightly change the conditions of our life , but it was difficult to imagine that the development of life in our part of the universe may depend crucially on the existence or nonexistence of a tiny vacuum energy density $|\rho_v| \geq 10^{-10^7} g.cm^{-3}$.

Of course, the events we are discussing here will occur in a very distant future (at $t \sim 10^{5.10^6}$ years), but it is very difficult not to be curious about our common fate as it can be understood at the present level of the development of science. Another possible application of these results would be to use them in order to obtain a strong anthropic bound $|\rho_v| < \rho_c \sim 10^{-10^7} g.cm^{-3}$, which would help us to solve the cosmological constant problem. At the first glance, such an idea certainly cannot work, since the existence of the vacuum energy $-10^{-29}g.cm^{-3} < \rho_v < 10^{-27}g.cm^{-3}$ does not prevent formation of galaxies and can hardly have any effect on the appearance of life in the universe [48]. Moreover, until very recently the general attitude of physicists to the Anthropic Principle was rather sceptical, to say the least. It was believed that the weak, strong and electromagnetic interactions are the same in all parts of our universe, that the fundamental constants of Nature are universal and it is meaningless to discuss the possibility to live in a universe of a different type. This attitude was somewhat changed first when it was understood that in accordance with the chaotic inflationary universe scenario the universe should contain an exponentially large (or maybe even infinitely large) number of exponentially large domains (mini-universes), in which all kinds of symmetry breaking and all kinds of compactification which are possible in a theory of a given type are actually realized [29]. In other words, the universe becomes divided into many exponentially large domains with different types of low-energy physics and maybe even with different dimensionality inside each of them. This made it possible to get a justification of a kind of Weak Anthropic Principle : If many exponentially large domains with the low-energy physics of our type do exist in the universe described by a given theory, then it is quite natural that we live in one of such domains rather than in a domain where life of our type is impossible [29].

To explain it in a more detailed way let us remember that in realistic theories of elementary particles there exist many different types of scalar fields $\phi_i$. The potential energy $V(\phi_i)$ often has many different local minima, in which the universe may live for an extremely long time, much greater than the $10^{10}$ years of our observable domain. For example, in the supersymmetric $SU(5)$ theory, $V(\phi_i)$ has several different minima of almost equal depth. Because the laws governing the interactions of elementary particles at the low energies at which we do experiments depend on the values of the classical fields $\phi_i$, each of these minima corresponds to a different low-energy physics. In one of them the $SU(5)$ symmetry between all types of interactions remains unbroken - that is, the scalar fields $\phi_i$ remain equal to zero. In other minima various symmetry breaking patterns are realized, and in only one of these minima is the broken symmetry of the weak, strong and electromagnetic interactions that which we in fact observe.

During inflation there are large-scale fluctuations of all the fields $\phi_i$. As a result, the inflationary universe becomes divided into an exponentially large number of inflationary mini-universes, with the scalar fields taking all possible values. As inflation ends in some mini-universes, these scalar fields roll down to all possible minima of $V(\phi_i)$. The universe becomes divided into many different exponentially large domains, realizing all possible types of symmetry breaking between the fundamental interactions. In some of these mini-universes the low-energy physics is quite different from our own. We cannot now see them because the size of our own domains is much greater than the size of its $10^{10}$-light-year observable portion. We could not live in these domains because our kind of life requires our kind of low-energy physics.

It is very important that in the inflationay universe there is lots of room for all possible types of symmetry breaking and for all possible types of life. There is, therefore, no longer any need to require that in the true theory the minimum of $V(\phi_i)$ corresponding to our type of symmetry breaking be the only one or the deepest one. This new cosmopolitan viewpoint may greatly simplify the task of building realistic models of the elementary particles.

The change of the values of the scalar fields $\phi_i$ – that is to say, the change of the vacuum state – is the simplest kind of "mutation" that may occur during inflation. Much more interesting possibilities appear if one considers chaotic inflation in the higher-dimensional Kaluza-Klein theories. In those domains in which the energy density of the field $\phi$ grows to the Planck density, quantum fluctuations of the metric at a length scale of $M_P^{-1}$ become of order unity. In such domains an inflationary $d$-dimensional universe can squeeze locally into a tube of smaller dimensionality $d-n$ (or vice versa). If this tube is also inflationary (in $d-n$ dimensions) and the initial length of the tube is greater than $M_P^{-1}$ (which is quite probable near the Planck density), then its further expansion proceeds independently of its prehistory and of the fate of its mother universe. In an eternally existing universe such processes should occur even if their probability is very small. In fact the probability of such processes is small only if $V(\phi)$ is far below $M_P^4$.

Thus the inflationary universe becomes divided into different mini-universes in which all possible types of compactification produce all sorts of dimensionalities [38]. By this argument we find ourselves inside a four-dimensional domain with our kind of low-energy physics not because other kinds of mini-universes are impossible or improbable, but simply because our kind of life cannot exist in other domains.

This may have important implication for the building of realistic Kaluza-Klein and superstring theories. For example, it is extremely complicated, if not impossible, to construct a theory in which only one type of compactification can occur, leading precisely to a four-dimensional inflationary universe with the low-energy particle physics of our experience. But from the point of view discussed here, there is no need to require that the results compactification and inflation have wrought in our realm be the only possible results, or the best. It is enough to find a theory in which such a compactification is possible. This problem is still difficult, but it is much easier than the one we have been trying to solve.

With the invention of the theory of many universes interacting with each other only globally [34], and especially with the development of the baby universe theory [19-21] the situation changed even more. As it will be argued below, in the context of the baby universe theory it is possible to justify the Strong Anthropic Principle as well.

According to the baby universe theory [19-21], we may live in different quantum states of the universe $|\alpha_i>$, corresponding to different choices of the "fundamental constants", such as the vacuum energy $\rho_v$ (cosmological constant $\Lambda = 8\pi\rho_v/M^2{}_p$), gravitational constant $G = M^{-2}{}_p$, fine structure constant $\alpha$, etc. By measuring the coupling constants we actually determine in which quantum state $|\alpha_i>$ we live now, and after that we cannot "jump" to another quantum state with other constants. That is why we are used to believe that we live in a classical universe with well-determined and unchanging laws of physics.

It would be very desirable to understand why do we live in a universe in a quantum state $|\alpha_i>$ with an extremely small (or even vanishing) cosmological constant, with $M_p \sim 10^{19}$ GeV, $\alpha = 1/137$, etc. A possible idea is to compute the wave function of the universe and then to prove that it is entirely concentrated near some particular values of the coupling constants [20, 23]. However, the validity of the euclidean approach to quantum cosmology used in most of the papers on this problem is far from being clear [29], and it is quite possible also that the wave function of the universe actually does not have any sharp peak at $\Lambda = 0$ [25, 26]. Moreover, it is not quite clear whether it is the wave function of the whole universe that is to be computed. It is known that in quantum mechanics only those statements can make sense that can be formulated in a concrete operational way. For example, at the classical level one can speak of the age of the universe $t$. However, the essence of the Wheeler-DeWitt equation for the wave function of the universe is that this wave function *does not depend on time*, since the total Hamiltonian of the universe, including the Hamiltonian of the gravitational field, vanishes identically [39]. Therefore if one would wish to describe the evolution of the universe with the help of its wave function, one would be in trouble. The resolution of this paradox is rather instructive [39]. The notion of evolution is not applicable to the universe as a whole since there is no external observer with respect to the universe, and there is no external clock as well which would not belong to the universe. However, we do not actually ask why the universe *as a whole* is evolving in the way we see it. We are just trying to understand our own experimental data. Thus, a more precisely formulated question is why *we see* the universe evolving in time in a given way. In order to answer this question one should first divide the universe into two main pieces: i) an observer with his clock and other measuring devices and ii) the rest of the universe. Then it can be shown that the wave function of the rest of the universe does depend on the state of the clock of the observer, i.e. on his "time" [39]. This time dependence in some sense is "objective", which means that the results obtained by different (macroscopic) observers living in the same quantum state of the universe and using sufficiently good (macroscopic) measuring apparatus agree with each other [49].

This example teaches us an important lesson. By an investigation of the wave function of the universe *as a whole* one sometimes gets information which has no direct relevance to the observational data, e.g. that the universe does not evolve in time. (Moreover, it is rather difficult to give any probabilistic interpretation to the wave function of the universe as a whole, since typically it is not normalizable.) In order to describe the universe *as we see it* one should divide the universe into several macroscopic pieces and calculate a conditional probability to observe it in a given state under an obvious condition that the observer and his measuring apparatus do exist. This simple condition, however, is very restrictive. In order to understand its meaning one should remember the famous Einstein-Podolsky-Rosen experiment, where two fermions (or photons) are produced by a decay of a spin zero particle. By a measurement of a spin of one of the particles an observer instantaneously determines the spin of another particle even if this particle and the observer are not in a causal contact at the moment of the measurement. This apparent paradox is resolved if one takes into account that prior to the measurement one should describe both particles by the wave function of the pair of particles rather than by separate wave functions of each of them. The knowledge of this wave function (the fact that it corresponds to a state with zero angular momentum) and the knowledge of the spin of one of these particles makes it possible to determine the spin of another particle without making any causal contact with it, just for the reason that there is a strong correlations between spins of particles in a system with a vanishing total angular momentum. For example, the conditional probability to find the second particle in a state with spin s = - 1/2 is equal to unity under the condition that the first particle has a spin s = + 1/2.

Similarly, any scientific exploration of the properties of the universe starts from the division of the universe into two subsystems (the observer with his measuring devices and the rest of the universe). Due to a strong correlation between the properties of these two subsystems, an investigation of our own properties, the properties of our environment and our measuring devices makes it possible to say a lot about the properties of the rest of the universe such as the vacuum energy, the values of some coupling constants etc.

It is important also, that in the EPR experiment one can choose different apparatus, e.g. the one which measures a linear polarization of the first photon, or the one which measures its circular polarization. After the measurment the second photon will be either linearly or circularly polarized, depending on our choice of measuring device. Similarly, many kinds of life and many types of measuring devices may exist, but what we are studying is the correlation of *our* kind of life, environment and measuring devices with the properties of the rest of the universe. But this means that in the context of the baby universe theory one can actually make sense of the Strong Anthropic Principle and try to explain why the electromagnetic coupling constant, the gravitational coupling constant and some other parameters of the theory have those particular values which we know: With other values of the coupling constants the existence of life *of our type* and of the measuring devices we use would be impossible. (Note, however, that in this interpretation the Strong Anthropic Principle has nothing in common with designing the universe for the benefit of human beings or with changing the whole universe by our own will, just as the determination of the properties of the second particle in the EPR experiment and the dependence of these properties on our choice of measuring device has nothing to do with the acausal action at a distance.)

If this idea is correct, then there is no need to compute the most probable values of the cosmological constant and of the gravitational constant by the investigation of the wave function of the universe as a whole, and it is not surprising that some of the results of the corresponding investigation look as counterintuitive as the statement that the universe as a whole does not depend on time. Rather one should investigate the probability to measure some particular eigenvalues of the operators corresponding to the "constants" $\Lambda, e, m_e, G$ in the quantum state of the universe in which we live now under the obvious condition that we can live and perform our measurements here. In some cases (e.g. in the case of the constants $e, m_e, G$) the corresponding investigation is rather simple and straightforward. For example, according to [50], an increase of $m_e$ by more than 2.5 times would make impossible the existence of atoms. An increase of $e$ by more than 30 % would lead to an instability of proton and nuclei. A change of the strong interaction constant by more than 10 % would lead to the absence of heavy nuclei and of hydrogen in the universe . An increase or decrease of the weak interaction constant by an order of magnitude would lead to the absence of complex elements or of hydrogen respectively. An increase of the gravitational constant G by an order of magnitude would make the lifetime of the Sun so small that no biological molecules would appear on the Earth. An even bigger increase of G would lead to an extremely efficient nuclearsynthesis and to the absence of hydrogen in the universe [50,51]. A smaller value of G would make expansion of the universe more slow. In such a universe the departure from thermal equilibrium during the process of baryosynthesis would be small, this process would be inefficient and the universe now would be practically empty.

This list can be easily continued, but we hope that the results discussed above clearly demonstrate that the correlation between the properties of observers and the properties of the quantum state of the rest of the universe is very strong indeed. Moreover, most of the results discussed above are not so much *anthropic*, they just show that a small deviation from the present values of coupling constants would make impossible the existence of all measuring devices we use. In the context of the baby universe theory this means that the distribution of the conditional probability to make measurements in a quantum state with given values of weak, strong, gravitational and some other "fundamental constants" proves to be sharply peaked near the actual values of these constants we have measured in the quantum state of the universe in which we live now. This peak in the probability distribution should be especially narrow in unified theories of all fundamental interactions where the values of different coupling constants are strongly correlated due to the underlying symmetry of the theory. This indicates that if one wishes to get an order-of-magnitude estimate of the coupling constants, in some cases one can do it without a computation of these constants in the context of the "big fix" paradigm [23], and if one still wishes to make a computation (which, of course, is a good idea), then one should compute the amplitudes corresponding to the actual experimental environment rather than the amplitudes corresponding to the metaphysical "universe in itself" investigated by an imaginary external observer. (The discussion of the time-independence of the wave function of the universe shows that in the context of quantum cosmology the subtle difference between the amplitudes of these two types can be crucially important.)

Now let us return to the discussion of the cosmological constant problem. The possibility to explain the small value of the vacuum energy density $|\rho_v| < 10^{-29} g.cm^{-3}$ with the help of the Anthropic Principle is much less clear, but is not hopeless. From the results of ref. [48] it follows that formation of galaxies and appearance of life of our type is possible only if $-10^{-29} g.cm^{-3} < \rho_v < 10^{-27} g.cm^{-3}$. In the context of the baby universe theory this means that the conditional probability to measure the vacuum energy density $\rho_v$ is concentrated in the interval between $-10^{-29} g.cm^{-3}$ and $10^{-27} g.cm^{-3}$ under the condition that the measurement is made by an observer of our type. This result could be strengthened up to the desirable constraint $|\rho_v| < 10^{-29} g.cm^{-3}$ if it would be possible to prove that life on the Earth has an extragalactic origin. (Any communication between galaxies would be hampered by the exponential expansion of the universe with $\rho_v > 10^{-29} g.cm^{-3}$.) Unfortunately, at the moment we have no evidence in favour of this extravagant hypothesis. (Of course, one can turn this argument upside down and say that the vanishing of the cosmological constant can be considered as an argument in favour of the extragalactic origin of life. However, this will be true only if no other solution of the cosmological constant problem will be found.)

Another possibility is related to the results obtained in this paper. It may be not unreasonable to assume that the conditional probability to be born in the universe in a quantum state with a given value of $\rho_v$ is proportional to the total volume of those parts of the universe with this value of $\rho_v$ where life of our type can appear. From the investigation performed above it follows that in the quantum states with $|\rho_v| < 10^{-29} g.cm^{-3}$ the total volume of the parts of the universe where life can exist is much bigger than in the states with $|\rho_v| > 10^{-29} g.cm^{-3}$, and it may be especially big in the states with $|\rho_v| < 10^{-10^7} g.cm^{-3}$. If there is some relation between the total volume of the parts of the universe where life can exist and the volume where life of our type can appear, then one can argue that the probability that we measure the vacuum energy density $\rho_v$ in our quantum state of the universe $|\alpha_i>$ is peaked in the region $|\rho_v| < 10^{-10^7} g.cm^{-3}$. This would be more than enough to solve the cosmological constant problem.

Unfortunately, this line of reasoning, which may lead to a justification of the so-called Final Anthropic Principle [51] in its strongest form, is still far from being well elaborated. There are many obvious and less obvious objections which can be raised against the arguments suggested above. It is quite possible that the cosmological constant problem eventually will be solved in quite a different way, e.g. by some modification of ideas proposed in [52, 20, 23]. However, the efficiency of the Anthropic Principle for the determination of many other coupling constants in the context of the baby universe theory and the existence of a strong correlation between the value of the cosmological constant and the lifetime of mankind makes it very tempting to continue an investigation of the possibilities discussed above. At the very least, what we have obtained already is a surprising result that the existence of the vacuum energy density as small as $10^{-10^7} g.cm^{-3}$ would be fatal for the future evolution of life in our part of the universe. This is the first indication of the potential practical importance of recent attempts to investigate the cosmological constant problem. Unfortunately (or, may be, fortunately), it may take as much as $10^{5.10^6}$ years until the significance of the current work on this problem will be fully appreciated.

# References


[1] Ya.B. Zeldovich and M.Yu. Khlopov, Phys. Lett. **79B** (1978) 239.
J.P. Preskill, Phys. Rev. Lett. **43** (1979) 1365.

[2] Ya.B. Zeldovich, M.-Yu. Kobzarev and L.B. Okun, Phys. Lett. **50B** (1974) 340.
S. Parke and S.-Y. Pi, Phys. Lett. **121B** (1981) 313.
G. Lazarides, Q. Shafi and T.F. Walsh, Nucl. Phys. **B195**(1982) 157.
P. Sikivie, Phys. Rev. Lett. **48** (1982) 1156.

[3] J. Ellis, A.D. Linde and D.V. Nanopoulos, Phys. Lett. **116B** (1982) 59.
M.Yu. Khlopov and A.D. Linde, Phys. Lett. **138B** (1984) 265.

[4] G.D. Coughlan, W. Fischler, E.W. Kolb, S. Raby and G.G. Ross, Phys. Lett. **131B** (1983) 59.

[5] A.S. Goncharov, A.D. Linde and M.I. Vysotsky, Phys. Lett. **147B** (1984) 279.

[6] S.M. Barr, K. Choi and J.E. Kim, Nucl. Phys. **B283** (1987) 591.

[7] I. Antoniadis, C. Bachas, J. Ellis and D.V. Nanopoulos, Phys. Lett.**B 211** (1988) 393.

[8] E.B. Gliner, Sov. Phys. JETP **22** (1965) 378; Dokl. Akad. Nauk SSSR **192** (1970) 771.
E.B. Gliner and I.G. Dymnikova, Pis. Astron. Zh. **1** (1975) 7.
I.E. Gurevich, Astrophys. Space Sci. **38** (1975) 67.

[9] A.A. Starobinsky, JETP Lett. **30** (1979) 682; Phys. Lett. **91B** (1980) 99.

[10] A.H. Guth, Phys. Rev. **D23** (1981) 347.

[11] A.D. Linde, Phys. Lett. **108B** (1982); **114B** (1982) 431; **116B** (1982) 335, 340.
A. Albrecht and P.J. Steinhardt, Phys. Rev. Lett. **48** (1982) 1220.

[12] A.D. Linde, Phys. Lett. **129B** (1983) 177.

[13] A.D. Linde, Phys. Lett. **201B** (1988) 437.

[14] A.D. Linde, Phys. Lett. **175B** (1986) 395; Physica Scripta **T15** (1987) 169; Physics Today **40** (1987) 61.

[15] A.S. Goncharov, A.D. Linde and V.F. Mukhanov, Int. J. Mod. Phys. **A2** (1987) 561.

[16] S.W. Hawking, Phys. Lett. **B 195** (1987) 277; Phys. Rev. **D37** (1988) 904.

[17] G.V. Lavrelashvili, V.A. Rubakov and P.G. Tinyakov, JETP Lett. **46** (1987) 167; Nucl. Phys. **B299** (1988) 757.

[18] S.B. Giddings and A. Strominger, Nucl. Phys. **B306** (1988) 890.

[19] S. Coleman, Nucl. Phys.**B 307** (1988) 867.

[20] T. Banks, Nucl. Phys. **B 309** (1988) 493.

[21] S. Giddings and A. Strominger, Nucl. Phys.**B 307** (1988) 854.

[22] S.W. Hawking and R. Laflamme, Phys. Lett. **209B** (1988) 39.

[23] S. Coleman, Nucl. Phys. **B 310** (1989) 643.

[24] S. Giddings and A. Strominger, Santa Barbara preprint (1988).

[25] J. Polchinski, Phys. Lett., to be published.

[26] W. Fischler, I. Klebanov, J. Polchinski and L. Susskind, Stanford University preprint, (1989).

[27] A.D. Linde, Phys.Lett. **B 211** (1988) 29.

[28] A.D. Linde, Phys. Lett. **B 162** (1985) 281.

[29] A.D. Linde, **Particle Physics and Inflationary Cosmology** (Gordon and Breach, New York, in preparation).

[30] A.D. Sakharov, JETP Lett. 5 (1967) 24;
V.A. Kuzmin, JETP Lett. **12** (19709 335;
M. Yoshimura, Phys. Rev. Lett. **41** (1978) 281;
A.Yu. Ignatiev, N.V. Krasnikov, V.A. Kuzmin and A.N. Tavkhelidze, Phys. Lett. **76B** (1978) 436;
S. Weinberg, Phys. Rev. Lett. **42** (1979) 850.

[31] A.D. Dolgov and A.D. Linde, Phys. Lett. **116B** (1982) 329.

[32] V.A. Kuzmin, V.A. Rubakov and M.E. Shaposhnikov, Phys. Lett. **155B** (1985) 83.

[33] I. Affleck and M. Dine, Nucl. Phys. B **249** (1985) 361.

[34] A.D. Linde, Phys. Lett. **160B** (1985) 243.

[35] B. Campbell, J. Ellis, D.V. Nanopoulos and K.A. Olive, Mod. Phys. Lett. **A1** (1986) 389;
J. Ellis, K. Enquist, D.V. Nanopoulos and K.A. Olive, Phys. Lett. **191B** (1987) 343.

[36] M. Fukugita and T. Yanagida, Phys. Lett. **174B** (1986) 45;
S. Dimopoulos and L.J. Hall, Phys. Lett. **196B** (1987) 135;
A. Dannenberg, L.J. Hall, LBL preprint 23728 (1987);
A.G. Cohen and D.B. Kaplan, Phys. Lett. **199B** (1987) 251

[37] V.F. Mukhanov and G.V. Chibisov, JETP Lett. **33** (1981) 523.
S.W. Hawking, Phys. Lett. **115B** (1982) 339.
A.A. Starobinsky, Phys. Lett. **117B** (1982) 175.
A.H. Guth and S.-Y. Pi, Phys. Lett. **49** (1982) 1110.
J. Bardeen, P.J. Steinhardt and M. Turner, Phys. Rev. **D28** (1983) 679.
R. Brandenberger, Rev. Mod. Phys. **57** (1985) 1.
V.F. Mukhanov, JETP Lett. **41** (1985) 493;
V.F. Mukhanov, L.A. Kofman and D.Yu. Pogosyan, Phys. Lett. **193B** (1987) 427.

[38] A.D. Linde and M.I. Zelnikov, Phys. Lett.**B 215** (1988) 59.

[39] J.A. Wheeler, in: **Relativity, Groups and Topology** eds. C.M. DeWitt and J.A. Wheeler (Benjamin, New York, 1968);
B.S. DeWitt, Phys. Rev. **160** (1967) 1113.

[40] J.B. Hartle and S.W. Hawking, Phys. Rev. **D28** (1983) 2960.

[41] A.D. Linde, JETP **60** (1984) 211; Lett. Nuovo Cim. **39** (1984) 401.

[42] Ya.B. Zeldovich and A.A. Starobinsky, Sov. Astron. Lett. **10** (1984) 135.

[43] V.A. Rubakov, Phys. Lett. **148B** (1984) 280.

[44] A. Vilenkin, Phys. Rev. **D30** (1984) 549.

[45] A.A. Starobinsky, in: **Fundamental Interactions** (MGPI Press, Moscow, 1984), p. 55; in: **Current Topics in Field Theory, Quantum Gravity and Strings**, Lecture Notes in Physics, eds. H.J. de Vega and N. Sanchez (Springer, Heidelberg 1986) **206**, p. 107.

[46] A.S. Goncharov and A.D. Linde, Sov. J. Part. Nucl. **17** (1986) 369;
S.J. Rey, Nucl. Phys. **B284** (1987) 706.

[47] A.Vilenkin, Phys. Rev. **D27** (1983) 2848.

[48] S. Weinberg, Phys. Rev. Lett. **59** (1987) 2607; Rev. Mod. Phys. **61** (1989) 1.

[49] H. Everett, Rev. Mod. Phys.,**29** (1957) 454; B. S. DeWitt and N. Graham, The Many-Worlds Interpretation of Quantum Mechanics (Princeton University Press, Princeton, 1973).

[50] I. L. Rosental,**Big Bang, Big Bounce: How Particles and Fields Drive Cosmic Evolution** (Springer, Berlin, 1988).

[51] J.D. Barrow and F.J. Tipler, **The Anthropic Cosmological Principle** (Oxford University Press, Oxford, 1986).

[52] A.D. Linde, Phys. Lett. **200B** (1988) 272.

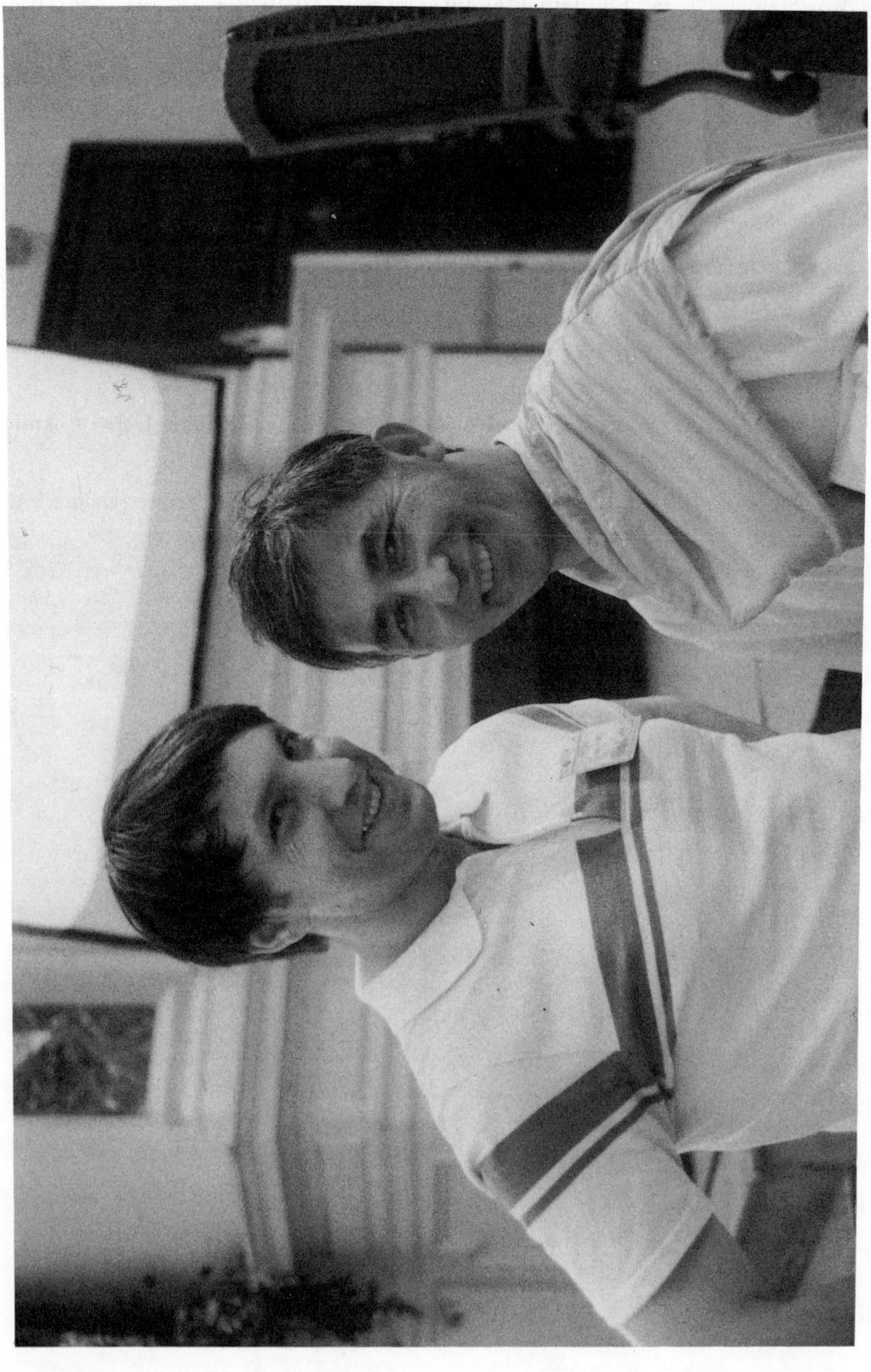

*Andrei Linde and Jean Tran Thanh Van.*

# Conclusions and Perspectives

*Lincoln Wolfenstein.*

# CP VIOLATION: WHERE WE STAND
# 25 YEARS LATER

LINCOLN WOLFENSTEIN

CARNEGIE MELLON UNIVERSITY

PITTSBURGH, PA 15213

ABSTRACT

Present information on CP violation is summarized. Experimental results are limited to the two parameters $\varepsilon$ and $\varepsilon'$ in $K^\circ$ decay. Results are consistent with the standard Kobayashi Maskawa model. The most clearcut prediction of that model is large CP violation in certain $B^\circ$ decays. CP violation phenomena provide interesting possibilities for exploring physics beyond the standard model. As an example the electric dipole moment of the neutron is discussed.

## 1. Introduction

The fundamental physics of the twentieth century has been dominated by symmetry principles. At this conference we have been concerned with three discrete symmetries:

P, symmetry with respect to inversion of spatial coordinates

T, symmetry with respect to reversal of time

C, symmetry with respect to exchange of particles and anti-particles.

These symmetries did not arise from theoretical postulates; rather they were discovered as summarized in the elegant introductory lecture of Professor Pais. The P and T symmetries were discovered as symmetries of the Schrodinger equation explaining observed regularities in atomic physics. C, P, and T were all found to be symmetries of quantum electrodynamics, the quantum version of the Maxwell-Einstein theory. If any postulate was involved, it was the implicit postulate that these symmetries held for the new interactions, strong and weak, needed for nuclear physics.

Particle physics of the last thirty-five years has been dominated by something more intriguing and perhaps more beautiful than perfect symmetry: it is called broken symmetry. There is a subtlety and a fascination about a broken symmetry that can keep one trapped for years--even twenty-five years.

Broken symmetry provides the foundation of the standard electroweak theory. The violation of parity gives the theory its chiral (V-A) structure while the breaking of the gauge symmetry allows for the masses of the gauge bosons and of the quarks and leptons. The starting point was the discovery of P violation following the observation by Lee and Yang that the P symmetry had never been tested in the weak interaction. The (V-A) theory that soon emerged involved a maximal violation of the P and C symmetries with CP remaining as a good symmetry. Very general principles of quantum field theory yield the result that CPT is a good symmetry so that CP symmetry implies T symmetry.

The discovery we honor at this meeting demonstrated that CP was not a good symmetry in the decay of the K° meson. I shall not go into the details of that discovery which is well-known to all of you. What I shall try to do in this talk is first to summarize what we know experimentally about CP violation; unfortunately, that will not take very long. Then I will explain how we look theoretically at CP violation from the point of view of gauge theories. Finally, I will discuss the experimental frontier. I will not try to summarize the many interesting talks presented here.

## 2. K° Decay

Shortly after the discovery of CP violation in the decay $K^\circ \to 2\pi$ I wrote a paper suggesting that this could be explained by a new very weak interaction, usually referred to now as the superweak interaction. The idea was that because the mass difference between $K_L$ and $K_S$ was so small, being second-order weak, the mass matrix was extremely sensitive to a new interaction provided it allowed $\Delta S = 2$ and violated CP. Thus the only effect of this CP violation would be in mixing so that

$$K_L = (K_2 + \varepsilon\, K_1)/(1 + |\varepsilon|^2)$$

where $K_2$ is CP-odd and $K_1$ is CP-even. After I submitted my paper to Physical Review Letters I received a reluctant acceptance from the referee who objected that my paper made no predictions. What he really meant was that the superweak theory predicted nothing; that is, nothing else would be found beyond the parameter $\epsilon$ in the K° system. Unfortunately, this prediction has proven all too true.

The parameters that measure CP violation in the K° system are:

$$\eta_{+-} = \frac{A(K_L \to \pi^+\pi^-)}{A(K_S \to \pi^+\pi^-)} = |\eta_{+-}| \, e^{i\varphi_{+-}}$$

$$\eta_{00} = \frac{A(K_L \to \pi^\circ\pi^\circ)}{A(K_S \to \pi^\circ\pi^\circ)} = |\eta_{00}| \, e^{i\varphi_{00}}$$

$$\delta = \frac{\Gamma(K_L \to \ell^+ \nu \pi^-) - \Gamma(K_L \to \ell^- \bar{\nu} \pi^+)}{\text{sum}}$$

The experimental results as they appeared a couple of years ago could be summarized (numerical values are taken from the Particle Data group).

$$|\eta_{+-}| = (2.27 \pm .02) \times 10^{-3}$$

$$\varphi_{+-} = (44.6 \pm 1.2)^\circ$$

$$|\eta_{00}| \approx |\eta_{+-}| \text{ within } 1\,\%$$

$$(\varphi_{00} - \varphi_{+-}) < 20^\circ$$

$$\delta = (3.30 \pm 0.12) \times 10^{-3}$$

To analyze these results we use the Wu-Yang variables $\varepsilon$ and $\varepsilon'$:

$$\eta_{+-} = \varepsilon + \varepsilon' \tag{1a}$$

$$\eta_{00} = \varepsilon - 2\varepsilon' \tag{1b}$$

Then from CPT and unitarity one can show

$$\phi(\varepsilon) \simeq \tan^{-1} (2\, \Delta m/\Delta\Gamma) = (43.67 \pm 0.14)^\circ \tag{2a}$$

$$\phi(\varepsilon') = \delta_2 - \delta_0 + \pi/2 \approx 48 \pm 8^\circ \tag{2b}$$

where $\delta_I$ is the final-state $\pi$-$\pi$ phase shift with isospin I. While the $\pi$-$\pi$ phase shifts are not well known the main point is that $\phi(\varepsilon')$ is close to $\phi(\varepsilon)$; exactly how close doesn't really matter. Finally, if we also assume the $\Delta Q = \Delta S$ rule we find

$$\delta = 2 \,\mathrm{Re}\, \varepsilon = 2\, |\varepsilon| \cos \phi(\varepsilon) \tag{3}$$

With these very minimal assumptions the five experimental numbers are reduced to two: $|\varepsilon|$ and $\varepsilon'/\varepsilon$. Since the latter is very small, $|\varepsilon|$ is essentially given by $|\eta_{+-}|$.

There exists great interest in $\varepsilon'/\varepsilon$. One reason is that this would represent the first new independent non-zero measure of CP violation beyond the value $|\eta_{+-}|$ measured in 1964. Secondly a non-zero value would rule out the superweak model and show that CP violation exists in weak decay amplitudes and is not confined to the mass matrix. Last year the CERN group published their results of the NA 31 experiment presented here by Peach giving

$$\varepsilon'/\varepsilon = (33 \pm 11) \times 10^{-4}$$

What is really measured is $|\eta_{00}/\eta_{+-}|$ but we interpret it as $\varepsilon'/\varepsilon$ using the phases given by Eqs.(2); note that the sign is significant indicating $|\eta_{00}| < |\eta_{+-}|$. As discussed in this conference the group has additional data being analyzed and is now running with some improvements. They expect to end up with three times the original data and with an error reduced to $7 \times 10^{-4}$.

A very different experiment at Fermilab E731 has also been measuring $|\eta_{00}/\eta_{+-}|$. Winstein gave to this conference a very preliminary result

$$\varepsilon'/\varepsilon = [(-10 \text{ to } 10) \pm 15] \times 10^{-4}$$

This result does not confirm the non-zero result of NA 31 nor does it significantly disagree with it. The present analysis of E731 includes only 20% of the data; when they are finished they expect to have reduced the error also to $7 \times 10^{-4}$.

When these two experiments are done each will have measured a CP-violating parameter $\varepsilon'$ to an accuracy of $\pm 1.5 \times 10^{-6}$. This is truly a remarkable achievement and both groups deserve high praise for their ingenuity and perseverance. What are we to conclude from these experiments? The most important conclusion is that they must be continued to still higher accuracy. The point is not to find the exact value of $\varepsilon'$; the point is to make absolutely sure that $\varepsilon'$ is non-zero. The NA 31 experiment has wounded the superweak theory. The time has come to really kill it. I remember in 1968 in Moscow discussing the death of the superweak theory based on an experiment (kindly omitted by Jack Steinberger in his review) that gave $\eta_{00} = 2\ \eta_{+-}$. The superweak theory does not die easily.

There are other experiments discussed in this conference that measure the same physics as $\varepsilon'$, that is CP violation in $\Delta S = 1$ decay amplitudes. Among these are CP violation in hyperon decays and $K^\circ \to 3\pi$ decays. It seems impossible that these experiments can compete in accuracy with $\varepsilon'$. A possible exception is the rare decay $K_L \to \pi^\circ + e^+ + e^-$ discussed here by Gilman. Analyses suggest that the branching ratio of this decay may be between $10^{-11}$ and $10^{-12}$ and that in the long-term future such a branching ratio can be measured. The problem is that there are three contributions to the decay:

(1) a CP-violating $\Delta S = 1$ amplitude like $\varepsilon'$, (2) a CP-violating effect from $K_S \to \pi^\circ + e^+ + e^-$, proportional to $\varepsilon$, and (3) a CP-conserving amplitude associated with two-photon exchange. These are expected to be roughly equal in order of magnitude. To identify the $\varepsilon'$ type term it is necessary to make a detailed analysis of this very rare decay; for example to study the charge asymmetry as discussed by Sehgal. Clearly, this is a program that will take a very long time.

Returning to the $K \to 2\pi$ decays there is the question of possible CPT violation. The most interesting place to look for such a violation is in the mass matrix because of its sensitivity to small effects. In the $K_1 - K_2$ representation the mass matrix may be written

$$\begin{bmatrix} m_1 & im' + \delta' \\ -im' + \delta' & m_2 \end{bmatrix}$$

where $m'$ is the usual CP-odd term which is also odd under T and even under CPT, whereas $\delta$ is CP-odd and T even and so violates CPT. The mixing parameter $\varepsilon$ is then given by

$$\varepsilon = \frac{im' + \delta'}{\Delta m + i\,\Delta\Gamma/2} \tag{4}$$

The effect of $\delta'$ is to shift the phase of $\varepsilon$ away from the value given by Eq (2a). The experimental result for $\eta_{+-}$ says that $\delta'$ must be small compared to $m'$ which itself is $2 \times 10^{-3}$ times the very small value $\Delta m$. It is easy to see in the $K^\circ$-$\bar{K}^\circ$ representation that the limit on $\delta'$ gives a limit on the mass difference

$$\left| m(K^\circ) - m(\bar{K}^\circ) \right| < 3 \times 10^{-10} \text{ ev}$$

Because this is the most accurate test of CPT invariance it would be good to improve it. Note that the result reported here by Gibson for NA 31, $\varphi_{+-} = 46.8 \pm 1.4 \pm 0.6 \pm 0.8$ (where the last error comes from the uncertainty in $\tau_S$), is almost 2 s.d. away from Eq. (2a). Since direct measurements of $\phi_{+-}$ are very difficult it is possible that ultimately the most accurate value will come from measurements of the charge asymmetry $\delta$ via Eq. (3).

Another test of CPT is the phase difference $\phi_{\infty} - \phi_{+-}$. Given the small value of $\varepsilon'$ it follows from Eq. (2) that this difference should be extremely close to zero. A non-zero value would require that the phase of $\varepsilon'$ be very different from that given by Eq. (2b) so that $\varepsilon'$ would have a large CPT-violating piece. It was gratifying at this conference to have two new relatively accurate results

NA 31 $\quad \phi_{\infty} - \phi_{+-} = 0.3° \pm 2.6° \pm 1.1°$

E731 $\quad \phi_{\infty} - \phi_{+-} = 2.8° \pm 2° \pm 0.5°$

Thus there is no sign of CPT violation. A dedicated experiment E773 at Fermilab discussed by Gollin aims at reducing the error to ± 0.5°.

## 3. Gauge Theories

Let me now turn to CP violation in gauge theories. All gauge theories have the following schematic Lagrangean

$$L = G\,(W,\psi,\varphi) + H(\varphi) + Y(\psi,\varphi)$$

where W are gauge boson fields, $\varphi$ are Higgs boson fields, and $\psi$ are fermion fields. G stands for the kinetic energy terms plus all the interactions involving W required by the gauge principle. H is the Higgs potential. Y is the Yukawa interaction between the Higgs bosons and the fermions. The major point is that G is automatically CP invariant just because of its gauge form; a pure gauge theory does not have CP violation. (A possible exception is the $\Theta$ term in QCD which we discuss later.) As a result there are three possible sources of CP violation:

(1) Complex coefficients in H

(2) Complex coefficients in Y

(3) L itself is CP invariant but the vacuum is not; that is, there are relative complex factors in the vacuum expectation values of the $\varphi$'s; this is called spontaneous CP violation.

Thus CP violation is tied to the Higgs sector of the theory, the part of the standard model which we understand and trust the least. One of the reasons CP violation is so important from this point of view is because it is so tied in with the aspect of gauge theories that we are striving to really understand. This may also be the reason progress on this subject has been so slow.

In the standard electroweak theory as originally formulated with two generations of quarks and one Higgs doublet there was no possibility of CP violation. With one Higgs $H(\varphi)$ is too simple to allow for complex coefficents or complex vacuum expectation values. The complex coefficients in Y eventually show up as complex terms in the quark mass matrix and hence as phases in the quark mixing matrix. With only two generations these phases can be removed by redefining the phase of quark fields. Thus, something had to be added. The simplest thing turns out to be the addition of a third generation of quarks, as proposed by Kobayashi and Maskawa (KM). With the discovery of the b quark, this has become the standard model of CP violation. Meanwhile many alternative gauge models were discussed, some of them superweak. Among the most-discussed non-superweak models have been (1) the Weinberg model containing three doublets of Higgs bosons in which CP violation is associated with Higgs boson exchange and (2) the $SU(2)_L \times SU(2)_R$ model involving additional gauge bosons the exchange of which is the source of CP violation.

It is customary to ask which of these various models is <u>the model</u> of CP violation. It is possibly more reasonable to think about the role of different mechanisms of CP violation. Since there are three generations and since we know CP is violated, the KM mechanism will be operative unless some additional symmetry makes the KM phase vanish. On the other hand it may turn out that the magnitudes of the parameters of the quark-mixing (CKM) matrix are not large enough to explain observed values of $\varepsilon$ and/or $\varepsilon'$. In that case there must be contributions from other mechanisms. If there is new physics beyond the standard model, in accordance with our hopes and dreams, then we expect this new physics to show CP violation also and provide additional CP-violating mechanisms. In particular, for those CP-violating observables which happen to be small in the standard model, the exploration of CP violation may prove to be an excellent way to probe physics beyond the standard model. I'll return to this point when I discuss the electric dipole moment of the neutron.

Let me turn now to the standard KM model. The effective weak interaction of the quarks is given by

$$H_{wk} \sim \bar{U}_{Lj} U_{ji} \gamma^{\lambda} D_{Li} W_{\lambda}$$

where $U = (u,c,t)$, $D = (d,s,b)$; and $U_{ji}$ is the CKM quark-mixing matrix; for example, $U_{us} = \sin\Theta_c$, where $\Theta_c$ is the Cabibbo angle. It turns out experimentally that there is a hierarchy among the $U_{ji}$ which allows the matrix to be conveniently expanded in powers of $\lambda \equiv \sin\Theta_c$. To order $\lambda^3$

$$U = \begin{bmatrix} 1-\frac{\lambda^2}{2} & \lambda & A\lambda^3(\rho-i\eta) \\ -\lambda & 1-\frac{\lambda^2}{2} & A\lambda^2 \\ A\lambda^3(1-\rho-i\eta) & -A\lambda^2 & 1 \end{bmatrix}$$

From experiment, as discussed in the talk of Harari, I put

$$\lambda = 0.22$$
$$A = 1 \pm 0.2$$
$$(\rho^2+\eta^2)^{\frac{1}{2}} \le 0.7 \qquad (5)$$

In the notation of Harari what I call $(\rho-i\eta)$ is written $(q/\lambda)e^{-i\delta}$. To fit the experimental result for $\varepsilon$ one needs a value of $\eta$ of the order of 0.3 within a factor of two or so, although this value goes down as $m_t$ goes up.

There are four features or predictions of the KM model I would like to emphasize:

(1) The value of $\varepsilon$ is of order $A^2\lambda^4 = 2.3 \times 10^{-3}$. This follows from Eq.(4) (with $\delta' = 0$) because the CP-violating part of the box diagram giving $m'$ is necessarily proportional to $A^2\lambda^6\eta$ (this corresponds to the J parameter of Jarlskog) whereas the mass difference $\Delta m$ is proportional to $\lambda^2$. Thus the small value of $\varepsilon$ in the standard model arises from the hierarchy in magnitudes among the elements of the CKM matrix.

(2) Most of the CP violation is in the mass matrix. The CP-violating amplitude needed for a non-zero value of $\varepsilon'$ comes from penguin diagrams. The theoretical value is very uncertain because of problems in "hadronic engineering"

$$(\varepsilon'/\varepsilon)_{th} \sim 6 \times 10^{-4} \text{ to } 6 \times 10^{-3}$$

The theoretical value goes down as $m_t$ goes up.

(3) The CP-violating phase is not small. Given the estimate of $\eta$ and the limit on $\rho$ from Eq. (5) the phase $\delta$ of $U_{ub}$ is likely to be greater than 15°.

(4) There is a unique place to look for large CP violating effects, namely in the $B°(b\bar{d})$ system. The reason is that, using the same calculation for $\varepsilon$ as in the K° system, $m'$ is still of order $A^2\lambda^6\eta$ but now $\Delta m$ is also of the order $\lambda^6$. Thus $\varepsilon$ for the B° system is proportional to $\lambda^0$ instead of $\lambda^4$. The reason for $\lambda^6$ is that b and d are extreme generations so that any leg of the box that connects them (whether via u,c, or t)--has a factor $\lambda^3$.

## 4. B° Decay

The experiment that would most clearly show CP violation in the B° system is a close analogue of the 1964 K° experiment. It involves B° decay into a CP eigenstate, such as $\psi\, K_S$ or $\pi\pi$. The difference is that the eigenstates of the $B^\circ$-$\bar{B}^\circ$ system (the analogues of $K_L$ and $K_S$) are expected to have almost the same lifetime so they cannot be separated. Instead one compares the decays from a $B^\circ$ beam to that from a $\bar{B}^\circ$. In the ideal experiment one produces $B\bar{B}$ pairs and at t = 0 tags a $B(\bar{B})$ on one side so as to assure that the other side starts out as a $\bar{B}(B)$. One then finds that

$$r = \frac{[B^\circ \to \psi Ks] - [\bar{B}^\circ \to \psi Ks]}{\text{Sum}}$$

$$= \frac{\Delta m/\Gamma}{1 + (\Delta m/\Gamma)^2} \sin 2\Theta_{td}$$

$$= 0.47 \sin 2\Theta_{td} = 0.94\ \eta\ (1-\rho)/[(1-\rho)^2+\eta^2] \qquad (6)$$

where we have inserted the experimental result $(\Delta m/\Gamma) \simeq 0.7$. The result in Eq.(6) is reduced from the maximum possible only by the factor $\sin 2\Theta_{td}$, for which a reasonable guess is 0.4. Note that the result depends only on the entries in the CKM matrix with no "hadronic engineering" involved.

One might ask whether this effect is due to CP violation in $B$-$\bar{B}$ mixing or in the decay amplitude. The answer depends almost entirely on the phase convention. In the phase convention we have chosen this effect (for final states like $\psi\ K_S$) appears to be entirely due to the difference between the transition (B to $\bar{B}$) and that of ($\bar{B}$ to B). It is possible to have this type of CP violation in a superweak model although there is no reason to expect a large value of r as there is in the KM model. In order to identify unambiguously CP violation in the decay amplitude it is necessary to measure values of r for two different final states corresponding to different quark transitions such as $\psi K_S$ and $\pi\pi$. More details are given in the talk of Sanda.

In spite of the large values expected for r the detection of CP violation in the B system is very difficult because of the low branching ratios for the final states of interest. Estimates indicate $10^7$ to $10^9$ $B\bar{B}$ pairs are needed. Note the number of pairs needed goes as $r^2$ and r is uncertain by at least a factor 5. Factors of 10 are easy for theoreticians to throw around but they have an enormous effect on the requirements for accelerators, not to mention the power bills.

We heard here about the intensive studies carried out during the last year directed at the design of a B meson factory. The obvious first idea is $e^+e^-$ colliding at the upsilon 4s state, CESR upgraded by a further factor of 20. There is, however, a problem. It is impossible to tag close to t = 0 because the slow moving B's decay too soon. Thus one must integrate over the tagging time, which corresponds to integrating over the relative time $(t-\bar{t})$ where t is the time of the "tagging" decay and $\bar{t}$ is the time the other B decays. At the 4s the state produced is $B\bar{B}-\bar{B}B$ (L = 1, C = −, corresponding to the virtual photon). As a result the observable is an odd function of $(t-\bar{t})$ and vanishes upon integration. Thus there has been a great interest in an asymmetric collider producing a moving 4s by colliding 7 GeV $e^-$ with 4 GeV $e^+$. It is gratifying to see the accelerator designers bowing before the power of a minus sign in a quantum-mechanical equation. Larger numbers of $B\bar{B}$ pairs could be produced at a hadron collider but signal to background is much worse; this possibility is also under active study.

## 5. Electric Dipole Moment of the Neutron

Let me turn now to direct tests of time-reversal violation. Of these the most sensitive appears to be the search for the electric dipole moment of the neutron $D_n$. The effective interaction $D_n \vec{\sigma} \cdot \vec{E}$ violates both the P and T symmetries. An experimental value reported some time ago by the Leningrad group was $D_n = -14 \pm 6 \times 10^{-26}$ e - cm. The latest result from the Grenoble group reported here by K. Smith is

$$D_n = (-6 \pm 4 \pm 2) \times 10^{-26} \text{ e-cm}$$

where the first error is a carefully-evaluated systematic error and the second statistical. At the present time the experiments must be interpreted as giving an upper limit of the order $2 \times 10^{-25}$ but planned improvements should explore values down to $10^{-26}$.

What makes these experiments so important is that the values that are being explored are qualitatively comparable to $\varepsilon'$ in their sensitivity to CP violation of the milliweak variety. As a rough guide one can write

$$D_n/\mu_n \sim 10^{-6} \times \varepsilon' \tag{7}$$

where $\mu_n$ is the magnetic moment of the neutron, $10^{-6}$ is the price paid for P violation, and $\varepsilon'$ the additional price for T violation. This suggests that present limits on $D_n$ correspond to $\varepsilon' \approx 10^{-5}$ and prospective experiments to $10^{-6}$. In fact various milliweak models (such as the Weinberg three Higgs model and left-right models) predict values of $D_n$ of the order of or greater than $10^{-26}$ e-cm together with $\varepsilon'/\varepsilon$ between $10^{-3}$ and $10^{-2}$. The actual results are sensitive to choices of parameters and hadronic engineering. In contrast, in the KM model $D_n$ vanishes in the lowest order; to estimate $D_n$ it is necessary to go through a virtual $|S| = 1$ state using the weak interaction twice, once with CP violation and once without. The result is a value of $D_n$ of $10^{-31}$ to $10^{-32}$ e-cm, much too small to be detected.

It is of interest to ask what conclusion one might reach if a non-zero value of $D_n$ in the range $10^{-25}$ to $10^{-26}$ were established. There is a tendency to say that this would show the KM model is wrong and one of the other models referred to is right. I prefer a more conservative approach.

Let me assume

(1) the standard model is a good effective theory for "low energy" physics but we expect there is new physics at a higher, perhaps much higher, mass scale

(2) the value of $\epsilon'$ when it is finally determined can be explained by the KM mechanism of CP violation.

We then write as the effective Lagrangian

$$L = L_{st} + L_{\Theta} + L'$$

where $L_{st}$ is the standard SU(3) × SU(2) × U(1) model and we discuss the other two terms below.

The term $L_{\Theta}$ represents the strong CP violation discussed by Barr and Sikivie. It is proportional to the parameter $\Theta_{QCD}$. A value of $D_n$ between $10^{-25}$ and $10^{-26}$ can be explained by a value of $\Theta_{QCD}$ between $10^{-9}$ and $10^{-10}$. The small value of $\Theta_{QCD}$ mandated by the present limit on $D_n$ has led to various schemes discussed by Sikivie to make its value essentially zero. On the other hand we here assume $\Theta_{QCD}$ is an effective parameter which may have a small non-zero value as a result of new physics at a higher mass scale. (Note that $L_{st}$ has some small parameters ($m_e/m_t < 10^{-5}$) that we hope will eventually be explained.) An interesting example is the model of Nelson. She assumes a model involving heavy fermions in which CP is violated spontaneously so that $\Theta_{QCD}$ is calculable. The effective theory at low energy is just the standard KM model plus $L_{\Theta}$ with a value of $\Theta_{QCD}$ which is small but greater than $10^{-11}$. In fact she invented the model with the aim of deriving the

KM theory and only afterward realized its relation to the strong CP problem. From a phenomenological point of view the only consequence of $L_\Theta$ is the electric dipole moment of the neutron; it is irrelevant for flavor-changing processes such as K decay.

The term L´ represents a sum of effective interactions proportional to $M^{-n}$ where M is the mass scale of new physics. A simple example of such a term is

$$(1/M_{SW}^2)(\bar{s}\ d\ \bar{s}\ de^{i\delta} + h.c.)$$

This, of course, is just the superweak interaction which would make a significant contribution to $\epsilon$ if $M_{SW}$ is of the order $10^4$ to $10^5$ times $m_W$. Of interest in the present context is a term

$$\left(\frac{e}{4\pi^2}\right)\left(1/M_D^2\right)\bar{d}_L\ \sigma_{\mu\nu}\ \gamma_5 d_R\ \phi\ F^{\mu\nu}$$

where $\phi$ is the Higgs doublet field required to make the term SU(2) invariant. When $\phi$ acquires the vacuum expectation value v we get

$$\frac{D_n}{\mu_n} \simeq \frac{vM_n}{\pi^2 M_D^2}$$

This term may be thought of as coming from a loop diagram at the mass scale $M_D$. (A detailed model is left as an exercise.) Values of $M_D$ of the order $10^4\ m_W$ could then be responsible for values of $D_n$ now being explored.

The main point of this approach is to indicate that while a value of $D_n$ of the order $10^{-25}$ to $10^{-26}$ implies new physics it may be that this new physics is at a much higher scale, far beyond the SSC, for example. On the other hand, in this way of looking at it, there is nothing special about the values $10^{-25}$ to $10^{-26}$ as there would be if

$D_n$ were related to $\varepsilon'$ as indicated by Eq.(7).

## 6. Conclusion

(1) C,P, and T are exact symmetries of unbroken gauge theories such as QCD and QED. None of these symmetries is required in the standard broken electroweak gauge theory. In particular CP and T violation are associated with the Higgs interactions.

(2) The only independent experimental numbers that measure CP violation (assuming CPT invariance) are $\varepsilon$ and $\varepsilon'$ in K° decay. The CERN measurement of $\varepsilon'$ (if confirmed) would disprove the superweak theory and is consistent with the standard Kobayashi-Maskawa (KM) model.

(3) The most clearcut prediction of the KM model is large CP violation in the B° system.

(4) The most sensitive test for T violation is a non-zero value of $D_n$, the electric dipole moment of the neutron. A non-zero value of $D_n$ in the present experiments would be a signal of new physics, but this might be at a very high mass scale.

(5) Given our limited experimental information and the lack of any real theoretical insight it is quite likely that the next step forward in this area will come from an unexpected direction.

It is natural to be disappointed that 25 years later no signal of CP or T violation has been found in any system other than K°. There is a natural frustration in the feeling that the fundamental origin of CP violation is still a mystery. There are among my theoretical colleagues those who wish to discover the theory of everything before they are too old to understand it. My own sentiments are closer to those of Freeman Dyson in his most recent book Infinite in All Directions. The title implies that as we explore further into the world of elementary particles or of cosmology we will continue always to make new discoveries. Therefore, I would like to say, and I think I can speak for Jim Cronin, Val Fitch, Jim Christenson, and Rene Turlay, we rejoice in leaving this intriguing problem of CP for the younger generation to solve. We urge the younger physicists here to attack the problem with all the experimental ingenuity and theoretical power they can muster. We hope you enjoy it as much as we have!

There is left now only the most pleasant task. On behalf of all the participants I want to thank Tran Thanh Van and all the others who have made this such a wonderful conference. Merci beaucoup. Thank you very much.

This work was supported in part by the U. S. Department of Energy contract number DE-AC02-76ER03066.

*Reception at the Château de Chambord.*

# LIST OF PARTICIPANTS

| | |
|---|---|
| ABASHIAN Alexander | Virginia Polytech. and State University<br>Dept of Physics<br>BLACKSBURG VA 24061<br>USA |
| ADAIR Robert | Yale University<br>Dept of Astronomy<br>NEW HAVEN CT 06511<br>USA |
| ALBRIGHT Carl H. | Northern Illinois University<br>Dept. of Physics<br>DEKALB IL 60115<br>USA |
| ASTBURY Alan | TRIUMF<br><br>VANCOUVER BC V6T 2A3<br>CANADA |
| AUBERT Bernard | IN2P3 - LAPP<br>Lab. de Physique des Particules<br>F- 74019 ANNECY LE VIEUX Cedex<br>FRANCE |
| AUDOUZE Jean | Institut d'Astrophysique<br>98 Bd Arago<br>F- 75014 PARIS<br>FRANCE |
| AUGE Etienne | Université Paris Sud<br>LAL<br>F- 914O5 ORSAY Cedex<br>FRANCE |
| BACKENSTOSS Gerhard | Universität Basel<br>Physik. Inst. Experimentalle Kernphysik<br>CH- 4056 BASEL<br>SWITZERLAND |
| BADIER Jean | Ecole Polytechnique<br>LPNHE<br>F- 91128 PALAISEAU Cedex<br>FRANCE |
| BARR Giles | CERN<br>Division EP<br>CH- 1211 GENEVE 23<br>SWITZERLAND |
| BARR Stephen | University of Delaware<br>Bartol Research Foundation<br>NEWARK DE 19716<br>USA |

| | |
|---|---|
| BASSOMPIERRE Gabriel | IN2P3 - LAPP<br>Lab. de Physique des Particules<br>F- 74019 ANNECY LE VIEUX Cedex<br>FRANCE |
| BECH NIELSEN Holger | Niels Bohr Institute<br><br>DK- 2100 COPENHAGEN O<br>DENMARK |
| BEREZHIANI Zurab | Academy of Science of the Georgian SSR<br>Institute of Physics<br>380 077 TBILISI<br>U S S R |
| BERNABEU José | Universidad de Valencia<br>Dept de Fisica Teorica<br>E- VALENCIA<br>SPAIN |
| BILENKY Samoti M. | Joint Inst. for Nuclear Research<br>Dubna Head Post Office<br>101 000 MOSCOW<br>U S S R |
| BISELLO Dario | Universita di Padova<br>INFN - Sezione di Padova<br>I- 35100 PADOVA<br>ITALY |
| BISELLO Dario | Universita di Padova<br>INFN - Sezione di Padova<br>I- 35100 PADOVA<br>ITALY |
| BLOCH Philippe | CERN<br>Division EP<br>CH- 1211 GENEVE 23<br>SWITZERLAND |
| BOCK Greg | FERMILAB<br>Dept of Physics<br>BATAVIA IL 60510<br>USA |
| BORDES Gisèle | Collège de France<br>Lab. de Physique Corpusculaire<br>F- 75231 PARIS Cedex 05<br>FRANCE |
| BOTELLA Francisco J. | Universidad de Valencia<br>Fac. de Ciencias Fisicas<br>E- BURJASOT Valencia<br>SPAIN |

| | |
|---|---|
| BRISSON Violette | Ecole Polytechnique<br>LPNHE<br>F- 91128 PALAISEAU Cedex<br>FRANCE |
| BURGESS Clifford | McGill University<br>Rutherford Physics Bldg - High Energy<br>MONTREAL Québec H3A 2T8<br>CANADA |
| CHARDIN Gabriel | CEN Saclay<br>DPhPE<br>F- 91191 GIF sur YVETTE Cedex<br>FRANCE |
| CHRISTENSON James | FERMILAB<br>Dept of Physics<br>BATAVIA IL 60510<br>USA |
| CLINE David | University of California<br>Dept of Physics<br>LOS ANGELES CA 90024<br>USA |
| COC Alain | Laboratoire René Bernas<br>CSNSM - Bâtiment 108<br>F- 91406 ORSAY Cedex<br>FRANCE |
| CRONIN James W. | University of Chicago<br>Enrico Fermi Inst. - Dept of Physics<br>CHICAGO IL 60637<br>USA |
| CUNDY Donald D. | CERN<br>Division EP<br>CH- 1211 GENEVE 23<br>SWITZERLAND |
| D'ESPAGNAT Bernard | Université Paris Sud<br>LPTHE<br>F- 91405 ORSAY Cedex<br>FRANCE |
| DALITZ Richard | University of Oxford<br>Dept of Theoretical Physics<br>OXFORD OX1 3NP<br>UNITED KINGDOM |
| DATTA Amitava | University of Jadavpur<br>High Energy Physics Division<br>700 032 CALCUTTA<br>INDIA |

| | |
|---|---|
| DAVIER Michel | Université Paris Sud<br>LAL<br>F- 914O5 ORSAY Cedex<br>FRANCE |
| DE RAFAEL Edouardo | CNRS<br>Centre de Physique Théorique<br>F- 13288 MARSEILLE Cedex 9<br>FRANCE |
| DEBU Pascal | CEN Saclay<br>DPhPE<br>F- 91191 GIF sur YVETTE Cedex<br>FRANCE |
| DOBLE Niels | CERN<br>Division EP<br>CH- 1211 GENEVE 23<br>SWITZERLAND |
| DOMINGUEZ Cesareo A. | Cape Town University<br>Dept of Astronomy<br>RONDEBOSCH CAPE 7700<br>SOUTH AFRICA |
| DRELL Persis | Cornell University<br>Newman Lab. of Nuclear Studies<br>ITHACA NY 14853<br>USA |
| DUCLOS Jean | DPhNHE<br>Orme des Merisiers<br>F- 91191 GIF-sur-YVETTE Cedex<br>FRANCE |
| EEG Jan-Olav | University of Oslo<br>Dept. of Physics<br>N- 1000 OSLO 3<br>NORWAY |
| EISELE Franz | DESY<br>Notkestrasse 85<br>D- 2000 HAMBURG 52<br>FEDERAL REP. OF GERMANY |
| ERNE Frederik | NIKHEF-H<br>Kruislaan 409<br>NL- 1009 DB AMSTERDAM<br>NETHERLANDS |
| ERNWEIN Jean | CEN Saclay<br>DPhPE<br>F- 91191 GIF sur YVETTE Cedex<br>FRANCE |

| | |
|---|---|
| FACKLER Orrin | University of California<br>Lawrence Livermore Nat. Lab.<br>LIVERMORE CA 94550<br>USA |
| FAURE Jean Louis | CEN Saclay<br>DPhPE<br>F- 91191 GIF sur YVETTE Cedex<br>FRANCE |
| FEINBERG Frank | Massachusetts Institute of Technology<br>Dept of Physics<br>CAMBRIDGE MA 02139<br>USA |
| FELDER Christoph | Universität Basel<br>Physik. Inst. Experimentalle Kernphysik<br>CH- 4056 BASEL<br>SWITZERLAND |
| FITCH Val | Princeton University<br>Dept of Physics<br>PRINCETON NJ 08540<br>USA |
| FLEURY Patrick | Ecole Polytechnique<br>Centre de Physique Théorique<br>F- 91128 PALAISEAU Cedex<br>FRANCE |
| FOURNIER Daniel | Université Paris Sud<br>LAL<br>F- 914O5 ORSAY Cedex<br>FRANCE |
| GABATHULER Erwin | University of Liverpool<br>Department of Physics<br>LIVERPOOL L69 3BX<br>UNITED KINGDOM |
| GARRETA Denis | CEN Saclay<br>DPhPE<br>F- 91191 GIF sur YVETTE Cedex<br>FRANCE |
| GERBER Hans-Jürg | ETH-ZURICH<br>Inst. for Interm. Energy Physics<br>CH 5234 VILLIGEN<br>SWITZERLAND |
| GIBSON Valérie | CERN<br>Division EP<br>CH- 1211 GENEVE 23<br>SWITZERLAND |

| | |
|---|---|
| GILMAN Fred | Stanford University<br>Stanford Linear Acc. Center (SLAC)<br>STANFORD CA 94305<br>USA |
| GOLLIN George | Princeton University<br>Dept of Physics<br>PRINCETON NJ 08540<br>USA |
| GOLOWICH Eugène | University of Massachusetts<br>Dept of Physics<br>AMHERST MA 01002<br>USA |
| GRECO Mario | Laboratorio Nazionale dell'INFN<br>Via E. Fermi 40<br>I- 00044 FRASCATI<br>ITALY |
| GRONAU Michael | Max Planck Institut<br>Physik und Astrophysik<br>D- 8000 MUNCHEN 40<br>FEDERAL REP. OF GERMANY |
| GROSSETETE Bernard | Université P. & M. Curie<br>LPNHE - Tour 32<br>F- 75230 PARIS Cedex 05<br>FRANCE |
| GUMPLINGER Peter | CERN<br>Division EP<br>CH- 1211 GENEVE 23<br>SWITZERLAND |
| GUZZO Marcelo | Inter. Scuola for Advanced Studies<br>Strada Costeria 11<br>I- 34100 TRIESTE<br>ITALY |
| HAGELIN John | Maharishi International University<br>Dept of Theoretical Physics<br>FAIRFIELD IOWA 52556<br>USA |
| HAGIWARA Kaoru | University of Durham<br>Sciences Labs - Dept of Mathematics<br>DURHAM DH1 3LE<br>UNITED KINGDOM |
| HARARI Haim | Weizmann Institut of Science<br>Dept of Nuclear Physics<br>76100 REHOVOT<br>ISRAEL |

| | |
|---|---|
| HARRISON Paul Fraser | University College London<br>Dept of Physics and Astronomy<br>LONDON WCIE 6BT<br>UNITED KINGDOM |
| HARRUS Ilana | Université Paris Sud<br>LAL<br>F- 914O5 ORSAY Cedex<br>FRANCE |
| HITLIN David | California Institute of Technology<br>Lauritsen Lab. of Physics<br>PASADENA CA 91 125<br>USA |
| HOFFMAN Cyrus | Los Alamos National Laboratory<br>Dept of Physics<br>LOS ALAMOS NM 87545<br>USA |
| HOLDER Martin | Universität Gesamthochschule Siegen<br>Fachbereich 7 - Physik<br>D- 5900 SIEGEN<br>FEDERAL REP. OF GERMANY |
| ICONOMIDOU-FAYARD Lydia | Université Paris Sud<br>LAL<br>F- 914O5 ORSAY Cedex<br>FRANCE |
| JARLSKOG Cecilia | University of Stockholm<br>Institut of Physics<br>S- 113 46 STOCKHOLM<br>SWEDEN |
| JARLSKOG Goran | University of Lund<br>Dept of Physics<br>S- 223 62 LUND<br>SWEDEN |
| JAUNEAU Louis | Université Paris Sud<br>LAL<br>F- 914O5 ORSAY Cedex<br>FRANCE |
| KABIR Prabaham | University of Virginia<br>Dept of Physics<br>CHARLOTTESVILLE VA 22901<br>USA |
| KANE Gordon | University of Michigan<br>Dept of Physics<br>ANN ARBOR MI 48109<br>USA |

KAPLAN Daniel
FERMILAB
Dept of Physics
BATAVIA IL 60510
USA

KASHA Henry
Yale University
Sloane Lab. - Dept of Physics
NEW HAVEN CT 06520
USA

KEPHART Thomas
International Centre for Theor. Physics
Strada Costeria 11
I- 34 100 TRIESTE
ITALY

KOLB Edward
FERMILAB
Dept of Physics
BATAVIA IL 60510
USA

KONIG Adriaan C.
University of Nijmegen
Fysisch Laboratorium
NL- 6525 ED NIJMEGEN
NETHERLANDS

LAMOREAUX Steve
University of Washington
Dept of Physics
SEATTLE WA 98195
USA

LEDERMAN Leon M.
FERMILAB
Dept of Physics
BATAVIA IL 60510
USA

LEHMAN Pierre
IN2P3
Inst. Nat. Phys. Nucl. et Phys. Partic.
F- 75640 PARIS Cedex 13
FRANCE

LEIPUNER Lawrence
Brookhaven Nat. Laboratory
Dept of Physics
UPTON NY 11973
USA

LI Ling-Fong
Carnegie-Mellon University
Dept of Physics
PITTSBURGH PA 15213
USA

LINDE Andrei
CERN
TH Division
CH- 1211 GENEVE 23
SWITZERLAND

| | |
|---|---|
| LIPKIN Harry | Weizmann Institut of Science<br>Dept of Nuclear Physics<br>76100 REHOVOT<br>ISRAEL |
| LUTZ Anne Marie | Université Paris Sud<br>LAL<br>F- 91405 ORSAY Cedex<br>FRANCE |
| MACHET Bruno | Université P. & M. Curie (Paris VI)<br>LPTHE - Tour 16<br>F- 75230 PARIS Cedex 05<br>FRANCE |
| MANNELLI Italo | Scuola Normale Superiore<br>Istituto di Fisica<br>I- 56100 PISA<br>ITALY |
| MAREL Gérard | CEN Saclay<br>DPhPE<br>F- 91191 GIF sur YVETTE Cedex<br>FRANCE |
| MARTIN Alan D. | University of Durham<br>Sciences Labs - Dept of Mathematics<br>DURHAM DH1 3LE<br>UNITED KINGDOM |
| MARTIN André | CERN<br>TH Division<br>CH- 1211 GENEVE 23<br>SWITZERLAND |
| MARTINELLI Guido | CERN<br>TH Division<br>CH- 1211 GENEVE 23<br>SWITZERLAND |
| Mc LERRAN Larry | University of Minnesota<br>Dept of Physics and Astronomy<br>MINNEAPOLIS MN 55455<br>USA |
| MERMOD Ronald | Université de Geneve<br>Dept de Physique Nucl. et Corpusculaire<br>CH- 1211 GENEVE 4<br>SWITZERLAND |
| MEYER Philippe | Ecole Normale Supérieure<br>Lab. de Physique Théorique<br>F- 75231 PARIS Cedex 05<br>FRANCE |

| | |
|---|---|
| MILLER James | University of Boston<br>Dept of Physics<br>BOSTON MA 02215<br>USA |
| MONTANET François | Faculté des Sciences<br>Centre de Phys. des Particules<br>F- 13288 MARSEILLE Cedex 2<br>FRANCE |
| MONTANET Lucien | CERN<br>Division EP<br>CH- 1211 GENEVE 23<br>SWITZERLAND |
| MORALES Angel | Universidad de Zaragoza<br>Depto de Fisica Teorica<br>E- 50009 ZARAGOZA<br>SPAIN |
| MOREL André | CEN Saclay<br>DPhT<br>F- 91191 GIF sur YVETTE Cedex<br>FRANCE |
| MORELLET Daniel | Université Paris Sud<br>LAL<br>F- 914O5 ORSAY Cedex<br>FRANCE |
| MORGENSTERN Brigitte | GSI<br>Ges. f. Schwerionenforschung<br>D- 6100 DARMSTADT 11<br>FEDERAL REP. OF GERMANY |
| MUGGE Marshall | University of California<br>Lawrence Livermore Nat. Lab.<br>LIVERMORE CA 94550<br>USA |
| MURPHY C. Thornton | FERMILAB<br>Dept of Physics<br>BATAVIA IL 60510<br>USA |
| NAPPI Aniello | CERN<br>Division EP<br>CH- 1211 GENEVE 23<br>SWITZERLAND |
| NARDI Enrico | Université Scie. et Techn. du Languedoc<br>Dept de Physique Mathématique<br>F- 34060 MONTPELLIER<br>FRANCE |

| | |
|---|---|
| NGUYEN-KHAC Ung | Ecole Polytechnique<br>LPNHE<br>F- 91128 PALAISEAU Cedex<br>FRANCE |
| OEHME Reinhard | University of Chicago<br>Enrico Fermi Inst. - Dept of Physics<br>CHICAGO IL 60637<br>USA |
| OLIVER Luis | Université Paris Sud<br>LPTHE<br>F- 91405 ORSAY Cedex<br>FRANCE |
| PAIS Abraham | Rockefeller University<br>Dept of Physics - Exp. Group<br>NEW YORK NY 10021<br>USA |
| PASCHOS Emmanuel | Universität Dortmund<br>Lehrstuhl f. Exp. und Theor. Phys.<br>D- 4600 DORTMUND 50<br>FEDERAL REP. OF GERMANY |
| PATTERSON Ritchie | University of Chicago<br>Enrico Fermi Inst. - Dept of Physics<br>CHICAGO IL 60637<br>USA |
| PEACH Ken | University of Edinburgh<br>High Energy Phys. Group - Physics Dept.<br>EDINBURGH EH9 3JZ<br>UNITED KINGDOM |
| PECCEI Roberto | University of California<br>Dept of Physics<br>LOS ANGELES CA 90024<br>USA |
| PENE Olivier | Université Paris Sud<br>LPTHE<br>F- 91405 ORSAY Cedex<br>FRANCE |
| PETIAU Pierre | Ministère de la Recher. et de la Techn.<br>Dept. de Physique<br>F- 75005 PARIS<br>FRANCE |
| PEYAUD Bernard | CEN Saclay<br>DPhPE<br>F- 91191 GIF sur YVETTE Cedex<br>FRANCE |

| | |
|---|---|
| PHAM Trinang | Ecole Polytechnique<br>Centre de Physique Théorique<br>F- 91128 PALAISEAU Cedex<br>FRANCE |
| PHAM QUANG HUNG | University of Virginia<br>Dept of Physics<br>CHARLOTTESVILLE VA 22901<br>USA |
| PICCIOTTO Charles | University of Victoria<br>Dept of Physics and Astrophysics<br>VICTORIA BC V8W 2Y2<br>CANADA |
| PICEK Ivica | Rudjer Boskovic Institute<br>Dept of Theor. Physics<br>41 001 ZAGREB<br>YUGOSLAVIA |
| PIERAZZIMI Giuseppe M. | Universita di Pisa<br>Istituto di Fisica<br>I- 56100 PISA<br>ITALY |
| RAU Ronald | DESY<br>Notkestrasse 85<br>D- 2000 HAMBURG 52<br>FEDERAL REP. OF GERMANY |
| RICKENBACH René | Universität Basel<br>Physik. Inst. Experimentalle Kernphysik<br>CH- 4056 BASEL<br>SWITZERLAND |
| RIORDAN Michael | Stanford University<br>Stanford Linear Acc. Center (SLAC)<br>STANFORD CA 94305<br>USA |
| RITCHIE Jack | University of Texas<br>Dept. of Physics<br>AUSTIN TX 787121081<br>USA |
| ROLDAN Jesus | Universidad de Valencia<br>Fac. de Ciencias Fisicas<br>E- BURJASOT Valencia<br>SPAIN |
| ROST Michael | Universität Gesamthochschule Siegen<br>Fachbereich 7 - Physik<br>D- 5900 SIEGEN<br>FEDERAL REP. OF GERMANY |

| | |
|---|---|
| SACHS Robert | University of Chicago<br>Enrico Fermi Institut<br>CHICAGO IL 60637<br>USA |
| SANCHIS Miguel A. | Universidad de Valencia<br>Fac. de Ciencias Fisicas<br>E- BURJASOT Valencia<br>SPAIN |
| SANDA Anthony | Rockefeller University<br>Dept of Physics - Exp. Group<br>NEW YORK NY 10021<br>USA |
| SCHAFFER Arthur | Université Paris Sud<br>LAL<br>F- 914O5 ORSAY Cedex<br>FRANCE |
| SCHALLER Lukas | Université de Fribourg<br>Institut de Physique Expérimentale<br>CH- 1700 FRIBOURG<br>SWITZERLAND |
| SCHMIDT Michael | Yale University<br>Dept of Physics<br>NEW HAVEN CT 06511<br>USA |
| SCHOPPER Andrews | CERN<br>Division EP<br>CH- 1211 GENEVE 23<br>SWITZERLAND |
| SCHRÖDER Henning | DESY<br>Notkestrasse 85<br>D- 2000 HAMBURG 52<br>FEDERAL REP. OF GERMANY |
| SCHUBERT Klaus | UNIVERSITAT KARLSRUHE<br>Inst. fur Experimentelle Kernphysik<br>D 7500 KARLSRUHE 1<br>FEDERAL REP. OF GERMANY |
| SEHGAL Lalit | Rhein Westf. Tech. Hochschule - RWTH<br>Institut für Theor. Physik E.<br>D- 51OO AACHEN<br>FEDERAL REP. OF GERMANY |
| SIKIVIE Pierre | University of Florida<br>Dept of Physics<br>GAINESVILLE FL 32601<br>USA |

| | |
|---|---|
| SKRINSKY Alexander | USSR Academy of Sciences<br>Siberian Division<br>630 090 NOVOSIBIRSK<br>U S S R |
| SMITH Kenneth | University of Sussex<br>Dept of Physics<br>BRIGHTON BN1 9QH<br>UNITED KINGDOM |
| SMITH Stewart | Princeton University<br>Dept of Physics<br>PRINCETON NJ 08540<br>USA |
| SOERGEL Volker | DESY<br>Notkestrasse 85<br>D- 2000 HAMBURG 52<br>FEDERAL REP. OF GERMANY |
| STEFFEN Peter | DESY<br>Notkestrasse 85<br>D- 2000 HAMBURG 52<br>FEDERAL REP. OF GERMANY |
| STEINBERGER Jack | CERN<br>Division EP<br>CH- 1211 GENEVE 23<br>SWITZERLAND |
| STIPPEL Susanne | Universität Wien<br>Institut für Theoretische Physik<br>A- 1090 WIEN<br>AUSTRIA |
| STROVINK Mark | Lawrence Berkeley Laboratory<br>Dept of Physics<br>BERKELEY CA 94720<br>USA |
| SVIRIDOV Victor A. | Joint Inst. for Nuclear Research<br>Dubna Head Post Office<br>101 000 MOSCOW<br>U S S R |
| TANIMOTO Morimitsu | Ehime University<br>Faculty of Education<br>790 MATSUYAMA<br>JAPAN |
| TAUSCHER Ludwig | Universität Basel<br>Physik. Inst. Experimentalle Kernphysik<br>CH- 4056 BASEL<br>SWITZERLAND |

TEIGE Scott — Rutgers University
Dept of Physics
NEW BRUNSWICK NJ 08903
USA

THOMPSON Julia — University of Pittsburgh
Dept. of Physics
PITTSBURGH PA 15260
USA

TRAN THANH VAN Jean — Université Paris Sud
LPTHE
F- 91405 ORSAY Cedex
FRANCE

TRUEMAN Larry — Brookhaven Nat. Laboratory
Dept of Physics
UPTON NY 11973
USA

TRUONG Nguyen Tran — Ecole Polytechnique
Centre de Physique Théorique
F- 91128 PALAISEAU Cedex
FRANCE

TURLAY René — CEN Saclay
DPhPE
F- 91191 GIF sur YVETTE Cedex
FRANCE

TUTS Michael — Columbia University
Dept of Physics
NEW YORK NY 10027
USA

VALLAGE Bertrand — CEN Saclay
DPhPE
F- 91191 GIF sur YVETTE Cedex
FRANCE

VAN BEVEREN Eef — Universidade de Coimbra
Depto. de Fisica
P- 3000 COIMBRA
PORTUGAL

VAN DIJK Adriaan — CERN
Division EP
CH- 1211 GENEVE 23
SWITZERLAND

WAHL Heinrich — CERN
Division EP
CH- 1211 GENEVE 23
SWITZERLAND

| | |
|---|---|
| WALI Kameshwar | University of Syracuse<br>Dept of Physics<br>SYRACUSE NY 13210<br>USA |
| WATTENBERG Albert | University of Illinois<br>Loomis Lab. - Dept of Physics<br>URBANA IL 61801<br>USA |
| WINSTEIN Bruce | University of Chicago<br>Enrico Fermi Inst. - Dept of Physics<br>CHICAGO IL 60637<br>USA |
| WITZIG Christoph | CERN<br>Division EP<br>CH- 1211 GENEVE 23<br>SWITZERLAND |
| WOLFENSTEIN Lincoln | Carnegie-Mellon University<br>Dept of Physics<br>PITTSBURGH PA 15213<br>USA |
| WORMSER Guy | Université Paris Sud<br>LAL<br>F- 914O5 ORSAY Cedex<br>FRANCE |
| YAMAGUCHI Yoshio | University of Tokai<br>Faculty of Science - Dept of Physics<br>259-12 KANAGAWA<br>JAPAN |
| ZAIMIDOROGA Oleg A. | Joint Inst. for Nuclear Research<br>Dubna Head Post Office<br>101 000 MOSCOW<br>U S S R |
| ZAVRTANIK Danilo | Institut J. Stefan<br>Jamova 39<br>61111 LJUBLJANA<br>YUGOSLAVIA |
| ZECH Günter | Universität Gesamthochschule Siegen<br>Fachbereich 7 - Physik<br>D- 5900 SIEGEN<br>FEDERAL REP. OF GERMANY |
| ZYLBERSTEIN Armand | CEN Saclay<br>DPhPE<br>F- 91191 GIF sur YVETTE Cedex<br>FRANCE |